THE INVERTEBRATES

An Illustrated Glossary

THE INVERTEBRATES

An Illustrated Glossary

Michael Stachowitsch
Department of Marine Biology
Institute of Zoology
University of Vienna
Vienna, Austria

Sylvie Proidl
Illustrations

WILEY-LISS

A JOHN WILEY & SONS, INC. PUBLICATION

New York • Chichester • Brisbane • Toronto • Singapore

Address All Inquiries to the Publisher
Wiley-Liss, Inc., 605 Third Avenue, New York, NY 10158-0012

© **1992 Wiley-Liss, Inc.**

Printed in the United States of America

Recognizing the importance of preserving what has been written, it is a policy of John Wiley & Sons, Inc., to have books of enduring value published in the United States printed on acid-free paper, and we exert our best efforts to that end.

Library of Congress Cataloging-in-Publication Data
Stachowitsch, Michael.
 The invertebrates : an illustrated glossary / Michael Stachowitsch; Sylvie Proidl, illustrations.
  p. cm.
 Includes index.
 ISBN 0-471-83294-4
 ISBN 0-471-56192-4 (paperback)
 1. Invertebrates—Terminology. 2. Invertebrates—Pictorial works. 3. Invertebrates—Classification. I. Proidl, Sylvie. II. Title.
QL362.S82 1991
592—dc20   91-21129
                    CIP

To Mother Nature
and all those dedicated
to maintaining her diversity

CONTENTS

PREFACE

The diversity of life on this planet is a never-ending source of fascination. Detecting and understanding order in this diversity has been one of the primary goals of biologists for centuries. A key lies in the realization that only a limited number of basic body plans exists and that a particular group is characterized by a limited number of anatomical features: the seemingly unending variety is based on modifications of these structures and reflects little more than "variations on a theme." This glossary is therefore divided into two sections. The first presents—group by group—the anatomical features of these body plans. The second part presents terms used in describing the modifications of these features.

This book has a bilingual accent. German was chosen because many terms were coined and many descriptions published at a time when German was a preeminent language in the scientific community. This is reflected, for example, in the repeated referral to two German treatises—Kükenthal-Krumbach's *Handbuch der Zoologie* and Bronn's *Klassen und Ordnungen des Tierreichs*—in an English-language standard work, *The Invertebrates*, by L. Hyman. These references and a wealth of more modern literature were consulted in compiling the present glossary.

The arrangement in the first, systematic section is conservative. It parallels that found in most textbooks and is not meant to favor any particular school of thought. A total of 77 living invertebrate groups—most on a phylum or class level—are presented. The somewhat finer differentiation among the crustaceans reflects the great variety of this group. Each chapter is meant to be autonomous; each definition should stand on its own. A certain degree of repetition was deemed less frustrating than using a generalized terminology for specialized groups. Generally, no taxonomic units other than those of the chapter headings are provided. Indeed, the names and relative ranks of even the largest taxonomic units are often a matter of controversy. Designating smaller subgroups within the text or illustrations was thought to detract more than add. Certain invertebrate groups have been omitted. These include single-celled forms (protozoans) and terrestrial arthropod groups (insects, mites, spiders, myriapods, etc.). The former goes beyond the format of the glossary, while specialized dictionaries exist for many of the latter.

The number of terms per chapter varies from 55 to 410. This reflects the different scope of the groups, differing levels of organization or complexity, and in some cases various states of knowledge. Unless specified otherwise, the definitions refer to adult forms. Not all characters are included. The number of references to musculature, for example, has been kept to a minimum, and those relating to features on the cellular or ultrastructural level are generally omitted. Commonly cited dimensions are presented under "standard measurements."

Each definition is a compromise between detail and brevity. The text structure comprises the main English entry (in its singular form), English equivalents or accepted spellings in parentheses, German translation(s) in brackets, the definition itself, adjectives describing variations (in parentheses), and cross-references to related terms not included in the definition (under "see also"). The number of adjectives after a definition reflects the importance (justified or not) attached to that feature in taxonomic work. If abundant and referring to different aspects, they may be separated by a semicolon. A number of descriptive terms commonly found in identification keys have been omitted. These include designations such as large/small, long/short, reduced/fully developed, present/absent. The former are relative, imprecise terms of little use to the non-specialist, while inclusion of the latter would unduly burden each definition: virtually no feature is present or fully developed in all representatives of a group (i.e., parasitic forms).

The second section of this book lists alphabetically and defines descriptive terms applied to the anatomical characters in the first section. It includes adjectives designating shape, ornamentation, position or orientation, growth type, function, or other condition. Adjectives directly derived from anatomical features (scutum/scutal plate) or from taxonomic names are generally omitted, as are ecological terms. Certain frequently used, self-explanatory adjectives are listed in order to provide for German translations (bell-shaped/*glockenförmig*) or to guide the reader to more technical equivalents (ear-shaped/auriform).

The illustrations are a fundamental component of the glossary and were conceived for rapid orientation within a group. The schematic line drawings reflect the same approach as the text: a generalized body plan supplemented by important or interesting variations. It is beyond the scope of either the illustrations or the text to specify which character combinations are realized by members of the group. A "lateral" perspective, i.e., showing one half of the organism from a longitudinal plane through the body, is generally provided. A certain degree of uniformity in the representation of internal organs ("cellular" digestive tract, solid nerve system, stippled gonads) compensates for varying perspectives dictated by diverse body plans and symmetries. Major differences between males and females are indicated when appropriate.

Not every definition can be ingenious, not every illustration a revelation. Nevertheless, the glossary should enable the reader to decipher the often highly technical diagnoses in the scientific literature and, in effect, provides a "key to the keys." It should also enable an intelligent description of collected specimens when more detailed monographs are not at hand, and can thus serve as a practical tool for those confronted with the astounding diversity of the invertebrates.

ACKNOWLEDGMENTS

This glossary contains more than 10,000 entries and 1,100 figures compiled in 79 full-page illustrations. Errors and oversimplifications are inevitable. In order to minimize these, most chapters were sent to specialists. I am very grateful to the following colleagues for corrections and advice: N. Boero, R. Brandstätter, D. Danielopol, T. Dunagan, P. Dworschak, P. Emschermann, E. Gaviria, S. Gaviria, R. Gibson, R. Higgins, H. Hilgers, H. Hinaidy, R. Kikinger, T. Klepal, R. Kristensen, H. Mehlhorn, D. Miller, R. Novak, J. Ott, G. Poinar, H.-H. Reichenbach-Klinke, R. Riedl, R. Rieger, H. Ruhberg, E. Ruppert, A. Ruttner-Kolisko, K. Rützler, L. v. Salvini-Plawen, A. Scheltema, F. Schram, W. Senz, G. Steiner, W. Sterrer, W. Sudhaus, N. Wawra, and K. Wittmann.

I am especially indebted to Luitfried v. Salvini-Plawen for graciously reviewing a large number of chapters and for discussions on the many finer points of invertebrate morphology. Thanks are also due to my wife Sylvie for her effort in designing the illustrations and for coping with the frustrations involved in seeing this endeavor to completion. This book would not have been possible without the continued support of Prof. Rupert Riedl.

M.S.

PLACOZOA

placozoan [G *Plakozoe*]

PLACOZOA

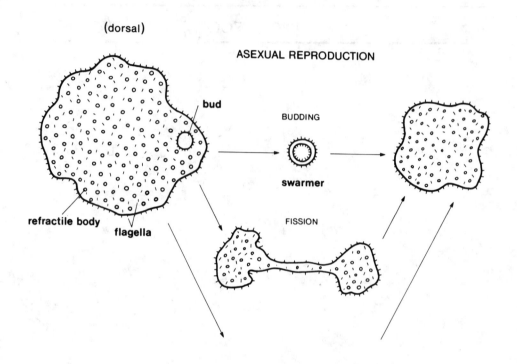

(dorsal)

ASEXUAL REPRODUCTION

bud

BUDDING

swarmer

FISSION

refractile body

flagella

SEXUAL REPRODUCTION

(lateral)

bacterium   egg   refractile body   vesicle
nucleus
nucleolus
dorsal
epithelium   concrement
vacuole

osmiophilic plate
intermediate   vesicle
layer   fiber cell

degenerating cell
gland cell

ventral   ciliated cell
epithelium
flagellum

substratum

PLACOZOA

placozoan [G *Plakozoe*]

bud [G *Knospe*]: In asexual reproduction, small, spherical, multicellular outgrowth; gives rise to swarmer.

ciliary pit [G *Wimpergrube*]: In cells of dorsal and ventral epitheliam, eccentrically located invagination from which flagellum projects.

ciliated cell [G *Wimperzelle, Geißelzelle*]: Flagellum-bearing (monociliate) cell of dorsal or ventral epithelium; typically used to differentiate flagellum-bearing cells from gland cells in ventral epithelium.

cilium: flagellum

concrement vacuole [G *Konkrementvakuole*]: Relatively large, deposit-filled vacuole in fiber cell or between nucleus and ventral surface of ciliated cells of ventral epithelium.

cortical granule [G *Cortex-Granula*]: Small, dense, membrane-enclosed bodies within egg; as egg matures, cortical granules concentrate on outer margin and play a role in formation of fertilization membrane.

degenerating cell [G *degenerierende Zelle*]: One of three cell types (ciliated cell, degenerating cell, gland cell) of ventral epithelium; bears osmiophilic deposits and lacks nucleus and cytoplasm.

desmosome [G *Desmosom*]: Dense material concentrated along internal margins of cells of dorsal epithelium.

dorsal epithelium [G *Dorsalepithel*]: Layer of flat, polygonal cells forming dorsal surface. Each cell typically bears single, eccentrically located flagellum arising from ciliary pit; ventral section of cell with nucleus projects into intermediate layer. (with/without flagella) (See also ventral epithelium)

egg [G *Ei*]: In sexual reproduction, ♀ reproductive body arising in ventral epithelium and maturing in intermediate layer; closely associated with fiber cells of intermediate layer and surrounded by fertilization membrane.

epithelium: dorsal epithelium, ventral epithelium

fertilization membrane [G *Befruchtungsmembran*]: Two-layer membrane surrounding mature egg; outer layer apparently formed by cortical granules.

fiber cell (stellate cell) [G *Faserzelle*]: Highly branched, contractile type of cell with tetraploid nucleus forming network within intermediate layer; often with bacteria-filled vesicles. Responsible for changes in body shape.

flagellum [G *Geißel*]: Elongate, motile projection arising from eccentrically located ciliary pit in cells of dorsal and ventral epithelia.

gland cell [G *Drüsenzelle*]: One of three cell types (ciliated cell, degenerating cell, gland cell) of ventral epithelium; less numerous than ciliated cells and differing from latter in being cone-shaped rather than cylindrical.

globule: refractile body

hood [G *Haube*]: One of up to three hood-shaped modifications on certain flagella of dorsal and ventral epithelia.

intercellular bacterium [G *interzelluläres Bakterium*]: One of numerous bacteria enclosed in vesicles within fiber cells of intermediate layer.

intermediate layer (middle layer) [G *Zwischenschicht*]: Layer of body between dorsal and ventral epithelia; consists of fluid-filled cavity containing network of contractile fiber cells.

marginal bulge [G *Randzone, Randwulst*]: Outer rim of body consisting of region where flat cells of dorsal epithelium join cylindrical cells of ventral epithelium; may contain separate space and, in life, be elevated above substrate.

"microvillus" [G *"Mikrovillus," Zotte*]: Term occasionally applied to one of numerous short projections of cells of ventral (more rarely dorsal) epithelium to exterior.

middle layer: intermediate layer

osmiophilic plate [G *osmiophile Platte*]: Dense demarcation between two fiber cells in intermediate layer; adjoined to each side by vesicle.

radiating filament [G *Radiärfilament*]: One of several fine filaments connecting flagellum of ciliated cell to margin of ciliary pit.

refractile body (globule) [G *Glanzkugel*]: One of numerous large, round lipid globules located directly below intercellular gaps in dorsal epithelium; bears thin layer of cytoplasmic material ventrally.

sperm [G *Spermium*]: Term applied to nonflagellated cell type in intermediate layer which may be ♂ reproductive body.

stellate cell: fiber cell

swarmer [G *Schwärmer*]: In asexual reproduction, free-swimming stage produced by budding; consists of outer layer of ciliated dorsal epithelium and inner layer of ciliated ventral epithelium.

ventral epithelium [G *Ventralepithel*]: Ventral layer of body consisting of cylindrical, flagellum-bearing cells (ciliated cells), club-shaped gland cells, and degenerating cells; nuclei of ciliated cells and gland cells located proximally. (See also dorsal epithelium)

vesicle [G *Vesikel*]: Small, round protuberance on dorsal epithelium or on fiber cell in intermediate layer; also intracellular structure enclosing intercellular bacteria or located on each side of osmiophilic plate at junction of two fiber cells.

MESOZOA

mesozoan [G *Mesozoe*]

MESOZOA

CEPHALOPOD INHABITANT (dicyemid)

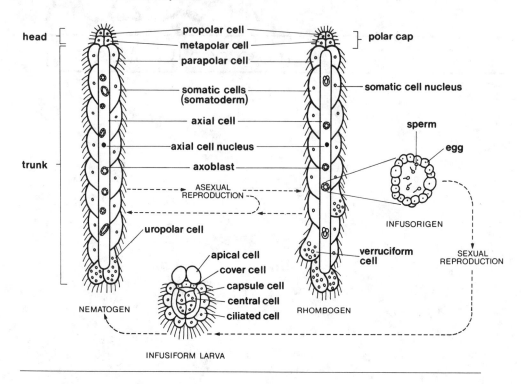

head

propolar cell
metapolar cell
parapolar cell

polar cap

somatic cells
(somatoderm)

somatic cell nucleus

axial cell

sperm

axial cell nucleus

egg

trunk

axoblast

ASEXUAL
REPRODUCTION

INFUSORIGEN

uropolar cell

apical cell
cover cell
capsule cell
central cell
ciliated cell

verruciform
cell

SEXUAL
REPRODUCTION

NEMATOGEN

RHOMBOGEN

INFUSIFORM LARVA

NON-CEPHALOPOD INHABITANT (orthonectid)

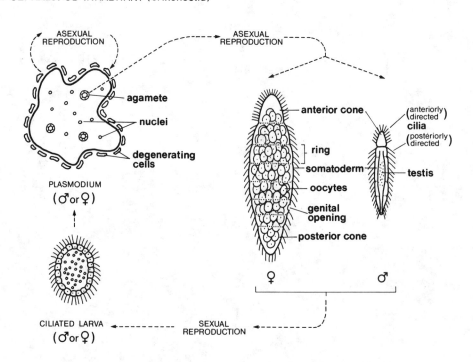

ASEXUAL
REPRODUCTION

ASEXUAL
REPRODUCTION

agamete

nuclei

degenerating
cells

anterior cone

(anteriorly)
(directed)
cilia
(posteriorly)
(directed)

ring
somatoderm
oocytes
genital
opening
posterior cone

testis

PLASMODIUM
(♂or♀)

♀

♂

CILIATED LARVA
(♂or♀)

SEXUAL
REPRODUCTION

MESOZOA

mesozoan [G *Mesozoe*]

agamete [G *Agamet*]: Either axoblast or one of numerous diploid reproductive cells in plasmodium of noncephalopod-inhabiting mesozoan. Gives rise to free-swimming sexual stage.

agamont [G *Agamont*]: General term for asexually reproducing parasitic stage. (See also plasmodium)

anterior cone [G *Vorder-Konus*]: In free-swimming sexual stage of noncephalopod-inhabiting mesozoan, typically well-delimited anteriormost ring or rings of somatoderm cells; as opposed to remainder of somatoderm, cilia are directed anteriorly. (See also posterior cone)

apical cell [G *Polzelle, Blasenzelle*]: In infusiform larva, one of two large unciliated anteriormost cells, each containing a large, dense refringent body.

axial cell [G *Axialzelle*]: In nematogen or rhombogen, single elongate inner cell surrounded by somatoderm; incorporates axoblasts.

axoblast (germinal cell, agamete) [G *Axoblast, Agamet*]: One of numerous diploid cells within axial cell; develops asexually into nematogen, rhombogen, or infusorigen in cephalopod-inhabiting mesozoan.

calotte: polar cap

calotte cell: polar cell

capsule cell [G *Kapselzelle*]: In infusiform larva, one of two granule-filled, curved cells enclosing central cells and surrounded externally by cover of ciliated cells.

central cell (urn cell) [G *Urnenzelle*]: In infusiform larva, one of four binucleate, granule-filled cells forming interior; each contains a germinal cell. Enclosed laterally by capsule cells.

cover cell [G *Deckzelle*]: In infusiform larva, one of four cells which, along with two ciliated cells, underlie apical cells and form anterior border of urn.

dense body: refringent body

dispersal larva: infusiform larva

egg: pseudoegg

embryo [G *Embryo*]: One of numerous young vermiform nematogens developing from agametes within axial cell of adult nematogen. Refers also to ciliated embryo developing inside free-swimming ♀ sexual stage of noncephalopod-inhabiting mesozoan.

genital opening [G *Genitalporus, Geschlechtsöffnung*]: Pore through somatoderm in posterior half of free-swimming ♀ sexual stage of noncephalopod-inhabiting mesozoan.

germinal cell [G *Axoblast, generative Zelle*]: Reproductive cell in axial cell and one in each central cell of infusiform larva.

head [G *Kopf*]: In nematogen or rhombogen, anterior and smaller of two major external divisions of body (head, trunk); consists of eight to nine polar, metapolar, and parapolar cells forming polar cap or calotte.

hermaphroditic gonad: infusorigen

infusoriform: infusoriform larva

infusoriform larva (swarmer, dispersal larva) [G *Infusoriforme-Larve, Schwärmlarve*]: Small, oval, free-swimming larva developing from infusorigen in axial cell of rhombogen; consists of apical cells, external cover of ciliated cells, and an urn formed by central cells, capsule cells, and cover cells. Released into seawater along with host urine.

infusorigen (hermaphroditic gonad) [G *Infusorigen, Zwittergonade*]: One of numerous hermaphroditic gonads (previously regarded as minute hermaphroditic individuals) within axial cell of rhombogen; consists of large inner cell containing sperm and outer layer of pseudoeggs. Gives rise to infusiform larvae after fertilization.

jacket cell: somatic cell

metapolar cell [G *metapolare Zelle*]: In nematogen or rhombogen, one of four or five cells forming posterior tier of polar cap. (See also propolar cell)

nematogen [G *Nematogen*]: In basic life cycle of cephalopod-inhabiting mesozoan (nematogen, rhombogen, infusiform larva), adult asexually reproducing stage found at low densities in juvenile host; consists of outer somatoderm and inner axial cell. One may distinguish a primary and a secondary nematogen. (See also stem nematogen)

oocyte [G *Oozyte*]: One of a number of immature diploid ♀ reproductive cells developing from oogonia and forming outer layer of infusorigen; gives rise to pseudoegg. Also one of numerous variously arranged interior cells of free-swimming ♀ sexual stage of noncephalopod-inhabiting mesozoan.

oogonium [G *Oogonium*]: One of a number of immature diploid ♀ reproductive cells giving rise to oocytes and pseudoeggs forming outer layer of infusorigen. (See also progenitor of oogonia)

ovocyte: oocyte

parapolar cell [G *Parapolarzelle*]: In nematogen or rhombogen, one of typically two cells constituting first trunk cells immediately posterior to polar cap; intermediate in size between polar and remaining trunk cells.

peripheral cell: somatic cell

plasmodium [G *Plasmodium, Agamont*]: In noncephalopod-inhabiting mesozoan, amoeboid, multinucleate, asexually reproducing stage in host.

polar cap (calotte, polar calotte) [G *Polarzellen-Kappe*]: In nematogen or rhombogen, expanded anterior section of body formed by polar cells; succeeded posteriorly by parapolar cells. Cilia of polar cap serve in attachment to host.

polar cell (calotte cell) [G *Polarzelle*]: In nematogen or rhombogen, one of eight or nine relatively small, modified somatic cells of head; typically in form of two groups of four cells forming polar cap. First group (propolar cells) and second group (metapolar cells) may be arranged bilaterally or radially, in the latter case each propolar cell either in line with or alternating with metapolar cell.

pole cell [G *Polzelle*]: apical cell

posterior cone [G *End-Konus*]: In free-swimming sexual stage of noncephalopod-inhabiting mesozoan, typically well-delimited posteriormost ring or rings of somatoderm cells; with posteriorly directed cilia.

primary nematogen [G *primäres Nematogen*]: Typical nematogen produced asexually by stem nematogen or by other subsequently produced nematogens. (See also secondary nematogen)

primary rhombogen [G *primäres Rhombogen*]: Type of rhombogen produced asexually by last generation of agametes within axial cell of last generation of nematogens. (See also secondary rhombogen)

primordial oogonium: progenitor of oogonia

progenitor of oogonia (primordial oogonium) [G *prospektive Eizelle*]: In sexual reproduction leading to infusorigen (= hermaphroditic gonad) within axial cell of rhombogen, smaller of two cells formed by second asymmetrical division of agamete. Continues to divide until larger cell is surrounded.

progenitor of somatic cells

[G *prospektive Somatodermzelle*]: In asexual reproduction leading to new nematogen within axial cell of adult nematogen, smaller of two cells formed by first asymmetrical division of agamete. Continues to divide until larger cell (new axial cell) is surrounded by somatic cells.

progenitor of spermatogonia

[G *prospektives Spermium*]: In sexual reproduction leading to infusorigen (= hermaphroditic gonad) and subsequently to infusiform larva within axial cell of rhombogen, smaller of two cells formed by third asymmetrical division of agamete. Progenitor of spermatogonia is then engulfed by large cell and continues to divide mitotically into spermatogonia, then meiotically into sperm.

propolar cell

[G *propolare Zelle*]: In nematogen or rhombogen, one of four cells forming anterior tier of polar cap. (See also metapolar cell, parapolar cell, polar cell)

pseudoegg

[G *Pseudo-Eizelle*]: In axial cell of rhombogen, mature egg developing from oocytes surrounding outer surface of infusorigen (= hermaphroditic gonad).

refringent body (dense body)

[G *lichtbrechender Körper*]: Large, refringent deposits in each of two apical cells of infusiform larva.

rhombogen

[G *Rhombogen*]: In basic life cycle of cephalopod-inhabiting mesozoan (nematogen, rhombogen, infusiform larva), final sexually reproducing stage found at high densities in mature host; differs from nematogen in type of cell inclusions and in production of infusorigen (= hermaphroditic gonad) rather than further nematogens. One may distinguish a primary and a secondary rhombogen.

ring

[G *Ring*]: In free-swimming sexual stage of noncephalopod-inhabiting mesozoan, one of up to more than 50 annulations formed by circular grooves between cells of somatoderm. (equal, unequal; ciliated, unciliated)

secondary nematogen

[G *sekundäres Nematogen*]: Term applied to nematogen resulting from reversal of sexually reproducing rhombogen into asexually reproducing nematogen. (See also primary nematogen)

secondary rhombogen

[G *sekundäres Rhombogen*]: Type of rhombogen arising by direct modification of nematogen. (See also primary rhombogen)

somatic cell

[G *somatische Zelle, Somatodermzelle*]: One of a number of variously differentiated, typically ciliated cells forming outer cell layer or somatoderm. (ciliated, nonciliated; separate, syncytial)

somatoderm

[G *Somatoderm*]: In nematogen or rhombogen, outer layer of somatic cells enclosing axial cell and forming polar cap. In free-swimming sexual stage of noncephalopod-inhabiting mesozoan, typically annulated outer layer of somatic cells enclosing reproductive cells.

sperm

[G *Sperm*]: In axial cell of rhombogen, haploid ♂ reproductive cell within infusorigen (= hermaphroditic gonad). Fertilizes layer of eggs forming outer cover of hermaphroditic gonad, giving rise to infusiform larva. In free-swimming ♂ of noncephalopod-inhabiting mesozoan, haploid reproductive cells in testis.

spermatogonium

[G *Spermatogonium*]: One of a number of diploid ♂ reproductive cells in infusorigen (= hermaphroditic gonad) within axial cell of rhombogen; gives rise to sperm. Also one of numerous immature reproductive cells in testis of free-swimming stage of noncephalopod-inhabiting mesozoan.

stem nematogen (stem vermiform)

[G *Stammnematogen*]: In life cycle of cephalopod-inhabiting mesozoan (nematogen, rhombogen, infusiform larva), first stage found after infection in young host; agametes in axial cells give rise to primary nematogens.

swarmer: infusiform larva

testis [G *Hoden*]: In free-swimming sexual stage of noncephalopod-inhabiting mesozoan, ♂ reproductive organ containing sperm; located anterior to oocytes in hermaphroditic form.

trunk [G *Rumpf*]: In nematogen or rhombogen, posterior, more elongate of two external divisions of body (head, trunk).

urn [G *Urne*]: Cavity in infusiform larva. Surrounded by cover cells and two ciliated cells anteriorly, by four central cells posteriorly. Opens to exterior by means of pore.

urn cell: central cell

uropolar cell [G *Uropolarzelle*]: In nematogen or rhombogen, one of two inclusion-rich posteriormost cells of trunk.

vermiform [G *wurmförmige Generation*]: Term applied to nematogen and rhombogen in order to distinguish these worm-shaped adult forms from oval infusiform larvae.

verruciform cell [G *verruciforme Zelle*]: One of several enlarged, inclusion-rich, wart-like somatic cells projecting from trunk of rhombogen.

Wagener's larva [G *Wagener-Larve*]: Type of larva produced by certain cephalopod-inhabiting mesozoans; bears both anteriorly and posteriorly directed cilia.

PORIFERA
(PARAZOA)

sponge [G *Schwamm*]

PORIFERA

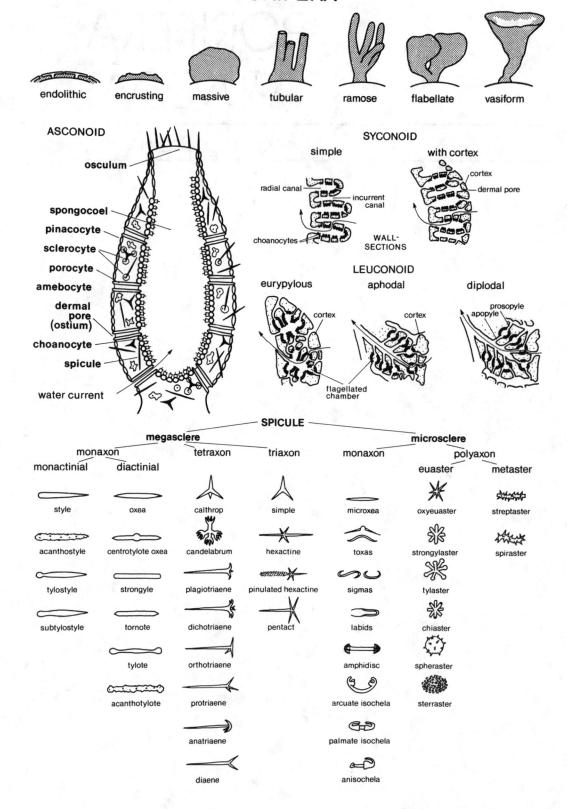

endolithic encrusting massive tubular ramose flabellate vasiform

ASCONOID

osculum

spongocoel
pinacocyte
sclerocyte
porocyte
amebocyte
dermal pore (ostium)
choanocyte
spicule

water current

SYCONOID

simple

with cortex

radial canal
cortex
dermal pore
incurrent canal

choanocytes

WALL-SECTIONS

LEUCONOID

eurypylous aphodal diplodal

cortex cortex prosopyle
apopyle

flagellated chamber

SPICULE

megasclere microsclere

monaxon tetraxon triaxon monaxon polyaxon

monactinial diactinial euaster metaster

style oxea calthrop simple microxea oxyeuaster streptaster

acanthostyle centrotylote oxea candelabrum hexactine toxas strongylaster spiraster

tylostyle strongyle plagiotriaene pinulated hexactine sigmas tylaster

subtylostyle tornote dichotriaene pentact labids chiaster

tylote orthotriaene amphidisc spheraster

acanthotylote protriaene arcuate isochela sterraster

anatriaene palmate isochela

diaene anisochela

PORIFERA (PARAZOA)

sponge [G *Schwamm*]

acantho [G *akantho-*]: Prefix added when any part of a spicule is spinose (i.e., acanthoxea, acanthostyle, acanthotornote, acanthostrongyle, acanthotylote, acanthotylostyle).

alpha stage [G *Alpha-Stadium*]: First growth stage of boring sponge in which sponge is under surface of substrate, exposing only incurrent and excurrent pores. (See also beta stage, gamma stage)

amebocyte **(amoebocyte)** [G *Amöbozyte, Amoebocyte*]: Mobile ameboid cell in mesohyl with phagocytic function; may also be applied to any type of ameboid cell (i.e., archeocyte, collencyte, sclerocyte).

amphidisc [G *Amphidisk*]: Type of spicule: microsclere with both ends expanded in umbrella-like fashion.

anatriaene [G *Anatriaen*]: Type of spicule (megasclere): triaene with one long ray and three short rays curved in direction of longer ray.

anisochela [G *Anisochela*]: Type of spicule (microsclere): chela with ends differing in size and shape.

apopyle [G *Apopyle, abführende Kammerpore*]: Opening through which water leaves flagellated chamber in leuconoid sponge; leads into excurrent canal. (See also prosopyle)

archeocyte **(archaeocyte)** [G *Archaeozyte, Archaeocyte*]: Large ameboid cell in mesohyl; undifferentiated and capable of developing into other types of cells.

ascon [G *Ascontypus*]: Simplest type of sponge structure in which incurrent pores (ostia) in body wall lead directly into central spongocoel; very small, radial forms. (See also leucon, sycon)

aster [G *Aster*]: Type of spicule: general term for polyaxial microsclere with numerous rays emanating from a common center (euaster) or from an elongated axis (metaster) (i.e., oxyaster, chiaster, spheraster, sterraster, tylaster).

atrium: spongocoel

beta stage [G *Beta-Stadium*]: Second growth stage of boring sponge in which sponge also grows on substrate surface as an encrustation. (See also alpha stage, gamma stage)

bipocillum [G *Bipocille*]: Type of spicule (microsclere): highly modified, irregular chela.

birotule (birotulate): Type of spicule (microsclere): highly modified chela.

calcoblast: Term occasionally applied to type of amebocyte forming calcium carbonate spicules in calcareous sponge. (See also scleroblast)

13

calthrop [G *Chelotrop*]: Type of spicule (megasclere): simple tetraxon, all four rays equal in length.

centro- [G *centro-*]: Prefix added when head of spicule is not terminal (i.e., centrotylote).

chela [G *Chela*]: Type of spicule: general term for microsclere whose ends have dissimilar (anisochela) or similar (isochela) shovel-like structures, the latter with either more highly developed shaft (isochela arcuata = arcuate chela) or more highly developed shovels (isochela palmata = palmate chela).

chiaster [G *Chiaster*]: Type of spicule (microsclere): euaster with cylindrical, truncate rays.

choanocyte (collar cell) [G *Choanozyte, Choanocyte, Kragengeiselzelle*]: Current-producing cell lining spongocoel (asconoid sponge), radial canals (syconoid sponge), or flagellated chambers (leuconoid sponge); bears a flagellum surrounded by a "collar" of filaments which trap food particles.

choanocyte chamber: flagellated chamber

choanosome: endosome

collar [G *Kragen*]: Contractile, mucus-lined, collar-shaped structure surrounding flagellum of choanocyte; formed by a circle of parallel filaments which trap food particles on their outer surface.

collar cell: choanocyte

collencyte [G *Collenzyte, Collencyte*]: Fixed ameboid cell attached to neighboring cells in mesohyl by slender, contractile pseudopodia.

conule [G *Conulus, Zipfel*]: Tent-like projection on surface of certain sponges.

cortex [G *Cortex, Rinde*]: Term applied to thickened ectosome in certain sponges.

dermal pore [G *Dermalpore*]: Intercellular pore on outer surface through which water enters incurrent canal in more complex sponge.

diactine [G *Diactine*]: Type of spicule (megasclere): general term for monaxon with two similar ends. (See also monactine)

dichotriaene [G *Dichotriaen*]: Type of spicule (megasclere): triaene with one long ray and three short, forked rays.

ectosome [G *Ektosoma*]: Thin or thick (cortex: reinforced by distinctive skeleton) outer layer. (See also endosome)

endosome (choanosome) [G *Endosoma*]: Tissue lying beneath ectosome in more complex (= leuconoid) sponge and containing choanocyte chambers.

euaster [G *Euaster*]: Type of spicule (microsclere): general term for star-shaped asters with rays emanating from a common center (vs. metaster).

excurrent canal [G *ausführender Kanal*]: Canal in more complex (= leuconoid) sponge leading water to exterior after it has left choanocyte chamber via apopyle; becomes progressively larger toward exterior as it merges with other canals.

fistule [G *Fistel*]: Tube-like elevation on surface of certain sponges bearing terminal osculum.

flagellated canal: radial canal

flagellated chamber (choanocyte chamber) [G *Geißelkammer, Choanocyten-Kammer, Choanocytenraum*]: Small, choanocyte-lined cavities in leuconoid sponge; water enters flagellated chamber via incurrent canal through apopyle and exits through prosopyle via excurrent canal.

flagellum [G *Geißel*]: Flagellum of choanocyte; its spiral beating creates a water current and draws suspended particles onto collar.

gamma stage [G *Gamma-Stadium*]: Third growth stage of boring sponge in which sponge is also elevated as a massive growth above the substrate surface. (See also alpha stage, beta stage)

gemmule [G *Gemmula*]: Reproductive structure, particularly common in freshwater sponges, consisting of a protective, spicule-reinforced outer cover and archeocyte-filled internal cavity; contents emerge under favorable conditions via micropyle.

head [G *Kopf*]: Swelling on a monaxon spicule; typically terminal, yet also nonterminal (centro-) or less pronounced (sub-). (See also shaft)

hexact (hexactine, hexactinal spicule) [G *Hexactin*]: Any type of spicule with six rays; typically refers to regular triaxon.

incurrent canal [G *zuführender Kanal*]: Canal through which water enters more complex sponge; formed in syconoid sponge by invaginations of exterior surface and equivalent to space between radial canals, in leuconoid sponge in form of branching channels eventually leading to choanocyte chambers.

incurrent pore: ostium

isochela [G *Isochela*]: Type of spicule (microsclere): chela whose ends bear similarly sized and shaped shovel-like structures; palmate isochela (isochela palmata) with more highly developed shovels, arcuate isochela (isochela arcuata) with more highly developed shaft.

leucon [G *Leucontypus*]: Most complex type of sponge structure, characteristic for most sponges, in which a high degree of folding leads to formation of choanocyte chambers and both incurrent and excurrent canals; permits large, irregular growth forms. (See also ascon, sycon)

megasclere [G *Megasklere, Megasklerit*]: Larger spicule (generally 60–2000 μm) functioning as chief supporting elements in all but the most complex sponges. (See also microsclere)

mesohyl [G *Mesohyl*]: Tissue between pinacoderm and choanoderm; typically containing spicules and ameboid cells.

metaster [G *Metaster*]: Type of spicule (microsclere): general term for asters with rays emanating from an elongated axis. (See also euaster)

micro- [G *mikro-*]: Prefix added to spicule designation when configuration resembles that of a megasclere, yet falls within size range of a microsclere (i.e., microxea, microstrongyle).

microsclere [G *Mikrosklerit*]: Small (generally 10–60 μm), more randomly distributed spicule type not functioning as a supporting element. (See also megasclere)

monactine [G *Monactine*]: Type of spicule (megasclere): general term for monaxon with two dissimilar ends. (See also diactine)

monaxon [G *Monaxon*]: Type of spicule: general term for skeletal element with a single straight or curved axis.

myocyte [G *Myozyte, Myocyte*]: Contractile cell in more complex sponge; concentrated around osculum or arranged parallel to surface, enabling slight contraction of body or constriction of osculum.

onychaete [G *Onychaet*]: Type of spicule: roughened/crenulate raphide.

orthotriaene [G *Orthotriaen*]: Type of spicule (megasclere): triaene with one long and three short rays directed away at right angles from longer ray.

osculum [G *Osculum*]: Excurrent opening; in simple sponge, a single distal opening of spongocoel, in complex sponge, openings formed by merger of excurrent canals.

ostium [G *Ostium*]: Any opening through which water enters sponge. Includes minute intracellular incurrent pores on external surface of less complex sponge (i.e., those formed by a porocyte). (See also dermal pore)

oxea [G *Ox*]: Type of spicule (megasclere): monaxon with both ends tapering to a point.

oxyaster [G *Oxyaster*]: Type of spicule (microsclere): euaster with thin, pointed rays.

paragaster: spongocoel

pinacocyte [G *Pinakozyte, Pinakocyte*]: Flattened, polygonal, contractile cell covering outer surface of simple sponge.

pinacoderm [G *Pinakoderm*]: One-cell-thick layer delimiting sponge from external milieu; formed by pinacocytes.

plagiotriaene [G *Plagiotriaen*]: Type of spicule (megasclere): triaene with one long ray and three short, curved rays.

polyaxon [G *Polyaxon*]: General term for any spicule having more than four centered axes (excluding hexacts). More or less synonymous with euaster.

porocyte [G *Porozyte, Porocyte*]: Tube-shaped cell in less complex sponge, derived from pinacocyte, extending from external surface to spongocoel. Lumen (ostium) serves as incurrent pore.

prosopyle [G *zuführende Kammerpore*]: Opening through which water enters radial canal or choanocyte chamber in complex (leuconoid) sponge.

protriaene [G *Protriaen*]: Type of spicule (megasclere): triaene with one long ray and three short rays directed forward from long ray.

radial canal [G *Radialtube*]: Choanocyte-lined cavity of finger-shaped fold extending outward from spongocoel in syconoid sponge.

raphide [G *Raphid*]: Type of spicule: very thin, hair-like microrhabd.

reduction body [G *Reduktionskörper*]: Packet including various ameboid cells and somewhat dedifferentiated choanocytes and enclosed by layer of pinacocytes; formed under adverse conditions and later giving rise to new colony.

rhabd [G *Rhabd*]: Type of spicule: monaxonic/diactinal megasclere. Also, longer ray of triaene.

scleroblast [G *Skleroblast*]: Ameboid cell giving rise to sclerocyte.

sclerocyte [G *Sklerozyte, Sklerocyte*]: Ameboid cell giving rise, either singly or in groups, to spicules.

shaft [G *Schaft*]: Long axis of monaxon spicule; may be smooth or spiny, straight or curved. (See also head)

sigma [G *Sigma*]: Type of spicule: C- or S-shaped microsclere.

skeleton [G *Skelett*]: Supporting framework consisting of calcareous spicules, siliceous spicules, spongin fibers, or a combination of the latter two.

spheraster (sphaeraster) [G *Sphaeraster*]: Type of spicule (microsclere): euaster with conical rays emanating from spherical center.

spicule [G *Spikel, Spiculum, Sklerit, Nadel*]: Variously shaped and sized calcareous or siliceous skeletal elements produced by ameboid cells; occurring either singly or associated to varying degrees.

spiraster [G *Spiraster*]: Type of spicule (microsclere): metaster with spinose, somewhat spiral-shaped axis.

spongin [G *Spongin*]: Elastic, proteinaceous material reinforcing mesohyle (spongin A fibrils) and forming highly branched fiber network (spongin B); important skeletal element in complex leuconoid sponge.

spongocoel (atrium, paragaster) [G *Spongocoel*]: Central cavity; lined by choanocytes in asconoid sponge, giving rise to radial canals in syconoid sponge, and reduced to numerous excurrent canals in leuconoid sponge.

sterraster [G *Sterraster*]: Type of spicule (microsclere): euaster in which very numerous short rays emanate from large central sphere.

streptaster [G *Streptaster*]: Type of spicule (microsclere): metaster with spinose axis.

strongyle [G *Strongyl*]: Type of spicule (megasclere): monaxon with both ends rounded.

style [G *Styl*]: Type of spicule (megasclere): monaxon with one end rounded, the other pointed.

subtylostyle [G *Subtylostyl*]: Type of spicule (megasclere): monaxon resembling tylostyle, yet with less pronounced head.

sycon [G *Sycontypus*]: Intermediate stage in structural complexity of sponge in which body wall evaginates to form radial canals; radial symmetry maintained. (See also ascon, leucon)

tetraxon [G *Tetractine, Vierstrahler*]: Type of spicule: general term for megasclere with four rays emanating from common center (calthrop, triaene).

tornote [G *Tornote*]: Type of spicule (megasclere): monaxon with both ends shaped like a lance.

toxa [G *Toxe*]: Type of spicule: bow-shaped microsclere.

triaene [G *Triaen*]: Type of spicule (megasclere): general term for tetraxon with one long and three short, equal rays (ana-, dicho-, ortho-, plagio-, protriaene).

triaxon (hexactine) [G *Dreistrahler*]: Type of spicule: three- (simple triaxon) or six-ray (hexactinial spicule) megasclere.

tylaster [G *Tylaster*]: Type of spicule (microsclere): euaster with rays having knobbed ends.

tylostyle [G *Tylostyl*]: Type of spicule (megasclere): monaxon with one end pointed, the other knobbed (head).

tylote [G *Tyl, Doppelkeule)*]: Type of spicule (megasclere): monaxon with both ends knobbed.

SCYPHOZOA

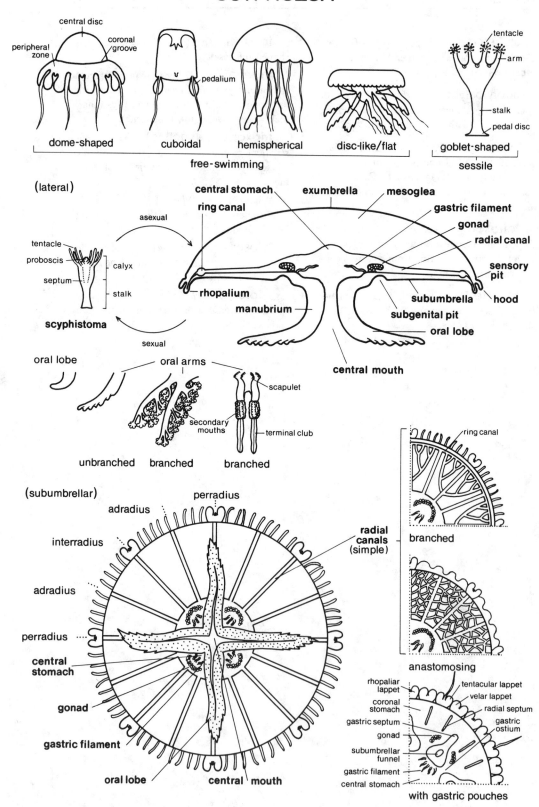

dome-shaped · central disc · peripheral zone · coronal groove

cuboidal · pedalium

hemispherical

disc-like/flat

goblet-shaped · tentacle · arm · stalk · pedal disc

free-swimming · sessile

(lateral)

asexual

tentacle · proboscis · septum · calyx · stalk

scyphistoma

sexual

central stomach · exumbrella · mesoglea · gastric filament · gonad · radial canal · sensory pit · hood

ring canal

rhopalium · manubrium · subumbrella · subgenital pit · oral lobe

central mouth

oral lobe

oral arms · scapulet · secondary mouths · terminal club

unbranched · branched · branched

(subumbrellar)

perradius · adradius · interradius · adradius · perradius

central stomach · gonad · gastric filament · oral lobe · central mouth

radial canals (simple)

ring canal

branched

anastomosing

rhopaliar lappet · coronal stomach · gastric septum · gonad · subumbrellar funnel · gastric filament · central stomach

tentacular lappet · velar lappet · radial septum · gastric ostium

with gastric pouches

CNIDARIA

SCYPHOZOA

scyphozoan [G *Scyphozoe*]

adhesive disc: pedal disc

adradius (adradial) [G *Adradius*]: One of eight radial axes halfway between each perradius and interradius; may correspond to projections (tentacles, rhopalia) of umbrellar margin.

anchor (colletocystophore) [G *Randanker, Colletocystophore*]: In sessile scyphozoan, well-developed, cushion-like rhopalioid containing mucus glands.

appendage [G *Anhang*]: One of numerous small projections of oral arms; bears nematocysts and mucus cells. (elongate, filiform, vesicular) (See also terminal club)

arm [G *Arm*]: In sessile scyphozoan, one of eight upwardly directed, adradial extensions of umbrellar margin; tip of each arm bears numerous minute, capitate tentacles. If short, may also be referred to as lobe. (simple, bifurcated) (See also oral arm)

arm canal (brachial canal) [G *Mundarmgefäß*]: Longitudinal gastrovascular canal within each oral arm; communicates with central stomach and opens to exterior via numerous suctorial mouths.

arm disc [G *Mundarmscheibe*]: Thick, disc-shaped, gelatinous projection of subumbrella. Mouth arms arise from lower side of arm disc.

bell: umbrella

brachial canal: arm canal

bud [G *Knospe*]: Lateral outgrowth of scyphistoma; develops asexually into new scyphistoma.

calyx [G *Calyx*]: Upper and wider of two divisions of scyphistoma (calyx, stalk); typically divided by longitudinal septa.

cathammatum: septal node

central stomach (stomach) [G *Zentralmagen*]: Undivided central part of gastrovascular system as distinguished from outer part which may be divided into gastric pouches or radial canals by septa. (See also coronal stomach)

ciliated groove [G *Mundrinne*]: Longitudinal, ciliated furrow running along middle of inner surface of oral arm.

circular canal: ring canal

claustrum [G *Claustrum*]: In sessile scyphozoan or medusa, partition parallel to umbrellar margin; divides each gastric pouch into inner and outer pocket.

collenchyme: mesoglea

colletocystophore: anchor

coronal groove [G *Coronalfurche, Ringfurche*]: Circular furrow extending around upper surface (exumbrella) of umbrella; clearly delimits marginal region from central region of bell.

coronal muscle [G *Ringmuskel*]: Circular band of muscle around periphery of umbrella; located in subumbrella and serves in swimming. (continuous, divided)

coronal sinus: coronal stomach

coronal stomach (coronary stomach, coronal sinus, gastrovascular sinus, ring sinus) [G *Coronaldarm, Kranzdarm, peripherische Darm*]: Outer part of gastrovascular system as distinguished from central part (central stomach); may be divided by septa.

coronary stomach: coronal stomach

delta muscle (deltoid muscle) [G *Deltamuskel*]: One in a series of radial muscle bands in subumbrella; spans from manubrium to circular coronal muscle or umbrellar margin and increases in width peripherally.

digitellus: gastric filament

ectoderm (epidermis) [G *Ektoderm*]: Outermost layer of body wall (ectoderm, mesoglea, endoderm) forming outer surface of exumbrella and subumbrella. Contains muscle, nerve, and gland cells as well as nematocysts.

endoderm (gastrodermis) [G *Entoderm*]: Flagellated innermost layer of body wall (ectoderm, mesoglea, endoderm). Lines manubrium and walls of gastrovascular system, forms gastric filaments, and gives rise to gonads.

ephyra [G *Ephyra*]: Young, free-swimming medusa stage typically developing from attached scyphistoma and occasionally directly from planula. Umbrellar margin typically characterized by eight bifurcated arms (primary lappets), each with medial rhopalium.

epidermis: ectoderm

excretory pore [G *Exkretionsporus*]: One in a series of small pores occasionally observed on subumbrellar surface at junction of radial canals with ring canal; radial canals open to exterior here.

exumbrella [G *Exumbrella*]: Upper (aboral) convex surface of bell (umbrella). (smooth, warty) (See also subumbrella)

eye [G *Auge, Linsenauge*]: Complex photosensitive organ consisting of cellular lens, retina-like arrangement of sensory cells, and epidermis cover. (See also ocellus)

festoon canal [G *Festonkanal*]: Narrow, ring-shaped canal running around umbrellar margin and extending into each marginal lappet.

filament: gastric filament

frenulum [G *Frenulum, Stützleiste*]: One of four perradial gelatinous folds supporting velarium and extending to subumbrellar surface.

gastric filament (filament, digitellus) [G *Gastralfilament, Digitellus, Digitulus*]: One of numerous tentacle-like structures projecting into central stomach either from free inner edge of each septum or from stomach floor. (See also phacella)

gastric ostium [G *Gastralostium*]: One of four openings in partitions separating central stomach from four gastric pouches. (See also septal ostium)

gastric pocket: gastric pouch

gastric pouch (gastric pocket) [G *Gastraltasche*]: One of four pocket-shaped extensions of central stomach; communicates with other gastric pouches via septal ostia, with central stomach via gastric ostia. (See also coronal stomach)

gastric septum: septum

gastrodermis: endoderm

gastrogenital membrane [G *Gastrogenitalmembran*]: Membrane separating central stomach from subgenital porticus. Consists of ectoderm, mesoglea, and endoderm.

gastrovascular cavity: central stomach

gastrovascular sinus: coronal stomach

genital sinus [G *Genitalsinus*]: Cavity between gonad and gastrogenital membrane.

gonad [G *Gonade*]: Aggregation of ♂ or ♀ reproductive cells; located in septate scyphozoan along sides of septa, in those lacking septa in groups on stomach floor.

(crescentic, horseshoe-shaped, oval, round, U-shaped, W-shaped)

gullet: manubrium

hood [G *Decklappen, Deckplatte*]: Small, hood-like extension of exumbrella over rhopalium; may bear outer and inner sensory pits basally on outer (exumbrellar) and inner (subumbrellar) sides.

hypostome: proboscis

interradius (interradial) [G *Interradius*]: One of four radial axes, 90° apart, lying between perradii; corresponds to position of septa and subumbrellar funnels as well as to projections of umbrellar margin (rhopalia, tentacles). (See also adradius)

lappet (marginal lappet) [G *Lappen*]: One in a series of lobe-like extensions around umbrellar margin; according to whether sensory structures (rhopalia) or tentacles are located between lappets one may distinguish rhopalial and tentacular lappets. (See also primary lappet, velar lappet)

lip: oral lip

lobe: arm, oral lobe

manubrium (gullet, pharynx) [G *Manubrium, Mundrohr, Magenstiel*]: Median quadrangular projection of subumbrella bearing terminal mouth; leads to central stomach. If well-developed, one may distinguish fused basal section (oral tube) and distal, elongate lobes (oral arms).

marginal lappet: lappet

mesoglea (mesogloea, collenchyme, mesolamella) [G *Mesogloea, Stützlamelle, Mesolamelle, Enchym, Pleurom, Collenchym*]: Middle layer of body wall between ectoderm and endoderm; typically thick, containing fibers and ameboid cells. Occasionally referred to as collenchyme because of presence of latter.

mesolamella: mesoglea

mouth [G *Mund*]: Four-cornered opening of central stomach at end of manubrium. May give rise to oral lobes or oral arms. (See also suctorial mouth)

mouth arm: oral arm

nematocyst: see Hydrozoa

nematocyst wart [G *Warze*]: One of numerous wart-like, nematocyst-bearing projections on upper surface (exumbrella) or manubrium of medusa.

nerve ring [G *Nervenring*]: Ring of nervous tissue around margin of umbrella; loops upward perradially to connect with rhopalia.

ocellus (pigment-spot ocellus, pigment-cup ocellus) [G *Ocellus, Becherauge*]: Small, simple photosensitive organ consisting either of pigment spot or pigment cup; typically associated with basal section of rhopalium at umbrellar margin. (See also eye)

olfactory pit: sensory pit

oral arm (mouth arm) [G *Mundarm, Mundfahne*]: One of four or eight elongate lobes with frilled edges extending from manubrium; bears nematocysts and numerous suctorial mouths. (branched, unbranched, three-winged, two-winged)

oral disc [G *Mundscheibe, Peristom*]: In polypoid stage (scyphistoma), flat oral surface with central, elevated mouth opening (proboscis) and marginal tentacles.

oral lip [G *Mundkrause*]: Folded, fringed margin of suctorial mouth on oral arm.

oral lobe (perradial lobe) [G *Mundlappen*]: One of four short, lobe-like projections at each corner of mouth or manubrium. (See also oral arm)

oral tube: manubrium

ostium: gastric ostium, septal ostium

pedal cyst: podocyst

pedal disc (adhesive disc) [G *Haftscheibe*]: Expanded, disc-shaped base of sessile scyphozoan or scyphistoma; contains gland cells and serves in attachment to substratum. May also produce podocysts in scyphistoma.

pedalium [G *Pedalium*]: Gelatinous basal expansion of tentacle at margin of umbrella.

peduncle: stalk

peristome: oral disc

peristomial pit [G *Peristomtrichter, Septaltrichter*]: On oral disc of scyphistoma, one of four shallow depressions associated with insertion of each longitudinal muscle. (See also subgenital pit, subumbrellar funnel)

perradial lobe: oral lobe

perradius (perradial) [G *Perradius*]: One of four main radial axes, 90° apart; correspond in position to oral lobes, but typically also to other major structures such as gastric pouches, radial canals, and structures along umbrellar margin. (See also adradius, interradius)

phacella (phacellus) [G *Phacella*]: Row of gastric filaments, one along each side of free inner edge of septum.

pharynx: manubrium

planula [G *Planula*]: Ciliated larva; attaches to substratum and develops into tentacle-bearing, polypoid stage (scyphistoma). (hollow, solid)

podocyst (pedal cyst) [G *Podozyste*]: At base of scyphistoma, chitinous cyst produced by pedal disc and containing ciliated larva.

primary lappet [G *Stammlappen*]: In ephyra, one of eight bifurcated arms projecting from umbrellar margin; each has median rhopalium and develops into rhopalial lappet of adult.

proboscis [G *Mundkegel, Hypostom, Proboscis*]: At center of oral disc of scyphistoma, cone-shaped projection bearing terminal mouth.

radial canal [G *Radialkanal*]: In scyphozoan lacking gastric pouches, one in a series of narrow canals extending from central stomach to rim of umbrella or to ring canal if present; may be associated with so-called excretory pore. Termed radial channel if larger in diameter. (anastomosed, branched, simple)

radial channel: radial canal

radial septum: septum

rhopalar canal [G *Rhopalienkanal*]: Small diverticulum of gastrovascular system extending in to rhopalium.

rhopalar lappet: rhopalial lappet

rhopalial lappet (rhopalar lappet) [G *rhopalarer Lappen*]: Pair of small, rudimentary lappets flanking rhopalium on umbrellar margin. (See also tentacular lappet, velar lappet)

rhopalioid [G *Rhopalioid*]: In sessile scyphozoan, short perradial and interradial projection (reduced tentacle) of exumbrellar margin between drawn-out adradial arm. (lobe-like, tentacle-like) (See also anchor)

rhopalium [G *Rhopalium, Randsinnesorgan*]: Small, hollow, club-shaped sensory structure projecting from umbrellar margin; flanked by rhopalial lappet and covered by hood. Contains statoliths distally.

ring canal [G *Ringkanal*]: In scyphozoan with reduced gastric pouches, continuous circular tube around umbrellar margin; connected to central stomach via radial canals.

ring sinus: coronal stomach

scapulet (epaulette, shoulder ruffle) [G *Scapulette, Schulterkrause*]: On proximal outer surface of oral arm, fringed outgrowth bearing numerous suctorial mouths.

scyphistoma [G *Scyphistoma, Scyphopolyp*]: Tentacle-bearing polypoid stage developing after attachment of planula to substratum; consists of pedal disc, stalk, peristome, and calyx with tentacles. Gastrovascular cavity typically divided by four longitudinal septa. (See also strobila)

scyphomedusa [G *Scyphomeduse*]: General term for medusa stage of scyphozoan. (See also scyphopolyp)

scyphopolyp [G *Scyphopolyp*]: General term for polyp stage of scyphozoan. (See also scyphistoma, scyphomedusa, strobila)

secondary mouth: suctorial mouth

sensory pit (olfactory pit) [G *Sinnesgrube*]: Small sensory depression associated with base of rhopalium; according to position one may distinguish an outer (exumbrellar) and inner (subumbrellar) sensory pit.

septal muscle [G *Septalmuskel*]: Longitudinal (oral-aboral axis) muscle band in septum.

septal node (cathammatum, cathammal plate) [G *Septalknoten, Cathamme, Kathammalleiste, Kathammalplatte*]: One of four small interradial areas of fusion between exumbrella and subumbrella; separates central stomach from coronal stomach and represents reduced gastric septum.

septal ostium [G *Septalostium*]: Circular opening in peripheral region of septum; permits gastric pouches to communicate with one another.

septum [G *Septum, Täniole*]: Partition spanning between exumbrella and subumbrella. Typically refers to one of four interradial partitions dividing gastrovascular system into central stomach and four marginal gastric pouches; may bear opening (septal ostium) and be penetrated by subumbrellar funnel. May be termed gastric septum to differentiate from more numerous radial septa in outer part of gastrovascular system (coronal stomach).

shoulder ruffle: scapulet

stalk (peduncle) [G *Stiel, Pedunculus*]: In sessile scyphozoan, slender aboral division of body; attached to substratum by pedal disc. May also refer to stalk of scyphistoma.

statocyst [G *Statozyste*]: Term applied to mass of cells, each bearing statolith, at distal end of rhopalium.

statolith [G *Statolith*]: Minute, solid, movable body within one of numerous cells at distal end of rhopalium.

stolon [G *Stolo*]: Hollow, tube-like projection from stalk or base of scyphistoma; gives rise asexually to new scyphistomae.

stomach: central stomach

strobila [G *Strobila*]: Stage of sessile polypoid form (scyphistoma) in which series of young medusae (ephyrae) are being produced simultaneously (polydisc strobilation) or singly (monodisc strobilation) by transverse constriction(s).

subgenital ostium: subgenital porticus

subgenital pit [G *Subgenitaltasche, Trichterhöhle, Trichtergrube*]: One of four shallow depressions on interradius of subumbrella. May fuse to form subgenital porticus. (See also subumbrellar funnel)

subgenital porticus [G *Subgenitalportikus, Subgenitalraum*]: Cross-shaped cavity formed by fusion of four subgenital pits. Located on subumbrellar surface between central stomach and bases of oral arms (original mouth degenerates); four original openings (subgenital ostia) remain.

subumbrella [G *Subumbrella*]: Lower (oral) concave surface of medusa body (umbrella).

subumbrellar funnel (peristomial pit) [G *Septaltrichter, Trichter*]: One of four deep, funnel-shaped pits in subumbrella; located on interradius and therefore projecting into septum dividing gastrovascular system. (See also subgenital pit)

subumbrellar sac [G *Umbralsack*]: Sac-like enlargement of proximal section of gastric pouch; hangs down into subumbrellar cavity.

suctorial mouth (secondary mouth) [G *Nebenmundöffnung, Haarröhrchenmund, Saugöffnung, Sekundärostium*]: On oral arm, one of numerous communications between longitudinal groove along inner surface of arm and internal arm canal. Opening to exterior surrounded by oral lips.

tentacle [G *Tentakel*]: One in a series of elongate processes around umbrellar margin, either on or between lappets or projecting from subumbrellar surface. In sessile scyphozoan, one of numerous minute, capitate tentacles at tip of each arm. (hollow, solid; capitate, filiform; single, clustered)

tentacular lappet [G *tentakulärer Lappen*]: Pair of lappets flanking tentacle on umbrellar margin. (See also rhopalial lappet, velar lappet)

terminal club (terminal appendage) [G *Endzapfen*]: Long, club-shaped appendage hanging from each oral arm. (See also appendage)

umbrella (bell) [G *Umbrella, Schirm, Glocke*]: Main body or bell of medusa; upper surface termed exumbrella, lower surface subumbrella. Typically with thick mesoglea and marginal lappets. (conical, cuboidal, disc-like/flattened, dome-shaped, hemispherical, pyramidal, saucer-shaped)

velarium (velum) [G *Velarium*]: Horizontal fold projecting inwardly from umbrella margin; consists only of subumbrellar tissue and may be supported by four frenula.

velar lappet [G *Randlappen*]: In scyphomeduse lacking marginal tentacles, type of lappet on umbrellar margin between rhopalial lappets. (See also tentacular lappet)

velum: velarium

wart: nematocyst wart

CNIDARIA

HYDROZOA

hydrozoan (hydropolyp, hydromedusa)
[G *Hydrozoe (Hydropolyp, Hydromeduse)*]

HYDROZOA (hydropolyp)

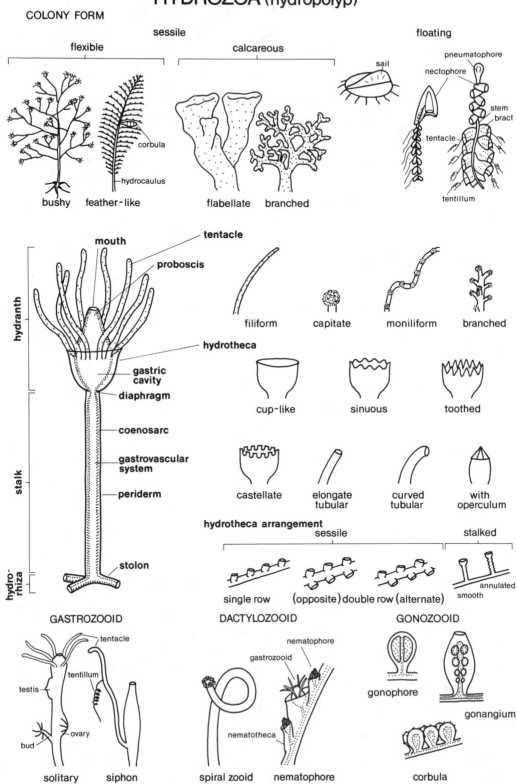

COLONY FORM

sessile

flexible

calcareous

floating

bushy   feather-like

corbula

hydrocaulus

flabellate   branched

sail

pneumatophore

nectophore

stem

bract

tentacle

tentillum

mouth

tentacle

proboscis

hydrotheca

gastric cavity

diaphragm

coenosarc

gastrovascular system

periderm

stolon

hydranth

stalk

hydrorhiza

filiform   capitate   moniliform   branched

cup-like   sinuous   toothed

castellate   elongate tubular   curved tubular   with operculum

hydrotheca arrangement

sessile

stalked

single row   (opposite) double row (alternate)

annulated

smooth

GASTROZOOID

tentacle

tentillum

testis

ovary

bud

solitary   siphon

DACTYLOZOOID

nematophore

gastrozooid

nematotheca

spiral zooid   nematophore

GONOZOOID

gonophore

gonangium

corbula

HYDROZOA (hydromedusa)

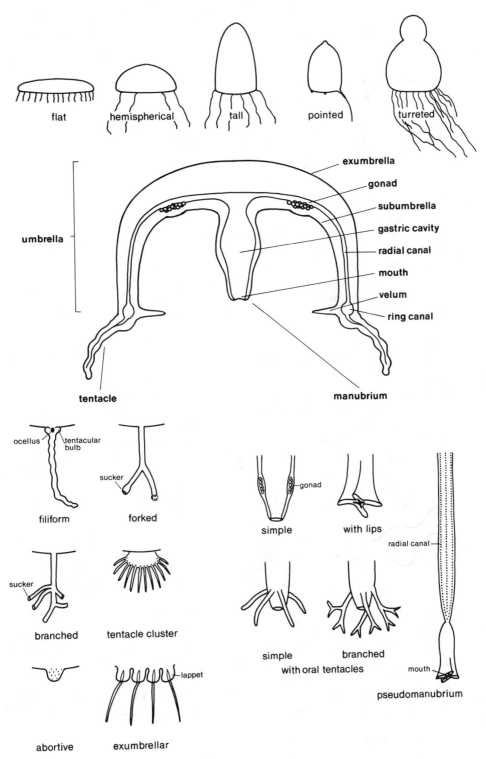

flat   hemispherical   tall   pointed   turreted

exumbrella

gonad

subumbrella

gastric cavity

radial canal

mouth

velum

ring canal

umbrella

tentacle

manubrium

ocellus — tentacular bulb

sucker

filiform   forked

sucker

branched   tentacle cluster

abortive   exumbrellar

lappet

gonad

simple   with lips

radial canal

simple   branched

with oral tentacles

mouth

pseudomanubrium

CNIDARIA

HYDROZOA

hydrozoan (hydropolyp, hydromedusa)
[G *Hydrozoe (Hydropolyp, Hydromeduse)*]

acrocyst [G *Marsupium*]: Brood chamber extending from gonophore.

acrophore [G *Acrophore*]: Type of nematocyst: rhopaloneme whose club-shaped tubule bears apical projection. (See also anacrophore)

actinula [G *Actinula, Actinula-stadium*]: Tentacle-bearing larva that creeps along substratum before attaching. (See also planula)

adhesive pad (sucker) [G *Klebkisse*]: In medusa, adhesive structure lacking nematocysts near tip of tentacle; tentacle may bend here to form "knee."

adradius [G *Adradius*]: In medusa, one of eight radial axes or sectors lying between each perradius and interradius.

air sac: pneumatosaccus

amastigophore [G *Amastigophore*]: Type of nematocyst: rhabdoid whose tubule does not continue beyond butt. (macrobasic, microbasic) (See also mastigophore)

ampulla [G *Ampulle*]: In hydrozoan colony with massive calcareous skeleton, rounded chamber in outer surface (coenosteum) containing gonophore; gives rise to medusa. Also refers to nematocyst-free sac at end of either spiral-shaped cnidoband or tentacle branch (tentillum) in floating colony.

anacrophore [G *Anacrophore*]: Type of nematocyst: rhopaloneme whose club-shaped tubule lacks apical projection. (See also acrophore)

anisorhiza [G *Anisorhize*]: Type of nematocyst: haploneme whose tubule is slightly dilated toward base. Tip of tubule open (stomocnidae). (atrichous, heterotrichous, homotrichous)

annulus [G *Ringelung*]: One in a series of rings in periderm; typically in groups directly below hydranth or at point of branching of stalk.

anthomedusa [G *Anthomeduse*]: Medusa stage of athecate hydrozoan; characterized by position of gonads along manubrium and absence of statocysts on umbrellar margin. Umbrella typically tall. (See also leptomedusa)

apical chamber [G *apikaler Kammer*]: In medusa, small aboral chamber at apex of gastrovascular cavity. (bead-like, pointed bead, spinous bulb, stalked bulb)

aspirotele [G *Aspirotele*]: Type of nematocyst: microbasic eurytele in which butt bears three strong spines and tubule does not continue beyond butt. (See also spirotele)

astomocnida [G *Astomocnide*]: One of two major categories of nematocyst; type in which tip of tubule is closed. (See also stomocnida)

aurophore [G *Aurophor*]: In floating colony, constricted-off section of float (pneumatophore); resembles and often closely adjoins swimming bells (nectophores).

base [G *Polypenbasis*]: Proximal division of polyp (base, stalk, hydranth); forms glandular site of attachment (pedal disc) in solitary hydrozoan and, in colonial forms, gives rise to root-like stolons.

basket: corbula

battery [G *Nesselbatterie*]: Aggregation of various types of nematocysts within single ectoderm cell on tentacle of polyp; often in groups.

bell: umbrella

birhopaloid [G *Birhopaloid*]: Type of nematocyst (heteroneme): butt of unequal diameter at distal and proximal ends.

blastostyle [G *Blastostyl*]: In sessile colony, modified gastrozooid bearing series of gonophores.

bract (hydrophyllium, phyllozooid) [G *Deckblatt*]: In floating colony, one of numerous thick, gelatinous zooids with gastrovascular canal; considered to provide buoyancy and protect colony. (helmet-shaped, leaf-like, prismatic)

bud [G *Knospe*]: Outgrowth of body wall (coenosarc) in stalk or stolon; develops asexually into new polyp or medusa (medusa bud).

butt (shaft) [G *Schaft*]: Enlarged basal portion of tubule of nematocyst; may bear stylet and/or spines. (diameter: uniform, not uniform)

capsule [G *Kapsel, Kapselwand*]: Capsule wall of nematocyst; contains tubule and bears distal operculum. Thick and double-walled in nematocyst, thin and consisting of single layer in spirocyst. (elongated, oval, spherical, pyriform)

caulome: rhizocaulome

centripetal canal [G *Zentripetalkanal*]: In medusa, one in a series of blind canals extending toward summit of umbrella from ring canal.

clasper [G *Halter, Blastostyl*]: Elongate projection on body wall; bears terminal sucker and grasps eggs emitted from gonophores.

cnidoband [G *Nesselknopf*]: In floating colony, structure on tentacle branch (tentillum); bears dense aggregations of several different types of nematocysts and gives rise to end filament. (coiled/spiral-shaped, oval)

cnidoblast [G *Cnidoblast*]: Developing cnidocyte; often used synonymously with cnidocyte.

cnidocil [G *Cnidocil*]: Bristle-shaped projection adjacent to operculum at distal end of cnidocyte; serves as trigger to discharge tubule.

cnidocyte (nematocyte) [G *Nesselzelle, Cnidozyte, Nematozyte*]: Specialized cell type in ectoderm. Consists of basal nucleus and distal cnidocil and contains nematocyst internally; cell wall contains supporting rods. Typically concentrated on tentacles.

cnidome [G *Cnidom*]: Entire complement of nematocyst types in one species.

cnidophore [G *Cnidophore, Nesselbatterie*]: In medusa, one in a series of nematocyst-filled structures attached to tentacles by elongate, contractile stalks.

cnidothylacium [G *Cnidothylacie*]: Cluster of nematocysts enclosed in endoderm canals in umbrella of medusa.

coenosarc [G *Coenosark*]: Three-layer tissue (ectoderm, mesolamella, endoderm) forming hollow tube (gastrovascular cavity) in stolon, stalk, or hydrocaulus; typically covered by periderm. May also refer to stem of floating colony.

coenosteum [G *Coenosteum*]: Surface of hydrozoan colony with massive calcareous skeleton; bears pores for gastrozooids (gastropores) and dactylozooids (dactylophores).

column [G *Polypenkörper*]: Body of solitary polyp; may be subdivided into distal tentacle-bearing hypostome, gastric region, stalk, and proximal pedal disc.

conaria [G *Conaria*]: Early larval stage of floating colony with disc-shaped float (pneumatophore); consists of hollow sphere with aboral thickening. (See also rataria)

coppinia [G *Coppinia*]: Dense aggregation of gonangia and dactylozooids.

corbula (basket) [G *Corbula*]: Highly developed type of phylactocarp in which series of leaf-like hydrocladium branches arise to each side of gonangia. (open, closed)

cordylus (sensory club) [G *Randkolbe*]: In medusa, one in a series of small, club-shaped sensory organs between tentacle bases along umbrella margin; composed of large endoderm cells surrounded by thin ectoderm and does not contain statolith. (See also lithostyle)

cormidium [G *Cormidium*]: In floating colony, one in a series of units containing different types of zooids attached to stem; each unit typically consists of bract, trophozooid (siphon), and gonophore.

craspedon: velum

cryptomedusoid [G *Cryptomedusoid*]: In four recognized stages of medusoid reduction, type of medusoid lacking tentacles, mouth, manubrium, velar invagination, and radial canals. (See also eumedusoid, heteromedusoid, styloid)

cyclosystem [G *Zyklosystem*]: In surface layer (coenosteum) of hydrozoan colony with massive calcareous skeleton, arrangement of pores in which several dactylopores surround central gastropore; may be covered by lid.

dactylopore [G *Dactyloporus*]: In surface layer (coenosteum) of hydrozoan colony with massive calcareous skeleton, smaller pore housing dactylozooid; may be borne on projection. (See also gastropore)

dactylozooid (tentaculozooid, machozooid, protective polyp) [G *Dactylozooid, Wehrpolyp*]: In colonial hydrozoan, type of individual serving to pro-

tect colony or stun prey; equipped with nematocysts, yet mouth, tentacles, and gastric cavity are reduced or lost. (See also nematophore, palpon, spiral zooid)

desmoneme (volvent) [G *Desmoneme*]: Type of nematocyst: haploneme whose tubule forms corkscrew-like coils. Tip of tubule closed (astomocnidae). (pyriform, spherical)

diaphragm [G *Diaphragma*]: Inwardly projecting chitinous shelf at base of hydrotheca; serves as site of attachment of hydranth and hydrotheca. (circular, one-sided)

ectoderm (epidermis) [G *Ektoderm*]: Outermost cellular layer of body wall (ectoderm, mesoglea, endoderm). Contains muscle, interstitial, nerve, and mucus-secreting cells as well as nematocysts. May secrete periderm in polypoid forms; constitutes outer layer of exumbrella and subumbrella in medusa.

embryonic theca [G *Embryothek, Oothek*]: Shell-like structure secreted by embryo developing on column of solitary polyp. (smooth, spiny; spherical, planoconvex)

end filament [G *Endfilament*]: In floating colony, slender, nematocyst-bearing projection hanging from cnidoband on tentacle branch (tentillum).

endoderm (gastrodermis) [G *Endoderm, Entoderm*]: Innermost flagellated cellular layer of body (ectoderm, mesoglea, endoderm). In polyp, lines gastrovascular cavity and either lines or forms core of tentacles; in medusa, lines manubrium and forms wall of radial and ring canals.

entocodon [G *Glockenkern*]: Cavity formed by invagination of ectoderm at distal end of gonophore during production of medusa or more complex type of medusoid; develops into subumbrella.

epidermis: ectoderm

eudoxid (eudoxome) [G *Eudoxid*]: Free-swimming unit of different types of zooids (cormidium) that has detached from stem of floating colony.

eudoxome: eudoxid

eumedusoid [G *Eumedusoid*]: In four recognized stages of medusoid reduction, type of medusoid most closely resembling medusa; endoderm forms four radial tubes, yet tentacles, mouth, and occasionally manubrium are lacking. Sex cells on manubrium or radial canals. (See also cryptomedusoid, heteromedusoid, styloid)

eurytele [G *Eurytele*]: Type of nematocyst (rhopaloid): butt dilated distally. Tip of tubule either open (stomocnidae) or closed (astomocnidae). (homotrichous microbasic, heterotrichous microbasic, holotrichous macrobasic, merotrichous macrobasic, microbasic, semiophoric microbasic, telotrichous macrobasic) (See also stenotele)

exumbrella [G *Exumbrella*]: In medusa, upper (aboral) convex surface of umbrella. (See also subumbrella)

feeler: palpon

festoon canal [G *Festonkanal, peroniales Ringkanalsystem*]: In medusa, term applied to reduced ring canal extending into each lappet of umbrellar margin.

fishing tentacle: tentacle

float: pneumatophore

frustule [G *Frustel*]: Nonciliated, planula-like larva; constricts off from polyp, creeps along substratum, and develops into polyp.

funnel (trichter) [G *Trichter*]: In floating colony, expanded chamber at bottom of gas-filled air sac (pneumatosaccus) in float; ectoderm lining is modified into gas gland.

gas gland [G *Gasdrüse*]: In floating colony, modified ectoderm lining lower chamber (funnel) of gas-filled air sac (pneumatosaccus) in float; may be branched, with branches terminating in giant cells. Secretes gas into pneumatosaccus.

gastral region [G *Gastralregion*]: Expanded basal section of hydranth (proboscis, gastral region) bearing gastrovascular cavity; may be joined to stalk by neck.

gastric pocket: gastric pouch

gastric pouch (gastric pocket) [G *Magentasche*]: In medusa, one of four (or multiple thereof) enlarged pockets of gastrovascular cavity.

gastrodermis: endoderm

gastropore [G *Gastroporus*]: In surface layer (coenosteum) of hydrozoan colony with massive calcareous skeleton, larger pore housing gastrozooid. (See also dactylopore)

gastrovascular cavity (gastric cavity, stomach) [G *Gastralraum, Magenraum*]: Simple (undivided) enlargement of gastrovascular system; located in lower region of hydranth or, in medusa, in or at aboral region of manubrium.

gastrozooid: trophozooid

glutinant: isorhiza

gonad [G *Gonade*]: ♂ or ♀ reproductive organs. In polyp, typically located near gastral region, also as various form of medusoid, or contained in modified zooids (gonozooids). In medusa, along radial canals (leptomedusa) or manubrium (anthomedusa).

gonangium [G *Gonangium*]: Reproductive unit consisting of outer gonotheca and enclosed blastostyle bearing series of gonophores.

gonodendrum [G *Gonophorentraube, Geschlechtstraube*]: Type of gonozooid in floating colony. Consists of branched stalk bearing clusters of gonophores; may be associated with gonopalpons.

gonopalpon [G *Genitaltaster*]: In floating colony, tentacle-like dactylozooid (palpon) associated with gonophores.

gonophore [G *Gonophor*]: In colonial hydrozoan, asexually developing reproductive bud arising either from hydranth, hydrocaulus, stalk, stolon, or gonozooid; gives rise to medusa, medusoid, or sex cells. May be enclosed by periderm (gonotheca).

gonosome [G *Gonosom*]: Collective term for all reproductive structures of colonial hydrozoan. (See also trophosome)

gonotheca [G *Gonothek*]: Periderm case around blastostyle; bears distal opening. (smooth, ribbed, spined)

gonozoid [G *Gonozooid, Geschlechts polyp*]: In colonial hydrozoan, type of individual giving rise to reproductive bodies (gonophores). (male, female)

haploneme [G *Haploneme*]: Type of nematocyst: general term for type whose tubule lacks dilated basal section (butt). Tip of tubule may be opened (stomocnidae) and include isorhizas and anisorhizas or be closed (astomocnidae) and include desmonemes. (See also heteroneme)

heteromedusoid [G *Heteromedusoid*]: In four recognized stages of medusoid reduction (eumedusoid, cryptomedusoid, heteromedusoid, styloid), type of medusoid in which endoderm retains simple form of original invagination, yet cavity (entocodon) is present.

heteroneme [G *Heteroneme*]: Type of nematocyst: general term for type whose tubule is differentiated into dilated basal section (butt) and thin distal section. Tip of tubule may be opened (stomocnidae) and include mastigophores, amastigophores, euryteles, and stenoteles or be closed (astomocnidae) and include euryteles, spiroteles, and aspiroteles. (See also haploneme)

holdfast [G *Wurzelhaare*]: One of many filiform structures that project from base of stalk and serve to anchor solitary polyp in sediment.

hydranth [G *Hydranth*]: Distal division of polyp (base, stalk, hydranth) bearing tentacles, terminal mouth, and gastrovascular cavity. (bottle-shaped, cylindrical, elongated, vase-like)

hydrocaulus [G *Hydrocaulus*]: Main stalk of sessile colony; typically bears branches with hydranths. (straight, zigzag; monosiphonic, polysiphonic/fascicled)

hydrocladium [G *Seitenachse*]: Lateral hydranth-bearing branch of hydrocaulus.

hydroecium [G *Hydroecium*]: In floating colony, sheath-like extension of swimming bell (nectophore) enclosing uppermost section of stem.

hydromedusa (medusa) [G *Hydromeduse, Meduse*]: Free-swimming sexual hydrozoan individual (as opposed to polyp) typically consisting of umbrella, manubrium, and tentacles; medusa of thecate hydrozoan termed leptomedusa, that of athecate forms anthomedusa.

hydrophyllium: bract

hydropolyp (polyp) [G *Hydropolyp, Polyp*]: General term for sessile hydrozoan individual (as opposed to hydromedusa). (See also zooid)

hydrorhiza [G *Hydrorhiza*]: Network of stolons serving to attach sessile colonies to substratum.

hydrotheca [G *Hydrothek*]: Expanded distal section of periderm partially or entirely surrounding hydranth; attached to base of hydranth by diaphragm. May bear operculum. (sessile, stalked; with/without operculum; elongate tubular, curved tubular, cylindrical, cup-like; rim: smooth, sinuous, toothed, castellate, crenulate)

internode [G *Internodium*]: Section of stalk between successive buds, zooids, or branches in sessile colony. In floating colony, section of stem between units of different zooids (cormidia).

interradius [G *Interradius*]: In medusa, one of four radial axes, 90° apart, lying between perradii; may also refer to entire sector between perradius. (See also adradius)

involucre [G *Involucrum*]: In floating colony, protective cap at origin of spiral-shaped type of cnidoband on tentacle branch (tentillum).

isorhiza (glutinant) [G *Isorhiza, Glutinant*]: Type of nematocyst: haploneme whose tubule is of same diameter throughout.

Tip of tubule open (stomocnidae). (apotrichous, atrichous, basitrichous, holotrichous, merotrichous) (See also anisorhiza)

lappet [G *Flügellappen*]: In medusa, one in a series of lobe-like extensions around umbrellar margin.

leptomedusa [G *Leptomeduse*]: Medusa stage of thecate hydrozoan; characterized by position of gonads along radial canals, presence of statocysts, and often numerous tentacles. Umbrella typically flat. (See also anthomedusa)

lid: operculum

lip [G *Lippe*]: One of typically four or eight lobe-like extensions of manubrium surrounding mouth in medusa. (See also oral tentacle)

lithocyst: statocyst

lithocyte [G *Lithozyte*]: In medusa, cell type containing statolith; located in either statocyst or lithostyle; closely associated with sensory cells.

lithostyle (tentaculocyst, sense club) [G *Lithostyl, Statolithenorgan*]: In medusa, one in a series of minute, club-shaped organs of equilibrium suspended from umbrellar margin between tentacle bases. Contains one to several lithocytes, each enclosing a statolith. (See also cordylus)

liver [G *Leber*]: In floating colony with disc-shaped float (pneumatophore), term occasionally applied to system of endoderm canals above main, central trophozooid.

machozooid: dactylozooid

manubrium [G *Manubrium*]: In medusa, median tubular projection of subumbrella bearing terminal mouth and either leading into or containing gastrovascular cavity. May also refer to proboscis in polyp. (cylindrical, quadrangular; simple, with lips, with oral tentacles)

mastigophore [G *Mastigophore*]: Type of nematocyst: rhabdoid whose tubule continues beyond butt. (macrobasic, microbasic) (See also amastigophore)

medusa: hydromedusa

medusoid [G *Medusoid*]: Reproductive bud (gonophore) that does not become a free-swimming medusa, but remains attached. According to degree of reduction of medusa features, one may distinguish eumedusoid, cryptomedusoid, heteromedusoid, or styloid.

mesoglea (mesogloea) [G *Mesogloea*]: Middle noncellular layer of body wall. In polyp, thin, lamella-like layer (mesolamella); in medusa, thicker gelatinous layer.

mesolamella [G *Stützlamelle*]: In polyp, thin, lamella-like mesoglea; in medusa, lamella-like layer separating gelatinous mesoglea from both endoderm and ectoderm.

mouth [G *Mund, Mundöffnung*]: Opening of gastrovascular system to exterior. In polyp, at end of proboscis; in medusa, at end of manubrium. Serves also as anus.

mouth tentacle: oral tentacle

neck [G *Hals*]: Constricted section of polyp between hydranth and stalk; also narrowed lower section of hydrotheca.

nectocalyx: nectophore

nectophore (swimming bell, nectocalyx) [G *Nectophor, Schwimmglocke*]: In floating colony, one of numerous muscular, medusoid zooids with bell, velum, radial canals, and ring canal yet without mouth, manubrium, tentacles, or sense organs. Serves in propulsion of colony.

nectosome [G *Nectosom*]: In floating colony, upper section of stem attached to float (pneumatophore) and bearing series of swimming bells (nectophores). (See also siphonosome)

nematocyst [G *Nematozyste, Cnidozyste, Cnide, Nesselkapsel*]: Organelle within cnidocyte; consists of thick, double-walled capsule with coiled tubule and distal operculum. Nematocysts basically classified according to whether end of tubule is open (stomocnidae) or closed (astomocnidae), whether discharged tubule is regionated into dilated proximal butt and tubule proper (heteroneme) or of equal diameter through-out (haploneme), and according to armature of butt and tubule. (See also acrophore, amastigophore, anacrophore, anisorhiza, as-pirotele, birhopaloid, desmoneme, eurytele, isorhiza, mastigophore, rhabdoid, rhopaloid, rhopaloneme, spirotele, stenotele)

nematocyte: cnidocyte

nematophore   (sarcostyle) [G *Nematophor*]: One of typically three small dactylozooids surrounded by theca (nematotheca) and closely adjoining each trophozooid. (capitate, club-shaped)

nematotheca [G *Nematothek, Nematotheca*]: Periderm case surrounding nematophore; typically closely adjoining case (hydrotheca) of trophozooid.

nerve ring [G *Nervenring*]: In medusa, nerve ring around umbrellar mar-gin: one above, a second below attachment point of velum. Upper ring connected to statocysts, lower ring responsible for pulsa-tion.

nettle ring [G *Nesselring*]: In medusa, dense band of nematocysts encir-cling umbrella near margin.

node [G *Nodium*]: In sessile colony, constricted section of stalk marking junction of two internodes.

nutritive polyp: trophozooid

ocellar bulb: tentacular bulb

ocellus [G *Ocellus, Ocelle*]: In medusa, series of small, simple photosensi-tive organs, one in tentacular bulb at base of each tentacle. (black, brown, red)

operculum (lid) [G *Operculum, Deckel, Deckelapparat*]: Lid-like structure of hydrotheca; consists of one to several ele-ments and closes over contracted hydranth. May also refer to lid covering opening of nematocyst.

oral cone: proboscis

oral tentacle (mouth tentacle) [G *Mundtentakel, Mundarm, oraler Tenta-kel*]: In medusa, one of typically four tenta-cle-like projections of manubrium surrounding mouth. (simple, branched) (See also lip)

otocyst: statocyst

otoporpa [G *Otoporpa*]: In medusa, short tract of nematocysts extending upward from each lithostyle at umbrella margin.

ovary [G *Ovarium, Ovar*]: In solitary polyp, large ectodermal bulge bearing egg; typically located in proximal section of gas-tral region.

palpon (feeler, taster) [G *Palpon, Taster*]: Dactylozooid of floating colony; ten-tacle-shaped or resembling trophozooid (si-phon) yet bearing single, unbranched tentacle. (See also gonopalpon)

pedal disc [G *Fuscheibe, Haft-scheibe*]: Adhesive base of solitary polyp.

pedicel: stalk

peduncle: pseudomanubrium

penetrant: stenotele

periderm (perisarc) [G *Periderm, Perisark*]: Nonliving, typically thin and transparent chitinous outermost layer sur-rounding stalk and hydrocaulus and forming gonotheca and hydrotheca; secreted by ecto-derm. Degree to which periderm surrounds hydranth serves as basis for distinguishing thecate from athecate hydrozoans.

perisarc: periderm

peronium [G *Peronium, Spange*]: In medusa, continuation of ectoderm from inner side of tentacle (when this is positioned on exumbrella) down to umbrella margin; con-tains nematocysts, muscle, and nerves. (See also tentacular root)

perradius [G *Perradius*]: In medusa, one of four major radial axes; located at right angles from one another and corresponding to main structures (radial canals, yet typi-cally also tentacles and sense organs). (See also adradius, interradius)

person: zooid

phorocyte [G *Phorozyte*]: In medusa, nurse cell in parent medusa within which larval development takes place.

phylactocarp [G *Phylactocarp*]: In sessile colony, modified branch (hydrocladium) or accessory branchlets of hydrocladium serving to protect gonangia by covering them and bearing nematophores. (See also corbula)

phyllozooid: bract

planula [G *Planula*]: Ciliated, free-swimming larva; lacks mouth yet contains differentiated cells and, in older stage, may contain gastrovascular cavity. Attaches to substratum by broader anterior end. (See also actinula, conaria, rataria)

pneumatocodon [G *Luftschirm*]: In floating colony, outer (exumbrellar) wall of float (pneumatophore); separated from inner wall (pnematosaccus) by gastrovascular space.

pneumatophore (float) [G *Pneumatophor*]: In floating colony, highly modified, gas-filled zooid serving as float; consists of outer (pneumatocodon) and inner (pneumatosaccus) walls separated by gastrovascular chamber.

pneumatopore [G *Porus, apikaler Porus*]: In floating colony, small opening of gas-filled air sac (pneumatosaccus) to exterior at apex of float.

pneumatosaccus (air sac) [G *Luftsack*]: In floating colony, inner (subumbrellar) wall of float (pneumatophore); separated from outer wall (pneumatocodon) by gastrovascular space and typically lined by chitinous layer.

polyp: zooid, hydropolyp

proboscis (hypostome, oral cone, manubrium) [G *Proboscis, Hypostom, Mundkegel*]: Elongate, often narrowed distal section of hydranth bearing terminal mouth. (conical, cylindrical, globular)

protective polyp: dactylozooid

pseudomanubrium (peduncle, peduncle-manubrium) [G *Magenstiel*]: In medusa, elongate median extension of subumbrella bearing radial canals internally and manubrium terminally.

radial canal [G *Radiärkanal, Radialkanal*]: In medusa, one of typically four (occasionally 6, 8, 12, or 16) tubes extending from central gastrovascular cavity to ring canal at rim of umbrella; may bear gonads. Position of four canals corresponds to perradii. (simple, branched)

rataria [G *Rataria*]: Later larval stage of floating colony with disc-shaped float (pneumatophore); consists of more elongate body surrounded by rim of growing disc. (See also conaria)

rhabdoid [G *Rabdoid*]: Type of nematocyst (heteroneme): general term for type of tubule whose dilated basal section (butt) is cylindrical (of same diameter throughout). Tip of tubule open (stomocnidae). Includes mastigophores and amastigophores. (See also rhopaloid)

rhizocaulome (caulome) [G *Rhizocaulom*]: In sessile colony, vertical, stalk-like structure formed by bundle of parallel stolons.

rhizome: stolon

rhopaloid [G *Rhopaloid*]: Type of nematocyst (heteroneme): general term for type of tubule whose dilated basal section (butt) is of unequal diameter. Tip of tubule may be open (stomocnidae) and include euryteles, spiroteles, and aspiroteles or be closed (astomocnidae) and include euryteles and stenoteles. (See also rhabdoid)

rhopaloneme [G *Rhopaloneme*]: Type of nematocyst: tubule modified into elongated sac and much greater in volume than capsule. Tip of tubule closed (astomocnidae). Includes acrophores and anacrophores.

ring canal [G *Ringkanal*]: In medusa, circular tube around umbrella margin; connected to central gastrovascular cavity via four radial canals.

sail [G *Segel*]: Thin, erect, sail-like structure extending across upper surface of disc-shaped float (pneumatophore) in floating colony.

sarcostyle: nematophore

sensory club: cordylus

shaft: butt

siphon [G *Siphon, Gasterozooid, Freßpolyp*]: Trophozooid of floating colony; gives rise at its base to single hollow, contractile tentacle with lateral branches (tentilla).

siphonosome [G *Siphosom*]: In floating colony, lower section of stem bearing series of cormidia. (See also nectosome)

somatocyst [G *Somatozyste, Saftbehälter*]: In floating colony, canal at apical end of colony; corresponds to origin of stem and may contain oil droplet. (simple, branched)

spadix [G *Spadix*]: In more reduced type of medusoid (sporosac), central core (reduced manubrium) on whose surface sex cells ripen.

spine [G *Stachel*]: One of numerous chitinous spines projecting from dense, periderm-covered network of stolons (hydrorhiza) in encrusting colony. Refers also to one in a series of minute spines (spinules) on tubule of nematocyst. (See also stylet)

spiralzooid [G *Spiralzooid*]: Type of protective polyp (dactylozooid) bearing terminal aggregation of short, capitate tentacles and often curved or coiled distally.

spironeme [G *Spironeme*]: Type of nematocyst: tubule typically forms spiral coil distally. Tip of tubule closed (astomocnidae). Includes haplonemes and heteronemes. (See also rhopaloneme)

spirotele [G *Spirotele*]: Type of nematocyst: microbasic eurytele in which butt bears three strong spines and tubule forms spiral coil distally. (See also aspirotele)

sporosac [G *Sporosac*]: Collective term for more reduced types of medusoid including heteromedusoid, styloid, and type in which sex cells ripen directly on sides of blastostyle.

stalk (stem, pedicel) [G *Stiel*]: Tubular section of polyp between base and hydranth. Main stalk of sessile colony termed hydrocaulus.

statocyst (lithocyst, otocyst, statoblast) [G *Statozyste, Statoblast*]: In medusa, one in a series of sensory structures (organ of equilibrium) between tentacle bases around umbrella margin; consists of pit or vesicle with specialized cells (lithocytes) containing solid statoliths. (closed, open)

statolith [G *Statolith*]: In medusa, minute, solid, movable body within lithocyte. Lithocytes may be found in both statocysts and lithostyles.

stem (coenosarc) [G *Stamm*]: In floating colony, section from which all zooid types bud; begins at apex of colony as somatocyst or is continuous with gastrovascular cavity of float. (tubular, disc-shaped) (See also stalk)

stenotele (penetrant) [G *Stenotele*]: Type of nematocyst (rhopaloid): butt is dilated proximally and bears three strong spines (stylets) distally as well as three rows of smaller spines which continue along tubule. Tip of tubule open (stomocnidae). (See also eurytele)

stolon (rhizome) [G *Stolo*]: In sessile colony, hollow, tube-like projection from base of polyp; serves to attach colony to substratum and may give rise to additional polyps. Network of stolons termed hydrorhiza.

stomach: gastrovascular cavity

stomocnida [G *Stomocnide*]: One of two major categories of nematocyst; type in which tip of tubule is open. (See also astomocnida)

style [G *Gasterostyl*]: In colony with massive calcareous skeleton, upright spine at base of gastropore, occasionally also dactylopore. (pointed, rounded, toothed)

stylet [G *Stilett*]: In stenotele type of nematocyst, one of three large spines on dilated basal portion of tubule (butt); each forms the first element in a series of spines on butt.

styloid [G *styloider Gonophor*]: In four recognized stages of medusoid reduction (eumedusoid, cryptomedusoid, heteromedusoid, styloid), most reduced form in which endoderm retains simple configuration of original evagination and cavity (entocodon) is absent.

subepidermal nerve plexus [G *subepidermaler Nervenplexus, Nervennetz*]: Network of nerve cells beneath ectoderm. In polyp, concentrated around mouth; in medusa, concentrated along radial canals and connected to nerve ring.

subumbrella [G *Subumbrella*]: In medusa, lower (oral) concave surface of umbrella; manubrium projects from center of subumbrella. (See also exumbrella)

sucker [G *Haftballen*]: Adhesive structure in medusa; located either at tip of certain branches in forms with branched tentacles, near tip of tentacle (adhesive pad), or at tip of stalk-like structure (stalked sucker).

tabula [G *Zwischenboden*]: In colony with massive calcareous skeleton, one in a series of horizontal plates traversing colony; polyps extend only down to topmost tabula.

tactile comb: In medusa, sensory projection on umbrella margin; bears long, stiff hairs.

taster: palpon

tentacle [G *Tentakel*]: One in a series of variously arranged, nematocyst-bearing processes around hydranth in polypoid hydrozoans or, in medusa, extending either from umbrella margin or exumbrella. Tentacle-like extensions of manubrium around mouth of medusa are termed oral tentacles; extremely elongated tentacle of dactylozooid in floating colony, fishing tentacle. (solid, hollow; filiform, capitate, branched, moniliform)

tentacular bulb (ocellar bulb) [G *Tentakelbasis*]: In medusa, expanded base of tentacle; functions in digestion and nematocyst formation and may bear ocellus.

tentacular root [G *Tentakelwurzel*]: In medusa, continuation of endoderm core of tentacle (when this is positioned on exumbrella) into umbrella. (See also peronium)

tentaculocyst: lithostyle

tentaculozooid: dactylozooid

tentillum [G *Tentille*]: In floating colony, one of several slender lateral branches of single tentacle of feeding polyp (siphon); bears expanded terminal aggregation of nematocysts (cnidoband).

testis [G *Hoden*]: In solitary polyp, one of several conical ectoderm bulges containing sperm; restricted to gastral region in dioecious species, more distally in hermaphroditic species.

theca: periderm, gonotheca, hydrotheca, embryonic theca

thread: tubule

trichter: funnel

trophosome [G *Trophosom*]: Collective term for all nonreproductive structures of colonial hydrozoan. (See also gonosome)

trophozooid (gastrozooid, nutritive polyp) [G *Trophozooid, Gasterozooid, Freßpolyp*]: In colonial hydrozoan, tentacle- and mouth-bearing individual whose function is to capture and ingest food. In floating colony, termed siphon. (See also dactylozooid, gonozooid)

tube: tubule

tubule (tube, thread) [G *Schlauch, Faden*]: In nematocyst, long, thin, hollow tube coiled inside capsule. Turns inside out when discharged. Discharged tubule may be differentiated into dilated proximal section (butt) and thinner distal section (tubule proper or thread). Shape and armature (as well as whether tip is open or closed) form main criteria for nematocyst classification.

umbrella (bell) [G *Umbrella, Schirm, Glocke*]: Main body or swimming bell of medusa; upper surface termed exumbrella, lower surface subumbrella. (bell-shaped, bowl-shaped, dome-shaped, flat, hemispherical, pointed, saucer-shaped, tall, turreted)

velum (craspedon) [G *Velum, Craspedon*]: Horizontal fold projecting inward from umbrella margin; consists of three layers: subumbrellar ectoderm (with muscle fibers), mesolamella, and exumbrellar ectoderm. Serves in propulsion.

volvent: desmoneme

zooid [G *Zooid*]: In colonial hydrozoan, any one of several types of individual (e.g., dactylozooid, gonozooid, trophozooid).

CNIDARIA

ANTHOZOA

anthozoan [G *Anthozoe, Blumentier*]

ANTHOZOA

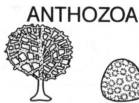

feather-shaped   capitate   reticulate   massive   branched   solitary

colonial   solitary

tentacle

papilliform   simple   capitate   pinnate   branched

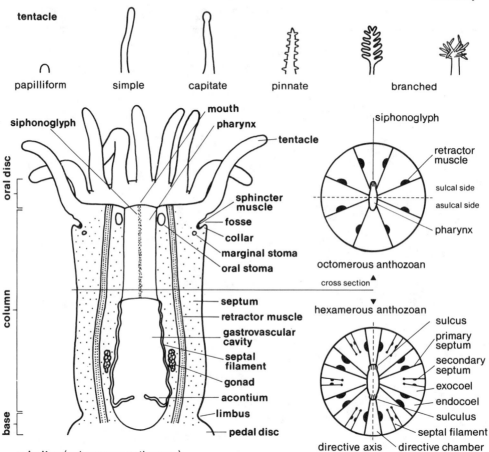

oral disc

siphonoglyph
mouth
pharynx
tentacle

sphincter muscle
fosse
collar
marginal stoma
oral stoma

column

septum
retractor muscle
gastrovascular cavity
septal filament
gonad
acontium
limbus
pedal disc

base

siphonoglyph
retractor muscle
sulcal side
asulcal side
pharynx

octomerous anthozoan

cross section

hexamerous anthozoan

sulcus
primary septum
secondary septum
exocoel
endocoel
sulculus
septal filament
directive chamber
directive axis

sclerites (octomerous anthozoan)

needle   rod   three-flanged rod   spindle   scaphoid   crutch   balloon club   wart club

leaf club   double club   triradiate   cross   six radiate   seven radiate   eight radiate (capstan)   tuberculate spheroid (ball)

double sphere (barrel)   rosette   double disc   double head (dumbbell)   double cone   finger biscuit   scale   thorn scale

CNIDARIA

ANTHOZOA

anthozoan [G *Anthozoe, Blumentier*]

octomerous anthozoan (octocoral, octocorallian, alcyonarian)	[G *achtstrahlige Koralle (Octokoralle)*]
soft coral	*Lederkoralle, Weichkoralle*
blue coral	*blaue Koralle*
organ-pipe coral	*Orgelkoralle*
sea pen	*Seefeder*
horny coral (gorgonian coral)	*Hornkoralle*
sea fan	*Seefächer*
sea whip	*Seepeitsche*
red coral	*Edelkoralle*
hexamerous anthozoan	*sechsstrahlige Koralle (Hexakoralle)*
sea anemone	*Seeanemone, Seerose*
stony coral (scleractinian, madreporarian coral)	*Steinkoralle, Riffkoralle*
black coral	*schwarze Koralle*
ceriantharian anemone	*Zylinderrose*
zoanthid anemone	*Krustenanemone*

accessory sclerite [G *Nebensklerit*]: In octomerous anthozoan, small, not chevroned sclerite at base of polyp between sclerites of anthocodia and those of supporting bundle. (See also crown, intermediate sclerite, point, secondary sclerite)

acontioid [G *Akontioide*]: In hexamerous anthozoan, acontium-like thread on lower section of septum; contains mucus glands yet lacks nematocysts. (simple, branched)

acontium [G *Akontie*]: Thread-like extension of lower end of septal filament into gastrovascular cavity; bears nematocysts and may be protruded through cinclides of body wall (column) to exterior.

acrorhagus [G *Akrorhagus*]: In hexamerous anthozoan, one in a series of hollow, nematocyst-covered projections forming a circle either around junction of oral disc and column (margin), around collar, or on pseudotentacles.

actinopharynx: pharynx

ameboid cell [G *amoeboide Zelle*]: Type of undifferentiated cell with stellate processes in mesoglea; may develop into scleroblast or cnidoblast.

anthocaulus [G *Anthocaulus*]: In solitary, exoskeleton-producing hexamerous anthozoan, stalk-shaped aboral end which produces and casts off oral discs (anthocyathis).

anthocodia [G *Anthocodie*]: In octomerous anthozoan, ectoderm-covered distal end of polyp bearing mouth and tentacles and projecting from surface of colony. May be withdrawn into proximal section (anthostele). (retractile, nonretractile)

anthocyathus [G *Anthocyathus*]: In solitary, exoskeleton-producing hexamerous anthozoan, disc-shaped oral end which is cast off from aboral anthocaulus.

anthostele [G *Anthostele, Polypenkelch*]: In octomerous anthozoan, thicker and rigid proximal section of polyp into which distal end (anthocodia) may be withdrawn; if heavily reinforced by sclerites, may be termed calyx.

autozooid [G *Autozooid, Trophozooid, Freßpolyp*]: In octomerous anthozoan, larger, normally developed polyp in dimorphic (autozooid, siphonozooid) species. (See also mesozooid)

axial polyp [G *Axialpolyp, Achsenpolyp*]: In octomerous anthozoan colony, topmost polyp whose elongate body gives rise to lateral secondary polyps. (See also founder polyp)

axial rod: axis

axial sheath [G *Achsenscheide*]: Layer of tissue immediately surrounding axis; typically delimited from overlying coenenchyme by longitudinal endodermal canals (boundary canals).

axial skeleton: axis

axis (axial rod, axial skeleton) [G *Achsenskelett, Achse*]: Central, longitudinal supporting structures forming main axis of erect octomerous anthozoan colony; consists either of inner central chord and outer horny cortex or inner medulla and outer coenenchyme.

axis cortex: cortex

axis epithelium (axial epithelium) [G *Achsenepithel*]: Thin layer of ectoderm surrounding and producing axis.

ball: spheroid

balloon club [G *Ballonkeule, tütenförmige Spicula*]: Type of sclerite in octomerous anthozoan: club with large, smooth head.

barrel: double sphere

basal plate [G *Basalplatte*]: Bottom of skeletal cup (corallite) enclosing polyp.

base [G *Basis, Basalscheibe, Fußscheibe*]: In solitary hexamerous anthozoan, one of three divisions of body (oral disc, column, base); typically formed into pedal disc or physa. Lower surface may bear one or more pores.

birotulate: double disc

botruchnid [G *Botruchnide*]: In hexamerous anthozoan, typically in larval form, cluster of rounded projections (cnidorhagi) on septal filament.

boundary canal [G *Kranz von Längskanälen*]: One in a series of longitudinal canals separating medulla of axis from coenenchyme.

branched spindle [G *verzweigte Spindel*]: Type of sclerite in octomerous anthozoan: spindle with elongated, branched processes projecting from all sides.

bulb [G *Bulbus*]: Expanded section of erect octomerous anthozoan colony located on stalk (peduncle) near border of peduncle and enlarged, polyp-bearing rachis. (See also end bulb)

calicoblastic layer [G *Calicoblastenschicht*]: In exoskeleton-producing hexamerous anthozoan, modified basal layer of ectoderm; secretes skeleton.

calyx [G *Kalyx, Kelch, Polypenkelch*]: In octomerous anthozoan, rigid, sclerite-reinforced proximal section (anthostele) of polyp. May also refer to skeletal cup (corallite) of hexamerous anthozoan.

capitulum [G *Capitulum*]: In octomerous anthozoan colony, general term for expanded, polyp-bearing distalmost of two divisions (peduncle, capitulum). (disc-shaped, hemispherical) (See also rachis). In hexamerous anthozoan, short, thin-walled distal section of column above sphincter. (See also scapus)

capstan [G *Gürtelstab*]: Type of sclerite in octomerous anthozoan: rod with upper and lower circlets of tubercular or wart-like processes as well as terminal projection. (foliate) (See also eight-radiate)

carcinoecium [G *Carcinoecium*]: Term applied to coenenchyme of certain anthozoans symbiotic with hermit crabs; may dissolve or extend occupied gastropod shell and therefore directly envelop crab.

caterpillar: unilaterally spinose spindle

central chord (core) [G *Zentralstrang, Achsenstrang*]: Central part of axis; contains loculi and is surrounded by cortex. (horny, horny-calcareous, hollow and cross-chambered) (See also medulla)

ciliated band: flagellated band

cinclide [G *Cinclide*]: One of numerous small pores in column, occasionally on papillae and arranged in rows, through which acontia may be discharged.

club [G *Keule*]: Type of sclerite in octomerous anthozoan: monaxial form with one enlarged end (head) and one tapered end (handle). (See also balloon club, double club, leaf club, rooted head, torch, wart club)

cnidoglandular band [G *Cnidoglandulartrakt, Drüsenwulst, Drüsenstreifen*]: In oral section of septal filament, central of three longitudinal ridges bearing nematocysts and gland cells. (See also flagellated band, intermediate tract)

cnidorhagus [G *Cnidorhagus*]: In hexamerous anthozoan, typically in larval form, one in a cluster (botruchnid) of numerous round projections on septal filaments; contains nematocysts.

coelenteron: gastrovascular cavity

coenenchyme [G *Coenenchym*]: Tissue of colony between polyps; continuous with polyp tissue and contains gastrovascular space continuous with gastrovascular cavity of polyp. In octomerous anthozoan, mesoglea of coenenchyme contains skeletal elements (sclerites) and is penetrated by network of endodermal tubes (solenia). In hexamerous anthozoan, lower surface of coenenchyme secretes skeleton (coenosteum) between individual skeletal cups (corallites). May also be referred to here as coenosarc, in which case secreted skeleton is termed coenenchyme.

coenosarc: coenenchyme

coenosteum [G *Coenosteum*]: In exoskeleton-producing hexamerous anthozoan, part of skeleton (corallum) between walls (thecae) of individual skeletal cups (corallites); secreted by coenenchyme.

collar (parapet) [G *Margo, Parapet*]: On column of hexamerous anthozoan, upwardly projecting fold of scapus at junction of scapus and capitulum. (See also fosse)

collaret: crown

columella [G *Columella*]: In exoskeleton-producing hexamerous anthozoan, column-shaped skeletal projection of basal plate in middle of skeletal cup (corallite). (See also pseudocolumella)

column [G *Mauerblatt*]: Cylindrical section of hexamerous anthozoan between pedal disc and oral disc; may be differentiated into several regions (capitulum, collar, fosse, scapulus, scapus) and be variously ornamented. (smooth, longitudinally grooved, warty)

conchula [G *Conchula*]: Spout-shaped projection of siphonoglyph onto oral disc.

corallite [G *Einzelpolypar*]: In exoskeleton-producing hexamerous anthozoan, skeletal structure of single polyp in colony; typically consists of skeletal cup with radially arranged, vertical ridges (sclerosepta). (See also basal plate, columella, costa, epitheca, theca)

corallum [G *Gesamtpolypar, Polyparium*]: In exoskeleton-producing hexamerous anthozoan, entire skeletal structure of colony.

core: central chord

cortex (cortical layer) [G *Cortex, Rinde*]: In octomerous anthozoan colony, layer surrounding central part of axis; refers either to coenenchyme surrounding medulla or to horny layer surrounding central chord.

costa [G *Costa*]: In exoskeleton-producing hexamerous anthozoan, radially arranged continuation of scleroseptum between inner (theca) and outer wall (epitheca) of skeletal cup (corallite) surrounding polyp.

couple [G *Zwillingsmesenterium, Zwillingsseptum*]: Two septa in which one member is located on one side of gastrovascular cavity and the second member, of the same age, is on opposite side of symmetry axis. (See also pair)

crescent [G *halbmondförmiger Sklerit*]: Type of sclerite in octomerous anthozoan; curved spindle with processes projecting from all sides, those of convex side typically longer. (See also scaphoid)

cross [G *Kreuz*]: Type of sclerite in octomerous anthozoan; consists of four rays in one plane.

crown (collaret) [G *Krone, Kranz transversaler Spindeln*]: In octomerous anthozoan, circlet of transversely oriented, usually bow-shaped sclerites below tentacles of polyp. (See also point)

crown spine [G *Kronstachel*]: Type of sclerite in octomerous anthozoan: spindle located circumorally in polyp, with one end modified into long, smooth spine.

crutch [G *krückenförmiger Sklerit*]: Type of sclerite in octomerous anthozoan: spindle bifurcated at one end.

cuticle [G *Kutikula*]: Noncellular outermost layer on column and pedal disc of certain anthozoans.

cycle [G *Zyklus*]: One in a series of concentric circles of tentacles around oral disc; tentacles of adjoining circles alternate. May also refer to cycle of septa (primary, secondary, tertiary septa).

directive [G *Richtungsseptum, Richtungsmesenterium*]: In hexamerous anthozoan, one of two pairs of complete septa, one at each narrow end of pharynx; attached to siphonoglyph(s) and distinguished from other septa in that they bear retractor muscle bands facing in opposite directions (exocoelic retractor muscles).

directive axis [G *Richtungsebene*]: In hexamerous anthozoan, long axis through narrowed ends of ellipsoid pharynx; defined by directives, siphonoglyphs, and/or (addition of) couples. (See also symmetry axis)

directive chamber [G *Richtungsfach*]: In hexamerous anthozoan, space between pair of directive septa (directive); opposite directive chambers define directive axis. (See also lateral chamber)

disc: oral disc

disc-spindle [G *Scheibenträger*]: Type of sclerite in octomerous anthozoan: spindle with tubercles of four or more circlets more or less fused into discs. (See also double disc)

dissepiment [G *Dissepiment*]: In exoskeleton-producing hexamerous anthozoan, thin, horizontal skeletal plate between sclerosepta of skeletal cup (corralite) surrounding polyp. (See also synapticulum, tabula)

dorsal track (prorachis) [G *Dorsalfeld, dorsales nacktes Kielfeld*]: In octomerous anthozoan colony, elongate, median, polyp-free strip along proximal division (rachis) of colony; corresponds to asulcal side of primary polyp. (See also ventral track)

double club (opera glass) [G *Doppelkeule*]: Type of sclerite in octomerous anthozoan: modified six-radiate form resembling two parallel clubs.

double cone [G *Doppelkegel*]: Type of sclerite in octomerous anthozoan: spindle with ornamentation-free midregion (waist).

double cup (rosette, spiny rosette) [G *kelchförmiger Sklerit, Rosette, Stachelrosette*]: Type of sclerite in octomerous anthozoan: cup- or funnel-shaped at one end, rounded and multituberculate at other end.

double disc (double wheel, birotulate) [G *Doppelscheibe, Doppelrad*]: Type of sclerite in octomerous anthozoan: modified capstan with tubercles of two circlets fused into discs. (See also disc-spindle)

double head [G *doppelköpfiger Sklerit*]: Type of sclerite in octomerous anthozoan: general term for symmetrical forms with narrow, smooth midregion (waist) and processes at each end not arranged in circlets.

double sphere (barrel) [G *Fäßchen*]: Type of sclerite in octomerous anthozoan: form with warty processes at both ends and short, ornamentation-free midregion (waist). (See also dumbbell)

double star [G *Doppelstern*]: Type of sclerite in octomerous anthozoan: rod with circlets of irregularly shaped, spiny projections at both ends.

double wheel: double disc

dumbbell [G *Doppelkegel*]: Type of sclerite in octomerous anthozoan: form with warty processes at both ends and elongate, ornamentation-free midregion (waist). (See also double sphere)

ectoderm (epidermis) [G *Ektoderm*]: Outermost cellular layer of body wall (ectoderm, mesoglea, endoderm). Contains mucus cells, sensory cells, interstitial cells, and nematocysts and is underlain by muscle fibers.

eight-radiate (capstan) [G *Achtstrahler*]: Type of sclerite in octomerous anthozoan: capstan with one circlet of three tubercles around each end as well as terminal tubercles at each end.

end bulb [G *Endblase*]: Enlarged swelling at base of peduncle in octomerous anthozoan colony; serves in anchoring colony. May also refer to physa in solitary hexamerous anthozoan. (See also bulb)

endocoel (endocoelic chamber) [G *Innenfach, Endocoel*]: In anthozoan with paired septa, gastrovascular space between members of each pair. Retractor muscle bands of each member project into endocoel (except directives). (See also exocoel, pair)

endoderm (gastrodermis) [G *Entoderm*]: Innermost cellular layer of body wall (ectoderm, mesoglea, endoderm). Lines entire gastrovascular cavity, solenia, inner surface of tentacles, oral disc, and faces of septa; forms retractor muscles and often contains symbiotic algae.

epidermis: ectoderm

epitheca [G *Epithek, Epitheca*]: In exoskeleton-producing hexamerous anthozoan, second outer wall of skeletal cup (corallite) surrounding polyp; may be joined to inner wall (theca) by costae.

esophagus: pharynx

exocoel (exocoelic chamber) [G *Zwischenfach, Exocoel*]: In anthozoan with paired septa, gastrovascular space between pairs. (See also endocoel, pair)

external stoma: stoma

filament: septal filament

finger biscuit-like form [G *biskuitförmige Platte*]: Type of sclerite in octomerous anthozoan: minute, flattened rod.

flagellated band (ciliated band) [G *Flimmerwulst, Flimmerstreifen, Ciliarstreifen*]: In oral section of septal filament, one of two flagellated, longitudinal ridges along each side of cnidoglandular band.

fosse [G *Fossa*]: In hexamerous anthozoan, groove around proximal end of column; formed by fold (collar) at junction of scapus and capitulum. Bears sphincter muscle.

founder polyp (primary polyp) [G *Gründerpolyp, Primärpolyp*]: In octomerous anthozoan, first polyp of colony formed by planula larva; may develop into axial polyp.

frond: pseudoacrorhagus

gastric cavity: gastrovascular cavity

gastrodermal canal [G *Gastrodermalkanal*]: In octomerous anthozoan, relatively wide, endoderm-lined canal in coenenchyme; continuous with solenia.

gastrodermis: endoderm

gastrovascular cavity (coelenteron, gastric cavity) [G *Gastralraum, Magenraum*]: Spacious cavity of gastrovascular system within column; divided into interseptal spaces by longitudinal partitions (septa).

gonad [G *Gonade*]: ♂ or ♀ reproductive cells along septa in gastrovascular cavity.

gorgonin [G *Gorgonin*]: Horny material (protein) forming, together with sclerites, inner and/or outer layers of axis.

gullet: pharynx

hockey-stick spindle [G *Hockeyschläger*]: Type of sclerite in octomerous anthozoan: large spindle with longer and shorter sections forming an obtuse angle.

holdfast [G *Basis*]: Variously shaped basal section of octomerous anthozoan colony; serves to attach colony to or in substratum. (See also end bulb)

hyposulculus [G *Hyposulcus*]: Continuation of flagellum-bearing groove (siphonoglyph) of pharynx along edges of one septum pair ("dorsal" septum or sulculus).

intermediate sclerite [G *intermediärr Sklerit*]: In octomerous anthozoan, minute sclerite between rows of larger, chevroned sclerites (points) at proximal end of polyp.

intermediate tract [G *Zooxanthellenstreifen*]: Longitudinal band between cnidoglandular and both flagellated tracts along oral region of septal filament; typically containing zooxanthellae.

internal stoma: stoma

internode [G *Internodium*]: Rigid, often enlarged, calcareous segment of jointed axis in octomerous anthozoan colony. (See also node).

interseptal chamber: interseptal space

interseptal space (interseptal chamber) [G *Fach, Interseptalraum*]: Spaces of gastrovascular cavity between septa. (See also endocoel, exocoel; directive chamber, lateral chamber)

introvert: neck

labial tentacle: oral tentacle

lateral chamber [G *Seitenfach*]: In hexamerous anthozoan, one of two gastrovascular cavities (spaces) between the two directive chambers; includes exocoels and endocoels.

leaf: polyp leaf

leaf club [G *Blattkeule*]: Type of sclerite in octomerous anthozoan: club with enlarged end (head) bearing leaf-like processes, occasionally from one side only.

leaf spindle [G *Blattspindel*]: Type of sclerite in octomerous anthozoan: spindle with leaf-like processes projecting from one side (only).

limbus [G *Limbus*]: In hexamerous anthozoan, constriction marking border between column and pedal disc.

loculus [G *Loculus*]: In hexamerous anthozoan colony, one in a series of calcified or fiber-filled compartments in innermost part (central chord) of axis. In exoskeleton-producing hexamerous anthozoan, series of radially oriented pockets between septa of polyp and inwardly projecting sclerosepta of skeletal cup (corallite) surrounding polyp.

longitudinal canal [G *Längska-nal*]: In octomerous anthozoan colony, one of typically four large, longitudinal partitions of gastrovascular cavity of founder polyp; surrounds a central axis. Also general term for larger longitudinal canals.

macrocneme: macroseptum

macroseptum   (macrocneme) [G *Makroseptum, Makromesenterium*]: Term applied to complete septum if size difference between complete and incomplete septa is great. (See also microseptum)

margin [G *Margo*]: In hexamerous anthozoan, junction of oral disc and column; may be further elaborated into collar and groove (fosse).

marginal stoma: stoma

marginal tentacle [G *Marginaltentakel, Randtentakel*]: In hexamerous anthozoan with two distinct groups of tentacles, typically larger tentacle within the set at outer margin of oral disc. (See also oral tentacle)

medulla [G *Medulla, Markschicht, Markstrang* (central part only)]: Central supporting structure of certain octomerous anthozoan colonies; consists of dense aggregations of sclerites, gorgonin, and occasionally one or more longitudinal canals (boundary canals). Surrounded by coenenchyme. (See also axis)

mesenterial filament: septal filament

mesenterial perforation: stoma

mesentery: septum

mesoglea (mesogloea) [G *Mesogloea*]: Middle noncellular matrix of body wall (ectoderm, mesoglea, endoderm); contains ameboid cells. Well-developed in coenenchyme of octomerous anthozoan colonies, thin in distal part (anthocodia) of polyp.

mesozooid [G *Mesozooid*]: In octomerous anthozoan colony, type of polyp intermediate between autozooid and siphonozooid; considered to be siphonozooid with well-developed septa and serves as exhalent zooid of colony.

metaseptum [G *Metaseptum, Metamesenterium*]: In hexamerous anthozoan, all later-developing septa "ventral" to protosepta.

microseptum   (miscocneme) [G *Mikroseptum, Mikromesenterium*]: Term applied to incomplete septum if size difference between complete and incomplete septa is great.

mouth [G *Mundöffnung*]: Opening of gastrovascular cavity to exterior; enters cavity via pharynx. (oval, elongated, slit-like)

muscle banner: retractor muscle

muscle pennon: retractor muscle

myoseptum: septum

neck (introvert) [G *Halszone*]: In octomerous anthozoan, soft, thin-walled section of polyp below tentacles; lacks sclerites and allows introversion of distal part (anthocodia) of polyp into basal part (anthostele).

needle [G *Nadel*]: Type of sclerite in octomerous anthozoan: long, slender, smooth monaxial form. (three-flanged)

nemathybome [G *Nemathybome, Nesselhöcker*]: In hexamerous anthozoan, one in a series of nematocyst-bearing mesogleal sacs in column; projects from outer surface of column in longitudinal rows between lines of septum attachment.

nematocyst: see Hydrozoa. Includes atrichous isorhiza, basitrichous isorhiza, holotrichous isorhiza, microblastic mastigophore, microblastic amastigophore, and macrobasic amastigophore. (See also spirocyst)

nerve   plexus   (subepidermal nerve plexus) [G *Nervenplexus, subepidermaler Plexus*]: Network of nerve cells beneath ectoderm. Concentrated on oral disc around mouth, upper end of pharynx, and tentacle bases.

node [G *Nodium*]: Flexible, often more slender, horny segment of jointed axis in octomerous anthozoan colony. (See also internode)

opera glass: double club

operculum [G *Operculum, Deckel*]: In octomerous anthozoan, lid-like structure covering withdrawn tentacles; consists of eight triangular elements.

oral disc (disc) [G *Mundscheibe, Tentakelscheibe)*]: Somewhat expanded, flat distal division of body (oral disc, column, base); bears central mouth and tentacles. (For hexamerous anthozoan, see peristome)

oral sphincter: sphincter

oral stoma: stoma

oral tentacle (labial tentacle) [G *Labialtentakel*]: In hexamerous anthozoan with two distinct groups of tentacles, typically smaller tentacle within the set near mouth. (See also marginal tentacle)

ostium: stoma

oval (ovoid) [G *Ellipsoid, Oval*]: Type of sclerite in octomerous anthozoan: short, occasionally flattened rod with rounded outline.

ovoid: oval

pair [G *Paar*]: Pair of septa in which both members adjoin one another; space between pair termed endocoel, space outside termed exocoel. (See also couple)

palus [G *Palus*]: In exoskeleton-producing hexamerous anthozoan, one in a series of small, vertical ridges between central columella and sclerosepta of skeletal cup (corallite) surrounding polyp.

parapet: collar

pedal disc [G *Fußscheibe*]: In hexamerous anthozoan, disc-shaped expansion of proximal division of body (base); serves in attachment to substratum (may be modified into float). (See also base, physa)

pedicel [G *Pedicellus*]: In octomerous anthozoan colony, type of stalk stiffened externally by large sclerites.

peduncle (stalk) [G *Stiel*]: In octomerous anthozoan colony, proximal stalk-like division of colony (peduncle, rachis); lacks secondary polyps or polyp leaves.

peristome [G *Mundregion*]: In hexamerous anthozoan, region of polyp surrounding mouth; formed by basal parts of tentacles. (For octomerous anthozoan, see oral disc)

pharynx (actinopharynx, esophagus, gullet, stomodeum) [G *Pharynx, Schlundrohr, Actinopharynx*]: Tubular section of digestive system extending from mouth into gastrovascular cavity; may be attached to column by septa and typically bears one or two flagellated grooves (siphonoglyphs). (circular, oval; smooth, ridged)

physa [G *Physa, Fußscheibe*]: In hexamerous anthozoan, bulb-shaped expansion of proximal division (base) of body; serves in attachment to soft substratum. (See also pedal disc)

pinnule [G *Fieder, Pinnula*]: One in a series of short, hollow projections arranged in two opposite rows along each tentacle.

plate [G *Platte*]: Type of sclerite in octomerous anthozoan: flat, yet relatively large (> 0.05 mm) and thick form. (outline: circular, irregular, oval, polygonal, stellate) (See also platelet, scale)

platelet [G *Plättchen*]: Type of sclerite in octomerous anthozoan: small (typically < 0.05 mm), smooth plate.

platform [G *Platte*]: In hexamerous anthozoan colony, one in a series of horizontal plates uniting erect tubes; consists of fused sclerites.

point [G *konvergierende Doppelreihe, Kronenspitze*]: In octomerous anthozoan, one of eight longitudinal rows of chevroned sclerites around distal end of polyp. (See also crown)

polyp (zooid) [G *Polyp, Anthopolyp*]: Individual anthozoan; typically refers to octomerous forms, where one may distinguish autozooid, siphonozooid, and mesozooid. (For octomerous forms: contractile, retractile)

polyparium (polypary) [G *Polyparium, Polypar*]: In octomerous anthozoan, part of colony bearing anthocodiae.

polyp leaf (leaf) [G *Polypenträger, polyptragender Wulst, Blatt*]: In octomerous anthozoan colony, flattened lateral projection from upper division of colony (rachis); bears secondary polyps. (See also ray, siphonozooid plate, stipule)

primary polyp: founder polyp

primary septum [G *primäres Septum, Septum erster Ordnung*]: One of six pairs of complete septa in hexamerous anthozoan; entire cycle referred to as the primaries. (See also secondary septum, tertiary septum)

protoseptum [G *Protoseptum, Protomesenterium*]: In hexamerous anthozoan, one of three "dorsal" couples of septa first to develop embryologically; includes directives and two couples to either side. (See also metaseptum)

pseudoacrorhagus (frond) [G *Pseudoacrorhagus*]: Compound, enlarged acrorhagus with few or no nematocysts; forms wide, protruding ring around hexamerous anthozoan at junction of oral disc and column.

pseudocolumella [G *Pseudocolumella*]: In exoskeleton-producing hexamerous anthozoan, type of columella formed by merger of central ends of sclerosepta.

pseudoseptum [G *Pseudoseptum*]: Inwardly directed skeletal projection in calyx of octomerous anthozoan; bears no constant relationship with septa of polyp. (See also tooth)

pseudotentacle [G *Pseudotentakel*]: In hexamerous anthozoan, one in a series of four to eight branched, tentacle-like projections at junction of oral disc and column; bears acrorhagi.

pseudotheca [G *Pseudothek*]: In exoskeleton-producing hexamerous anthozoan, type of theca formed by fusion of outer ends of sclerosepta.

rachis (rhachis) [G *Rhachis*]: In octomerous anthozoan colony, distal division of colony (peduncle, rachis); bears secondary polyps or polyp leaves. May also refer to main part of tentacle which bears pinnules.

radiate [G *Strahler*]: Type of sclerite in octomerous anthozoan: general term for sclerite with processes radiating symmetrically in one or more planes. Includes crosses, six- and eight-radiates, and stellate plates.

ray [G *Hauptstrahl*]: In octomerous anthozoan colony, one in a series of bundled sclerites supporting polyp-bearing lateral projection (polyp leaf).

retractor muscle (muscle banner, muscle pennon) [G *Retraktor, Muskelfahne*]: Well-developed band of longitudinal muscle along one side of each septum; in octomerous anthozoan along sulcal sides, in hexamerous forms with paired septa, along sides facing each other (except on directives where they face in opposite directions).

rhachis: rachis

rhizoid [G *Rhizoid, wurzelförmiger Fortsatz*]: In octomerous anthozoan, root-like process extending from base of colony. (axial, coenenchymal) (See also stolon)

rind: cortex

rod [G *Stab*]: Type of sclerite in octomerous anthozoan: monaxial form blunt at both ends. (curved, straight) (See also capstan, double star, oval, seven-radiate, six-radiate)

rooted head [G *Wurzelkopf*]: Type of sclerite in octomerous anthozoan: form with one enlarged and rounded end and one branched end.

rooted leaf [G *Wurzelblatt*]: Type of sclerite in octomerous anthozoan: form with one flattened, disc-shaped end and one branched end.

rosette: double cup

scale [G *Schuppe*]: Type of sclerite in octomerous anthozoan: thin, flat form. According to position on polyp one may distinguish (from distal to proximal) opercular, buccal, medial, basal, and infrabasal scales.

scaphoid [G *Klammer*]: Type of sclerite in octomerous anthozoan: curved spindle with processes typically projecting only from concave side. (See also crescent)

scapulus [G *Scapulus*]: On column of hexamerous anthozoan, modified, short distal region of scapus; differs histologically and in general appearance.

scapus [G *Scapus*]: In hexamerous anthozoan, longer, thick-walled proximal section of column; may form collar. (See also capitulum)

sclerite (spicule) [G *Sklerit*]: In octomerous anthozoan, minute calcareous skeletal element secreted by scleroblast and lodged in anthocodium or coenenchyme.

scleroblast [G *Skleroblast*]: Type of cell, developing from ameboid cell in mesoglea, which produces sclerite.

scleroseptum [G *Skleroseptum*]: In exoskeleton-producing hexamerous anthozoan, one in a series of radially arranged, vertical ridges projecting from floor (basal plate) of skeletal cup (corallite) surrounding polyp. May be referred to as septum if term mesentery rather than septum is applied to longitudinal partitions of gastrovascular cavity of polyp. (endocoelic, exocoelic; spiny, thorny; upper edge: jagged, toothed)

secondary polyp [G *sekundärer Polyp, Sekundärpolyp*]: In octomerous anthozoan colony, series of lateral polyps arising from elongate axial polyp.

secondary septum [G *sekundäres Septum, Septum zweiter Ordnung*]: One of six pairs of septa between pairs of primary septa; entire cycle referred to as secondaries. (complete, incomplete)

septal filament (mesenterial filament) [G *Septalfilament, Mesenterialfilament*]: Thickened free inner edge of septum below pharynx. In octomerous anthozoan, two asulcal filaments may be heavily flagellated. In hexamerous anthozoan, filament may be trilobed in cross section orally, consisting of central cnidoglandular band and two lateral flagellated bands; only the former continues to aboral end. (sulcal, asulcal) (See also acontium)

septum (mesenterium) [G *Septum, Mesenterium, Sarcoseptum*]: In octom-

erous anthozoan, eight longitudinal partitions dividing gastrovascular cavity into interseptal spaces; in hexamerous forms, typically a multiple of six. Extends from wall of column to pharynx orally; free inner edge below pharynx bears thickening (septal filament). Each septum bears longitudinal muscle (retractor muscle). (complete, incomplete; macroseptum, microseptum; paired, coupled; fertile, sterile; primary, secondary, tertiary, quaternary)

seven-radiate [G *Siebenstrahler*]: Type of sclerite in octomerous anthozoan: rod with one circlet of three tubercles around each end, yet terminal tubercle at one end only.

siphonoglyph (sulcus) [G *Siphonoglyphe, Flimmerrinne, Wimperrinne, Schlundrinne*]: Flagellum-bearing longitudinal groove extending from mouth into gastrovascular cavity along pharynx; typically located at narrow end of oval pharynx; defines "ventral" side of directive axis. If two siphonoglyphs are present (diglyphic forms), one may distinguish a "ventral" sulcus and "dorsal" sulculus.

siphonozooid [G *Siphonozooid*]: In dimorphic octomerous anthozoan (autozooid, siphonzooid), smaller polyp with reduced tentacles, septa, septal filaments, and well-developed siphonoglyph. Serves to drive water through colony. (See also mesozooid, siphonozooid plate, stipule)

siphonozooid plate [G *Zooidplatte*]: In octomerous anthozoan colony, dense aggregation of siphonozooids forming band on underside of polyp leaves.

six-radiate [G *Sechsstrahler*]: Type of sclerite in octomerous anthozoan: rod with one circlet of three tubercles around each end.

skeleton [G *Skelett*]: Flexible or more rigid, internal or external structure (secretion, deposit) supporting individual polyp or entire colony.

solenium [G *Solenium*]: In octomerous anthozoan colony, narrow endodermal tube continuous with gastrovascular cavity and gastrodermal canal and therefore joining polyps with one another.

sphaerome [G *Sphaerom*]: Small, nematocyst-packed vesicle on or forming modified tentacle. (round, oval)

spheroid (ball) [G *Kugel*]: Type of sclerite in octomerous anthozoan: ball-shaped, typically with ornamentation. (foliate, spiny, tuberculate, unilaterally foliate, unilaterally spiny)

sphincter [G *Sphinkter*]: Well-developed band of circular muscle around proximal end of column; located either below junction (margin) of oral disc and column or, if column is regionated into scapus and capitulum, in groove (fosse) formed by projecting upper end (collar) of scapus.

spicule [G *Spikel, Spiculum, Nadel*]: Type of sclerite in octomerous anthozoan; monaxial form with pointed tips (spindles, needles). Occasionally (incorrectly) applied as general term for all calcareous skeletal elements.

spindle [G *Spindel*]: Type of sclerite in octomerous anthozoan: straight or curved monaxial form wider than needle and pointed at both ends; typically with ornamentation. (straight, curved, three-flanged) (See also caterpillar, crescent, scaphoid, and branched, disc-, hockey-stick, leaf, thorn, and unilaterally spinose spindle)

spiny rosette: double cup

spirocyst [G *Spirozyste, Klebkapsel*]: Type of cell organelle, related to nematocyst, also with double-walled capsule; unarmed tubule surrounded by spirally wound sticky material when discharged.

stalk [G *Stiel*]: In certain octomerous anthozoan colonies, proximal polyp-free section of colony. May also refer to narrow proximal part of nonretractile polyp. (See also peduncle, stem)

stem [G *Stamm, Hauptstamm*]: In certain octomerous anthozoan colonies, proximal thicker section of colony with or without branches. In other forms may also refer to polyp-bearing (polypiferous) section of colony usually giving rise to branches. (See also peduncle, stalk)

stipule [G *Stielwulst, Nebenblatt*]: In octomerous anthozoan colony, small, lobe-like aggregation of siphonzooids on polyp leaves.

stolon [G *Stolo*]: In octomerous anthozoan colony, basal expansion of tissue (coenenchyme) along substratum; gives rise to new polyps. (rounded, flattened, membranous) (See also rhizoid)

stoma (ostium, mesenterial perforation) [G *Septalostium, Stoma*]: Opening through distal end of septum through which adjoining interseptal spaces may communicate; according to location relative to mouth and column one may distinguish oral (or internal) and marginal (or external) stomata.

stomodeum: pharynx

sulculus: siphonoglyph

sulcus: siphonoglyph

supplementary sclerite [G *zusätzlicher Sklerit*]: In octomerous anthozoan, type of chevroned sclerite below crown. (See also accessory sclerite, intermediate sclerite)

supporting bundle [G *Stützbündel*]: In octomerous anthozoan, sheath of sclerites below anthocodia; may consist of single large spicule.

symmetry axis [G *Symmetrie-Achse*]: Either directive axis (hexamerous anthozoan) or, in octomerous anthozoan, long axis through narrowed ends of pharynx; defined by siphonoglyphs or paired arrangement of septa.

synapticulum [G *Synaptikel*]: In exoskeleton-producing hexamerous anthozoan, rod-shaped skeletal structure connecting adjacent sclerosepta in skeletal cup (corallite) surrounding polyp. (See also dissepiment)

tabula [G *Tabula, Querboden*]: In exoskeleton-producing hexamerous anthozoan, one in a series of thicker horizontal plates extending across entire skeletal cup (corallite) surrounding polyp. Deposition of new tabulae allows polyps to remain on surface of growing colony. (See also dissepiment)

tenaculus [G *Tenaculus*]: In hexamerous anthozoan, one in a series of glandular, papilla-shaped adhesive structures along outer wall of column; may bear cuticle cover.

tentacle [G *Tentakel*]: One in a series of variously arranged, motile and contractile, nematocyst-bearing processes around oral disc; internal cavity continuous with gastrovascular cavity. (simple, branched, pinnate, papilliform, laterally flattened, with sphaeromes) (see also marginal tentacle, oral tentacle)

terminal polyp [G *Terminalpolyp*]: In octomerous anthozoan colony, distal end of primary polyp.

tertiary septum [G *tertiäres Septum, Septum dritter Ordnung*]: One of 12 pairs of septa within exocoels formed by secondary septa.

theca [G *Theka*]: In exoskeleton-producing heamerous anthozoan, wall of skeletal cup (corallite) surrounding polyp. (See also epitheca)

thorn club [G *Stachelkeule*]: Type of sclerite in octomerous anthozoan: club with enlarged end (head) bearing sharp, thorn- or spine-like processes.

thornscale [G *Stachelplatte*]: Type of sclerite in octomerous anthozoan: flat form with spine-like or digitiform process projecting from margin or center.

thornspindle [G *Stachelspindel*]: Type of sclerite in octomerous anthozoan: spindle with thorn-like processes projecting from one side only.

thornstar [G *Stachelfuß*]: Type of sclerite in octomerous anthozoan: thorn scale with one or more thorn- or leaf-like processes projecting vertically from middle of divided, root-like base.

tooth [G *Zahn, Kelchzahn*]: In octomerous anthozoan, one of typically eight pointed, upwardly projecting lobes of calyx margin; typically stiffened by sclerites.

torch [G *Fackel*]: Type of sclerite in octomerous anthozoan: club with slanted, strongly incised, leaf-like processes projecting from enlarged end (head).

triradiate [G *Dreistrahler*]: Type of sclerite in octomerous anthozoan: consists of three rays in one plane.

unilaterally spinose spindle (caterpillar) [G *Raupe*]: Type of sclerite in octomerous anthozoan: spindle with spine-like processes projecting from one side only.

ventral track (metarachis) [G *ventrales nacktes Kielfeld*]: In octomerous anthozoan colony, elongate, median, polyp-free strip along proximal division (rachis) of colony; corresponds to sulcal side of primary polyp. (See also dorsal track)

verruca (wart) [G *Saugwarze*]: In hexamerous anthozoan, one in a series of muscular, hollow adhesive structures lined with gland cells; in longitudinal rows along outer surface of column between lines of septum attachment. (See also tenaculus)

vesicle [G *Vesikel, Blase*]: One of numerous hollow, nematocyst-covered projections along thick-walled proximal section (scapus) of column. (simple, compound)

wart: verruca

wart club [G *Warzenkeule*]: Type of sclerite in octomerous anthozoan: club with enlarged end (head) bearing wart-like processes.

zooid: polyp

CTENOPHORA (COLLARIA, ACNIDARIA)

ctenophore, comb jelly [G *Rippenqualle, Kammqualle*]

CTENOPHORA

ovate

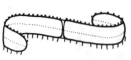

ribbon-shaped

dorsoventrally
flattened

tentaculate

cone-shaped
(laterally compressed)

atentaculate

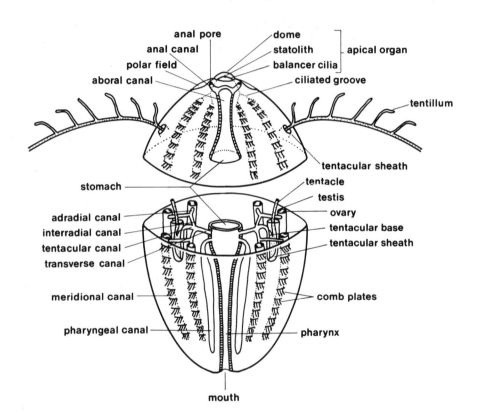

anal pore
anal canal
polar field
aboral canal

dome
statolith
balancer cilia
ciliated groove

} apical organ

tentillum

tentacular sheath

stomach
tentacle
testis
ovary
adradial canal
interradial canal
tentacular canal
transverse canal
tentacular base
tentacular sheath

meridional canal
comb plates

pharyngeal canal
pharynx

mouth

mouth

narrow

wide

comb rows
mouth
auricle
oral lobe

with oral lobes
(lobate ctenophores)

CTENOPHORA (COLLARIA, ACNIDARIA)

ctenophore, comb jelly [G *Rippenqualle, Kammqualle*]

aboral canal (infundibular canal, funnel [G *Aboralkanal, Trichtergefäß*]: Gastrovascular canal leading from stomach to aboral pole. May give rise to anal canals.

aboral crest [G *Schwertkiel*]: One of two elongate aboral projections of body wall. Comb rows may extend onto crest.

aboral pole [G *aboraler Pol, apikaler Pol, Scheitelpol*]: Region or end of body diametrically opposite mouth; bears apical organ.

adhesive spherule (granule) [G *Sekretgranulum, Klebkorn*]: One of numerous adhesive granules secreted by hemispherical head of colloblast.

adradial canal [G *adradialer Kanal*]: In gastrovascular system, one of two canals leading from each interradial canal to meridional canal.

anal canal (excretory canal) [G *Analkanal, Exkretionskanal*]: One of typically four short gastrovascular canals extending from aboral canal to aboral pole; two canals open to exterior near apical organ via anal pores.

anal pore (excretory pore) [G *Analporus, Exkretionsporus*]: One of two openings of gastrovascular system to exterior on aboral pole; joined to stomach via anal canal and aboral canal.

apical organ (statocyst) [G *Apikalorgan, Scheitelorgan, Statozyste*]: Sensory (equilibrium) organ on aboral pole; contains statolith and balancer cilia and is connected to comb rows via ciliated grooves.

auricle [G *Aurikel*]: One of four ciliated, finger-shaped projections near mouth of lobate ctenophore.

auricular groove [G *Aurikelfurche*]: Ciliated groove extending from tentacular base to base of auricles; bears series of short tentacles.

balancer cilia [G *Wimperfeder*]: One of four tufts of cilia on which statolith in apical organ rests; connected to comb rows via ciliary grooves.

bell: dome

brood chamber [G *Brutkammer*]: In dorsoventrally flattened ctenophore, pouches in aboral surface in which egg develops into cydippid larva.

cell rosette: rosette

chimney [G *Tentakelscheide, Tentakelrohr*]: Term applied to one of two upwardly directed, tube-like extensions of oral surface of ctenophore; includes tentacle sheath and is continuous with stomach.

ciliated furrow: ciliated groove

ciliated groove (ciliated furrow) [G *Wimperstrasse, Wimperschnur, Wimperband*]: One of four bands of cilia arising from balancer cilia and, after forking, extending through each comb row.

collenchyme: mesoglea

colloblast [G *Colloblast, Klebzelle*]: Adhesive, prey-capturing cell in ectoderm of tentacle; consists of hemispherical head with adhesive spherules, spiral filament, and straight filament.

comb plate [G *Wimperkamm, Wimperplatte*]: Transverse row of long, fused cilia; typically connected into a series along a ciliated groove to form a comb row.

comb row (plate row, rib, costa) [G *Rippe, Wimperrippe, Wimpermeridian*]: Locomotory organ of many ctenophores; typically eight in number and consisting of a series of comb plates overlying each meridional canal.

costa: comb row

ctene: comb plate

cupule: dome

cydippid larva [G *Cydippe-Stadium*]: Free-swimming larval stage of ctenophore; characterized by simple gastrovascular system and short comb rows.

dome (bell, cupole) [G *Glocke, Kuppel*]: Transparent, dome-shaped structure of apical organ formed by fused cilia and covering statolith.

ectoderm [G *Ektoderm*]: Outermost layer of body (ectoderm, mesoglea, endoderm); typically ciliated and containing gland cells, pigment cells, nerve cells, and colloblasts. Lines pharynx and tentacle sheath. (cellular, syncytial)

endoderm [G *Entoderm*]: Ciliated layer of body (ectoderm, mesoglea, endoderm) lining gastrovascular system (with the exception of pharynx).

esophagus (oesophagus) [G *Ösophagus*]: Relatively narrow section of gastrovascular system between pharynx and stomach.

excretory canal: anal canal

excretory pore: anal pore

funnel: stomach

gonopore [G *Gonoporus*]: One of a variable number of openings of ovaries or testes to exterior, typically at level of tentacular sheath.

granule: adhesive spherule

head [G *Klebeglocke*]: Outer hemispherical section of colloblast bearing adhesive spherules; attached to tentacle by spiral and straight filaments.

infundibulum: stomach

lateral filament: tentillum

meridional canal [G *Meridionalgefäß, Rippengefäß*]: One of eight gastrovascular canals underlying each comb row and connected to stomach via adradial, interradial, and transverse canals; wall of each meridional canal contains testes and ovaries. (simple, branched, anastomosing)

mesoglea (collenchyme) [G *Mesogloea*]: Thick layer of body between ectoderm and endoderm; contains fibers, muscles, nerves, and ameboid cells.

mouth [G *Mund, Mundöffnung*]: Opening of gastrovascular system to exterior at oral end of body; typically slit-shaped.

oral lobe [G *Mundlappen, Orallappen, Schwimmlappen*]: One of two large lobes parallel to tentacular plane and partially enveloping body.

oral pole [G *oraler Pol*]: Pole diametrically opposite apical organ; bears mouth. (See also aboral pole)

ovary [G *Ovar*]: Female gonad, forming one of two bands (ovary, testis) along wall of each meridional canal. May open to exterior via oviduct and gonopore.

oviduct [G *Ovidukt*]: Narrow canal leading from each ovary in meridional canal to exterior; opens via gonopore.

papilla [G *Papille*]: One of various short projections of outer surface of body, typically bordering polar fields or overlying meridional canals. (simple, branched)

paragastric canal: pharyngeal canal

perradial canal: transverse canal

pharyngeal canal (paragastric canal) [G *Schlundgefäß, Pharyngealgefäß*]: One of two gastrovascular canals extending from stomach to oral pole parallel to pharynx.

pharynx (stomodeum, stomodaeum) [G *Pharynx, Schlund*]: Ectoderm-lined section of gastrovascular system between mouth and esophagus or stomach. Typically flattened perpendicular to plane of tentacles (pharyngeal plane).

planula [G *Planula*]: Ciliated, free-swimming larva of certain parasitic ctenophores. (See also cydippid larva)

plate row: comb row

polar field (polar plate, pole plate) [G *Polfeld*]: One of two elongate, ciliated depressions extending from apical organ in pharyngeal plane.

polar plate: polar field

pole plate: polar field

rib: comb row

ring canal [G *Ringkanal*]: In certain atentaculate ctenophores, gastrovascular ring formed around mouth by interconnection of meridional and pharyngeal canals.

rosette (cell rosette) [G *Wimperrosette, Rosette*]: Structure consisting of two circlets of ciliated cells in lining of gastrodermal canal; one circlet beats toward lumen of canal, the other into mesoglea.

seminal receptacle [G *Receptaculum seminis*]: In certain dorsoventrally flattened ctenophores, small, invaginated ectoderm sacs containing sperm.

spiral filament [G *Spiralfaden*]: Contractile filament originating from hemispherical head of colloblast and coiling spirally around straight filament.

statocyst: apical organ

statolith [G *Statolith*]: Calcareous body supported by four tufts of balancer cilia in apical organ.

stomach (infundibulum) [G *Magen, Zentralmagen*]: Expanded section of gastrovascular system between pharynx or esophagus and aboral canal; gives rise to orally directed pharyngeal canals and to transverse canals, the latter branching to form interradial, adradial, and tentacular canals.

stomodeum (stomodaeum): pharynx

straight filament [G *Zentralfaden*]: Filament connecting hemispherical head of colloblast to tentacle; surrounded by spiral filament.

subepidermal nerve plexus [G *subepidermaler Nervenplexus, Nervennetz*]: Network of nerve fibers underlying ectoderm; typically more concentrated under comb row.

tentacle [G *Tentakel*]: One of two solid, purely ectodermal, muscular structures bearing colloblasts and typically originating within tentacle sheath. May also refer to series of short projections in auricular grooves or along oral edge of ribbon-shaped ctenophore. (simple, branched)

tentacular base (tentacle base) [G *Tentakelbasis*]: Enlarged base of tentacle lying alongside tentacular canal within tentacular sheath.

tentacular canal [G *Tentakelgefäß*]: One of two gastrovascular canals extending from transverse canal into tentacular sheath; adjoined by tentacular base.

tentacular sheath (tentacle sheath) [G *Tentakeltasche*]: Ectodermal invagination into which tentacle may be retracted; contains tentacular canal and adjoining tentacular base.

tentillum (lateral filament)

[G *Seitenfaden*] : One in a series of variously shaped branches of tentacle.

testis

[G *Hoden*]: Male gonad, typically forming one of two bands (ovary, testis) along wall of each meridional canal. May open to exterior via vas deferens and gonopore.

transverse canal (perradial canal)

[G *Transversalgefäß*]: One of two gastrovascular canals extending from stomach in plane of tentacles and, after branching, giving rise to interradial and tentacular canals.

vas deferens

[G *Vas deferens, Samenleiter*]: Narrow canal leading from each testis in meridional canal to exterior; opens via gonopore.

PLATYHELMINTHES

TURBELLARIA

turbellarian, free-living flatworm [G *Turbellar, Strudelwurm*]

TURBELLARIA

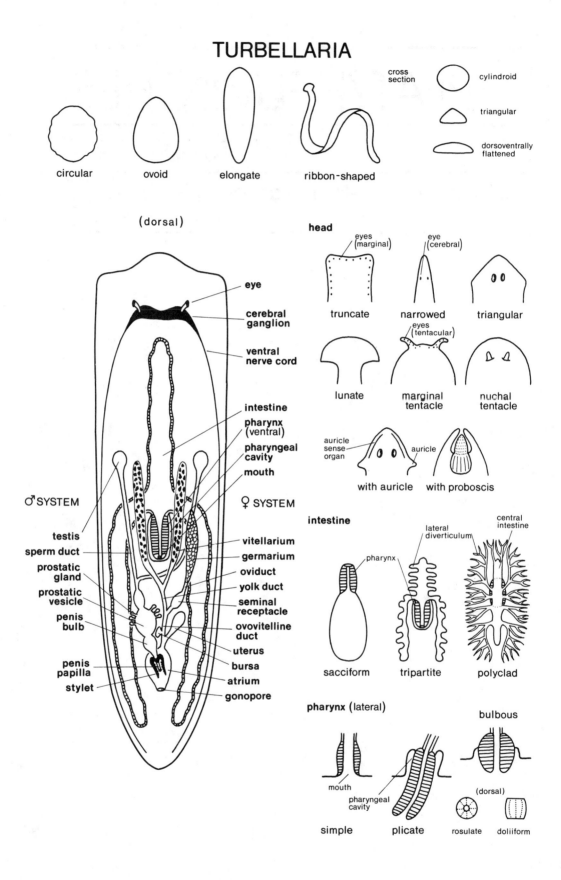

circular   ovoid   elongate   ribbon-shaped

cross section — cylindroid, triangular, dorsoventrally flattened

(dorsal)

eye
cerebral ganglion
ventral nerve cord
intestine
pharynx (ventral)
pharyngeal cavity
mouth

♂ SYSTEM

testis
sperm duct
prostatic gland
prostatic vesicle
penis bulb
penis papilla
stylet

♀ SYSTEM

vitellarium
germarium
oviduct
yolk duct
seminal receptacle
ovovitelline duct
uterus
bursa
atrium
gonopore

head

eyes (marginal) — truncate
eye (cerebral) — narrowed
triangular
lunate
eyes (tentacular) — marginal tentacle
nuchal tentacle
auricle sense organ — auricle — with auricle
with proboscis

intestine

lateral diverticulum — central intestine
pharynx
sacciform   tripartite   polyclad

pharynx (lateral)

bulbous
mouth
pharyngeal cavity
simple   plicate   (dorsal) rosulate doliiform

PLATYHELMINTHES

TURBELLARIA

turbellarian, free-living flatworm [G *Turbellar, Strudelwurm*]

---

acetabulum (sucker) [G *Saugnapf*]: True muscular sucker separated from parenchyma by septum and characterized by rim with central depression; occurs in longitudinal row along each margin. (sessile, stalked)

adenodactyl [G *männlicher Reizorgan*]: Prostatic vesicle-like structure associated with ♂ reproductive system or in wall of common ♂ and ♀ antrum.

adhesive disc [G *Saugnapf, Haftplatte*]: Adhesive structure lacking musculature; typically refers to expansion of marginal adhesive zone into discrete adhesive structure, yet often applied to wide range of glandulomuscular adhesive organs.

adhesive papilla [G *Haftpapille*]: Minute papilla formed by opening of two adhesive glands and specialized epidermal cells which can quickly adhere and release. Often in groups, forming finger-like projection.

anal pore [G *Analporus*]: Minute pore of intestinal branch to exterior; typically arranged in longitudinal series; sometimes a single pore on caudal end of straight gut. (dorsal, lateral)

antrum [G *Antrum masculinum, Antrum femininum*]: Expanded terminal section of reproductive system; may form common ♂ and ♀ opening (then termed atrium) or, if separate (then termed antrum),

enclosing penis in ♂ and forming enlarged section of vagina in ♀.

apioid organ: prostatoid

athrocyte (paranephrocyte) [G *Athrozyte, Paranephrozyte*]: Large cell type with conspicuous nucleus and vacuolated cytoplasm adjoining or enclosing nephridial tubule. (single, in clusters)

atrium: antrum

auricle [G *Aurikel, Öhrchen*]: Lateral extension of head; bears sensory cells (auricular groove) and gives head triangular appearance.

auricular groove (auricular sense organ) [G *Aurikularsinnesorgan*]: Sensory groove along margin of head or across base of auricles when present.

bowl-shaped body [G *schüsselförmiger Körper*]: One of four pairs of sensory structures connected to brain in head; each consists of refringent structure(s) forming bowl over clear vesicle.

brain: cerebral ganglion

buccal tube [G *Mundrohr*]: Elongated, tubular elaboration of pharyngeal cavity; termed buccogenital canal in certain turbellarians.

bursa [G *Bursa*]: General term for sperm storage organ in ♀ reproductive system; according to duration of storage one may distinguish copulatory bursa or seminal bursa. Opens either into seminal receptacle, indirectly (via bursal canal) or directly into atrium/vagina, or through separate bursal pore via copulation canal.

bursal canal [G *Bursastiel*]: In ♀ reproductive system, canal joining sperm storage organ (bursa) to antrum. (See also insemination canal)

bursal pore [G *Bursalporus*]: In ♀ reproductive system in which sperm storage organ (bursa) connects to antrum or atrium, opening of bursa (independent of genital pore) to exterior.

capillary [G *Kapillar*]: Slender branch of nephridial tubule; bears flame bulb terminally.

capsule: cocoon

cement gland [G Schalendrüse]: One of numerous glands opening into or near terminal section of ♀ reproductive system; secretes material joining eggs to one another or to substratum. (See also cement sac)

cement sac (glandular duct) [G *Drüsengang*]: Term applied to terminal section of ♀ reproductive system into which all glands involved in egg-laying process open.

cerebral ganglion (brain) [G *Zerebralganglion, Gehirn*]: Bilobed, band-shaped concentration of nerve tissue at anterior end (head); may be associated with statocyst and gives rise anteriorly to series of sensory nerves, posteriorly either to pair of large ventral nerve cords or to pairs of dorsal, lateral, and ventrolateral cords.

chondrocyst [G *Chondrozyste*]: Type of rhabdoid in terrestrial turbellarian; large rhabdite with granular surface.

ciliated pit [G *Wimpergrube*]: Small sensory depression in head; typically bears stiff cilia centrally. (single, unpaired; rounded, flask-shaped)

circular muscle [G *Ringmuskel*]: Outermost of typically three layers (circular, diagonal, longitudinal muscles) of subepidermal musculature of body wall. May also refer to circular muscles of pharynx or other inner organs.

cirrus [G *Zirrus, Cirrus*]: Terminal part of ♂ reproductive system in which ejaculatory duct is lined with spines or thorns, forming a protrusible penis papilla. During copulation, cirrus is everted and armature comes to lie on outer surface.

cocoon (capsule) [G *Kokon, Kapsel*]: Hard shell, formed in antrum or uterus, enclosing eggs.

commissure [G *Kommissur*]: Thin transverse nerve connection between longitudinal nerve cords (dorsal, lateral, ventrolateral cords).

concrement [G *Konkrement*]: Minute, crystal-like body; typically forms dense aggregations.

conorhynch: proboscis

copulation canal (ductus vaginalis) [G *Ductus vaginalis*]: In ♀ reproductive system in which sperm storage organ (bursa) does not open into regular gonopore, canal between bursa and separate bursal pore.

copulatory bursa (bursa copulatrix) [G *Bursa copulatrix*]: In ♀ reproductive system, sperm storage organ; often receives penis during copulation. Termed seminal bursa when sperm is stored over longer periods. (See also seminal receptacle)

creeping sole [G *Kriechsohle*]: On ventral surface, longitudinal strip(s) or ridge(s) of well-developed cilia; serves in locomotion.

cuticle [G *Kutikula*]: Rigid intracellular material coating tip of prostatoid or forming elements of false cuticle (e.g., stylet).

diagonal muscle [G *Diagonalmuskel*]: Middle of three layers (circular, diagonal, longitudinal muscles) of subepidermal musculature of body wall.

dorsoventral muscle [G *Dorsoventralmuskel*]: In parenchymal musculature, muscle spanning from dorsal to ventral side of body.

ductus spermaticus: insemination canal

ductus vaginalis: copulation canal

egg [G *Ei*]: ♀ reproductive body produced in ovary or germarium, in the latter case supplied with yolk by vitellarium, maturing in uterus, and encapsulated in antrum. (entolecithal, ectolecithal, subitaneous, dormant)

ejaculatory duct [G *Ductus ejaculatorius*]: In ♂ reproductive system, lumen of terminal section of common sperm duct.

enteron: intestine

epidermis [G *Epidermis*]: Outermost cell layer of body wall (epidermis, basement membrane, musculature); typically ciliated, at least on ventral surface, and containing gland cells and rhabdoid glands. (cellular, syncytial, insunk)

esophagus [G *Ösophagus*]: Short, narrow, nonmuscular section of digestive tract between pharynx and intestine.

excretory beaker [G *Exkretionsbecher*]: Cavity between mouth and pharyngeal cavity into which nephridial tubule opens.

eye (pigment-cup ocellus) [G *Auge, Pigmentbecherocellus*]: Term applied to one of variously arranged photosensitive structures in mesenchyme. May also refer to epidermal pigment spots. (epidermal, cerebral, marginal, tentacular)

flame bulb [G *Terminalorgan, Cyrtozyte*]: In protonephridium, series of osmoregulatory structures along nephridial canal; contains tuft of cilia and is attached to tubule by capillary. (See also lateral flame)

frontal gland [G *Kopfdrüse, Stirndrüse*]: Cluster of glands near brain in head; opens to exterior at anterior end of body via long necks.

frontal organ [G *Frontalorgan*]: Sensory and glandular complex formed in head by frontal gland and accompanying nerve fibers.

genital sucker: glandulomuscular adhesive organ

genitointestinal canal [G *Ductus genito-intestinalis, Canalis genito-intestinalis*]: Connection between intestine and ♀ reproductive system.

germarium [G *Germarium, Keimstock*]: In ♀ reproductive system, egg-producing part of ovary (if this is completely separated from vitellarium or yolk-producing part). (single, paired; club-shaped, cylindrical, lobed, oval, pyriform) (See also germovitellarium).

germovitellarium [G *Germovitellarium*]: In ♀ reproductive system, term applied to ovary if this is differentiated into yolk- and egg-producing regions (yet two regions not completely separated). (single, paired)

glandular duct: cement sac

glandular margin [G *Drüsenkante*]: In terrestrial turbellarian, strip of glands opening on ventral surface along each margin. (See also marginal adhesive zone)

glandulomuscular adhesive organ [G *Drüsenpolster, Drüsentasche, Haftgrube, Haftwulst, Sauggrube, Saugorgan, Kleborgan*]: General term for various adhesive organs consisting of muscular protrusions surrounding depression with openings of gland cells; differs from marginal adhesive glands through well-developed musculature. Typically single organ on anterior margin, on ventral side of head, or between ♂ and ♀ gonopore. (single, paired)

gonad: ovary, testis

gonopore [G *Gonoporus, Geschlechtsöffnung*]: Opening of ♂ or ♀ reproductive system to exterior; ♂ and ♀ may have separate or common gonopore. (See also antrum)

Götte's larva [G *Götte'sche Larve*]: Free-swimming larval stage bearing four lobes fringed by long cilia. (See also Müller's larva)

head [G *Kopf*]: Anterior end of turbellarian; may be delimited from remainder of body by neck-like constriction and bear brain, eyes, frontal organ, auricles, and adhesive organs.

insemination canal (ductus spermaticus) [G *Ductus spermaticus*]: In ♀ reproductive system, thin canal between sperm storage organ (bursa) and oviduct.

intestine (enteron, gut) [G *Mitteldarm, Darm*]: Main section of digestive tract, typically traversing entire body length; may bear lateral diverticula. Lacking in smaller microturbellarians. (polyclad, sac-like, three-branched/tripartite/triclad)

Lang's vesicle [G *Bursalorgan*]: In ♀ reproductive system, elongate modification of seminal bursa.

lateral diverticulum [G *Divertikel, Darmast*]: One in a series of lateral branches of intestine. (simple, branched, anastomosing)

lateral flame [G *Treibflamme, Treibwimperflamme*]: In protonephridium, tuft of cilia located within nephridial tubule itself.

lateral nerve cord (marginal nerve cord) [G *lateraler Nervenstamm*]: Pairs of longitudinal nerve cords arising from brain and extending to posterior end of body; joined to other longitudinal cords (dorsal, ventrolateral, ventral cords) by commissures.

lithocyte [G *Lithozyte*]: Cell type containing statolith; located within vesicle to form statocyst.

longitudinal muscle [G *Längsmuskel*]: Innermost of three layers (circular, diagonal, longitudinal muscles) of subepidermal musculature of body wall. May occur in bundles. Also refers to longitudinal muscles in proboscis, pharynx, and penis or in parenchymal musculature.

macroturbellarian: Category of turbellarian measuring more than one centimeter in length. (See also microturbellarian)

marginal adhesive gland [G *Haftdrüse*]: In aquatic turbellarian, series of glands encircling ventral surface of body; open to exterior via series of serial pores

(marginal adhesive zone). (See also adhesive disc, adhesive papilla, glandulomuscular adhesive organ)

marginal adhesive zone [G *Haftzellenring*]: In aquatic turbellarian, adhesive zone around periphery of ventral surface; formed by serial openings of marginal adhesive glands. (See also glandular margin)

mesenchyme: parenchyma

microturbellarian: Category of turbellarian measuring no more than a few millimeters. (See also macroturbellarian)

mouth [G *Mund*]: Opening of digestive tract; with the exception of turbellarians bearing anal pores, serves both for ingestion and egestion. Located along midline of ventral surface. Opens directly into pharynx simplex or parenchyma of simple turbellarians, into pharyngeal cavity of forms with more complex, protrusible pharynges. (circular, slit-shaped)

Müller's larva [G *Müller'sche Larve*]: Free-swimming larval stage bearing eight lobes fringed with long cilia.

musculature: parenchymal musculature, subepidermal musculature

nephridial tubule [G *Exkretionskanal*]: Elongate, tubular section of protonephridium between flame bulbs and nephridiopore; typically forms two to four longitudinal stems underlying epidermis. May be expanded into muscular bladder before opening to exterior. (dorsal, ventral; single, paired) (See also capillary)

nephridiopore [G *Exkretionsporus*]: Opening of protonephridium to exterior; may open into pharyngeal cavity or excretory beaker. If tubule bears numerous openings along its length, one may refer to accessory nephridiopores. (single, paired; ventral, dorsal; central, posterior)

ovary [G *Ovar, Ovarium*]: Section of ♀ reproductive system in which only entolecithal eggs are produced. (single, paired, numerous; tubular, sacciform)

oviduct [G *Ovidukt*]: Section of ♀ reproductive system between ovary and sperm storage organ (bursa, seminal receptacle). (single, paired)

oviductule [G *Kanälchen*]: In ♀ reproductive system characterized by multiple ovaries, series of minute ducts joining each ovary to oviduct.

ovovitelline duct: In ♀ reproductive system, term applied to oviduct if yolk-producing vitellaria discharge directly (and not via yolk ducts) into oviduct.

papilla: adhesive papilla, tubercle

paranephrocyte: athrocyte

parenchyma   (mesenchyme) [G *Parenchym, Mesenchym*]: Tissue between body wall and major organs; typically consists of net-like parenchymal cells and small free cells.

parenchymal   musculature [G *Körpermuskulatur*]: System of muscles traversing parenchyma; consists of dorsoventral, longitudinal, and transverse muscles. (See also subepidermal musculature)

penis [G *Penis*]: Terminal section of ♂ reproductive system; typically produced into muscular, conical projection (penis papilla). (unarmed, armed) (See also cirrus, stylet)

penis bulb [G *Penisbulbus*]: In terminal section of ♂ reproductive system, muscular, rounded structure supporting penis papilla; may incorporate seminal vesicle and/or prostatic vesicle.

penis papilla [G *Penispapille*]: Muscular, protrusible type of penis located in chamber (male antrum). May be armed with thorns, spines, or stylet. (short, long; truncate, conical)

pharyngeal cavity [G *Pharyngealtasche*]: In turbellarian with folded (plicate, pharynx plicatus) pharynx, cavity surrounding retracted pharynx and opening to exterior through mouth.

pharyngeal gland [G *Pharyngealdrüse*]: One of numerous gland cells located in pharynx wall or opening into pharynx via long necks.

pharynx [G *Pharynx, Schlund*]: Typically muscular, protrusible or eversible section of digestive tract; many-layered, including several muscle layers, nerve tissue, and gland cells. One may distinguish 1) pharynx O: simple mouth pore; 2) pharynx simplex: simple, invaginated, nonmuscular tube with terminal mouth; 3) pharynx plicatus: folded, muscular, cylindrical, cone-shaped, suspended in large pharyngeal cavity, protrusible from mouth; 4) pharynx bulbosus: short, thick, muscular pharynx with pharyngeal cavity and delimited from parenchyme—further divided into rosulate and doliiform types as well as variable pharynx. (bulbous, doliiform, flaring, plicate, rosulate, simple, tubular, variable)

pigment-cup ocellus: eye

proboscis (conorhynch) [G *Rüssel, Fangrüssel*]: Protrusible glandulomuscular adhesive organ in anterior tip of body; may be differentiated into anterior end cone and posterior muscle cone. Serves in capture of prey. (unarmed, armed; undivided, divided/bifurcated; eversible, protrusible; conical, bipartite)

proboscis pore [G *Rüsselscheidenöffnung*]: Opening of proboscis sheath to exterior at anterior tip of body.

proboscis sheath [G *Rüsselscheide*]: Cavity enclosing retracted proboscis.

prostatic duct: In ♂ reproductive system, short tubule connecting free prostatic vesicle to ejaculatory duct.

prostatic gland [G *Kornsekretdrüse*]: In terminal section of ♂ reproductive system, aggregation of gland cells opening either into ejaculatory duct, penis base, or prostatic vesicle.

prostatic vesicle [G *muskulöses Drüsenorgan, Kornsekretblase*]: In terminal section of ♂ reproductive system, muscular vesicle into which prostatic glands open. Either located between seminal vesicle and penis, incorporated with seminal vesicle, or connected to ejaculatory duct via prostatic duct. (free, interpolated; rounded, oval, tubular)

prostatoid (apioid organ, pyriform organ) [G *männlicher Reizorgan*]: Term applied to one of many prostatic vesicle-like structures when multiple such organs are located around male antrum or along ventral surface; consists of muscular, pyriform bodies with attached glands and pointed, hardened tip.

protonephridium [G *Protonephridium*]: Osmoregulatory and excretory organ typically consisting of two or four pairs of longitudinal nephridial tubules bearing series of flame bulbs; opens to exterior via nephridiopore. (single, paired)

pseudorhabdite [G *Pseudorhabdit*]: Type of rhabdoid: indefinite mass of slimy secretion inside epidermal cells.

pyriform organ: prostatoid

rhabdite [G *Rhabdit*]: Type of rhabdoid: rod-shaped body, shorter than height of epidermis cell, oriented perpendicular to surface of body. Typically more abundant dorsally and laterally. (adenal, epidermal; straight, curved) (See also rhammite, sagittocyst)

rhabdoid [G *Rhabdoid*]: General term for variously shaped, minute, rigid bodies (rhabdite, rhammite, chondrocyst, sagittocyst) in body wall; secreted by gland cells and typically consist of hard outer layer and more fluid inner core.

rhammite [G *Rhammit*]: Type of rhabdoid: long, slender. Formed by group of glands in mesenchyme, with visible rhammite tract extending to anterior tip of animal (part of frontal gland).

sagittocyst [G *Sagittozyste*]: Type of rhabdoid: pointed vesicle containing central protrusible rod or needle.

seminal bursa (bursa seminalis) [G *Bursa seminalis*]: In ♀ reproductive system, term applied to sperm storage organ (bursa, copulatory bursa) if storage takes place over longer period. (single, paired, numerous) (See also seminal receptacle)

seminal receptacle [G *Receptaculum seminis*]: In ♀ reproductive system, sperm storage organ; consists of expanded section of oviduct or follows bursa.

seminal vesicle [G *Samenblase, Vesicula seminalis*]: Expanded muscular section of ♂ reproductive system following sperm duct(s); serves to store sperm and may form complex together with penis and prostatic apparatus. (round, pyriform, tubular)

sensory margin [G *Sinneskante*]: In terrestrial turbellarian, concentration of sensory cells in longitudinal strip along each margin of body.

spermatophore [G *Spermatophore*]: Packet of sperm produced in seminal vesicle of ♂ and injected into ♀ during copulation, after which it may be found in bursa.

sperm duct (vas deferens) [G *Vas deferens*]: Tubular section of ♂ reproductive system between testis and seminal vesicle. May join sperm duct from second testis to form common duct; each duct or common duct may be enlarged to form spermiducal vesicle. (single, paired)

sperm ductule [G *Vas efferens*]: If multiple testes are present in ♂ reproductive system, minute duct connecting each testis to sperm duct.

spermiducal vesicle (false seminal vesicle) [G *innere Samenblase*]: In ♂ reproductive system, enlarged section of each sperm duct or common duct; serves to store ripe sperm. Leads into copulatory complex (prostatic vesicle, penis) or seminal vesicle if present. (sacciform, tubular; thin-walled, thick-walled)

spine [G *Stachel*]: In certain turbellarians, one in a series of minute spines along margin of body; each consists of hardened epidermal secretion overlying papilla.

statocyst [G *Statozyste*]: Sensory structure adjoining brain; consists of vesicle containing lithocyte with internal statolith.

statolith [G *Statolith*]: Minute, solid body within lithocyte.

stylet [G *Stilett*]: In ♂ reproductive system, rigid structure(s) enclosing penis or penis papilla; consists either of hollow tube or complex system of elements. (tubular: straight, curved; complex)

subepidermal musculature [G *Hautmuskelschlauch*]: Stratum of muscles underlying epidermis. Typically consists of outer circular layer, middle diagonal layer, and inner longitudinal layer, yet may consist of up to six layers. Better developed ventrally. (See also parenchymal musculature)

subepidermal plexus [G *subepidermaler Plexus, Hautnervenplexus*]: Delicate network of nerves directly underlying basement membrane of epidermis. (See also submuscular plexus)

submuscular plexus [G *submuskulärer Plexus*]: Main network of nerves located in parenchyma and typically consisting of brain, one to six pairs of longitudinal nerve cords, and numerous transverse connections (commissures). (See also subepidermal plexus)

sucker [G *Saugnapf*]: General term for sucker-like adhesive structures. When strictly applied, refers only to muscular sucker (acetabulum), yet frequently used for wide array of glandulomuscular adhesive organs.

tail filament [G *Schwanzfaden, Schwanzanhang*]: Slender, tail-like appendage at posterior end of body.

tentacle [G *Tentakel*]: One of typically two, yet up to 12 projections of head; may bear eyes. One may distinguish nuchal tentacle located over brain region or marginal tentacle along anterior margin.

testis [G *Hoden*]: Proximal section of ♂ reproductive system; opens to exterior via sperm duct, seminal vesicle, prostatic vesicle, and penis. (single, paired, numerous; oval, elongated, lobulated, branched; dorsal, lateral, ventral)

transverse muscle [G *Transversalmuskel*]: In parenchymal musculature, transverse muscle spanning through mesenchyme.

uterus [G *Uterus*]: Section of ♀ reproductive system serving to store ripe eggs or occasionally immature individuals; consists either of modified, expanded section of oviduct or separate organ extending from antrum. (single, paired; sac-like, tubular)

vagina [G *Vagina*]: Section of ♀ reproductive system between oviduct and gonopore; may be expanded into antrum.

vas deferens: sperm duct

ventral nerve cord [G *ventraler Nervenlängsstamm, ventraler Längsnerv*]: Pair of well-developed ventral nerve cords arising from brain and extending to posterior end; joined to other longitudinal cords, if present, via commissures. (See also lateral nerve cord, ventrolateral nerve cord)

ventral nerve plate [G *ventrale Nervenplatte, Nervenplatte*]: In terrestrial turbellarian, advanced submuscular nervous system consisting of thick ventral nerve layer; extends along entire ventral surface of body (brain is lacking).

ventrolateral nerve cord [G *ventrolateraler Längsnervenstamm*]: Pair of longitudinal nerve cords arising from brain and extending to posterior end of body; joined to other longitudinal cords (dorsal, lateral, ventral nerve cords) by commissures.

vitellarium (yolk gland) [G *Vitellarium, Vitellar, Dotterstock*]: In ♀ reproductive system, yolk-producing part of ovary (if this is completely separated from egg-producing germarium). (elongate, lobed, branched, follicular)

yolk duct [G *Vitellodukt*]: In ♀ reproductive system, duct between yolk gland (vitellarium) and oviduct. (See also ovovitelline duct)

yolk gland: vitellarium

zooid [G *Zooid, Tochterindividuum*]: In turbellarian reproducing asexually by transverse fission, one in a series of individuals attached to one another in a chain.

TREMATODA

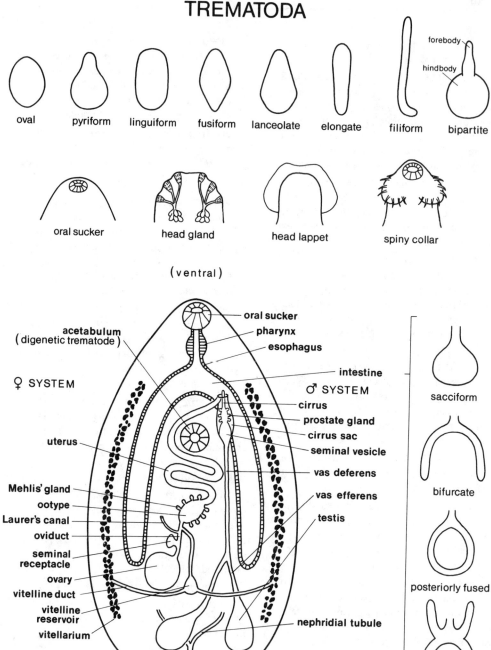

oval   pyriform   linguiform   fusiform   lanceolate   elongate   filiform   bipartite

forebody
hindbody

oral sucker   head gland   head lappet   spiny collar

(ventral)

oral sucker
pharynx
esophagus
acetabulum
(digenetic trematode)
♀ SYSTEM
intestine
♂ SYSTEM
cirrus
prostate gland
cirrus sac
seminal vesicle
uterus
vas deferens
vas efferens
Mehlis' gland
ootype
testis
Laurer's canal
oviduct
seminal
receptacle
ovary
vitelline duct
vitelline
reservoir
vitellarium
nephridial tubule
opisthaptor
(monogenetic trematode)
excretory bladder
nephridiopore

sacciform
bifurcate
posteriorly fused
H-shaped

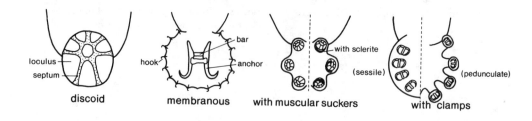

loculus
septum
discoid

hook   bar   anchor
membranous

with sclerite
(sessile)
with muscular suckers

(pedunculate)
with clamps

PLATYHELMINTHES

TREMATODA

trematode, fluke [G *Trematode, Saugwurm]*

acetabulum (ventral sucker) [G *Bauchsaugnapf, Acetabulum*]: True muscular sucker working primarily on vacuum principle (not primarily glandular or with rigid elements); typically refers to ventral sucker of digenetic trematode. (central, posterior; pedunculate, sessile)

adhesive disc [G *Saugscheibe*]: Greatly enlarged ventral adhesive organ extending along virtually entire length of body; bears elevated septa and suctorial depressions (loculi) or consists of longitudinal rows of suckers.

adolescaria [G *letztes Larvenstadium*]: Collective obsolete term for cercaria and metacercaria larval stages in digenetic trematode life cycle.

alveolus: loculus

amphistome [G *Amphistom*]: Digenetic trematode with ventral sucker (acetabulum) located at or near posterior end of body. (See also distome, echinostome, gasterostome, holostome, monostome)

anal pore [G *Anus, Analporus*]: Opening of intestinal cecum to exterior. (single, paired)

anchor [G *Anker, Ankerhaken*]: One of typically two large hooks on posterior adhesive organ (opisthaptor) of monogenetic trematode; often connected by bar. Occasionally termed claw or hook. (See also bar, hook)

anterior sucker [G *präoraler Saugnapf*]: Type of anterior adhesive organ (prohaptor) in monogenetic trematode; typically consists of pair of suckers located at anterior margin of body at some distance from mouth.

antrum (genital atrium, atrium) [G *Antrum, Atrium, Atrium genitale, Genitalatrium*]: Common terminal section of ♂ and ♀ reproductive systems. Typically forms chamber into which ♂ copulatory apparatus and uterus enter; opens to exterior via gonopore. (armed, unarmed)

anus: anal pore

atrium: antrum

bar [G *Klammer, Balken*]: In posterior adhesive organ (opisthaptor) of monogenetic trematode, rigid, transverse element joining base of two anchors. (See also hook)

bladder: excretory vesicle

bothrium [G *Bothrium*]: Type of anterior adhesive organ (prohaptor) in monogenetic trematode; consists of two lateral or ventral pits or grooves.

brain: cerebral ganglion

buccal sucker [G *innerer Saugnapf, Bukkaltasche*]: Type of anterior adhesive organ (prohaptor) in monogenetic trematode; consists of pair of small suckers opening into buccal tube.

buccal tube [G *Mundhöhle*]: General term for relatively short, nonmuscular section of digestive tract between mouth and pharynx; termed prepharynx in digenetic trematodes and may bear buccal suckers in monogenetic forms.

bursa: copulatory bursa

caecum: intestinal cecum

capillary [G *Kapillar*]: Slender branch of nephridial tubule; bears flame bulb terminally.

capsule: egg

cecum: intestinal cecum

cephalic gland [G *Kopfdrüse*]: Multicellular glands in head of monogenetic trematode; open via ducts at anterior margin of body to form head organ.

cercaria [G *Cercarie*]: In life cycle of digenetic trematode, fourth of typically five larval stages (miracidium, sporocyst, redia, cercaria, metacercaria); typically briefly free-swimming and characterized according to variously shaped tail or position of ventral sucker. (amphistome, distome, echinostome, gasterostome, holostome, monostome; mesostomate, stenostomate; lophocercous; oculate, nonoculate; apharyngeate, pharyngeate; in distome forms: brevifurcate, chaetomicrocercous, cotylocercous, cystocercous, cystophorous, furcocercous, longifurcate, macrocercous, microcercous, obscuromicrocercous, pleurolophocercous, rhopalocercous, sulcatomicrocercous, trichocercous, virgulate) (See also cercariaeum, rat-king cercaria, xiphidiocercaria)

cercariaeum [G *Cercariaeum*]: Type of cercaria lacking tail.

cerebral ganglion (brain) [G *Zerebralganglion, Kopfganglion, Gehirn*]: One of two adjoining concentrations of nerve tissue connected by broad commissure. Located dorsally at anterior end of body; in monogenetic trematodes above and behind pharynx, in digenetic forms between oral sucker and pharynx. Gives rise posteriorly to pair of larger ventral nerve cords and often pair of smaller dorsal and lateral cords.

circular muscle [G *Ringmuskel*]: Outermost of typically three layers (circular, longitudinal, diagonal; circular, diagonal, longitudinal muscles) of subcuticular musculature.

cirrus [G *Zirrus, Cirrus*]: ♂ copulatory organ consisting of eversible ejaculatory duct (as opposed to noneversible penis); typically enclosed in and continuous with cirrus sac. (armed, unarmed, spinous, nonspinous)

cirrus sac (cirrus pouch) [G *Zirrusbeutel*]: Terminal section of ♂ reproductive system; consists of muscular, pouch-like structure; contains cirrus and may enclose seminal vesicle, prostatic glands, and prostatic vesicle. (See also sinus sac)

clamp [G *Haftklappe*]: One in a series of jaw-like structures located either in posterior adhesive organ (opisthaptor) of monogenetic trematode or on posterior end of body itself. Each consists of bivalved structure of two membranous flaps with supporting framework of rigid elements (sclerites). (pedunculate, sessile; fire-tong-shaped, guitar-shaped, oblong, oval)

cloaca (uroproct) [G *Kloake*]: Common duct formed by junction of posterior ends of intestinal ceca with excretory vesicle; opens to exterior via cloacal pore.

cloacal pore [G *Kloakalporus*]: Opening of cloaca to exterior at posterior end of body.

collar spine (head spine) [G *Kopfstachel, "Halsstachel"*]: Single or double row of spines on expanded collar (head collar) of echinostome trematode. (See also spine)

commissure [G *Kommissur*]: Thin transverse nerve connection between longitudinal nerve cords (ventral, lateral, dorsal cords) or thick connective between two cerebral ganglia.

copulation canal: vagina

copulatory bursa [G *Bursa copulatrix*]: In digenetic trematode, pouch-like invagination enclosing genital cone and occasionally genital bulb.

cotylocidium [G *Cotylocidium*]: Larval stage of trematode whose adult stage is characterized by large ventral adhesive disc.

cuticle: tegument

cyclocoel [G *Cyclocoel*]: Type of intestine in which two intestinal ceca fuse posteriorly.

daughter sporocyst [G *Tochtersporozyste*]: Second generation of sporocysts in life cycle of digenetic trematode; produced within mother sporocyst and gives rise to cercaria.

diagonal muscle [G *Diagonalmuskel*]: Middle or innermost of typically three layers (circular, diagonal, longitudinal; circular, longitudinal, diagonal muscles) of subcuticular musculature.

diporpa larva [G *Diporpa*]: Type of larva of monogenetic trematode; characterized by ventral sucker and dorsal papilla and permanently attached to second individual.

distome [G *Distome*]: Digenetic trematode with ventral sucker (acetabulum) located at or anterior to center of body. (See also amphistome, echinostome, gasterostome, holostome, monostome)

diverticulum [G *Divertikel*]: One in a series of lateral branches of intestinal cecum. (lobed, branched, anastomosing)

dorsal nerve cord [G *dorsaler Längsnervenstrang, dorsaler Nervenstamm*]: Pair of longitudinal nerve cords arising from cerebral ganglia and extending to posterior end of body; joined to ventral and lateral cords by commissures.

echinostome [G *Echinostome*]: Digenetic trematode whose oral sucker is surrounded by spinous collar (head collar). (See also amphistome, distome, gasterostome, holostome, monostome)

ecsoma [G *Ecsoma*]: In certain digenetic trematodes, thin-walled, tail-like posterior end of body; may be withdrawn into anterior end (soma) like telescope.

egg [G *Ei*]: Reproductive body produced in ovary and supplied with yolk and capsule by vitelline gland. Capsule may be drawn out at one or both ends to form polar filaments. Operculate egg typical of digenetic trematode, nonopeculate egg of monogenetic trematode. (operculate, nonoperculate; embryonated, unembryonated; ectolecithal; oval, fusiform)

ejaculatory duct [G *Ductus ejaculatorius*]: In ♂ reproductive system, terminal, muscular, cuticle-lined section of common sperm duct (vas deferens).

esophageal bulb [G *Ösophagealbulbus*]: Muscular bulb at posterior end of esophagus.

esophagus [G *Ösophagus*]: Relatively short, cuticle-lined section of digestive tract between pharynx and intestine; typically slender, yet occasionally with muscles and receiving ducts of salivary glands.

excretory vesicle (bladder) [G *Exkretionsblase*]: In monogenetic trematode, small bladder preceding each anterior nephridiopore; in digenetic forms, single large bladder preceding posterior nephridiopore. (elongate, funnel-like, sac-like, saccular, tubular, V-shaped, Y-shaped, much branched)

eye (ocellus) [G *Auge, Augenfleck, Ocelle*]: Photosensitive organ, typically restricted to larval stages and adult monogenetic trematodes; consists of pigment spot or simple pigment cup, occasionally with lens-like structure.

false cirrus sac: sinus sac

fimbria (tentacle) [G *Fimbria, Tentakel*]: In digenetic trematode, one in a circlet of tentacle-like projections around oral sucker or rhynchus.

flame bulb [G *Terminalzelle*]: In protonephridium, series of excretory structures along each nephridial tubule; contains tuft of cilia and is attached to tubule by capillary.

forebody [G *Vorderkörper*]: In certain digenetic trematodes, anterior division of body. (concave, flattened, slender) (See also hindbody)

gasterostome [G *Gasterostome*]: Digenetic trematode whose mouth is located at center of body. (See also amphistome, distome, echinostome, holostome, monostome)

genital atrium: antrum

genital bulb [G *Bulbus*]: In digenetic trematode, muscular projection adjoining genital cone in copulatory bursa.

genital cone [G *Genitalkegel*]: Cone-shaped terminal end of reproductive system in digenetic trematode; traversed by hermaphroditic duct and surrounded by copulatory bursa.

genital sucker [G *Genitalsaugnapf*]: In digenetic trematode, muscular, suckerlike structure encircling gonopore.

genitointestinal canal [G *Genito-Intestinal-Kanal, Ductus genito-intestinalis*]: In monogenetic trematode, slender canal between oviduct and (typically right) intestinal cecum.

genitovitelline canal: vitellovaginal canal

germinal sac: sporocyst

gonopore [G *Gonoporus*]: Opening of ♂ and/or ♀ reproductive system (or of common antrum) to exterior. Typically located on ventral surface at anterior end of body. (anterior: ventral, lateral, dorsal; posterior: ventral)

gonotyl [G *Gonotyl*]: In digenetic trematode, protrusible, muscular structure within ventrogenital sac.

gynecophoric canal (gynaecophoric canal) [G *Canalis gynaecophorus*]: In sexually dimorphic digenetic trematode, ventral groove of ♂ in which ♀ is carried.

haptor [G *Haptor, Haftnapf, Klammerapparat*]: General term for adhesive organ in monogenetic trematode. One may distinguish anterior adhesive apparatus (prohaptor) and posterior apparatus (opisthaptor). (See also larval haptor)

head collar (spiny collar) [G *Kragen*]: Expanded collar surrounding oral sucker of echinostome trematode; bears row of spines (collar spines).

head lappet [G *Kopflappen*]: Type of anterior adhesive organ (prohaptor) in monogenetic trematode; consists of membranous lateral expansions at anterior end of body and may bear pseudosuckers.

head organ [G *Kopforgan*]: Type of anterior adhesive organ (prohaptor) in monogenetic trematode; consists of papilla-like outgrowths associated with aggregations of multicellular cephalic glands. (one, two, three pairs; several pairs; numerous and scattered)

head spine: collar spine

hermaphroditic duct [G *Zwittergang*]: In digenetic trematode, common duct formed by fusion of terminal portions of ♂ and ♀ reproductive canals; may be enclosed in sinus sac.

hindbody [G *Hinterkörper*]: In certain digenetic trematodes, posterior division of body. (narrow, cylindroid, rounded, saclike) (See also forebody)

holdfast: tribocytic organ

holostome [G *Holostome*]: Digenetic trematode whose body is divided into forebody and hindbody by constriction; may bear additional adhesive organ (tribocytic organ) besides oral and ventral suckers. (See also amphistome, distome, echinostome, gasterostome, monostome)

hook (hooklet, marginal hook) [G *Marginalhaken, Larvenhaken, lateraler Haken*]: One in a circlet of smaller hooks surrounding posterior adhesive organ (opisthaptor) of monogenetic trematode. (See also anchor, bar)

hump: In trematode characterized by ventral adhesive disc extending along virtually entire length of body, dorsal part of body containing internal organs.

intestinal cecum (cecum, caecum) [G *Darmschenkel*]: One of typically two posteriorly directed, blind branches of intestine; may be joined posteriorly to form cyclocoel, or each cecum may open into excretory vesicle or separately to exterior. (simple, with diverticula, H-shaped)

intestine [G *Darm, Mitteldarm*]: Elongate section of digestive tract posterior to esophagus; typically branches to form two posteriorly directed ceca. (sac-like, bifurcate)

larval haptor [G *larvaler Haftnapf*]: In monogenetic trematode, small larval adhesive disc retained on margin of secondarily developed, adult posterior adhesive organ (opisthaptor).

lateral nerve cord [G *lateraler Längsnervenstrang, lateraler Längsstamm*]: Pair of longitudinal nerve cords arising from cerebral ganglia and extending to posterior end of body; joined to ventral cord and, if present, dorsal cord by commissures.

Laurer's canal [G *Laurer'scher Kanal*]: In digenetic trematode, section of ♀ reproductive system leading from oviduct to dorsal surface of body, where it opens via pore. May also end blindly or open into excretory system.

loculus (alveolus) [G *Loculus, Alveolus*]: In posterior adhesive organ (opisthaptor) of monogenetic trematode or in ventral adhesive disc, one of several suctorial depressions delimited by septa. (central, intermediate, peripheral)

longitudinal muscle [G *Längsmuskel*]: Middle or innermost of typically three layers (circular, longitudinal, diagonal; circular, diagonal, longitudinal muscles) of subcuticular musculature.

lymphatic system [G *Lymphsystem*]: In certain digenetic trematodes, series of blind, branching, longitudinal vessels containing fluid with free cells.

marita [G *Marita*]: Adult (sexually mature) trematode

Mehlis' gland (shell gland) [G *Mehlis'sche Drüse, Schalendrüse*]: Collective term for cluster of unicellular gland cells opening into ootype; involved in assemblage of egg and yolk and in egg shell (capsule) formation.

mesenchyme: parenchyma

mesocercaria [G *Mesocercarie*]: In life cycle of digenetic trematode, additional larval stage between cercaria and metacercaria; lacks tail and is not encysted.

metacercaria (adolescaria) [G *Metacercarie*]: Fifth and final larval stage (miracidium, sporocyst, redia, cercaria, metacercaria) in life cycle of digenetic trematode; encysted in intermediate host and infecting final or definitive host after this consumes intermediate host.

metraterm [G *Uterusring*]: Modified, muscular terminal part of uterus; may be "cuticle"-lined and be provided with hooks and glands.

miracidium [G *Miracidium*]: In life cycle of digenetic trematode, first of typically five larval stages (miracidium, sporocyst, redia, cercaria, metacercaria). Free-swimming stage hatching from egg and characterized by ciliated epidermal plates, anterior papilla, and glands.

monostome [G *Monostome*]: Digenetic trematode with only one (typically oral) sucker or lacking suckers. (See also amphistome, distome, echinostome, gasterostome, holostome)

mouth [G *Mund*]: Opening of digestive tract to exterior; with the exception of trematodes with excretory pore or anus, serves both for ingestion and egestion. Typically located at anterior tip of body, although also ventrally at center of body in certain digenetic trematodes. Surrounded by oral sucker in most digenetic trematodes and leads into buccal tube or directly into pharynx.

mouth tube: buccal tube

nephridial tubule [G *Exkretionskanal, Nephridialkanal*]: On each side of body, typically elongate, tubular section of protonephridium between flame bulb (or capillary) and excretory vesicle or nephridiopore.

nephridiopore [G *Nephridioporus*]: Opening of protonephridium to exterior; typically preceded by excretory vesicle. (anterior, posterior; single, paired)

ocellus: eye

onchomiracidium (onco-miracidium) [G *Onchomiracidium*]: In monogenetic trematode, ciliated, free-swimming larval stage hatching from egg; resembles adult. (See also miracidium)

ootype [G *Ootyp*]: Somewhat expanded section of ♀ reproductive system between oviduct and uterus; considered to be modified region of oviduct and surrounded by Mehlis' gland.

operculum (lid) [G *Operculum, Deckel*]: Lid-like structure at one end of egg capsule. Young larva opens operculum at preformed line.

opisthaptor (opisthohaptor) [G *Opisthaptor*]: General term for variously elaborated principle adhesive organ at posterior end of monogenetic trematode; consists either of adhesive disc bearing anchors and hooks, of septa and loculi, or of a number of smaller, well-developed suckers or clamps. (symmetrical, asymmetrical; sessile, pedunculate; armed, unarmed; septate, aseptate) (See also prohaptor)

oral sucker [G *Mundsaugnapf*]: Cup-shaped, muscular sucker surrounding mouth at anterior end of body. In monogenetic trematode, less well-developed, sucker-like prohaptor surrounding mouth; in digenetic trematode, well-developed sucker surrounding mouth. (simple, lobed, with fimbria) (See also acetabulum, opisthaptor)

ovary [G *Ovar, Ovarium*]: Section of ♀ reproductive system in which eggs are produced; opens to exterior via oviduct, ootype, ovovitelline duct, uterus, and gonopore. (single, paired; pretesticular, posttesticular; intercecal, extracecal, lateral, median; bilobed, club-shaped, elongate, entire, follicular, lobed, ovoid, spherical, trilobed, tubular)

oviduct [G *Ovidukt*]: Relatively short, narrow section of ♀ reproductive system between ovary and ootype; vitelline duct, seminal receptacle, and Laurer's canal may open into oviduct.

ovovitelline duct [G *Ductus communis*]: In ♀ reproductive system, term applied either to section of oviduct after

entrance of vitelline duct or to duct leading from ootype (continuing as uterus).

paraprostate organ: Muscular sac opening via gonopore in certain digenetic trematodes; surrounded by gland cells.

parenchyma (mesenchyme) [G *Parenchym*]: Tissue between body wall and major organs; consists of either discrete cells or syncytium and free (mobile) cells. Gives rise to tegument.

parenchymal musculature [G *Parenchymmuskel*]: System of muscles traversing parenchyma; consists largely of dorsoventral muscles. (See also subcuticular musculature)

parthenita [G *pädogenetische Formen*]: Collective obsolete term for sporocyst and redial larval stages in digenetic trematode life cycle. (See also adolescaria, marita)

penis [G *Penis*]: Protrusible (as opposed to eversible cirrus) ♂ copulatory organ; typically produced into muscular projection (penis papilla). (muscular, fibrous; armed, unarmed)

penis bulb [G *Penisbulbus*]: In monogenetic trematode, term applied to muscular, rounded structure supporting penis papilla.

penis papilla [G *Penispapille*]: Muscular, elongate penis; may be armed with hooks or stylets. (armed, unarmed)

pharyngeal gland [G *Pharynxdrüse*]: Gland either contained within pharynx wall or located outside of pharynx and opening into pharynx lumen.

pharyngeal sac [G *Pharynxtasche*]: One of two muscular lateral pouches of pharynx.

pharynx [G *Pharynx, Schlund*]: Muscular, "cuticle"-lined section of digestive tract between mouth or buccal tube and esophagus. (elongated, globose, spherical, with median constriction)

plaque [G *Schild*]: In certain monogenetic trematodes, adhesive thickening of body wall adjoining opisthaptor at posterior end of body. (dorsal, lateral, ventral, spinous)

prepharynx [G *Präpharynx*]: Relatively short, nonmuscular, "cuticle"-lined section of digestive tract between mouth and pharynx in digenetic trematodes. (See also buccal tube)

prepuce [G *Präputium*]: In digenetic trematode, ring of tissue surrounding genital cone in copulatory bursa.

prohaptor [G *Prohaptor*]: General term for wide range of adhesive organs (bothrium, buccal sucker, lead lappet, head organ, oral sucker, pseudosucker) located at anterior end of monogenetic trematode. (See also opisthaptor)

prostate gland (prostatic gland) [G *Prostatadrüse*]: In terminal section of ♂ reproductive system, aggregation of gland cells opening into penis bulb or prostatic vesicle; may be contained within cirrus sac.

prostatic vesicle [G *Prostata*]: In terminal section of ♂ reproductive system, muscular vesicle into which prostatic glands enter; opens into penis bulb or is contained within cirrus sac. (single, paired)

protonephridium [G *Protonephridium*]: Osmoregulatory organ consisting of typically two longitudinal nephridial tubules bearing capillaries with flame bulbs; opens to exterior via excretory vesicle and/or nephridiopore.

pseudodisc (pseudohaptor) [G *Pseudohaptor*]: In monogenetic trematode, large, secondarily developed adhesive disc (opisthaptor) at posterior end of body; bears radial rows of spines and reduced true disc on posterior margin.

pseudosucker [G *Pseudosaugnapf*]: Type of anterior adhesive organ (prohaptor) in monogenetic trematode; consists of sucker-like depression, typically on membranous head lappet.

radial muscle [G *Radiärmuskel*]: Strong bundle of radial fibers in sucker.

rat-king cercaria [G *Rattenkönig Cercarie*]: Aggregation of cercariae radiating from a common center where they are attached to one another by their tail ends.

redia [G *Redie*]: In life cycle of digenetic trematode, third of typically five larval stages (miracidium, sporocyst, redia, cercaria, metacercaria). Characterized by muscular pharynx and sac-like intestine; gives rise to cercaria. If two generations of redia exist, one may distinguish mother and daughter redia. Occasionally interpreted as representing paedomorphic generation rather than being a larva.

reserve bladder [G *Pseudoblase*]: In certain digenetic trematodes, peripheral system of excretory vessels opening into protonephridium.

rhynchus [G *Rhynchus*]: In digenetic trematode with mouth located at center of body (gasterostome), muscular structure replacing oral sucker at anterior tip of body. (with spines, without spines, with fimbria)

salivary gland [G *Ösophagusdrüse*]: Gland opening into esophagus in certain digenetic trematodes.

sclerite [G *Haftspange*]: Rigid element associated with clamps or suckers of posterior adhesive organ (opisthaptor) in monogenetic trematode. (bifid, curved, trifid, U-shaped, X-shaped)

seminal receptacle [G *Receptaculum seminis*]: In ♀ reproductive system, sperm storage organ either opening into oviduct, Laurer's canal, or vagina.

seminal vesicle [G *Samenblase*]: Expanded muscular section of ♂ reproductive system following sperm duct(s); serves to store sperm and may be located either outside of (external seminal vesicle = vesicula seminalis externa) or within (internal seminal vesicle = vesicula seminalis interna) cirrus sac.

septum [G *Leiste*]: In posterior adhesive organ (opisthaptor) of monogenetic trematode or in ventral adhesive disc, slightly raised muscular ridge delimiting each suctorial depression (loculus). (radial, parallel to margin, fused, incomplete, bifurcated)

shell gland: Mehlis' gland

sinus sac (false cirrus sac)
[G *Genitalsinus*]: Pouch-like structure, resembling cirrus pouch, enclosing common, often protrusible hermaphroditic duct in digenetic trematode.

soma [G *"Hüllkörper"*]: In digenetic trematode, thick-walled anterior end of body into which posterior end (ecsoma) can be retracted like telescope.

sperm duct: vas deferens, vas efferens

spermiducal vesicle [G *sekundäre Samenblase*]: In ♂ reproductive system, enlarged section of each vas efferens or of vas deferens prior to entrance into seminal vesicle.

spine [G *Dorn, Stachel*]: General term referring to spine-like projection on head collar (collar spine), around oral sucker (circumoral spine), or in tegument.

spiny collar: head collar

sporocyst (germinal sac, mother sporocyst) [G *Sporozyste, Muttersporozyste*]: In life cycle of digenetic trematode, second of typically five larval stages (miracidium, sporocyst, redia, cercaria, metacercaria); characterized by unciliated surface and body cavity containing germinal bodies. Gives rise to daughter sporocyst or redia. Occasionally also interpreted as representing paedomorphic generation rather than being a larva. (branched, elongate, ovoid, unbranched, vermiform)

squamodisc [G *Häckchenapparat*]: In monogenetic trematode, modified tegument adjoining adhesive organ (opisthaptor) at posterior end of body; bears concentric circles of small spines, scales, or ridges.

stylet [G *Stylet*]: In monogenetic trematode, rigid structure enclosing penis; consists of hollow tube or complex system of elements.

stylet cercaria: xiphidiocercaria

subcuticular musculature [G *Hautmuskelschlauch*]: Stratum of muscles underlying tegument; typically consists of outer circular layer, middle longitudinal layer, and inner diagonal layer, or diagonal

layer lies between circular and longitudinal muscles. (See also parenchymal musculature)

submuscular plexus [G *submuskulärer Plexus*]: Network of nerves located in parenchyma. Typically consists of cerebral ganglia, three pairs of longitudinal cords (ventral, lateral, dorsal), and numerous transverse connections (commissures).

sucker [G *Saugnapf*]: General term for adhesive organ. In monogenetic trematode, typically refers to sucker-like elaboration of either anterior adhesive organ (prohaptor) or posterior adhesive organ (opisthaptor); in digenetic trematode, to anterior oral sucker and ventral sucker (acetabulum). (See also anterior sucker, buccal sucker, genital sucker, pseudosucker)

tail [G *Schwanz*]: Tail-like appendage of cercaria. Shape often used as basis for differentiation of cercaria types. (See also cercaria)

tegument [G *Tegument*]: Highly modified epidermis (nonciliated syncytium) forming outer layer of body wall. Previously referred to as cuticle. (annulated, smooth, spinous)

tentacle: fimbria

testis [G *Hoden*]: Proximal section of ♂ reproductive system. Located in posterior half of body and opening to exterior via vas efferens, vas deferens, seminal vesicle, and penis or cirrus. (single, paired, numerous; diagonal, opposite, tandem; branched, dendritic, elongate, entire, follicular, lobate, spherical)

tribocytic organ (holdfast) [G *tribozytisches Halteorgan*]: In digenetic trematode, gland-bearing, muscular accessory adhesive organ (in addition to oral sucker and acetabulum); located posterior to acetabulum. (bulbous, foliaceous, lobed)

uroproct: cloaca

uterus [G *Uterus*]: Section of ♀ reproductive system between ootype and gonopore. Extremely elongate and coiled in digenetic trematodes, where one may distinguish descending (posteriorly directed) loop or limb and ascending (anteriorly directed) loop. Terminal part may be modified to form metraterm.

vagina (copulation canal) [G *Vagina*]: In monogenetic trematode, canal leading from special exterior pore to oviduct, vitelline duct, or vitelline reservoir. (single, paired)

vas deferens [G *Vas deferens*]: In ♂ reproductive system, common sperm duct formed by junction of vasa efferentia; may be enlarged to form spermiducal vesicle and leads into seminal vesicle.

vas efferens [G *Vas efferens*]: Elongate, slender section of ♂ reproductive system between each testis and common sperm duct (vas deferens) or seminal vesicle. (sinuous, straight)

ventral nerve cord [G *ventraler Längsnervenstrang, ventraler Längsstamm*]: Pair of well-developed ventral nerve cords extending posteriorly from each cerebral ganglion. Joined to each other and to dorsal and lateral cords, if present, by commissures.

ventral sucker: acetabulum

ventrogenital sac [G *Ventralgrube*]: In digenetic trematode, invagination containing ventral sucker (acetabulum), gonopore, and occasionally muscular gonotyl.

vitellarium (vitelline gland, yolk gland) [G *Vitellarium, Dotterstock*]: Section of ♀ reproductive system producing shell material and possibly yolk for egg; opens into oviduct or ootype via vitelline duct and vitelline reservoir. (unilateral, bilateral; compact, follicular, tubular)

vitelline duct [G *Vitellodukt*]: Section of ♀ reproductive system between vitellarium and vitelline reservoir; may refer either to longitudinal duct along follicular type of vitellarium or to transverse duct between each longitudinal duct and vitelline reservoir.

vitelline gland: vitellarium

vitelline reservoir (yolk reservoir) [G *Dotterreservoir*]: In ♀ reproductive system, expanded junction of vitelline ducts; opens via short duct into oviduct or ootype.

vitellovaginal canal (genitovitelline canal) [G *Vitello-Vaginalkanal*]: In monogenetic trematode, common duct formed by junction of vitelline duct and vagina; leads to oviduct.

xiphidiocercaria (stylet cercaria) [G *Xiphidiocercarie*]: Type of cercaria with anterior boring stylet.

yolk gland: vitellarium

yolk reservoir: vitelline reservoir

CESTODA

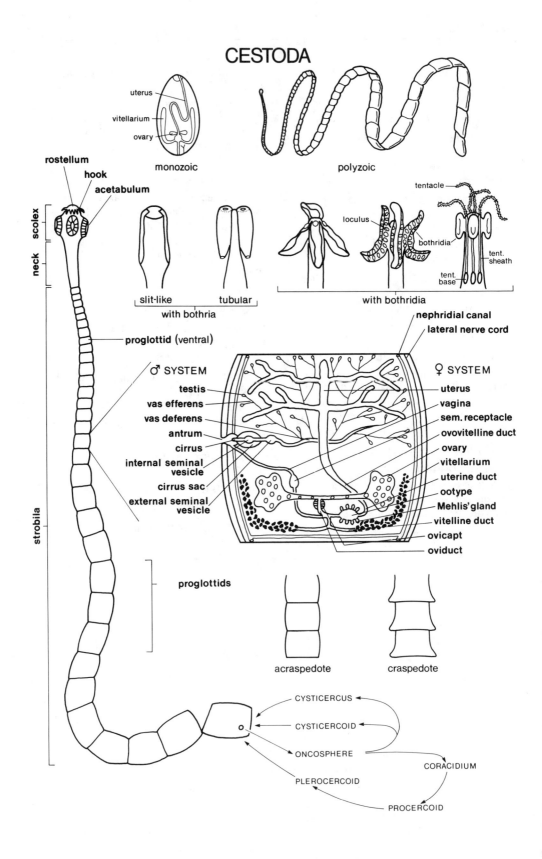

uterus
vitellarium
ovary

monozoic

polyzoic

rostellum
hook
acetabulum

scolex

neck

tentacle
loculus
bothridia
tent. sheath
tent. base

slit-like   tubular
with bothria

with bothridia

proglottid (ventral)

nephridial canal
lateral nerve cord

♂ SYSTEM   ♀ SYSTEM

testis   uterus
vas efferens   vagina
vas deferens   sem. receptacle
antrum   ovovitelline duct
cirrus   ovary
internal seminal   vitellarium
  vesicle   uterine duct
cirrus sac   ootype
external seminal   Mehlis' gland
  vesicle   vitelline duct
                            ovicapt
                            oviduct

strobila

proglottids

acraspedote   craspedote

CYSTICERCUS
CYSTICERCOID
ONCOSPHERE
CORACIDIUM
PLEROCERCOID
PROCERCOID

PLATYHELMINTHES

CESTODA

cestode, tapeworm [G *Cestode, Bandwurm*]

accessory lateral nerve cord [G *akzessorischer Längsnervenstrang*]: One of two longitudinal nerve cords arising from scolex and accompanying major lateral nerve cords to posterior end of body; joined to other longitudinal cords (lateral, dorsal, ventral cords) by ring commissures.

accessory sucker (apical sucker) [G *akzessorischer Saugnapf*]: Small sucker at tip of scolex or on anterior end of each bothridium.

acetabulum [G *Acetabulum, Saugnapf*]: One of four suckers symmetrically arranged on scolex; each consists of hemispherical depression with strong radial musculature. (armed, unarmed)

antrum (genital atrium, atrium) [G *Antrum, Atrium, Atrium genitale*]: Common terminal part of ♂ and ♀ reproductive systems; typically consists of chamber into which ♂ copulatory apparatus (cirrus, cirrus sac) and vagina enter. Opens to exterior via gonopore.

apical ring [G *apikaler Nervenring*]: Nerve ring anterior to rostellar ring of scolex.

areola: loculus

atrium: antrum

bladder (excretory bladder) [G *Blase*]: In protonephridium of anapolytic cestode, median excretory vesicle in last proglottid; opens to exterior via nephridiopore. May also refer to enlarged posterior end of cysticercus (bladder worm).

bladder worm: cysticercus

blastocyst [G *Blase*]: Bladder-like posterior end of plerocercus into which anterior end can withdraw.

bothridium (phyllidium) [G *Bothridium*]: One of typically four (occasionally only two) broad, leaf-like adhesive structures on scolex; distinguished from bothrium by well-developed musculature. May bear secondary depressions (loculi) or accessory sucker. (sessile, stalked; elongate, rounded; smooth, ruffled, with hooks)

bothrium [G *Bothrium, Sauggrube*]: One of two shallow, typically elongated adhesive structures on scolex; characterized by weak musculature. Margins may project, be ruffled, or be fused to form tube or cup. (elongated, fan-like, foliaceous, pit-like, slit-like, spherical) (See also acetabulum, bothridium)

brain [G *Gehirn*]: Main concentration of nerve tissue in scolex; consists of anterior expansions (cerebral ganglia) of two lateral longitudinal nerve cords, well-developed transverse commissure, and nerve ring.

brood capsule [G *Brutkapsel*]: Asexually produced vesicle on inner surface of hydatid or multilocular cyst; projects into central cavity of hydatid and buds protoscoleces internally.

calcareous corpuscle (calcareous body) [G *Kalkkörper*]: Minute body consisting of concentric layers of calcareous material. Located in large numbers in parenchyma and secreted within certain parenchymal cells. (round, oval, irregular)

capillary [G *Kapillar*]: In protonephridium, thin excretory tubule.

capsule: egg

cephalic peduncle (head stalk) [G *Kopfstiel*]: Elongate posterior section of scolex; typically refers to region posterior to adhesive organs and may contain tentacles internally.

cercomer [G *Cercomer*]: Tail-like posterior appendage of infectious stage after metamorphosis (metacestode); typically bears hooks of oncosphere.

cerebral ganglion: brain

chainette [G *Hakenreihe, Häkchenreihe*]: Longitudinal row of hooks on tentacle.

circular muscle [G *Ringmuskel*]: Outer layer of subtegumental musculature. (See also longitudinal muscle)

cirrus [G *Zirrus, Cirrus*]: ♂ copulatory organ consisting of protrusible terminal section of ejaculatory duct; enclosed in cirrus sac. (armed, unarmed)

cirrus pore [G *Zirrusöffnung*]: Opening through which cirrus is everted. May also refer to opening in cirrus through which sexual products are extruded.

cirrus pouch: cirrus sac

cirrus sac (cirrus pouch) [G *Zirrusbeutel*]: Terminal section of ♂ reproductive system; consists of muscular, pouch-like structure and contains ejaculatory duct, internal seminal vesicle, and cirrus. (cylindrical, oval, pyriform, round, tubular)

coenurus [G *Coenurus*]: Tissue cyst in which several protoscoleces (yet no daughter cysts or brood capsules) develop asexually.

collecting canal: nephridial tubule

comidial layer: tegument

commissure: ring commissure, transverse commissure

coracidium [G *Coracidium*]: Type of larva developing when first larval stage (oncosphere) is released into water. May be regarded as free-swimming oncosphere. Develops into procercoid in first intermediate host.

cortex [G *Cortex*]: Region of body exterior to mesenchymal longitudinal muscle bands. (See also medulla)

cuticle: tegument

cysticercoid [G *Cysticercoid*]: In abbreviated life cycle of certain polyzoic cestodes, stage of development following oncosphere; typically with tail and characterized by internal cavity permitting withdrawal (not invagination) of anterior scolex. Budding of scolex typically exogenous. (rostello-, arostello-, acanthorostello-, anacanthorostello-; caudate, acaudate, brevicaudate, circumcaudate, longicaudate) (See also cysticercus, metacestode)

cysticercus (bladder worm) [G *Cysticercus, Blasenwurm, Finne*]: In abbreviated life cycle of certain polyzoic cestodes, stage of development following oncosphere; characterized by lack of tail and presence of round, hollow bladder with depression into which anterior scolex is invaginated (not merely withdrawn). Budding of scoleces typically endogenous. (anacantho-, heteracantho-, homacantho-, multicephalo-, strobilo-) (See also cysticercoid, metacestode)

daughter cyst (daughter bladder) [G *Tochterblase*]: New, asexually produced hydatid; buds either from outer surface (exogenous budding) or inner surface (endogenous budding) of hydatid. (See also brood capsule)

decacanth (lycophore) [G *Decacanth, Lycophorala-Larve*]: In life cycle of most monozoic cestodes, larval stage bearing 10 hooks on posterior end. (See also hexacanth)

dithyridium: tetrathyridium

dorsal nerve cord [G *dorsaler Längsnervenstrang, dorsaler Nervenstrang*]:

Pair of minor longitudinal nerve cords arising from nerve ring of brain and extending to posterior end of body; joined to other longitudinal cords (lateral, accessory lateral, ventral cords) by ring commissures.

dorsoventral muscle [G *dorsoventraler Muskel*]: In parenchymal musculature, dorsoventral muscle fibers traversing parenchyme; typically restricted to margins of proglottid.

egg [G *Ei*]: ♀ reproductive body produced in ovary and supplied with yolk cells and capsule by vitellarium/ootype/Mehlis' gland complex. (operculate, nonoperculate; embryonated, unembryonated; oval, fusiform, spherical; thick-walled, thin-walled; with/without filaments)

ejaculatory duct [G *Ductus ejaculatorius*]: In ♂ reproductive system, muscular section of vas deferens within cirrus sac; may be enlarged to form internal seminal vesicle.

external seminal vesicle (spermiducal vesicle) [G *Samenblase, Vesicula seminalis externa*]: In ♂ reproductive system, enlarged section of vas deferens prior to entrance into cirrus sac. (See also internal seminal vesicle)

flame cell [G *Terminalzelle, Cyrtozyste*]: In protonephridium, excretory structure typically containing one ciliary flame; scattered throughout body and connected to longitudinal nephridial tubules via capillaries.

frontal gland [G *Frontaldrüse*]: One to several clusters of glands opening at apex of scolex, especially in developmental stages.

genital atrium: antrum

gonoduct: oviduct, vas deferens, vas efferens

gonopore (genital pore) [G *Gonoporus, Genitalporus, Geschlechtsöffnung*]: Opening of ♂ or ♀ reproductive system (or of common antrum) to exterior; typically located on lateral margin. (dorsal, lateral, marginal, preequatorial, submarginal, ventral; unilateral, irregularly alternating)

head: scolex

head stalk: cephalic peduncle

hexacanth [G *Oncosphaera, 6-Haken-Larve*]: In life cycle of polyzoic cestode, larval stage hatching from egg capsule and developing into either procercoid, cysticercoid, or cysticercus; characterized by six hooks. Often used synonymously with oncosphere.

homogeneous layer: tegument

hook [G *Haken*]: One in a circlet of hooks around rostellum at anterior end of scolex. May also refer to hooks at anterior end of bothridia or along tentacles. (simple, bifid, trifid, hammer-shaped, T-shaped; arrangement: heteracanthus, homeoacanthus, in circlets, in festoons)

hydatid [G *Hydatide*]: Complex tissue cyst with multilayered wall which asexually produces daughter cysts and brood capsules. (unilocular, multilocular)

internal seminal vesicle [G *Samenblase, Vesicula seminalis interna*]: In ♂ reproductive system, expanded section of vas deferens (ejaculatory duct) within cirrus bulb. (See also external seminal vesicle)

interproglottidal gland: Cluster of gland cells, often around sac-like structure; located in a series along posterior margin of proglottid in certain cestodes.

lateral nerve cord [G *lateraler Längsnervenstrang, lateraler Längsstamm*]: Pair of major longitudinal nerve cords arising from brain and extending to posterior end of body; joined to other longitudinal cords (accessory lateral, dorsal, ventral nerve cords) by ring commissures.

loculus [G *Loculus*]: One of several to numerous depressions on each bothridium.

longitudinal muscle [G *Längsmuskel*]: Inner layer of subtegumental musculature. (See also circular muscle)

lycophore: decacanth

medulla [G *Medulla*]: Region of body interior to parenchymal longitudinal muscle bands. (See also cortex)

Mehlis' gland (shell gland) [G *Mehl'sche Drüse, Schalendrüse*]: In ♀ reproductive system, collective term for cluster of unicellular gland cells opening into ootype; involved in egg capsule formation.

mesenchymal musculature: parenchymal musculature

mesenchyme: parenchyma

metacestode [G *Metacestode*]: General term for infectious stages following metamorphosis of larvae; includes procercoid, plerocercoid, neoplerocercoid, cysticercoid, cysticercus.

metascolex [G *Metascolex*]: Posterior region of divided scolex (proscolex, metascolex). May also refer to ruff on neck.

microthrix [G *Mikrothriche*]: Minute, spine-like processes of outer surface (distal cytoplasm) of tegument; each consists of distal shaft and expanded base.

myzorhynchus [G *Wurzelhals*]: Protrusible, muscular mass at anterior end of scolex; may bear suckers.

neck [G *Hals*]: Relatively short, slender, and unsegmented division of body between scolex and strobila; gives rise to proglottids.

neoplerocercoid [G *Neoplerocercoid*]: In life cycle of certain polyzoic cestodes, type of plerocercus which has lost expanded posterior end (blastocyst) during transfer between hosts.

nephridial tubule [G *Exkretionskanal, Nephridialkanal*]: One of typically four longitudinal (two ventrolateral, two dorsolateral) canals of protonephridium; ventral pair may be connected by transverse canal. (straight, sinuous, coiled)

nephridiopore [G *Nephridioporus*]: Opening of protonephridium to exterior; consists of either single terminal pore associated with bladder, separate terminal pores of each nephridial canal, or numerous accessory pores.

oncosphere [G *Oncosphaera*]: First stage in life cycle of polyzoic cestode; consists of encapsulated, oval embryo with six hooks (hexacanth). Develops in certain species into procercoid; in others into either cysticercoid, cysticercus, or hydatid. (See also coracidium)

oocapt: ovicapt

ootype [G *Ootyp*]: Expanded section of ♀ reproductive system between oviduct and ovovitelline duct or uterus; located next to entrance of vitelline duct into oviduct and surrounded by Mehlis' gland.

operculum (lid) [G *Operculum, Deckel*]: Lid-like structure at one end of egg capsule.

ovary [G *Ovar, Ovarium*]: Unpaired, bilobed section of ♀ reproductive system in which eggs are produced; opens to exterior via oviduct, ootype, uterus, and uterine pore. (single, double; bilobed, bialate, H-shaped, four-lobed, reticulate; cortical, medullary)

ovicapt (oocapt) [G *Schluckapparat*]: Muscular thickening of oviduct at border of ovary and oviduct. (bulbiform, funnel-shaped, pyriform)

oviduct [G *Ovidukt*]: Tubular section of ♀ reproductive system between ovary and ootype; typically bears ovicapt. Vagina and vitelline duct open into oviduct.

ovovitelline duct [G *Eidottergang*]: Section of ♀ reproductive system between ootype and uterus proper. Slender portion may be termed uterine duct.

parauterine organ (paruterine organ) [G *Paruterin-Organ*]: In ♀ reproductive system, sac-shaped, fibrous structure adjoining uterus and receiving and retaining embryos while uterus degenerates.

parenchyma (mesenchyme) [G *Parenchym*]: Tissue between body wall and organs; contains parenchymal musculature, cells, and calcareous corpuscles.

parenchymal musculature (mesenchymal musculature) [G *parenchymale Muskulatur*]: System of muscles in parenchyma; consists of longitudinal, dorsoventral, and transverse fibers. (See also sub-tegumental musculature)

pars bothridialis [G *Pars bothridialis*]: In cestode with tentacles, region of scolex from apex to posterior margin of bothridia.

pars bulbosa [G *Pars bulbosa*]: In cestode with tentacles, region of scolex occupied by tentacular bulbs.

pars postbulbosa [G *Pars postbulbosa*]: In cestode with tentacles, region of scolex posterior to that occupied by tentacular bulbs (pars bulbosa).

pars vaginalis [G *Pars vaginalis*]: In cestode with tentacles, region of scolex traversed by tentacular sheaths.

paruterine organ: parauterine organ

phyllidium: bothridium

plerocercoid [G *Plerocercoid*]: In life cycle of certain polyzoic cestodes, stage following procercoid; characterized by scolex and first traces of segmented strobila. May also refer to larval stage of monozoic cestode following procercoid. (acantho-, acetabulo-, bothridio-, bothrio-, culcito-, glando-, tentaculo-) (See also metacestode, sparganum)

plerocercus [G *Plerocercus*]: In life cycle of cestode with tentacles, type of plerocercoid characterized by bladder-like posterior structure (blastocyst). (See also neoplerocercoid)

polycercus [G *Polycercus*]: Modified cysticercus which asexually produces cysticercoids by budding into central cavity.

prebulbar organ: In cestode with tentacles, small, often red organ anterior to tentacular bulb.

proboscis: tentacle

proboscis bulb: tentacular bulb

proboscis sheath: tentacular sheath

procercoid [G *Procercoid*]: Stage of development between oncosphere and plerocercoid in life cycle of certain polyzoic cestodes; characterized by elongate shape, lack of both internal cavity and scolex, yet with six hooks on posterior end. May also refer to

larval stage of monozoic cestode between decacanth and plerocercus. (See also metacestode)

proglottid [G *Proglottide, Proglottis*]: In posteriormost division of polyzoic cestode (scolex, neck, strobila), one in a series of segments. (elongate, short and broad, quadratic, subquadratic; margin: smooth, lobed, fimbriated)

proscolex [G *Proscolex*]: Anterior region of divided scolex (proscolex, metascolex).

prostatic gland [G *Prostatadrüse*]: In terminal section of ♂ reproductive system, aggregation of gland cells opening into vas deferens or cirrus sac.

prostatic vesicle [G *Prostata*]: In ♂ reproductive system of certain cestodes, muscular vesicle opening into cirrus sac.

protonephridium [G *Protonephridium*]: Excretory organ typically consisting of flame cell connected with capillaries, longitudinal nephridial tubules, transverse canals, and often a bladder; opens to exterior via nephridiopore(s).

protoscolex [G *Protoscolex*]: Scolex of developmental stage (metacestode) of cestode.

pseudoscolex [G *Pseudoscolex*]: In those cestodes in which scolex is reduced or cast off, crest-like modification of neck and anterior strobila.

ring commissure [G *ringförmige Kommissur*]: One of several dorsal and ventral transverse nerves connecting longitudinal nerve cords in each proglottid.

rosette [G *Rosette*]: Ruffled, well-innervated structure surrounding funnel-like depression at posterior end of certain monozoic cestodes.

rostellar receptacle [G *Rostellumscheide*]: Cavity at apex of scolex into which rostellum can be withdrawn.

rostellar ring [G *rostellarer Nervenring*]: Nerve ring anterior to brain at base of rostellum; unites anterior nerves. (See also apical ring)

rostellum [G *Rostellum*]: Modification of anterior end of scolex into highly mobile, muscular cone; typically armed with hooks and retractile into rostellar receptacle. (armed, unarmed)

scolex [G *Scolex*]: Anteriormost of three divisions of body (scolex, neck, strobila); bears organs of attachment (acetabula, bothria, bothridia, tentacles, hooks) and contains brain. (armed, unarmed; cardiform, clavate, fan-like, globular, rounded, prismatic, pyriform, six-sided, with/without velum)

seminal receptacle [G *Receptaculum seminis*]: In ♀ reproductive system, expanded section of vagina; serves to store sperm.

seminal vesicle: external seminal vesicle, internal seminal vesicle

shell gland: Mehlis' gland

sparganum [G *Sparganum*]: Plerocercoid whose identity is unknown.

sperm duct: vas deferens, vas efferens

spermiducal vesicle: external seminal vesicle

strobila [G *Strobila*]: Posterior of three divisions (scolex, neck, strobila) of body; consists of a series of increasingly larger, flattened segments (proglottids). (acraspedote, craspedote; flattened, cylindroid)

strobilocercoid [G *Strobilocercoid*]: Modified cysticercoid with strobila.

strobilocercus [G *Strobilocercus*]: Modified cysticercus bearing immature strobila between scolex and posterior bladder.

subtegumental musculature [G *Hautmuskelschlauch, subtegumentale Muskulatur*]: Stratum of muscles underlying tegument; consists of outer circular layer and inner longitudinal layer. (See also parenchymal musculature)

sucker [G *Saugnapf*]: General term for sucking structures on scolex; includes acetabula, bothria, bothridia, and accessory suckers.

tegument [G *Tegument*]: Highly modified syncytial outer layer of body wall. Consists of outer zone (distal cytoplasm) forming microthriches and underlying cell bodies (tegumentary cytons). The two zones are separated by a basement layer (basal lamina + connective fibrils). Previously (erroneously) termed cuticle and considered to consist of outer cormidial layer and inner homogeneous layer.

tentacle (proboscis) [G *Tentakel, Rüssel*]: One of four elongate, eversible projections of scolex, each contained within tentacular sheath in retracted state; when everted, hooks and spines of tentacle come to lie on outer surface. (armed, unarmed)

tentacular bulb (proboscis bulb) [G *Tentakelbulbus*]: Oval, muscular cavity forming base of each tentacular sheath.

tentacular sheath (proboscis sheath) [G *Tentakelscheide*]: One of four elongate tubes enclosing tentacles and terminating posteriorly in muscular tentacular bulbs.

testis [G *Hoden*]: Proximal section of ♂ reproductive system; typically consists of numerous spherical bodies, each connected to vas deferens by vas efferens. Number, size, and distribution are of taxonomic importance. (evenly distributed, in lateral fields, central, cortical, medullary, preovarian)

tetrathyridium (dithyridium) [G *Tetrathyridium*]: Type of elongate cysticercoid whose enlarged anterior end contains scolex.

transverse canal [G *querverlaufender Exkretionskanal*]: In excretory system, transverse excretory vessel connecting ventrolateral pair of longitudinal nephridial tubules in posterior region of each proglottid.

transverse commissure [G *Scolexkommissur*]: In brain, relatively well-developed transverse nerve connection between expanded anterior ends (cerebral ganglia) of lateral longitudinal nerve cords. (See also ring commissure)

transverse muscle [G *Transversalmuskel*]: In parenchymal musculature, transverse muscle bands lying inside or interwoven with longitudinal fibers.

uterine capsule [G *Uteruskapsel*]: In cestode in which uterus breaks apart, one of many sac-shaped uterus fragments containing one to several embryos.

uterine duct [G *Uterusansatz*]: In ♀ reproductive system, slender portion of ovovitelline duct between ootype and expanded uterus.

uterine pore [G *Uterusöffnung*]: Opening of uterus to exterior; if present, typically located on ventral surface of polyzoic cestodes, at anterior end in monozoic forms.

uterovaginal canal [G *Uterovaginalgang*]: Common canal formed by merger of terminal section of uterus and vagina; typically opens separately from ♀ gonopore.

uterus [G *Uterus*]: Conspicuous section of ♀ reproductive system following ovovitelline duct and containing embryos; may open to exterior via uterine pore. (single, double; cortical, medullary; branching, horseshoe-shaped, lobulated, ring-shaped, reticulate, rosette-shaped, sacciform, tubular)

vagina [G *Vagina*]: In ♀ reproductive system, elongate canal leading from gonopore or common genital antrum to oviduct; typically bears expanded seminal receptacle.

vas deferens [G *Vas deferens*]: In ♂ reproductive system, main sperm duct into which vasa efferentia open; leads to cirrus and may bear expanded sections (seminal vesicles). (coiled, sinuous)

vas efferens [G *Vas efferens*]: In ♂ reproductive system, one of numerous thin canals joining each testis to vas deferens.

velum [G *Velum*]: Collar-like fold of posterior end of scolex; projects over beginning of neck. May also refer to posterior section of proglottid overlapping following segment. (smooth, lobed, scalloped, toothed)

ventral nerve cord [G *ventraler Längsnervenstrang, ventraler Längsstamm*]: Pair of minor longitudinal nerve cords arising from nerve ring of brain and extending to posterior end of body; joined to other longitudinal nerves (lateral, accessory lateral, dorsal cords) by ring commissures.

vitellarium (yolk gland) [G *Vitellarium, Dotterstock*]: Section of ♀ reproductive system supplying material for egg capsule formation; opens into oviduct near ootype via vitelline duct. (cortical, medullary; bilobed, follicular)

vitelline duct (yolk duct) [G *Vitellodukt*]: In ♀ reproductive system, canal leading from vitellarium to oviduct near ootype; may bear expanded vitelline reservoir.

vitelline reservoir (yolk reservoir) [G *Dotterreservoir*]: In ♀ reproductive system, expanded section of vitelline duct.

yolk gland: vitellarium

GNATHOSTOMULIDA

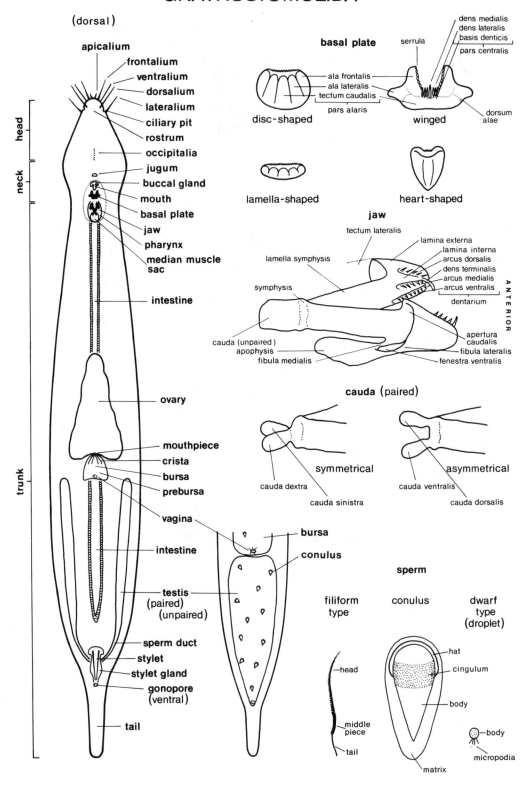

(dorsal)

apicalium
frontalium
ventralium
dorsalium
lateralium
ciliary pit
rostrum
occipitalia
jugum
buccal gland
mouth
basal plate
jaw
pharynx
median muscle sac
intestine
ovary
mouthpiece
crista
bursa
prebursa
vagina
intestine
testis (paired) (unpaired)
sperm duct
stylet
stylet gland
gonopore (ventral)
tail

head
neck
trunk

basal plate

dens medialis
dens lateralis
basis denticis
pars centralis
serrula

ala frontalis
ala lateralis
tectum caudalis
pars alaris

dorsum alae

disc-shaped

winged

lamella-shaped

heart-shaped

jaw

tectum lateralis
lamina externa
lamina interna
arcus dorsalis
dens terminalis
arcus medialis
arcus ventralis
dentarium

lamella symphysis
symphysis

cauda (unpaired)
apophysis
fibula medialis

apertura caudalis
fibula lateralis
fenestra ventralis

ANTERIOR

cauda (paired)

symmetrical

asymmetrical

cauda dextra
cauda sinistra

cauda ventralis
cauda dorsalis

bursa
conulus

sperm

filiform type

conulus

dwarf type (droplet)

head
middle piece
tail

hat
cingulum
body

body
micropodia

matrix

GNATHOSTOMULIDA

gnathostomulid [G *Gnathostomulide, Kiefermäuler*]

adhesive gland [G *Haftdrüse*]: Term applied to a variety of epidermal inclusions considered to have adhesive function; may be scattered, arranged in longitudinal rows, or concentrated in posterior region of body. (See also rhabdoid)

ala frontalis (frontal wing, rostral wing) [G *Ala frontalis*]: On basal plate, anteriormost pair of wing-shaped extensions of pars alaris, one to each side of pars centralis; frontomedial margin may bear serrulae.

ala lateralis (lateral wing) [G *Ala lateralis*]: On basal plate, lateral pair of wing-shaped extensions of pars alaris.

anus [G *After, Anus*]: Tissue connection found between intestine and dorsal epidermis and considered to represent posterior opening of digestive tract.

apertura caudalis [G *Apertura caudalis*]: Part of involucrum of jaw. Elliptical posterior opening of cone-shaped structure formed by involucrum.

apicalium [G *Apicalium*]: Elongate, single sensory cilia, two or four in number, at tip of rostrum; typically followed posteriorly by four pairs (frontalia, ventralia, dorsalia, lateralia) of elongate sensory bristles.

apophysis [G *Apophysis*]: Part of suspensorium of jaws. Rostral apophysis: lateral wing-shaped projection of jaws; may be attached to ventral sections of lamina interna and lamina externa by fibula medialis and fibula lateralis, respectively. Caudal apophysis: shorter lateral projection of jaws, between rostral apophysis and symphysis.

arcus dorsalis (dorsal row) [G *Arcus dorsalis*]: Part of dentarium of jaw. Dorsalmost of three longitudinal, tooth-bearing ridges (arcus dorsalis, arcus medialis, arcus ventralis) on lamina interna.

arcus medialis (median row) [G *Arcus medialis*]: Part of dentarium of jaw. Median of three longitudinal, tooth-bearing ridges (arcus dorsalis, arcus medialis, arcus ventralis) on lamina interna; bears dens terminalis anteriorly.

arcus ventralis (ventral row) [G *Arcus ventralis*]: Part of dentarium of jaw. Ventralmost of three longitudinal, tooth-bearing ridges (arcus dorsalis, arcus medialis, arcus ventralis) on lamina interna.

articularium [G *Articularium*]: One of four major parts of jaw (articularium, dentarium, involucrum, suspensorium); consists of lamellae symphysis joined by symphysis and serves to prevent twisting of jaw.

basal plate [G *Basalplatte*]: Variously elaborated, cuticular, plate-like structure secreted by ventral wall of pharynx and closely associated with jaw; in more complex gnathostomulids, consists of dentate (dens medialis, dentes laterales) central part (pars centralis), wing-shaped extensions (pars alaris) divided into five regions (ala frontalis, two alae laterales, tectum caudalis), and scaled frontomedial margin (serrula) of ala frontalis.

basis denticis [G *Basis denticis*]: On basal plate, thickened base of insertion of central tooth (dens medialis) and lateral teeth (dentes laterales) on posterior end of pars centralis.

body [G *Körper*]: Cone-shaped internal structure of conulus; consists of concentric lamellar layers.

brain [G *Gehirn*]: Main concentration of nerve tissue; typically basiepithelial, yet occasionally located subepithelially anterior to mouth and giving rise to longitudinal nerves, including pair of buccal nerves joining brain to buccal ganglion.

buccal cavity [G *Mundhöhle*]: Section of digestive tract between mouth and pharynx; associated with various buccal glands.

buccal ganglion [G *Schlundganglion*]: Concentration of nerve tissue at posterior end of pharynx; joined to brain by pair of buccal nerves.

buccal gland [G *Pharyngealdrüse*]: One of variously positioned, often paired glands associated with anterior section of digestive tract. Termed, according to position in different gnathostomulids, buccal, lateropharyngeal, prepharyngeal (preoral), or postoral glands.

buccal nerve [G *Schlundnerv*]: Pair of nerves joining brain and buccal ganglion.

bursa [G *Bursa*]: Variously elaborated sperm storage sac posterior to ovary in ♀ reproductive system; may bear longitudinal cristae forming mouthpiece and be adjoined posteriorly by prebursa. (cuticularized, soft)

bursa appendix: prebursa

cauda [G *Cauda*]: Part of suspensorium of jaw; paired or unpaired extension of symphysis. If paired, either symmetrical from the dorsal view (cauda sinistra, cauda dextra) or asymmetrical (cauda dorsalis, cauda ventralis).

caudal apophysis: apophysis

central tooth: dens medialis

ciliary pit [G *Sinnesgrübchen*]: Type of sensory structure consisting of ciliated depression laterally on rostrum. (See also spiral-ciliary organ)

cingulum [G *Cingulum*]: Multilayered girdle encircling blunt end of body of conulus.

circular muscle [G *Ringmuskel*]: Thin layer of fine circular muscle fibers between epidermis and intestine; overlying longitudinal muscles.

conulus [G *Conulus*]: One of at least three types of sperm (conulus, dwarf type, filiform type); relatively large, cone-shaped type of sperm consisting of cingulum, hat, and body surrounded by matrix.

crista [G *Crista*]: One of typically three longitudinal ridges formed by tripartite structure of bursa; cristae unite anteriorly to form mouthpiece.

dens lateralis (lateral tooth) [G *Dens lateralis*]: On basal plate, one in a series of smaller teeth to each side of large central tooth (dens medialis) on pars centralis; inserting, together with dens medialis, on basis denticis.

dens medialis (central tooth) [G *Dens medialis*]: On basal plate, longest and thickest unpaired central tooth of pars centralis; flanked on each side by lateral teeth (dentes laterales) and inserting on basis denticis.

dens terminalis (terminal tooth) [G *Dens terminalis*]: Part of dentarium of jaw. Largest tooth in jaw, located in anteriormost position on arcus medialis.

dentarium [G *Dentarium*]: One of four major parts of jaw (articularium, dentarium, involucrum, suspensorium); tooth-bearing distal part of jaw consisting of lamina interna, arcus dorsalis, arcus medialis, and arcus ventralis, dens terminalis, and incisura dorsalis.

dorsalium [G *Dorsalium*]: One member in a pair of elongate, compound sensory bristles; located on each side of rostrum, dorsal to ventralia and between frontalia and lateralia.

dorsum alae [G *Dorsum alae*]: On basal plate, ridge-shaped thickening of pars alaris; forms border between alae frontales, alae laterales, and tectum caudalis, dividing pars alaris into five regions.

dwarf type [G *Zwerg-Typus*]: One of at least three types of sperm (conulus, dwarf type, filiform type); small, round or drop-shaped sperm with micropodia.

egg [G *Ei*]: ♀ reproductive body in ovary; typically in a series, the most mature egg being most posterior.

epidermal inclusion [G *Hauteinschluß*]: Variously shaped bodies in epidermis; either scattered over entire surface, grouped in bundles, or arranged in rows. Certain types may serve as adhesive glands. (round, spindle-shaped)

epidermis [G *Epidermis*]: Outermost layer of body wall; consists of one layer of monociliated cells, occasionally arranged in a regular pattern, and bearing epidermal inclusions.

excretory organ (protonephridium) [G *Protonephridium*]: Paired groups of two or three organs in body wall to each side of pharynx, bursa, and penis; each organ consists of terminal cell and, according to interpretation, canal and/or outlet cell.

fenestra ventralis [G *Fenestra ventralis*]: Part of suspensorium of jaw. Window-like space in apophysis between fibula medialis and fibula lateralis; may be divided into more anterior fenestra ventrofrontalis and more posterior fenestra ventrocaudalis if traversed by fibula radialis.

fenestra ventrocaudalis: fenestra ventralis

fenestra ventrofrontalis: fenestra ventralis

fibula [G *Fibula*]: Part of suspensorium of jaw. One of several struts joining apophysis to lamina interna (fibula medialis), to lamina externa (fibula lateralis), or joining the two lamellae (fibula radialis).

fibula lateralis [G *Fibula lateralis*]: fibula

fibula medialis [G *Fibula medialis*]: fibula

fibula radialis [G *Fibula radialis*]: fibula

filiform type [G *Filiform-Typus*]: One of at least three types of sperm (conulus, dwarf type, filiform type); elongate, thin sperm consisting of head, middle piece, and tail.

frontalium (rostralium) [G *Frontalium*]: One member of anteriormost pair of typically four pairs (frontalia, ventralia, dorsalia, lateralia) of elongate, compound sensory bristles on each side of rostrum.

frontal wing: ala frontalis

gonopore [G *Gonoporus, Geschlechtsöffnung*]: Opening of reproductive system to exterior. In ♂, ventral opening at posterior end of trunk and often associated with penis or stylet. In ♀, see vagina.

hat [G *Hut*]: Thin, multilayered, electron-dense band capping blunt end of conulus.

head [G *Kopf*]: Anteriormost of three divisions of body (head, neck, trunk) tapering anteriorly into rostrum and/or bearing series of sensory projections (apicalia, frontalia, ventralia, dorsalia, lateralia); separated from trunk by neck.

incisura dorsalis [G *Incisura dorsalis*]: Part of dentarium of jaw. Dorsal fissure cutting posteriorly into lamina interna; formed by reduction of posterior extension of lamina interna.

incisura ventralis [G *Incisura ventralis*]: Part of involucrum of jaw. Ventral fissure formed by fusion of anterior end of apophysis and lamina externa.

infundibulum [G *Infundibulum*]: On basal plate, funnel-shaped structure formed by fusion of frontomedial margins of alae frontales into transverse crest under pars centralis.

intestine [G *Darm*]: Section of digestive tract posterior to pharynx; tissue connection between intestine and dorsal epidermis considered to be anal pore.

involucrum [G *Involucrum*]: One of four major parts of jaw (articularium, dentarium, involucrum, suspensorium). Consists of lamina externa, incisura ventralis, apertura caudalis, and tectum lateralis. Considered to be extension of lamina interna; forms anterior hollow, cone-shaped structure for muscle attachment.

jaw [G *Kiefer*]: Pair of movable, toothed cuticular structures closely associated with basal plate and secreted by ventral wall of pharynx; consists of articularium, dentarium, involucrum, and suspensorium.

jugum [G *Jugum*]: Cuticular, crescent-shaped skeletal structure anterior to mouth in wall of buccal cavity.

lamella symphysis [G *Lamella symphysis*]: Part of articularium of jaw. Longitudinal axis of each jaw half; joined posteriorly by symphysis.

lamina externa [G *Lamina externa*]: Part of involucrum of jaw. Lateral loop-shaped extension of lamina interna forming a cone for muscle attachment.

lamina interna [G *Lamina interna*]: Part of dentarium of jaw. Anterior inner surface of jaw bearing three longitudinal, thickened ridges with teeth (arcus dorsalis, arcus medialis, arcus ventralis).

lateralium [G *Lateralium*]: One member of posteriormost of typically four pairs (frontalia, ventralia, dorsalia, lateralia) of elongate, compound sensory bristles laterally on head.

lateral muscle sac [G *laterale Muskelbeutel der Kiefer*]: In tripartite elaboration of pharynx musculature, paired group of muscles to each side of jaw. (See also median muscle sac)

lateral wing: ala lateralis

lateropharyngeal gland [G *Lateropharyngealdrüse*]: Pair of buccal glands located lateral to pharynx.

longitudinal muscle [G *Längsmuskel*]: One of three to four groups of relatively thick, longitudinal muscle bands underlying thin circular muscles in body wall.

median lobe: pars centralis

median muscle sac [G *medianer Muskelbeutel der Kiefer*]: In tripartite elaboration of pharynx musculature, unpaired group of muscles inserting medially either at anterior end of jaw or at symphysis. (See also lateral muscle sac)

micropodia [G *Füßchen*]: In dwarf type of sperm, tubular protuberances of plasma membrane.

mouth [G *Mund*]: Anterior opening of digestive tract located ventrally behind head; associated with various skeletal elements (jugum, basal plate, jaws) and buccal glands.

mouthpiece [G *Mundstück*]: Constricted anterior section of bursa, formed by merger of cristae.

neck [G *Hals*]: Short, slightly constricted division of body between head and trunk; typically corresponds in position with mouth.

nodus [G *Nodus*]: Part of suspensorium of jaw. Knob-like structure, either at junction of fibula lateralis and lamina externa (nodus lateralis) or at junction of fibula medialis and lamina interna (nodus medialis).

nodus lateralis: nodus

nodus medialis: nodus

occipitalium [G *Occipitalium*]: Mediodorsal row of single, stiff, somewhat elongated cilia on head.

ovary [G *Ovar*]: Unpaired, elongate section of ♀ reproductive system dorsal to intestine in trunk; associated posteriorly with bursa. Eggs mature posteriorly in ovary.

parenchyma [G *Parenchym*]: Poorly developed mass of unspecialized cells between body wall and internal organs.

pars alaris [G *Pars alaris*]: Wing-shaped, expanded part of basal plate surrounding pars centralis. Typically consists of five regions: pair of frontal wings (alae frontales), pair of lateral wings (alae laterales), and median, unpaired, posterior tectum caudalis, all delimited from one another by ridges (dorsum alae).

pars centralis [G *Pars centralis*]: Median anterior part of basal plate; anterior margin may bear central tooth (dens medialis) and lateral teeth (dentes laterales) projecting from thickening (basis denticis). (See also pars alaris)

penis [G *Penis*]: Simple glandular or more complex muscular copulatory organ forming terminal section of ♂ reproductive system in posterior region of trunk; may be provided with stylet.

pharyngeal gland: buccal gland

pharynx [G *Pharynx, Schlund*]: Muscular section of digestive tract between buccal cavity and intestine; may exhibit tripartite elaboration with unpaired median and paired lateral muscle sacs. Ventral pharynx wall secretes jaws.

postlateralium [G *Postlateralium*]: One member of fifth and posteriormost pair of elongate, compound sensory bristles laterally on head.

postoral gland [G *postorale Drüse*]: Pair of buccal glands located posterior to mouth on each side of jaw.

prebursa (bursa appendix) [G *Präbursa*]: Section of ♀ reproductive system adjoining bursa posteriorly.

preoral gland: prepharyngeal gland

prepharyngeal gland (preoral gland) [G *Präpharyngealdrüse*]: Buccal gland located anterior to mouth. (paired, unpaired)

protonephridium: excretory organ

rhabdoid [G *Rhabdoid*]: Spindle-shaped epidermal inclusion; typically in bundles in posterior region of body and considered to have adhesive function.

rostral apophysis: apophysis

rostralium: frontalium

rostral wing: ala frontalis

rostrum [G *Rostrum*]: Variously shaped section of body anterior to mouth; may bear single apical cilia (apicalia), series of lateral compound sensory bristles (frontalia, ventralia, dorsalia, lateralia), and ciliary pits.

seminal vesicle [G *Vesicula seminalis*]: More or less muscular enlargement of sperm ducts in ♂ reproductive system.

serrula [G *Serrula*]: Serrated frontomedial margin of alae frontales of basal plate; formed by scale-like teeth (serrulae). May be differentiated into serrula infundibulum and serrula lateralis.

serrulae [G *Serrulae*]: Collective term applied to small, scale-like teeth along frontomedial margin of alae frontales.

serrula infundibulum [G *Serrula infundibulum*]: On basal plate, extension of scale-like dentition (serrulae) onto infundibulum.

serrula lateralis [G *Serrula lateralis*]: On basal plate, section of serrula restricted to frontomedial margins of alae frontales to each side of pars centralis. (See also serrula infundibulum)

sperm [G *Sperma*]: One of three or four types of ♂ reproductive body, including conulus, dwarf type, and filiform type.

sperm duct [G *Samenleiter, Vas deferens*]: Section of ♂ reproductive system between testes and gonopore; when two are present they may unite before entering penis.

spiral-ciliary organ [G *Spiralorgan*]: Presumable sensory organ consisting of one or two pairs of cells located in epidermis to each side of or anterior to mouth. Each cell encloses spiral cavity bearing one long, spirally coiled cilium.

standard measurements:
basal plate index: length of basal plate divided by its maximum width
body index: length of body through maximum width
pharynx index: length of pharynx divided by length of jaws

rostrum index: length of rostrum divided by its maximum width

U: relative scale of measurement (total length = 100 U, anterior tip = U 0)

stylet [G *Stylet*]: Tube-like structure surrounding penis and consisting of 8–10 parallel rods; may be enclosed by stylet sheath.

stylet gland [G *Styletdrüse*]: One of a number of glandular cells surrounding stylet and producing stylet sheath.

stylet sheath [G *Styletscheide*]: Sheath, produced by sheath gland cells and considered to be cuticular, enclosing stylet.

sulcus prepharyngealis [G *Sulcus präpharyngealis*]: Lateral constriction delimiting rostrum from body.

suspensorium [G *Suspensorium*]: One of four major parts of jaw (articularium, dentarium, involucrum, suspensorium). Suspends jaw within pharynx and consists of cauda, apophysis, fenestra ventralis, and fibula radialis.

symphysis [G *Symphysis*]: Part of articularium of jaw. Thickening connecting posterior ends of lamellae symphysis.

symphysis lamella: lamella symphysis

tail [G *Schwanz*]: Tapered posterior end of trunk.

tectum caudalis [G *Tectum caudalis*]: On basal plate, unpaired median posterior region of pars alaris.

tectum lateralis [G *Tectum lateralis*]: Part of involucrum of jaw. Thin posterolateral extension of lamina externa.

terminal cell [G *Terminalzelle*]: Proximal cell of excretory organ; bears single cilium.

testis [G *Testis, Hoden*]: Unpaired dorsal or paired lateral section of ♂ reproductive system located posteriorly in trunk; connected to gonopore via sperm ducts.

tooth [G *Zahn*]: Any one of numerous tooth-like projections of pars centralis of basal plate or of dentarium of jaw. (See also dens lateralis, dens medialis, dens terminalis)

trunk [G *Rumpf*]: Posteriormost and largest of three divisions of body (head, neck, trunk); may be produced posteriorly into tail.

vagina [G *Vagina*]: Opening of ♀ reproductive system to exterior; located dorsally, posterior to or above bursa near middle of trunk. Connected to bursa by short, wide canal.

ventralium [G *Ventralium*]: One member in a pair of elongate compound sensory bristles; located on each side of rostrum, ventral to dorsalia and between frontalia and lateralia.

NEMERTINI
(RHYNCHOCOELA,
NEMERTEA,
NEMERTINEA)

nemertine, rhynchocoel, nemertean, ribbon-worm [G *Nemertine, Schnurwurm*]

NEMERTINI

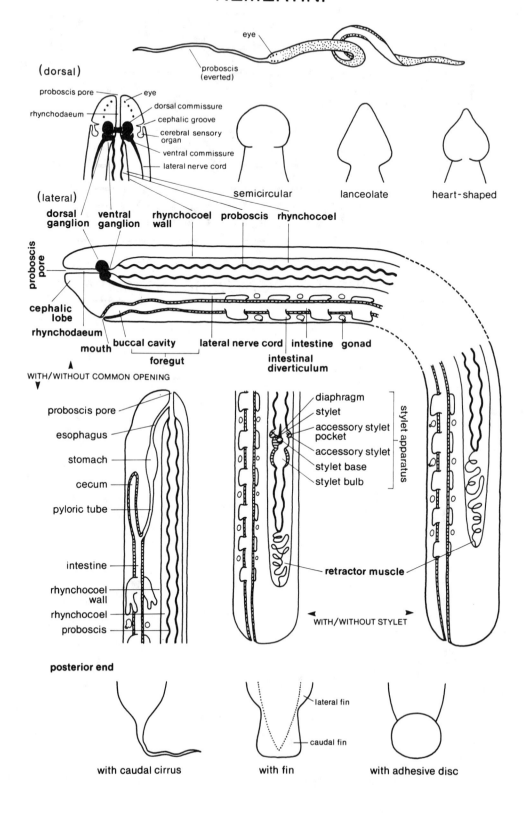

eye

proboscis (everted)

(dorsal)

proboscis pore — eye
rhynchodaeum
dorsal commissure
cephalic groove
cerebral sensory organ
ventral commissure
lateral nerve cord

semicircular   lanceolate   heart-shaped

(lateral)

dorsal ganglion **ventral ganglion** **rhynchocoel wall** **proboscis** **rhynchocoel**

proboscis pore

cephalic lobe
rhynchodaeum
mouth **buccal cavity** **foregut** **lateral nerve cord** **intestine** **gonad**

intestinal diverticulum

WITH/WITHOUT COMMON OPENING

proboscis pore
esophagus
stomach
cecum
pyloric tube

intestine

rhynchocoel wall
rhynchocoel
proboscis

diaphragm
stylet
accessory stylet pocket
accessory stylet
stylet base
stylet bulb

stylet apparatus

retractor muscle

WITH/WITHOUT STYLET

posterior end

with caudal cirrus   lateral fin   caudal fin   with fin   with adhesive disc

NEMERTINI (RHYNCHOCOELA, NEMERTEA, NEMERTINEA)

nemertine, rhynchocoel, nemertean, ribbon-worm [G *Nemertine, Schnurwurm*]

accessory stylet (reserve stylet) [G *Reservestylet, Nebenstylet*]: In armed type of proboscis, one of several variously shaped, developing stylets located in accessory stylet pockets lateral to stylet. Replaces main stylet when latter is damaged or lost.

accessory stylet pocket [G *Reservestylettasche*]: In armed type of proboscis, one of typically two, sometimes several, lateral pouches adjacent to stylet; houses a variable number of accessory stylets.

adhesive disc (sucker) [G *Saugscheibe*]: In certain nemertines, muscular, plate-shaped, ventral sucker at posterior end of body; contains gland cells, blood vessels, and ganglionic swellings of lateral nerve cords. Serves to attach commensal nemertine to host.

anal commissure [G *Analkommissur*]: Transverse nerve commissure joining lateral nerve cords at posterior end of body near anus.

anal lacuna [G *Anallakune*]: In circulatory system, transverse space usually below anus at posterior end of body; lacks definite walls. Serves as channel of communication between lateral blood vessels.

anus [G *After*]: Posterior opening of digestive tract; located terminally, occasionally subterminally and somewhat dorsal.

basement membrane: ground layer

brain [G *Gehirn*]: Main concentration of nerve tissue; typically four-lobed, consisting of two dorsal and two ventral ganglia. Surrounds rhynchocoelom or rhynchodaeum by means of dorsal and ventral commissures. Gives rise posteriorly to two large lateral nerve cords, foregut nerves, and middorsal nerve, anteriorly to cephalic nerves.

buccal cavity [G *Mundhöhle*]: Short, bell-shaped anteriormost section of foregut between mouth and esophagus; may be delimited from mouth opening by lips and

95

be surrounded anteriorly by ring of salivary glands.

bulb: stylet bulb

caecum: cecum

caudal cirrus (tail) [G *Schwänzchen*]: Short, slender, tail-like extension of posterior end of body; contains all layers of body wall as well as posteriormost sections of digestive tract, longitudinal blood vessels, and lateral nerve cords.

caudal fin (tail fin) [G *Afterflosse*]: In certain nemertines, dorsoventrally compressed, fin-like extension of posterior end of body. (simple, bilobed) (See also lateral fin)

cecum (intestinal cecum) [G *Blinddarm*]: Anteriorly directed, blind section of intestine lying parallel to and directly under foregut; entrance of pyloric tube into dorsal wall of intestine marks boundary between cecum and posteriorly directed intestine. May also bear anterior and/or lateral diverticula.

central stylet: stylet

cephalic groove (cephalic furrow) [G *Kopffurche*]: One, occasionally two pairs of shallow, transverse furrows in epidermis, one across each side of anterior end of body. Left and right cephalic grooves typically do not meet middorsally or midventrally. (See also cephalic slit)

cephalic lacuna [G *Kopfgefäß*]: In circulatory system, space at anterior tip of body serving as channel of communication between cephalic blood vessels; lacks definite walls. (See also anal lacuna)

cephalic lobe [G *Kopflappen*]: Variously shaped extension of anterior end of body. (heart-shaped, lancet-shaped, spatula-shaped, semicircular)

cephalic nerve [G *Kopfnerv*]: One of a number of smaller nerves arising from frontal side of dorsal and ventral ganglia of brain and extending anteriorly to eyes, cerebral organs, frontal organ, and other cephalic structures.

cephalic slit [G *Kopfspalte*]: Pair of deep, longitudinal furrows in epidermis, one

along each side at anterior end of body; extend from proboscis pore to mouth or brain.

cerebral canal (ciliated canal) [G *Zerebralkanal*]: In cerebral organ, elongated, ciliated canal leading from exterior to gland- and nerve cell-bearing inner end of organ; may be forked.

cerebral sensory organ [G *Zerebralorgan*]: Pair of invaginations or canals opening to exterior via cephalic grooves, cephalic slits, or separate lateral pores near brain; lined with elongate cilia and bearing gland and nerve cells at inner end. Connected to brain by cephalic nerves or closely adjoining brain at inner end.

circular muscle [G *Ringmuskel*]: Component of innermost layer of body wall (epidermis, dermis, musculature). If two-layered, musculature consists of outer circular and inner longitudinal muscles; if three-layered, two circular muscle layers may sandwich longitudinal muscle layer or single circular muscle may be sandwiched between two longitudinal muscle layers.

cirrus: tentacle (See also caudal cirrus)

cutis: dermis

dermis (cutis) [G *Kutis*]: Middle layer of body wall (epidermis, dermis, musculature); may be composed of either hyaline, gelatinous tissue (basement membrane or ground layer) or fibrous tissue (dermis or cutis in restricted sense). May be divisible into outer glandular and inner connective tissue zones. (fibrous, gelatinous)

Desor's larva [G *Desor'sche Larve*]: Oval, ciliated type of larva remaining within egg membrane and lacking apical sense organ, oral lobes, and ciliated oral band. (See also pilidium)

diagonal muscle [G *Diagonalmuskel*]: One of two relatively thin layers of muscle fibers, either between outer circular and inner longitudinal muscles (two-layer musculature) or between outer longitudinal and middle circular muscles (three-layer system).

diaphragm [G *Diaphragma*]: Muscular thickening in midregion of armed type of proboscis; secretes stylet base and bears stylet on anterior surface.

diverticulum: intestinal diverticulum

dorsal commissure (dorsal cerebral commissure, dorsal brain commissure) [G *dorsale Gehirnkommissur*]: Transverse band of nerve fibers passing across rhynchodaeum or rhynchocoel and connecting two dorsal ganglia of brain. (See also ventral commissure)

dorsal ganglion [G *dorsales Ganglion*]: In brain, one of two dorsal concentrations of nerve tissue on each side of rhynchodaeum or rhynchocoel; connected to each other by dorsal commissure and closely adjoining ventral ganglia.

dorsal peripheral nerve: peripheral nerve branch

dorsomedian nerve: middorsal nerve cord

epidermis [G *Epidermis, Epithel*]: Outermost layer of body wall (epidermis, dermis, musculature) consisting of tall, ciliated epithelial cells and bearing gland, interstitial, and sensory cells.

esophageal lacuna (foregut vascular plexus) [G *Schlundgefäß, Vorderdarmgefäß*]: In circulatory system, anterior network of channels formed by branches of each lateral blood vessel immediately posterior to ventral connective; serves foregut.

esophageal nerve: foregut nerve

esophagus (oesophagus) [G *Ösophagus*]: Section of foregut between buccal cavity and stomach; in contrast to remainder of foregut, epithelium of esophagus lacks glands. If digestive tract and proboscis apparatus share common opening at proboscis pore, esophagus forms anteriormost section of digestive tract.

eye [G *Auge*]: One of two or more symmetrically arranged pigment-cup ocelli located dorsally at anterior end of body, typically anterior to brain; connected to brain by cephalic nerves. (black, brown, red, blue)

flame bulb (flame cell) [G *Endkölbchen, Wimperkölbchen, Kölbchen*]: In protonephridium, one of up to several thousand variously elaborated excretory structures containing tuft of fused cilia and closely associated with lateral blood vessels. Opens to exterior via protonephridial tubule.

foregut [G *Vorderdarm*]: Anterior and shorter of two major divisions of digestive tract (foregut, midgut, rarely hindgut); may be further subdivided into buccal cavity, esophagus, stomach, and pyloric tube.

foregut nerve (esophageal nerve) [G *Schlundnerv*]: Single or pair of nerves arising from caudal side of either ventral ganglia or ventral commissure and extending posteriorly to foregut.

frontal gland (cephalic gland) [G *Kopfdrüse*]: Elongate series of closely adjoining glands opening together at anterior end of body above proboscis pore, often through frontal organ; sometimes extend posteriorly behind brain. May open independently via numerous improvised pores over head surface.

frontal organ [G *Frontalorgan*]: Single flask-shaped pit or, when protruded, mound-like elevation above proboscis pore at anterior tip of body; bears elongate cilia and may be penetrated by ducts of frontal glands. Term may also refer to three sensory papillae at anterior tip of body.

gonad: ovary, testis

gonopore (genital pore) [G *Genitalporus, Geschlechtsporus*]: Opening of ♂ or ♀ reproductive system to exterior; typically series of lateral openings communicating with gonads via short gonoducts, in ♂ occasionally at end of penes.

ground layer (basement membrane) [G *Grundschicht*]: Hyaline, gelatinous type of dermis.

hindgut [G *Enddarm*]: Term occasionally applied to short, posterior, diverticulum-free section of intestine before anus.

intestinal cecum: cecum

intestinal diverticulum [G *Darmtasche*]: Lateral, typically regularly arranged pairs of outpocketings of intestine; ovaries and testes typically positioned between diverticula. (simple, branched)

intestine (midgut) [G *Mitteldarm*]: Posterior and longer of two major divisions of digestive tract (foregut, midgut, rarely hindgut); usually bears lateral diverticula and may be divided into anteriorly directed cecum and posteriorly directed intestine proper. (straight, sinuous)

lateral blood vessel [G *Seitengefäß*]: One of two main longitudinal blood vessels running along sides of body; communicates anteriorly via cephalic lacuna and ventral connective, posteriorly via anal lacuna. Gives rise to rhynchocoel vessel and may communicate with middorsal vessel posteriorly via numerous transverse connectives.

lateral fin [G *lateraler Flossensaum*]: In certain nemertines, short, fin-like lateral extension of body; typically better developed posteriorly and merging with caudal fin.

lateral ganglion: ventral ganglion

lateral nerve cord [G *Seitenstamm, Seitennerv*]: One of two main nerve cords emanating from ventral ganglion in brain and extending laterally or ventrolaterally to anal commissure at posterior end of body; may be joined to one another ventrally and to middorsal nerve by transverse commissures, or give rise to series of peripheral nerve branches.

lateral organ [G *Seitenorgan*]: In certain nemertines, pair of pit-shaped or, when protruded, mound-shaped sensory organs adjacent to nephridiopore, one on each side at anterior end of body.

lateral peripheral nerve: peripheral nerve branch

lateral rhynchocoel blood vessel [G *Rhynchocoelomseitengefäß*]: Branch of each lateral blood vessel posterior to esophageal lacuna; runs alongside proboscis sheath. (See also rhynchcoel blood vessel)

lip [G *Lippe*]: Ring-like fold of epidermis extending into lumen of digestive tract between mouth and buccal cavity; differs histologically from epithelium of buccal cavity.

longitudinal muscle [G *Längsmuskel*]: Component of innermost layer of body wall (epidermis, dermis, musculature). If two-layered, musculature consists of outer circular and inner longitudinal muscles; if three-layered, single longitudinal muscle layer may be sandwiched between two circular muscle layers or two coats of longitudinal muscles may sandwich single circular layer.

lower dorsal nerve [G *untere Rückennerv*]: Longitudinal dorsal nerve emanating from and running parallel directly below middorsal nerve.

mesenchyme [G *Mesenchym, Parenchym*]: Undifferentiated tissue between body wall and digestive tract and surrounding blood vessels, nerves, gonads, and proboscis sheath; gives rise to gonads. May also refer to nonfibrous connective tissue in body wall.

middorsal blood vessel [G *Rückengefäß*]: Longitudinal blood vessel emanating at anterior end of body from ventral connective or branch of cephalic vessel and extending posteriorly to anal lacuna; typically connected to lateral blood vessels posteriorly by numerous transverse connectives. Usually penetrates rhynchocoel wall to form vascular plug or rhynchocoelic villus.

middorsal nerve cord [G *Rückennerv*]: Longitudinal nerve, typically arising from dorsal commissure and extending posteriorly to anal commissure. May give rise to lower dorsal nerve and be connected to lateral nerve cords by transverse commissures.

midgut: intestine

midventral nerve [G *Bauchnerv*]: Longitudinal nerve running ventrally in body wall musculature.

mouth [G *Mund*]: Anterior opening of digestive tract located ventrally below or posterior to brain. Digestive tract may open into rhynchodaeum and therefore share common opening with proboscis at proboscis pore. (round, elongate)

nephridial gland [G *Nephridialtasche*]: Modified proximal section of protonephridium; end of protonephridial tube gives off dense series of capillaries which form ridge penetrating lateral blood vessel.

nephridiopore [G *Exkretionsporus*]: Minute opening of excretory system to exterior; typically one on each side at level of foregut, yet also forming series of many thousand openings along entire body.

nephridium: protonephridium

ovary [G *Ovar, Ovarium*]: Sac-shaped ♀ reproductive organ, typically in a series between intestinal diverticula on both sides of intestine.

ovotestis [G *Ovotestis*]: In hermaphroditic nemertines, reproductive organ bearing both ovocytes and spermatocytes.

packet gland [G *Packetdrüse*]: Clusters of gland cells in epidermis or dermis opening to exterior via common duct.

parenchyma: mesenchyme

penis [G *Penis*]: In ♂ reproductive system of certain nemertines, finger-like extension of sperm duct; projects outward from body wall in two rows at anterior end of body.

peripheral nerve branch [G *peripherer Nerv*]: One in a series of branching nerves arising at regular intervals from each lateral nerve cord. Termed, according to position, dorsal, lateral, and ventral peripheral nerve branches.

pilidium [G *Pilidium*]: Helmet-shaped, free-swimming type of larva with apical sensory organ, oral lobes, and ciliated oral band. (See also Desor's larva)

proboscis [G *Rüssel*]: Elongate, often coiled, eversible, blind tube lying in rhynchocoel above digestive tract; may bear stylet. Similar to body wall in structure and attached to posterior end of rhynchocoel by retractor muscle. (armed, unarmed, simple, branched)

proboscis pore [G *Rüsselöffnung*]: Opening at anterior tip of body through which proboscis is everted; either located dorsal to mouth or forming common opening with digestive tract.

proboscis retractor muscle: retractor muscle

proboscis sheath [G *Rüsselscheide*]: Wall of rhynchocoel; in certain nemertines, bears paired lateral outpocketings, corresponding in position to intestinal diverticula, or unpaired dorsal or ventral pouches.

protonephridial tubule [G *Nephridialkanal, Protonephridialkanal*]: Section of protonephridium between flame bulb and nephridiopore; may be simple, branched, or regionally differentiated into thicker convoluted and thinner straight sections.

protonephridium (nephridium) [G *Protonephridium, Nephridium*]: Excretory organ, associated proximally with lateral blood vessel and consisting of flame bulb and protonephridial tubule opening to exterior via nephridiopore; typically paired, one on each side at level of foregut, yet also highly branched or forming series along entire body.

pyloric tube (pylorus) [G *Pylorusrohr*]: Narrow, tube-shaped section of digestive tract between stomach and intestine; enters dorsal wall of intestine, thus marking boundary of posteriorly directed intestine proper and anteriorly directed cecum.

reserve stylet: accessory stylet

retractor muscle (proboscis retractor muscle) [G *Retraktor, Retraktor des Rüssels*]: Muscle attaching posterior end of proboscis to wall of rhynchocoel.

rhabdite [G *Rhabdit*]: Small, hard calcareous structures in epidermis of body wall or epithelium of proboscis. (rod-shaped, sickle-shaped)

rhynchocoel [G *Rhynchocoelom*]: Closed, fluid-filled cavity enclosing proboscis; located dorsal and parallel to digestive tract and typically extending the length of the body. Wall of rhynchocoel is termed proboscis sheath.

rhynchocoel blood vessel [G *Rhynchocoelomgefäß*]: Branch of each lateral blood vessel posterior to esophageal lacuna; runs through proboscis sheath and rejoins lateral vessel. (See also lateral rhynchocoel blood vessel)

rhynchocoelic villus: Thin-walled, elongate protrusion of middorsal blood vessel into rhynchocoel lumen.

rhynchodaeum [G *Rhynchodaeum*]: In proboscis apparatus, short, tubular cavity between proboscis pore and insertion of proboscis at level of brain.

salivary gland [G *Speicheldrüse*]: One in a series of elongate gland cells forming ring around anterior region of buccal cavity.

sensory papilla [G *Frontalorgan*]: One of three small projections bearing elongate cilia at anterior tip of body; histologically similar to and often termed frontal organ.

sperm duct [G *Samenleiter, Genitalductus*]: Section of ♂ reproductive system between testis and gonopore; typically, short tube between each testis and exterior. In certain parasitic nemertines, also median longitudinal tube serving as common duct of numerous testes and opening into posterior end of intestine.

spermiducal vesicle: Posterior expanded section of median longitudinal type of sperm duct.

statocyst [G *Statozyste, Otolithenblase*]: One, occasionally two pairs of vesicles associated with ventral ganglia in brain; contains statolith.

statolith [G *Statolith, Otolith*]: Variously elaborated solid element or elements in statocyst. (dumbbell-shaped, spherical)

stomach [G *Magen, Magendarm*]: Enlarged gland-bearing posterior section of foregut (buccal cavity, esophagus, stomach); may open directly into intestine or be joined to intestine by pyloric tube.

stylet (central stylet) [G *Stachel, Giftstachel, Hauptstachel, Angriffsstilet*]: In armed type of proboscis, barb on stylet base on anterior surface of diaphragm; located in posterior region of retracted proboscis, at tip of everted proboscis. (curved, straight) (See also accessory stylet)

stylet base [G *Basis*]: In armed type of proboscis, conical, granular mass secreted by diaphragm and bearing either single stylet terminally or numerous stylets distributed over shield-like base.

stylet bulb (bulb) [G *Ballon, Blase*]: In armed type of proboscis, expanded muscular region immediately posterior to stylet-bearing diaphragm.

submuscular gland: One of numerous clusters of mucus-secreting glands in mesenchyme; typically open to exterior on ventral surface.

sucker: adhesive disc

tail: caudal cirrus

tail fin: caudal fin

tentacle (cirrus) [G *Zirrus, Cirrus*]: In certain nemertines, pair of elongate, muscular lateral projections of body wall at anterior end of body.

testis [G *Hoden*]: Sac-shaped ♂ reproductive organ, often in a series between intestinal diverticula on both sides of intestine; typically opens to exterior via short sperm duct, rarely by penis or common longitudinal sperm duct.

vascular plug: Bulbous protrusion of middorsal blood vessel through ventral rhynchocoel wall, usually close to cerebral ganglia.

ventral commissure (ventral brain commissure) [G *ventrale Gehirnkommissur*]: Transverse band of nerve fibers passing below rhynchodaeum and connecting two ventral ganglia of brain. (See also dorsal commissure)

ventral connective [G *ventrale Gefäßkommissur*]: In circulatory system, transverse blood vessel joining lateral blood vessels anteriorly below rhynchodaeum; mid-dorsal blood vessel sometimes arises from ventral connective.

ventral ganglion (lateral ganglion) [G *ventrales Ganglion*]: In brain, one of two ventral concentrations of nerve tissue on each side of rhynchodaeum or rhynchocoel; connected to each other by ventral commissure and closely adjoining dorsal ganglia.

ventral peripheral nerve: peripheral nerve branch

ENTOPROCTA

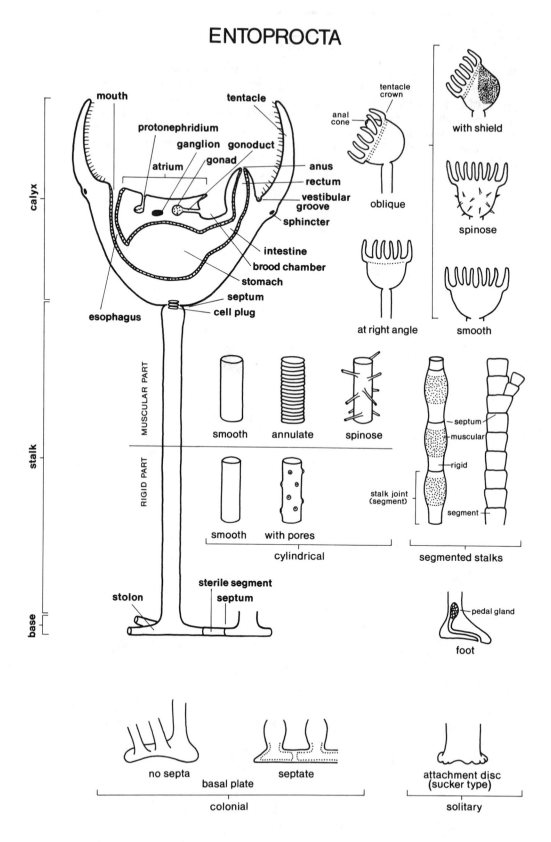

mouth
tentacle
protonephridium
ganglion
gonoduct
atrium
gonad
anus
rectum
vestibular groove
sphincter
intestine
brood chamber
stomach
septum
cell plug
esophagus

tentacle crown
anal cone
oblique
with shield
spinose
at right angle
smooth

calyx

MUSCULAR PART
smooth
annulate
spinose

RIGID PART
smooth
with pores
cylindrical

septum
muscular
rigid
stalk joint (segment)
segment
segmented stalks

stalk

stolon
sterile segment
septum

pedal gland
foot

base

no septa
basal plate
septate
colonial

attachment disc (sucker type)
solitary

ENTOPROCTA (KAMPTOZOA)

entoproct [G *Kamptozoon, Entoproct, Kelchwurm, Nicktier*]

anal cone [G *Analkonus*]: Barrel-shaped elevation on posterior end = aboral side of calyx; contains rectum and bears anus terminally.

anus [G *After*]: Posterior opening of U-shaped digestive tract within tentacle crown; typically located on elevated anal cone.

atrium (vestibule) [G *Atrium, Vestibulum*]: Within tentacle crown of calyx, concave surface between mouth and anus; section between gonopore and anus is termed genital recess and may serve as brood chamber.

attachment disc [G *Haftscheibe*]: In solitary entoproct, disc-shaped base of stalk; either a rudimentary foot or foot modified into a sucker without pedal gland. (See also basal plate, stolon)

basal plate [G *Fußplatte*]: Common attachment structure of several zooids. Either nonseptate due to nondetachment of buds from mother zooid or elongated and septate with single zooid stem originating from each segment (i.e., stolon rudiment).

base: attachment disc, basal plate, stolon

brood chamber [G *Brutraum*]: In ♀ or hermaphrodite, concave depression in atrium between gonopore and anus (genital recess); eggs or embryos brooded here. (See also embryophore)

bud [G *Knospe*]: Outgrowth of body wall (only ectoderm and mesenchyme); in solitary entoproct on oral side of calyx, in colonial forms orally or lateroorally on stalk base or muscular portions of stalk segments. Gives rise asexually to new zooid.

calyx (kalyx, calix) [G *Calyx, Kelch*]: Distal of two major divisions of entoproct (stalk, calyx); bears marginal tentacle crown, entire digestive tract, and all major organs. (smooth, spinous, with shield)

cell plug [G *Zellhaufen, Sternzellapparat*]: Stack of transversely oriented, star-like muscle cells at septum between calyx and stalk; serves as pumping organ to facilitate communication between calyx and stalk.

ciliary band (frontal ciliary band, lateral ciliary band) [G *Frontalcilien, Lateralcilien*]: On inner (atrial) surface of tentacle, one of three longitudinal bands of cilia; consists of two lateral bands of longer cilia (lateral cilia) responsible for water current through tentacular crown and median band of shorter cilia (frontal or medial cilia) leading to vestibular groove.

cuticle [G *Kutikula*]: Noncellular outermost layer of body wall (cuticle and underlying epidermis); covers entire surface and may form thick shield on one side of calyx.

cuticular shield: shield

dorsal shield: shield

embryophore [G *Embryophor*]: Wall of genital recess of atrium facing anal cone; eggs emerge from gonopore and then attach by a gelatinous envelope to embryophore in a series of increasingly mature embryos or larvae.

epidermis [G *Epidermis, Epithel, Körperwandepithel, Außenepithel*]: Single layer of cells below cuticle in body wall.

esophagus [G *Ösophagus, Schlund*]: Relatively narrow, tubular section of U-shaped digestive tract between mouth and stomach; constricted by series of loop-shaped constrictor muscles.

foot [G *Fuß*]: In solitary entoproct, expanded base of stalk bearing pedal gland and by which zooid is attached to substratum.

foot gland: pedal gland

frontal cilium: ciliary band

ganglion (subenteric ganglion) [G *Zentralganglion, Subösophagealganglion*]: Main concentration of nerve tissue; consists of single median bilobular ganglion located in calyx between atrium and stomach; gives rise to nerves for afferent and efferent innervation of tentacles, calyx, and stalk.

genital recess [G *Rezessus genitalis*]: Within tentacle crown of calyx, depression of atrium between gonopore and anus; serves as brood chamber in ♀ or hermaphrodite.

gland cell [G *Drüsenzelle*]: In epithelium of tentacle or calyx of certain entoprocts, bead-like unicellular or pluricellular mucus gland of unknown function.

gonad: ovary, testis

gonoduct [G *Gonodukt*]: General term for oviduct or sperm duct; typically associated with gland cells. In hermaphroditic entoproct, common duct formed by merger of sperm ducts and oviducts.

gonopore [G *Geschlechtsöffnung, Gonoporus, Geschlechtsporus*]: Opening of ♂ or ♀ reproductive system to exterior; located on vestibule at margin of genital recess (brood chamber).

intestine [G *Mitteldarm, Intestinum*]: Short section of U-shaped digestive tract between stomach and rectum; typically separated from rectum by constriction (sphincter).

lateral cilium: ciliary band

lateral sense organ [G *Sinnesorgan, Sinnespapille*]: Pair of sensory projections, one on each side of calyx under tentacle crown; consists of small, papilla-like structure with tuft of bristles.

lip [G *Lippe, Oberlippe, Unterlippe*]: Fold of body wall adjoining mouth; occasionally one may distinguish an upper lip anterior to mouth and lower lip posterior to mouth.

liver [G *Leber*]: Term occasionally applied to modified resorbing and excretory upper, subatrial area of stomach wall.

longitudinal muscle [G *Längsmuskel*]: Longitudinal muscle fibers along inner (oral, atrial) wall of tentacles and in stalk; run either continuously from calyx to stalk base (solitary entoproct) or are interrupted by an incomplete transverse septum in narrow connection between calyx and stalk (stalk neck).

lower lip: lip

mesenchyme [G *Mesenchym*]: Loose network of pluripotent cells filling body cavity.

mouth [G *Mund, Mundöffnung*]: Ciliated, funnel-shaped opening of U-shaped digestive tract at anterior end of tentacle crown; circular vestibular groove opens into each side of mouth.

muscular socket [G *Muskelsockel*]: In certain colonial entoprocts, thickened muscular base of zooid stalk; gives rise to stolonial budding of zooids and is covered by a thick, squamous, and flexible cuticle. May also refer to similar muscular swellings further up stalk (which, in turn, may serve as branching points of colony. (See also stalk joint)

nephridioduct [G *Nephrodukt*]: Section of protonephridium between flame bulb and nephridiopore; consists of channel surrounded by specialized ion- and osmoregulatory cells.

nephridiopore [G *Nephroporus*]: Opening of protonephridium to exterior. Typically paired openings posterior to mouth on atrium. May also refer to openings of osmoregulatory structures in stalk. (single, paired)

ovary [G *Ovar, Ovarium*]: Paired, sacciform ♀ reproductive organ located in calyx between atrium and stomach; each gives rise to oviduct. In hermaphroditic entoproct, located anterior to testis.

oviduct [G *Ovidukt*]: Short section of ♀ reproductive system leading from each ovary and merging to form single gonoduct. Opens to exterior via gonopore.

pedal gland (foot gland) [G *Kittdrüse, Fußdrüse, Klebdrüse*]: In solitary entoproct, large, multicellular gland at base of stalk; opens to exterior on foot either directly via pore or by groove.

pore (pore organ) [G *Pore, Stielpore, Porenorgan*]: Single-cell or multicellular glandular organ in epidermis of stalk. Located under pore opening (or under hollow spine) and considered to serve in ion regulation.

protonephridium [G *Protonephridium*]: Pair of small, ion- and osmoregulatory excretory organs located in oral region of calyx between stomach and atrium; each consists of flame bulb and intracellular canal (nephridioduct) opening to exterior via nephridiopore posterior to mouth. In freshwater entoproct, one of up to 30 flame bulbs opening via common nephridiopore on vestibule. May also refer to osmoregulatory structures in stalk.

rectum [G *Enddarm, Rektum*]: Terminal section of U-shaped digestive tract; typically separated from intestine by constriction (sphincter) and forming elevated anal cone.

resting bud [G *Dauerknospe*]: Protuberance of stolon with thick cuticle and enlarged mesenchyme cells; survives winter or other unfavorable conditions. Muscular sockets or muscular stalk joints may also serve as resting buds.

segment [G *Glied*]: General term for series of more or less distinct units of stolon or stalk. (See also stalk joint)

seminal vesicle [G *Samenblase*]: In ♂ reproductive system, enlarged section of common sperm duct.

septum [G *Septum, Diaphragma*]: Infolding of body wall at base of calyx forming incomplete constriction between calyx and stalk; may also refer to incomplete cuticular partition between stalk-bearing and sterile segments of stolon or similar structures in stalk.

shield (cuticular shield, dorsal shield) [G *Kelchschild, Schutzschild*]: Plate-like cuticular thickening on side of outer calyx wall close to anus. Typically directed upward due to position of calyx at an angle to stalk.

sperm duct [G *Vas deferens, Samenleiter*]: Short section of ♂ reproductive system leading from each testis and merging to form single gonoduct. May bear enlarged section (seminal vesicle) and opens to exterior via gonopore.

sphincter [G *Ringmuskel, Sphinkter*]: Band of circular muscle within tentacular membrane around rim of calyx. Allows membrane to cover tentacle crown when tentacles curl into atrium. May also refer to sphincter muscles around various sections of digestive tract.

spine [G *Stachel, Dorn*]: Spine-like cuticular protuberance on calyx or stalk; either solid thorns (in certain solitary entoprocts) or hollow spines in flexible cuticular portions of colonial forms (covering pore organs), being equivalent to cuticular pores in rigid cuticular portions.

stalk [G *Stiel*]: Proximal of two major divisions of entoproct (calyx, stalk); may be separated from calyx by septum. Attached to substratum in solitary entoproct by foot, in colonial forms by stolons or attachment zone of stalk base. (cylindrical, tapering; flexible, stiff; continuously muscular, segmented; branched, unbranched; annulated, with pores in stiff portion, with spines in flexible portion)

stalk joint [G *Stielglied, Glied*]: One in a series of segments of stalk. Each joint formed by stiff, muscle-free proximal portion and swollen, muscular distal portion; separated from adjacent joints by incomplete septum. May function as resting bud.

standard measurements:
total length
calyx: length, width, depth
reduncle: length, width, depth
ratio: calyx length/peduncle length
calyx length/width
peduncle width/depth

stolon [G *Stolo*]: In colonial entoproct, slender, tube-like connection between zooids; adheres to substratum. Formed by elongation between bud (formed at zooid base) and mother zooid; separated from zooids on either side by a septum (and thus merely serving as spacer).

stomach [G *Magen*]: Large, sacciform section of U-shaped digestive tract between esophagus and intestine; upper (ventral) surface underlying atrium is referred to as floor (liver), lower (dorsal) surface near stalk is referred to as roof.

tentacle [G *Tentakel*]: One of up to 30 nonretractile projections surrounding margin of calyx; each bears one band of frontal cilia and two bands of lateral cilia on surface facing atrium. Four oral tentacles are longer in one species.

tentacle crown [G *Tentakelkrone, Tentakelkranz*]: Circlet of tentacles around rim of calyx; basically horseshoe-shaped, with arms connecting aborally. (oval, elliptical)

tentacular membrane [G *Tentakelmembran, Epithelfalte*]: Tissue (epidermal fold) between base of tentacles. Forms rim of calyx and, because of sphincter muscle, covers tentacle crown when tentacles curl into atrium.

testis [G *Testis, Hoden*]: Paired, sacciform ♂ reproductive organ located in calyx between atrium and stomach; each gives rise to sperm duct, which unites with second sperm duct before opening to exterior via gonopore. In hermaphroditic entoproct, located posterior to ovary.

tube cell [G *Röhrenzelle*]: Term formerly applied to fusiform mesenchyme cell type in stalk.

upper lip: lip

vestibular groove [G *Atriumrinne*]: Horseshoe-shaped, ciliated groove extending around outer margin of atrium at base of tentacles; opens into each side of mouth. Cilia of tentacles merge into vestibular groove.

vestibular retractor (atrial retractor, vestibular depressor) [G *Retraktor des Vestibulums*]: Two groups of muscle fibers spanning from atrial floor to either side of calyx base.

vestibule: atrium

zooid [G *Zooid*]: Term applied to single individual of colony.

GASTROTRICHA

gastrotrich [G *Gastrotriche, Flaschentierchen*]

GASTROTRICHA

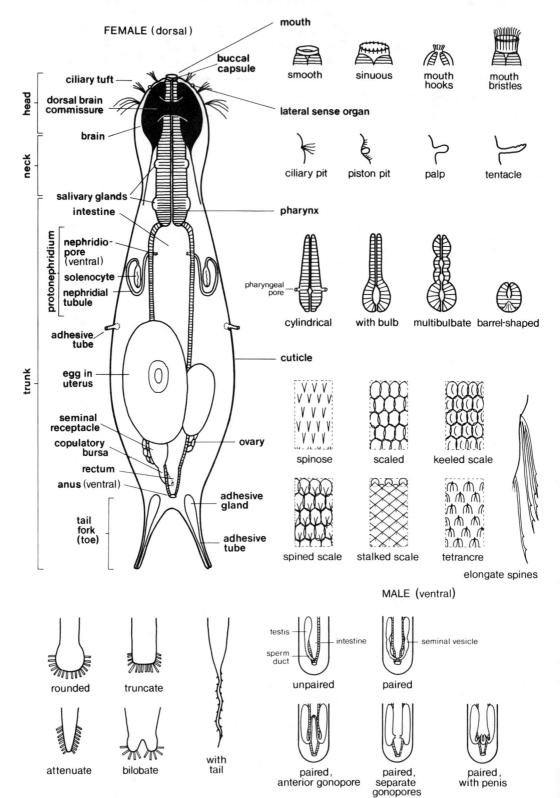

FEMALE (dorsal)

mouth

buccal capsule

head

ciliary tuft

dorsal brain commissure

brain

lateral sense organ

neck

salivary glands

intestine

protonephridium

nephridio-pore (ventral)

solenocyte

nephridial tubule

pharynx

adhesive tube

egg in uterus

trunk

seminal receptacle

copulatory bursa

rectum

anus (ventral)

tail fork (toe)

cuticle

ovary

adhesive gland

adhesive tube

smooth   sinuous   mouth hooks   mouth bristles

ciliary pit   piston pit   palp   tentacle

pharyngeal pore

cylindrical   with bulb   multibulbate   barrel-shaped

spinose   scaled   keeled scale

spined scale   stalked scale   tetrancre

elongate spines

MALE (ventral)

rounded   truncate   with tail

attenuate   bilobate

testis   intestine
sperm duct
unpaired   paired   seminal vesicle

paired, anterior gonopore   paired, separate gonopores   paired, with penis

GASTROTRICHA

gastrotrich [G *Gastrotrich, Flaschentierchen*]

adhesive gland (cement gland) [G *Klebdrüse*]: One of up to several hundred epidermal gland cells under cuticle producing an adhesive secretion; open laterally, dorsolaterally, and ventrolaterally in longitudinal rows or in aggregations on head and/or tail fork via adhesive tubes.

adhesive tube [G *Haftröhrchen*]: Cuticular, often independently movable tube serving as opening to exterior for adhesive gland. Can number up to several hundred; arranged in longitudinal rows laterally, dorsolaterally, and ventrolaterally on trunk as well as aggregated on head and/or tail fork.

anus [G *After*]: Posterior opening of digestive tract typically located either terminally between tail forks or ventrally, a short distance from end of trunk; may be cuticle-lined and bear sphincter muscle. (terminal, ventral)

brain (cerebral ganglion) [G *Gehirn*]: Main concentration of nerve tissue in form of two masses, one on each side of anterior section of pharynx, connected by dorsal brain commissure; each mass gives rise posteriorly to lateral nerve.

bristle: sensory bristle

buccal capsule [G *Mundröhre, Bukkalhöhle*]: Short, cuticle-lined section of digestive tract between mouth and pharynx; may bear longitudinal ridges or mouth bristles.

cement gland: adhesive gland

cephalion [G *Kephalion, Stirnkappe*]: Unpaired, smooth cuticular head plate anterior to subventral mouth and in contact with buccal capsule. (See also hypostomium, pleurion)

ciliary pit [G *Wimpergrube*]: Type of paired lateral sense organ, one on each side of head, consisting of ciliated depression in cuticle.

ciliary tuft [G *Wimperbüschel*]: Tufts of sensory cilia on head; sometimes forming three separate clusters—an anterior lateral, posterior lateral, and oral ciliary tuft—on each side of head.

cilium: locomotory cilium, sensory cilium

circular muscle [G *Ringmuskel*]: Delicate circular muscle fibers associated with wall of digestive tract and reproductive organs.

cirrus [G *Zirrus, Cirrus*]: Thick ambulatory structure formed by modification (fusion) of cilia on ventral surface of head or trunk.

cloaca [G *Kloake*]: Term applied to common opening of gonopore and anus in ♀.

copulatory bursa [G *Bursa copulatrix*]: Variously expanded posterior section of oviduct.

cuticle [G *Kutikula*]: Thin, noncellular layer lining pharynx and covering outer surface of body, where it is often elaborated into a wide range of scales and/or plates.

dorsal brain commissure (dorsal pharyngeal nerve commissure) [G *Faserring des Gehirns*]: Narrow or broad transverse band of nerve fibers passing dorsally across pharynx and connecting two lateral masses of brain.

dorsal gland [G *Rückendrüse*]: One of numerous variously elaborated dorsolateral epidermal glands opening via pores. (See also adhesive gland)

egg [G *Ei*]: One of two types of often conspicuous, oval ♀ reproductive bodies produced in ovary, stored in uterus, and passing through oviduct to exterior via gonopore (basically through body wall rupture). (smooth, spinous, tuberculate)

eye spot (ocellus) [G *Augenfleck, Ocelle*]: One of two or four dorsolateral clusters of red pigment in brain.

flame bulb: solenocyte

gonopore [G *Genitalöffnung, Geschlechtsöffnung*]: Opening of reproductive system to exterior. In ♀, single (dorsal, lateral, or ventral) pore near ripe egg. Eggs released by body wall rupture. In ♂, opening in common with anus or a single midventral or pair of ventrolateral pores in posterior third of trunk. (paired, unpaired)

gymnocoel: pseudocoel

head (head lobe) [G *Kopf*]: Anteriormost of three divisions of body (head, neck, trunk) bearing terminal or subventral mouth and varied array of cilia and sensory bristles; separated from trunk by constriction (neck), but often not clearly delimited.

head plate [G *Kopfplatte*]: Large cuticular thickening on head; typically includes unpaired cephalion anterior to mouth, unpaired hypostomium posterior to mouth, and pair of lateral pleurons.

hypostomium [G *Hypostomium*]: Cuticular head plate posterior to mouth. (See also cephalion, pleurion)

intestine [G *Intestinum*]: Entire digestive tract posterior to pharynx.

lateral nerve [G *Hauptnerv*]: Longitudinal nerve cord, one arising from each brain mass, extending laterally to posterior end of trunk.

lateral sense organ [G *Seitensinnesorgan*]: Variously elaborated, paired sense organ, one on each side of head and closely associated with brain. (See also ciliary pit, palp, piston pit, tentacle)

locomotory cilium [G *Ventralwimper*]: Coat of typically short cilia arranged in longitudinal bands or transverse rows along entire ventral surface. (See also sensory cilium)

longitudinal muscle [G *Längsmuskel*]: One of a varying number of typically isolated, delicate dorsal and ventrolateral muscle fibers inserting in mouth region and spanning to posterior end of trunk.

mesenteron: intestine

midgut: intestine

mouth [G *Mund*]: Variously elaborated and positioned anterior opening of digestive tract on head; may be cuticularized and bordered by numerous inwardly curved mouth hooks. (polygonal, rounded, subventral, terminal)

mouth bristle [G *Mundstachel*]: One of numerous long cuticular projections associated with buccal capsule and emerging from mouth. (See also mouth hook)

mouth hook [G *Häkchen*]: One of numerous short, distally inwardly curved, cuticular hooks bordering mouth. (See also mouth bristle)

neck [G *Hals*]: Short, slightly constricted, yet often poorly delimited division of body between head and trunk.

nephridial tubule [G *Kanalknäuel der Protonephridien*]: Thin, elongated, coiled section of protonephridium along each side of intestine between solenocytes and nephridiopore.

nephridiopore [G *Mündung des Protonephridiums*]: One of two openings of protonephridium in middle of trunk; located either laterally or between both ventral bands of locomotory cilia.

organ X: X-organ

ovary [G *Ovar*]: ♀ reproductive organ in form of cell bands or masses at posterior end of trunk. Eggs are released by body wall rupture.

oviduct [G *Ovidukt*]: Unpaired, poorly defined section of ♀ reproductive system between uterus and gonopore.

palp [G *Palporgan*]: Type of lateral sense organ characterized by lobate elaboration of piston.

penis [G *Penis*]: Hollow, papilla-like modification of terminal end of sperm duct.

pharyngeal bulb [G *Bulbus*]: Muscular enlargement of pharynx.

pharyngeal plug [G *Zapfen*]: Short, conical extension of posteriormost pharynx into intestine.

pharyngeal pore [G *Pharyngealporus*]: One of a pair of dorsolateral or ventrolateral openings of pharynx to exterior; typically positioned at posterior end of pharynx.

pharynx [G *Pharynx*]: Long, muscular, cuticle-lined section of digestive tract between mouth and intestine; triradiate in cross section and often with pharyngeal bulbs.

piston (pestel organ) [G *Stempel*]: Unciliated projection positioned centrally in depression of piston pit type of lateral sense organ.

piston pit (pestel organ) [G *Stempelgrube*]: Type of paired lateral sense organ, one on each side of head, consisting of cilia-bordered depression occupied by central unciliated projection (piston).

plate [G *Platte, Schiene*]: Large cuticular structure, either modified scale or separate structure on ventral surface or head. (See also head plate)

pleurion [G *Pleurion*]: One of two lateral cuticular head plates. Each pleurion on side of head may be subdivided into epipleurion and hypopleurion. (See also cephalion, hypostomium)

protonephridium [G *Protonephridium*]: Paired excretory organ, one along each side of anterior region of intestine; con-

sists of elongated solenocytes connected to nephridiopore via coiled nephridial tubule.

pseudocoel (gymnocoel) [G *Pseudocoel*]: Body compartments defined by muscle fibers and mesenchyme and occupied by reproductive organs and digestive tract.

rectum [G *Endarm*]: In certain gastrotrichs, short, cuticle-lined section of digestive tract between intestine and anus.

salivary gland (pharyngeal gland) [G *Speicheldrüse*]: One of two smaller and two larger modified ventrolateral cells in anterior and posterior regions of pharynx respectively; may also refer to numerous glandular cells in posterior region of pharynx.

scale [G *Schuppe*]: Type of cuticular cover consisting of flat or overlapping, variously elaborated small plates. (keeled, spined, stalked)

seminal receptacle [G *Receptaculum seminis*]: Variously elaborated, typically sperm-filled anterior section of oviduct. (See also copulatory bursa)

seminal vesicle [G *Vesicula seminis*]: Term applied to expanded section of sperm duct.

sensory bristle [G *Sinnesborste, Tastborste*]: Long, often rigid dorsolateral or lateral sensory projection; typically numerous on head and closely associated with brain as well as in symmetrical pairs on trunk.

sensory cilium [G *Tastwimper*]: Elongated cilium accompanying each lateral adhesive tube. May also refer to variously modified cilia on head. (See also ciliary pit, locomotory cilium)

solenocyte (flame bulb) [G *Solenozyte, Wimperkölbchen, Endkölbchen*]: In excretory system (protonephridium), specialized, tubular excretory cell containing a cilium. One to several pairs located along intestine; connected to lateral nephridiopores via nephridial tubules.

sperm duct (vas deferens) [G *Vas deferens*]: Elongate, relatively thin, typically posteriorly directed section of ♂ reproductive system between testes and gonopore.

spine [G *Stachel*]: Variously elaborated, solid or hollow cuticular projection inserting rigidly on a scale or movable and arising directly from cuticle.

tail [G *Schwanz*]: Term applied to single long, thin extension of posterior end of trunk; bears adhesive tubes.

tail fork [G *Gabelschwanz, Gabelfuß*]: Bifid, posteriorly directed extension of body at end of trunk; bears adhesive tubes. (See also toe)

tentacle [G *Tentakel*]: Type of lateral sense organ characterized by tentacle-like elaboration of piston. May also refer to more posterior, tentacle-like projection on head.

testis [G *Hoden*]: When present, single or paired elongate section of ♂ reproductive system located laterally along intestine; connected to gonopore(s) via sperm duct(s).

toe [G *Zehe*]: Term applied to each half of tail fork if it is well developed and bears adhesive gland with terminal adhesive tube.

trunk [G *Rumpf*]: Posteriormost and largest of three divisions of ventrally flattened body (head, neck, trunk); with variously elaborated posterior end.

uterus [G *Uterus, Eilager*]: Poorly delimited section of ♀ reproductive system between ovary and oviduct containing ripe eggs; typically anterior to ovary and dorsal to intestine in posterior half of trunk.

vas deferens: sperm duct

ventral gland [G *Ventraldrüse*]: One or more pairs of small, granular masses near border of pharynx and intestine and opening ventrolaterally.

X-organ (organ X) [G *Organ X*]: Rounded or bilobate section of ♀ reproductive system corresponding to oviduct or copulatory bursa; located ventrally under rectum and opening to exterior via ventral pore.

NEMATODA

roundworm, nematode [G *Rundwurm, Fadenwurm, Spulwurm, Nematode*]

NEMATODA

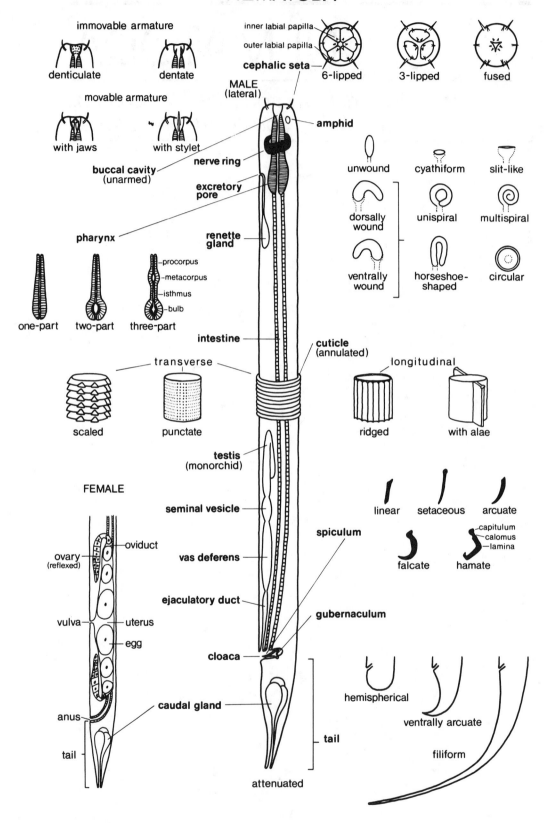

NEMATODA

roundworm, nematode [G *Rundwurm, Fadenwurm, Spulwurm, Nematode*]

adhesion tube [G *Haftröhrchen*]: ambulatory bristle

ala (wing) [G *Lateralmembran*]: Lateral or sublateral wing-shaped thickening or projection of cuticle (cervical ala, longitudinal or lateral ala, and caudal ala = bursa).

ambulatory bristle (stilt seta) [G *Kriechborste, Stelzborste*]: One in a series of elongate ventral setae located either anterior (preanal ambulatory bristle) or posterior (postanal ambulatory bristle) to anus and aiding in locomotion; when hollow and tube-like also termed adhesion tube.

amphid [G *Seitenorgan, Lateralorgan*]: Paired cuticular invaginations in a lateral position on or near head and having sensory function. (unwound, dorsally wound, ventrally wound; circular, cyathiform, horseshoe-shaped, multispiral, oval, porelike, shepherd's crook, spiral, unispiral)

ampulla [G *Ampulle*]: Small vesicle at end of excretory duct, immediately before excretory pore. Also refers to terminal swelling of duct of pharyngeal gland.

annule [G *Ringel*]: Type of cuticular ornamentation (light microscope magnification); one in a series of elevated rings of cuticle.

anus [G *After*]: Posterior opening of digestive tract on ventral surface.

buccal cavity (stoma) [G *Mundhöhle*]: Section of digestive tract between mouth and pharynx, often armed with teeth or stylet; when strongly cuticularized also termed buccal capsule.

buccal ridge [G *Zahnleiste*]: Elevated, cuticularized, and toothed ridge in buccal cavity.

bulb (pharyngeal bulb) [G *Bulbus, Pharyngealbulbus*]: Muscular enlargement of pharynx, often containing glands and provided with valve.

bursa (caudal ala) [G *Bursa*]: Lateral wing-shaped modification of cuticle in tail region of certain ♂ nematodes; serves to hold nematode in position during copulation. (aphelenchoid, arakoderan, leptoderan, peloderan, rhabditoid, tylenchoid)

calomus [G *Calomus*]: "Shaft" or midregion of spiculum between capitulum and lamina.

capitulum [G *Capitulum*]: "Head" or proximal, often set-off section of spiculum (also termed manubrium). Also refers to section of gubernaculum flanging cuneus and running parallel to corpus.

cardia [G *Cardia*]: Triradiate, cuticularized constriction between pharynx and intestine; acts as a valve (esophagointestinal valve).

caudal ala: bursa

caudal gland [G *Schwanzdrüse, Klebdrüse, Kittdrüse*]: Three (rarely two) cells in or near tail, emptying by separate ducts into a common opening (spinneret) and having an adhesive function.

cephalic capsule (helmet) [G *Kopfkapsel, Kopfpanzer*]: Any thickened cuticular, cap-like structure on head region.

cephalic seta [G *Kopfborste*]: Sensory setae belonging to second or third circlet of sensilla on head (first circlet of cephalic sensory organs = labial papillae); often refers only to third circlet, whose setae are typically longer and four in number.

cephalic shield (amphidial plaque) [G *Kopfplatte*]: Cuticular plate on head region and on which amphids may be located.

cephalid [G *Cephalid*]: Prominent nerve fibers connecting longitudinal nerves; of taxonomic importance in certain nematode groups.

cervical capsule [G *Halskapsel*]: Thickened section of cuticle in "neck" region behind head.

cervical papilla: deirid

cervical region [G *zervikale Region*]: "Neck" (= esophageal or pharyngeal) region.

circumesophageal commissure: nerve ring

cloaca [G *Kloake*]: Terminal section of intestine in ♂ into which ejaculatory duct opens and whose posterior wall typically bears two spicules. Equivalent to rectum in ♀.

coelomocyte [G *Coelomozyte*]: Term applied to large cell in body cavity; usually isolated or in pairs, often with specific position.

corpus [G *Corpus*]: Anterior section of three-part cylindrical pharynx (consisting of corpus, isthmus, and bulb); may be subdivided into procorpus and metacorpus. Also refers to main plate-like portion of gubernaculum.

crura [G *Crura*]: Lateral reinforcement of corpus on gubernaculum.

cuneus [G *Cuneus*]: Ventral medial extension of corpus on gubernaculum.

cuticle [G *Kutikula*]: Noncellular, multilayered, variously ornamented layer covering outer surface of body and lining both pharynx and rectum. (annulated, punctate, ridged, scaled, smooth, with alae)

deirid (cervical papilla) [G *Deiride*]: Paired lateral sensory papillae in certain nematodes; usually located in cervical region.

demanian system [G *de Man'sches Organ*]: Variously elaborated, tubular system in certain ♀ nematodes; connects intestine to uterus and also opens to exterior. Contains sperm and therefore considered to function as seminal receptacle. (See also osmosium, uvette)

denticle [G *Zähnchen*]: Small teeth, often in rows, in buccal cavity (as opposed to large teeth or jaws).

diverticulum [G *Darmblindsack*]: Outpocketing or cecum of anterior region of intestine.

egg [G *Ei*]: Reproductive body produced by ♀ in ovary, passing through oviduct, fertilized in uterus (or spermatheca), and emerging from vagina; often prominent, filling oviduct in a series of variously developed stages.

ejaculatory duct [G *Ductus ejaculatorius*]: Posterior, often muscular section of vas deferens in ♂ reproductive system; connects seminal vesicle to cloaca.

ejaculatory gland (cement gland) [G *Zementdrüse*]: Gland or glands surrounding ejaculatory duct and said to produce an adhesive material which helps hold ♂ to ♀ during copulation.

esophageal bulb [G *Ösophageal-bulbus*]: bulb

esophagus [G *Ösophagus*]: pharynx

excretory pore [G *Exkretions-porus*]: Opening of excretory system to exterior; usually located along anterior midventral line close to nerve ring.

gonopore [G *Geschlechtsöffnung*]: vulva

gubernaculum [G *Gubernaculum, akzessorisches Stück*]: Cuticular thickening formed from walls of spicular pouch and serving to guide spicules through cloacal chamber upon protrusion. May consist of apophysis, corpus, crura, and cuneus.

head [G *Kopf*]: Term occasionally applied to anterior section of body, especially when this is well set-off by a constriction or displays heavy cuticularization. A morphologically distinct head does not exist.

helmet: cephalic capsule

hemizonid [G *Hemizonid*]: Prominent (refractive) nerve fibers connecting nerve ring to ventral nerve cord in certain nematodes.

H-system [G *H-förmiges Seitenkanal-system*]: Characteristic type of tubular excretory system in which the longitudinal canals are connected by a transverse canal (which is further connected to an excretory pore by a median duct).

integument: cuticle

intestine [G *Mitteldarm*]: Section of digestive tract between pharynx and rectum; usually simple, straight, and tubular (at times reduced).

isthmus [G *Isthmus*]: Narrow section of three-part cylindrical pharynx between anterior corpus and posterior bulb.

jaw (mandible) [G *Kiefer*]: Movable biting or grasping organs formed from thickening of cuticle which lines buccal cavity; basically consists of three elements.

labial papilla [G *Lippenpapille*]: First circlet of sensilla on lips, usually six in number; when setiform, termed labial seta.

labium: lip

lamina [G *Lamina*]: "Blade" or distal section of spiculum.

lateral chord [G *Seitenwulst*]: Lateral thickening of hypodermis (two of the four hypodermal chords); often shimmering through.

lateral field [G *Lateralfeld, Seiten-feld*]: Differentiated cuticular region on lateral side of body.

lip (labium) [G *Lippe*]: Lip-like lobes bordering mouth, fundamentally six in number although often reduced through fusion to three; usually bearing circlet of labial papillae.

mandible: jaw

manubrium: capitulum

mesostome [G *Mesostom*]: Term occasionally applied to posterior part of buccal cavity when it appears subdivided by cuticular structures. (See also prostome)

metacorpus [G *Metacorpus*]: In three-part cylindrical pharynx (corpus, isthmus, bulb), posterior swollen section of corpus (forming first bulb). (See also procorpus)

metaneme [G *Metanem*]: Subcuticular, serially arranged, filiform organs either parallel or at an angle to the longitudinal axis in certain nematodes; interpreted as stretch receptors.

mouth [G *Mund*]: Anteriormost section of digestive tract; essentially a six-lip structure. (See also lip)

nerve ring (circumesophageal commissure) [G *Nervenring, periösophagealer Nervenring*]: Main concentration of nerve tissue ("brain") forming ring around pharynx; gives rise to ventral nerve cord.

ocellus [G *Ocellus*]: Simple photoreceptor with lens-like structures and pigment material, often in pharynx region.

odontium [G *Odontium*]: Tooth type arising from the anterior part of buccal cavity. (See also onchium)

onchium [G *Onchium*]: Tooth type arising from the posterior part of buccal cavity and associated with pharynx. (See also odontium)

osmosium [G *Osmosium*]: Section of demanian system connecting main tube of this organ to intestine.

ovary [G *Ovar, Eierstock*]: Section of ♀ reproductive system in which eggs are produced; opens into oviduct. (outstretched, reflexed; monovarial, diovarial, polyovarial; hologonic, telogonic)

oviduct [G *Ovidukt, Eileiter*]: Section of ♀ reproductive system between ovary and uterus; may take up greatest portion of reproductive system.

pharyngeal gland (esophageal gland) [G *Pharynxdrüse, Ösophagusdrüse*]: Glands, usually three in number (one dorsal, two ventrolateral), connected to lumen of pharynx or buccal cavity by means of cuticularized ducts.

pharynx (esophagus) [G *Pharynx, Ösophagus*]: Cuticle-lined, often regionated section of digestive tract between buccal cavity and intestine; triradiate in cross section. Synonymous with esophagus; the former being morphologically more correct, the latter generally accepted through common usage. (one-part: conoid, cylindrical, crenate, moniliform = multibulbar; two-part: mermithoid, spiruroid; three-part: aphelenchoid, diplogasteroid, rhabditoid, tylenchoid)

phasmid [G *Schwanzpapillendrüse, Phasmide*]: Pair of postanal lateral sensory organs; serve as a taxonomic character in the major division of nematodes.

plaque [G *Platte*]: Type of cuticular ornamentation.

preanal supplement: supplement

procorpus [G *Procorpus*]: In three-part cylindrical pharynx (corpus, isthmus, bulb), anterior section of corpus. (See also metacorpus)

proctodeum: rectum

prostome [G *Prostom*]: Term occasionally applied to anterior part of buccal cavity when it appears subdivided by cuticular structures. (See also mesostome)

rectal gland [G *Rektaldrüse*]: Gland cells, varying in number according to species and sex, missing entirely in certain nematodes; enter rectum.

rectum (proctodeum) [G *Enddarm*]: Cuticle-lined section of digestive tract between intestine and anus. Simple tube in most ♀; joins with reproductive system in ♂ to form cloaca which bears spicules and other copulatory structures.

renette gland (ventral gland) [G *Ventraldrüse*]: Single large gland cell located ventrally in pseudocoelomic cavity of cervical region and connected to an "excretory" pore; considered to be a primitive type of excretory system, although some authors prefer the more neutral terms cervical gland and cervical pore.

scale [G *Schuppe*]: Type of cuticular ornamentation; formed by pronounced outgrowth of annulation and considered to represent modification of spine.

seminal receptacle: spermatheca

seminal vesicle [G *Samenblase*]: Section of ♂ reproductive system between testes and vas deferens; usually dilated and serving as sperm storage organ.

sensory formula: Abbreviated notation of number and position of sensory projections (papillae, setae) in the circlets of sensilla on head.

seta [G *Borste*]: Bristle-shaped cuticular projection with sensory or locomotory function; one distinguishes cephalic (head) setae, cervical (neck) setae, and somatic (body) setae.

spermatheca (seminal receptacle) [G *Spermathek*]: Sperm storage organ at junction of oviduct and uterus in ♀.

sperm duct: vas deferens

spicular pouch [G *Spikulumscheide*]: Cuticle-lined pouch in cloaca; encloses spicules.

spiculum (spicule) [G *Spikulum*]: Protrusible, cuticular, spine-shaped structure lodged in invaginations of cloaca and aiding in copulation; usually two in number. May consist of capitulum, calomus, lamina, and velum. (arcuate, cuneiform, falcate, fusiform, hamate, L-shaped, linear, setaceous)

spine [G *Stachel*]: Type of cuticular ornamentation; usually associated with annules and closely related to scales.

spinneret [G *Haftröhrchen*]: Cuticular structure associated with common duct of caudal gland.

standard measurements:

a = body length/maximum diameter
b = body length/pharynx length
c = body length/tail length

	pharynx end	middle ♂ vulva ♀	anus
distance from head to corresponding	0	260	760	944
diameter	12	36	40	28
total length				1180 um

stilt seta [G *Stelzborste*]: ambulatory bristle

stoma: buccal cavity

stomodeum [G *Vorderdarm*]: Anterior cuticle-lined section of digestive tract; includes lips, mouth, buccal cavity, and pharynx.

stylet (spear, spur) [G *Stilett*]: Variously elaborated, cuticular, often hollow, spear-shaped structure in buccal cavity.

stylet knob [G *Styletkopf*]: Knob or swelling at base of stylet; serving for attachment of muscles.

subcephalic seta [G *subkephale Borste*]: Setae immediately posterior to cephalic setae (yet before cervical setae).

supplement [G *Hilfsorgan*]: In ♂, structures often projecting from cuticle in a series and usually positioned anterior to anus (preanal supplements); aids in copulation.

tail [G *Schwanz*]: Posterior section of nematode extending from anus to posterior tip of body. (arcuate, attenuated, conoid, digitate, filiform, hemispherical, mucronate, multidigitate, subdigitate))

tooth [G *Zahn*]: Cuticular, solid or hollow, tooth-like structure in buccal cavity. (See also odontium, onchium)

telamon [G *Telamon*]: Thickened immovable plate of ventral cloacal wall in certain nematodes; turns spiculum upon protrusion.

testis [G *Hoden*]: Section of ♂ reproductive system in which germ cells or spermatocytes are produced; opens into sperm duct. (hologonic, telogonic; diorchic, monorchic, proorchic)

U-system [G *U-System*]: Type of tubular excretory system in which the longitudinal canals do not extend far beyond the transverse connecting canal. (See also H-system)

uterus [G *Uterus*]: Section of ♀ reproductive system lying between oviduct and vagina. (amphidelphic, opisthodelphic, prodelphic; monodelphic, didelphic, tetradelphic, polydelphic)

uvette [G *Uvette*]: Glandular section of demanian system leading to exterior by way of duct.

vagina [G *Vagina*]: In ♀ reproductive system, muscular, cuticle-lined section following uterus and opening to exterior via vulva.

valve [G *Klappe*]: Structure either in pharynx (esophagointestinal valve) preventing the regurgitation of food or at junction of intestine and rectum (intestinorectal valve).

vas deferens [G *Vas deferens*]: Section of ♂ reproductive system between seminal vesicle and cloaca; usually with anterior glandular region and posterior muscular region (ejaculatory duct).

vas efferens [G *Vas efferens*]: Section of ♂ reproductive system connecting testes to seminal vesicle.

velum [G *Velum*]: Delicate wing-like extension on distal section (lamina) of spiculum.

ventral gland [G *Ventraldrüse*]: renette gland

ventral nerve cord [G *ventraler Markstrang*]: Longitudinal nerve cord arising from nerve ring and extending along ventral surface to posterior end of body.

vestibule [G *Vestibulum*]: In the absence of a well-developed buccal cavity, short section of digestive tract between mouth and pharynx.

vulva [G *Vulva*]: Opening of ♀ reproductive system to exterior, usually on midventral line in midregion of body.

wing: ala

NEMATOMORPHA

nematomorph, gordian worm, hairworm, horsehair worm [G *Nematomorphe, Saitenwurm*]

NEMATOMORPHA

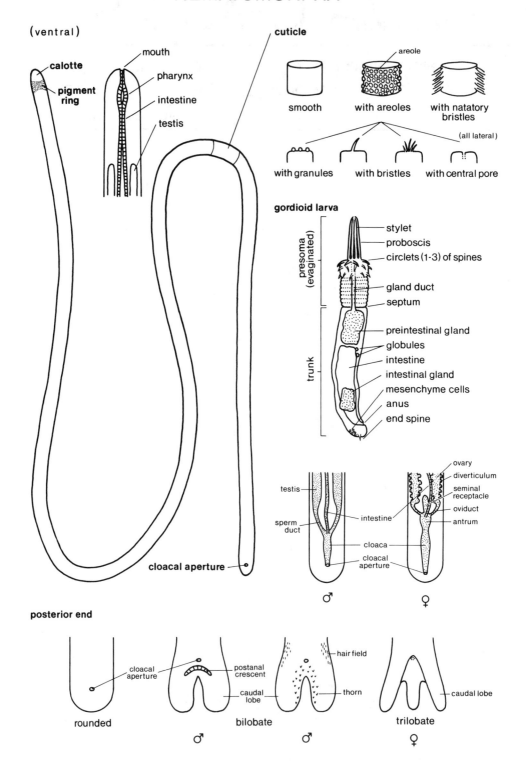

(ventral)

calotte
pigment ring

mouth
pharynx
intestine
testis

cuticle

areole

smooth
with areoles
with natatory bristles

(all lateral)

with granules
with bristles
with central pore

gordioid larva

presoma (evaginated)

stylet
proboscis
circlets (1-3) of spines
gland duct
septum

trunk

preintestinal gland
globules
intestine
intestinal gland
mesenchyme cells
anus
end spine

testis
sperm duct
intestine

ovary
diverticulum
seminal receptacle
oviduct
antrum

cloaca
cloacal aperture

♂   ♀

cloacal aperture

posterior end

cloacal aperture

postanal crescent

hair field

caudal lobe

thorn

caudal lobe

rounded

bilobate
♂

♂

trilobate
♀

NEMATOMORPHA

nematomorph, gordian worm, hairworm, horsehair worm [G *Nematomorphe, Saitenwurm*]

adhesive wart [G *Greifwarze*]: Variously elaborated modification of cuticle in cloacal region of ♂; arranged in bands, typically in addition to thorns and bristles. (elongate, reniform, rounded)

antrum [G *Atrium*]: Expanded glandular section of ♀ reproductive system into which oviducts open; adjoined posteriorly by cuticle-lined cloaca. Seminal receptacle extends anteriorly from antrum.

anus [G *After*]: In gordioid larva, opening of intestine to exterior. (In adult, see cloaca)

areole [G *Areole*]: Variously elaborated, minute elevation of outer homogeneous layer of cuticle covering surface of body; bears granules or bristles or is penetrated by a pore. Mosaic pattern of areoles separated by interareolar furrows. (polygonal, rounded)

brain [G *Gehirn*]: Main concentration of nerve tissue, forming ring around digestive tract in calotte; connected to eye by dorsal nerve, to posterior end by ventral cord.

bristle [G *Borste*]: Term applied to variously shaped cuticular projections arising either singly or in groups from areoles, distributed in interareolar furrows or in tracts in cloacal region of ♂. (See also interareolar bristle)

calotte [G *Kopfkalotte*]: Anterior tip of body, typically lighter in color than posteriorly adjoining pigment ring; contains brain,

eye, and pharynx, as well as terminal or ventral mouth.

capsule: eye capsule

caudal lobe [G *Lappen, postkloakaler Lappen*]: One of two or three lobes at posterior end of body; may bear fields of thorns.

cephalic nerve [G *Kopfnerv*]: Either of two lateral tracts of ventral nerve cord extending anteriorly from brain; each gives rise to two ventral nerves extending into ventral epidermis of calotte.

cloaca [G *Kloake*]: Posteriormost cuticle-lined section of digestive tract and/or reproductive system; located terminally or ventrally at posterior end of body and either flanked by tracts of thorns or bristles or closely adjoined by postanal crescent. (subterminal, terminal, ventral)

cloacal ganglion [G *Kloakenganglion*]: Thickening of ventral nerve immediately anterior to cloaca; gives rise in ♀ to pair of anterior and posterior cloacal nerves supplying cloaca, in ♂ to dorsal nerves extending into caudal lobes and small fibers supplying sperm ducts.

cloacal nerve [G *hinterer Bauchmarkast*]: In ♀, one of four nerves arising from cloacal ganglion and supplying cloaca; in ♂, pair of nerves arising from cloacal ganglion and extending into caudal lobes.

cuticle [G *Kutikula*]: Relatively thick, noncellular layer covering outer surface of body and lining both pharynx and cloaca; consists of thinner outer homogeneous layer and inner lamellate fibrous layer. Outer layer modified into areoles and interareolar furrows.

diverticulum [G *Ovarialdivertikel, Ovarialblase*]: One of up to several thousand lateral outpocketings of each ovary in which eggs ripen before reentering ovary or uterus.

dorsal epidermal chord [G *Rückenwulst*]: Longitudinal dorsal thickening of epidermis; interrupts longitudinal body wall muscles. (See also ventral epidermal chord)

dorsal nerve [G *Dorsalnerv*]: One of two dorsal nerves arising from brain and extending to eye.

egg [G *Ei*]: ♀ reproductive body produced in ovary, maturing in diverticula, reentering ovary or uterus, and passing through oviducts, antrum, and cloaca to exterior.

end spine (tail spine) [G *Endstachel*]: In gordioid larva, terminal cuticular spine at posterior end of trunk.

epidermis [G *Epidermis*]: Layer of body wall between cuticle and muscle; consists of one cell layer. Thickened ventrally and occasionally dorsally to form ventral and dorsal epidermal chords.

eye [G *Auge*]: Sac, located in calotte, containing coagulated fluid and retinal cells. Bordered anteriorly by modified, transparent epidermis of calotte, laterally and posteriorly by eye capsule. Joined to brain by dorsal nerves.

eye capsule [G *Augenkapsel*]: Modified mesenchyme forming capsule enclosing eye laterally and posteriorly.

fibrous layer [G *faserige Schicht*]: Thicker inner layer of cuticle consisting of up to 45 strata. (See also homogeneous layer)

giant nerve cell [G *Riesenzelle*]: One of several large nerve cells at junction of brain and ventral nerve cord.

gland [G *Drüse, Speicheldrüse*]: In gordioid larva, large glandular structure posterior to septum separating presoma and

trunk; opens to exterior at tip of proboscis via gland duct.

gland duct [G *Speichelkanal*]: In gordioid larva, long canal serving as outlet of gland; penetrates septum and opens at tip of proboscis.

gordioid [G *Larve*]: Larva of nematomorph. Encompasses period between hatching from egg and assumption of parasitic habit.

granule [G *Körnchen*]: One of a varying number of small, rounded cuticular elevations on surface of areole.

homogeneous layer [G *homogene Schicht*]: Thinner outer layer of cuticle; modified into variously elaborated areoles and interareolar furrows. (See also fibrous layer)

interareolar bristle [G *Interareolarborste*]: One in a series of bristles along interareolar furrows.

interareolar furrow [G *Interareolarfurche, interareolarer Feld*]: Network of furrows separating areoles in outer homogeneous layer of cuticle; bears characteristic array of pores, warts, or bristles.

intestine (midgut) [G *Mitteldarm*]: Section of digestive tract between pharynx and cloaca in pseudocoel; typically thin and degenerate, not always connecting with pharynx and cloaca.

longitudinal muscle [G *Längsmuskel*]: Layer of longitudinal muscle fibers underlying epidermis; interrupted ventrally and occasionally dorsally by ventral and dorsal epidermal chords.

mesenchyme [G *Parenchym*]: Undifferentiated cells more or less filling pseudocoel, enclosing gonads, and occasionally forming more clearly delimited partitions. Reduced in marine nematomorph.

midgut: intestine

midventral groove [G *Bauchrinne*]: Longitudinal depression along midventral line of body; corresponds to position of ventral nerve cord.

mouth [G *Mund*]: Anterior opening of digestive tract on calotte. If present, opens into pharynx. (subterminal, terminal, ventral)

natatory bristle [G *Schwebeborste*]: In marine nematomorph, one of a large number of elongate bristles arranged in a double row along both ventral and dorsal epidermal chords; absent at anterior and posterior ends of body. Serves in swimming and appears to be lateral due to body torsion.

nervous lamella [G *Neurallamelle*]: Longitudinal lamella joining ventral epidermal chord with ventral nerve cord; absent in marine nematomorph, where ventral nerve cord is included in epidermis.

ovary [G *Ovar*]: Elongate section of ♀ reproductive system; extends length of body and bears lateral diverticula. Occasionally, the term ovary is applied to lateral diverticula only, while longitudinal, cylindrical section is termed uterus. (paired, unpaired)

oviduct [G *Ovidukt*]: Thin, tubular section of ♀ reproductive system between each ovary and antrum.

pharyngeal bulb [G *Bulbus*]: Posterior enlargement of pharynx.

pharynx [G *Pharynx, Vorderdarm, Ösophagus*]: Section of digestive tract between mouth and intestine; may be a slender cuticular tube, be provided with posterior pharyngeal bulb, be solid, or be reduced.

pigment ring [G *Halsring*]: Ring of dark pigment encircling body posterior to calotte.

pore [G *Porus*]: Term applied to central opening in large areoles or to series of pores in interareolar furrows.

postanal crescent [G *postkloakale Hautfalte, postkloakale Kutikularfalte*]: Ventral crescentic cuticular fold between cloaca and caudal lobes in ♂.

presoma [G *Präsoma*]: In gordioid larva, anterior of two divisions of body (presoma, trunk) bearing proboscis; separated from trunk by septum.

proboscis [G *Proboscis, Rüssel*]: In gordioid larva, evaginable section of presoma bearing, when protruded, three circlets of proboscis spines laterally and three parallel proboscis stylets distally.

proboscis spine [G *Stachel*]: In gordioid larva, one of three, four or six, and seven cuticular spines in first, second, and third circlet of spines around evaginable proboscis.

proboscis stylet [G *Stilett*]: In gordioid larva, one of three elongate, closely adjoining, and parallel cuticular shafts located distally in evaginated proboscis.

pseudocoel [G *Leibeshöhle*]: Body cavity between digestive tract and body wall; more or less filled with mesenchyme in freshwater nematomorph, spacious and divided by septum into small anterior and large posterior sections in marine nematomorph.

retinal cell [G *Retinazelle*]: One of numerous fusiform cells within eye.

retractor muscle [G *Rüssel-Retraktor*]: In gordioid larva, longitudinal muscle fibers in presoma; serve to invaginate proboscis.

seminal receptacle [G *Receptaculum seminis*]: Elongate, unpaired section of ♀ reproductive system extending anteriorly from antrum.

seminal vesicle [G *Vesicula seminalis*]: Term occasionally applied to posterior section of testis.

septum [G *Septum*]: Transverse partition at level of brain separating small anterior section of pseudocoel from large posterior section in adult marine nematomorph. In gordioid larva, partition separating presoma from trunk.

sperm duct [G *Samenleiter, Vas deferens*]: Short, thin section of ♂ reproductive system between testis and cloaca. (paired, unpaired)

spine: end spine, proboscis spine

stylet: proboscis stylet

testis [G *Testis, Hoden*]: Cylindrical section of ♂ reproductive system extending length of body; opens into cloaca via sperm duct. (paired, unpaired)

thorn [G *Stachel*]: Cuticular modification, typically forming fields of thorn-shaped projections posterior to cloaca and on caudal lobes.

trunk [G *Rumpf*]: In gordioid larva, posterior of two divisions of body (presoma, trunk) bearing gland and intestine; separated from presoma by septum.

uterus [G *Uterus*]: In ♀ reproductive system, term applied to main longitudinal ovarian tube or to anterior section of antrum.

ventral epidermal chord [G *Bauchwulst*]: Longitudinal ventral thickening of epidermis; interrupts longitudinal body wall muscles. Encloses ventral nerve cord or is attached to it by nervous lamella. (See also dorsal epidermal chord)

ventral nerve cord [G *Bauchmark, Bauchmarkstrang*]: Longitudinal midventral nerve cord, subdivided into three tracts by glial partitions, extending from brain to cloaca; located either in ventral epidermal chord or joined to it by nervous lamella.

KINORHYNCHA

kinorhynch [G *Kinorhynche*]

KINORHYNCHA

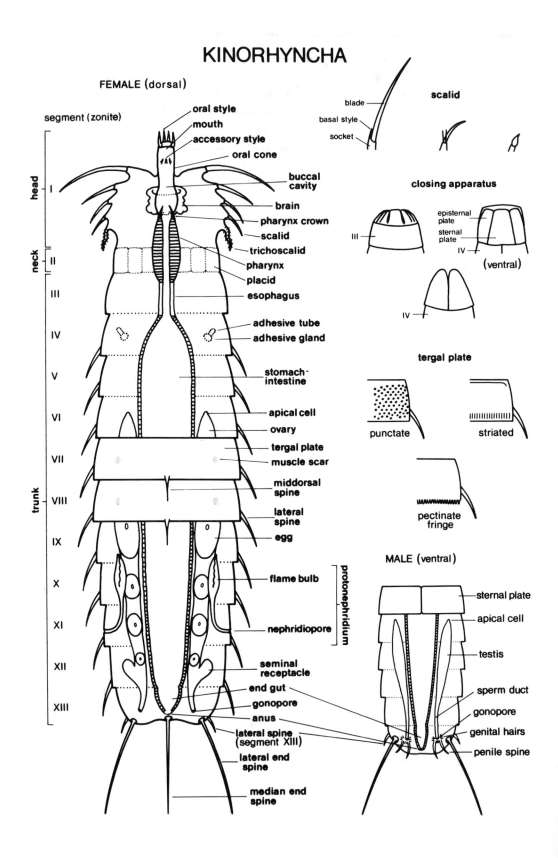

FEMALE (dorsal)

segment (zonite)

oral style
mouth
accessory style
oral cone
buccal cavity
brain
pharynx crown
scalid
trichoscalid
pharynx
placid
esophagus
adhesive tube
adhesive gland
stomach-intestine
apical cell
ovary
tergal plate
muscle scar
middorsal spine
lateral spine
egg
flame bulb
nephridiopore
seminal receptacle
end gut
gonopore
anus
lateral spine (segment XIII)
lateral end spine
median end spine

head
neck
trunk

I
II
III
IV
V
VI
VII
VIII
IX
X
XI
XII
XIII

protonephridium

scalid

blade
basal style
socket

closing apparatus

episternal plate
sternal plate

III
IV

(ventral)

IV

tergal plate

punctate
striated
pectinate fringe

MALE (ventral)

sternal plate
apical cell
testis
sperm duct
gonopore
genital hairs
penile spine

KINORHYNCHA

kinorhynch [G *Kinorhynch*]

accessory style [G *akzessorischer Stylus*]: An element in two closely adjoining circlets of small, cuticularized projections in buccal cavity.

acetabulum [G *Gelenkzapfen*]: Peg of peg-and-socket arrangement if tergal and sternal plates articulate.

adhesive gland [G *Klebdrüse*]: One of a pair of large glands located ventrolaterally in fourth segment (zonite); opens to exterior via adhesive tube.

adhesive tube [G *Klebröhre*]: Cuticularized tubes emerging ventrally from sternal plates on fourth segment (zonite) or other segments; serves as opening of adhesive gland. (apiculate, flattened)

amebocyte (amoebocyte) [G *Amöbozyte*]: Mobile cell with pseudopodia in fluid of body cavity (pseudocoel).

anus [G *After*]: Posterior opening of digestive tract located terminally between tergal and sternal plates of last segment (zonite)

apical cell [G *Apikalzelle*]: Distinct anteriormost cell in each ovary; gives rise to all other cells of gonad.

basal style [G *Schutzstachel*]: One of occasionally several small spines projecting from socket of scalid.

blade [G *Stachel*]: Distal, curved, pointed element of scalid; separated from basal socket by circular fold.

brain (nerve ring, circumenteric nerve ring) [G *Gehirn*]: Main concentration of nerve tissue forming a ring around base of mouth cone or anterior part of pharynx. Gives rise posteriorly to ventral cord.

buccal cavity [G *Mundhöhle*]: Anteriormost section of digestive tract between mouth and pharynx; formed by cavity of mouth cone. Pharynx crown projects into posterior end of buccal cavity.

closing apparatus [G *Verschlußapparat*]: Term applied to cuticular plates covering retracted head; consists of placids and three ventral plates (one sternal and two episternal) of third segment (zonite).

condyle [G *Gelenknapf*]: Socket of peg-and-socket arrangement if tergal and sternal plates articulate. (See also acetabulum)

copulatory spicule: penial spicule

cuticle [G *Kutikula*]: Noncellular layer covering outer surface of body in form of variously ornamented plates (tergal and sternal), placids, and scalids, as well as lining mouth cone, pharynx, esophagus, and posterior section of rectum. (denticulate, hairy, porous, spinous)

diagonal muscle band [G *schräge Seitenmuskel*]: Segmentally arranged series of muscle bands in lateral trunk region; span diagonally from anterior edge of one segment (zonite) to anterior edge of preceding zonite.

dorsal cord: middorsal cord

dorsal plate: tergal plate

dorsolateral muscle [G *dorsales Längsmuskelband*]: One of two dorsolateral, segmented, longitudinal muscle bands running along tergal plates in trunk. (See also ventrolateral muscle)

dorsoventral muscle [G *Dorsoventralmuskel*]: Segmentally arranged series of muscle bands, one on each side of every trunk segment (zonite); spans from tergal to sternal plates.

end gut (hindgut) [G *Enddarm*]: Posteriormost section of digestive tract between stomach-intestine and anus. Separated from stomach-intestine by a sphincter, with a second sphincter posteriorly at narrowed end of end gut; region from second sphincter to anus is cuticle-lined.

end spine: terminal spine

episternal plate [G *Episternalplatte*]: One of two ventral plates flanking single sternal plate on third segment (zonite); serves as element in closing apparatus.

esophagus (oesophagus) [G *Ösophagus*]: Short, cuticle-lined section of digestive tract between pharynx and stomach-intestine; four salivary glands open into esophagus anteriorly, two or more pancreatic glands posteriorly.

eyespot (ocellus, phaeosome) [G *Augenfleck, Auge*]: One of two red, dorsolateral pigment-cup ocelli in epidermis anterior to brain at base of anteriormost scalids; connected to brain by nerve tissue.

flame bulb [G *Terminalorgan*]: In protonephridium, pair of excretory structures, one on each side of stomach-intestine in 10th zonite. Bears long flagellum or both long and short flagella and opens to exterior via nephridiopore.

genital bristle [G *Genitalborste*]: One of a varying number of short, cuticularized projections accompanying penial spicules around ♂ gonopore; may surround gonopore in form of a dense crown.

gonopore (genital pore) [G *Genitalporus, Geschlechtsöffnung*]: Opening of reproductive system to exterior. In ♀, pair of pores either ventrolaterally on 13th segment (zonite) or laterally between zonites 12 and 13. In ♂, pair of pores either ventrolater-ally on 13th zonite or laterally between zonites 12 and 13; typically accompanied by penial spicules and genital bristles.

head [G *Kopf*]: Anteriormost of three divisions of body (head, neck, trunk) bearing terminal mouth and surrounded by scalids; corresponds to first segment (zonite) and is retractable into trunk.

head retractor muscle: inner head retractor, outer head retractor

hindgut: endgut

inner head retractor [G *innerer Retraktor*]: One in a series of longitudinal retractor muscle bands spanning from furrow around mouth cone on head to tergal and sternal plates on successive trunk segments (zonites); passes between brain and pharynx. (See also outer head retractor)

intestine: stomach-intestine

lateral cord [G *lateraler Längsstrang, Seitenstrang*]: One of two more highly developed longitudinal thickenings of epidermis along sides of body; swelling of lateral cord in each trunk segment (zonite) typically associated with sensory structures. (See also middorsal cord)

lateral end spine [G *Seitenendstachel*]: One of two conspicuously elongate, hollow, epidermis-filled lateral cuticular projections on tergal plate of last segment (zonite); jointed into socket in tergal plate and movable by special muscles.

lateral spine [G *Seitenstachel*]: Hollow, epidermis-filled lateral cuticular projection on each tergal plate of trunk; jointed into tergal plate but lacking musculature and therefore not independently movable (compare lateral end spine). (See also middorsal spine)

longitudinal muscle: dorsolateral muscle, ventrolateral muscle

median end spine [G *Mittelendstachel, medianer Endstachel*]: Conspicuous, elongate, hollow, epidermis-filled median cuticularized projection on tergal plate of last segment (zonite) between lateral end spines;

jointed into tergal plate and movable by special muscles.

middorsal cord (dorsal cord) [G *dorsomedianer Längsstrang, Rückenstrang*]: Middorsal longitudinal thickening of epidermis; swelling of cord in each trunk segment (zonite) typically associated with sensory structures. (See also lateral cord)

middorsal spine [G *Rückenstachel*]: Hollow, epidermis-filled middorsal cuticularized projection on each tergal plate of trunk; jointed into tergal plate but lacking musculature and therefore not independently movable. (See also lateral spine)

mouth [G *Mund*]: Anterior opening of digestive tract located in a central terminal position on head; situated on protrusible mouth cone and surrounded by oral styles. Leads into buccal cavity.

mouth cone [G *Mundkegel*]: Protrusible cone bearing terminal mouth and surrounded by circlet of oral styles and occasionally additional rings of smaller accessory styles; delimited from remainder of head by furrow.

neck [G *Hals*]: Division of body between head and trunk; consists of second segment (zonite).

nephridiopore [G *Exkretionsporus*]: One of two sieve-like openings of paired protonephridia to exterior; located laterally on tergal plate of 11th segment (zonite).

ocellus: eyespot

oesophagus: esophagus

oral style [G *Munddolch, Stylus*]: One element in a circlet of variously elaborated, hollow cuticular spines projecting anteriorly from rim of mouth cone.

outer head retractor [G *äußerer Retraktor*]: One in a series of longitudinal retractor muscles spanning from base of inner scalids on head to tergal and sternal plates of successive trunk segments (zonites); passes to the outside of brain. (See also inner head retractor)

ovary [G *Ovar*]: Paired, elongate section of ♀ reproductive system along both sides of stomach-intestine in trunk; each is connected to separate gonopore via oviduct.

oviduct [G *Ovidukt*]: Short, tubular section of ♀ reproductive system between each ovary and gonopore; bears dorsal seminal receptacle.

pachycyclus [G *Pachycyclus*]: Variously elaborated thickening of tergal or sternal plates; restricted to anterior margin of plates.

pancreatic gland [G *Pankreasdrüse, hintere Ösophagusdrüse*]: As distinguished by early authors, one of two or more glands entering digestive tract at junction of esophagus and stomach-intestine. Not demonstrable in recent studies. (See also salivary gland)

penial spicule (penial spine, copulatory spicule) [G *Penisstachel, Penisborste*]: One of two or three ventral cuticular projections associated with each ♂ gonopore at junction of 12th and 13th segments (zonites).

pharynx [G *Pharynx*]: Muscular, cuticle-lined section of digestive tract between buccal cavity and esophagus; pharynx crown at anterior end of pharynx projects into buccal cavity.

pharynx crown [G *Pharynxkrone*]: Thick cuticular ring forming anterior end of pharynx and projecting into buccal cavity.

placid [G *Plakid*]: Typically, well-developed cuticular plates on second segment (zonite) (neck) covering head when head is retracted. (trapezoidal, triangular)

protonephridium [G *Protonephridium*]: Paired excretory organ, one along each side of stomach-intestine in 10th segment (zonite); consists of modified flame bulb and opens to exterior via nephridiopore on tergal plate of zonite 11.

pseudocoel [G *Leibeshöhle*]: Term applied to fluid-filled body cavity between digestive tract and body wall containing numerous amebocytes.

rectum: endgut

salivary gland [G *Speicheldrüse*]: As distinguished by earlier authors, one of two dorsolateral and two ventrolateral glands opening into anterior section of esophagus. (See also pancreatic gland)

scalid [G *Skalide, Kopfstachel, Stachelskalide*]: Hollow, epidermis-filled, posteriorly directed cuticular projection consisting of basal socket and distal blade; in four to seven circlets around head. Size of scalids, typically numbering 10–20 per ring, decreases in successive circlets.

segment: zonite

seminal receptacle [G *Receptaculum seminis*]: Short dorsal diverticulum of oviduct in ♀.

sensory bristle (sensory seta) [G *Sinnesborste*]: Short, cuticularized, plasma-filled sensory projection on outer surface of body; typically associated with longitudinal thickenings of the epidermis (see lateral cord, middorsal cord) and therefore concentrated in up to ten longitudinal rows along trunk, each segment (zonite) thus bearing up to ten sensory bristles.

socket [G *Sockel*]: Basal element of scalid; separated from distal blade by circular fold and typically bearing one or more small basal styles.

sperm duct [G *Vas deferens*]: Short section of ♂ reproductive system between testes and gonopore.

spine [G *Stachel*]: Long, hollow, epidermis-filled cuticular projection attached to trunk by means of joint; typically forming two lateral and one middorsal series and including various independently movable terminal spines. (See also lateral end spine, lateral spine, median end spine, middorsal spine)

standard measurements:
    trunk length: anterior margin of third segment (first trunk segment) to posterior margin of segment 13 (measured along midline, exclusive of spines)
    standard width: sternal width at segment 12 (measured at anteroventral margin of 12th sternal plates)
    maximum sternal width: measured at anteroventral margin of widest pair of sternal plates as first encountered in measuring each segment from anterior to posterior
    lateral terminal spine length (often expressed in percent of trunk length)

sternal plate (ventral plate) [G *Ventralplatte*]: One of two thickened cuticular plates covering ventral surface of each segment (zonite) of trunk; typically flat, sternal plates of third or fourth zonite often bear pair of adhesive tubes. Overlapping and attached to succeeding sternal plate by thin, flexible cuticle, to tergal plate by thin cuticle and/or peg and socket.

stomach-intestine [G *Magendarm, Mitteldarm*]: Long, tube-like section of digestive tract between esophagus and end gut; lacks cilia and cuticular lining.

tergal plate (dorsal plate) [G *Tergalplatte*]: Thickened cuticular plate covering dorsal surface of each segment (zonite) of trunk; typically arched and bearing middorsal and lateral spines, last tergal plate also often with lateral and median end spines. Overlapping and attached to succeeding tergal plate by thin, flexible cuticle, to sternal plates by thin cuticle and/or peg and socket.

terminal spine [G *Endstachel*]: Term applied to movable lateral end and median end spines on tergal plate of last segment (zonite).

testis [G *Hoden*]: Paired, elongate section of ♂ reproductive system along both sides of stomach-intestine in trunk; each is connected to separate gonopore via short sperm duct.

trichoscalid [G *Trichoskalide*]: Modified scalid located in or forming posteriormost circlet of scalids; borne on cuticular plate and distinguished from scalid by weak cuticularization, coat of fine bristles, and no differentiation into socket and blade.

trunk [G *Rumpf*]: Posteriormost and largest of three divisions of body (head, neck, trunk), typically consisting of 11 segments (zonites).

ventral nerve cord [G *Bauch-strang, ventromedianer Nervenstrang*]: Two adjacent longitudinal strands of nerve tissue originating at brain and extending posteriorly along midventral line at junction of sternal plates; forms one ganglion pair per segment (zonite).

ventral plate: sternal plate

ventrolateral muscle [G *ventrales Längsmuskelband*]: One of two ventrolateral, segmented, longitudinal muscle bands running along sternal plates in trunk.

zonite (segment) [G *Zonit*]: One in a series of 13 divisions of cuticle into clearly defined segments. First zonite constitutes head; the second, the neck; and the remainder, the trunk, each of the latter characterized externally by dorsal tergal plate and two ventral sternal plates (or forming complete ring).

LORICIFERA

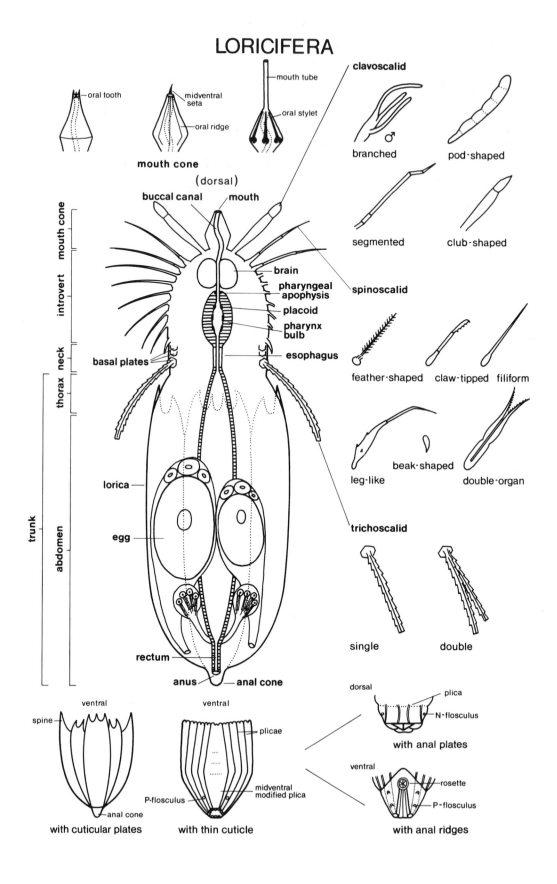

oral tooth

midventral seta

oral ridge

mouth tube

oral stylet

mouth cone

(dorsal)

clavoscalid

branched

pod-shaped

segmented

club-shaped

buccal canal — **mouth**

mouth cone

introvert

neck

thorax

abdomen

trunk

brain

pharyngeal apophysis

placoid

pharynx bulb

basal plates — **esophagus**

spinoscalid

feather-shaped   claw-tipped   filiform

leg-like   beak-shaped   double-organ

lorica

egg

trichoscalid

single   double

rectum

anus — **anal cone**

ventral

spine

anal cone

with cuticular plates

ventral

plicae

P-flosculus

midventral modified plica

anal cone

with thin cuticle

dorsal

plica

N-flosculus

with anal plates

ventral

rosette

P-flosculus

with anal ridges

LORICIFERA

loriciferan [G *Loricifer*]

abdomen [G *Abdomen*]: Posterior-most of five divisions of body (mouth cone, introvert, neck, thorax, abdomen); encased in lorica and bearing midgut and gonads internally.

accessory basal plate [G *akzessorische Basalplatte*]: One plate in either one or two circlets of plates (upper and middle basal plates) immediately anterior to trichoscalid-bearing basal plate (lower basal plate) on neck; may bear trichoscalid plate spine.

accessory stylet [G *akzessorisches Stilett*]: One of two smaller stylets supporting buccal tube ventrally in introvert; attached posteriorly to pharynx bulb and may open into buccal cavity anteriorly.

anal cone [G *Analconus, After-conus*]: On posteriormost region of lorica, strongly cuticularized, cone-shaped structure associated with anus.

anal field [G *Analfeld*]: On posteriormost region of lorica, larger modified cuticular region associated with anus and often bearing pores.

anal plate [G *Analplatte*]: On posteriormost region of lorica, cuticular plate associated with anus (especially in larva).

anal ridge [G *Analstreifen*]: On lorica, one of several cuticular ridges extending between rosette structure and posterior end on midventral plica.

anterolateral seta [G *Anterolateralborste*]: In larva, pair of thin lateral projections on anterior rim of lorica; probably only with sensory function. (with/without hairs) (See also anteroventral seta, lorical seta)

anteroventral seta [G *Anteroventralborste*]: In larva, pair of thin ventral projections on anterior rim of lorica; probably only with sensory function. (with/without hairs) (See also anterolateral seta, lorical seta)

anus [G *After*]: Posterior opening of digestive tract; located terminally on lorica and typically associated with variously modified cuticular surfaces (anal cone, anal field, anal plate, anal ridge). (dorsoterminal, ventroterminal)

basal plate [G *Basalplatte, Nackenplatte*]: One in a circlet of plates around neck from which trichoscalids project; trichoscalid-bearing plate (lower basal plate) may be adjoined anteriorly by one or two circlets of accessory basal plates (upper and middle basal plates). May also refer to plate at base of toe in larva.

brain [G *Gehirn*]: Main dorsal concentration of nerve tissue in introvert; innervates scalids and is closely adjoined by circumoral ganglia.

buccal canal [G *Bukkalkanal*]: In digestive tract, long, thin, cuticular tube in mouth cone between mouth and pharynx bulb; may be extrusible. (straight, curved)

buccal canal gland [G *Bukkalkanaldrüse*]: In mouth cone, three pairs of glands opening into buccal canal.

buccal tooth [G *Bukkalzahn*]: One in a series of minute teeth within mouth tube.

buccal tube: mouth tube

caudal ganglion [G *Kaudalganglion*]: Large ventral ganglion in abdomen; innervates flosculae.

circumoral ganglion [G *Zirkumoralganglion*]: One of eight radially arranged, small ganglia closely adjoining brain and innervating clavoscalids.

clavoscalid [G *Clavoskalid*]: One in a series of originally eight, often anteriorly directed scalids forming first circlet around introvert; if branched, one may distinguish primary, secondary, and tertiary clavoscalids. (branched, unbranched; segmented, unsegmented; club-shaped, pod-shaped, spinose)

closing apparatus: neck

collar: neck

collar pore [G *Halsporus*]: In larva, one in a circlet of several small pores around collar; may bear minute hook at anterior margin.

diaphragm (ligament) [G *Septum*]: In larva, transverse partition separating thorax and abdomen.

double organ [G *Doppelorgan*]: Strongly cuticularized projection formed by modification and fusion of two spinoscalids on midventral line of second spinoscalid ring around introvert.

esophagus (oesophagus) [G *Ösophagus*]: Short section of digestive tract between pharynx bulb and midgut.

flosculus [G *Flosculus*]: Sensory structure consisting of either cilium with collar of modified microvilli (N-flosculus) or small, round papilla with very thin cuticle (P-flosculus); typically present in one or two pairs ventrally near posterior end of lorica. (anterior, posterior)

gonad: ovary, testis

gonopore [G *Gonoporus, Geschlechtsöffnung*]: One of two lateral openings of ♂ or ♀ reproductive system to exterior at posterior end of lorica.

head: introvert

Higgins larva [G *Higgins-Larve*]: Larval stages two through five of loriciferan; differ most notably from adult in bearing toes, lorical setae, and collar-like closing apparatus.

introvert (head) [G *Introvert, Kopf*]: One of five divisions of body (mouth cone, introvert, neck, thorax, abdomen); eversible, bearing nine circlets of appendages (scalids).

introvert retractor muscle [G *Introvert-Retraktor*]: Longitudinal muscle spanning from introvert, through abdomen, to posterior end of lorica near anal cone.

larva: Higgins larva

ligament: diaphragm

locomotory spine (ventral spine) [G *Kriechborste*]: In larva, one of three pairs of ventral spines between thorax and abdomen at anterior rim of lorica; probably with locomotory and sensory function. (See also anterolateral seta, anteroventral seta, lorical seta)

lorica [G *Lorica*]: Cuticular encasement of abdomen; consists of either six heavy plates or many longitudinal plicae. Larval lorica characterized by various setae or spines. (sculptured, unsculptured)

lorical seta [G *Lorica-Borste*]: In larva, one of two or three pairs of thin, movable sensory projections on posterior end of lorica; termed, according to position on lorica, posterodorsal, posterolateral, and posteroterminal setae. (See also anterolateral seta, anteroventral seta, locomotory spine)

lower basal plate: basal plate

middle basal plate: basal plate

midgut [G *Mitteldarm*]: Long, straight section of digestive tract between esophagus and rectum.

midventral seta [G *midventrale Borste*]: Midventral seta projecting from mouth cone.

mouth [G *Mund*]: Anterior opening of digestive tract at end of mouth tube; joined to pharynx bulb via thin buccal canal.

mouth cone [G *Mundconus*]: Anteriormost of five divisions of body (mouth cone, introvert, neck, thorax, abdomen); typically tripartite, with terminal mouth at end of mouth tube. May bear oral stylets or oral ridges.

mouth tube [G *Mundrohr*]: Long tube bearing buccal canal and terminal mouth; may be telescopically protrusile.

mucro [G *Mucro*]: In larva, one in a series of leaf-like structures along lateral margins of toe; may all change direction at the same time.

neck [G *Nackenregion*]: One of five divisions of body (mouth cone, introvert, neck, thorax, abdomen); located posterior to last row of spinoscalids and bearing basal plates and trichoscalids. May form collar or closing apparatus in larva.

nephridiopore [G *Nephridioporus*]: One of two openings of protonephridium at posterior end of lorica.

N-flosculus: flosculus

oesophagus: esophagus

oral ridge [G *Mundconus-Furche*]: One in a series of cuticular longitudinal ridges on mouth cone.

oral stylet [G *Mundstilett, Oralstilett*]: One in a series of rigid, basally expanded, protrusile stylets oriented longitudinally in wall of mouth cone; each has separate anterior opening. (single, double)

oral tooth [G *Zahn*]: One in a series of small, tooth-like projections surrounding mouth opening.

ovary [G *Ovar*]: Paired, saccate section of ♀ reproductive system located in abdomen.

oviduct [G *Ovidukt*]: Section of ♀ reproductive system between each ovary and gonopores; may bear seminal receptacle.

penile spine [G *Penisstachel*]: One of two short laterocaudal spines with small basal pore on posterior end of lorica in ♂; possible accessory reproductive structure.

P-flosculus: flosculus

pharynx apophysis [G *Pharynx-Apophyse*]: Thickening at junction of buccal canal and pharynx bulb.

pharynx bulb [G *Pharynx, Bulbus*]: Expanded muscular section of digestive tract between buccal tube and esophagus; may bear placoids.

placoid [G *Placoid*]: In pharynx bulb, one in a series of several cuticular plates located at points of radial muscle attachment.

plica [G *Plica*]: One in a series of longitudinal elements making up lorica. Each plica bears a central secondary double ridge and is bordered by a primary double ridge on each side; midventral plica may be wider and bear rosette structure.

posterodorsal seta: lorical seta

posterolateral seta: lorical seta

posteroterminal seta: lorical seta

primary clavoscalid: clavoscalid

primary trichoscalid [G *primärer Trichoscalid*]: In trichoscalid type consisting of two elements, longer, lower element. (See also secondary trichoscalid)

protonephridium [G *Protonephridium*]: Paired excretory organ; opens to exterior via nephridiopore.

protoscalid [G *Protoskalid*]: Because of difficulty in comparing adult and larval appendages, term applied to certain short larval scalids on head.

rectum [G *Enddarm*]: Posteriormost cuticle-lined section of digestive tract between midgut and anus.

rosette structure [G *Rosette-Struktur*]: Rosette-shaped structure of unknown function located posteriorly on midventral plica of lorica; consists of six cells with central pore.

ruff [G *Krause*]: Series of longitudinal wrinkles in cuticle between mouth cone and introvert; formed by thin fibers.

salivary gland [G *Speicheldrüse*]: In introvert, pair of large glands embedded near brain.

scalid [G *Skalid*]: One of variously elaborated appendages forming up to 10 circlets around introvert and neck. (See also clavoscalid, spinoscalid, trichoscalid)

secondary clavoscalid: clavoscalid

secondary trichoscalid [G *sekundärer Trichoskalid*]: In trichoscalid type consisting of two elements, shorter element attached to upper surface of primary trichoscalid near its base.

seminal receptacle [G *Receptaculum seminis*]: In ♀ reproductive system, sperm storage organ associated with each oviduct.

seminal vesicle [G *Samenblase, Vesicula seminalis*]: In ♂ reproductive system, expanded section of each sperm duct.

seta: anterolateral seta, anteroventral seta, lorical seta

sperm duct [G *Samenleiter*]: In ♂ reproductive system, duct between each testis and gonopores; may bear seminal vesicle.

spinoscalid [G *Spinoskalid*]: One in a series of typically spine-shaped scalids forming second through ninth circlets around introvert. (segmented, unsegmented; beak-shaped, claw-tipped, filiform, leg-shaped, spine-shaped)

stylet: accessory stylet, oral stylet

tertiary clavoscalid: clavoscalid

testis [G *Hoden*]: Large, paired section of ♂ reproductive system located dorsally in abdomen; opens posteriorly via sperm duct.

thorax [G *Thorax*]: One of five divisions of body (mouth cone, introvert, neck, thorax, abdomen); located between trichoscalids and loricated abdomen.

toe [G *Zehe*]: In larva, pair of large caudal appendages, typically articulating with abdomen in a ball-and-socket joint. Either adhesive in function and spine-shaped or used in swimming and expanded by series of leaf-like structures (mucros) along each lateral margin.

toe gland [G *Zehendrüse*]: In larva with adhesive type of toe, pair of large glands located posteriorly in abdomen and opening via toe gland pore at base of toe.

toe gland pore [G *Zehendrüsenöffnung*]: In larva, opening of toe gland to exterior; located at base of adhesive type of toe.

trichoscalid [G *Trichoskalid, Nackenschuppe*]: One in a series of highly modified scalids forming posteriormost circlet of appendages; located on basal plates in neck region. (single, double; feather-shaped, serrated)

trichoscalid plate: basal plate

trichoscalid plate spine [*Basalplatten-Stachel*]: Short spine on accessory basal plate.

trunk [G *Rumpf*]: Term applied to posterior region of body encompassing thorax and abdomen; posterior section encased in lorica.

upper basal plate: basal plate

ventral ganglion [G *Bauchganglion*]: Large ventral ganglion in thorax; innervates trichoscalids.

ventral spine: locomotory spine

PRIAPULIDA

priapulid [G *Priapulide, Priapswurm*]

PRIAPULIDA

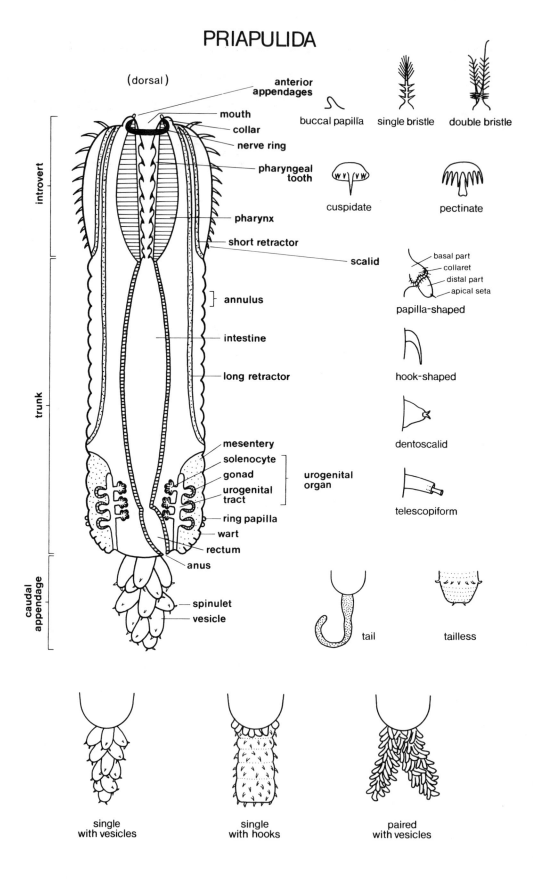

(dorsal)

anterior appendages
mouth
collar
nerve ring
pharyngeal tooth
pharynx
short retractor
scalid
annulus
intestine
long retractor
mesentery
solenocyte
gonad
urogenital tract
urogenital organ
ring papilla
wart
rectum
anus
spinulet
vesicle

introvert
trunk
caudal appendage

buccal papilla
single bristle
double bristle

cuspidate
pectinate

basal part
collaret
distal part
apical seta
papilla-shaped

hook-shaped

dentoscalid

telescopiform

tail
tailless

single with vesicles
single with hooks
paired with vesicles

PRIAPULIDA

priapulid [G *Priapulide, Priapswurm*]

abdomen: trunk

annulus (annule) [G *Annulus, Ringwulst*]: One in a series of elevated rings of body wall of trunk; annulation is superficial (corresponding to circular muscle bands) and does not reflect segmentation.

anus [G *After*]: Posterior opening of digestive tract; located terminally or, if caudal appendage is present, somewhat laterally on posterior end of trunk.

basis [G *Basis*]: In pharynx of interstitial priapulid, elevation containing tooth gland and enclosing basal section (manubrium) of tooth; if well developed, one may also refer to stalked tooth. Term also applied to base of tubulus or flosculus.

bullula [G *Bullula*]: In ♂ interstitial priapulid, microscopic, rounded elevation of cuticle (smaller than tumulus) on ventral surface of trunk.

caudal appendage [G *Schwanzanhang*]: One or two hollow, stalk-like appendages on posterior end of trunk; typically bears hollow, spherical vesicles. (See also tail)

caudal gland [G *Caudaldrüse*]: One of numerous glands opening via minute pores on warts at posterior end of trunk in adults, via tube-like structures at posterior end of lorica in larvae.

circular muscle [G *Ringmuskel*]: One in a series of circular muscle bands; conspicuous on trunk, where they correspond to annuli.

circumenteric nerve ring: nerve ring

circumoral region [G *Mundfeld*]: Area surrounding mouth and extending posteriorly to collar or, if this is absent, to row of scalids on introvert. Beset with geometrically arranged cuticular projections.

clavula [G *Clavula*]: In interstitial priapulid, club-shaped projection of cuticle anterior to urogenital pore on ventral surface of trunk.

collar [G *Kragen*]: In certain priapulids, circular band between spinous circumoral region and scalid-bearing region of introvert; corresponds in position to nerve ring.

cuticle [G *Kutikula*]: Noncellular, multilayered, variously ornamented layer covering outer surface of body and also lining both pharynx and rectum.

dentoscalid [G *Dentoscalid*]: Type of scalid provided with one or two strongly cuticularized distal teeth.

esophagus (oesophagus) [G *Ösophagus*]: In interstitial priapulid, short, narrow section of digestive tract between pharynx and intestine; terminates in polythyridium.

fimbrilla [G *Fimbrille*]: One in a row of minute hairs along cuticular ridge in posterior section of pharynx in interstitial priapulid.

flosculus [G *Flosculus*]: One of several small, flower-shaped sense organs projecting from cuticle of neck, introvert, and trunk; consists of basis, accessory seta, and flosculus proper.

gonad [G *Gonade*]: ♂ (testis) or ♀ (ovary) reproductive portion of urogenital organ on each side of intestine; opens on posterior end of trunk via urogenital duct.

intestine (midgut) [G *Mitteldarm*]: Section of digestive tract between pharynx and rectum; usually straight and tubular.

introvert (presoma) [G *Introvert, Präsoma*]: Anterior introversible division of body (introvert, trunk) with terminal mouth; characterized externally by spinous circumoral region, collar, and long rows of scalids, internally by pharynx.

long retractor muscle [G *langer Retraktor*]: One in a series of long muscle bands spanning from introvert to trunk wall; serve to retract introvert. (See also short retractor muscle)

lorica [G *Lorica*]: In larva, thick, rigid elaboration of trunk cuticle forming two or more longitudinal plates; typically bears longitudinal ridges and tubuli. (circular, dorsoventrally flattened)

manubrium [G *Griff*]: In pectinate pharyngeal tooth of interstitial priapulid, proximal section of tooth embedded in elevation (basis) of pharynx cuticle.

midventral nerve cord: ventral nerve cord

mouth [G *Mund*]: Anterior opening of digestive tract at anterior end of introvert; typically invaginated into introvert and surrounded by cuticular projections in geometric arrangement.

neck [G *Hals*]: One of four divisions of body (introvert, neck, trunk, caudal appendage) in larva, where it may form closing apparatus. Typically absent in adult.

nephridiopore: urogenital pore

nerve ring (circumenteric nerve ring, peribuccal nerve ring) [G *Schlundring*]: Main concentration of nerve tissue; intraepithelial and forming ring around anterior end of pharynx in introvert. Gives rise ventrally to ventral nerve cord.

ovary: gonad

papilla [G *Papille*]: Type of ornamentation on outer surface of trunk. (See also ring papilla)

pecten [G *Zahnkamm*]: In pectinate pharyngeal tooth of interstitial priapulid, distal section of tooth bearing row of approximately 12 spines. (See also manubrium)

pentagon [G *pentaradiärer Zahnkreis*]: One of several circlets of five teeth in anterior section of pharynx.

peribuccal nerve ring: nerve ring

pharyngeal tooth [G *Zahn*]: One in a series of cuticular projections of pharynx; each tooth typically consists of basis, median cusp, and several lateral cusps. Tooth size decreases posteriorly. (cuspidate, pectinate)

pharynx [G *Pharynx*]: Muscular, cuticle-lined, tooth-bearing section of digestive tract between mouth and intestine. Protruded during feeding, otherwise invaginated into intestine.

polythyridium [G *Polythyridium*]: Cuticular organ at junction of esophagus and intestine in interstitial priapulid; consists of anterior and posterior circlets of differently elaborated valvulae.

presoma: introvert

prickle: In interstitial priapulid, conical papilla with long, thin distal spine in posterior region of pharynx.

"proboscis": Equivalent to introvert, although occasionally used to denote main part of introvert exclusive of mouth, circumoral region, and collar.

protonephridial tubule: urogenital duct

rectum (end gut) [G *Enddarm*]: Short, terminal, cuticle-lined section of digestive tract between intestine and anus.

ring papilla [G *Ringpapille*]: One element in ring of special papillae around one or more annuli at posterior end of trunk.

scalid [G *Skalide*]: One in a series of posteriorly directed cuticular projections in numerous longitudinal rows along proboscis; typically decreases in size posteriorly. May consist of basal and distal sections. (bifid, papilla-shaped, pinnate, spine-shaped)

short retractor muscle [G *kurzer Retraktor*]: One in a series of short muscle bands spanning from introvert to junction of introvert and trunk. (See also long retractor muscle)

solenocyte [G *Solenozyte*]: Proximal cell of protonephridial organ for excretory ultrafiltration; aggregated into clusters (solenocyte tree), located opposite gonads, and opening into urogenital duct.

spine: scalid

spinulet [G *Dorn*]: Minute spine on vesicle of caudal appendage.

tail [G *Schwanzanhang*]: Term applied to single tubular appendage at posterior end of trunk in interstitial priapulid; may also be used as synonym for caudal appendage.

testis: gonad

tooth: pharyngeal tooth

tooth gland [G *Zahndrüse*]: In anterior section of pharynx of interstitial priapulid, cluster of epidermal cells apparently functioning in formation of pectinate pharyngeal teeth.

trunk (abdomen) [G *Rumpf*]: Posterior division of body (introvert, trunk); typically bearing caudal appendage and, in larger priapulids, characterized externally by superficial annulation.

tubulus [G *Tubulus*]: One of several short, tube-like structures projecting from lo-rica of larva or trunk of interstitial priapulid; positioned on basis and adjoined by accessory seta.

tumulus [G *Tumulus*]: One of numerous minute, conical elevations with ridges on cuticle of trunk.

twin-chaeta [G *gefiederte Doppelborste*]: Pinnate, tentacle-shaped elaboration of first ring of scalids on introvert.

urogenital duct (protonephridial tubule) [G *Urogenitalgang*]: Tubular structure on each side of intestine into which both products of gonads and ultrafiltrate of solenocytes are released.

urogenital organ [G *Urogenitalorgan*]: Organs on each side of intestine; each consists of urogenital duct bearing bulges of gonads on outer side and clusters of solenocytes on inner side.

urogenital pore (nephridiopore) [G *Urogenitalporus*]: One of two openings of urogenital ducts at posterior end of trunk.

valvula [G *Valvula*]: One element in several circlets of cuticular plates in polythyridium; inwardly directed margins of valvulae of anterior circlets bear tubercles, those of posterior circlets bear hairs.

ventral nerve cord (midventral nerve cord) [G *Nervenlängsstamm, ventraler Markstrang*]: Main longitudinal nerve extending midventrally within epidermis from circumenteric nerve ring to posterior end of trunk; slightly enlarged at each trunk annulus.

ROTIFERA

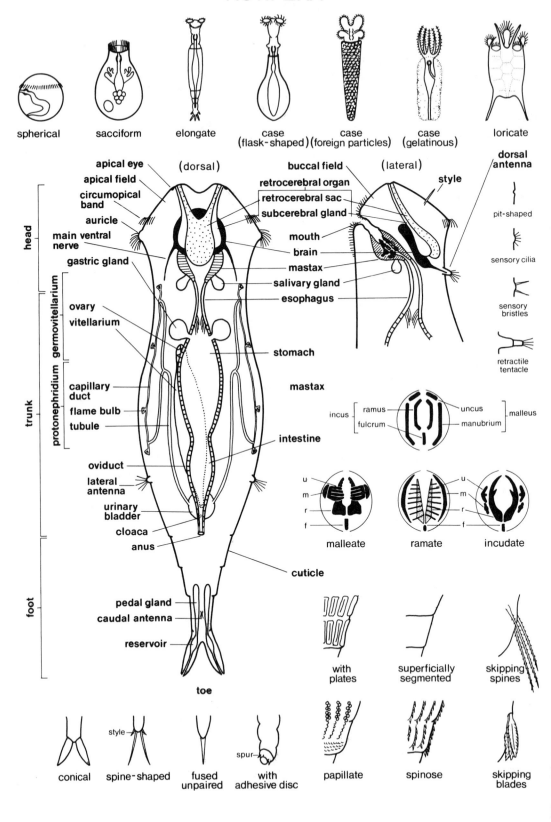

spherical

sacciform

elongate

case (flask-shaped)

case (foreign particles)

case (gelatinous)

loricate

(dorsal)

(lateral)

head

apical eye
apical field
circumopical band
auricle
main ventral nerve
gastric gland

germovitellarium

ovary
vitellarium

protonephridium

capillary duct
flame bulb
tubule

trunk

oviduct
lateral antenna
urinary bladder
cloaca
anus

buccal field
retrocerebral organ
retrocerebral sac
subcerebral gland
mouth
brain
mastax
salivary gland
esophagus

style

dorsal antenna

pit-shaped

sensory cilia

sensory bristles

retractile tentacle

stomach

mastax

intestine

incus ⟨ ramus / fulcrum ⟩ ⟨ uncus / manubrium ⟩ malleus

cuticle

foot

pedal gland
caudal antenna
reservoir

u
m
r
f
malleate

u
m
r
f
ramate

u
m
r
f
incudate

with plates

superficially segmented

skipping spines

toe

conical

style
spine-shaped

fused unpaired

spur
with adhesive disc

papillate

spinose

skipping blades

ROTIFERA (ROTATORIA)

rotifer, wheel-animalcule [G *Rädertier*]

adhesive disk [G *Haftscheibe*]: Disc-shaped adhesive structure forming end of foot.

alula [G *Alula*]: In mastax, posterior projection of rami; serves as attachment site for muscles extending to fulcrum.

amebocyte **(amoebocyte)** [G *Amöbozyte*]: Mobile cell with pseudopodia forming a loose network in fluid of body cavity (pseudocoel).

antenna: dorsal antenna, lateral antenna

anterior ganglion [G *Ganglion anterius*]: Small, paired ganglion, one on each of the two main ventral nerve cords; located anteriorly between brain and geniculate ganglion.

anus [G *After*]: Posterior opening of digestive tract; typically located dorsally on posterior half of trunk near trunk/tail border. Absent in certain pelagic rotifers and in certain ♂♂.

apical eye (frontal eye) [G *Apikalauge, Frontalauge*]: Pair of small, multinucleate, red pigment-spot ocelli, typically with lens-like structure; located in apical field or on rostrum.

apical field [G *Apikalfeld*]: Central region of head surrounded by corona and typically bearing openings of retrocerebral organ, apical eyes, and an array of sensory projections.

arm [G *Arm*]: One of six movable extensions of trunk bearing numerous setose bristles distally; with locomotory function.

auricle [G *Wimperohr, Aurikel*]: Pair of conspicuous, retractile projections of corona, one on each side of head, bearing terminal tuft of cilia; with locomotory function.

bladder: cloacal bladder, urinary bladder

brain (cerebral ganglion) [G *Gehirn, Zerebralganglion*]: Main concentration of nerve tissue dorsal to mastax. Gives rise to array of nerves including sensory nerves anteriorly and pair of main ventral nerves laterally. (bilobed, quadrangular, rounded, triangular)

buccal cavity [G *Bukkaltrichter, Mundhöhle*]: Term occasionally applied to anterior section of buccal tube.

buccal field [G *Bukkalfeld*]: Ventral ciliated area surrounding mouth.

buccal tube [G *Schlundrohr*]: Variously shaped, ciliated section of digestive tract between mouth and pharynx; one may occasionally distinguish an anterior buccal cavity and posterior pharyngeal tube. (funnel-shaped, tube-shaped)

capillary duct [G *Kapillarrohr*]: In excretory system, thin-walled section of each protonephridial tubule; bears delicate cilia and connects flame bulbs to one another. (See also tubule)

case [G *Hülle, Gehäuse*]: Variously formed, gelatinous or rigid housing secreted by certain rotifers; may incorporate specially consolidated and symmetrically arranged foreign particles.

cauda [G *Cauda*]: In mastax, pointed posterior end of each manubrium.

caudal antenna [G *Kaudaltaster*]: Unpaired dorsal sensory structure at base of toes on foot; consists of either small pit, small papilla, or two to four bristles.

caudal ganglion (pedal ganglion) [G *Kaudalganglion, Fußganglion*]: Unpaired, often not clearly delimited ganglion at posterior end of main ventral nerve cords; located in foot or, when this is reduced, in posterior end of trunk. (See also caudovesicular ganglion)

caudovesicular ganglion [G *Kaudovesikularganglion*]: Term applied to ganglion formed by fusion of caudal and vesicular ganglia.

cement gland: pedal gland

cerebral eye [G *Zerebralauge, Zervikalauge*]: Small, typically unpaired, red pigment-cup ocellus consisting of one cell and located dorsally, occasionally ventrally, on posterior part of brain. (paired, unpaired) (See also apical eye, lateral eye)

cerebral ganglion: brain

chin [G *Kinn*]: Term occasionally applied to ciliated, posteriorly directed projection of buccal field.

ciliated pit [G *Wimpergrube*]: Type of apical sensory structure consisting of several often symmetrically arranged, ciliated depressions. May also refer to ciliated pit under lower lip; this bears pit glands and is used to consolidate foreign particles for construction of case.

cingulum [G *Cingulum*]: Postoral circlet of membranelles on head; formed by modified cilia of circumapical band posterior to mouth. (See also paracingulum, paratrochus, trochus)

circumapical band [G *Zirkumapikalband*]: Band of cilia extending around head from buccal field; may be modified into a circlet anterior (trochus) and posterior (cingulum) to mouth.

cirrus [G *Zirrus*]: Elongated projection(s), typically on head, formed by fusion of cilia.

cloaca [G *Kloake*]: Posteriormost nonciliated, occasionally cuticularized section of digestive tract between intestine and anus into which protonephridial tubules and oviduct open. Absent in certain pelagic rotifers.

cloacal bladder [G *Kloakalblase, Rektalblase, Harnblase*]: Type of muscular cloaca, clearly delimited from intestine, into which both protonephridial tubules open via common duct. (See also urinary bladder)

corona [G *Räderorgan, Korona*]: Conspicuous, highly variable circlet of often modified cilia on head formed by cilia of circumapical band and/or buccal field. (bilobed, circular, cordate, horseshoe-shaped, lobed, oval)

coronal lobe [G *Coronallobus, Koronallappen*]: One of variously shaped, cilia-bordered lobes on head; formed by expansion of corona. (blunt, capitate, elongated, setose)

coronal matrix [G *Hypodermiswulst, Hypodermispolster*]: Conspicuous, thickened epidermal cushion underlying corona and containing basal bodies and roots of coronal cilia.

coronal sphincter [G *Kronenschließmuskel, Sphincter coronae*]: Well-developed circular muscle posterior to corona on head; composed of a number of bands and serving to seal neck when corona is retracted.

cutaneovisceral muscle [G *Haut-Eingeweidemuskel, Kutaneoviszeralmuskel*]: Muscle spanning from body wall to inner organs, especially digestive tract. One may distinguish cutaneopharyngeal, cutaneointestinal, and cutaneogenital muscles.

cuticle [G *Kutikula*]: Variously elaborated, noncellular layer covering outer surface of body as well as lining buccal cavity, pharynx, and in some cases cloaca; may be superficially segmented or thickened to form lorica. (papillate, ridged, smooth, spinose)

diaphragm [G *Diaphragma*]: In funnel formed by coronal lobes, slightly protruding ring forming border between upper infundibulum and lower vestibule; bears cilia ventrally and pharyngeal sense organs laterally at border of ciliated and nonciliated sections.

dorsal antenna [G *Dorsaltaster, Nackentaster*]: Variously shaped, typically unpaired and movable middorsal projection on anterior third of trunk; provided with sensory nerve cells basally, sensory cilia distally, and is innervated by short nerve from brain. (biarticulate, papilla-shaped, pit-shaped, tentacle-shaped, tubular; paired, unpaired) (See also lateral antenna)

dorsoventral muscle [G *Dorsoventralmuskel*]: Muscle spanning from ventral to dorsal body wall of loricate rotifers.

egg [G *Ei*]: ♀ reproductive body produced in ovary, provided with yolk by vitellarium, and passing through oviduct to exterior via cloaca or gonopore. (smooth, spinose, thick-shelled, thin-shelled, tuberculate)

epipharyngeal ganglion [G *Epipharynxganglion*]: Small, paired ganglion, one on each side of epipharynx; connected to brain by epipharyngeal nerve.

epipharyngeal nerve [G *Nervus epipharyngeus*]: Pair of nerves, arising either directly from brain or branching from pharyngeal nerve, connecting brain with epipharyngeal ganglion.

epipharynx [G *Epipharynx*]: One or more variously shaped, cuticular plates reinforcing anterior dorsal wall of virgate or cardate sucking type of mastax; may be protrusible.

esophagus [G *Ösophagus*]: Long or short, relatively narrow, thin-walled section of digestive tract between mastax and stomach. (See also postesophagus and preesophagus)

eye [G *Auge*]: One of variously elaborated and positioned photosensitive structures on head. (See also apical eye, cerebral eye, lateral eye)

flame bulb [G *Wimperkolben, Terminalorgan, Wimperflamme*]: Excretory structure containing membranelle of fused cilia and forming proximal section of protonephridium.

foot [G *Fuß*]: Posteriormost of three divisions (head, trunk, foot) of body; typically with ringed cuticle and either gradually tapering or considerably narrower than trunk. May bear pedal glands. (See also foot stalk, spur, style, tail)

foot stalk [G *Stiel, Haftstiel, Fußstiel, Pedunculus*]: Shorter or more elongated stalk-like structure produced by pedal glands in foot; used for attachment to substratum in certain sessile rotifers.

frontal eye: apical eye

fulcrum [G *Fulcrum*]: Median, unpaired posteriormost masticatory element (trophus) in mastax; typically a thin plate in sagittal plane. Forms incus together with rami.

gastric gland [G *Magendrüse, Mitteldarmdrüse*]: One of typically two variously shaped glands at junction of esophagus and stomach; each opens via separate pore into stomach. (band-shaped, bilobate, clavate, elongate sacciform, fusiform, ovate, spherical, stalked)

geniculate ganglion [G *Ganglion genu, Ganglion geniculatus*]: Small, paired ganglion, one on each main ventral nerve cord posterior to anterior ganglion. Gives rise to a number of nerves.

germovitellarium [G *Keimdotterstock*]: Conspicuous section of ♀ reproductive system formed by close association of unpaired ovary and vitellarium; located ventral to digestive tract. (band-shaped, horseshoe-shaped, multilobate, ovate, spherical; paired, unpaired)

gonopore [G *Genitalöffnung, Genitalporus*]: Opening of ♂ reproductive system to exterior; typically bearing tuft of cilia and located dorsally at posterior end of trunk. (See also cloaca, penis)

head [G *Kopf*]: Variously shaped, often poorly delimited anteriormost of three divisions (head, trunk, foot) of body; bears mouth, unciliated apical field, and array of sensory projections.

head shield [G *Kopfschild*]: Thickened cuticle covering neck region in loricate rotifers; formed by one to several plates and often seals off anterior opening of lorica when head is retracted.

Huxley's anastomosis [G *Huxley'-sche Anastomose*]: Transverse commissure of excretory system dorsal to brain; connects capillary canal of right and left protonephridium and may bear flame bulbs.

hypopharyngeal plate [G *Hypopharyngealplatte*]: One of several accessory cuticular plates adjacent to hypopharynx in rotifers with virgate mastax.

hypopharynx (piston) [G *Hypopharynx*]: Piston-shaped muscular mass creating suction in rotifers with virgate mastax.

incus [G *Incus*]: In mastax, term applied to combined fulcrum and rami, especially when these form a functional unit. (See also malleus)

infundibulum [G *Infundibulum*]: Outer region of funnel formed by coronal lobes; separated from inner mouth-bearing vestibule by diaphragm.

intestine [G *Intestinum, Mitteldarm*]: Variously elaborated, thin-walled, typically ciliated section of digestive tract between stomach and cloaca; either not clearly delimited from stomach or set off by pyloric constriction. (bladder-like, tubular)

intramalleus [G *Intramalleus*]: One of several small accessory masticatory elements occasionally located between unci and mallei.

lateral antenna [G *Lateraltaster, Seitentaster*]: One of two variously shaped, distally ciliated lateral projections on trunk;

provided with only single sensory nerve cell basally. (capitate, papilla-shaped, pit-shaped, tentacle-shaped, tubular) (See also dorsal antenna)

lateral eye [G *Lateralauge*]: Pair of small, multinucleate, red eyespots, typically with lens-like structure; located on each side of corona near lateral sense organs.

lip [G *Lippe*]: Ciliated projection ventral to mouth; formed by cingulum. May also refer to variously elaborated protuberances ventral (lower lip) or dorsal (upper lip) to mouth.

longitudinal muscle [G *Längsmuskel*]: One in a series of variously elaborated longitudinal muscles either running along body wall or extending through pseudocoel; with nuclei. (See also retractor muscle)

lorica [G *Lorica, Panzer*]: Variously ornamented outer layer of body formed by thickened cuticle; typically encloses only trunk and consists either of one piece with or without longitudinal suture or of several dorsal and ventral plates. (See also head shield)

lorical spine [G *Panzerstachel*]: Rigid, nonmovable cuticular projection, typically of anterior or posterior margin of lorica.

main ventral nerve [G *ventraler Hauptnerv*]: Pair of ventrolateral longitudinal nerve cords arising laterally from brain and extending to caudal ganglion. Gives rise to numerous nerves and bears anterior and geniculate ganglia.

malleus [G *Malleus*]: In mastax, term applied to combined uncus and manubrium, especially when these form a functional unit. (See also incus)

manubrium [G *Manubrium*]: In mastax, paired, elongated masticatory elements (trophi) adjoining outer ends of unci and extending posteriorly, parallel to fulcrum and rami. Forms malleus together with uncus.

mastax [G *Mastax*]: Ventral region of pharynx bearing trophi. Incorrectly used as synonym for pharynx.

mastax ganglion [G *Mastaxganglion*]: Unpaired ganglion located ventrally in midventral wall of mastax; connected to brain via pair of pharyngeal nerves.

mastax muscle [G *Mastaxmuskel*]: Any one of numerous muscles between individual masticatory elements (trophi) and between these trophi and pharynx wall.

membranelle [G *Membranelle*]: One in a series of modified cilia of corona, often forming projections of anterior (trochus) and posterior (cingulum) circlets on head.

mouth [G *Mund*]: Anterior opening of digestive tract located ventrally on head and typically surrounded by circumapical band of corona; leads either into buccal tube or directly into pharynx. (funnel-shaped, rounded, slit-like, triangular)

neck [G *Hals*]: Term occasionally applied to constricted region between head and trunk; may be retractile or covered with cuticular plates. (See also head shield)

ocellus: eye

oral plate [G *Oralplatte*]: Term occasionally applied to one of several cuticular plates which, as opposed to epipharynx covering dorsal wall of pharynx, cover ventral or ventrolateral surface of pharynx near mouth.

ovary [G *Ovar, Ovarium, Keimstock*]: Section of ♀ reproductive system which, together with immediately adjoining vitellarium, forms germovitellarium; Located ventral to digestive tract and connected to cloaca via oviduct. (paired, unpaired)

oviduct [G *Ovidukt*]: Section of ♀ reproductive system between vitellarium (germovitellarium) and cloaca. (paired, unpaired)

palp [G *Palporgan*]: Type of apical organ; consists of one or more unciliated, movable, retractile projections. (conical, finger-shaped)

papilla [G *Wimperzapfen*]: Type of apical sensory structure; typically two pairs of conical projections bearing distal cilia, one

pair adjacent to each opening of retrocerebral organ.

paracingulum [G *Paracingulum*]: Postoral circlet of membranelles on head; formed by modified cilia of both circumapical band and posterior margin of buccal field. (See also cingulum)

paratrochus [G *Paratrochus*]: Preoral circlet of membranelles on head; formed by modified cilia of both circumapical band and anterior margin of buccal field.

pedal ganglion: caudal ganglion

pedal gland [G *Klebdrüse*]: One of several to numerous often symmetrically arranged glands in foot; connected to openings at tips of toes via ducts. Secretes adhesive material or mucus.

pedicel [G *Trochusstiel, Stiel, Pedicelle*]: Retractile stalk of each trochal disc.

penis [G *Penis*]: Copulatory structure formed by modified terminal section of sperm duct; either an independent, cuticularized tube or formed by everted, cuticle-lined posterior section of sperm duct.

pharyngeal nerve [G *Pharyngealnerv, Nervus pharyngeus*]: Pair of nerves encircling digestive tract and connecting brain to mastax ganglion.

pharyngeal sense organ [G *Pharyngealsinnesorgan*]: In funnel formed by coronal lobes, one or two tufts of sensory cilia at border of dorsal nonciliated and ventral ciliated sections of diaphragm separating outer infundibulum from inner vestibule.

pharyngeal tube [G *Pharyngealrohr*]: Term occasionally applied to posterior section of buccal tube.

pharynx [G *Pharynx*]: Variously elaborated, cuticle-lined, muscular section of digestive tract between mouth or buccal tube and esophagus; bears ventral mastax with trophi. Often incorrectly used as synonym for mastax. (cylindrical, elongate oval, spherical, trilobate)

piston: hypopharynx

pleural rod [G *Pleuralstab*]: Pair of variously shaped cuticular rods adjacent or posterior to rami in lateral wall of mastax.

postesophagus [G *Postösophagus*]: Term applied to posterior region of esophagus if this is ciliated and anterior region is cuticle-lined; dorsal and ventral inner surfaces often with tuft of long cilia directed into stomach. (See also preesophagus)

preesophagus [G *Präösophagus*]: Term applied to anterior region of esophagus if this is cuticle-lined and posterior region is ciliated. (See also postesophagus)

prostatic gland [G *Prostatadrüse*]: One of typically two relatively large glands opening laterally into ♂ reproductive system near border of testes and sperm duct.

protonephridial tubule [G *Protonephridialkanal*]: Long, looped or coiled section of protonephridium between flame bulbs and cloaca; typically divided into capillary duct connecting flame bulbs with one another and tubule proper. Serves in fluid resorption.

protonephridium [G *Protonephridium*]: Paired excretory organ, one along each side of digestive tract; consists of several flame bulbs opening into common cloaca via protonephridial tubule.

proventriculus [G *Proventriculus*]: Term applied to modified, voluminous, sacciform pharynx associated with uncinate mastax.

pseudocoel [G *Leibeshöhle*]: Well-developed, fluid-filled body cavity between digestive tract and body wall; contains amebocytes.

pseudotroch [G *Pseudotrochus*]: Row of stiff cirri on head; formed by modification of marginal cilia of buccal field anterior to mouth.

pseuduncus [G *Pseuduncus*]: Pair of protrusile, hook-like elaborations of epipharynx.

pyloric sphincter [G *Ringfurche*]: Muscular constriction forming border between stomach and intestine.

ramus [G *Ramus*]: In mastax, paired, typically thicker, triangular masticatory element (trophus) anteriorly adjoining fulcrum. Forms incus together with fulcrum.

reservoir [G *Reservoir*]: Expanded section of duct of pedal glands.

retractor muscle [G *Retraktormuskel*]: One of a number of modified longitudinal muscles serving to retract head or foot; characterized by detachment from body wall, division into anterior and posterior bands, and terminal bifurcations.

retrocerebral organ [G *Retrozerebralorgan*]: Structure dorsal and posterior to brain in head; consists of unpaired median retrocerebral sac and paired lateral subcerebral glands, all opening into apical field.

retrocerebral sac [G *Retrozerebralsack*]: Unpaired median sac dorsal and posterior to brain in head and flanked by subcerebral glands; duct of retrocerebral sac is forked and opens onto apical field via paired openings, often on elevated papilla in close association with openings of subcerebral glands.

ring muscle [G *Ringmuskel*]: One in a series of muscle bands forming complete or incomplete rings under body wall; lack nuclei. (See also coronal sphincter)

rostral lamella [G *Rüssellamelle*]: One or two thin cuticular plates covering rostrum dorsally.

rostral sense organ [G *Rüsseltaster*]: Term referring either to sensory cilia at tip of rostrum or to small median apical sense organ with distal tuft of cilia.

rostrum [G *Rostrum, Rüssel*]: Mid-dorsal projection on head with array of variously modified cilia distally and covered dorsally by rostral lamellae; may bear apical eyes and openings of retrocerebral organ.

salivary gland [G *Speicheldrüse*]: One of up to seven variously shaped glands opening into anterior region of mastax or between mastax and mouth; typically one pair of well-developed ventral glands.

segment (joint) [G *Segment*]: One of a varying number of superficial segments of body formed by constrictions of cuticle; often more distinctly developed in neck and foot regions.

skipping blade [G *Flosse*]: One of several pairs of long, movable, sword-shaped cuticular structures with serrate margins projecting from anterior margin of trunk; with locomotory function.

skipping spine [G *Schwimmborste, Springborste*]: One of two long, movable bristles projecting from anterior region of trunk; with locomotory function.

sperm duct [G *Vas deferens*]: Relatively short, ciliated section of ♂ reproductive system between testes and gonopore; may be modified terminally to form penis.

spur [G *Spore*]: One of typically two thorn- or hook-shaped projections anterior and dorsal to toes on foot, often on penultimate foot "segment."

stomach [G *Magen*]: Large, ciliated, thick-walled section of digestive tract between esophagus and intestine. (sacciform, tube-shaped, U-shaped)

stomach-intestine [G *Magendarm*]: Term applied to section of digestive tract comprising stomach and intestine when the former is not clearly delimited from the latter.

style [G *Stylus*]: One of several accessory cuticular spines adjacent to toe. May also refer to type of apical sensory structure consisting of a number of elongate, rigid, symmetrically arranged projections formed by fusion of cilia and in close association with brain.

subcerebral gland [G *Subzerebraldrüse*]: Pair of glands, one on each side of retrocerebral sac; open onto apical field in close association with openings of retrocerebral sac. (paired, unpaired)

subuncus [G *Subuncus*]: In mastax, small accessory masticatory elements (trophi) occasionally located between unci and rami.

supra-anal sense organ [G *Supraanalorgan*]: Paired sensory structure immediately anterior to anus, each consisting of short, comb-like row of ciliary projections.

tail [G *Fuß*]: Term occasionally applied to foot if this is slender, elongate, and clearly delimited from trunk.

testis [G *Hoden*]: Relatively large section of ♂ reproductive system ventral to digestive tract if the latter is present; connected to gonopore via sperm duct. (paired, unpaired)

toe [G *Zehe*]: One of up to four variously shaped, movable projections at end of foot; distal end typically serves as opening for pedal glands. (conical, spine-shaped)

trochal disc [G *Trochalscheibe, Trochusscheibe*]: One of two discs formed by division of apical field, each bearing membranelles of trochus along margins; elevated on retractile pedicel.

trochus [G *Trochus*]: Preoral circlet of membranelles on head; formed by modified cilia of circumapical band anterior to mouth. (See also cingulum, paracingulum, paratrochus)

trophus [G *Trophus*]: One of seven variously elaborated and positioned main masticatory elements (fulcrum, manubria, rami, unci) in pharynx. Often used as synonym for mastax. (cardate, forcipate, fulcrate, incudate, malleate, ramate, submalleate, uncinate; symmetrical, asymmetrical)

trunk [G *Rumpf*]: Largest of three divisions (head, trunk, foot) of body bearing most internal organs and, in certain rotifers, with stiff integument (lorica).

tubule [G *Drüsenkanal*]: Relatively thick-walled, often coiled, looped, or forked, syncytial section of each protonephridium between capillary duct and cloaca; may bear long, driving cilia internally.

uncus [G *Uncus*]: In mastax, paired masticatory elements (trophi) anteriorly adjoining rami and typically at right angles to these. Form malleus together with manubrium.

urinary bladder [G *Protonephridial-blase, Harnblase*]: Unpaired contractile sac, ventral to cloaca, into which each protone-phridial tubule opens separately; empties into cloaca. (See also cloacal bladder)

vas deferens: sperm duct

ventral nerve: main ventral nerve

vesicular ganglion [G *Blasen-ganglion*]: Unpaired ventral ganglion adjacent to urinary bladder; located anterior to caudal ganglion at posterior end of main ventral nerve cords. (See also caudovesicular ganglion)

vestibule [G *Vestibulum*]: Inner region of funnel formed by coronal lobes; bears mouth at its base and is separated from upper infundibulum by diaphragm.

visceral muscle [G *Viszeralmuskel*]: Various muscles, either with or without nucleus, in wall of inner organ.

vitellarium [G *Vitellarium, Dotterstock*]: Large yolk-producing section of ♀ reproductive system between ovary and oviduct; located ventral to digestive tract and, together with ovary, forms germovitellarium. (paired, unpaired)

ACANTHOCEPHALA

acanthocephalan, spiny-headed worm
[G *Acanthocephale, Kratzer*]

ACANTHOCEPHALA

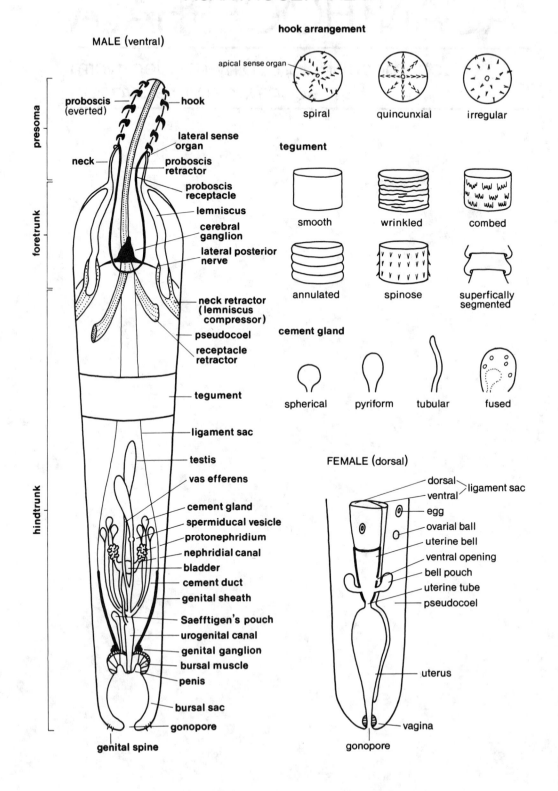

MALE (ventral)

presoma

- proboscis (everted)
- hook
- neck
- lateral sense organ
- proboscis retractor
- proboscis receptacle
- lemniscus
- cerebral ganglion
- lateral posterior nerve

foretrunk

- neck retractor (lemniscus compressor)
- pseudocoel
- receptacle retractor
- tegument

hindtrunk

- ligament sac
- testis
- vas efferens
- cement gland
- spermiducal vesicle
- protonephridium
- nephridial canal
- bladder
- cement duct
- genital sheath
- Saefftigen's pouch
- urogenital canal
- genital ganglion
- bursal muscle
- penis
- bursal sac
- gonopore
- genital spine

hook arrangement

apical sense organ

spiral   quincunxial   irregular

tegument

smooth   wrinkled   combed

annulated   spinose   superfically segmented

cement gland

spherical   pyriform   tubular   fused

FEMALE (dorsal)

- dorsal ligament sac
- ventral
- egg
- ovarial ball
- uterine bell
- ventral opening
- bell pouch
- uterine tube
- pseudocoel
- uterus
- vagina
- gonopore

ACANTHOCEPHALA

acanthocephalan, spiny-headed worm
[G *Acanthocephale, Kratzer*]

acanthella [G *Acanthella*]: Larval stage attained after acanthor larva passes through gut of intermediate host and begins to differentiate.

acanthor [G *Acanthor, Hakenlarve*]: First larval stage of acanthocephalan characterized by anterior rostellum and spinose body. This stage is maintained until larva passes through gut of intermediate host and may also refer to larva while still within egg. (See also acanthella)

aclid: Boring device on anterior end of acanthella.

apical sense organ [G *apikale Sinnespapille*]: Small pit or papilla located centrally at tip of proboscis; innervated by nerves running anteriorly from cerebral ganglion. (See also lateral sense organ)

basement layer: dermis

bell pouch (bell pocket) [G *Glockentasche*]: One of two anteriorly directed, pouch-like structures of efferent duct system in ♀. Ripe eggs pass from bell pouches into uterus; immature eggs move to ventral ligament sac through ventral opening of uterine bell.

binding layer: dermis

bladder (urinary bladder) [G *Blase, Nephridialsack*]: In certain acanthocephalans, expanded section of excretory system at junction of ligament sac and genital sheath in ♂ and dorsal outer rim of uterine bell in ♀; opens into uterus in ♀ and common sperm duct in ♂.

brain: cerebral ganglion

bursa [G *Bursa copulatrix, Bursa*]: Posteriormost muscular, eversible section of ♂ reproductive system surrounding penis at end of trunk; possesses sensory bulbs innervated by genital ganglion.

bursal muscle (muscular cap) [G *Bursalkappe, Muskelkappe*]: Thickened muscular layer capping anterior section of bursa in ♂; attached to posterior end of genital sheath and often providing bursa wall with muscular rays.

cement duct [G *Zementdrüsengang*]: Canal leading from each cement gland to common sperm duct; often united into one or two main ducts before entering ♂ reproductive system. If cement glands are fused, one duct leads to cement reservoir, two from reservoir to ♂ reproductive system.

cement gland [G *Zementdrüse*]: One of typically six or eight variously shaped unicellular glands in posterior third of trunk in ♂; open into sperm duct via cement duct. Cement glands may also be fused into one large mass. (clavate, fused, multinucleate, pyriform, rounded, tubular, uninucleate)

cement reservoir [G *Zementreservoir*]: In acanthocephalan with fused cement glands, expanded reservoir joined to gland mass by duct and itself connected to sperm duct via two ducts.

cerebral ganglion (ganglion, brain) [G *Proboscisganglion, Gehirn, Ganglion cephalicum*]: Main concentration of nerve tissue located along inner ventral wall of proboscis receptacle. Gives rise to array of nerves including lateral posterior nerves and those innervating apical and lateral sense organs. (See also genital ganglion)

circular muscle [G *Ringmuskel*]: Relatively thin layer of circular muscle fibers between longitudinal muscles and dermis in body wall (cuticle, epidermis, dermis, circular muscles, longitudinal muscles); may be divided into distinct rings if longitudinal muscles attach directly to epidermis in superficially segmented acanthocephalan. (See also tegument)

compressor of the lemnisci [G *Compressor lemniscorum*]: Those muscles of neck retractors encircling lemnisci over part of their length.

cuticle [G *Kutikula*]: Variously elaborated, thin, noncellular outermost layer of body wall (cuticle, epidermis, dermis, circular muscles, longitudinal muscles). (annulate, smooth, spinose, wrinkled) (See also tegument)

dermis (basement layer, binding layer) [G *Bindesubstanz*]: Relatively thin layer between epidermis and circular muscle layer of body wall (cuticle, epidermis, dermis, circular muscles, longitudinal muscles). (See also tegument)

efferent duct system [G *efferentes System*]: In ♀ reproductive system, collective term for all passages through which eggs move in going to outside of acanthocephalan.

egg [G *Ei*]: ♀ reproductive body arising from ovarian balls, developing in pseudocoel, and passing to exterior via uterine bell, bell pockets, uterine tube, uterus, and vagina. Ripe egg has several membranes, including shell, and contains acanthor.

epidermis (hypodermis) [G *Hypodermis*]: Relatively thick, differentiated layer between cuticle and dermis in body wall (cuticle, epidermis, dermis, circular muscles, longitudinal muscles); its inner layer contains lacunar system. (See also tegument)

flame bulb [G *Wimperkolben, Terminalorgan*]: One of several hundred variously arranged excretory structures containing row of cilia and forming proximal section of protonephridium.

foretrunk [G *Vorderrumpf*]: In certain acanthocephalans, enlarged anterior region of trunk. (See also hindtrunk)

genital ganglion [G *Genitalganglion*]: One of two smaller ganglia at base of penis; connected by commissure and innervating sensory bulbs of penis and bursa. (See also cerebral ganglion)

genital sheath [G *Genitalscheide, Ductus ejaculatorius, Leitungsschlauch*]: Muscular sheath in ♂ surrounding cement gland ducts, sperm ducts, and nephridial canals; continuous with dorsal ligament sac anteriorly and terminating posteriorly on bursal musculature.

genital spine [G *Rumpfstachel im Umkreis der Geschlechtsöffnung*]: In certain acanthocephalans, one in a series of spines surrounding gonopore.

gonopore [G *Geschlechtsöffnung, Genitalöffnung*]: Unpaired opening of reproductive system to exterior at or near posterior tip of trunk. Formed in ♂ by penis or rim of bursa, in ♀ by external opening of vagina.

hindtrunk [G *Hinterrumpf*]: In certain acanthocephalans, more slender posterior region of trunk. (See also foretrunk)

hook [G *Haken*]: One in an array of variously sized and distributed, curved, cuticle-covered elements of proboscis armature arising from dermis. May only refer to those larger projections with roots. (See also spine)

hypodermis: epidermis

invaginator muscle: proboscis retractor muscle

invertor muscle: proboscis retractor muscle

lacunar system [G *Lakunensystem, Hautgefäßsystem*]: System of fluid-filled channels, located in epidermis of body wall, with main longitudinal channels and either net-shaped or regularly spaced transverse channels.

lateral posterior nerve (main lateral posterior nerve) [G *lateraler Nervenstrang, Nervus lateralis posterior*]: Pair of longitudinal nerve cords; arise from posterior end of cerebral ganglion, pass through proboscis receptacle, and, enclosed in retinacula, run to lateral body wall where they continue to posterior end of trunk.

lateral sense organ [G *lateraler Sinnespapille, Halspapille*]: One of two small lateral organs located immediately posterior to last row of hooks on neck; innervated by nerves running anteriorly from cerebral ganglion. (See also apical sense organ)

lemniscus [G *Lemniskus, Lemnisk*]: Pair of elongate bodies suspended in pseudocoel of trunk. Formed by invagination of tegument between neck and proboscis; do not communicate with lacunar system of body wall.

ligament sac [G *Ligamentsack*]: In ♂, a single large, elongate, longitudinal sac; in ♀, a single or pair of sacs (dorsal and ventral ligament sacs) extending through trunk; attached anteriorly to end of proboscis receptacle, posteriorly to reproductive organs, which one sac largely envelops. (ephemeral, persistent)

ligament strand [G *Ligamentstrang*]: Nucleated strand between two ligament sacs or along ventral surface of single ligament sac; serves as site of attachment of reproductive organs.

longitudinal muscle [G *Längsmuskel*]: Relatively thin layer of longitudinal muscles underlying circular muscles and forming innermost layer of body wall (cuticle, epidermis, dermis, circular muscles, longitudinal muscles).

muscular cap: bursal muscle

neck [G *Hals*]: Variously elaborated posterior region (proboscis, neck) of presoma lacking hooks; retractile along with proboscis into trunk.

neck bulb [G *Halsbulbus*]: In certain acanthocephalans, bulbous enlargement of neck.

neck retractor muscle [G *Halsretraktor, Retraktormuskel des Halses*]: Longitudinal muscles spanning from posterior region of neck to trunk; lie outside proboscis retractor and receptacle protrusor muscles and envelop lemnisci. (See also compressor of the lemnisci)

nephridial canal (protonephridial duct) [G *Exkretionskanal, Nephridialkanal*]: Tubular section of each protonephridium between flame bulbs and bladder; may also refer to common canal formed by union of two protonephridia or to single canal leading from bladder to reproductive system.

ovarian ball [G *Ovarialball*]: One of numerous ♀ reproductive bodies floating in ligament sac or pseudocoel; produced by early fragmentation of ovary and giving rise to eggs.

penis [G *Penis*]: Conical, muscular projection of posterior end of urogenital canal; enclosed by bursa and bearing sensory bulbs.

presoma (praesoma) [G *Präsoma*]: Anterior and smaller of two major divisions (presoma, trunk) of body; consists externally of hook-bearing proboscis and smooth neck, internally of proboscis receptacle, lemnisci, and cerebral ganglion.

proboscis [G *Proboscis*]: Variously elaborated, spinose anterior section (proboscis, neck) of presoma; retractile along with neck into trunk.

proboscis receptacle (receptacle) [G *Proboscis-Receptaculum, Receptaculum*]: Muscular sac suspended into pseudocoel at anterior end of trunk; serves as receptacle for invaginated proboscis and neck and bears cerebral ganglion on inner ventral wall.

proboscis retractor muscle (invaginator muscle, invertor muscle) [G *Invaginatormuskel der Proboscis, Proboscisretraktor*]: Muscle or group of muscles spanning from tip of proboscis to proboscis receptacle along inner proboscis wall; passes in part through receptacle wall to form receptacle retractor muscles.

protonephridium [G *Protonephridium*]: In certain acanthocephalans, paired excretory organ on dorsal side of reproductive system. Each consists of variously arranged flame bulbs, nephridial canal, and either a common canal or bladder opening into posterior end of reproductive system. (branching, sacciform)

protonephridial duct: nephridial canal

pseudocoel [G *Leibeshöhle, Pseudocoel*]: Fluid-filled body cavity between ligament sac(s) or genital sheath and body wall in trunk and extending into presoma; may contain large number of eggs in ♀.

receptacle: proboscis receptacle

receptacle protrusor muscle [G *Protractor receptaculum*]: Longitudinal muscle spanning from wall of neck to posterior region of proboscis receptacle. (dorsal, lateral, ventral)

receptacle retractor muscle [G *Retraktor receptaculum*]: Muscle or group of muscles, forming continuation of proboscis retractor muscles, spanning from wall of proboscis receptacle to wall of trunk. (dorsal, ventral)

retinaculum [G *Retinaculum*]: One of two muscle sheaths extending from proboscis receptacle to body wall of trunk and containing lateral posterior nerves.

retractor muscle: neck retractor muscle, proboscis retractor muscle, receptacle retractor muscle.

root [G *Hakenwurzel*]: Enlarged section of hook embedded in proboscis wall.

rostellum [G *Rostellum*]: Anterior end of acanthor larva consisting of a crown of three pairs of larval hooks.

Saefftigan's pouch [G *Markbeutel*]: Conspicuous, muscular, fluid-filled, elongated sac within genital sheath in ♂; communicates with cavities in bursal cap and may serve to evert bursa.

seminal vesicle (spermiducal vesicle) [G *Samenblase, Vesicula seminis*]: Expanded section of sperm duct (vas deferens, vas efferens) in ♂ reproductive system.

sensory bulb [G *Sinnespapille*]: One of a number of bulbous nerve endings on rim of bursa or in penis.

sperm duct: vas deferens, vas efferens

spermiducal vesicle: seminal vesicle

spine [G *Stachel*]: Smaller curved, cuticle-covered projection of trunk; lacks root. Often incorrectly used as synonym for hook.

tegument [G *Tegument*]: Preferred term for outer syncytial layer of body wall. Includes, from outer to inner layer, glycocalyx/epicuticle, cuticle, epidermis, dermis.

testis [G *Testis, Hoden*]: Paired, asymmetrically arranged (in tandem: typically one more anterior, one more posterior) section of ♂ reproductive system; each is connected to common sperm duct or seminal vesicle via separate sperm duct. (diorchic, monorchic)

trunk [G *Rumpf*]: Posterior and larger of two major divisions (presoma, trunk) of body; may be differentiated into broader foretrunk and more slender hindtrunk and ornamented with spines.

urinary bladder: bladder

urogenital canal [G *Urogenitalgang*]: In ♂, common canal formed in genital sheath by entrance of nephridial canals and cement ducts into common sperm duct; in ♀, section of reproductive system posterior to junction of nephridial canals and uterine tube.

uterine bell [G *Uterusglocke*]: Muscular, funnel- or tube-shaped proximal section of ♀ reproductive system continuous with ligament sac. Captures eggs and passes them to selector apparatus; immature eggs pass through posterior ventral opening of uterine bell into ventral ligament sac or pseudocoel.

uterine tube [G *Uterusgang*]: Narrowed section of ♀ reproductive system between uterine bell and uterus and bearing bell pouches; nephridial canals, when present, enter uterine tube.

uterus [G *Uterus*]: Elongate, muscular section of ♀ reproductive system between uterine tube and vagina.

vagina [G *Vagina*]: Short, nonmuscular section of ♀ reproductive system following uterus and opening to exterior via gonopore.

vas deferens [G *Vas deferens*]: In ♂ reproductive system, common sperm duct formed by merger of two vasa efferentia. Enclosed in genital sheath.

vas efferens [G *Vas efferens*]: In ♂ reproductive system, duct leading from each testis; merges with vas efferens of second testis to form common sperm duct (vas deferens).

ECHIURIDA

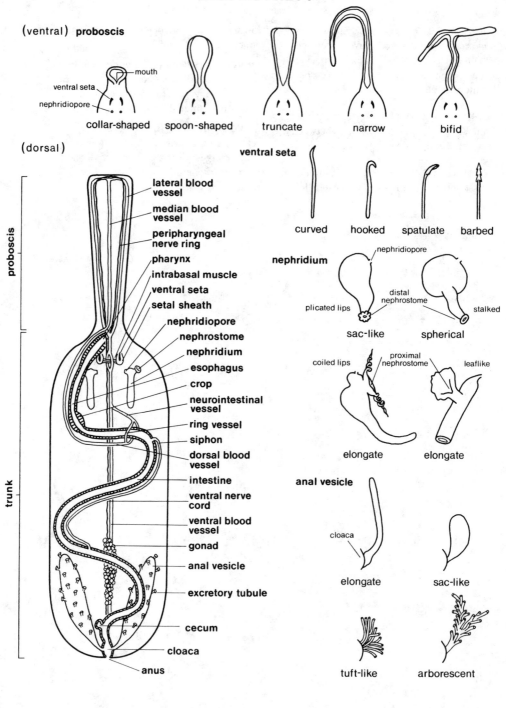

(ventral) **proboscis**

mouth
ventral seta
nephridiopore

collar-shaped   spoon-shaped   truncate   narrow   bifid

(dorsal)

ventral seta

curved   hooked   spatulate   barbed

proboscis

lateral blood vessel
median blood vessel
peripharyngeal nerve ring
pharynx
intrabasal muscle
ventral seta
setal sheath
nephridiopore
nephrostome
nephridium
esophagus
crop
neurointestinal vessel
ring vessel
siphon
dorsal blood vessel
intestine
ventral nerve cord
ventral blood vessel
gonad
anal vesicle
excretory tubule
cecum
cloaca
anus

trunk

nephridium

nephridiopore
distal nephrostome
plicated lips
stalked
sac-like   spherical

coiled lips
proximal nephrostome
leaflike
elongate   elongate

anal vesicle

cloaca
elongate   sac-like

tuft-like   arborescent

smooth

with papillae

with anal setae

ECHIURIDA

echiuran, spoon-worm [G *Echiuride, Igelwurm, Stachelschwanz*]

anal seta [G *Analborste*]: One or two rings of setae surrounding anus at posterior end of trunk in certain echiurans.

anal vesicle [G *Analschlauch, Rektalsäckchen*]: Excretory organ consisting of a pair of variously shaped structures opening to exterior via cloaca; typically bears excretory tubules. (arborescent, bush-like, elongate, sac-like)

androecium [G *Androecium*]: Specially developed section of nephridium in ♀ of certain echiurans containing dwarf ♂.

anterior seta: ventral seta

anus [G *After*]: Posterior opening of digestive tract; located terminally on trunk.

cecum(caecum) [G *Caecum, Coecum*]: Blind outpocketing of intestine anterior to cloaca.

ciliated funnel [G *Wimpertrichter*]: One of numerous ciliated, funnel-shaped structures projecting into body cavity from excretory tubules on anal vesicles.

ciliated groove [G *Flimmerrinne*]: Ciliated groove extending along ventral surface of intestine as far as cecum; may be elevated or associated with siphon.

circular muscle [G *Ringmuskel*]: Outermost layer of muscle of trunk. (See also longitudinal muscle, oblique muscle)

cloaca (rectum) [G *Enddarm, Rektum*]: Terminal section of digestive tract into which anal vesicles open.

collateral intestine: siphon

crop [G *Kropf*]: Section of digestive tract between gizzard and stomach; typically bears longitudinal markings and is surrounded by ring vessel.

diaphragm [G *Diaphragma*]: Thin membrane anterior to sheath of ventral setae; incompletely separates an anterior coelom from general body cavity in certain echiurans.

diverticulum: cecum

dorsal blood vessel [G *Rückengefäß*]: Dorsal contractile blood vessel often running parallel to foregut; joined to ventral blood vessel by neurointestinal vessel and extending into proboscis via median blood vessel.

esophagus [G *Ösophagus*]: Section of digestive tract between pharynx and gizzard.

excretory tubule [G *Trichterstiel*]: One of numerous small, variously arranged tubules on anal vesicles; bears ciliated funnel distally.

fascicle [G *Bündel*]: Elevated bundle of oblique musculature on inner wall of trunk in certain echiurans; usually in a longitudinal series.

foregut [G *Vorderdarm*]: Anterior section of digestive tract; typically consists of pharynx, esophagus, gizzard, stomach or crop. (See also hindgut, midgut)

genital groove [G *Genitalfurche*]: Groove extending from nephridiopore(s) to mouth on outer ventral surface of trunk in certain echiurans. (with/without setae)

gizzard [G *Kaumagen*]: Muscular, often bulbous and ringed section of digestive tract between esophagus and stomach.

gonad [G *Gonade*]: Aggregation of ♂ or ♀ reproductive cells at posterior end of trunk; located along midventral membrane.

gonoduct: nephridium

gutter [G *Rinne*]: Trough formed by ventrally upturned edges of proboscis.

heart: dorsal blood vessel

hindgut: cloaca

intestine [G *Mitteldarm*]: Long, highly coiled section of digestive tract between stomach and cloaca; one may distinguish a presiphonal, siphonal, and postsiphonal region according to length of accompanying siphon.

interbasal muscle [G *interbasaler Borstenmuskel*]: Short muscle connecting sheaths of the two ventral setae in certain echiurans.

lappet [G *Gabelast*]: Term applied to each lateral arm of a bifid proboscis if these are short.

lateral blood vessel [G *laterales Blutgefäß, Randgefäß*]: Blood vessels running along lateral margins of proboscis; arising from bifurcation of median blood vessel and merging with ventral blood vessel in mouth region.

longitudinal muscle [G *Längsmuskel*]: Intermediate layer of muscle in trunk; may be concentrated in bands visible on exterior. (See also circular muscle, oblique muscle)

median blood vessel [G *medianes Kopflappengefäß, Mediangefäß*]: Extension of dorsal blood vessel into proboscis; bifurcates distally to form lateral blood vessels.

midgut: intestine

mouth [G *Mund*]: Anterior opening of digestive tract near junction of proboscis and trunk.

muscular pad [G *Muskelkissen*]: Muscle concentration(s) in which ventral setae of certain echiurans are embedded.

nephridiopore [G *Geschlechtsöffnung*]: Opening of nephridium to exterior on anterior ventral surface of trunk; often indistinct.

nephridium [G *Metanephridium, Nephridium*]: One of typically two variously elaborated vessels in anterior end of trunk; opens to exterior via ventral nephriodiopore and into body cavity via nephrostome. Temporarily stores sperm and eggs.

nephrostomal lip [G *Trichterlappen*]: Variously shaped tissue surrounding nephrostome.

nephrostome [G *Coelomtrichter*]: Variously elaborated (nephrostomal lips) and often elevated opening of nephridium into body cavity; entrance into nephridium for sperm or eggs. (basal = proximal, terminal = distal)

nerve cord [G *Bauchnervenstrang, Ventralnerv*]: Single longitudinal, unsegmented ventromedian nerve cord parallel to ventral blood vessel in trunk; bifurcates at base of proboscis and runs along its margins.

net gland: slime gland

neurointestinal blood vessel [G *Darmgefäß*]: Blood vessel branching from ventral blood vessel posterior to ventral setae and connected either directly or indirectly to dorsal blood vessel.

oblique muscle [G *Schrägmuskel, Diagonalmuskel*]: Innermost layer of muscle in trunk; may form fascicles between longitudinal muscle bands in certain echiurans. (See also circular muscle, longitudinal muscle)

papilla [G *Papille*]: Type of ornamentation on outer surface of trunk; variously shaped, often associated with glandular cells, and often most prominent in anterior and posterior regions of trunk.

pharynx [G *Pharynx*]: Anteriormost section of digestive tract; located between mouth and esophagus.

proboscis [G *Kopflappen, Prostomium*]: Narrowed, variously elaborated, contractile anterior division of body (proboscis, trunk); typically bears a ventral groove leading to mouth at junction of proboscis and trunk.

reserve seta [G *Ersatzborste*]: Fresh seta, complete with sheath, embedded in trunk and emerging to replace functioning ventral seta.

ring vessel [G *Darmgefäßring*]: Blood vessel formed by bifurcation of neurointestinal vessel; encircles crop and merges with dorsal blood vessel.

segmental organ: nephridium

seta: anal seta, ventral seta

sheath (of seta) [G *Borstentasche, Borstenfollikel*]: Muscular sheath enclosing base of each ventral seta.

siphon (collateral intestine) [G *Nebendarm*]: Narrow tube accompanying intestine over a considerable part of its length.

slime gland (net gland): Glands on anterior part of trunk in certain echiurans; produce a mucus net used in feeding.

stomach [G *Magen*]: Posteriormost section of foregut.

trunk [G *Rumpf*]: Broad posterior division (proboscis, trunk) of body.

ventral blood vessel [G *Rückengefäß*]: Longitudinal blood vessel accompanying ventral nerve cord; branches anteriorly near ventral setae to form neurointestinal vessel.

ventral groove [G *Wimperstreifen*]: Ciliated groove along ventral surface of proboscis; leads to mouth.

ventral seta [G *ventrale Borste, Bauchborste*]: One of typically two movable setae emerging posterior to mouth on ventral surface of trunk. (See also sheath)

SIPUNCULIDA

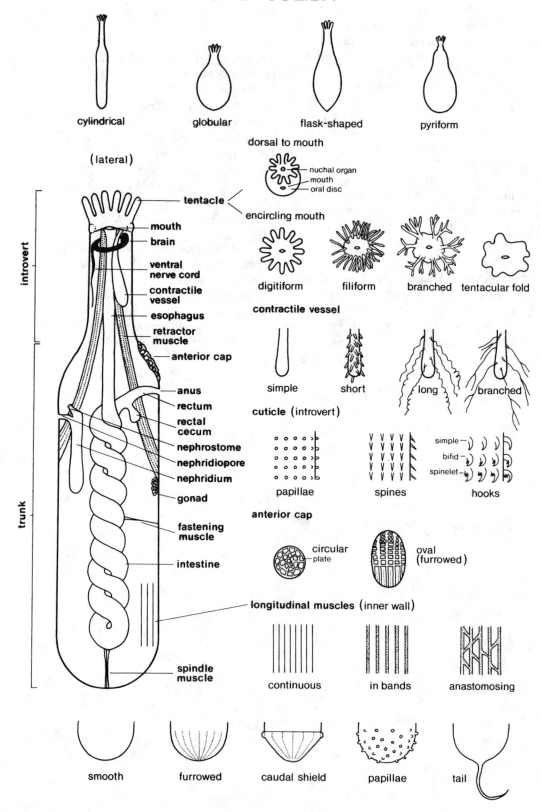

cylindrical

globular

flask-shaped

pyriform

(lateral)

dorsal to mouth

nuchal organ
mouth
oral disc

tentacle

encircling mouth

mouth

brain

ventral
nerve cord

contractile
vessel

esophagus

retractor
muscle

anterior cap

digitiform

filiform

branched

tentacular fold

contractile vessel

anus

rectum

rectal
cecum

nephrostome

nephridiopore

nephridium

gonad

simple

short

long

branched

cuticle (introvert)

papillae

spines

simple
bifid
spinelet

hooks

anterior cap

fastening
muscle

intestine

circular
plate

oval
(furrowed)

longitudinal muscles (inner wall)

spindle
muscle

continuous

in bands

anastomosing

smooth

furrowed

caudal shield

papillae

tail

introvert

trunk

SIPUNCULIDA

sipunculan, sipunculid [G *Sipunculide, Spritzwurm*]

ampullary base [G *Tentakelbasis*]: Enlarged proximal section of tentacle.

anal gland: racemose gland

anal shield: anterior cap

anterior cap (anal shield) [G *Afterschild, Analschild, Vorderschild*]: Thick, hardened structure on outer surface of trunk in certain sipunculans. Located either somewhat laterally, anterior to anus, or terminally; in the former, introvert emerges ventral to cap, in the latter through middle of cap. (circular, oval; furrowed) (See also caudal shield)

anus [G *After*]: Posterior opening of digestive tract; typically located on anterior dorsal surface of trunk, occasionally on introvert.

benthosphaera [G *Benthosphaera*]: Benthic form of secondary larva. (See also pelagosphaera)

brain [G *Gehirn*]: Dorsal bilobed main concentration of nerve tissue in introvert posterior to mouth; connected to single median ventral nerve cord by pair of circumesophageal commissures. (See also cerebral organ, digitate process)

caudal shield [G *Caudalschild, Schwanzschild*]: Thick, hardened structure on posterior end of trunk. (flat, subconical) (See also anterior cap)

cecum: rectal cecum

cephalic collar [G *Unterlippe, Hautfalte*]: collar

cephalic tube [G *Zerebraltubus, Zerebralgang*]: Epidermis-lined tube leading from dorsal side of oral disc to cerebral organ of brain. (median, unpaired; lateral, paired)

cerebral organ [G *Zerebralorgan, dorsales Sinnesorgan*]: Anterior neurosecretory modification between dorsofrontal epidermis and brain; typically associated with cephalic tube. (paired, unpaired)

cervical collar: collar

chloragogue cell (chloragen cell) [G *Chloragogen-Zelle*]: Modified cell type coating outer surface of intestine or contractile vessels. (cylindrical, pyruvate)

ciliated groove [G *Ventralrinne*]: Ciliated groove extending ventrally along greater part of digestive tract. Also refers to ciliated groove on tentacle surface facing mouth.

circular muscle [G *Ringmuskel*]: Outermost muscle layer of body wall (cuticle, epidermis, dermis; circular, oblique, longitudinal muscles); may be visible on exterior as distinct rings.

circumenteric connective: circumesophageal commissure

circumesophageal commissure (circumenteric connective) [G *Schlundkommissur, Schlundring*]: One of two strands of nerve tissue joining ventral nerve cord to brain and forming ring around esophagus.

coelomic canal [G *Hautkanal, Coelomkanal, Kutiskanal*]: Network of canals or spaces of body wall below epidermis; communicates with secondary body cavity via pores.

coelomic papilla [G *Coelothel-Papille*]: One in a field of small, leaf-like processes on inner wall of trunk, typically anterior to nephridia.

collar [G *Kragen*]: Smooth zone around anterior end of introvert behind tentacles; may bear fold (cephalic collar) anteriorly and posteriorly (cervical collar).

compensatory sac: contractile vessel

contractile tubule (diverticulum, villus) [G *Poli'scher Divertikel, Villus*]: One in a series of branches of contractile vessel. (short, long, branched)

contractile vessel (compensatory sac, polian vessel) [G *Poli'scher Schlauch, kontraktile Blase, Ausgleichs-Gefäß*]: Single (dorsal) or paired (lateral), fluid-filled tubes attached to esophagus; communicates with hollow tentacles anteriorly and is blind posteriorly. (See also contractile tubule)

cuticle [G *Kutikula*]: Noncellular outermost layer of body wall (cuticle, epidermis, dermis; circular, oblique, longitudinal muscles); devoid of chitin. Thicker on trunk, thinner on introvert and tentacles.

dermis [G *Grundsubstanz, peripheres Bindegewebe*]: Layer of body wall between epidermis and circular muscles; may contain coelomic canals.

digitate process (papilliform process) [G *Frontallappen, Zerebrallappen, fingerförmiger Sinnesorgan*]: One in a series of projections from anterior frontal surface of brain. (finger-shaped, leaf-shaped)

dissepiment [G *Dissepiment, Falte*]: In certain sipunculans, one in a series of transverse folds of inner trunk wall.

dorsal vessel: contractile vessel

epidermis [G *Epidermis*]: Layer of body wall between cuticle and dermis; secretes cuticle, bears numerous glands, and is involved in formation of cuticular projections.

esophagus (oesophagus) [G *Ösophagus*]: Section of digestive tract between pharynx and intestine.

eye (pigment-cup ocellus) [G *Pigmentfleck, Augenfleck*]: One of typically two pigment spots at base of ocular tube in brain.

fastening muscle (fixing muscle) [G *Muskelfädchen*]: Thin, short muscle filament attaching digestive tract (e.g., esophagus, posterior intestine) to body wall.

fixed urn: urn

free urn: urn

frontal organ: digitate process

glans [G *Eichel*]: In certain sipunculans, acorn-shaped posterior end of trunk.

gonad [G *Gonade*]: Cluster of ♂ or ♀ reproductive cells at base of retractor muscles on body wall of trunk.

holdfast (adhesive papilla, attaching papilla) [G *Haftpapille, Drüsenpapille*]: Hardened, often horseshoe-shaped, modified papilla on outer surface of trunk; serves to anchor sipunculan.

hook [G *Haken*]: Curved, typically hollow type of cuticular ornamentation on outer surface of introvert or trunk. (single tip, bifid tip, with spinelets)

intestine [G *Mitteldarm, Intestinum*]: Prominent, U-shaped, and spirally wound section of digestive tract between esophagus and rectum.

introvert [G *Introvert, "Rüssel"*]: Narrowed, invaginable, often ornamented anterior division of body (introvert, trunk); bears mouth and tentacles distally.

longitudinal muscle [G *Längsmuskel*]: Innermost muscle layer of body wall (circular, oblique, longitudinal muscles); may be visible on exterior as distinct bands. (anastomosing, continuous, in bands)

nephridiopore [G *Nephroporus*]: Opening of each nephridium to exterior; usually located ventrally at anterior end of trunk.

nephridium (metanephridium) [G *Nephridium, Metanephridium, Segmentalorgan*]: One of typically two tubular excretory organs located ventrally at anterior end of trunk; each opens to exterior via separate nephridiopore, into body cavity via nephrostome. (attached, free).

nephrostome [G *Nephridialtrichter*]: Ciliated opening of nephridium into body cavity; typically located at anterior end of nephridium and closely adjoining body wall.

nerve cord: ventral nerve cord

nuchal organ [G *Nuchalorgan, Nackenorgan*]: Richly innervated, ciliated protuberance; located on oral disc, typically dorsal to and immediately adjoining tentacles or, if these are displaced dorsally, within tentacle crown. (bilobed, quadrilobed)

oblique muscle [G *Schrägmuskel, Diagonalmuskel*]: Diagonal muscle layer between circular and longitudinal muscles of body wall.

ocular tube [G *Augentubus*]: Tubular depression in brain containing vitreous body and photosensitive pigment basally (eye); extends from oral disc to brain or associated with cephalic tube. (paired, unpaired)

oesophagus: esophagus

oral disc [G *Mundscheibe, Tentakelscheibe*]: Anteriormost section of introvert bearing mouth, tentacles, and nuchal organ.

papilla [G *Papille*]: Type of surface ornamentation on trunk or introvert; often associated with glandular and sensory cells and most prominent on anterior and posterior regions of trunk. (See also holdfast, hook, spine)

papilliform process: digitate process

paraneural muscle [G *paraneuraler Muskel*]: Pair of longitudinal muscles along each side of ventral nerve cord anteriorly.

pelagosphaera [G *Pelagosphaera*]: Pelagic form of secondary larva. (See also benthosphaera)

pharynx [G *Pharynx*]: Muscular section of digestive tract between mouth and esophagus in introvert.

pigment-cup ocellus: eye, ocular tube

polian tubule: contractile tubule

polian vessel: contractile vessel

postesophageal loop [G *vordere Darmschleife*]: Additional loop of digestive tract between esophagus and major intestinal coils.

primary tentacle: tentacle

protractor muscle [G *Introvert-Protraktor*]: Occasionally present, additional (third) pair of retractor muscle bands spanning from introvert near brain to anterior wall of trunk.

racemose gland (rectal gland) [G *Analdrüse, büschelförmiger Körper*]: Group of bushy glandular structures on each side of rectum.

rectal cecum (rectal caecum) [G *Rektalcaecum, Blindfortsatz*]: Blind outpocketing of rectum at end of ciliated groove.

rectum [G *Enddarm*]: Posteriormost section of digestive tract between intestine and anus; may bear rectal cecum and/or racemose gland.

retractor muscle [G *Retraktor, Rückziehmuskel*]: Single, one pair (ventral), two pairs (dorsal and ventral), or even three pairs of longitudinal muscle bands spanning from anterior end of introvert to inner trunk wall; serves to invaginate introvert. Posterior point of attachment of ventral muscles is site of gonads.

secondary tentacle: tentacle

skin body [G *Hautkörper, Papille*]: General term for groups of glandular cells (glandular papilla) on outer surface of trunk.

spindle muscle [G *Spindelmuskel*]: Long, thin muscle spanning from anterior to posterior end of trunk; runs along center of intestinal coils and supports digestive tract.

spine [G *Stachel*]: Straight, typically hollow, spine-shaped type of cuticular ornamentation on outer surface of introvert or trunk. (See also hook, papilla)

spinelet [G *Nebenzahn*]: One in a series of several small spines projecting from base of hook.

stomach [G *Magen*]: Term occasionally applied to section of digestive tract between esophagus and intestine bearing internal longitudinal ridges.

tentacle [G *Tentakel*]: Variously shaped and arranged hollow projections located at anterior end of introvert (oral disc) and bearing ciliated groove. If arranged in concentric circlets, one may distinguish primary (innermost), secondary, and tertiary (outermost) tentacles. (dendritically branched, palmately branched, conical, digitiform, filiform, leaf-like, lobate)

tentacular fold [G *Tentakularmembran, Tentakelfalte*]: Fused, membrane-like elaboration of tentacles surrounding oral disc. (foliaceous, scalloped)

terminal organ [G *Terminalorgan*]: Small, innervated invagination with glands at posterior end of trunk in benthosphaera and pelagosphaera larva as well as occasionally in adult.

tertiary tentacle: tentacle

trunk [G *Rumpf*]: Posterior broader division of body (introvert, trunk) into which introvert may be invaginated.

urn [G *Urne, Wimperurne*]: Vase-shaped cluster of cells bearing prominent terminal ciliated cell; either attached to inner coelomic wall (fixed urns) or detached in body cavity (free urns).

ventral nerve cord (nerve cord) [G *Ventralnervenstrang, Bauchmark*]: Single unsegmented ventromedian nerve cord extending to posterior end of trunk; connected to brain by pair of circumesophageal commissures.

villus: contractile tubule

MOLLUSCA

CAUDOFOVEATA
(CHAETODERMOMORPHA)

caudofoveate [G   *Caudofoveat, Schildfüßer*]

CAUDOFOVEATA

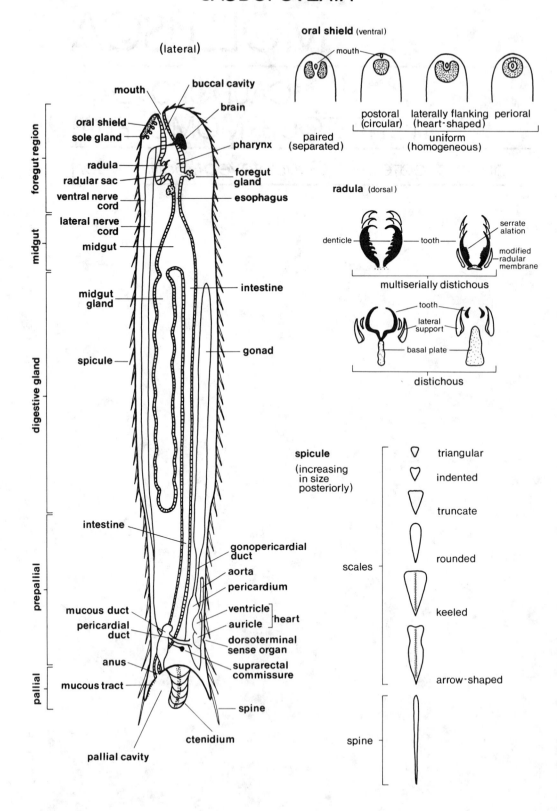

(lateral)

foregut region

midgut

digestive gland

prepallial

pallial

mouth
buccal cavity
brain
oral shield
sole gland
pharynx
radula
foregut gland
radular sac
ventral nerve cord
esophagus
lateral nerve cord
midgut
midgut gland
intestine
spicule
gonad
intestine
gonopericardial duct
aorta
pericardium
mucous duct
ventricle
auricle
heart
pericardial duct
dorsoterminal sense organ
anus
suprarectal commissure
mucous tract
spine
ctenidium
pallial cavity

oral shield (ventral)

mouth

paired (separated)

postoral (circular)

laterally flanking (heart-shaped)

perioral

uniform (homogeneous)

radula (dorsal)

denticle — tooth
serrate alation
modified radular membrane

multiserially distichous

tooth
lateral support
basal plate

distichous

spicule
(increasing in size posteriorly)

scales

triangular
indented
truncate
rounded
keeled
arrow-shaped

spine

MOLLUSCA

CAUDOFOVEATA (CHAETODERMOMORPHA)

caudofoveate [G *Caudofoveat, Schildfüßer*]

anal chamber: pallial cavity

anterior constriction: Ring-like constriction of body at anterior end between foregut and midgut regions (between neck and anterior trunk in alternate designation of body regions).

anterior retractor [G *vorderer Retraktormuskel, Rückzieher der Fußschildregion*]: One of several pairs of muscle bands arising from longitudinal muscle layer of body wall and extending close to oral shield. (See also gill retractor)

anterior trunk: In alternate designation of body regions (anterium, neck, anterior trunk, posterior trunk, posterium), external region from anterior constriction to anterior end of digestive (midgut) gland/gonad region. (See also body regions)

anterium: In alternate designation of body regions (anterium, neck, anterior trunk, posterior trunk, posterium), anteriormost, protrusible external region bearing oral shield; usually bears few spicules and is often retracted. (See also body regions)

anus [G *After, Anus*]: Posterior opening of digestive tract; located below ctenidia in lower region of pallial cavity.

aorta [G *Aorta*]: Blood vessel leading from anterior chamber (ventricle) of heart forward into anterior region of body; may be separated from ventricle by distinct sphincter or aortal bulb.

auricle (atrium) [G *Atrium, Aurikel, Vorkammer*]: One of typically two muscular posterior chambers of heart; receives blood from ctenidia and pumps it anteriorly to ventricle via each auriculoventricular opening. (single, paired)

auriculoventricular opening (atrioventricular opening) [G *Atrioventricularöffnung*]: In heart, opening leading from posterior auricle(s) to anterior ventricle; may be equipped with sphincter. (single, paired)

basal cone (basal plate) [G *Basalplatte*]: In certain caudofoveates, thick, cone-shaped chitinous support for reduced single pair of radular teeth.

body regions: Two systems of body region designations have been introduced. One is based on internal organs and recognizes a foregut, midgut, digestive gland, prepallial, and pallial region. The second is based on body wall musculature and recognizes an anterium, neck, anterior trunk, posterior trunk, and posterium, as well as an anterior constriction between neck and anterior trunk. The anterium of the latter system has no counterpart in the former. Approximate correspondence of other regions: neck = foregut region (may include part of stomach), anterior trunk = midgut region in part, posterior trunk = digestive gland/gonad region, posterium = prepallial and pallial regions.

bolster [G *Radulapolster*]: In radular apparatus, one of two oval bodies laterally underlying radular sheath. Serves as internal support and as site of muscle attachment.

brain (cerebral ganglion, cephalic ganglion, supraenteric ganglion) [G *Gehirn, Zerebralganglion, Oberschlundganglion*]: One of a pair of main concentrations of nerve tissue above pharynx; bears precerebral ganglia anteriorly and gives rise posteriorly to pair of buccal connectives and pair of lateral and ventral nerve cords. (separated, fused-bilobed) (See also cerebrobuccal connective, cerebrolateral connective, and cerebroventral connective)

buccal cavity [G *Mundhöhle*]: Cuticle-lined section of digestive tract between mouth and pharynx; may be protrusible.

buccal connective: cerebrobuccal connective

buccal ganglion [G *Bukkalganglion*]: Ganglionic swelling at end of each cerebrobuccal connective; closely associated with radular apparatus. (See also lateral ganglion, ventral ganglion)

buccal plate: oral shield

buccal spatula: jaw

cerebral ganglion: brain

cerebral nerve [G *Gehirnnerv, Zerebralnerv*]: One in a series of anteriorly directed nerves arising from precerebral ganglia and innervating oral shield and oral region.

cerebrobuccal connective (buccal connective) [G *Bukkalkonnektiv*]: One of two relatively small strands of nerve tissue originating at brain (typically from common root with cerebrolateral and cerebroventral connectives); each gives rise to buccal ganglion and may be joined to a

second cerebrobuccal connective by commissure.

cerebrolateral connective (lateral connective) [G *Lateralkonnektiv, Zerebrolateralkonnektiv*]: Short part of each lateral nerve cord extending laterally from brain; typically has common root with cerebrobuccal and cerebroventral connectives. Continues posteriorly as lateral nerve cord proper.

cerebropedal connective: cerebroventral connective

cerebroventral connective (cerebropedal connective, ventral connective) [G *Ventralkonnektiv, Zerebroventralkonnektiv*]: Part of each ventral nerve cord extending ventrally from brain (around foregut); typically has common root with cerebrobuccal and cerebrolateral connectives. Ends in ventral ganglion and continues posteriorly as ventral cord proper.

circular muscle [G *Ringmuskel*]: Outermost muscle layer of body wall (cuticle, epidermis; circular, diagonal, longitudinal muscles).

cloaca (cloacal chamber, cloacal cavity): pallial cavity

ctenidium (gill) [G *Kieme, Ctenidiumkieme*]: One of two well-developed respiratory organs in pallial cavity; consists of central axis and alternating series of lamellae.

ctenidium retractor: gill retractor

cuticle [G *Kutikula*]: Outermost noncellular layer of body wall (cuticle, epidermis; circular, diagonal, longitudinal muscles); encloses calcareous spicules on exterior. Secreted by epidermis.

denticle [G *Zähnchen, Dentikel, Nebenhaken*]: One in a series of posteriorly directed projections of each radular tooth; may also refer to small remaining elements in reduced type of radula.

dermis [G *Bindegewebe*]: Layer of connective tissue occasionally present between epidermis and musculature.

diagonal muscle [G *Diagonalmuskel*]: Layer of two sets of muscles at right angles to each other between outer circular and inner longitudinal muscles of body wall (cuticle, epidermis; circular, diagonal, longitudinal muscles).

digestive gland: midgut gland

digestive gland region [G *Mitteldarmdrüsenregion*]: Typically widest and longest externally visible region of body (foregut, midgut, digestive gland, prepallial, and pallial regions); corresponds internally with expanded midgut gland. (See also body regions)

dome-shaped membrane: lateral support

dorsocaudal sensory groove: dorsoterminal sense organ

dorsoterminal sense organ (terminal sense organ, dorsocaudal sensory groove, osphradial sense organ) [G *Dorsoterminalorgan, dorsoterminales Sinnesorgan; osphradiales Sinnesorgan*]: Longitudinal sensory depression at posterior end of body; located middorsally, at border of pallial cavity. Lacks cuticle or spicules and is innervated by suprarectal commissure.

dorsoventral muscle (laterodorsal-ventral muscle, oblique muscle) [G *Dorsoventralmuskel*]: In certain caudofoveates, poorly developed muscle bands spanning from laterodorsal body wall to ventral portion in anterior region of body.

epidermal papilla (giant cell) [G *Epidermalpapille, Riesenzelle, Papille*]: In mantle, vesicle of fused epidermal cells extending into cuticle and eventually reaching surface.

epidermis [G *Epidermis*]: Outermost cell layer of body wall (cuticle, epidermis; circular, diagonal, longitudinal muscles); forms mantle, contains gland cells and papillae, and secretes both cuticle and spicules.

esophagus (oesophagus) [G *Ösophagus*]: Relatively narrow section of digestive tract between pharynx (radula) and midgut; may be separated from midgut by sphincter.

foot shield: oral shield

foregut [G *Vorderdarm*]: Anterior, often cuticle-lined region of digestive tract (foregut, midgut, hindgut) consisting of buccal cavity, pharynx, radula, and esophagus; ventral foregut glands typically present. May be eversible due to well-developed radial and anterior retractor muscles.

foregut gland (salivary gland) [G *Vorderdarmdrüse*]: Aggregation of pair of glands opening ventrally and/or dorsally into pharynx close to radula.

foregut region [G *Vorderdarmregion*]: Anteriormost region of body (foregut, midgut, digestive gland, prepallial, and pallial regions); corresponds internally with foregut and visible on exterior due to slender diameter. (See also body regions)

gametoduct: lower gametoduct, upper gametoduct

gametopore: Opening of each gametoduct (= pericardioduct + mucus duct) to exterior in pallial cavity. Closed by sphincter; in ♂ often on papilla.

gastric shield [G *Magenschild*]: In digestive tract of certain caudofoveates, cuticular plate between midgut and intestine.

gill: ctenidium

gill retractor (ctenidium retractor) [G *Kiemenratrektor*]: One of six or more bands of muscle arising from longitudinal muscle layer of body wall and extending to ctenidia. (See also anterior retractor)

glandular duct: mucous duct

gonad [G *Gonade*]: ♂ (testis) or ♀ (ovary) reproductive organ. Consists of elongate tube dorsal to intestine and opens into pericardium via short gonoducts (gonopericardial ducts). Seminal products travel through pericardium, pericardioduct, mucous duct, and, in ♀, onto mucous tract in pallial cavity. (single. paired)

gonoduct: gonopericardial duct

gonopericardial duct (gonopericardial canal) [G *Gonoperikardialgang*]: Short, somewhat narrowed section of reproductive system between gonad and pericardium. (single, paired)

heart [G *Herz*]: In circulatory system, contractile organ within dorsal pericardium; consists of anterior ventricle and posterior auricle(s).

hindgut [G *Enddarm*]: Posteriormost of three regions of digestive tract (foregut, midgut, hindgut); also referred to as intestine (including rectum).

horizontal muscle [G *Horizontalmuskel*]: Transverse sheet of horizontal muscle forming dorsal boundary of ventral sinus.

intestine (hindgut, rectum) [G *Enddarm, Intestinum*]: Narrow, elongate section of digestive tract between midgut and anus; bends ventrally in front of pallial cavity.

jaw (mandible, buccal spatula) [G *Kieferspange, Bukkalspange, Mandibel*]: In certain caudofoveates, cuticular and chitinized element on each side of pharynx anterior to radula.

kidney: pericardioduct

lamella [G *Lamelle*]: One in a series of several to many alternating, flattened projections on each side of ctenidium.

lateral connective: cerebrolateral connective

lateral ganglion [G *Lateralganglion*]: Ganglionic swelling near origin of lateral nerve cord at brain. (See also buccal ganglion, ventral ganglion)

lateral nerve cord [G *lateraler Längsnervenstrang*]: Pair of longitudinal nerve cords extending from brain to suprarectal commissure after merging with ventral nerve cord; connected in anterior section to ventral cord by lateral commissures. Anteriormost section of nerve surrounding foregut and before lateral ganglion is termed cerebrolateral connective.

lateral support [G *Lateralplatte*]: In radular apparatus with basal cone, thickened lateral element of cuticle along outer side of each bolster.

laterodorsal-ventral muscle: dorsoventral muscle

liver: midgut gland

lobus impar [G *Lobus impar*]: Median posterior lobe of brain.

longitudinal muscle [G *Längsmuskel*]: Innermost muscle layer of body wall (cuticle, epidermis; circular, diagonal, longitudinal muscles); may be divided into four bundles and gives rise anteriorly and posteriorly to retractors of oral shield and ctenidia. Well developed ventrally in certain caudofoveates in order to allow the animal to roll up.

lower gametoduct: mucous duct

mandible: jaw

mantle [G *Mantel*]: Epidermis covering external body; secretes spiculose cuticular coat.

mantle cavity: pallial cavity

midgut (stomach) [G *Mitteldarm*]: Expanded section of digestive tract between esophagus and intestine; gives rise posteriorly to large midgut gland.

midgut cecum: midgut gland

midgut gland (digestive gland, midgut sac, liver) [G *Mitteldarmsack, Mitteldarmdrüse*]: Elongate, blind sac lying beneath gonad and intestine and emptying into posterior end of midgut (stomach); contains two types of digestive cells.

midgut region [G *Mitteldarmregion*]: Region of body between foregut region and more expanded digestive tract region; corresponds internally with midgut. (See also body regions)

mouth [G *Mundöffnung*]: Anterior opening of digestive tract; located anteroventrally and partially or entirely surrounded by oral shield. Opens into buccal cavity.

mouth shield: oral shield

mucus duct (lower gametoduct, glandular duct) [G *Schleimgang,*

Schleimdrüse]: One of two enlarged ducts between pericardioducts and pallial cavity; may bear blind anterior part and is delimited from pallial cavity by sphincter muscle.

mucus tract (pallial groove, spawning groove) [G *Schleimrinne*]: One of two ventral mucus tracts in pallial cavity. Typically better developed in ♀.

neck: In alternate designation of body regions (anterium, neck, anterior trunk, posterior trunk, posterium), external region between anterium and trunk, from which it is set off by ring-like anterior constriction.

nidamental gland: mucus duct

oblique muscle: dorsoventral muscle

odontoblast [G *Odontoblast*]: Modified cell type in radular sac; secretes teeth.

oral shield (pedal shield, foot shield, mouth shield, buccal plate) [G *Mundschild, Fußschild, Schild*]: Smooth, cuticular, plate-like structure with thick extracellular outer layer adjoining mouth; supplied with thick retractor muscles and mucous gland (oral shield gland). (homogeneous = fused = simple, paired = not fused = divided; perioral, postoral, preorally fused; circular, heart-shaped, oval)

oral shield gland (sole gland) [G *Mundschilddrüse, Fußschilddrüse, Sohlendrüse*]: One of numerous unicellular glands concentrated along edges of oral shield; produces mucus.

ovary: gonad

pallial cavity (mantle cavity, anal chamber, cloacal cavity) [G *Pallialraum, Mantelhöhle, Analhöhle*]: Terminal external cavity of body bordered by mantle; typically surrounded by elongate spicules (spines) and contains ctenidia, anus, openings (gametopores) of mucus ducts, and mucous tracts.

pallial region [G *Pallialregion*]: Posteriormost region of body (foregut, midgut, digestive gland, prepallial, and pallial regions); consists of pallial cavity surrounded by elongate spicules (spines). (See also body regions)

pedal ganglion: ventral ganglion

pedal nerve cord: ventral nerve cord

pedal shield: oral shield

pedal sinus: ventral sinus

pericardial horn [G *perikardiales Horn*]: One of two posteriorly directed extensions of pericardium; continuous with pericardioduct.

pericardioduct (coelomoduct, kidney) [G *Perikardiodukt, Coelomodukt*]: Pair of anteroventrally directed ducts leading from posterior side of pericardium to each mucous duct.

pericardium [G *Perikard, Herzbeutel*]: Sac located dorsal to intestine at posterior end of body; contains heart, receives products of gonads, and gives rise to pair of pericardioducts.

pharynx [G *Schlund*]: Muscular, often cuticle-lined section of digestive tract between buccal cavity and esophagus; well-developed radial and retractor muscles span between pharynx and body wall. Bears radular apparatus posteroventrally.

posterior trunk: In alternate designation of body regions (anterium, neck, anterior trunk, posterior trunk, posterium), external region between anterior and posterior ends of midgut gland/gonad. (See also body regions)

posterium: In alternate designation of body regions (anterium, neck, anterior trunk, posterior trunk, posterium), external region from posterior end of midgut gland/gonad to posterior end of body. (See also body regions)

precerebral ganglion [G *Präzerebralganglion*]: One of three or more pairs of ganglia along anterior margin of brain; innervates oral shield and oral region via cerebral nerves.

precloacal organ: mucous duct

prepallial region [G *Präpallialregion*]: Region of body (foregut, midgut, digestive gland, prepallial, and pallial regions) between expanded digestive gland region and terminal pallial region; corresponds internally with gonopericardial ducts. (See also body regions)

protostyle [G *Protostyl*]: In type of caudofoveate characterized by stomach with gastric shield, elaboration of mechanically rotating column of ferment secretions in stomach.

radula [G *Radula*]: Specialized feeding structure in ventral pocket of pharynx; refers to chitinized teeth on a chitinized ribbon. (bipartite = biserial = distichous; smooth, serrate, denticulated)

radular apparatus (radular eminence, tongue) [G *Radulaapparat, Zunge*]: Entire feeding structure consisting of radular sac, supporting muscles and bolsters, radular membrane bearing teeth (radula), and occasional lateral supports; movable.

radular membrane (radular ribbon) [G *Radulamembran*]: Elongate membrane to which teeth are attached.

radular pouch: radular sac

radular sac (radular sheath, radular pouch) [G *Radulascheide, Radulasack, Radulatasche*]: Pouch-like depression in ventral surface of digestive tract between pharynx and esophagus; wall of sac may be cuticularized. Odontoblasts at base of sac produce teeth of radula.

rectum: hindgut

salivary gland: foregut gland

scale [G *Schuppe*]: One element in a dense coat of scale-like calcareous structures (spicules) covering most of body. Producted by epidermis of mantle and typically changing in shape as well as increasing in size posteriorly. (lanceolate, squamiform; keeled, nonkeeled; base: indented, rounded, truncate)

sole gland: oral shield gland

spawning groove: mucous tract

spicule [G *Spikel, Spiculum*]: General term for one of numerous calcareous elements covering outer surface of body except for oral shield, pallial cavity, and dorsoterminal sense organ. Termed scales or spines according to shape and size.

spine [G *Spikulum, Stachel*]: Type of large, elongate spicule surrounding pallial cavity.

stomach: midgut

style: protostyle

subpharyngeal commissure [G *Subpharyngealkommissur*]: Commissure underlying or nerve ring surrounding pharynx; arises from cerebrobuccal connective and may bear ganglion (subpharyngeal ganglion).

subpharyngeal ganglion [G *Subpharyngealganglion*]: Ganglion of subpharyngeal commissure.

support: lateral support

suprarectal commissure [G *Suprarektalkommissur*]: Concentration of nerve tissue dorsal to intestine at posterior end of body; consists of ganglionate transverse commissure joining merged lateral and ventral nerve cords. Innervates ctenidia and dorsoterminal sense organ.

suture [G *Ventralnaht*]: External midventral line along anterior region of certain caudofoveates; represents midventral line of fusion of mantle.

testis: gonad

tongue: radular apparatus

tooth [G *Radulazahn, Dentikel*]: Paired cuticular element of radula; produced by odontoblasts in radular sac and may bear denticles or serrations. Obsolete term for basal cone.

trunk: anterior trunk, posterior trunk

upper gametoduct: pericardioduct

ventral connective: cerebroventral connective

ventral ganglion (pedal ganglion) [G *Ventralganglion*]: Ganglionic swelling near origin of each ventral nerve cord at brain. (See also buccal ganglion, lateral ganglion)

ventral nerve cord (pedal nerve cord) [G *ventraler Längsnervenstrang*]: Pair of longitudinal nerve cords extending from brain to suprarectal commissure after merging with lateral cord; anteriorly connected to each other and to lateral cords via transverse commissures. Anteriormost section of nerve surrounding foregut and before ventral ganglion is termed cerebroventral connective.

ventral sinus (ventromedian sinus, ventral blood sinus, pedal sinus) [G *ventraler Sinus, ventraler Blutsinus*]: In circulatory system, medioventral blood channel through which blood flows from anterior region of body to ctenidia.

ventricle [G *Ventrikel, Hauptkammer*]: Muscular anterior chamber of heart; receives blood from auricle(s) via auriculoventricular openings and pumps it anteriorly via aorta.

SOLENOGASTRES

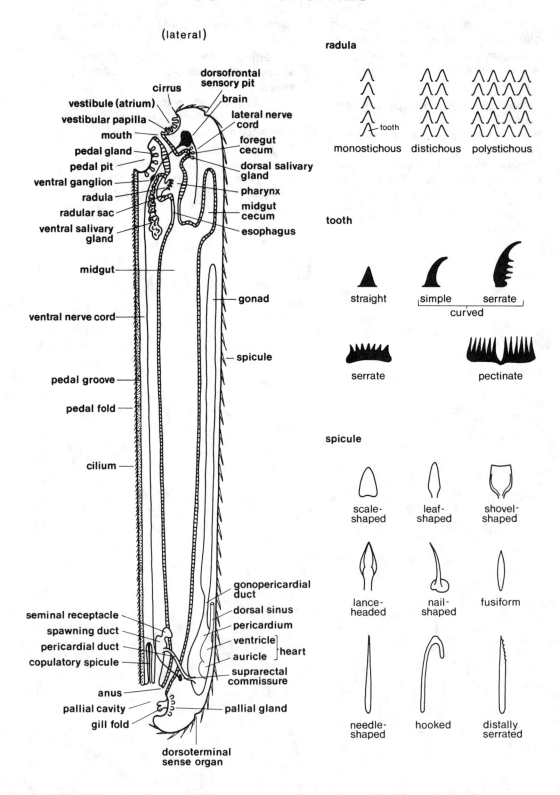

(lateral)

cirrus
dorsofrontal sensory pit
brain
vestibule (atrium)
lateral nerve cord
vestibular papilla
mouth
foregut cecum
pedal gland
dorsal salivary gland
pedal pit
ventral ganglion
pharynx
radula
midgut cecum
radular sac
ventral salivary gland
esophagus

midgut

gonad

ventral nerve cord

spicule

pedal groove

pedal fold

cilium

gonopericardial duct
seminal receptacle
dorsal sinus
spawning duct
pericardium
pericardial duct
ventricle
copulatory spicule
auricle
heart
suprarectal commissure
anus
pallial cavity
pallial gland
gill fold

dorsoterminal sense organ

radula

tooth

monostichous   distichous   polystichous

tooth

straight   simple   serrate
curved

serrate   pectinate

spicule

scale-shaped   leaf-shaped   shovel-shaped

lance-headed   nail-shaped   fusiform

needle-shaped   hooked   distally serrated

MOLLUSCA

SOLENOGASTRES

solenogaster [G *Furchenfüßer*]

accessory gland: foregut gland

anal chamber: pallial cavity

anterior pedal gland: pedal gland

anus [G *After, Anus*]: Posterior opening of digestive tract; located in upper region of pallial cavity.

atrial sense organ: vestibule

atriobuccal cavity [G *Atriobukkalraum*]: Common chamber of vestibule (atrium) and buccal cavity.

atrium: vestibule

auricle [G *Atrium, Vorkammer*]: One of typically two muscular posterior chambers of heart; receives blood from respiratory organs and pumps it anteriorly to ventricle via each auriculoventricular opening. (single, paired)

auriculoventricular opening [G *Atrioventrikularöffnung*]: In heart, opening leading from auricle to ventricle; may be equipped with sphincter. (single, paired)

bolster (radular bolster) [G *Radulapolster*]: In radular apparatus, muscular support underlying radula; occasionally with large, clear chondroid cells. Serves as site of muscle attachment.

brain (cerebral ganglion, cephalic ganglion, supraenteric ganglion) [G *Gehirn, Zerebralganglion, Oberschlundganglion*]: Main concentration of nerve tissue above pharynx; gives rise anteriorly to cerebral nerves supplying vestibule and oral region, posteriorly to buccal connectives and lateral and ventral nerve cords.

buccal cavity [G *Mundhöhle*]: Posterior portion of atriobuccal cavity leading to mouth or, in solenogasters where vestibule is more clearly separated from mouth opening, cuticle-lined section of digestive system between mouth and pharynx.

buccal ganglion [G *Bukkalganglion*]: Ganglionic swelling of each cerebrobuccal connective; closely associated with radular apparatus. (See also lateral ganglion, ventral ganglion)

cerebral nerve [G *Gehirnnerv, Zerebralnerv*]: One in a series of six or more anteriorly directed nerves arising from brain (occasionally with basal swelling) and innervating vestibule and oral region.

cerebrobuccal connective (buccal connective) [G *Zerebrobukkalkonnektiv, Bukkalkonnektiv*]: One of two relatively small strands of nerve tissue originating at brain; each gives rise to buccal ganglion. (See also cerebrolateral connective, cerebroventral connective)

cerebrolateral connective (lateral connective) [G *Zerebrolateralkonnektiv, Lateralkonnektiv*]: Part of each lateral nerve cord extending from brain around foregut (typically not having common root with cerebrobuccal and cerebroventral connectives); forms lateral ganglion.

cerebropedal connective: cerebroventral connective

cerebroventral connective (cerebropedal connective, ventral connective) [G *Zerebroventralkonnektiv, Ventralkonnektiv*]: Part of each ventral nerve cord extending from brain around foregut (typically not having common root with cerebrobuccal and cerebrolateral connectives); forms ventral ganglion.

chondroid cell: bolster

circular muscle [G *Ringmuskel*]: Outermost layer of body wall muscle (cuticle, epidermis; circular, diagonal, longitudinal muscles); an inner layer of circular muscle may also be present.

cirrus [G *Zirrus, Cirrus*]: Unit of many closely packed cilia. Refers to 1) one of several bundles of modified sensory stereocilia at periphery of vestibule or atriobuccal cavity (stereocirrus, "seta"); 2) one of many bundles of kinocilia, often elaborated as "driving wheel" within pedal pit (macrocilium, polycilium); 3) incorrectly applied to vestibular papilla.

cloaca (cloacal chamber, cloacal cavity): pallial cavity

copulatory spicule [G *Kopulationsstachel, Kopulationsstilett*]: One, two, or several paired, elongate spicules, each of pair contained in tubule underlying mucous ducts; may be protruded from pallial cavity.

crest: keel

cuticle [G *Kutikula*]: Outermost noncellular layer of body wall (cuticle, epidermis; circular, diagonal, longitudinal muscles). Secreted by epidermis.

denticle [G *Dentikel, Nebenhaken*]: Projection of tooth.

dermis (ground substance) [G *Bindegewebe, Grundsubstanz*]: In certain solenogasters, layer of collagene substance with connective tissue below epidermis; occasionally incorporates musculature and may be thickened to form keel.

diagonal muscle [G *Diagonalmuskel*]: Layer of two sets of muscles at right angles to each other between outer circular and inner longitudinal muscles of body wall (cuticle, epidermis; circular, diagonal, longitudinal muscles).

diffuse gland: foregut gland

dorsal sinus [G *Dorsalsinus*]: In circulatory system, mediodorsal blood channel through which blood flows from anterior chamber (ventricle) of heart forward into anterior region of body. (See also ventral sinus)

dorsocaudal sensory pit: dorsoterminal sense organ

dorsofrontal sensory pit [G *dorsofrontales Sinnesorgan*]: In certain solenogasters, one or more small sensory depressions, similar to dorsoterminal sense organ, located at anterior end of body above atriobuccal opening.

dorsoterminal sense organ (terminal sense organ, dorsocaudal sensory pit, osphradial sense organ) [G *Dorsoterminalorgan, dorsoterminales Sinnesorgan, osphradiales Sinnesorgan*]: Protrusible sensory depression at posterior end of body; located middorsally, anterior to pallial cavity. May be surrounded by small scales and is innervated by suprarectal commissure.

dorsoventral muscle (laterodorsal-ventral muscle, oblique muscle) [G *Dorsoventralmuskel*]: Well-developed, serially arranged pairs of muscles spanning on each side from laterodorsal body wall to pedal groove. May result in constrictions of midgut and therefore previously termed septal muscle.

epidermal papilla [G *Epidermalpapille, Papille*]: Single or multicellular granular cells of mantle epidermis extending into cuticle; may have stalk-like basal part and distal vesicle. Vesicle eventually reaches surface, leaving depression in cuticle.

epidermis [G *Epidermis*]: Outermost cellular layer of body, either differentiated into mantle or pedal groove.

esophagus [G *Ösophagus*]: Relatively narrow section of digestive tract with poorly developed musculature located between pharynx (radula) and midgut; may bear foregut (pharyngeal) glands.

foot (pedal fold) [G *Fußleiste, Pedalfalte, Längsfalte*]: One or more ciliated ridges lying within pedal groove; serves in locomotion.

foregut [G *Vorderdarm*]: Anterior, often cuticle-lined region of digestive tract (foregut, midgut, hindgut) consisting of buccal cavity or vestibule/mouth complex, pharynx, radula, and esophagus; typically well supplied with foregut glands.

foregut cecum [G *dorsale Drüsentasche*]: Outpocketing of foregut associated with foregut glands (dorsal salivary glands).

foregut gland [G *Vorderdarmdrüse*]: One of variously positioned and shaped glands opening into foregut. One may distinguish 1) pharyngeal (diffuse or accessory) glands opening anterior to radula, 2) dorsal salivary glands opening into foregut cecum, and 3) ventral salivary glands opening on or near radular apparatus (arranged in clusters, tubes, or pouches).

gametoduct: lower gametoduct, upper gametoduct

gametopore: gonopore

gill fold [G *Kiemenfalte*]: One of up to several longitudinal respiratory folds in pallial cavity.

gonad [G *Gonade*]: Hermaphroditic reproductive organ consisting of elongate pair of tubes dorsal to intestine; ovary located on medial wall, testis on lateral wall of each gonad. Gonads generally give rise to pair of short gonopericardial ducts leading into pericardium.

gonoduct [G *Gonodukt*]: In solenogasters in which gonads do not lead into pericardium via gonopericardial ducts, one of two ducts leading directly into spawning duct or pallial cavity.

gonopericardial duct [G *Gonoperikardialgang*]: Short, somewhat narrowed section of gonoduct leading from each gonad to pericardium.

gonopore (gametopore) [G *Gonoporus, Geschlechtsöffnung*]: Opening of gonoduct or spawning duct into pallial cavity. (single, paired)

heart [G *Herz*]: In circulatory system, contractile dorsal invagination of pericardium; consists of anterior ventricle and posterior auricle(s).

hindgut (rectum) [G *Enddarm, Rektum*]: Posteriormost of three regions of digestive tract (foregut, midgut, hindgut).

horizontal muscle [G *Horizontalmuskel*]: Transverse portion of dorsoventral muscles forming dorsal boundary of ventral sinus.

intestine: hindgut

keel (crest, carina) [G *Kiel*]: One or more longitudinal dorsal ridges formed by thickened dermis, cuticle, and/or mantle spicules; may also refer to medial ridge of spicule.

lateral ganglion [G *Lateralganglion*]: First ganglionic swelling of lateral nerve cord. (See also buccal ganglion, ventral ganglion)

lateral nerve cord [G *lateraler Nervenstrang, lateraler Längsnervenstrang*]: Pair of longitudinal nerve cords extending from brain to supraractal commissure; bears series of more or less distinct ganglionic swellings and is joined to ventral nerve cord by connectives. (See also cerebrolateral connective)

laterodorsal-ventral muscle: dorsoventral muscle

longitudinal muscle [G *Längsmuskel*]: Innermost muscle layer of body wall (cuticle, epidermis; circular, diagonal, longitudinal muscles); may be well developed ventrally or be present in the form of separate bundles enabling the solenogaster to roll up.

lower gametoduct: spawning duct

mantle [G *Mantel*]: Dorsal and lateral epidermis of body which produces cover of cuticle and spicules; borders pedal groove, atriobuccal opening, and pallial cavity.

midgut [G *Mitteldarm*]: Large, expanded section of digestive tract between esophagus and hindgut; gives rise to anteriorly projecting dorsal midgut cecum. (with/ without lateral pouches or diverticula) (See also foregut, hindgut)

midgut cecum [G *vorderer Darmblindsack*]: Anteriorly directed, blind section of midgut lying parallel to and directly above foregut. (simple, paired)

mouth [G *Mund, Mundöffnung*]: Anterior opening of digestive tract; located anteroventrally and either opens into buccal cavity or is associated with vestibule.

mucous duct: spawning duct

needle: spicule

oblique muscle: dorsoventral muscle

odontoblast [G *Odontoblast*]: Modified cell type in radular sac; secretes teeth.

ovary: gonad

pallial cavity (mantle cavity, anal chamber, cloaca, cloacal cavity [G *Pallialraum, Mantelhöhle, Analhöhle*]: Subterminal cavity of body; contains simple respiratory structures (gill folds, papillae) and anus, as well as openings of spawning ducts and sheaths of copulatory spicules.

pallial gland (preanal gland) [G *Präanaldrüse*]: One of numerous glands opening into pallial cavity.

papilla: epidermal papilla, respiratory papilla, vestibular papilla

pedal fold: foot

pedal gland [G *Fußdrüse*]: One in an aggregation of multicellular, mucus-producing glands associated with pedal pit (also termed anterior pedal gland). (See also sole gland)

pedal ganglion: ventral ganglion

pedal groove (ventral groove, foot groove) [G *Fußfurche, Ventralfurche, Bauchfurche*]: Longitudinal medioventral groove extending the length of the body from pedal pit to pallial cavity; lacks spicules, contains one to many longitudinal pedal folds, and receives outlets of sole glands.

pedal nerve cord: ventral nerve cord

pedal pit [G *Flimmergrube*]: Protrusible ventral depression in anterior body behind buccal cavity; supplied by pedal glands and often provided with polyciliate cirri; followed posteriorly by pedal groove.

pedal sinus: ventral sinus

penis (copulatory organ) [G *Penis, Genitalkegel, Kopulationsorgan*]: Term occasionally applied to muscular, conical elevation in pallial cavity; bears single opening of fused spawning ducts and may be associated with copulatory spicules.

pericardial horn [G *perikardiales Horn*]: One of two posteriorly directed extensions of pericardium; continuous with pericardioduct.

pericardioduct (upper gametoduct) [G *Perikardiodukt, Coelomodukt*]: Pair of anteroventrally directed ducts leading from posterior end of pericardium to spawning duct. Proximal section of each pericardioduct may bear seminal vesicle; distal section may bear seminal receptacle.

pericardium [G *Perikard, Herzbeutel*]: Chamber located dorsal to intestine at posterior end of body; contains heart, receives products of gonads, and gives rise to pair of pericardioducts.

pharyngeal gland (diffuse gland, accessory gland): foregut gland

pharynx [G *Schlund*]: Muscular, cuticle-lined section of digestive tract between buccal cavity or vestibule and esophagus; bears radular apparatus posteroventrally and is typically well supplied with foregut (pharyngeal) glands.

posterior pedal gland: sole gland

preanal gland: pallial gland

proboscis [G *Rüssel*]: Modification of preradular section of foregut into protrusible structure, often with sheath.

radula [G *Radula*]: Specialized feeding structure in ventral pocket of pharynx; refers to variously shaped, chitinized teeth on a chitinized ribbon. (distichous = biserial, monostichous = monoserial, polystichous = polyserial)

radular apparatus (radular eminence, tongue) [G *Radulaapparat, Zunge*]: Entire feeding structure consisting of radular sac, bolster with complex musculature, radular membrane bearing teeth, and typically internal supports (chondroid cells). Ducts of certain foregut glands (ventral salivary glands) open on or anterior to radular apparatus.

radular membrane (radular cuticle) [G *Basalmembran*]: Elongate layer of cuticle underlying teeth of radula.

radular pouch: radular sac

radular sac (radular sheath, radular pouch) [G *Radulascheide, Radulasack, Radulatasche*]: Simple pouch-like depression or well-developed diverticle in ventral surface of pharynx; odonotoblasts at base of sac produce teeth of radula.

rectum: hindgut

respiratory papilla [G *Atempapille*]: One of several respiratory protrusions in posterior pallial cavity.

scale [G *Schuppe*]: Flattened type of spicule. (round, oval, elongate, triangular; proximally reinforced)

seminal receptacle [G *Receptaculum seminis*]: Expanded proximal section of spawning duct or occasionally of adjacent pericardioduct; serves as organ for storage of partner's sperm. (sac-like, pedunculate; single, in clusters)

seminal vesicle [G *Vesicula seminalis, Samenblase*]: Expanded section of pericardioduct near pericardium; serves as organ for storage of own sperm.

seta [G *Tastborste*]: Term applied to series of modified sensory stereocilia surrounding atriobuccal cavity. (See also cirrus)

shell gland: spawning duct

sole gland [G *Sohlendrüse*]: One of many subepithelial gland cells arranged symmetrically above pedal groove and opening into it (also termed posterior pedal glands). (See also pedal gland)

spawning duct (mucus duct, lower gametoduct, glandular duct, nidamental duct, precloacal organ, shell gland) [G *Laichgang, Laichdrüse, Schalendrüse, präkloakales Organ, Nidamentaldrüse*]: One of two highly glandular, expanded ducts between pericardioducts and pallial cavity; usually fused terminally and bearing seminal receptacles at anterior portion. May be associated with copulatory spicules.

spicule [G *Spikel, Spiculum*]: General term for one of numerous calcareous elements produced by mantle; flattened spicules are termed scales, elongate forms are termed spines or (hollow) needles. (distally hooked, distally serrated, fusiform, lance-headed, leaf-like, shovel-like) (See also copulatory spicule)

spine: spicule

suprarectal commissure [G *Suprarektalkommissur*]: Commissure dorsal to hindgut; joins lateral nerve cords to each other and innervates pallial cavity and dorsoterminal sense organ(s).

testis: gonad

tooth [G *Zahn*]: One in a series of cuticular elements of radula; produced by odonoblasts in radular sac. (curved, straight; acute pectinate, serrate; alternate terminology distinguishes distichous denticulate bars, distichous denticulate hooks, distichous pectinate bars, monostichous denticles fused/not fused, monostichous pectinate bars, polystichous)

transverse muscle: dorsoventral muscle

upper gametoduct: pericardioduct

ventral ganglion (pedal ganglion) [G *Ventralganglion*]: Large ganglionic swelling near origin of each ventral nerve cord at brain; both ventral ganglia are connected to each other by commissures and to lateral nerve cord by connectives.

ventral groove: pedal groove

ventral nerve cord (pedal nerve cord) [G *ventraler Nervenstrang, ventraler Längsnervenstrang*]: Pair of longitudinal nerve cords extending from brain to pallial cavity; bears series of more or less distinct ganglionic swellings and innervates pedal groove via ventral nerves. Connected to each other by commissures and to lateral nerve cord via connectives.

ventral salivary gland: foregut gland

ventral sinus (ventromedian sinus, ventral blood sinus) [G *ventraler Sinus, ventraler Blutsinus*]: In circulatory system, medioventral blood channel through which blood flows from anterior region of body to respiratory organs; typically delimited by horizontal muscles.

ventricle [G *Ventrikel*]: Muscular anterior chamber of heart; receives blood from posterior auricle(s) via auriculoventricular opening(s) and pumps it anteriorly into dorsal sinus.

vestibular papilla ("cirrus") [G *Atrialpapille, "Cirrus"*]: One of numerous projections within vestibule. (simple, forked, branched)

vestibule (atrium) [G *Atrium, atriales Sinnesorgan*]: Relatively large cavity at anterior end of body, either confluent with or partially separated from buccal cavity; innervated by brain and bears numerous vestibular papillae. May be surrounded by setae (cirri).

MOLLUSCA

MONOPLACOPHORA (TRYBLIDIIDA, GALEROCONCHA)

monoplacophoran, monoplacophore [G *Monoplacophore, Tryblidiide, Galeroconche*]

MONOPLACOPHORA

(lateral)

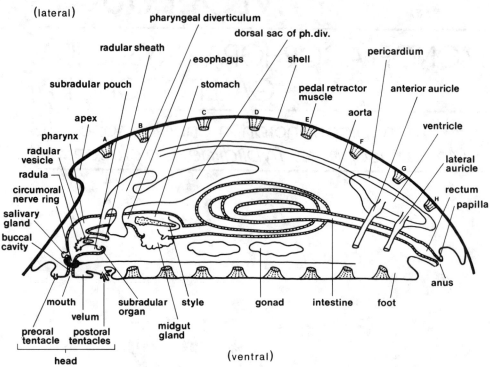

pharyngeal diverticulum
dorsal sac of ph.div.
radular sheath
esophagus
shell
pericardium
subradular pouch
stomach
pedal retractor muscle
anterior auricle
apex
aorta
ventricle
pharynx
radular vesicle
lateral auricle
radula
rectum
circumoral nerve ring
papilla
salivary gland
buccal cavity
anus
mouth
velum
subradular organ
style
gonad
intestine
foot
preoral tentacle
postoral tentacles
midgut gland

head

(ventral)

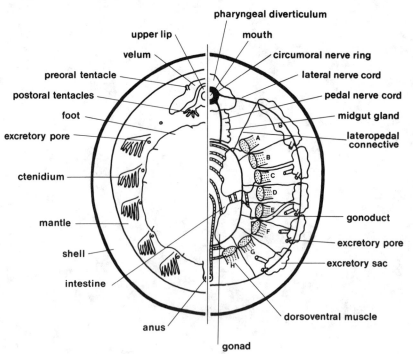

pharyngeal diverticulum
upper lip
mouth
velum
circumoral nerve ring
preoral tentacle
lateral nerve cord
postoral tentacles
pedal nerve cord
foot
midgut gland
excretory pore
lateropedal connective
ctenidium
gonoduct
excretory pore
mantle
excretory sac
shell
intestine
dorsoventral muscle
anus
gonad

MOLLUSCA

MONOPLACOPHORA (TRYBLIDIIDA, GALEROCONCHA)

monoplacophoran, monoplacophore [G *Monoplacophore, Tryblidiide, Galeroconche*]

anterior lip: upper lip

anus [G *After*]: Posterior opening of digestive tract; located on papilla in pallial groove behind foot.

aorta [G *Aorta*]: Dorsal blood vessel leading from heart to anterior end of body; initially paired (one branch arising from each ventricle), then fused into single vessel.

apex [G *Apex*]: Anteriorly directed, earliest formed point of shell. (relation to anterior shell margin: projecting beyond, immediately over, set back)

auricle [G *Atrium, Vorkammer*]: In heart, one of two tubular chambers opening into each ventricle. Blood from anterior ctenidia pairs enters heart via anterior auricle, that from last pair via lateral auricle.

blood sinus [G *Blutsinus, Sinus*]: One of a number of blood channels or spaces in head, above pallial groove, surrounding digestive tract and enclosing pedal nerve cords.

bolster [G *Radulapolster*]: In radular apparatus, one of two pairs of more rigid bodies (so-called cartilages) associated with anterior end of a pair of fluid-filled vesicles; serves as internal support and site of muscle attachment.

brain: cerebral ganglion

buccal cavity [G *Mundhöhle*]: Somewhat expanded, partially cuticle-lined section of digestive tract between mouth and pharynx; continuous posteriorly with subradular pouch.

buccal ganglion [G *Bukkalganglion*]: One of two ganglia connected to circumoral nerve ring by pair of buccal connectives.

cartilage: bolster

cerebral ganglion (brain) [G *Zerebralganglion*]: Concentration of nerve tissue, one on each side of circumoral nerve ring; each gives rise to buccal connective (to buccal ganglion), subradular nerve (to subradular ganglion), lateral nerve cord, and pedal nerve cord.

circular foot muscle [G *palliopedaler Ringmuskel, Musculus circularis*]: Band of muscle running around periphery of foot between two branches of each pedal retractor muscle. Additional circular muscles are associated with pallial groove.

circumoral nerve ring (circumenteric nerve ring) [G *circumoraler Nervenring*]: Ring of nerve tissue surrounding buccal cavity; bears cerebral ganglion on each side. Anterior part of ring may be termed cerebral commissure, posterior part termed subenteric commissure.

crystalline style: style

ctenidium (gill) [G *Kieme, Ctenidium*]: One in a series of five to six pairs of respiratory organs suspended from roof of pallial groove. Each ctenidium consists of an axis with a series of up to eight triangular, thin, ciliated lamellae at posterior face.

dorsal coelomic sac: pharyngeal diverticulum

dorsoventral muscle (pedal retractor muscle, shell muscle) [G *Dorsoventral-Muskel, Schalenmuskel*]: One of eight pairs of muscle bundles arranged serially between shell and ventral surface (foot, head). Numbered A–H or I–VIII.

esophageal gland: pharyngeal diverticulum

esophagus (posterior esophagus) [G *(hinterer) Ösophagus*]: Relatively short, narrow section of digestive tract between pharynx (entrance of pharyngeal diverticula) and stomach. (See also pharynx)

excretory organ (emunctorium, kidney, nephridium) [G *Exkretionsorgan, Niere, Emunktorium*]: One of six or seven pairs of serially arranged excretory organs. Each consists of a highly lobulated excretory sac opening into pallial cavity via excretory pore; posterior organs apparently have contact with pericardium (renopericardial pore).

excretory pore (nephridiopore) [G *Exkretionsöffnung, Nephroporus*]: Opening of each excretory organ to exterior; located in pallial groove adjacent to base of each ctenidium. Pore of anteriormost excretory organ either lacking or in front of first pair of ctenidia.

excretory sac (nephridial sac) [G *Exkretionssack*]: Highly lobulated section of each excretory organ; opens to exterior in pallial groove via excretory pore. Gonoducts open into sacs of second and third pairs of excretory organs.

foot [G *Fuß, Kriechfuß*]: Muscular creeping sole forming central ventral part of body. Surrounded by pallial groove and separated from head by transverse groove. Peripherally provided with bundles of fibers of dorsoventral muscles.

gill: ctenidium

gonad [G *Gonade*]: One pair of ♀ (ovary) or two pairs of lobulated ♂ (testis) reproductive organs. Located below intestine in region of second and third ctenidia and open via gonoducts into adjoining nephridial sacs.

gonoduct [G *Gonodukt*]: In reproductive system, ♂ (sperm duct) or ♀ (oviduct) duct leading from each gonad to adjoining excretory sac. Apparent discovery of third pair of gonoducts may indicate presence of third pair of gonads in certain species.

head [G *Kopf*]: On ventral side, anterior region of body bearing mouth, velum, and postoral tentacles; surrounded by pallial groove and separated from foot by transverse groove.

heart [G *Herz*]: In circulatory system, contractile organ located within pericardium at posterior end of body; consists of paired ventricle and two pairs of auricles. Pumps blood anteriorly via aorta.

interpedal commissure [G *pedale Kommissur*]: Nerve commissure joining two pedal nerve cords at anterior end of foot.

intestine [G *Mitteldarm*]: Elongate, highly coiled section of digestive tract between stomach and rectum; originates from ventrolateral wall of stomach.

jaw [G *Kiefer*]: U-shaped cuticular thickening of oral margin of upper lip.

lateral nerve cord (visceral nerve cord) [G *Lateralstrang, lateraler Markstrang*]: Pair of longitudinal nerve cords originating from cerebral ganglia on circumoral nerve ring and extending to posterior end of body where they unite below rectum (subrectal commissure). Lies dorsal to pallial groove and innervates each ctenidium. Connected to pedal nerve cord by lateropedal connectives.

lateral tooth [G *Lateralzahn*]: In radula, one of three teeth to each side of central rachidian tooth. First lateral tooth is small and slender; the second and third are the strongest teeth and have large blades.

lateropedal connective [G *lateropedaler Konnektiv*]: One in a series of transverse nerves joining lateral and pedal nerve cords.

lip: lower lip, upper lip

liver: midgut gland

lower lip (posterior lip) [G *Unterlippe*]: Lip-like structure bordering mouth posteriorly. Adjoined posteriorly by postoral tentacles.

mantle [G *Mantel*]: Dorsal epidermis of body which secretes shell; ventromarginally delimited by mantle fold (pallial fold).

marginal tooth [G *Randzahn*]: In radula, one of outermost two teeth on each side of transverse tooth row. First (inner) marginal tooth is comb-shaped (fringed), the second (outer) is triangular.

midgut gland (liver) [G *Mitteldarmdrüse, Leber*]: Relatively large digestive organ broadly attached to ventral side of stomach. Gland is unpaired, yet consists of two main lobes, each with highly lobulated margins.

mouth [G *Mund*]: Anterior opening of digestive tract in front of foot on ventral surface of head. Bordered by anterior and posterior lips and surrounded anteriorly by velum, posteriorly by postoral tentacles. Opens into buccal cavity.

nacreous layer [G *Perlmutterschicht*]: Relatively thin, lamellate lower (innermost) layer of shell (periostracum, prismatic layer, lamellate nacreous layer).

nephridial sac: excretory sac

nephridiopore: excretory pore

nephridium: excretory organ

ovary: gonad

oviduct: gonoduct

pallial groove [G *Mantelrinne, Kiemenrinne, Pallialraum*]: On ventral side of body, circular groove between foot and rim of mantle. Along each side of body, pallial groove bears five to six ctenidia and adjoining excretory pores.

pedal nerve cord [G *Pedalstrang, Pedal-Markstrang*]: Pair of longitudinal nerve cords originating at cerebral ganglia on circumoral nerve ring and extending to posterior end of foot. Connected to each other anteriorly by interpedal commissure and to lateral nerve cord by series of lateropedal connectives. Lies within blood sinus and is united posteriorly.

pedal retractor muscle: dorsoventral muscle

pericardium [G *Perikard*]: In circulatory system, posterodorsal chamber enclosing heart. Posterior and perhaps also anterior excretory organ have a connection with pericardium.

periostracum [G *Periostracum*]: Thin, organic, horny outermost layer of shell (periostracum, prismatic layer, lamellate nacreous layer).

pharyngeal diverticulum (esophageal gland) [G *Ösophagealdrüse, Vorderdarmdrüse*]: Pair of large digestive glands opening into posterior region of pharynx. Each is elaborated into a number of sacs; large dorsal sacs were originally incorrectly interpreted as "dorsal coelomic sacs."

pharyngeal gland: salivary gland

pharynx (anterior esophagus) [G *Schlund, Pharynx, vorderes Ösophagus*]: Muscular section of digestive tract between buccal cavity and esophagus. Receives unpaired salivary gland anteriorly, paired pharyngeal diverticula posteriorly, and is associated ventroposteriorly with radular apparatus.

posterior lip: lower lip

postoral commissure (subenteric commissure) [G *Schlundkommis-*

sur]: Term applied to section of circumoral nerve ring posterior to buccal cavity.

postoral tentacle [G *postoraler Tentakel*]: One of two tufts of tentacles bordering mouth on posterior margin of head. (simple/unbranched, multiple) (See also preoral tentacle)

preoral tentacle [G *präoraler Tentakel*]: On lateral flaps of velum bordering mouth, pair of short, claviform projections. (See also postoral tentacle)

prismatic layer [G *Prismenschichte*]: Thick calcareous layer of shell between periostracum and inner lamellate nacreous layer.

rachidian tooth (rhachidian tooth, central tooth, median tooth) [G *Mittelzahn*]: Small, slender central element in each transverse row of teeth in radula.

radula [G *Radula*]: Specialized, protrusile armature of ventral pharynx. Refers to full complement of chitinized teeth on radular membrane (ribbon).

radular apparatus [G *Radulaapparat*]: Protrusile anterior region of digestive tract consisting of radula (teeth), radular membrane (ribbon), subradular membrane, supporting muscles, and bolsters with radular vesicles.

radular membrane: ribbon

radular sheath (radular sac) [G *Radulascheide*]: Elongate sheath producing and enclosing radula. (See also subradular pouch)

radular vesicle [G *Radulavesikel, Radulaknorpel*]: In radular apparatus, one of two fluid-filled, fusiform or ovoid bodies underlying teeth; associated with bolsters and serves as internal support and site of muscle attachment.

rectum [G *Enddarm, Rektum*]: Short section of digestive tract between intestine and anus. Flanked on each side by ventricle of heart.

ribbon (radular membrane)

[G *Radulamembran*]: Membrane in radular sheath into which radula teeth insert.

salivary gland (pharyngeal gland) [G *Speicheldrüse*]: Relatively small gland(s) consisting of pouch-like to globular diverticula and opening into beginning of pharynx.

shell [G *Schale*]: Single shell element secreted by mantle and covering dorsal surface of body. Apex positioned anteriorly. Composed of outer periostracum, prismatic layer, and inner lamellate nacreous layer. (flattened/shield-like, patelliform; smooth, sculptured)

sperm duct: gonoduct

statocyst [G *Statozyste*]: Pair of equilibrium sensory organs, one located on each side of mantle in region delimited by second and third lateropedal connectives and lateral and pedal nerve cords. Each statocyst opens via long duct into pallial groove.

stomach [G *Magen*]: Enlarged section of digestive tract between esophagus and intestine; contains style sac and style and is broadly connected to midgut gland.

style (crystalline style) [G *Kristallstiel, Fermentstiel*]: Cone-shaped mucoid body extending along roof of stomach; enclosed posteriorly in style sac.

style sac [G *Stielsack, Kristallstiel-Divertikel*]: In posterior end of stomach, sac-like structure enclosing pointed posterior end of style.

subradular ganglion [G *Subradularganglion*]: Unpaired ganglion connected to circumoral nerve ring by pair of subradular connectives. Lies dorsal to and innervates subradular organ.

subradular membrane [G *Subradularmembran*]: In anterior radular sheath, membrane underlying tooth-bearing radular membrane; serves as site of muscle attachment.

subradular organ [G *Subradularorgan*]: Unpaired, cushion-shaped sensory organ in roof of subradular pouch.

subradular pouch (subradular sac) [G *Subradulartasche*]: In digestive tract, unpaired, flat sac extending posteriorly from buccal cavity below radula; bears subradular organ.

testis: gonad

tooth [G *Zahn*]: One of 11 cuticular elements in each transverse row of teeth in radula. According to position one may distinguish rachidian (central) tooth, three lateral teeth, and two marginal teeth.

upper lip (anterior lip) [G *Oberlippe*]: Lip-like structure bordering mouth anteriorly. Thickening of cuticle of upper lip is termed jaw. (See also lower lip)

velum [G *Velum, Mundlappen*]: Fleshy fold of tissue bordering mouth anteriorly and laterally; expanded laterally to form flaps which may bear preoral tentacles.

ventricle [G *Ventrikel*]: Paired, muscular median chamber of heart, one lying to each side of rectum. Each ventricle receives blood via two auricles and pumps it anteriorly via aorta.

visceral nerve cord: lateral nerve cord

POLYPLACOPHORA

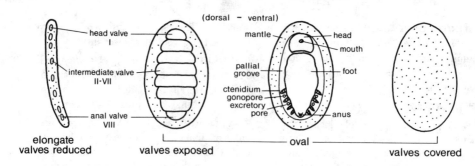

(dorsal — ventral)

head valve I	mantle — head	
	— mouth	
intermediate valve II-VII	pallial groove — foot	
	ctenidium	
	gonopore —	
	excretory	
anal valve VIII	pore — anus	

elongate
valves reduced   valves exposed   ⌊——— oval ———⌋   valves covered

(lateral)

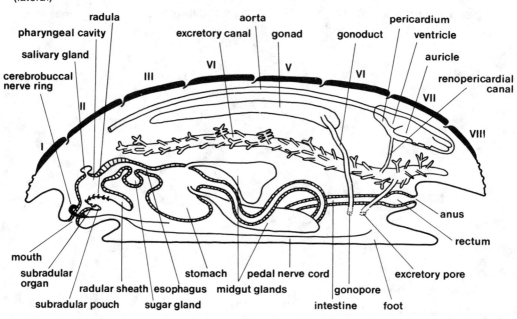

radula
pharyngeal cavity
salivary gland
cerebrobuccal nerve ring
aorta
excretory canal   gonad
gonoduct
pericardium
ventricle
auricle
renopericardial canal

III   VI   V   VI   VII   VIII

II

I

anus
rectum

mouth
subradular organ
radular sheath   esophagus
subradular pouch   sugar gland
stomach   pedal nerve cord
midgut glands
gonopore
intestine   foot
excretory pore

girdle

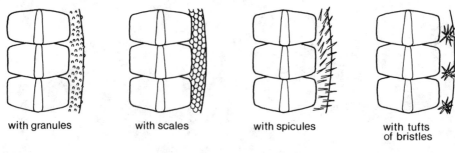

with granules   with scales   with spicules   with tufts of bristles

intermediate valve
(dorsal)

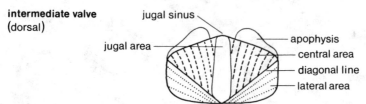

jugal sinus
jugal area
apophysis
central area
diagonal line
lateral area

MOLLUSCA

POLYPLACOPHORA

chiton [G *Chiton, Käferschnecke, Placophore*]

aesthete: esthete

anal valve (anal plate, tail valve) [G *Endplatte, letzte Schalenplatte, Platte VIII*]: Last of eight plates covering dorsal surface; typically semicircular in shape and characterized by elevated mucro.

anterior intestine: intestine

anus [G *After*]: Posterior opening of digestive tract; located behind foot, usually on papilla.

aorta [G *Aorta*]: Dorsal blood vessel leading from median chamber (ventricle) of heart to head sinus at anterior end of body. Gives rise to channels leading to gonad(s) and valve muscles.

apophysis (sutural plate, sutural lamina) [G *Apophyse, Suturalplatte*]: One of two thin, anteriorly directed processes of each valve except head valve; composed of hypostracum (termed articulamentum in apophysis) and underlies preceding valve.

articulamentum [G *Articulamentum*]: Term applied to layer of valve (hypostracum) which projects anteriorly to form apophyses.

atrium: auricle

auricle (atrium) [G *Atrium, Vorkammer, Vorhof*]: One of two lateral muscular chambers of heart. Receives blood from ctenidia via auricular pores and conveys it into median ventricle via auriculoventricular ostium.

auricular pore [G *Atrialpore*]: In heart, one of several pores by which blood passes from branchial sinuses into auricles.

auriculoventricular ostium [G *atrioventrikulare Öffnung*]: In heart, one of typically one to three pairs of openings leading from lateral auricles into median ventricle.

blade [G *Schneide*]: Thickened distal cutting edge of radular tooth.

bolster [G *Radulapolster*]: In radular apparatus, pair of oval bodies (so-called cartilages) underlying anterior radula and associated with pair of fluid-filled vesicles; serves as internal support and as site of muscle attachment.

branchial sinus [G *Kiemensinus, Kiemenarterie, Kiemenvene*]: In circulatory system, blood channel lying in roof of pallial groove and associated with ctenidia. One may distinguish afferent and efferent branchial sinuses.

bristle [G *Borste*]: Type of ornamentation of girdle; modified spicule in which cuticularized base and sheath form greater part of shaft.

buccal cavity [G *Mundhöhle*]: Somewhat expanded, cuticle-lined section of digestive tract between buccal tube and pharynx; bears buccal glands and is continuous with subradular pouch.

buccal commissure [G *Bukkalkommissur*]: Strand of nerve tissue joining two buccal ganglia. One may distinguish dorsal and ventral buccal commissures.

buccal connective [G *Bukkalkonnektiv*]: One of two strands of nerve tissue extending medially from cerebrobuccal nerve ring to buccal ganglia.

buccal ganglion [G *Bukkalganglion*]: One of two relatively small ganglia innervating anterior digestive tract; connected to each other by buccal commissure, to cerebrobuccal ring by buccal connectives.

buccal gland: salivary gland

buccal tube [G *Mundrohr, Bukkalrohr*]: Short, narrow section of digestive tract between mouth and buccal cavity; cuticle-lined and bearing longitudinal folds.

central area (central triangle, median triangle) [G *Zentralfeld*]: In intermediate valve, median triangular area demarcated from lateral areas by two diagonal lines (ribs); may bear median jugal tract.

central tooth: rachidian tooth of radula

cephalic plate: head valve

cerebrobuccal nerve ring [G *Schlundring*]: Ring of nerve tissue surrounding buccal cavity. Gives rise to posteriorly directed pair of lateral and pedal nerve cords as well as to medially directed pair of buccal and subradular connectives.

cirrus [G *Zirrus, Cirrus*]: One in a series of fleshy projections along anterior margin of mantle.

ctenidium (gill) [G *Kieme*]: One in a series of respiratory organs suspended from roof of pallial groove and joined to circulatory system by branchial sinuses. Each gill consists of an axis with two alternating series of oval, thin, ciliated lamellae. (abanal, adanal, holobranchial, merobranchial)

diagonal line (rib) [G *Diagonallinie*]: One of two elevated, diagonal ridges of a valve. It divides intermediate valves into central and two lateral areas, anal valve into antemucronal and postmucronal areas.

diaphragma [G *Diaphragma, Septum*]: Vertical partition of connective tissue at anterior end of body; separates head sinus from visceral sinus.

digestive gland: midgut gland

dorsoventral muscle [G *Dorsoventral-Muskel*]: One of 16 pairs of serially arranged muscles (two pairs per valve) spanning between valve and foot. Serves to adhere chiton to substratum.

emunctorium: excretory organ

esophageal gland: sugar gland

esophagus [G *Ösophagus*]: Narrow section of digestive tract between pharynx and stomach. Pair of large sugar glands open into digestive tract at junction of pharynx and esophagus.

esthete (aesthete) [G *Aesthet*]: Sensory structure, including photoreceptor, in tegmentum of valve. According to their relative size one may distinguish larger megalesthetes or macresthetes (in megalopores) and smaller micresthetes (in micropores). Innervated by lateral nerve cords and typically arranged in definite pattern. (See also ocellus)

excretory canal (nephridial canal) [G *Exkretionskanal, Nephridialkanal*]: In excretory system, main canal of each excretory organ; extends from anterior to posterior end of foot and bears numerous branching diverticula.

excretory duct (ureter, emunctorial duct) [G *Exkretionsgang, Ureter, Emunktorialgang*]: Term applied to short section of excretory system between expanded excretory sac and excretory pore.

excretory organ (nephridium, kidney, metanephridium, emunctorium) [G *Exkretionsorgan, Niere, Nephridium, Metanephridium, Emunktorium*]: Paired organ between renopericardial pore and excretory pore; consists of excretory canal with excretory sac and excretory duct.

excretory pore (nephridiopore)
[G *Exkretionsporus, Nephroporus*]: One of two openings of excretory system to exterior. Located at posterior end of pallial groove (posterior to gonopores) near largest ctenidium.

excretory sac (nephridial sac)
[G *Exkretionssack*]: In excretory system, expanded section of main excretory canal; opens to exterior via excretory duct and excretory pore.

exhalent chamber [G *Ausström-rinne*]:Term applied to inner region of pallial groove through which water flows after passing through ctenidia; separated from inhalent chamber by ctenidia.

eye: ocellus

fiber cord [G *Faserstrang*]: Epidermis-coated fibers extending through valve and giving rise to esthetes.

foot [G *Fuß, Kriechfuß*]: Creeping sole (pedal sole) forming central ventral part of body; provided with fibers of dorsoventral muscles. Surrounded by pallial groove and separated from head by transverse groove.

gill: ctenidium

girdle (perinotum) [G *Gürtel, Perinotum*]: Well-developed part of mantle peripheral to valves; forms margin of body. Dorsal surface is typically armed, while ventral surface often lacks ornamentation. (naked, granular, scaly, bristly)

girdle muscle (lateral longitudinal muscle) [G *Gürtel-Längsmuskel, Lateralmuskel, Einrollmuskel*]: In girdle, muscle band running along both sides from anterior to posterior end; located under outer margin of valves. Serves to roll chiton up for protection.

gonad [G *Gonade*]: ♂ (testis) or ♀ (ovary) reproductive organ; typically a single sac-shaped structure anterior to pericardium beneath valves III–VI. Gives rise to pair of gonoducts. (single, paired)

gonoduct [G *Gonodukt*]: In reproductive system, paired ♂ (sperm ducts) or ♀ (oviducts) ducts leading from dorsal surface of posterior end of gonad to gonopore in pallial groove. May bear enlargement or outpocketing.

gonopore [G *Gonoporus, Geschlechtsöffnung*]: One of two openings of reproductive system to exterior. Located in pallial groove anterior to excretory pores at junction of valves VI and VII.

gutter: esophagus

head [G *Kopf*]: On ventral surface, anterior region of body bearing mouth; bordered by mantle groove and separated from foot by transverse groove.

head sinus [G *Kopfblutraum, Kopfsinus*]: In circulatory system, anterior blood cavity into which aorta opens; separated by diaphragm from visceral sinus and gives rise to several longitudinal sinuses.

head valve (cephalic plate)
[G *erste Schalenplatte, Platte I*]: First of eight plates covering dorsal surface; lacks apophyses and is typically semicircular. (crescentic, hemispherical) (See also anal valve, intermediate valve)

heart [G *Herz*]: In circulatory system, contractile organ located within pericardial cavity beneath last two valves (VII and VIII); consists of median ventricle and pair of lateral auricles.

hindgut: rectum

hypostracum [G *Hypostracum*]: Relatively thick, lower layer of valves (periostracum, tegmentum, hypostracum); projects forward to form apophyses, where it is referred to as articulamentum and is locally specialized into myostracum.

inhalent chamber [G *Einström-rinne*]: Term applied to outer region of pallial groove through which water flows before passing through ctenidia; separated from exhalent chamber by ctenidia.

insertion plate [G *Insertionsplatte*]: Lateral extension of lower layer (hypostracum) of valve; margin typically divided into insertion teeth by radiating slit rays. Serves to anchor valve in mantle.

insertion tooth [G *Insertionszahn, Zahn*]: In valve, one in a series of tooth-like structures formed along outer margin of insertion plates by slit rays. (pectinated, undivided)

intermediate valve [G *Mittelplatte*]: One of six plates between anterior head valve and posterior anal valve; may be regionated by two diagonal lines into one central and two lateral areas.

intestine [G *Mitteldarm*]: Elongate, variously coiled section of digestive tract between stomach and rectum; according to lining of intestine one may distinguish an anterior and posterior intestine (separated by valve). Left or larger of two midgut glands is entwined with intestine.

jugal lamina [G *Jugallamina*]: In intermediate valves and anal valve, short extension of anterior margin (jugal sinus) between apophyses; composed of articulamentum. (smooth, slitted)

jugal sinus [G *Jugalsinus, Sinus jugalis*]: Anterior margin of each intermediate valve or anal valve between apophyses. (smooth, denticulate)

jugal tract (jugal area) [G *Jugalfeld, Area jugalis*]: In intermediate valve, narrow middorsal strip differing in sculpture from rest of central area.

kidney: excretory organ

lateral area (lateral triangle) [G *Seitenfeld, Area lateralis*]: In intermediate valve, triangular area separated from median area by two diagonal lines (ribs).

lateral fold: pallial fold

lateral longitudinal muscle: girdle muscle

lateral muscle: girdle muscle, lateropedal muscle

lateral nerve cord (palliovisceral nerve cord) [G *Viszeropallialstrang, Lateralstrang, Pleuroviszeralstrang*]: Pair of longitudinal nerve cords extending from cerebrobuccal nerve ring to posterior end of body, where they are joined by suprarectal commissure. Lies dorsal to pallial groove and innervates each ctenidium. May be joined to pedal nerve cords by lateropedal connectives and is enclosed in neurolateral sinus.

lateral tooth [G *Zwischenzahn, Hauptzahn*]: In radula, one of two teeth to each side of central rachidian tooth. Inner smaller tooth is referred to either as first, intermediate, or minor lateral tooth. Outer tooth is referred to either as second, hooked, or major lateral tooth; it is the largest tooth and bears a long shaft and typically tridentate blade.

lateral triangle: lateral area

lateropedal connective (pedolateral commissure) [G *Lateropedal-Konnektiv, Palliopedalkommissur*]: One in a series of transverse nerves joining pedal and lateral nerve cords.

lateropedal muscle (lateral muscle) [G *lateraler Dorsoventral-Muskel*]: Part of dorsoventral musculature spanning from lateral margins of valve to foot. One may distinguish anterior and posterior lateral muscles.

lateropleural area [G *Lateropleuralfeld*]: In chitons whose intermediate valve lacks diagonal line (rib), area to each side of median jugal tract.

macresthete: esthete

mantle [G *Mantel*]: Dorsal epithelium of body which secretes cuticle and calcareous material (valves, scales, spicules, granules); peripheral part forming margin of body is termed girdle and may partially overgrow valves.

mantle cavity: pallial groove

mantle muscle [G *Mantelmuskel*]: Muscle band spanning from valve into mantle. One may distinguish inner and outer mantle muscles.

marginal tooth [G *äußerer Marginalzahn, Randplatte*]: In radula, outermost three plate-like teeth (or outermost six teeth, i.e., including uncinal teeth) on each side of tooth row.

median sinus: ventral sinus

median triangle: central area

megalesthete: esthete

megalopore [G *Megalopore*]: One of numerous larger pores in outer surface of valve; each megalopore corresponds to end of a canal traversing valve. Bears esthete (megalesthete). (See also micropore)

metanephridium: excretory organ

micresthete: esthete

micropore [G *Mikropore*]: One of numerous smaller pores in outer surface of valve; each micropore is associated with a megalopore and corresponds to end of a canal traversing valve. Bears esthete (micresthete).

midgut gland (liver) [G *Mitteldarmdrüse, Leber*]: Pair of glands opening into posterior end of stomach; right gland is typically smaller and lies alongside stomach, left gland is very large and entwined with intestine. (opening into stomach: separate, fused)

mouth [G *Mund*]: Anterior opening of digestive tract in front of foot on ventral surface of head; opens into buccal tube.

mucus tract [G *Schleimkrause, Schleimrinne*]: In posterior part of pallial groove, epithelium bearing tall mucous cells. Arranged throughout groove or in two or three longitudinal bands (pedal, neural, pallial bands)

mucro [G *Mukro*]: Highest point of anal valve; typically located centrally, yet occasionally also anteriorly or posteriorly. Diagonal lines radiating from mucro divide tail valve into antemucronal and postmucronal areas. May also refer to apex of intermediate valve.

myostracum [G *Myostracum*]: In valve, term applied to structurally specialized areas of hypostracum at insertion sites of dorsoventral musculature.

nephridial canal: excretory canal

nephridial sac: excretory sac

nephridiopore: excretory pore

nephridium: excretory organ

nephrostome: renopericardial pore

neurolateral sinus [G *Lateralsinus*]: In circulatory system, pair of longitudinal blood channels extending above pallial groove and enclosing lateral nerve cords.

neuropedal sinus [G *Lateropedalsinus*]: In circulatory system, pair of longitudinal blood channels originating at head sinus and lying to each side of ventral sinus; encloses pedal nerve cord.

oblique muscle [G *Musculus obliquus*]: Pair of muscle bands spanning obliquely (beneath diagonal line) from anterior margin (jugal sinus) of one valve to body wall beneath preceding valve.

ocellus (eye, shell eye) [G *Ocellus, Schalenauge*]: Complex, pigmented photosensitive structure formed by modified megalesthete in outer layer of valve. Consists of cornea, lens, fibrillar area, cup, and pigment cells. Ocelli may be arranged in definite pattern.

odontophore: radular apparatus

osphradial sense organ (osphradium) [G *osphradiales Sinnesorgan, Osphradium*]: One of two short, chemoreceptive sensory tracts lateral to anus.

ostium: auriculoventricular ostium

ovary: gonad

oviduct: gonoduct

pallial fold (lateral fold) [G *Lateralleiste, Mantelsaum, Mantelfalte*]: On ventral surface of girdle, mantle fold projecting into pallial groove. May be enlarged to form lappet posteriorly. (smooth, fimbriated)

pallial groove (mantle cavity) [G *Mantelrinne, Kiemenrinne, Pallialraum*]: On ventral side of body, groove between foot and marginal mantle (girdle). Along each side of body, pallial groove bears ctenidia; posteriorly it bears pair of excretory pores, gonopores, and mucus tracts.

pallioviseral nerve cord: lateral nerve cord

pedal commissure (ventral commissure) [G *Pedalkommissur*]: One in a series of transverse nerves joining paired pedal nerve cords.

pedal nerve cord (ventral nerve cord) [G *Pedalstrang, Ventralstamm*]: Pair of longitudinal nerve cords extending from cerebrobuccal nerve ring to posterior end of body. Lies dorsally in median region of foot. Connected to each other by pedal commissures, to lateral nerve cords by lateropedal connectives. Enclosed by neuropedal sinus.

pedolateral commissure: lateropedal connective

pedal sinus [G *Pedalsinus*]: In circulatory system, pair of longitudinal blood channels originating at head sinus and lying lateral to neuropedal sinuses.

pericardium [G *Perikard*]: In circulatory system, dorsoventrally flattened, fluid-filled chamber located beneath last two valves (VII and VIII); contains heart and is connected to excretory system via two renopericardial pores.

perinotum: girdle

periostracum [G *Periostracum*]: Relatively thin, organic, horny outermost layer of valve (periostracum, tegmentum, hypostracum).

pharyngeal gland: sugar gland

pharynx [G *Schlund, Pharynx*]: Muscular, cuticle-lined section of digestive tract between buccal cavity and esophagus; may be expanded laterally to form two diverticula.

plate: valve

posterior intestine: intestine

rachidian tooth (rhachidian tooth, central tooth) [G *Mittelzahn, Mittelplatte*]: Central element in each transverse row of teeth in radula; typically small, elongate, with narrow blade.

radula [G *Radula*]: Specialized, protrusile armature of anterior region of digestive tract. Refers to full complement of chitinized teeth.

radular apparatus (odontophore) [G *Radulaapparat*]: Protrusile anterior region of digestive tract consisting of radula (teeth), ribbon, supporting muscles, and bolsters.

radular sheath (radular sac) [G *Radulascheide*]: Elongate sheath enclosing radula and ribbon.

rectum [G *Enddarm*]: Short, occasionally somewhat expanded posteriormost section of digestive tract between intestine and anus.

rectus muscle [G *Musculus rectus*]: Pair of parallel muscle bands spanning straight from anterior margin (jugal sinus) of one valve to body wall beneath preceding valve.

renopericardial pore [G *Renoperikardialporus*]: One of two connections between pericardium and excretory system; continues via renopericardial canal into main excretory canal.

renopericardial canal (pericardioduct) [G *Renoperikardialgang, Perikardiodukt*]: In excretory system, canal leading from renopericardial pore (at pericardium) to main excretory canal.

rhachidian tooth: rachidian tooth

ribbon (radular membrane) [G *Radulamembran*]: Membrane in radular sheath into which radula teeth insert.

salivary gland (buccal gland) [G *Bukkaldrüse*]: Pair of glands near junction of buccal cavity and pharynx; associated with mucous gutter dorsally and ciliated groove ventrally (simple, arborescent)

scale [G *Schuppe*]: Type of ornamentation of girdle; flat, scale-like structures secreted by mantle epidermis. (rounded, cuboidal; smooth, sculptured)

shell: valve

sinus: jugal sinus, ventral sinus

slime sac [G *Oviduktdrüse, Schleimsac*]: Outpocketing of ♀ gonoduct (oviduct); produces slimy material of egg strings.

slit ray [G *Naht*]: One in a series of radiating grooves in lower layer (hypostracum) of valve; may consist of a row of pores and forms insertion teeth at margin of insertion plate.

sperm duct: gonoduct

spicule [G *Stachel*]: Type of ornamentation of girdle; relatively small projection secreted by mantle epidermis and typically surrounded basally by cuticularized sheath. (straight, curved; cylindrical, styliform, fluted)

spine: bristle, spicule

stomach [G *Magen*]: Expanded section of digestive tract between esophagus and intestine; separated from esophagus by sphincter. Pair of large midgut glands opens into stomach.

subradular connective [G *Subradularkonnektiv*]: One of two strands of nerve tissue extending medially from cerebrobuccal nerve ring to subradular ganglia.

subradular ganglion [G *Subradularganglion*]: One of two relatively small, adjoining ganglia connected to cerebrobuccal nerve ring by subradular connectives. Lies dorsal to and innervates subradular organ.

subradular organ [G *Subradularorgan*]: One of two adjoining, cushion-shaped sensory organs on roof of subradular pouch; may be everted to mouth opening.

subradular pouch [G *Subradulartasche*]: In digestive tract, blind sac extending posteriorly from buccal cavity; bears pair of subradular organs.

sugar gland (esophageal gland, pharyngeal gland) [G *Ösophagealdrüse, Zuckerdrüse, Schlundsack*]: Pair of large glands opening into digestive tract at junction of pharynx and esophagus.

suprarectal commissure [G *Suprarektalkommissur*]: Transverse commissure joining two lateral nerve cords at posterior end of body; innervates osphradial sense organs.

sutural lamina: apophysis

sutural plate: apophysis

tail valve: anal valve

tegmentum [G *Tegmentum*]: Calcified organic layer of each valve between upper periostracum and lower hypostracum; encloses esthetes and may turn under posterior margin of valve.

testis: gonad

tooth [G *Zahn, Platte*]: One of 17 chitinized cuticular elements in each transverse row of teeth in radula. According to position one may distinguish rachidian (central), minor and major lateral, uncinal, and marginal teeth.

transverse muscle [G *Musculus transversalis, Quermuskel*]: Muscle fibers spanning dorsoventrally from anterior margin (apophyses) of one valve to posterior margin of preceding (overlapping) valve.

trochophore (pseudotrochophore) [G *Trochophor-Larve, Pseudotrochophora*]: Larval stage of chiton; characterized externally by apical ciliary tuft and prototroch.

uncinal tooth [G *innerer Marginalzahn*]: One of three teeth adjoining major lateral tooth in each tooth row of radula. The first two are plate-like and may be termed uncinal plates, the third bears an inwardly curving blade and may be termed spatulate uncinal. (See also marginal tooth)

ureter: excretory duct

valve (plate, mantle plate) [G *Platte, Schalenplatte, Mantelplatte*]: One in a series of eight arched plates partially covering dorsal surface (mantle) of body. Consists of several layers (periostracum, tegmentum, hypostracum) and is embedded to various degrees in mantle. One may distinguish an anterior head valve, six intermediate valves, and a posterior anal valve. (in contact, separated; with apophyses, sutures, insertion plate; smooth, granulated, mucronate, ridged)

ventral nerve cord: pedal nerve cord

ventral sinus (median sinus) [G *Ventralsinus*]: In circulatory system, large, longitudinal blood channel extending from head sinus in foot to posterior end of body; gives rise to branchial sinuses.

ventricle [G *Ventrikel*]: Muscular median chamber of heart; receives blood from lateral auricles via auriculoventricular ostia and pumps it anteriorly via aorta.

vesicle [G *Papille, Bläschen*]: Type of ornamentation of girdle; stalked epidermal structure with terminal enlargement.

visceral sinus [G *Viszeralsinus, Körpersinus*]: General body cavity containing greater part of digestive tract and serving as blood sinus; separated from head sinus by diaphragma.

MOLLUSCA

GASTROPODA

gastropod, snail [G *Gastropode, Schnecke*]

GASTROPODA

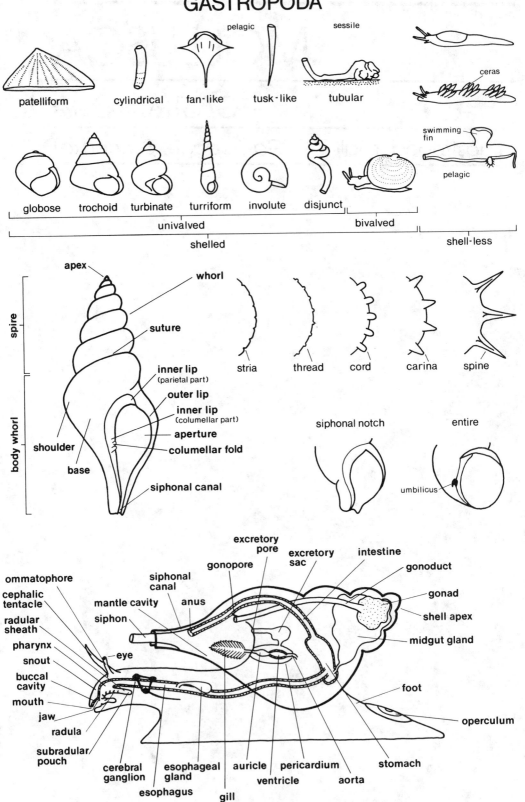

pelagic · sessile

patelliform · cylindrical · fan-like · tusk-like · tubular · ceras

swimming fin · pelagic

globose · trochoid · turbinate · turriform · involute · disjunct

univalved · bivalved

shelled · shell-less

apex · whorl

spire

suture

inner lip (parietal part)

outer lip

inner lip (columellar part)

aperture

columellar fold

body whorl

shoulder

base

siphonal canal

stria · thread · cord · carina · spine

siphonal notch · entire

umbilicus

ommatophore · excretory pore · excretory sac · intestine

cephalic tentacle · gonopore · gonoduct

radular sheath · siphonal canal · gonad

pharynx · mantle cavity · anus · shell apex

snout · siphon · midgut gland

buccal cavity · eye

mouth · foot

jaw · operculum

radula

subradular pouch

cerebral ganglion · esophageal gland · auricle · pericardium · stomach

esophagus · ventricle · aorta

gill

MOLLUSCA

GASTROPODA

gastropod, snail [G *Gastropode, Schnecke*]

albumen gland [G *Eiweißdrüse, Albumindrüse*]: In ♀ reproductive system, innermost expanded section of oviduct or distinct gland opening into oviduct; in hermaphroditic gonad (ovotestis), large gland opening into carrefour. Secretes nutritive material surrounding egg. (See also capsule gland)

anal canal: posterior canal

anal gland [G *Analdrüse*]: Dark-colored gland opening into rectum. (simple, branched)

anterior canal: siphonal canal

anus [G *After*]: Posterior opening of digestive tract; opens in mantle cavity or at its vestigial area, in typical (torted) snail anteriorly along with gonopore and excretory pore.

aorta [G *Aorta*]: Major blood vessel leading from ventricle of heart; typically divides into anterior and posterior aortas supplying the major organs.

aperture [G *Mündung*]: Opening of shell through which head and foot may be protruded. Characterized by outer and inner lips and may be closed by means of operculum. (effuse, entire, holostomatous, patulous, siphonostomatous)

apex [G *Apex, Spitze*]: Part of shell that was formed first; typically pointed, consisting of protoconch and smallest whorls of spire. (entire, eroded)

arm [G *Bukkalarm, Fangarm*]: In pelagic snail, one of up to several elongate extensions of buccal tube; bears suckers and serves in capturing prey.

ascus (saccus) [G *Saccus, Ascus*]: Specialized, blind pouch of pharynx below radular apparatus (subradular pouch) in which used and worn teeth accumulate.

astragal [G *Reifen*]: Type of spiral shell sculpture: very large, round-topped, steep-sided elevation around whorls.

auricle [G *Atrium, Vorkammer*]: Relatively thin-walled chamber of heart receiving blood from ctenidium or respiratory surface and pumping it via ventricle into aorta. Reduction of one ctenidium generally leads to loss of one auricle. (single, paired)

axis [G *Achse*]: In shell, imaginary line around which whorls coil, i.e., line extending from apex to siphonal canal; corresponds to columella.

base [G *Basis*]: In typical (conispiral) shell, part or surface of last whorl (body whorl) farthest from apex, i.e., abapical from widest circumference (periphery) of body whorl.

body whorl (last whorl) [G *Endwindung*]: Last-formed and largest complete whorl of shell; growth margin of body whorl forms outer lip.

bolster [G *Radulapolster*]: In odontophore of radular apparatus, one of typically two pairs of masses of turgescent cells ("cartilages") underlying subradular membrane;

serves as internal support and site of muscle attachment.

boss [G *Knoten*]: Type of shell sculpture; relatively large, rounded elevation on surface of whorl. Larger than tubercle.

brain: cerebral ganglion

branchial cordon [G *Kiemenkranz*]: In snail with patelliform shell, secondary external gill in pallial groove between mantle edge and foot; consists of series of leaf-like structures.

buccal cavity [G *Mundhöhle*]: Somewhat expanded, cuticle-lined section of digestive tract between mouth and pharynx; typically bears pair of jaws laterally. (See also buccal tube)

buccal cone [G *Bukkalkonus, Bukkalkegel*]: In pelagic snail, one element in up to three pairs of cone-shaped projections of buccal cavity. Protruded to exterior in order to capture prey.

buccal ganglion [G *Bukkalganglion*]: One of two concentrations of nerve tissue adjoining radular apparatus and innervating anterior section of digestive tract; joined to each other by buccal commissure and to cerebral ganglia by cerebrobuccal commissures.

buccal gland (oral gland, anterior gland, anterior salivary gland) [G *Bukkaldrüse, vordere Speicheldrüse*]: Gland often adjoining pharynx or esophagus yet opening into buccal cavity via long duct. (one pair, two pairs) (See also pharyngeal gland)

buccal tube (oral tube) [G *Mundrohr*]: Tube-shaped buccal cavity.

bursa copulatrix [G *Bursa copulatrix*]: In ♀ reproductive system, blind pouch whose opening closely adjoins ♀ gonopore; duct leading from pore to bursa is termed vagina. Receives sperm after copulation and passes them on to seminal receptacle.

callus [G *Kallus*]: In shell, thickened layer (inductura) either on parietal region (parietal callus), extending across part of body whorl, or covering umbilicus.

canal: posterior canal, siphonal canal

capsule gland (membrane gland) [G *Eikapseldrüse*]: In ♀ reproductive system, conspicuous, thick-walled gland opening into oviduct after albumen gland. Secretes capsules enclosing eggs.

cardiac region: stomach

carina [G *Carina*]: Type of spiral shell sculpture: prominent ridge or keel.

carrefour [G *postampullärer Zwittergang*]: In hermaphroditic gonad (ovotestis), section of gonoduct between hermaphroditic duct and separate oviduct and sperm duct; albumen gland and fertilization pouch open into carrefour.

caryophyllid: One in a series of sensory structures projecting from dorsal surface of certain shell-less snails; supported by circle of upright spicules and bears a terminal sensory knob in a terminal depression.

cecum (caecum) [G *Caecum, Spiralcaecum*]: Blind pouch of stomach. Food particles pass from stomach into cecum along ridge (typhlosole) and are then sorted. (simple, coiled)

cephalic shield [G *Kopfschild*]: Expanded thickening of dorsal surface of head corresponding to cephalic tentacle. May be used in burrowing. (quadrangular, posteriorly bifid)

cephalic tentacle (frontal tentacle) [G *Kopftentakel, Tentakel*]: Pair of large tentacles on head; may be associated with ommatophores and thus closely adjoin or be fused with eye, or may be modified into cephalic shield. Innervated by cerebral ganglion. (smooth, fringed) (See also rhinophore, tentacle)

cephalic veil: frontal sail

ceras (pl.: cerata) [G *Ceras, Rückenanhang*]: One in a series or bundle of respiratory projections on dorsal surface of snail with reduced or no shell. (in series, in clusters, uniformly distributed; branched, cylindrical, flattened, fusiform, serrated)

cerebral commissure [G *Zerebralkommisur*]: Nerve commissure passing dorsal to digestive tract and joining two cerebral ganglia. (See also subcerebral commissure)

cerebral ganglion [G *Zerebralganglion*]: One of two concentrations of nerve tissue located above anterior section of digestive tract and representing nervous center; connected dorsally to each other by cerebral commissure, occasionally also ventrally by subcerebral commissure. Gives rise to nerves innervating head region and associated sense organs; may more closely adjoin additional ganglia to form circumenteric nerve ring.

cerebropedal connective [G *Zerebropedalkonnektiv*]: Nerve interconnection joining each cerebral ganglion to corresponding pedal ganglion.

cerebropleural connective [G *Zerebropleuralkonnektiv*]: Nerve interconnection joining each cerebral ganglion to corresponding pleural ganglion.

cervical lappet: nuchal lobe

circumenteric nerve ring [G *Schlundring, zircumpharyngealer Ganglienring*]: Concentration of nerve tissue surrounding anterior section of digestive tract; consists of closely adjoining cerebral, pleural, and pedal ganglia, not infrequently also of parietal and visceral ganglia.

clausilium [G *Clausilium*]: Spoon-shaped plate attached to columella inside shell of certain terrestrial snails. Serves as operculum.

cloacal siphon (anal siphon) [G *Mantelsipho*]: Tubular structure formed by posterolateral corners of mantle. Serves as excurrent opening of mantle cavity and encloses anus.

columella (axis) [G *Columella, Spindel*]: Calcareous structure forming central axis of shell; shell whorls form and coil around columella. (solid, hollow; smooth, with columellar folds)

columellar fold (columellar plica) [G *Spindelfalte*]: In shell, one or more (parallel) elevated ridges winding around columella; typically visible as folds or "teeth" on inner lip (columellar lip).

columellar lip [G *Spindellippe, Innenlippe*]: In shell, part of inner lip formed by terminal section of columella or consisting of thickened columellar material encroaching onto body whorl. (reflected) (See also parietal lip)

columellar muscle (spindle muscle, shell muscle) [G *Spindelmuskel*]: Dorsoventral bundles fused into one large muscle originating on columella and spanning through body into foot; forms scar on operculum and serves to retract snail into shell. (See also shell muscle)

conchiolin (conchin) [G *Conchin*]: Organic (horny) material forming outermost layer (periostracum) of shell.

cord [G *Reifen*]: Type of spiral or axial shell sculpture: relatively large, round-topped elevation.

costa (rib) [G *Rippe*]: Type of axial shell sculpture: prominent, round-topped elevation.

crop [G *Kropf*]: Thin-walled, expanded posterior modification of esophagus.

crystalline style (style) [G *Krystallstiel, Fermentstiel*]: In stomach, rod-shaped, concentrically layered body within style sac; end of style directed toward esophagus is in contact with gastric shield.

ctenidium (gill) [G *Kieme, Fiederkieme*]: Respiratory organ, typically located in mantle cavity and forming part of pallial complex; consists of axis (septum) bearing series of triangular, ciliated leaflets alternating along both sides. (external, internal; monopectinate = unipectinate, bipectinate, finger-like, plicate) (See also gill)

dart (love dart) [G *Liebespfeil*]: Pointed calcareous body secreted in dart sac; may be shot into partner during copulation; typically with complex cross section.

dart sac [G *Pfeilsack, Liebespfeilsack*]: Relatively large, muscular pouch opening into vagina; secretes calcareous dart which may be shot into partner during copulation. (single, paired)

digitiform gland [G *fingerförmige Drüse*]: Pair of branched glands opening into vagina next to dart sac.

epiphallus [G *Epiphallus*]: Muscular portion of sperm duct prior to penis sheath; produces spermatophores in conjunction with flagellum.

epipodial tentacle [G *Epipodialtentakel*]: One of a pair or series of tentacles projecting from epipodium. (ciliated, fringed) (See also tentacle)

epipodium [G *Epipodium, Epipodialfalte*]: Horizontal fold of body wall along upper border of foot. (fringed, papillate, tentaculate)

esophageal bulb (pyriform organ, crop, jabot, pharynx of Leiblein) [G *Jabot, Blättermagen*]: Highly folded expansion of digestive tract at anterior end of esophagus.

esophageal gland (esophageal pouch) [G *Schlundtasche, Schlunddrüse*]: Pair of glandular pouches opening into lateral part of beginning of esophagus. May be modified into gland of Leiblein.

esophagus (oesophagus) [G *Ösophagus*]: Elongate section of digestive tract between pharynx and stomach; may bear esophageal pouch, esophageal bulb, crop, or gland of Leiblein, and is characterized internally by longitudinal folds.

excretory canal (ureter) [G *Exkretionsgang, Ureter*]: In excretory organ, canal leading from excretory sac to excretory pore; may also serve as gonoduct.

excretory organ (nephridium, metanephridium, kidney) [G *Exkretionsorgan, Niere*]: Primitively paired organ closely adjoining pericardium into which it opens via renopericardial pore; consists of renopericardial canal, excretory sac, excretory canal (ureter), and excretory pore. (single, paired)

excretory pore (nephridiopore) [G *Exkretionsöffnung, Nephroporus*]: Ciliated, muscular opening of excretory system to exterior; typically located in mantle cavity, forming part of pallial complex, yet also opening onto body surface or into rectum. (single, paired)

excretory sac (nephridial sac) [G *Exkretionssack, Nierensack*]: Expanded, often regionated and internally folded section of excretory system; connected to pericardium via renopericardial canal, to excretory pore via excretory canal (ureter). (simple, lobulated, branched)

eye [G *Auge*]: One of two photosensitive organs on head or one of several photoreceptors on notum; often located on projections (ommatophores or tentacles) and consisting of cornea, retina, and lens.

fasciole [G *Fasciole*]: Type of shell sculpture; band formed by series of elevations, each representing an irregularity in successive growth lines (typically in outer lip). According to position, one may refer to anal, basal, or siphonal fascioles.

fertilization pouch [G *Befruchtungstasche*]: In hermaphroditic gonad (ovotestis), small pouch opening into carrefour next to albumen gland.

flagellum [G *Flagellum*]: In ♂ section of hermaphroditic gonad (ovotestis), elongate, narrow sac opening into penial complex. Produces spermatophores together with epiphallus. May also refer to terminal filament of penis.

foot [G *Fuß, Kriechfuß*]: Part of body, typically in the form of a muscular creeping sole separated to varying degrees from head and visceral mass and which may be retracted into shell. May be regionated and bear operculum on posterior end. (See also epipodium, propodium)

forestomach: proventriculus

frontal sail (velum) [G *Stirnsegel*]: Relatively large, erect, transverse extension or fold on anterior end of head. (edge: simple, branched, scalloped, tuberculate)

gastric shield [G *Magenschild*]: Heavily cuticularized area of stomach positioned near entrance of esophagus; in contact with crystalline style.

gill [G *Kieme*]: Respiratory organ, primitively a ctenidium yet including other outgrowths and respiratory appendages.

gizzard [G *Kaumagen*]: Muscular, cuticle-lined modification of posterior stomach.

gland of Leiblein [G *Leiblein'sche Drüse, unpaare Vorderdarmdrüse*]: Large, unpaired, and often lobulated, modified esophageal gland entering digestive tract at various levels of esophagus.

gonad: ovary, ovotestis, testis

gonoduct: hermaphroditic duct, oviduct, sperm duct, spermoviduct

gonopericardial duct [G *Gonoperikardialdukt*]: In ♀ reproductive system, duct joining expanded section (uterus) of oviduct to pericardium.

gonopore [G *Gonoporus, Geschlechtsöffnung*]: Opening of ♂ or ♀ reproductive system to exterior; typically located in mantle cavity where it forms part of pallial complex.

growth line [G *Zuwachsstreifen, Anwachsstreifen*]: In shell, thin axial line across surface of whorl indicating former position of outer lip; termed growth rugum if somewhat elevated and irregular in shape, varix if strongly developed. (opisthocline, prosocline, opisthocyrt, prosocyrt)

growth rugum [G *Zuwachsstreifen*]: In shell, axial undulation or wrinkled ridge across surface of whorl indicating former position of outer lip; represents thickening of outer lip formed during pause in growth of snail. Termed growth line if thin, varix if strongly developed.

head [G *Kopf*]: Anterior part of body typically bearing pair of eyes, tentacles, and mouth; separated to varying degrees from foot.

heart [G *Herz*]: In circulatory system, contractile organ located within pericardium; consists of one or two auricles and one ventricle. Blood, coming from gill or respira-

tory surface, is pumped via auricle(s) and ventricle into aorta.

helicocone [G *Kegelspirale*]: Term applied to typical snail shell, i.e., coiled tube whose diameter increases with each turn.

hermaphroditic duct [G *Zwittergang*]: Common duct leading from follicles comprising hermaphroditic gonad (ovotestis); continues into carrefour, after which it branches into separate oviduct and sperm duct or forms spermoviduct.

hindgut: rectum

hypobranchial gland [G *Hypobranchialdrüse*]: Glandular area in roof of mantle cavity; may be developed as mere thickening or as series of folds or lamellae. Composed of several types of gland cells. (single, paired)

hypostracum: nacreous layer

inductura [G *Schwiele, Parietalkallus*]: In parietal region of aperture, smooth, shelly outermost layer secreted by mantle. May extend over inner lip and across part of body whorl; if very thick, termed callus or parietal callus.

infrapallial lobe (mantle cecum) [G *Pallial-Caecum*]: In snails with head shield, posteriorly directed lobe formed by unattached posterior part of mantle; may bear tentacle.

inner lip (labium) [G *Innenlippe*]: In shell, wall of aperture opposite outer lip; often thickened, consisting of columellar material (columellar lip) and/or parietal region (parietal lip).

intestine [G *Mitteldarm, Intestinum*]: Section of digestive tract between stomach and rectum; in typical (torted) snail, intestine proceeds anteriorly along or through pericardium in direction of mantle cavity. (straight, coiled, looped)

jaw [G *Kiefer*]: Pair of separate or fused cuticular elements in lateral wall of buccal cavity; movable by special muscles. Considered to form border between buccal cavity and pharynx.

keel [G *Kiel*]: Type of spiral shell sculpture; prominent ridge around whorls.

labial palp [G *Mundlappen*]: On head, pair of flattened, fleshy extensions of snout; each palp originates at entrance of mouth. Tentacle-like projections bordering mouth are termed labial tentacles.

labial veil [G *Lippensegel*]: Lobe-like flap overhanging mouth.

labium: inner lip

labrum: outer lip

lamella [G *Lamelle, Falte*]: Thin plate, typically referring to series of projections in aperture of certain terrestrial snails. According to position one may distinguish columellar, infraparietal, interparietal, and parietal lamellae.

lamellate layer [G *Lamellenschicht*]: Layer of shell underlying prismatic layer (periostracum, prismatic layer, lamellate layer, nacreous layer); consists of sheets (lamellae) of calcium carbonate (typically aragonite).

last whorl: body whorl

lateral tooth [G *Lateralzahn*]: In radula, one of typically several teeth to each side of central or rachidian tooth; differs in size and shape from outer marginal teeth.

lip [G *Lippe*]: Modified body wall surrounding mouth; pair of protrusions at entrance of buccal cavity are termed inner lips. May also refer to shell structure, i.e., inner and outer lips of aperture.

lira: Type of shell sculpture; fine, elevated line on whorl or on inside of outer lip, in the latter case corresponding to depressions on shell exterior.

liver: midgut gland

mantle [G *Mantel*]: Fleshy fold of body wall. In shell-bearing snails, circular fold lining body whorl and secreting shell. Typically forms space (mantle cavity) either enclosing gill or forming respiratory surface itself; may also be elongated to form siphon or mantle tentacles.

mantle cavity (pallial cavity) [G *Mantelhöhle*]: Space between mantle and rest of body. Located in body whorl of snails with coiled shells and either encloses gill and pallial complex or forms pulmonary sac.

mantle tentacle [G *Manteltentakel*]: Elongate, tentacle-like projection of mantle margin.

marginal tooth (uncinus) [G *Randzahn*]: In radula, outermost elements on each side of transverse tooth row; typically differ in size and shape from lateral teeth.

mentum [G *Mentum*]: In certain shell-bearing snails, enlarged upper labial fold.

metanephridium: excretory organ

metapodium [G *Metapodium*]: Posterior section of regionated type of foot; may bear lobe-like parapodia and be innervated by metapodial ganglion. (See also propodium)

midgut gland (liver) [G *Mitteldarmdrüse, Leber*]: Large, typically unpaired digestive gland opening into stomach via duct(s); may form greater part of visceral mass. (single, paired)

mouth [G *Mund*]: Anterior opening of digestive tract; may be positioned on prolongation of head (snout, proboscis) and be surrounded by various labial projections. Opens into buccal cavity. (circular, slit-like, T-shaped)

mucus gland [G *Mucusdrüse*]: In ♀ reproductive system, outermost glandular, swollen section of oviduct following capsule gland.

nacreous layer (hypostracum) [G *Perlmutterschicht, Hypostracum*]: Innermost layer of shell (periostracum, prismatic layer, lamellate layer, nacreous layer); consists of sheets (lamellae) of calcium carbonate (typically aragonite) alternating with thin layers of conchiolin.

nephridiopore: excretory pore

nephridium: excretory organ

nephrostome: renopericardial pore

nidamental gland [G *Eischalendrüse*]: In ♀ reproductive system, term applied either to mucus gland or to entity including albumen, capsule, and mucus glands.

node [G *Knoten*]: Type of shell sculpture; knob-like elevation on whorl, either in a series or formed by crossing of axial and spiral sculpturing. Small nodes are termed nodules.

notum [G *Notum*]: In shell-less marine snail, dorsal surface of flattened visceral body.

nuchal lobe (cervical lappet) [G *Nackenlappen*]: In certain freshwater snails, pair of leaf-like projections behind head; either left or right lobe may be modified to serve as siphon (pseudosiphon).

odontophore [G *Odontophor*]: In radular apparatus, entire complex of supporting bolsters and muscles.

ommatophore (eye stalk) [G *Augententakel, Augenstiel, Ommatophor*]: On head, one of two tentacle-like projections bearing eyes; may be separate structure or more or less fused with cephalic tentacles.

ootheca [G *Eikapsel*]: Capsule surrounding egg; produced by ♀ in capsule gland which opens into oviduct.

operculum [G *Operculum, Deckel*]: Lid-like structure serving to close aperture when snail retracts into shell; borne on posterior end of foot and composed of organic material (similar to periostracum) strengthened to varying degrees by calcareous material. (horny, calcareous; round, oval, elongate; flat, hemispherical; smooth, with bristles; growth lines: spiral, paucispiral, concentric; apex: central, marginal)

organ of Hancock [G *Hancock'-sches Sinnesorgan*]: Elongate sensory organ located in groove between cephalic shield and foot; innervated by cerebral ganglion. (folded, bipectinate)

ornament: shell sculpture

osphradium [G *Osphradium*]: Chemosensory structure adjoining gill in roof of mantle cavity; innervated by osphradial ganglion. (single, paired; bipectinate, ridge-like, wart-like)

outer lip (labrum) [G *Außenlippe*]: In shell, wall of aperture opposite inner lip; corresponds to growth margin of body whorl and may be variously thickened during pause in growth of snail. (emarginate, everted, explanate, inflected, marginate, reflected)

ovary [G *Ovar, Ovarium, Eierstock*]: Expanded section of ♀ reproductive system in which eggs are produced; closely adjoins midgut gland in visceral mass and opens to exterior via oviduct. (See also ovotestis)

oviduct [G *Ovidukt, Eileiter*]: Section of ♀ reproductive system between ovary and ♀ gonopore. May be expanded to form uterus or vagina, be connected to pericardium by gonopericardial duct, and bear seminal receptacles as well as various glands for egg capsule production. In hermaphroditic gonad (ovotestis), separate oviduct branches from hermaphroditic duct.

ovotestis [G *Zwitterdrüse*]: In reproductive system, hermaphroditic gonad producing both eggs and sperm; closely adjoins midgut gland in visceral mass and consists of few too many follicles.

palisade layer: prismatic layer

pallial cavity: mantle cavity

pallial complex [G *Mantelkomplex*]: Term applied to group of structures located on roof of mantle cavity; includes gill(s), osphradium, hypobranchial gland, and terminal parts of excretory and reproductive systems.

pallial ganglion: pleural ganglion

palmette [G *Palmette*]: Pair of flattened projections between cephalic tentacles on head; considered to be remnant of epipodium.

papillary sac (papillated sac) [G *Papillensack*]: Term applied to left, typically small excretory organ when this bears large number of hollow papillae projecting into its lumen.

parapedal commissure [G *Parapedalkommissur*]: Additional nerve commissure (accompanying pedal commissure) joining two pedal ganglia.

parapodium [G *Parapodium*]: One of two lateral lobes of posterior division (metapodium) of foot; may be used to cover shell or as swimming organ. (edge: smooth, scalloped, lobulated)

parietal callus: callus

parietal ganglion (intestinal ganglion) [G *Parietalganglion*]: One of two ganglia between pleural ganglia and visceral ganglia; connected to these by pleurovisceral connectives. Because of torsion of digestive tract, one may distinguish a supraintestinal ganglion lying above gut and a subintestinal ganglion lying below.

parietal lip [G *Parietallippe*]: On parietal region of shell, part of inner lip not consisting of columellar material. May be termed parietal callus if highly thickened. (See also columellar lip)

parietal region [G *Parietalregion*]: In shell, general area of aperture opposite outer lip, i.e., on columellar side of aperture; generally termed inner lip, especially if this is thickened. May also refer only to section of inner lip not consisting of columellar material.

pedal commissure [G *Pedalkommissur*]: Nerve commissure passing ventral to digestive tract and joining two pedal ganglia. (See also parapedal commissure)

pedal ganglion [G *Pedalganglion*]: One of two concentrations of nerve tissue below anterior section of digestive tract. Each give rise to posteriorly directed pedal nerve cord or to propodial and metapodial ganglia. May be connected to each other by pedal and parapedal commissures and may more closely adjoin other ganglia to form circumenteric nerve ring.

pedal gland [G *Fußdrüse*]: One of various glands associated with anterior end of foot, either as transverse cleft along anterior margin or opening via pore in midventral anterior region. Also refers to gland opening between pedal tentacles of sessile, tube-dwelling snails.

pedal nerve cord [G *Pedalstrang, Pedal-Markstrang*]: Pair of longitudinal nerve cords originating at pedal ganglia and extending to posterior end of foot.

pedal tentacle [G *Pedaltentakel*]: In sessile, tube-dwelling snail, pair of tentacles between operculum and cephalic tentacles.

penis (verge) [G *Penis*]: ♂ copulatory structure often located on head; either separate structure or associated with tentacle. May be enclosed in penis sheath, associated with accessory copulatory glands and organs, or armed with stylet or spines. (armed, unarmed; conical, flattened, foliaceous, tubular)

penis sheath (male antrum) [G *Penisscheide*]: Invagination or fold enclosing penis.

pericardium (pericardial cavity) [G *Perikard, Herzbeutel*]: Fluid-filled chamber enclosing heart and connected to excretory system via renopericardial canal.

pericardial gland [G *Perikardialdrüse*]: Modified inner wall of pericardium, either forming folds in pericardium itself or in the form of protuberances on auricle(s).

periphery (shoulder) [G *Peripherie*]: In typical (conispiral) shell, part of shell (i.e., part of body whorl or any other whorl) projecting or bulging farthest from shell axis.

periostracum [G *Periostracum*]: Relatively thin outermost layer of shell (periostracum, prismatic layer, lamellate layer, nacerous layer). Composed of horny organic material termed conchiolin. (smooth, rough, hairy)

pharyngeal gland (posterior salivary gland, dorsal foregut gland) [G *Speicheldrüse, dorsale Vorderdarmdrüse, Pharyngealdrüse*]: Gland opening into dorsal part of postradular pharynx. (single, paired; band-like, sacciform, tubular) (See also buccal gland)

pharynx [G *Pharynx, Schlund*]: Muscular section of digestive tract between buccal cavity and esophagus. Receives various glands and is associated with radular apparatus.

pleural ganglion (pallial ganglion) [G *Pleuralganglion, Pallialganglion*]: Concentration of nerve tissue, one to each side of digestive tract between more dorsal cerebral and more ventral pedal ganglia. May adjoin these ganglia more closely to form circumenteric nerve ring, Innervates mantle and gives rise to pleurovisceral connectives leading to parietal and visceral ganglia.

pleural tooth [G *Pleuralzahn*]: In snails in which lateral teeth of radula are indistinguishable from marginal teeth, any tooth lateral to central or rachidian tooth.

pleurovisceral   connective [G *Pleuroviszeralkonnektiv, Pleurointestinalkonnektiv*]: Strand of nerve tissue, one extending from each pleural ganglion to visceral ganglion via parietal ganglion. In typical (torted) snails, pleurovisceral connectives cross each other to form loop.

pneumostome [G *Atemloch, Pneumostom*]: In snails lacking gills, contractile opening of pulmonary sac to exterior.

posterior canal (anal canal) [G *hintere Mündungsrinne*]: In shell, posteriorly (adapically) directed, tubular or trough-like extension of aperture; termed posterior notch if not elongate. (See also siphonal canal)

prepuce [G *Präputium*]: In copulatory apparatus of ♂ with hermaphroditic gonad (ovotestis), continuation of penis sheath to exterior; may bear sarcobelum, preputial gland, or preputial organ.

prismatic layer (palisade layer) [G *Prismenschicht*]: Thick layer of shell underlying outer periostracum (periostracum, prismatic layer, lamellate layer, nacreous

layer); consists of vertical crystals of calcium carbonate (typically calcite).

proboscis [G *Rüssel*]: Retractile or invaginable, tubular, highly elongated type of snout; bears terminal mouth and serves in capture of prey. (acrembolic, pleurembolic)

proboscis sheath [G *Rüsselscheide*]: Cylindrical fold of body wall enclosing retractile (pleurembolic) type of proboscis.

propodium [G *Propodium*]: Anterior section of regionated type of foot; may be developed as flap- or hood-shaped structure and even be developed as swimming organ. Typically   delimited   from   posterior metapodium by constriction or groove and innervated by propodial ganglion.

prostate (prostate gland) [G *Prostata*]: In ♂ reproductive system, either enlarged glandular section of sperm duct or separate gland opening into sperm duct.

protoconch (nuclear whorls) [G *Embryonalgewinde, Protoconch*]: At apex of shell, whorl or whorls formed by larval snail; if not eroded away, typically distinguished from other whorls of spire by different (or lacking) sculpture. (anastrophic, deviated, heterostrophic, homeostrophic)

proventriculus   (forestomach) [G *Vormagen*]: Short, constricted, anterior section of stomach.

pseudoconch [G *Pseudoconch*]: Translucent, cartilaginous internal shell of certain pelagic snails.

pseudotrochophore   (preveliger) [G *Pseudotrochophora, Präveliger*]: Free-swimming larva of more conservative marine snails; characterized by ovoid body with preoral girdle of cilia (trochus). (See also veliger)

pseudoumbilicus (false umbilicus) [G *Pseudoumbilicus, falscher Nabel*]: Cavity or depression at base of shell; affects only body whorl and does not extend into columella. (See also umbilicus)

pulmonary sac [G *Atemhöhle, Lungenhöhle*]: Highly vascularized chamber formed by fusion of mantle cavity to body of snail; serves as respiratory surface in snails lacking gill(s) and opens to exterior via pneumostome.

punctum [G *Pore*]: Type of shell sculpture: minute pit on surface of whorl.

pustule [G *Höcker*]: Type of shell sculpture: small, rounded elevation on surface of whorl.

pyloric region: stomach

rachidian tooth (rhachidian tooth) [G *Mittelzahn, Rachis-Zahn*]: Central, typically symmetrical element in each transverse row of teeth in radula. (See also lateral tooth, marginal tooth, tooth)

radula [G *Radula*]: Specialized, protrusile armature of anterior region of digestive tract. Refers to full complement of chitinized teeth along radular membrane. (doccoglossate, hystrichoglossate, ptenoglossate, rachiglossate, rhipidoglossate, taenioglossate, toxoglossate)

radular apparatus [G *Radulaapparat*]: Protrusible anterior region of digestive tract consisting of radula (teeth), radular membrane, subradular membrane, and odontophore (supporting muscles and bolsters).

radular membrane (ribbon) [G *Radulamembran*]: In radular apparatus, membrane upon which radula teeth insert; underlain by subradular membrane.

radular sheath (radular sac) [G *Radulascheide*]: Elongate sheath producing and enclosing radula and radular membrane. Term radular sac may refer only to posterior, somewhat expanded section of radular sheath. (short, elongate, looped, coiled, posteriorly bifurcated, V-shaped)

raphe (ciliary tract) [G *Wimperband*]: One of two (dorsal, ventral) ciliated bands extending from edge of mantle posteriorly through mantle and mantle cecum; maintains water current into mantle cavity.

rectum (hindgut) [G *Enddarm, Rektum*]: Terminal expanded and cuticle-lined section of digestive tract between intestine and anus; may receive opening of anal gland.

reinforcement sac (crystal sac) [G *Kristallsack*]: In ♀ reproductive system, blind pouch whose opening closely adjoins ♀ gonopore; contains foreign bodies which are added to outer capsule layer of eggs.

renopericardial canal [G *Renoperikardialgang*]: In excretory organ, canal leading from pericardium to excretory sac.

renopericardial pore (nephrostome) [G *Nephrostom*]: Opening of excretory organ into pericardium; leads via renopericardial canal into excretory sac.

rhinophore [G *Rhinophor, Riechtentakel*]: Pair of tentacles posterior to cephalic tentacles; forms sole pair of projections on head of certain shell-less snails. Innervated by cerebral ganglion and may be retractable into sheath. (retractile, nonretractile; simple, bifurcated; smooth, papillate)

ribbon [G *Reifen*]: Type of spiral shell sculpture: flat elevation around whorls. Also refers to radular membrane.

salivary gland: buccal gland, pharyngeal gland

sarcobelum [G *Sarcobelum, Reizkörper*]: In copulatory apparatus of ♂ with hermaphroditic gonad (ovotestis), muscular or glandular diaphragm projecting into prepuce from junction of penis sheath and prepuce.

sculpture: shell sculpture

selenizone (slit band) [G *Schlitzband*]: Type of spiral shell sculpture; notch in outer lip prevents formation of shell there; this may result in long slit which may or may not be filled in by series of crescentic growth lines (slit bands).

seminal receptacle (spermatheca) [G *Receptaculum seminis*]: In ♀ reproductive system, somewhat expanded section of oviduct or distinct sperm storage organ opening into oviduct; may receive foreign sperm (allosperm) indirectly via bursa copulatrix. (sac-shaped, branched)

seminal vesicle (ampulla) [G *Samenblase, Vesicula seminalis, Ampulle*]: In ♂ reproductive system, enlarged, often convoluted section of sperm duct; serves in storing autosperm (own sperm). In hermaphroditic gonad (ovotestis), series of small diverticula along hermaphroditic duct.

septum [G *Lamelle*]: On interior of certain patelliform snail shells, lamella- or cup-shaped structure serving as support for body.

shell (concha) [G *Schale, Concha*]: Single shell element secreted by mantle; typically consists of outer periostracum, prismatic layer, lamellate layer, and inner nacreous layer. Head and foot of snail generally retractable into shell. (basic structure: patelliform, plate-like, tubular, bivalved, coiled; type of coiling: dextral, sinistral, paucispiral, multispiral, conispiral, planispiral, disjunct; shape: bulloid, cylindrical, fusiform, lenticular, globose, obovate, obconical, ovate, trochoid = trochiform, turbinate = turbiniform, turriculate = turriform; operculate, nonoperculate; umbilicate = omphalous, anomphalous; smooth, sculptured)

shell muscle [G *Schalenmuskel, Dorsoventralmuskel*]: In certain snails (i.e., limpets, lacking columella), horseshoe-shaped arrangement (either continuous or separate bundles) of dorsoventral muscle fibers from shell to foot. Modified in other snails into columellar muscle.

shell sculpture (ornament) [G *Skulptur*]: General term applied to any type of elevation or depression on exterior of shell; includes both series of individual structures and continuous surface features. According to direction of markings one may distinguish axial (or transverse, i.e., parallel to axis of shell) and spiral (winding around whorls) sculpture. These may meet to form cancellated or reticulated pattern. In order of increasing prominence:
1. individual structures
    depression: punctum, pit
    elevation: granule, pustule, tubercle, boss, nodule, node, spine
2. continuous features
    axial: (depression) stria, groove; (elevation) thread, cord, costella, costa, varix
    spiral: thread, ribbon, cord, carina, astragal

Above arrangement is only approximate and does not take shape into consideration; certain terms are preferentially used for specific taxonomic groups.

shoulder: periphery

sinus [G *Sinus*]: In circulatory system, one of numerous distinct and permanent spaces in which blood collects before returning to gills and heart. In contrast to blood vessels (e.g., aorta), lacks mesothelial wall. According to position one may distinguish, among others, abdominal, basibranchial, cephalic, cephalopedal, pedal, periradular, perivisceral, and subrenal sinuses.

siphon [G *Sipho*]: Anteriorly directed, trough-like or tubular extension of mantle margin; typically partially enclosed in siphonal canal of shell. Serves to direct water into pulmonary sac or mantle cavity.

siphonal canal (anterior canal) [G *Siphonalkanal, Siphonalrinne*]: In shell, anteriorly (abapically) directed tubular, trough-like or notch-like (siphonal notch) extension of aperture. Serves as groove for protruded siphon. (See also posterior canal)

slit band: selenizone

snout [G *Schnauze*]: Anterior noninvaginable extension of head bearing terminal mouth; may be drawn out into pair of labial palps. (See also proboscis)

spermatheca [G *Spermathek*]: Term variously applied to organ storing foreign sperm (allosperm), i.e., either to fertilization chamber or to seminal receptacle.

spermatophore [G *Spermatophor*]: Packet of sperm enclosed in membrane; produced at end of sperm duct either by prostate or by epiphallus and flagellum. Typically regionated or provided with thorn-like projections.

sperm duct (vas deferens)
[G *Vas deferens*]: Section of ♂ reproductive system between testis and penis; typically modified to form seminal vesicle and prostate. In hermaphroditic gonad (ovotestis), separate sperm duct branches from hermaphroditic duct.

spermoviduct [G *Spermoviduct*]: In hermaphroditic gonad (ovotestis), gonoduct arrangement consisting of closely adjoining or partially fused spermduct and oviduct following carrefour. (See also hermaphroditic duct)

spindle muscle: columellar muscle

spine [G *Stachel*]: Type of shell sculpture: elongate, pointed projection.

spire [G *Gewinde*]: In shell, visible part of all whorls except last whorl (body whorl).

standard measurements: As applicable to typical (conispiral) shells:
shell height (length): greatest length of shell from apex to abapical extremity (typically siphonal canal or aperture margin); measured along axis or parallel to it if two ends are not directly aligned.
shell width (diameter): greatest width of shell; measured as distance between two planes parallel to shell axis, each touching one side of shell.
spire angle: in shell whose diameter increases at a constant rate, less common measurement consisting of angle between two straight lines which touch all whorls on opposite sides.

statoconia [G *Statoconia, Otoconia*]: One of few to many smaller bodies within fluid-filled cavity of statocyst. (See also statolith)

statocyst [G *Statozyste*]: Pair of sensory (equilibrium) organs, one closely adjoining each pedal ganglion; consists of fluid-filled vesicle containing single large statolith or larger number of small statoconia. Innervated by cerebral ganglion. (round, oval, planoconvex)

statolith [G *Statolith, Otolith*]: Single solid body within fluid-filled cavity of statocyst. (See also statoconus)

stomach [G *Magen*]: Weakly muscularized section of digestive tract between esophagus and intestine; receives midgut gland. In typical (torted) snail, located at point in which digestive tract turns anteriorly. May be regionated into anterior cardiac area (into which esophagus and ducts of midgut gland open and which bears ciliated ridges = typhlosoles) and posterior pyloric area. Gives off stomach cecum and contains crystalline style.

stria [G *Ritze*]: Type of shell sculpture; very fine groove.

style: crytalline style

style sac [G *Stielsack, Magenstiel*]: In stomach, sac-like part adjoining intestine and typically containing crystalline style.

stylet [G *Stylet*]: Hollow, pointed, rigid structure (jaw or radular tooth) located within stylet tube in proboscis; projected from opening dorsal to mouth and used to pierce and suck out prey tissue. May also refer to copulatory structure associated with tip of penis.

subcerebral commissure (sublingual commissure) [G *Subzerebralkommissur, Unterschlundkommissur*]: Nerve commissure passing ventral to digestive tract and joining two cerebral ganglia. (See also cerebral commissure)

subintestinal ganglion: parietal ganglion

subradular membrane [G *Subradularmembran*]: In anterior radular sheath, multilayered membrane underlying tooth-bearing radular membrane; serves, together with bolsters, as site of muscle attachment.

subradular organ [G *Subradularorgan*]: Sensory cushion at roof of subradular pouch; may be folded and may project into pharyngeal cavity.

subradular pouch (subradular pocket) [G *Subradulartasche*]: In digestive tract, flat sac extending posteriorly below radula; roof of subradular pouch may include subradular organ.

sucker [G *Saugnapf*]: One of up to several hundred muscular, disc-shaped adhesive organs along buccal arms of pelagic snail. May also refer to sucker-like structure encircling mouth at tip of proboscis or on foot of pelagic snail.

supraintestinal ganglion: parietal ganglion

suture [G *Naht*]: In shell, border between adjoining whorls; typically in form of a line coiling around shell. (adpressed, canaliculate, channeled, flush, impressed)

teleoconch [G *Teleoconch*]: All shell whorls exclusive of protoconch.

tentacle [G *Tentakel*]: One of several types of elongate projections of body. One may distinguish the typically paired cephalic tentacles and rhinophoral tentacles on head, epipodial tentacles along foot, mantle tentacles along mantle margin, pedal tentacles of sessile, tube-dwelling snails, or labial tentacles around mouth. (smooth, fringed, grooved, papillate; simple, forked; cylindrical, flattened)

testis [G *Hoden, Testis*]: Expanded section of ♂ reproductive system in which sperm are produced; closely adjoins midgut gland in visceral mass. Opens to exterior via sperm duct and either excretory pore or penis. (See also ovotestis)

thread [G *Streifen*]: Type of spiral shell sculpture: fine elevation around whorls.

tooth [G *Zahn*]: One in a series of elements of radula (rachidian tooth, lateral tooth, marginal tooth). May also refer to tooth-like structure on columella (columellar fold). (tip of radula tooth: smooth, cuspidate, comb-like, harpoon-like, serrate, with bristles)

trochophore: pseudotrochophore

tubercle [G *Tuberkel*]: Type of shell sculpture: relatively small, rounded elevation on surface of whorl.

typhlosole [G *Typhlosolis*]: In stomach, one of two (major and minor typhlosoles) longitudinal ridges bordering a ciliated furrow; major typhlosole may continue into cecum.

umbilicus [G *Nabel, Umbilicus*]: In shell, depression or cavity at base of body whorl; aligned with axis and formed when whorls are loosely coiled and do not fuse axially to form solid columella. (anomphalous, cryptomphalous, hemiomphalous, phaneromphalous)

uncinus: marginal tooth

ureter: excretory canal

uterus (pallial oviduct) [G *Uterus, pallialer Ovidukt*]: Expanded, histologically differentiated terminal section of oviduct; may be connected to pericardium via gonopericardial duct and opens to exterior through ♀ gonopore.

vagina [G *Vagina*]: Terminal section of ♀ reproductive system. Refers either to terminal section of oviduct, to section between uterus and ♀ gonopore, or to duct leading from exterior to bursa copulatrix. May be associated with various sperm-receiving organs or with dart sac.

varix [G *Varize*]: In shell, strongly developed axial elevation indicating former position of outer lip; represents thickening of outer lip formed during pause in growth of snail. (See also growth line).

vas deferens: sperm duct

veliger [G *Veliger, Veligerlarve*]: Free-swimming larva of more advanced snails; characterized by lobed extensions of preoral trochus (velum), larval shell (protoconch), and foot rudiment. (See also pseudotrochophore)

velum: frontal sail (See also veliger)

ventricle [G *Ventrikel*]: Muscular chamber of heart receiving blood from auricle(s) and pumping it into aorta; typically separated from auricle by auriculoventricular valve.

verge: penis

visceral ganglion (abdominal ganglion) [G *Viszeralganglion, Abdominalganglion*]: Concentration of nerve tissue joined to pleural ganglia by pleurovisceral connectives. Innervates visceral mass. (single, paired)

visceral mass (visceral hump) [G *Eingeweidesack*]: Region of dorsoposterior body, generally well separated from head/foot, containing most of digestive, excretory, reproductive, and circulatory (heart) systems. Midgut gland may form greater part of visceral mass. (See also foot, head)

whorl [G *Umgang*]: In shell, any complete (360°) coil or exposed surface of complete coil. First-formed and smallest whorls begin at apex and increase in size until last whorl (body whorl). Line forming border between two adjoining whorls is termed suture. (adpressed, disjunct, imbricate, immersed)

wing [G *Mündungsflügel*]: In shell, flattened, outwardly directed expansion of outer lip.

MOLLUSCA

SCAPHOPODA (SOLENOCONCHA)

scaphopod, tusk shell, tooth shell
[G *Scaphopode, Kahnfüßer, Elefantenzahn*]

SCAPHOPODA

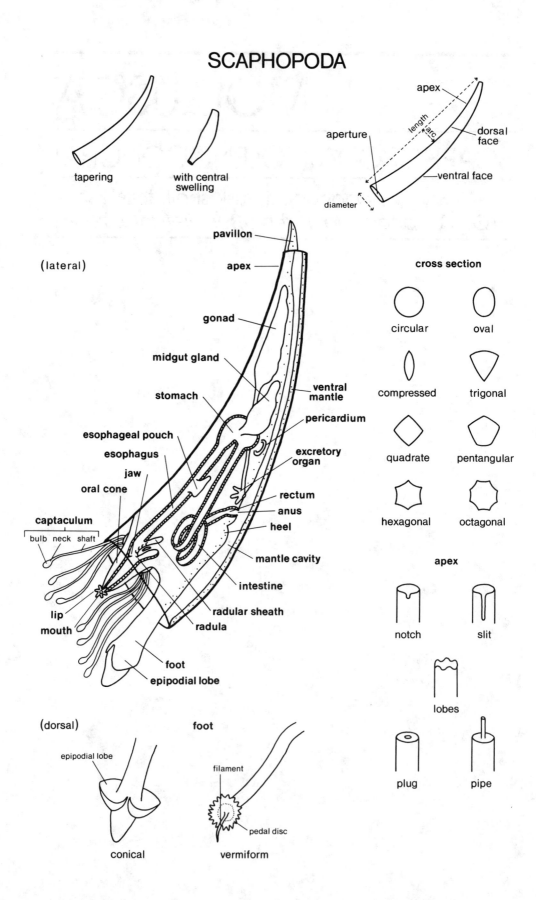

tapering

with central
swelling

apex

length

arc

aperture

dorsal
face

ventral face

diameter

(lateral)

pavillon

apex

gonad

midgut gland

stomach

esophageal pouch

esophagus

jaw

oral cone

captaculum

bulb neck shaft

lip

mouth

ventral
mantle

pericardium

excretory
organ

rectum

anus

heel

mantle cavity

intestine

radular sheath

radula

foot

epipodial lobe

cross section

circular   oval

compressed   trigonal

quadrate   pentangular

hexagonal   octagonal

apex

notch   slit

lobes

plug   pipe

(dorsal)

foot

epipodial lobe

filament

pedal disc

conical

vermiform

MOLLUSCA

SCAPHOPODA (SOLENOCONCHA)

scaphopod, tusk shell, tooth shell
[G *Scaphopode, Kahnfüßer, Elefantenzahn*]

anal orifice: apical orifice

anus [G *After*]: Posterior opening of digestive tract in mantle cavity between excretory pores. Waste is transported in a posterior direction by water currents and exits via apical orifice.

aperture (anterior aperture) [G *vordere Schalenöffnung*]: Larger anterior opening of shell through which both foot and captacula may be protruded. (constricted, not constricted; oblique, not oblique; for cross sections, see shell)

apex [G *Apex, Spitze*]: Posterior end of shell; characterized by small opening (apical orifice) with variously shaped rim. Serves for both inhalent and exhalent water currents. (simple, lobed, notched, with slit, with pipe, with plug; for cross sections, see shell)

apical orifice (anal orifice) [G *hintere Schalenöffnung*]: Relatively small posterior opening at apex of shell. Part of mantle may be protruded from orifice.

bolster [G *Radulapolster*]: In radular apparatus, pair of more rigid bodies underlying subradular membrane; serves as internal support and site of muscle attachment.

buccal cavity (buccal tube) [G *Mundhöhle, Mundrohr*]: Term occasionally applied to section of digestive tract in oral cone; may bear pair of lateral glandular pouches.

buccal ganglion [G *Bukkalganglion*]: One of several pairs of nerve concentrations connected to cerebral ganglia and surrounding anterior end of digestive tract. (See also stomatogastric complex)

callus: preapical callus

captaculum [G *Captaculum, Captakel, Fangfäden*]: One of numerous contractile, thread-like structures originating from lobe to each side of oral cone; each captaculum consists of terminal bulb with ciliated depression, neck, and shaft. Serves in capturing prey.

central tooth [G *Mittelzahn*]: Central tooth in each transverse tooth row of radula; flanked on each side by lateral and marginal teeth. (smooth, denticulated; almost square, wider than high, higher than wide)

cerebral ganglion [G *Zerebralganglion*]: One of two adjoining concentrations of nerve tissue above buccal cavity; joined to immediately following pleural ganglion by very short cerebropleural connective and to pedal ganglion by cerebropedal connective. Innervates captacula, oral cone, statocyst, and part of mantle. (See also stomatogastric complex)

cerebropedal connective
[G *Zerebropedalkonnektiv*]: Nerve interconnection joining each cerebral ganglion to corresponding pedal ganglion; forms common strand with pleuropedal connective until shortly before cerebral ganglion.

cerebropleural connective
[G *Zerebropleuralkonnektiv*]: Very short nerve interconnection joining each cerebral ganglion to corresponding, posteriorly adjoining pleural ganglion.

digestive gland: midgut gland

dorsal face [G *dorsale Schalenseite*]: Dorsal (concave) side of shell.

emunctorium: excretory organ

epipodial collar: epipodial lobe

epipodial lobe [G *Epipodiallappen, Lobus*]: In scaphopod with conical foot, one of two lateral lobes on foot; may be bent outward to help anchor foot in sediment. Dorsally interrupted, collar-like structure formed by both lobes may be termed epipodial collar or epipodial sheath.

epipodial sheath: epipodial lobe

esophageal pouch [G *Ösophagusdrüse*]: Pair of relatively large, glandular pouches, one broadly entering each side of esophagus.

esophagus [G *Ösophagus*]: Straight section of digestive tract between pharynx and stomach; bears ventral and dorsal longitudinal ciliary tracts internally (dorsal food channel) and receives pair of esophageal pouches.

excretory organ (nephridium, metanephridium, renal organ, kidney, emunctorium) [G *Exkretionsorgan, Niere, Nephridium*]: Paired organ, one on each side of rectum; each excretory organ opens to exterior through excretory pore in mantle cavity. Gonoduct enters right excretory organ.

excretory pore (nephridiopore)
[G *Exkretionsporus, Nephroporus*]: Opening of each excretory organ to exterior; located in mantle cavity posterior and lateral

to anus. May be closed by sphincter. Right excretory pore serves as gonopore.

filament: median filament

food channel [G *Nahrungsrinne*]: Longitudinal groove along dorsal wall of esophagus; formed by two parallel ciliated folds projecting into lumen of esophagus.

foot [G *Fuß*]: Muscular part of body which may project from aperture of shell; bears statocysts and pedal ganglia internally and serves as burrowing organ. Developed either as a cone with two epipodial lobes or as a cylindrical structure with pedal disc and associated median filament. (conical, vermiform)

foot disc: pedal disc

foot retractor muscle [G *Fußretraktor, Retraktormuskel des Fußes*]: Pair of longitudinal muscle bands spanning from dorsal midline around digestive tract and into foot; may divide into numerous thinner bundles.

gastric cecum [G *Caecum*]: Small ventrolateral outpocketing of stomach.

gastric shield [G *Magenschild*]: Thickened cuticle on ventral side of stomach; located next to sorting area and gastric cecum.

gonad [G *Gonade*]: Relatively large, unpaired ♂ (testis) or ♀ (ovary) reproductive organ; located behind stomach and extends dorsally to posterior end of body. Opens to exterior via right excretory organ.

head [G *Kopf*]: Anterior part of body corresponding to oral cone with internal buccal cavity and radular apparatus as well as captacula and both cerebral and pleural ganglia.

heart [G *Herz*]: In circulatory system, poorly developed contractile organ formed by dorsal invagination of, and lying within, pericardium. Consists of only a single chamber (considered to be rudimentary ventricle). Blood enters heart anteriorly and emerges posteriorly.

heel [G *Ferse*]: Posteriormost region of foot; may protrude somewhat into mantle cavity directly before anus.

hemocoel: sinus

intestine [G *Mitteldarm*]: Relatively narrow section of digestive tract between stomach and rectum; extends far into anterior region of body from stomach, coils three to five times, and curves back before opening into mantle cavity.

jaw [G *Kiefer*]: Single horseshoe-shaped cuticular structure in roof of digestive tract near junction of buccal cavity and pharynx.

labial flap: lip

lamellate layer [G *Lamellenschicht*]: Layer of shell underlying prismatic layer (periostracum, prismatic layer, lamellate layer, nacreous layer); consists of sheets (lamellae) of calcium carbonate (aragonite).

lateral tooth [G *Seitenzahn*]: In radula, relatively large tooth to each side of central tooth; flanked by marginal tooth.

lip (labial flap) [G *Lippe, Mundlappen*]: One in a series of lip-like folds surrounding mouth at tip of oral cone. (crenulated, smooth)

lobe: epipodial lobe

mantle [G *Mantel*]: Fleshy fold of body wall which secretes shell. Fused ventrally to form mantle cavity and thus lines entire tube-shaped shell. May bear one or two glandular fields anteriorly and one posteriorly.

mantle cavity (pallial cavity) [G *Mantelhöhle*]: Space between mantle and rest of body. Forms longitudinal cavity extending ventrally from aperture to apex and serves as respiratory surface. Anus and excretory pores open into mantle cavity.

marginal tooth [G *Marginalzahn, Randzahn*]: In radula, outermost element on each side of transverse tooth row. (See also central tooth, lateral tooth)

median filament (terminal filament, filament) [G *Zentralfilament,*

fingerförmiger Fortsatz, Endanhang]: In scaphopod with thread-like foot, one or more filaments extending from expanded terminal pedal disc.

midgut gland (digestive gland, liver) [G *Mitteldarmdrüse, Leber*]: Large, lobulated gland opening through one or two pores into proximal region of stomach. (single, paired)

mouth [G *Mund*]: Anterior opening of digestive tract at tip of oral cone; surrounded by lips and captacula. Opens into buccal cavity.

nacreous layer (hypostracum) [G *Perlmutterschicht, Hypostracum*]: Innermost layer of shell (periostracum, prismatic layer, lamellate layer, nacreous layer); consists of sheets (lamellae) of calcium carbonate (aragonite) within conchiolin matrix.

nephridiopore: excretory pore

nephridium: excretory organ

notch [G *Einkerbung, Kerbe*]: In shell, relatively short indentation of apex rim. (dorsal, ventral; rounded, V-shaped) (See also slit)

oral cone (proboscis) [G *Schnauze, Schnauzenkegel, Bukkalkegel*]: Contractile, conical snout located dorsally between foot and shell margin; bears terminal mouth and contains buccal cavity.

ovary: gonad

pavillon (sipho) [G *Pavillon*]: Short posteriormost section of mantle; may be protruded from apical orifice and serves as funnel for both incurrent and excurrent water flow. Bears sensory fields.

pedal disc (foot disc, terminal disc) [G *Endscheibe*]: In scaphopod with thread-like foot, disc-like terminal expansion of foot; typically with crenulated margin and occasionally with median filament. Considered to be form of epipodial collar.

pedal ganglion [G *Pedalganglion*]: One of two adjoining concentrations of nerve tissue in foot next to statocysts. Joined to cerebral ganglia by cerebropedal connectives, to pleural ganglia by pleuropedal connectives, two connectives forming a common strand over most of their length.

pericalymma (test-cell larva) [G *Pericalymma, Hüllglockenlarve*]: Later larval stage with three to five girdles of cilia and apical tuft. Bears additional layer of cells (testa) over prospective adult body. (See also pseudotrochophore)

pericardium [G *Perikard*]: In circulatory system, small, fluid-filled chamber enclosing heart; located ventrally under stomach, behind esophagus.

periostracum [G *Periostracum*]: Outermost layer of shell (periostracum, prismatic layer, lamellate layer, nacreous layer). Composed of horny organic material termed conchiolin. Existence uncertain.

pharynx [G *Pharynx, Schlund*]: Term applied to section of digestive tract which is somewhat expanded due to radular apparatus; follows buccal cavity and continues posteriorly as esophagus. Bears jaw anteriorly.

pipe (terminal pipe) [G *Siphonalröhre*]: In shell, thin, hollow tube projecting from plug closing apex; encloses and forms extension of pore in plug.

pleural ganglion [G *Pleuralganglion*]: Concentration of nerve tissue, one posteriorly adjoining each cerebral ganglion (or connected to the latter by very short cerebropleural connective). Gives rise to pleurovisceral connective leading posteriorly to corresponding visceral ganglion.

pleuropedal connective [G *Pleuropedalkonnektiv*]: Nerve interconnection joining each pleural ganglion to corresponding pedal ganglion; forms common strand with cerebropedal connective until shortly before pleural ganglion.

pleurovisceral connective [G *Pleuroviszeralkonnektiv*]: Nerve interconnection joining each pleural ganglion with corresponding visceral ganglion.

plug [G *"verengte Öffnung"*]: In apex of shell, solid septum across apical orifice; perforated centrally by pore or bearing thin pipe.

preapical callus (callus) [G *Kallus*]: Thickening of inner shell wall shortly before apex.

prismatic layer [G *Prismenschicht*]: Layer of shell underlying outer periostracum (periostracum, prismatic layer, lamellate layer, nacreous layer); consists of vertical crystals of calcium carbonate (aragonite).

proboscis: oral cone

pseudotrochophore [G *Pseudotrochophora*]: Early larval stage characterized by three girdles of cilia and apical tuft. (See also pericalymma)

radula [G *Radula*]: Specialized armature of anterior region of digestive tract. Refers to full complement of chitinized teeth along radular membrane.

radular apparatus [G *Radulaapparat*]: Anterior expanded region of digestive tract consisting of radula (teeth), radular membrane, subradular membrane, bolsters, and supporting muscles.

radular membrane (ribbon) [G *Radulamembran*]: In radular apparatus, membrane upon which radula teeth insert; underlain by subradular membrane.

radular sheath (radular sac) [G *Radulascheide*]: Elongate sheath producing and enclosing radula and radular membrane.

rectal gland [G *Rektaldrüse*]: Unpaired gland opening into rectum; considered to have excretory function.

rectum [G *Enddarm, Rektum*]: Enlarged terminal end of intestine; receives opening of rectal gland and opens into mantle cavity via anus.

renal organ: excretory organ

rib [G *Rippe*]: Type of shell structure; one in a series of large, longitudinal elevations. Less prominent ribs may be termed riblets. (rounded, angular; smooth, serrate)

ribbon: radular membrane

shell [G *Schale*]: Single tubular shell with large anterior opening (aperture) and smaller posterior opening (apical orifice); secreted by mantle and consisting of outer periostracum, prismatic layer, lamellate layer, and inner nacreous layer. (tapering, with swelling near middle; almost straight, slightly curved, strongly curved/strongly arcuate; cross section: circular, oval, compressed, trigonal, quadrate, pentagonal, hexagonal, octagonal, subpolygonal; sculpture: longitudinal, transverse/annular, smooth, striated, ribbed)

sinus [G *Sinus, Blutsinus*]: In circulatory system, one of numerous spaces in which blood collects. According to position one may distinguish abdominal, anal, cephalic, intestinal, and pedal sinuses; separated by muscular septa.

sipho: pavillon

slit [G *Schlitz*]: In shell, relatively long and narrow incision in apex rim. (dorsal, ventral) (See also notch)

sorting area [G *Sortierfeld*]: In digestive tract, ciliated area of stomach; located near opening of midgut gland. (See also gastric shield)

standard measurements:
shell length: length from dorsal side of aperture to dorsal side of apex, i.e., total length
shell diameter: greatest width of aperture or apex
arc: measure of degree of curvature, i.e., ratio of shell length to arc length

statocyst [G *Statozyste*]: Pair of sensory (equilibrium) organs, one adjoining each pedal ganglion in foot. Innervated by cerebral ganglion and consists of vesicle containing numerous statoliths.

statolith [G *Statolith*]: One of numerous small, solid calcareous bodies in each statocyst.

stomach [G *Magen*]: Expanded section of digestive tract between esophagus and intestine. Marks point at which digestive tract turns anteriorly. Typical regionation includes gastric shield and sorting area; paired or unpaired midgut gland opens into proximal and distal ends of stomach.

stomatogastric complex [G *stomatogastrisches Nervensystem*]: Term applied to closely adjoining nerve complex comprising subradular ganglia and several pairs of buccal ganglia.

stria [G *Stria, Streifen*]: Type of shell sculpture; very fine groove.

subradular membrane [G *Subradularmembran*]: In anterior radular sheath, multilayered membrane underlying tooth-bearing radular membrane. Serves, together with bolsters, as site of muscle attachment.

subradular organ [G *Subradularorgan*]: Sensory organ formed by modified floor of subradular pouch.

subradular pouch [G *Subradulartasche*]: In digestive tract, sac extending posteriorly below radula; modified floor of subradular pouch may be termed subradular sense organ.

terminal pipe: pipe

testis: gonad

thread [G *Streifen*]: Type of shell sculpture; fine longitudinal or transverse elevation.

tooth [G *Zahn*]: One of five (in one case, seven) plate-like, chitinized elements in each of the approximately 20 transverse rows of teeth in radula. According to position one may distinguish a central tooth flanked on each side by one lateral tooth and one marginal tooth.

trochophore: pseudotrochophore

veliger: pericalymma

ventral face [G *ventrale Schalen-seite, Bauchseite*]: Ventral (convex) side of shell.

ventricle: heart

visceral ganglion [G *Viszeralganglion*]: Concentration of nerve tissue, one on each side of anus. Joined to corresponding pleural ganglion by pleurovisceral connective and gives rise to posteriorly directed nerve.

water pore [G *Wasserpore*]: Pair of minute pores opening into mantle cavity, one next to each excretory pore. Considered to serve as pressure valve for body fluids.

MOLLUSCA

BIVALVIA (LAMELLIBRANCHIA, PELYCYPODA)

bivalve [G *Bivalve, Muschel*]

BIVALVIA

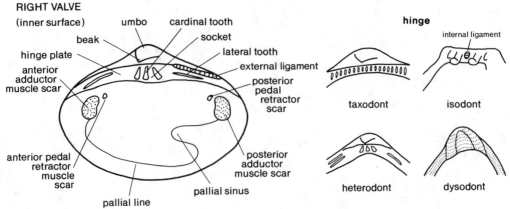

orbicular trigonal subquadrate elongate - elliptical auriculate (pectiniform)

cylindrical sword-shaped (ensiform) fan-shaped alate (pteriform) vermiform

RIGHT VALVE
(left gill removed)

crystalline style
gastric shield
digestive gland
style sac
aorta
pericardium
ventricle
stomach
esophagus
cerebral ganglion
anterior adductor muscle
mouth
labial palp
pedal ganglion
foot
auricle
rectum
excurrent siphon
anus
incurrent siphon
posterior adductor muscle
visceral ganglion
excretory organ
excretory pore
gonad
gill
gonopore
mantle lobe

RIGHT VALVE
(inner surface)

umbo
cardinal tooth
beak
socket
hinge plate
lateral tooth
anterior adductor muscle scar
external ligament
posterior pedal retractor scar
anterior pedal retractor muscle scar
posterior adductor muscle scar
pallial sinus
pallial line

hinge
internal ligament

taxodont
isodont
heterodont
dysodont

MOLLUSCA

BIVALVIA (LAMELLIBRANCHIA, PELYCYPODA)

bivalve [G *Bivalve, Muschel*]

abdominal sense organ [G *abdominales Sinnesorgan*]: In certain bivalves, one of two knob-like mechanoreceptive organs close to anus (in addition to and more posterior than osphradium). Innervated by visceral ganglion.

accessory plate [G *akzessorische Platte, Akzessorium, Schalenschaltstück*]: In certain wood- or stone-boring bivalves, one of several plates covering parts of animal remaining exposed when valves are closed (gape). According to position in front of, across, or behind umbo, one may distinguish a protoplax, mesoplax, and metaplax dorsally and a hypoplax ventrally. (paired, unpaired; calcareous, chitinous) (See also callum, siphonoplax)

adductor muscle [G *Schließmuskel*]: Large muscle spanning between valves; one may typically distinguish an anterior and posterior adductor muscle, each leaving a muscle scar. Anterior muscle often reduced, posterior one rarely reduced. Serves to draw valves together and functions as antagonist to ligament. May consist of both quick (phasic, vitreous, striated) and catch (tonic, nacreous, smooth) muscle bundles. (dimyarian, monomyarian; homomyarian = isomyarian, anisomyarian = heteromyarian)

adductor scar [G *vorderer Muskeleindruck, Schließmuskeleindruck*]: muscle scar

anterior lateral tooth: lateral tooth

anteroposterior axis: oroanal axis

anus [G *After*]: Posterior opening of digestive tract into mantle cavity; located above and behind posterior adductor muscle. Fecal pellets reach exterior through excurrent siphon.

aorta [G *Aorta*]: Single (anterior) or paired (anterior and posterior) major blood vessels leading from ventricle of heart to major organs; anterior aorta may be variously expanded into bulbus arteriosus.

apophysis [G *Apophyse*]: On inner wall of valve of certain wood- or stone-boring bivalves, calcareous projection of hinge region serving for attachment site of pedal muscle.

atrium: auricle

auricle (atrium) [G *Atrium, Vorhof*]: In heart, one of two relatively thin-walled chambers leading into ventricle; receives blood from ctenidia and other respiratory surfaces. Separated from ventricle by atrioventricular opening with valve.

auricle (ear, hinge ear) [G *Ohr*]: Triangular, relatively compressed extension of valve anterior and/or posterior to beak; typically clearly delimited from main body or disc of shell by furrow (auricular sulcus). Byssal notch or sinus in anterior auricle permits passage of byssus. (unequal, nearly equal)

bead [G *Knopf*]: Type of valve sculpture: relatively small, rounded protuberance, typically formed at point of intersection of radial and concentric (commarginal) sculpture.

beak [G *Wirbelspitze, Apex des Wirbels, Embryonalschale*]: Small, beak-like part of valve along or above hinge; represents earliest part of shell and continues to form outwardly curving umbo. Often used as synonym for umbo.

bladder [G *Exkretionssack, Exkretionsblase*]: Term applied to expanded and relatively thin-walled upper (distal) limb of each excretory organ; opens into mantle cavity via excretory pore.

blood sinus (lacuna) [G *Blutsinus, Sinus, Lakune*]: In circulatory system, one of numerous spaces in which blood collects. Largest sinus, located below and associated with excretory organ, is termed sinus venosus.

branchia: ctenidium

brood chamber [G *Bruttasche, Marsupium*]: Modified space, typically between gill leaflets yet occasionally also in free mantle cavity, in which fertilized eggs may be brooded.

bulbus arteriosus: aorta

byssal disc pit [G *Trichter*]: Distal depression of byssal groove in which tip of each byssus thread is formed.

byssal foramen [G *Byssusloch*]: In bivalve permanently attached to hard substrate on flattened right valve, opening in right valve for passage of calcified byssus.

byssal gape: gape

byssal gland [G *Byssusdrüse*]: On posterior end of (often reduced) foot, internally folded gland secreting byssus; opens via pore which is continuous with byssal groove and ends in byssal disc pit.

byssal groove [G *Byssusrinne*]: Thin groove extending along ventral edge of foot from opening of byssal gland to byssal disc pit; serves as mold for individual byssus threads.

byssal notch [G *Byssuseinschnitt*]: In right valve of byssiferous bivalve, notch between anterior auricle and main body or disc of shell; smaller notch in corresponding position of left valve is termed byssal sinus.

byssus [G *Byssus*]: Bundle of strong, hair-like threads originating from posterior end of (often reduced) foot. Produced by byssal gland in conjunction with byssal groove and byssal disc pit and extended from shell to attach to hard substrates.

callum [G *Kallusumschlag*]: In certain wood- or stone-boring bivalves, calcareous anterior extension of shell; closes pedal gape.

cardinal area [G *Ligamentfläche*]: When viewing valve from dorsal side, triangular surface between beak and hinge; partly or wholly occupied by ligament. (flat, concave)

cardinal platform: hinge plate

cardinal tooth [G *Hauptzahn*]: In valve, one of two basic categories of tooth (cardinal teeth, lateral teeth). One of up to several relatively short, ventrally radiating teeth below beak. Separated from lateral teeth by toothless space. (smooth, denticulate; opisthocline, orthocline, prosocline) (See also pseudocardinal tooth)

carina (keel, ridge) [G *Carina, Kiel*]: Type of valve sculpture: prominent radial or concentric (commarginal) ridge or keel.

cartilage [G *Knorpel*]: internal ligament

cecum [G *Caecum*]: Any outpocketing of stomach; may contain part of sorting area, serve for storage, or receive openings of digestive gland.

cerebral ganglion [G *Zerebralganglion*]: One of two concentrations of nerve tissue behind anterior adductor muscle and above esophagus; connected dorsally to each other by cerebral commissure. Innervates labial palps. Joined to pedal ganglion by cerebropedal connective. Generally fused posteriorly with pleural ganglion and thus termed cerebropleural ganglion.

cerebropedal connective [G *Zerebropedalkonnektiv*]: Interconnection extending ventrally to join each cerebral ganglion to corresponding pedal ganglion; initially separate from, then merging with pleural connective.

cerebropleural ganglion [G *Zerebropleuralganglion*]: One of two concentrations of nerve tissue formed by fusion of cerebral and pleural ganglia; located behind anterior adductor muscle and above esophagus. Gives rise to two eventually merging interconnections (cerebropedal and pleuropedal connectives) leading to pedal ganglion and connected to visceral ganglion by cerebrovisceral connective. Innervates anterior mantle, anterior adductor muscle, labial palps, and statocysts.

cerebrovisceral connective: pleurovisceral connective

chondrophore [G *Chondrophor, Knorpelträger*]: In valve, shelly projection with hollowed-out surface on hinge plate; serves for attachment of internal ligament. (spoon-shaped) (See also resilifer)

conchiolin: periostracum

costa (rib) [G *Rippe*]: Type of valve sculpture: moderately broad and prominent radial or concentric (commarginal) elevation. According to relative size on valve, one may distinguish primary, secondary, or tertiary ribs. (rounded, flat-topped)

costella (riblet) [G *Rippchen, Costula*]: Type of valve sculpture: more narrow costa.

cruciform muscle [G *Spangenmuskel, Musculus cruciformis*]: In certain bivalves, relatively small muscle spanning posteroventrally between valves (in front of incurrent siphon); intersect in midbody to form cross.

crystalline layer (inner/outer crystalline layer): lamellate layer, nacreous layer, prismatic layer

crystalline style (style) [G *Kristallstiel, Fermentstiel, Gallertstiel*]: In stomach, relatively soft, concentrically layered, rod-shaped body secreted by and partially contained within style sac. One end (head) projects into stomach, where it rotates against gastric shield and partially dissolves. (straight, curved)

ctenidium (gill, branchia) [G *Kieme, Ctenidium*]: Respiratory organ, mostly combined with feeding function, suspended in posterior region of mantle cavity to each side of median visceral mass. Each ctenidium consists of a longitudinally directed axis with series of variously shaped, heavily ciliated filaments. Inhalent, anteriorly directed water current passes gills ventrally, exhalent (posteriorly directed) current dorsally. (ctenidobranchiate, eulamellibranchiate, protobranchiate, pseudolamellibranchiate, septibranchiate; heterorhabdic, homorhabdic) (See also demibranch)

demibranch [G *äußere/innere Kieme, Kiemenblatt*]: One half of each gill. In more advanced bivalve, typically consists of series of recurved filaments extending from main axis of gill, thus being V-shaped (set closer to visceral mass may be termed inner demibranch, those closer to mantle and valve, outer demibranch). Limb attached to axis is descending limb, while upturned section is ascending (typically shorter) limb. (heterorhabdic, homorhabdic; reflected, not reflected)

denticle [G *Zähnchen, Knötchen*]: Type of marginal sculpture along inner surface of valve; small, rounded, tooth-like protuberance.

dentition [G *Bezahnung*]: Collective term for full complement of hinge teeth and sockets in both valves.

digestive gland (midgut gland, liver) [G *Mitteldarmdrüse, Verdauungsdrüse*]: Basically paired gland opening into left and right sides of stomach; consists of highly branched system of ducts and digestive diverticula.

dissoconch [G *Dissoconch, Adultschale*]: Term applied to shell secreted after metamorphosis of larva. Typically differs from larval shell (prodissoconch) in sculpture and shape; separated from latter by distinct line.

diverticulum [G *Divertikel, Mitteldarmdivertikel*]: digestive gland

dorsal hood (dorsal pouch) [G *dorsale Magentasche*]: Anterodorsal sac-like outpocketing of stomach above entrance of esophagus.

ear: auricle

escutcheon [G *Area*]: In valve, flattened, structurally differentiated dorsal surface posterior to beak; represents successive growth layers of hinge structures. When entire shell is viewed from dorsal side, escutcheon typically forms elongate depression bordered by nymph and split longitudinally by hinge and ligament.

esophagus [G *Ösophagus, Vorderdarm*]: Relatively short, tubular or somewhat dorsoventrally depressed section of digestive tract between mouth and stomach. Lacks larger glands or muscular swellings.

excretory organ (nephridium, kidney, organ of Bojanus, emunctorium) [G *Exkretionsorgan, Niere, Nephridium, Bojanus'sches Organ, Emunktorium*]: Paired organ lying below pericardium; each organ consists basically of U-shaped pericardioduct with longitudinally oriented limbs and bend next to posterior adductor muscle. Lower (proximal) limb is attached to pericardium via renopericardial pore, upper (distal) limb opens into mantle cavity via excretory pore.

excretory pore (nephridiopore) [G *Exkretionsöffnung, Nephroprokt*]: Opening of each excretory organ (upper, distal limb) into mantle cavity. Located at base of gill between foot and posterior adductor muscle. May also serve as gonopore.

external ligament [G *äußeres Ligament*]: ligament

eye (photoreceptor, ocellus) [G *Auge, Photoreceptor, Ocellus*]: Any one of variously developed and arranged photosensitive organs, either on mantle edge, siphon, gill, or pallial tentacles. May be well developed, even consisting of cornea, lens, and retina, or be aggregated to form compound eye.

fold [G *Falte*]: Type of valve sculpture; broad radial or concentric (commarginal) undulation.

food channel [G *Nahrungsrinne*]: In more primitive bivalve, dorsal groove in esophagus; bordered by ciliated, longitudinal folds.

foot [G *Fuß*]: Relatively large, muscular ventral part of body continuous with visceral mass and suspended in mantle cavity; protruded from shell and primitively serves in plowing or burrowing. Innervated by pedal ganglia and may be modified to accommodate byssal apparatus.

fossette: socket

fulcrum: chondrophore

gape [G *Spalt, Klaffspalt*]: Opening remaining in shell margin even after valves are closed. According to position or part of exposed soft body one may distinguish a siphonal (posterior, for siphons), pedal (anterior, for foot), and variously positioned byssal gapes.

gastric shield [G *Magenschild*]: Heavily cuticularized dorsal area of stomach; head of crystalline style rotates against gastric shield.

gastric tooth [G *Magenzahn*]: Hook- or tooth-like protuberance of gastric shield.

genital pore: gonopore

gill: ctenidium

glochidium [G *Glochidium*]: In freshwater bivalve, larval form characterized by bivalved shell, with or without pair of hooks, and long adhesive thread (tentacle).

Lives as temporary parasite on fish gills or fins. Modified glochidium in certain species is termed lasidium before attachment, haustorium after attachment.

gonad [G *Gonade*]: Paired ♂ (testis) or ♀ (ovary) reproductive organ located to either side of and intermingled with intestinal loops; opens into mantle cavity either via excretory organ or via gonoduct and gonopore.

gonoduct [G *Gonodukt*]: In reproductive system, shorter ♂ (sperm duct) or ♀ (oviduct) duct leading from each gonad into mantle cavity; either enters excretory organ and opens via excretory pore (renogenital pore) or leads directly into mantle cavity where it opens via gonopore.

gonopericardial duct [G *Gonoperikardialgang*]: In more primitive bivalve, duct leading from pericardium to gonad.

gonopore (genital pore) [G *Gonoporus, Geschlechtsöffnung*]: Opening of reproductive system into mantle cavity; located at base of gill between foot and posterior adductor muscle. May be termed renogenital pore if gonad and excretory organ open through common pore.

groove: sulcus

growth line (incremental line) [G *Anwachsstreifen, Zuwachsstreifen*]: On outer surface of valve, one in a series of concentric (commarginal) thin lines indicating former positions of margin. Termed growth ruga if more elevated and irregular in shape.

haustorium: glochidium

heart [G *Herz*]: In circulatory system, contractile organ located within pericardium; consists of two auricles and one ventricle. Blood enters ventricle through auricles and is pumped into one or two (anterior and posterior) aortas.

hinge [G *Schloß*]: In valve, collective term for all structures (hinge teeth, sockets, ligament, chondrophore, etc.) along dorsal margin which function during opening and closing of shell. (actinodont, anodont = toothless = edentulous, anomalodesmatic,

ctenodont = pseudotaxodont, desmodont, diagenodont, dysodont, hemidapedont, heterodont, isodont, orthodont, pachydont, prionodont, pseudotaxodont, schizodont, taxodont)

hinge axis (cardinal axis, hinge line) [G *Schloßachse*]: Imaginary straight line about which valves open; located dorsally in hinge region and passes through ligament. May be used as an orientation for certain standard measurements. Angle between hinge axis and oroanal axis indicates orientation of soft body in shell.

hinge plate [G *Schloßplatte*]: In dorsal (hinge) region of valve, prominent shelly plate below beak; bears hinge teeth.

hinge tooth: tooth

hypobranchial gland (mucus tract) [G *Hypobranchialdrüse, Schleimkrause*]: In more primitive bivalve, tall glandular epithelium lining posterior mantle cavity.

hypoplax: accessory plate

hypostracum: nacreous layer

imprint [G *Eindruck*]: On inner surface of valve, impression left by muscle (muscle scar) or gill.

internal ligament [G *inneres Ligament*]: ligament

interspace (interval) [G *Zwischenraum*]: In valve sculpture, depression between adjacent radial or concentric (commarginal) elevations.

intestine (midgut) [G *Intestinum, Mitteldarm*]: Elongate, typically highly coiled section of digestive tract between stomach and rectum. Anterior intestine may be closely associated with style sac, posterior part surrounded by ventricle and pericardium. Forms fecal pellets.

isthmus (mantle isthmus) [G *Mantel-Isthmus*]: Dorsomedian part of mantle underlying hinge region and connecting mantle lobes lining valves.

keel: carina

kidney: excretory organ

labial palp [G *Labialpalpe, Mund-palpe, Mundsegel*]: One of two pairs (anterior and posterior) of elongate, triangular flaps on each side of mouth. Primatively continuous with esophageal fold. Smooth on outer side, with ciliated ridges on inner side. Serves as sorting organ for filtered particles; edible particles are transported to mouth via oral groove.

lacuna: blood sinus

lamella [G *Lamelle*]: Type of valve sculpture: thin, plate-like concentric (commarginal) elevation. May also refer to filaments of ctenidium.

lamellate layer (inner crystalline layer, hypostracum) [G *Lamellen-schicht, Kreuzlamellenschicht*]: Layer of valve underlying prismatic layer (periostracum, prismatic layer, lamellate layer); consists of sheets (lamellae) of crystals of calcium carbonate (typically aragonite) and usually forms crossed lamellar structure. Replaces nacreous layer in advanced bivalve.

larva [G *Larve*]: Generally an early pelagic stage of development characterized by one to three girdles of cilia (trochi) and apical tuft. Barrel-shaped with three trochi = lecithotroph pericalymma or test-cell larva; globular to pear-shaped with one prominent trochus = pseudotrochophora or rotiger; with enlarged trochus as well as valves = veliconcha, advanced also with foot = pediveliger. (See also glochidium)

lasidium [G *Lasidium*]: glochidium

lateral tooth [G *Seitenzahn*]: In valve, one of two basic categories of hinge tooth (cardinal teeth, lateral teeth). One of up to several relatively long teeth extending more or less parallel to dorsal shell margin. Separated from cardinal teeth by toothless space. Lateral teeth anterior to beak are termed anterior lateral teeth, those posterior to beak, posterior lateral teeth. (smooth, denticulate) (See also pseudolateral tooth)

ligament [G *Ligament, Schloßband*]: Elastic, multilayered structure joining two valves of shell dorsally, typically posterior to beak. According to its position either dorsal

(external) to or within hinge one may distinguish an external and internal ligament. Serves as antagonist of adductor muscles. (single, bipartite; amphidetic, opisthodetic; alivincular, duplivincular, multivincular, paravincular) (See also resilium)

lithodesma (ossiculum) [G *Li-thodesma*]: In hinge region of valve, small shelly plate reinforcing internal ligament.

lunule [G *Lunula*]: In valve, flattened, structurally differentiated dorsal surface anterior to beak; represents successive growth layers of hinge structures. When entire shell is viewed from above, both halves of lunule typically form a heart-shaped depression (split longitudinally by hinge). (See also escutcheon)

mantle [G *Mantel*]: Entire dorsal body wall; according to laterally compressed body, consists of two lobes, one lining and secreting each valve. Joined dorsally by mantle isthmus. Ventral margin may be fused posteriorly to form siphon(s).

mantle cavity (mantle chamber, pallial cavity) [G *Mantelhöhle, Mantel-raum, Pallialraum*]: Large space between midbody and two mantle lobes lining valves. Main body of bivalve and gills are suspended in mantle cavity, which may be somewhat regionated to separate incurrent and excurrent water flow.

mantle fold [G *Mantelfalte*]: One of typically three (inner, middle, outer) folds forming margin of mantle (lobes). Outer fold functions in secretion of shell, middle fold is sensory, and inner fold contains muscles (attachment line on valve termed pallial line).

mantle lobe [G *Mantellappen*]: One of two lateral extensions of mantle forming flattened lobes and directly underlying secreting valve; margin bears two or three mantle folds. Joined dorsally by mantle isthmus.

mesoplax: accessory plate

metaplax: accessory plate

mouth [G *Mund, Mundöffnung*]: Anterior opening of digestive tract behind anterior adductor muscle at base of foot; typically surrounded by labial palps. Food enters mouth via gills, labial palps, and oral grooves and then proceeds into esophagus.

muscle scar [G *Muskeleindruck, Muskelansatzstelle*]: On inner wall of valve, impression left by attachment of muscles. Largest and most conspicuous muscle scars are formed by muscles spanning between valves, where one may often distinguish an anterior and posterior adductor muscle scar. Additional muscle scars include those formed by pallial muscles (pallial line), siphonal retractor muscles (pallial sinus), and pedal retractor muscles. (anisomyarian = heteromyarian, dimyarian, isomyarian = homomyarian, monomyarian)

myostracum [G *Myostracum*]: Restricted areas of inner crystalline layer of valve secreted by muscle attachments; according to muscles involved, one may distinguish a pallial myostracum (pallial line), adductor myostracum (adductor muscle scars), etc.

nacreous layer (inner crystalline layer, hypostracum) [G *Perlmutterschicht*]: Innermost layer of valve in more primitive bivalve (periostracum, prismatic layer, nacreous layer); consists of sheets (lamellae) of calcium carbonate (aragonite) alternating with thin layers of conchiolin. In advanced bivalve, replaced by lamellate layer. A similar arrangement of calcite in valve is termed foliated structure.

nephridiopore: excretory pore

nephridium: excretory organ

nymph [G *Ligamentleiste*]: In valve, narrow platform extending posteriorly from beak along dorsal margin; serves as attachment surface for external ligament.

oral groove [G *Mundrinne*]: Ciliated groove between each pair of labial palps; food particles from ctenidia are transported along oral grooves into mouth and esophageal food channel.

organ of Bojanus: excretory organ

oroanal axis (anteroposterior axis) [G *Körperlängsachse*]: Axis of soft body passing through mouth and anus; generally equivalent to anteroposterior axis. Angle between hinge axis and oroanal axis indicates orientation of soft body in shell.

osphradium [G *Osphradium*]: Paired sensory modification of body wall in posterior region of body; located close to visceral ganglia between base of foot and anus near attachment of ctenidia. Innervated by visceral ganglia, occasionally with fibers from pleural ganglia.

ossiculum: lithodesma

ostium [G *Wasserpore*]: In outer surface of gills characterized by fusion of adjacent filaments, one in a series of small holes allowing incurrent water to flow through gills (into suprabranchial chamber).

ostracum: lamellate layer, prismatic layer

otocyst: statocyst

otolith: statolith

ovary: gonad

pallet [G *Palette*]: In wood-boring bivalve, one of two small calcareous structures serving to plug burrow when siphons are retracted. Consists of stalk and expanded distal part and is operated by set of muscles.

pallial line [G *Palliallinie, Mantellinie*]: On inner wall of valve, thin, elongate impression made by attachment of mantle muscle. Typically extends from anterior to posterior adductor muscle scars and runs parallel to ventral margin of valve. May curve inward posteriorly to form pallial sinus. (disjunct, integripalliate, sinupalliate = entire)

pallial muscle [G *Pallialmuskel*]: Collective term for any muscle of mantle. Radial pallial muscles are attached along margin of valve to form pallial line; serve to withdraw mantle edge into shell before valves are closed.

pallial nerve [G *Pallialnerv*]: Nerve, one running from each pleural ganglion along mantle margin to corresponding visceral ganglion in posterior region of body.

pallial tentacle [G *Mantelrandtentakel*]: One in a series of tentacle-like projections of middle mantle fold; sensory in function, often with eyes.

pallial sinus (sinus) [G *Pallialsinus, Mantelbucht*]: On inner wall of valve, inwardly directed, U-shaped curve of pallial line. Located in posterior region of valve and representing muscle scar made by attachment of siphonal retractor muscles.

palp proboscis (palp appendage) [G *Mundlappenanhang*]: In certain bivalves, one of two tentacle-like projections associated with palps; one appendage originates from posterior end of each upper labial palp. Protruded from shell during feeding.

pedal elevator muscle [G *Fußheber, Levator pedis*]: pedal muscle

pedal ganglion [G *Pedalganglion*]: One of two closely adjoining concentrations of nerve tissue in foot. Joined to cerebropleural ganglia via cerebropedal and pleuropedal (separate, then merging) connectives. Innervates foot and byssus retractor muscle. Positioned close to (but not innervating) statocysts.

pedal gape: gape

pedal levator muscle: pedal muscle

pedal muscle [G *Pedalmuskel, Fußmuskel*]: One of several muscles either within foot or extending from foot to valve (dorsoventral musculature). Includes pedal elevator, pedal protractor, and pedal retractor (levator) muscles. The latter are the largest and may form muscle scars on inner wall of valve; one may often distinguish anterior and posterior pedal retractors, the former usually behind anterior adductor muscle scar, the latter above posterior adductor scar. May be modified in byssiferous bivalve to form byssal-pedal retractor.

pericardial gland [G *Perikardialdrüse*]: Glandular tissue associated with wall of pericardium or heart auricle; involved in excretion (ultrafiltration).

pericardium (pericardial cavity) [G *Perikard, Herzbeutel*]: In circulatory system, fluid-filled chamber enclosing heart and part of posterior intestine (rectum); located under dorsal margin of shell and connected to excretory system via renopericardial pore.

periostracum [G *Periostracum*]: Outermost, relatively thin layer of shell (periostracum, prismatic layer, lamellate or nacreous layer). Composed of horny organic material termed conchiolin and secreted by periostracal glands in mantle folds.

pleural ganglion [G *Pleuralganglion*]: Concentration of nerve tissue closely adjoining cerebral ganglion posteriorly or fused with latter to form cerebropleural ganglion. Joined to visceral ganglion by pleurovisceral (cerebrovisceral) connective. Gives rise to pallial nerve.

pleuropedal connective [G *Pleuropedalkonnektiv*]: Nerve interconnection extending ventrally to join each pleural ganglion to corresponding pedal ganglion; initially separate from, then merging with cerebropedal connective.

pleurovisceral connective [G *Pleuroviszeralkonnektiv*]: Nerve interconnection extending posteriorly to join each pleural ganglion to corresponding visceral ganglion. May be termed cerebrovisceral connective if cerebral and pleural ganglia are fused to form cerebropleural ganglion.

plica [G *Plica, Falte*]: Type of valve sculpture; fold or costa involving entire thickness of valve.

posterior lateral tooth: lateral tooth

prismatic layer (outer crystalline layer, ostracum) [G *Prismenschicht*]: Layer of valve underlying periostracum (periostracum, prismatic layer, lamellate or nacreous layer); consists of vertical, columnar crystals of calcium carbonate (calcite or aragonite).

proboscis: palp proboscis

prodissoconch [G *Prodissoconcha*]: Bivalved shell formed by larva prior to metamorphosis; according to shell-secreting organ, one may distinguish an earlier and smaller prodissoconch I and a later and larger prodissoconch II (enclosing entire animal). (See also dissoconch, provinculum)

protoplax: accessory plate

protostyle (ergatulum, mucus rod) [G *Protostyl*]: In stomach of certain more primitive bivalves, rod-shaped body resembling crystalline style yet composed of mucus and nutritive material.

provinculum [G *Provinculum*]: Precursor to adult hinge on larval shell (prodissoconch); consists of small teeth or crenulations along median part of hinge margin.

pseudocardinal tooth: Term applied to cardinal tooth in certain bivalves; cardinal tooth is not separated from lateral tooth on hinge (pseudolateral teeth extend up to or overlap pseudocardinal teeth) and is somewhat irregular.

pseudolateral tooth: Term applied to lateral tooth in certain bivalves; lateral tooth is not separated from cardinal tooth on hinge (pseudolateral teeth extend up to or overlap pseudocardinal teeth).

punctum [G *Nadelstich*]: Type of valve sculpture: minute pit, typically in a series in interspace between radial or concentric (commarginal) elevations.

pustule [G *Pustel*]: Type of valve sculpture: small, rounded protuberance.

rectum [G *Enddarm, Rektum*]: Terminal section of digestive tract between intestine and anus; may be surrounded by ventricle and pericardium.

renopericardial pore [G *Renoperikardialöffnung, Nephrostom*]: Opening of pericardium into excretory organ (lower, proximal limb).

resilifer [G *Resilifer*]: In valve, any type of structure on hinge plate for attachment of internal ligament (resilium). (See also chondrophore)

resilium [G *Resilium, Schließknorpel*]: Synonym for internal ligament; occasionally also applied to inner layer of ligament. (See also resilifer)

rib: costa

riblet: costella

rotiger: larva

scale [G *Schuppe*]: Type of valve sculpture: thin, flat projection. Commonly on radial sculpture and arising at crossing point of concentric (commarginal) lamellae and radial ribs.

scar: muscle scar

shell [G *Schale*]: Entire calcareous structure, consisting of two valves, more or less enclosing soft body.

sinus: blood sinus, pallial sinus

siphon [G *Sipho*]: Tube-like, muscular extension of mantle formed by fusion of mantle margins. Typically present in the form of two posteriorly directed tubes, the lower (more ventral) siphon for inhalent current and the upper one for exhalent current. (separate, fused)

siphonal gape: gape

siphonal retractor muscle [G *Siphonalretraktor, Siphonen-Rückziehmuskel*]: Modified muscles of mantle margin (pallial muscles) serving to retract siphons. Attachment line forms pallial sinus on inner wall of valve.

siphonoplax [G *Siphonoplax*]: In certain wood- or stone-boring bivalves, calcareous posterior extension of shell; forms tube enclosing proximal portion of siphons.

socket (fossette, fosset) [G *Zahngrube, Grube*]: In valve, one of typically several depressions along dorsal margin or in hinge plate. Corresponds with tooth on opposing valve.

soft body [G *Weichkörper*]: Collective term for all parts of body exclusive of valves; includes visceral mass, foot, gills, mantle, and muscles.

sorting area [G *Sortierfeld*]: Ciliated and ridged ventral region in stomach.

spine [G *Stachel*]: Type of valve sculpture: elongate, pointed projection. (straight, hooked)

standard measurements: Shell measurements tend to vary considerably even with only slight changes in orientation; shell should therefore be aligned so that hinge axis is horizontal.

shell length: greatest length from anterior to posterior end; measured parallel to hinge axis.

shell height: greatest dimension perpendicular to length; measured from highest dorsal point (umbo tip, hinge margin) to most ventral margin or parallel to this if both extremities are not directly aligned.

shell width (inflation): greatest width measured as distance between two planes parallel to hinge axis and touching outermost part of each valve.

thickness: distance between inner and outer surfaces of a valve (at a specified point).

statoconia (otolith) [G *Statolith, Statoconia*]: One of few to many smaller bodies within statocyst. (See also statolith)

statocyst (otocyst) [G *Statozyste, Otozyste*]: Pair of sensory (equilibrium) organs in foot; located along connectives (cerebropedal and pleuropedal) between cerebropleural and pedal ganglia, often near or adjoining the latter. Innervated by cerebropleural ganglion and typically containing either single large statolith or more numerous and smaller statoconiae. (open, closed)

statolith (otolith) [G *Statolith, Otolith*]: Single, relatively large, solid body within statocyst. (See also statoconia)

stomach [G *Magen*]: Relatively thin-walled section of digestive tract between esophagus and intestine; receives openings of digestive glands. Generally characterized internally by gastric shield and sorting area, as well as crystalline style in style sac; gives off cecum.

stria [G *Stria, Streife, Linienstreife*]: Type of valve sculpture: very fine radial or concentric (commarginal) groove. Occasionally also applied to fine elevation.

style: crystalline style

style sac [G *Kristallstielscheide, Fermentstielsack*]: In stomach, posterior sac-like or elongate outpocketing or separation more or less closely associated with intestine and typically containing crystalline style.

sulcus [G *Furche, Sulcus*]: Type of valve sculpture; radial groove.

suprabranchial chamber (exhalent mantle chamber) [G *Suprabranchialraum*]: More or less well-defined space in upper part of mantle cavity, one to each side of visceral mass. Delimited below by ctenidia, which thus also serve in separating inhalent and exhalent water currents.

test-cell larva: larva

testis: gonad

thread [G *Faden*]: Type of valve sculpture: fine radial or concentric (commarginal) elevation.

tooth (hinge tooth) [G *Zahn, Schloßzahn*]: In valve, one of typically several projections along dorsal margin or on hinge plate. According to position one may basically distinguish cardinal and lateral teeth. Teeth fit into corresponding sockets on opposing valve and function in alignment and stability of shell. (See also pseudocardinal tooth, pseudolateral tooth)

trochophore (pseudotrochophore): larva

tubercle [G *Tuberkel*]: Type of valve sculpture: relatively small, rounded protuberance.

umbo [G *Umbo, Wirbel*]: In valve, strongly curving dorsal region following beak. Represents early growth stages of shell (radial and concentric valve sculpture most tightly spaced here. (opisthogyrate, orthogyrate = straight, prosogyrate, spirogyrate) (See also beak)

umbonal cavity [G *Wirbelhöhle*]: On inner side of valve, space within umbo.

valve [G *Klappe, Schalenklappe*]: One of two calcareous shell elements secreted by mantle and more or less covering left and right sides of soft body (therefore, left and right valves). Consists of outer periostracum followed by two calcareous crystalline layers (basically prismatic and nacreous or lamellate layers). Two valves are joined dorsally by hinge and ligament. (relation of valves to one another: closed, gaping, equal = equivalved, unequal = inequivalved; symmetry of single valve: equilateral, inequilateral; outer surface: smooth, sculptured; inner surface: anisomyarian = heteromyarian, dimyarian, isomyarian = homomyarian, monomyarian, integripalliate, sinupalliate; lateral outline: circular = orbicular, elongate, elongate-elliptical, ovate = oval, quadrangular, rhomboidal, subcircular = suborbicular, subelliptical, suboval, subquadrate, subrectangular, subtrigonal, trigonal = triangular, wedge-shaped, alate, bialate, auriculate. For lateral outline, terms incorporating genus names are also frequently employed, e.g., donaciform, ensiform, modioliform, nuculaniform, pectiniform, pteriform) (See also accessory plate, valve sculpture)

valve sculpture (ornament) [G *Skulptur*]: General term applied to any type of elevation on or depression in outer surface of valve; includes both series of individual structures or continuous surface features. According to direction of markings one may distinguish concentric or commarginal (running parallel to valve margins) and radial (running from beak to margins) sculpture. In order of increasing prominence:

1. individual structures
     depression: punctum, pit
     protuberance: granule, pustule, bead, tubercle, scale, spine

2. continuous features
     depression: stria, groove
     elevation: thread, costella (riblet), costa (rib), carina (keel), lamella, fold (plica)

Above arrangement is only approximate and does not take shape into consideration; certain terms are preferentially used for specific taxonomic groups.

veliger (veliconcha, pediveliger): larva

ventricle [G *Ventrikel*]: Muscular chamber of heart receiving blood from auricles and pumping it into aorta(s). May surround rectum.

visceral ganglion [G *Viszeralganglion, Viszeroparietalganglion*]: One of two concentrations of nerve tissue beneath posterior adductor muscles. Joined to (cerebro)-pleural ganglion by pleurovisceral (cerebropleurovisceral) connective. Innervates posterior mantle, siphons, posterior adductor muscles, gills, osphradia, abdominal sense organs, excretory system, heart, gonads, and gut.

visceral mass [G *Viszeralmasse*]: Region of body containing most of digestive, excretory, circulatory (heart), and nervous systems. Distinguished from other major parts of soft body such as muscles, foot, gills, and mantle.

wing [G *Flügel*]: Extension of dorsal region of valve; similar to auricle, yet extending beyond main body of shell.

CEPHALOPODA

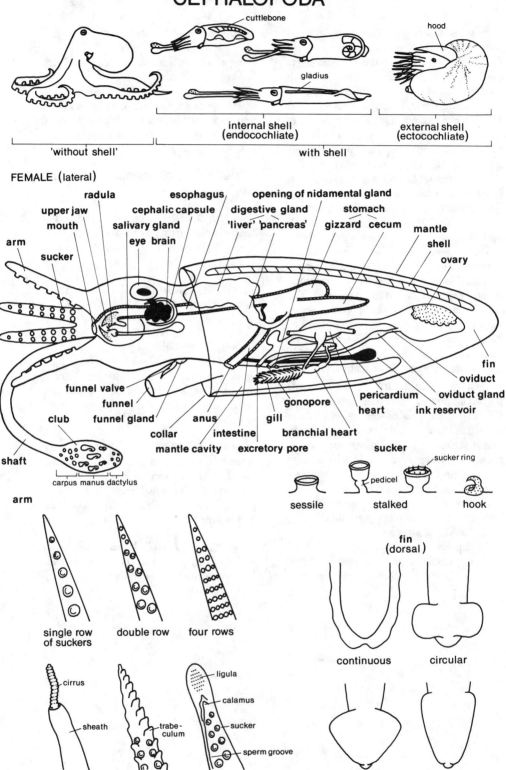

cuttlebone

gladius

hood

internal shell
(endocochliate)

external shell
(ectocochliate)

'without shell'

with shell

FEMALE (lateral)

radula · esophagus · opening of nidamental gland

upper jaw · cephalic capsule · digestive gland · stomach

mouth · salivary gland · 'liver' 'pancreas' · gizzard cecum · mantle

arm · eye brain · shell

sucker · ovary

fin

oviduct

funnel valve · gonopore · pericardium · oviduct gland

funnel · heart · ink reservoir

club · funnel gland · anus · gill · branchial heart

collar · intestine · sucker

shaft · mantle cavity · excretory pore

carpus manus dactylus

sucker ring

sessile · stalked · hook

arm · pedicel

single row · double row · four rows
of suckers

cirrus · ligula

sheath · calamus

trabe-culum · sucker

sperm groove

bipartite · hectocotylized

fin
(dorsal)

continuous · circular

triangular · elongate

MOLLUSCA

CEPHALOPODA (SIPHONOPODA)

cephalopod [G *Cephalopode, Kopffüßer*]

The terms anterior, posterior, dorsal, and ventral in text refer to orientation of animal in illustration.

anal flap (anal leaflet) [G *Anallappen*]: Small, fleshy projection, one on each side of anus.

annulus [G *Annulus*]: In ectocochliate cephalopod, ring-shaped attachment band of soft body on inner wall of final shell chamber; passes through attachment points of retractor muscles.

anus [G *After*]: Posterior opening of digestive tract located anteriorly in mantle cavity behind funnel; may be positioned at tip of papilla-like protuberance and flanked by pair of anal flaps.

aorta [G *Aorta*]: Major blood vessel leading from ventricle of heart; one may basically distinguish an anterior aorta (cephalic aorta) extending into head and a posterior aorta (aorta abdominalis) extending to end of visceral mass.

appendix [G *Appendix, Umkehrsack*]: In ♂ gonoducal system (spermatophoric organ), narrow loop or sac at interconnection of finishing gland and Needham's sac.

aquiferous duct: gonopericardial duct

aquiferous pore: water pore

arm [G *Arm*]: One in a circle of 8 (octobrachiate or octopod cephalopod), 10 (decabrachiate or decapod cephalopod), or approximately 90 (in two circlets—ectoco-chliate cephalopod) elongate, muscular projections surrounding mouth. Represents perioral tentacles of other molluscs. Arms bear suckers along surface facing mouth and may be more or less connected to each other by web. One or more arms of ♂ may be modified (hectocotylized) for copulation. (annulated, with/without suckers, circular, flattened) (See also cirrus, tentacle)

auricle (atrium) [G *Atrium, Vorkammer*]: One of two or four (depending on gill number) relatively thin-walled chambers of heart. Blood enters auricles from gills and passes into ventricle, where it is pumped into body through aortas.

baculum (rod) [G *Baculum, Stab, Sattel*]: One of two rod-shaped remnants of shell in octobrachiate cephalopod.

bolster (odontophore, cartilage) [G *Radulapolster, Odontophor, Radulaknorpel*]: In radular apparatus, rigid mass of tissue and turgescent cells underlying teeth and radular membrane; serves as support and site of muscle attachment.

brachial ganglion [G *Brachialganglion*]: Concentration of nerve tissue anterior to pedal ganglion. May be separate or more or less fused to pedal ganglion; gives rise to large nerves (brachial nerves) extending into each arm/tentacle.

brain [G *Gehirn*]: Main concentration of nerve tissue in head. Consists, in ectocochiliate cephalopod, of circumesophageal ring with closely adjoining cerebral, pedal, pleurovisceral, and buccal ganglia. In endocochliate cephalopod, these ganglia are typically fused into a more uniform mass bearing numerous lobes. More or less enclosed in cephalic cartilage.

branchial ganglion [G *Branchialganglion*]: Relatively small concentration of nerve tissue, one innervating each ctenidium. Joined to pleurovisceral ganglion by connective.

branchial gland [G *Branchialdrüse, Kiemenbanddrüse*]: Glandular tissue along line of fusion of ctenidium and mantle cavity.

branchial heart (gill heart) [G *Kiemenherz*]: In endocochliate and shell-less cephalopods, accessory pumping organ in circulatory system; consists of muscular, thick-walled expansion at base of each of the two ctenidia. Pumps deoxygenated blood from body (coming via vena cava) into ctenidium. Continuous with pericardial gland.

buccal cavity [G *Mundhöhle, Pharyngealhöhle*]: Expanded section of digestive tract between mouth and esophagus; bears radular apparatus ventrally and pair of jaws anteriorly and receives openings of various salivary glands. (See also buccal mass)

buccal ganglion [G *Bukkalganglion*]: One of two relatively small pairs (superior and inferior buccal ganglia) of ganglia located behind buccal mass. Superior buccal ganglia either adjoin cerebral ganglia anteriorly or are joined to latter by connectives. Superior and inferior buccal ganglia joined by interbuccal connectives.

buccal gland (ventral buccal gland, submandibular gland, submaxillary salivary gland, sublingual gland) [G *Bukkaldrüse, Submandibulardrüse*]: In endocochiate and shell-less cephalopods, relatively small glandular mass below radular apparatus in buccal

cavity. Dorsal buccal gland may also be present. (See also salivary gland)

buccal lappet [G *Bukkalzipfel, Mundarm, Bukkallappen*]: One of six to eight triangular extensions of buccal membrane surrounding mouth; attached to base of arms by buccal membrane connectives.

buccal mass [G *Schlundkopf, Bukkalmasse*]: Highly muscular, well-delimited section of digestive tract between mouth and esophagus; functions as a unit and consists of jaws, radular apparatus, and associated glandular organs. Lumen delimited by buccal mass is termed buccal cavity.

buccal membrane [G *Bukkalmembran, Bukkaltrichter*]: In decabrachiate cephalopod, thin membrane encircling mouth; may be drawn out aborally into buccal lappets, bear minute buccal suckers, and is attached to base of tentacles by buccal membrane connectives. In ♀ may bear fold serving as seminal receptacle.

buccal membrane connective [G *Bukkalpfeiler*]: Muscular band connecting buccal membrane to bases of arms. (dorsally attached, ventrally attached)

buccal sucker [G *Bukkalsaugnapf*]: One of several minute suckers on buccal membrane or buccal lappets surrounding mouth.

calamus [G *Calamus*]: In octobrachiate cephalopod, conical, papilla-like projection at end of modified arm (hectocotylus) of ♂; located after last sucker, at end of sperm groove.

carpal cluster (carpal pad) [G *Karpal-Gruppe*]: On proximal region (carpus) of tentacular club, clearly delimited group of small suckers (carpal suckers) and knobs (carpal knobs); adheres to carpal club of second tentacular club during capture of prey. (See also fixing apparatus)

carpal knob [G *Haftknopf*]: On proximal region (carpus) of tentacular club, one of numerous small, rounded, hemispherical protuberances; typically grouped with small suckers (carpal suckers) to form carpal cluster. Each carpal knob adheres to carpal sucker of second tentacular club during capture of prey. (See also fixing apparatus)

carpal pad: carpal cluster

carpal sucker [G *karpaler Saugnapf*]: On proximal region (carpus) of tentacular club, one of numerous small suckers; typically grouped with small protuberances (carpal knobs) to form carpal cluster. Each carpal sucker adheres to carpal knob of second tentacular club during capture of prey. (See also fixing apparatus)

carpus [G *Karpalteil*]: Most proximal of three regions (carpus, manus, dactylus) of expanded terminal section (club) of tentacle; Bears small suckers (carpal suckers) and knobs (carpal knobs); these adhesive organs may be clustered (carpal cluster) and interlock with corresponding elements of second tentacular club during capture of prey.

cartilage: cephalic cartilage, funnel-locking cartilage, mantle-locking cartilage, neck-locking cartilage

"cartilaginous" scale [G *Schuppenblase*]: In certain decabrachiate cephalopods, one of numerous minute, rigid elements in outer surface of body. (knob-like, papilla-like, scale-like and overlapping)

cecal sac [G *Appendix des Caecums, Caecum-Anhang, Spiraldarm-Sack*]: Elongate, posteriorly tapering extension of ventral chamber (cecum) of stomach; extends to end of visceral mass.

cecum [G *Caecum, Blindsack, Spiraldarm*]: Thin-walled ventral chamber of stomach (gizzard, cecum); inner surface of cecum bears numerous ciliated folds (leaflets). Receives opening of digestive gland and is separated from gizzard by sphincter. (straight, spiraled)

cephalic cartilage [G *Kopfkapsel, Knorpelkapsel*]: Elastic capsule more or less enclosing brain; may serve as site of muscle attachment.

cerebral ganglion [G *Zerebralganglion, Gehirnganglion, Supraösophagealmasse*]: Concentration of nerve tissue above

esophagus. In ectocochliate cephalopod, cerebral ganglion forms circumesophageal ring along with pedal and pleurovisceral ganglia; in endocochliate cephalopod, it is fused with additional ganglia to form more uniform mass (brain) with numerous lobes. Joined to more anteriorly positioned buccal ganglia by connectives.

chromatophore [G *Chromatophor*]: Pigment-filled, sac-shaped cell in outer surface of body; contraction and expansion of chromatophores result in lighter or darker coloration and color patterns. Expansion (by associated radial muscles) regulated by subesophageal part of brain.

cirrus [G *Zirrus, Cirrus*]: 1) In ectocochliate cephalopod, slender, annulated distal portion of each (of approximately 90) arm; retractable into thicker basal portion (sheath). 2) One in a series of elongate, finger-like projections along lateral edges of arms. 3) Warty protuberance above each eye in certain octobrachiate cephalopods.

club: tentacular club

collar [G *Mantelvorderrand*]: Rim-like anterior margin of mantle, especially free ventral edge. Water enters mantle cavity around edge of collar and is expelled through funnel.

cone (conus) [G *Endkegel, Phragmoconus*]: At posterior end of elongate internal shell (gladius, cuttlebone), spoon- or cup-like conical expansion; may bear spike-like extension (rostrum).

crop (prestomach) [G *Kropf, Vormagen*]: Expanded posterior region of esophagus; adjoins stomach and serves to store food after feeding.

ctenidium (gill) [G *Kieme, Ctenidium*]: Respiratory organ suspended in mantle cavity; consists of axis and alternating series of flattened projections (lamellae). Either one pair supported by a pair of accessory pumping organs (branchial heart) or doubled to two pairs of gills.

cuttlebone [G *Schulp*]: Rigid, chambered, elongate inner shell of certain cephalopods.

dactylus [G *Finger*]: Most distal of three regions (carpus, manus, dactylus) of expanded terminal section of tentacle (tentacular club); bears small suckers.

digestive gland (midgut gland, hepatopancreas) [G *Mitteldarmdrüse*]: Large gland opening into ventral chamber (cecum) of stomach; typically subdivided into "liver" and "pancreas."

diverticulum: finishing gland

epistellar body [G *Epistellarkörper*]: Pair of glands, one attached to each stellate ganglion by neurosecretory processes.

esophagus [G *Ösophagus, Schlundrohr*]: Relatively narrow, muscular section of digestive tract between buccal cavity and crop or stomach.

excretory organ (kidney, nephridium, emunctorium) [G *Exkretionsorgan, Niere*]: Paired (in ectocochliate cephalopod, two pairs) organ located in visceral mass, associated with pericardium, and opening into mantle cavity. Consists of renopericardial canal, renal sac, and excretory pore.

excretory pore (nephridiopore, renal pore) [G *Exkretionsöffnung*]: Opening of each renal sac into mantle cavity, typically on an elevation (renal papilla).

eye [G *Auge*]: Well-developed photosensitive and image-perceiving organ on each side of head; most complex type consists of eyelid, outer cornea, iris, lens, and retina. Innervated by large optic lobe of brain and enclosed by cartilage plates. (See also orbital pore, orbital sinus)

fin (lateral fin) [G *Flosse, Seitenflosse*]: Typically one pair of muscular dorsolateral extensions of tube- or sac-like mantle; serves in locomotion, steering, and stabilization. (circular, elongate, triangular, fused)

finishing gland (prostate, diverticulum) [G *Rangierdrüse, Prostata, ak-zessorische Drüse*]: In ♂ reproductive system, terminal part of sperm duct in which cap and thread of spermatophore are secreted; leads into Needham's sac.

fin lobe [G *vorderer Flossenlappen*]: Part of each fin extending anteriorly beyond most anterior point of attachment of fin and mantle.

fixing apparatus [G *Haftapparat*]: Term applied to entire complement of knobs and suckers in carpal region of both tentacular clubs. Serves to interlock both clubs during capture of prey.

foot: funnel

foregut gland: salivary gland

foveola [G *Foveola*]: Transverse membranous fold forming pocket in funnel groove along underside of head. (See also side pocket)

funnel (siphon) [G *Trichter*]: Muscular, funnel-shaped structure extending from mantle cavity below neck; either uniform tube or consisting of two adjoining folds. Water expelled from mantle cavity through movable funnel enables rapid jet swimming. Represents ventrally rolled up foot.

funnel groove [G *Trichtergrube*]: Elongate depression in underside of head in which anterior portion of funnel lies.

funnel-locking cartilage [G *Knorpelgrube des ventralen Schließapparates, Trichterknorpel, Trichterhaft*]: One of two cartilaginous recesses (groove, pit, pocket, depression) located ventrolaterally on outer surface of funnel. Interlocks with corresponding cartilaginous projection (mantle-locking cartilage) on mantle during swimming in order to direct water out through funnel (and not through mantle opening between head and collar).

funnel organ (organ of Verrill) [G *Trichterorgan, Trichterdrüse, Fußdrüse*]: Glandular structure on inner surface of funnel; either W- or V-shaped with two additional oblong components.

funnel valve [G *Trichterventil, Trichterklappe*]: Muscular flap in dorsal wall of funnel; projects into distal end of funnel and is anteriorly directed. Prevents water flow from exterior into mantle cavity through funnel (it must pass through mantle opening between head and collar).

giant fiber [G *Riesenfaser*]: One element in a system of large nerve fibers extending through mantle and serving to enable bilaterally symmetrical contractions of mantle. According to location of neurons and fibers one may distinguish first-, second-, and third-order giant neurons/fibers.

gill: ctenidium

gill heart: branchial heart

gizzard [G *Magensack, Muskelmagen*]: Muscular, cuticle-lined dorsal chamber of stomach; joined to ventral cecum by sphincter.

gladius (pen) [G *Gladius, Rückenfeder*]: Thin, flexible, elongate inner shell of certain cephalopods; lacks chambers. Consists of thickened central axis (rachis), thinner lateral expansions (vanes), and conical posterior section (cone) with spike-like tip (rostrum).

gonad: ovary, testis

gonoduct: oviduct, sperm duct

gonopericardial duct (aquiferous duct) [G *Gonoperikardialkanal*]: Long, narrow duct joining gonad (gonocoel) with each remnant of pericardium around pericardial glands; each gonopericardial duct opens to exterior by water pore.

gonopore [G *Geschlechtsöffnung*]: Opening of ♂ or ♀ reproductive system into mantle cavity next to anus. Paired, yet generally only left gonopore retained in ♂.

head [G *Kopf*]: Anterior region of body bearing mouth, perioral arms, buccal mass, brain, and eye.

head-foot [G *Kopffuß, Cephalopodium*]: Original ventral portion of molluscan body, later separated into head (see arm) and foot (see funnel). Arms and tentacles were formerly considered to represent part of modified foot, both regions therefore occasionally being subsumed under the term head-foot.

heart (systemic heart) [G *Herz*]: In circulatory system, contractile organ typically located within pericardium; consists of two or four auricles (depending on gill number) and one ventricle. Blood, coming from gills, is pumped via auricles and ventricle into aortas. (See also branchial heart)

hectocotylus [G *Hectocotylus, Begattungsarm*]: In ♂, modified arm(s) serving in copulation. Suckers, sucker stalks, protective membranes, and trabeculae may be modified; characterized, in octobrachiate cephalopod, by terminal calamus and ligula. Transfers spermatophores from funnel or Needham's sac and deposits them on or into ♀.

hepatopancreas: digestive gland

hood [G *Kopfkappe*]: In ectocochliate cephalopod, fleshy, hood-like structure above head; represents modified (expanded and fused) tentacle bases and serves to cover animal when it withdraws into shell.

hook [G *Haken*]: On arm or expanded terminal end (club) of tentacle, chitinous, claw-like structure; represents modified chitinous ring of sucker.

ink gland: ink sac

ink sac (ink gland) [G *Tintenbeutel, Tintendrüse*]: In visceral mass, posteroventral organ consisting of ink-producing gland, reservoir, duct, sphincter, and ampulla. Opens into rectum; ink released into mantle cavity is expelled to exterior via funnel.

intestine [G *Intestinum, ausführender Mitteldarm*]: Elongate, muscular section of digestive tract between stomach and rectum; proceeds in anterior direction to open in mantle cavity. Regionated proximally in ectocochliate cephalopod to reflect different internal structure of incoming gizzard and cecum. (straight, coiled)

iridocyte [G *Iridozyte, Flitterzelle*]: Cell type underlying chromatophores in outer surface of body; each iridocyte contains minute iridescent plates that reflect incoming light and may produce color patterns in combination with chromatophores.

jaw [G *Kiefer, Schnabel*]: In anterior section of buccal cavity, pair of beak-like biting and tearing organs. Consists of upper and lower jaws, the lower element encompassing and jutting more anteriorly than upper element. Both upper and lower jaws bear relatively large, posteriorly directed lamellae serving as muscle attachments.

keel (swimming membrane) [G *Schwimmsaum*]: On arm, flattened muscular extension along surface directed away from mouth. May also refer to one or two expanded muscular membranes on terminal end (club) of tentacle.

kidney: excretory organ

kidney sac: renal sac

Kölliker's canal [G *Ductus koellikeri, Kölliker'scher Gang*]: In ectocochliate cephalopod, narrow duct leading from each statocyst to exterior. In endocochliate cephalopod, retained as small blind diverticulum from statocyst (as a remnant of embryological invagination).

Kölliker's organ (Kölliker's bristles) [G *Kölliker'sches Büschel, Kölliker'-sches Organ*]: In juvenile octobrachiate cephalopod, bundle of bristles used in hatching.

lateral lobe [G *lateraler Bukkallappen*]: Fleshy lobe projecting into buccal cavity, one on each side of radular apparatus; bears opening of anterior salivary gland on inner face and may be armed with denticles.

lateral tooth: tooth

light organ: photophore

ligula [G *Ligula, Greifplatte*]: In octobrachiate cephalopod, spatulate to spoon-shaped, transversely structured portion at end of modified arm (hectocotylus) of ♂.

lip [G *Lippe*]: In ectocochliate cephalopod, fold or set of folds of body wall around mouth; located in front of jaws at entrance of buccal cavity. (smooth, papillated) (See also prelingual process)

"liver" (distal digestive gland) [G *"Leber"*]: Relatively large distal or anterior subdivision of digestive gland (liver, pancreas); connected to pancreas by pair of narrow ducts.

lobe (lobus) [G *Gehirnlappen, Lobus*]: One of numerous more or less distinctly delimited centers in more highly developed type of cephalopod brain; according to position and function one may distinguish, for example, anterior basal, posterior basal, lateral basal, inferior frontal, superior frontal, verticalis, subverticalis, and optic lobes.

mantle [G *Mantel*]: Muscular, sac- or tube-shaped part of body; encloses cavity (mantle cavity) containing gills and, by contracting, expels water through funnel (respiration, locomotion). Mantle may secrete shell which either houses animal (ectocochliate cephalopod) or becomes enclosed by mantle (endocochliate cephalopod).

mantle cavity (pallial cavity) [G *Mantelhöhle, Pallialraum*]: Cavity formed by mantle; contains gills and openings of ink sac as well as of digestive, reproductive, and excretory systems. Water enters mantle cavity around edge of collar and is expelled through funnel.

mantle-locking cartilage [G *Knorpelkopf des ventralen Schließapparates, Trichterhaftknorpel*]: One of two cartilaginous projections (ridge, knob, swelling) on inner surface of mantle; located ventrolaterally near anterior rim (collar) of mantle. Interlocks with corresponding cartilaginous recess (funnel-locking cartilage) on funnel during swimming in order to direct water out through funnel (and not through mantle opening). (See also neck-locking cartilage)

manus [G *Handteil*]: Central of three regions (carpus, manus, dactylus) of expanded terminal section (club) of tentacle; bears large suckers and hooks.

median tooth: tooth

midgut gland: digestive gland

mouth [G *Mund, Mundöffnung*]: Anterior opening of digestive tract; surrounded by circlet of arms and tentacles, lips, and, in decabrachiate cephalopod, by buccal membrane. Opens into buccal cavity.

mucilaginous gland: spermatophore gland

nacreous layer [G *Perlmutterschicht, Hypostracum*]: In ectocochliate cephalopod, inner of two layers of shell (nacreous layer, porcelaneous layer).

neck [G *Hals, Nackenregion*]: Term occasionally applied to somewhat narrowed region of body between head and visceral mass.

neck-locking cartilage (dorsal cartilage) [G *dorsaler Schließapparat, Nackenknorpel, Mantelhaft*]: One of two dorsal cartilaginous formations enabling neck and mantle to interlock.

Needham's sac (spermatophore sac, spermatophoric sac) [G *Spermatophorensack, Needham'sche Tasche*]: In ♂ reproductive system, large, elongate sac serving to store finished spermatophores. Located at end of sperm duct and surrounds its distal glandular section; opens into mantle cavity via gonopore.

nephridiopore: excretory pore

nephridium: excretory organ

nidamental gland [G *Nidamentaldrüse*]: In ♀ reproductive system, one or two pairs of sac- or groove-like glands opening into mantle cavity next to gonopore. Secretes outermost cover of eggs or gelatinous substance forming egg mass. Term incorrectly applied to accessory gland (accessory nidamental gland) containing luminescent symbiotic bacteria and functioning as a photophore.

ocellus [G *Augenfleck*]: In certain cephalopods, pigmented patch usually consisting of central chromatophore spot surrounded by one or more concentric rings of chromatophores.

odontophore: bolster

olfactory papilla (rhinophore, postocular tubercle) [G *Rhinophor*]: Protuberance on each side of head behind eye; considered to be olfactory in nature. (bump-like, finger-like)

olfactory pit [G *Riechgrube*]: Sensory depression on each side of head behind or beneath eye.

optic lobe [G *Lobus opticus*]: Large, conspicuous lobe of brain, one extending to each eye; may be located between brain and eye, joined to former by stalk-like nerve.

orbital pore (eye pore) [G *Cornea-Porus*]: Minute pore in cornea covering eye; exposes lens to sea water.

orbital sinus [G *Lidsinus*]: Anteriorly directed indentation of eyelid.

organ of Verrill: funnel organ

osphradium [G *Osphradium*]: Sensory organ in ectocochliate cephalopod; consists of modified epithelium near gills in mantle cavity.

ovary [G *Ovar, Ovarium, Eierstock*]: Expanded, unpaired section of ♀ reproductive system in which eggs are produced. Located medially in posterior region of body and continuous with pericardium; opens into mantle cavity via oviduct(s), oviducal gland, and gonopore.

oviducal gland [G *Eileiterdrüse*]: In ♀ reproductive system, glandular distal region of oviduct. Secretes outer cover (membrane/capsule) of egg. (See also nidamental gland)

oviduct [G *Ovidukt, Eileiter*]: Typically elongate, narrow, and winding section of ♀ reproductive system leading from ovary into mantle cavity; expanded distally to form seminal receptacle and oviducal gland. (single, paired)

"pancreas" (proximal digestive appendage) [G *"Pankreas"*]: Relatively small proximal or posterior subdivision of digestive gland (liver, pancreas); connected to liver by pair of narrow ducts.

pedal ganglion [G *Pedalganglion*]: One of two closely adjoining concentrations of nerve tissue ventral to esophagus. Either joined to cerebral ganglion to form antero-ventral band of circumesophageal ring (posterior band = pleurovisceral ganglion) or fused with posterior pleurovisceral and cerebral ganglia to form more uniform brain. Connected or continuous with anterior brachial ganglion.

pedicel (stalk) [G *Stiel*]: Short, tubular stalk of sucker; contains nerve.

pen: gladius

penis [G *Penis*]: Term occasionally applied to muscular, somewhat extended section of terminal ♂ genital duct (spermatophore, Needham's sac) protruding into mantle cavity; may bear diverticulum containing spermatophores.

pericardial gland [G *Perikardial-drüse, Kiemenherzanhang*]: Glandular, internally highly folded organ adjoining each branchial heart; continuous with both pericardial cavity and branchial heart. Serves in excretion.

pericardial pore [G *vordere Exkretionsöffnung*]: In ectocochliate cephalopod, one of two openings of pericardium via second (anterior) excretory organ into mantle cavity in front of excretory pore.

pericardium (pericardial chamber) [G *Perikard, Herzbeutel*]: In circulatory system, fluid-filled chamber enclosing heart; opens into mantle cavity either separately or through renopericardial duct and excretory pore. May be reduced to two smaller chambers surrounding pericardial gland/branchial heart complex.

photophore (light organ) [G *Leuchtorgan*]: One of variously constructed light-producing organs. Bioluminescence is produced by either cephalopod (self-generated, intrinsic) or symbiotic bacteria (extrinsic). Luminescent material may be ejected into water.

pleurovisceral ganglion (visceral ganglion) [G *Pleuroviszeralganglion*]: One of two closely adjoining concentrations of nerve tissue ventral to

esophagus. Either joined to cerebral ganglia to form posteroventral band of circumesophageal ring (anterior band = pedal ganglion) or fused with anterior pedal and cerebral ganglia to form more uniform brain. Gives rise to posteriorly directed nerves (visceral or pallial nerves) extending to stellate ganglia, visceral mass, and branchial ganglia.

pocket [G *Tentakeltasche, Hauttasche*]: One of two depressions on underside of head; each pocket receives one retractile tentacle when animal is not feeding.

poison gland (posterior salivary gland) [G *Giftdrüse, hintere Speicheldrüse*]: Relatively large pair of glands located at anterior end of midgut gland and opening into buccal cavity on papilla in front of radular apparatus.

porcelaneous layer [G *Prismenschicht, Ostracum*]: In ectocochliate cephalopod, outer of two layers of shell (nacreous layer, porcelaneous layer); consists of prisms of calcium carbonate in an organic matrix.

prelingual process (ventral buccal lobe) [G *ventraler Bukkallappen*]: In ectocochliate cephalopod, one of two (anterior and posterior prelingual process) fleshy processes inside buccal cavity between lower jaw and radula.

prestomach: crop

prostate: finishing gland

protective membrane [G *Schutzsaum*]: Thin, web-like integument along arm or expanded terminal section (club) of tentacle. Located to each side of suckers along margins facing mouth; supported by muscular rods (trabeculae).

rachidian tooth: tooth

rachis [G *Achse*]: Thickened central axis of flexible, elongate inner shell (gladius); thinner lateral expansions along rachis are termed vanes.

radula [G *Radula*]: Specialized armature of anterior region of digestive tract (floor of buccal cavity). Refers to full complement of teeth along radular membrane.

radular membrane (ribbon) [G *Radulamembran*]: In radular apparatus, membrane upon which radula teeth insert; underlain by subradular membrane.

radular apparatus [G *Radulaapparat*]: Modified anterior region of digestive tract (floor of buccal cavity) consisting of radular sac, radula (teeth), radular membrane (ribbon), subradular membrane, bolster, and supporting muscles.

radular sac [G *Scheide, Radulasack*]: Posteriorly directed buccal diverticulum in which radula and radular membrane (ribbon) are differentiated.

rectum [G *Rektum, Enddarm*]: Section of digestive tract between intestine and anus; receives duct of ink gland.

renal appendage [G *Venenanhang, renaler Appendix*]: One in a series of branched projections of veins (venae cavae) returning to gills; extends into renal sac and serves in excretion.

renal sac (kidney sac) [G *Exkretionssack, Nierensack, Renalsack*]: Large, expanded section of excretory organ; connected to pericardium via renopericardial canal and opens into mantle cavity via excretory pore. Veins (venae cavae) returning to gills are in contact with and send projections (renal appendages) into renal sac. (two pairs, one pair: fused, separate)

renopericardial canal [G *Renoperikardialgang*]: In excretory system, canal leading from pericardium to renal sac.

resisting apparatus [G *Verschlußapparat*]: Term applied to full complement of cartilages serving to temporarily attach anterior rim of mantle (collar) to funnel; consists of projecting elements (mantle-locking cartilages) which interlock with corresponding cartilaginous recesses (funnel-locking cartilages). Serves to direct water flow from mantle cavity to exterior through funnel (and not through mantle opening).

retractor muscle (head retractor, funnel retractor, pallial retractor) [G *Schalenmuskel, Kopfretraktor, Trichterretraktor, Musculus depressor infundibuli*]: Muscle bundle spanning from each side of final shell chamber, of shell rudiment, or of corresponding posterior site of mantle to cephalic cartilage and funnel (depressor infundibuli muscle). Serves to withdraw head and arms into shell or mantle.

rhinophore: olfactory papilla

rostrum (spine) [G *Rostrum*]: Spike-like extension of cone at posterior end of flexible, elongate inner shell (gladius).

salivary gland (foregut gland) [G *Speicheldrüse*]: One of two pairs of glands opening into buccal cavity. One may distinguish 1) anterior salivary glands located in posterior region of buccal mass and opening on lateral lobes and 2) posterior salivary glands (poison glands) located at anterior end of midgut gland and opening on papilla in front of radular apparatus.

secondary shell: shell

secondary web [G *Nebenhaut*]: Narrow membrane connecting web (primary web) to arms.

seminal receptacle [G *Receptaculum seminis*]: In ♀ reproductive system, expanded distal section of oviduct serving as sperm storage organ. May also refer to sperm-receiving area located below mouth in a fold of buccal membrane.

seminal vesicle [G *Samenblase, Vesicula seminalis*]: In ♂ reproductive system, glandular, somewhat expanded section of sperm duct. First phase of spermatophore production may occur with sperm collected in seminal vesicle. Leads into spermatophore organ.

septum [G *Septum, Scheidewand, Kammerwand*]: One in a series of transverse partitions forming chambers in shell. Secreted by mantle and, in coiled shells, perforated by siphuncle.

shell [G *Schale*]: Variously developed supporting structure secreted by mantle. Either 1) external, coiled, with septa, chambers, and siphuncle (ectocochliate cephalopod) or 2) internal, coiled or elongate, in the latter case either flexible (gladius) or rigid (cuttle-bone) (endocochliate cephalopod). External shell consists of outer procelaneous and inner nacreous layers. May also refer to thin, unchambered, coiled secondary shell secreted by arms of certain species to protect eggs.

side pocket [G *Trichtertasche*]: In funnel groove along underside of head, one of several smaller membranous folds forming pockets lateral to foveola.

sinus (blood sinus, lacuna) [G *Sinus, Lakune*]: In circulatory system, space in which blood collects. Largest (visceral) sinus surrounds digestive tract.

siphon: funnel

siphuncle [G *Schalensipho, Siphonalrohr*]: In shell, slender calcareous tube perforating septa and extending to earliest chambers; contains narrow extension of soft body (siphuncular cord).

spadix [G *Spadix*]: In ectocochliate cephalopod, conical, muscular copulatory organ of ♂; formed by fusion of four arms from inner perioral circle of arms.

spermatophore [G *Spermatophore*]: Elongate packet of sperm produced at distal end of sperm duct (seminal vesicle, spermatophore gland, mucilaginous gland, finishing gland), stored in spermatophore sac (Needham's sac), and transferred onto or into ♀ by modified arm (hectocotylus) of ♂. Regionated, consisting of sperm mass, cement body, and coiled ejaculatory organ enclosed in one or more sac-like layers (tunic, capsule) and topped by a cap and thread.

spermatophore gland [G *Spermatophorendrüse*]: In ♂ reproductive system, region of two glandular sections (mucilaginous gland, tunic gland) of lower sperm duct. Secretes inner parts of spermatophores.

spermatophoric organ [G *Spermatophorenorgan*]: Region or regions of lower sperm duct and accessory organs in which various stages of spermatophore production take place. May refer to section of sperm duct proper, to spermatophore glands, to finishing glands in which spermatophores

are hardened, to Needham's sac, or to a terminally associated appendix.

spermatophore pad [G *Pharetra*]: Fleshy patch of tissue in mantle cavity of ♀; serves as attachment site of spermatophores.

spermatophore sac: Needham's sac

sperm duct (vas deferens) [G *Vas deferens, Samenleiter*]: Elongate, narrow, winding section of ♂ reproductive system leading from testis into mantle cavity; variously modified into seminal vesicle and spermatophore-producing regions. (single, paired)

sperm groove [G *Samenrinne*]: In ♂, fold extending along arm modified for copulation (hectocotylus); serves to transfer spermatophore to tip of arm.

spine: rostrum

standard measurements:
arm length: distance from tip of arm to point of origin on head
mantle length: distance from collar to tip of tail or, in octobrachiate cephalopod, from eye to end of mantle
tail length: distance from posteriormost point of attachment of fin to tip of tail
fin length: distance from anterior to posterior margin of fin
fin angle: angle between longitudinal axis of mantle and posterior border of fin

statoconia [G *Statoconia*]: One of numerous smaller bodies in statocyst. (See also statolith)

statocyst [G *Statozyste*]: Pair of sensory (equilibrium) organs located in ventral part of capsule (cephalic cartilage) enclosing brain. May be internally regionated into plate(s) (maculae) and folds (cristae); contains statolith and/or statoconiae. In primitive (ectocochliate) cephalopod, statocyst still opens to exterior via duct.

statolith (otolith) [G *Statolith, Otolith*]: Single, relatively large, solid body within statocyst. (See also statoconia)

stellate ganglion [G *Stellargang-lion, Mantelganglion*]: Large, star-shaped ganglion, one on each side of mantle. Joined to brain by visceral or pallial nerves and giving rise to numerous nerves innervating mantle musculature; contains giant neurons.

stomach [G *Magen*]: Section of digestive tract between esophagus (or crop) and intestine. Consists of two chambers, a muscular, cuticle-lined dorsal gizzard and thin-walled ventral cecum. The latter receives digestive glands.

sublingual gland: buccal gland

submandibular gland: buccal gland

submaxillary salivary gland: buccal gland

subradular ganglion [G *Sub-radularganglion*]: Relatively small pair of ganglia located below radula and innervating both subradular organ (if present) and salivary glands.

subradular membrane [G *Sub-radularmembran*]: In radular apparatus, membrane underlying radular membrane; serves, together with bolsters, as site of muscle attachment.

subradular organ [G *Subradular-organ*]: In roof of subradular pouch, modified section serving as chemoreceptive sense organ in ectocochliate cephalopod.

subradular pouch (subradular sac) [G *Subradulartasche*]: In digestive tract, flat sac extending posteriorly below radula; roof of subradular pouch may be modified into subradular organ.

sucker [G *Saugnapf*]: One in a series of variously structured and arranged, muscular adhesive structures along arm or concentrated on terminal expansion (club) of tentacle. May also refer to small suckers (buccal suckers) on buccal membrane or to sucker-like structure on mantle (for attachment to algae in tidal pools). (sessile, stalked; rim: smooth, with sucker ring) (See also hook)

sucker ring [G *Hornring*]: Rigid chitinous ring around rim of sucker. (denticulated, serrate)

supraradular pouch (radular diverticulum) [G *Supraradulardivertikel*]: Flat dorsal pouch of anterior radular sac; separates the latter from esophagus.

swimming membrane: keel

systemic heart: heart

tail [G *Schwanz*]: Posterior, tapering, and elongate extension of mantle; fins may extend along tail.

tentacle [G *Tentakel*]: Term occasionally used as synonym for arm, yet more precisely applied to highly elongate arm used in capture of prey. Consists of more slender stalk and terminal sucker-bearing expansion (tentacular club). (contractile, retractile) (See also pocket)

tentacular club (club) [G *Tentakelkeule, Endkeule*]: Terminal expanded section of tentacle. May be regionated into proximal carpus with small suckers and knobs, central manus with larger suckers and hooks, and distal dactylus with smaller suckers. Serves in capturing prey.

testis [G *Hoden, Testis*]: Expanded, unpaired section of ♂ reproductive system in which sperm are produced. Located medially in posterior region of body and continuous with pericardial cavity; opens into mantle cavity via sperm duct(s), seminal vesicle, spermatophore-producing and -storing organs, and gonopore.

tongue [G *Zunge*]: In buccal cavity, refers to radula or radular apparatus.

tooth (radula tooth) [G *Zahn, Radulazahn*]: One of 5 to 13 cuticular elements in each transverse row of teeth in radula. According to position one may distinguish median (rachidian), lateral, and marginal teeth.

trabecula [G *Trabekel*]: One in a series of muscular rods supporting protective membranes along arm or expanded terminal section (club) of tentacle.

tunic gland: spermatophore gland

valve: funnel valve

vane [G *Fieder*]: In flexible, elongate inner shell (gladius), relatively thin lateral expansion of central axis (rachis).

vas deferens: sperm duct

vena cava [G *Vena cava*]: Major vein returning deoxygenated blood to ctenidia either directly or via branchial hearts. Venae cavae are in contact with and send branched projections into renal sac.

ventricle [G *Ventrikel, Hauptkammer*]: Muscular, typically elongate chamber of heart receiving blood from auricles and pumping it into body through aortas. In octobrachiate cephalopod, more or less subdivided by longitudinal septum.

visceral ganglion: pleurovisceral ganglion

visceral mass [G *Eingeweidesack*]: Tubular to sac-shaped region of body covered by mantle and containing most of digestive, excretory, reproductive, and circulatory systems. Continuous anteriorly with head and ventrally with funnel.

water pore (aquiferous pore) [G *Wasserpore*]: Small opening of each gonopericardial duct on head.

web (primary web) [G *Haut, Velarhaut*]: Thin membrane spanning from arm to arm in certain octobrachiate cephalopods; if well developed, may present umbrella-like appearance when arms are spread out. (See also secondary web)

ANNELIDA

POLYCHAETA

polychaete [G *Polychaet*]

POLYCHAETA

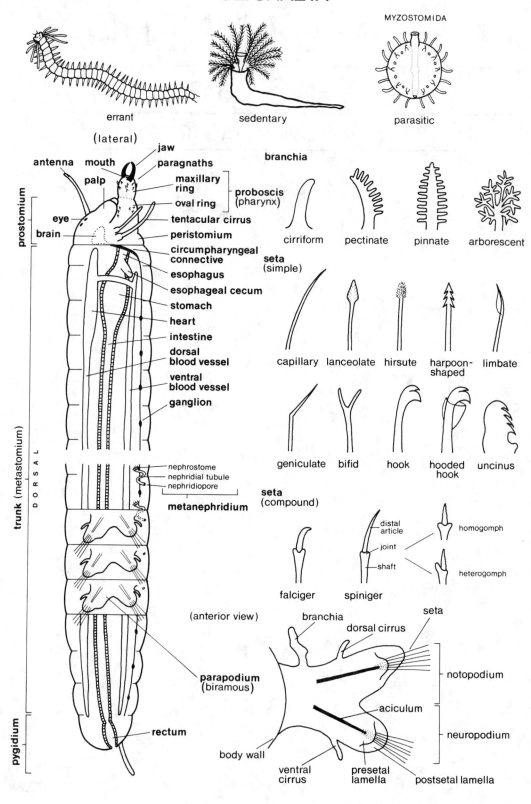

errant

sedentary

MYZOSTOMIDA

parasitic

(lateral)

antenna — jaw
mouth
palp — paragnaths
maxillary ring
oval ring
eye — tentacular cirrus
brain — peristomium
circumpharyngeal connective
esophagus
esophageal cecum
stomach
heart
intestine
dorsal blood vessel
ventral blood vessel
ganglion

prostomium

trunk (metastomium)

DORSAL

pygidium

nephrostome
nephridial tubule
nephridiopore

metanephridium

parapodium (biramous)

rectum

proboscis (pharynx)

branchia

cirriform   pectinate   pinnate   arborescent

seta (simple)

capillary   lanceolate   hirsute   harpoon-shaped   limbate

geniculate   bifid   hook   hooded hook   uncinus

seta (compound)

distal article
joint
shaft

homogomph

heterogomph

falciger   spiniger

(anterior view)

branchia   dorsal cirrus   seta

notopodium

aciculum

neuropodium

body wall

ventral cirrus   presetal lamella   postsetal lamella

ANNELIDA

POLYCHAETA

polychaete [G *Polychaet*]

abdomen [G *Abdomen*]: In regionated type of trunk (metastomium), posterior of two regions (thorax, abdomen); gonads or branchiae may be concentrated here. Abdomen is occasionally followed by tail.

aciculum [G *Acicula, Acikel*]: Thick internal seta supporting dorsal (notopodium) or ventral (neuropodium) branch of parapodium.

aileron [G *Aileron*]: Accessory jaw plate (rod-shaped, wing-shaped)

anal cirrus [G *Analzirrus*]: Projection of pygidium. (filiform, leaf-shaped, ovate, spherical)

anal cone [G *Analkonus*]: Cone-like projection of modified pygidium (anal plaque); bears anus.

anal funnel [G *Analtrichter*]: Funnel-like elaboration of modified pygidium (anal plaque).

anal plaque [G *Analplatte, Analscheibe*]: Flattened, plate-like elaboration of pygidium. (symmetrical, asymmetrical; smooth-rimmed, crenulate, with marginal anal cirri)

antenna [G *Antenne*]: Elongate sensory projection of prostomium; innervated by single nerve from anterior part of brain. (paired, unpaired; dorsal, frontal, lateral, occipital; smooth, annulated, articulated, biarticulate)

antennular membrane: cephalic veil

anus [G *After*]: Posterior opening of digestive tract on pygidium. (dorsal, ventral, terminal)

atoke [G *atoker Polychaet*]: Nonsexual stage or unmodified sexual reproductive individual. (See also epitoke)

bacillary seta (supernumerary seta) [G *Drüsenborste*]: Long, very thin seta emerging from thread gland.

brain [G *Gehirn*]: Main concentration of nerve tissue in prostomium; joined to longitudinal ventral nerve cords via circumesophageal connective and innervates palps, antennae, eyes, and nuchal organs.

branchia (gill, ctenidium) [G *Kieme*]: Any highly vascularized projection of body wall; typically associated with dorsal branch (notopodium) of each parapodium, yet also restricted to anterior parapodia or projecting from dorsal surface. (arborescent, cirriform, globular, lanceolate, papillose, pectinate, pinnate, spirally branching)

branchial crown: crown

branchial lobe [G *Tentakelträger*]: In sedentary polychaete, one of two lobes on head bearing radioles.

branchial vesicle: One of numerous club-shaped papillae with respiratory function on parapodium.

buccal cavity [G *Mundhöhle*]: Cuticle-lined section of digestive tract between mouth and pharynx.

buccal cirrus: buccal tentacle

buccal segment [G *Bukkalsegment*]: Anterior segment formed by fusion of peristomium (in the strict sense) and anterior trunk segment(s); may be fused to prostomium to form head.

buccal tentacle (buccal cirrus) [G *Peristomialzirrus, Tentakel*]: One of typically numerous elongate appendages in or around mouth. Whether they are completely withdrawable into mouth or not is of taxonomic importance. (smooth, papillose)

capillary [G *kapillare Borste, Kapillarborste, Haarborste*]: Hair-like seta; also general term for long, slender seta tapering to a fine point. May be used as an adjective.

caruncle [G *Karunkel*]: Posterior extension of prostomium, occasionally covering several segments; with sensory function.

cecum: esophageal cecum, intestinal cecum

cephalic cage [G *Schutzkorb*]: Series of elongate, anteriorly directed setae of first segments; form protective structure into which head can be retracted.

cephalic plaque (cephalic plate) [G *Kopfscheibe, Kopfplatte, Nackenplatte*]: Flattened, plate-like elaboration of anterior end (dorsal surface of head); may bear slit-like nuchal organs. (round, oval, elongate, keeled)

cephalic plate: cephalic plaque

cephalic rim [G *Saum*]: Raised margin of flattened, plate-like elaboration of head (cephalic plaque). (smooth, lobed, notched, scalloped)

cephalic veil (antennular membrane) [G *Tentakelmembran*]: Hood-like, curved membrane between strong anterior setae (paleae) and buccal tentacles. (margin: smooth, fringed, cirrate)

cephalon: head

ceratophore [G *Ceratophor*]: Basal part of jointed antenna. (See also ceratostyle)

ceratostyle [G *Ceratostyl, Styl*]: Distal part of jointed antenna. (See also ceratophore)

chaeta: seta

chevron [G *Chevron*]: One in a series of V-shaped chitinous structures along each side at base of proboscis.

chloragogen tissue [G *Chloragogengewebe, Chloragoggewebe*]: Metabolically active, often distinctly colored tissue developing from peritoneum on intestine or blood vessels.

chromophil gland [G *chromophile Drüse*]: In certain pelagic polychaetes, gland in distally expanded section of ventral branch (neuropodium) of parapodium. (See also hyaline gland, rosette gland, spur gland)

circular muscle [G *Ringmuskel*]: Relatively thin outermost muscle layer of body wall (cuticle, epidermis; circular, diagonal, longitudinal muscles; peritoneum).

circumesophageal connective: circumpharyngeal connective

circumpharyngeal connective (circumesophageal connective) [G *Schlundkonnektiv*]: One of two strands of nerve tissue surrounding digestive tract and joining dorsal brain to ventral nerve cord; may give rise to nerves innervating palps.

cirratostyle [G *Cirratostyl*]: Distal part of jointed cirrus. (See also cirrophore)

cirrophore [G *Cirrophor*]: Basal part of jointed cirrus. (See also cirratostyle)

cirrus: dorsal cirrus, interramal cirrus, tentacular cirrus, ventral cirrus

collar [G *Kragen*]: In tubiculous polychaete, collar-like fold around base of each branchial lobe; may be united ventrally. Serves in tube construction.

companion seta [G *Begleitborste*]: Small, simple seta accompanying or alternating with larger seta (hook or incinus).

copragogue [G *Fäkalfurche, Kotrinne*]: In certain tubiculous polychaetes, ciliated groove extending from posterior to anterior end of body, typically running ventrally along abdomen and then dorsally along thorax; serves to void feces.

crochet [G *Hakenborste*]: Type of seta with long shaft and typically one or more distal teeth; often used synonymously with hook.

crown (branchial crown, tentacle crown) [G *Tentakelkrone, Kiemenkrone*]: In tubiculous polychaete, series of radioles forming filter-feeding and respiratory apparatus; radioles originate from two branchial lobes on head, thus forming two semicircles or two spirals.

ctenidium [G *Ctenidium*]: In elytron-bearing polychaete, small, ciliated, cushion-like respiratory structure located in concave region between notopod and branchia.

cuticle [G *Kutikula*]: Multilayered, noncellular outer layer of body wall (cuticle, epidermis; circular, diagonal, longitudinal muscles; peritoneum); relatively thin and secreted by epidermis.

denticle: paragnath

diagonal muscle [G *Diagonalmuskel*]: Thin layer of muscle between outer circular and inner longitudinal muscle layers of body wall (cuticle, epidermis; circular, diagonal, longitudinal muscles; peritoneum).

diaphragm [G *Diaphragma*]: In certain sedentary polychaetes, one of few remaining septa located in anterior region of trunk (thorax).

dorsal blood vessel [G *Dorsalgefäß*]: In circulatory system, contractile longitudinal vessel lying dorsal to intestine. Receives blood from parapodia and propels it anteriorly. (See also ventral blood vessel)

dorsal cirrus [G *Dorsalzirrus*]: Projection from dorsal side of uniramous parapodium or, if parapodium is biramous, projection from upper ramus (notopodium). May be modified into branchia. (annulated,

conical, filiform, leaf-shaped, lobe-like, spherical) (See also ventral cirrus)

dorsal feltage: felt

dorsal organ [G *dorsales Organ*]: Two longitudinal, ciliated sensory grooves or ridges extending from prostomium across several anterior segments; considered to be modified nuchal organ.

dorsal tubercle [G *Dorsaltuberkel*]: In elytron-bearing polychaete, tubercle-shaped projection occupying same position as elytrophore on segments without elytra.

elytron [G *Elytron*]: On parapodium, scale-like modification of distal part (cirratostyle) of dorsal cirrus; basal, stalk-like part termed elytrophore. May cover entire dorsal surface of body. (circular, oval, reniform, heart-shaped; smooth, with tubercules, with papillae, with smooth margin, with fringed margin)

elytrophore [G *Elytrophor*]: On parapodium, modified basal part (cirrophore) of dorsal cirrus; bears elytron.

epidermis [G *Epidermis*]: Outermost cellular layer of body wall (cuticle, epidermis; circular, diagonal, longitudinal muscles; peritoneum); secretes cuticle.

epitoke [G *epitoker Polychaet*]: Modified, sexually reproductive individual; characterized by altered structure of head, parapodia, or segments. Develops either by direct transformation or asexually by budding. (See also heteronereid, stolon)

esophageal cecum [G *Blindsack*]: One of two lateral outpocketings of esophagus.

esophagus [G *Ösophagus*]: Section of digestive tract between pharynx and stomach; may bear esophageal ceca.

excretory sac [G *Blase*]: In excretory system of sedentary polychaete with only one pair of metanephridia, expanded section of each nephridial tubule before merger into common canal.

eye [G *Auge*]: One of typically several pairs of photosensitive organs on protostomium; may also refer to less well-developed photosensitive structures on trunk or crown of sedentary polychaete. (sessile, stalked)

falciger [G *falcigere Borste*]: Seta, typically of the compound type, whose distal part is stout, hook-shaped, and generally blunt. (See also spiniger)

false operculum [G *Pseudooperculum*]: In tubiculous polychaete, second, incompletely developed reserve operculum.

felt (dorsal feltage) [G *Borstenfilz*]: In certain elytron-bearing polychaetes, dense mat of thin setae (capillaries) formed by dorsal branch (notopodium) of parapodium. May cover elytra.

genital spine [G *Genitalhaken*]: One of several thick, modified setae associated with gonopore on dorsal or ventral surface of ♂ or ♀; typically located on posterior segment(s) of thorax.

gill: branchia

glandular shield (glandular pads, ventral pads) [G *Drüsenschild, Bauchplatte, Ventralschild*]: In tubiculous polychaete, pair of cushion-shaped glandular elevations typically located on ventral surface of each segment. (ventral, dorsal)

gonad [G *Gonade*]: ♂ or ♀ reproductive organ (ovary, testis); consists of mass of cells in peritoneum of segment walls or in parapodia of certain pelagic species. Eggs and sperm are typically released to exterior via nephridial tubules.

gonoduct [G *Gonodukt*]: In those polychaetes in which eggs and sperm are not released via nephridial tubules or by rupture of body wall, special duct leading to exterior; opening may be associated with genital spines.

gonopore [G *Genitalöffnung, Gonoporus*]: In those polychaetes in which eggs and sperm are not released via nephridiopores or by rupture of body wall, opening of gonoduct to exterior; typically located on posterior region of thorax and associated with genital spines.

head (cephalon) [G *Kopf*]: Anterior or presegmental division of body (head, trunk, pygidium); consists of prostomium and peristomium along with their appendages, although it may include fused segments of trunk.

heart [G *Herz*]: Any expanded, contractile section of circulatory system; typically refers to dorsal blood vessel.

heteronereid [G *Heteronereis*]: Term applied to extreme type of epitokous polychaete in which anterior and posterior ends of body differ markedly in appearance.

hood (guard) [G *Kaputze, Scheide*]: Delicate, hood-like membrane partially surrounding tip of seta; occasionally one may distinguish a primary and secondary hood.

hook [G *Haken*]: General term for thick, blunt, distally curved seta. (See also crochet, uncinus)

horn: prostomial horn

hyaline gland [G *hyaline Drüse*]: In certain pelagic polychaetes, gland in distally expanded section of dorsal branch (notopodium) of parapodium. (See also chromophil gland, rosette gland)

interramal cirrus (intermediate cirrus) [G *interramaler Zirrus*]: Variously shaped cirrus projecting between dorsal branch (notopodium) and ventral branch (neuropodium) of parapodium; may be modified as branchia.

intestinal cecum [G *Blindsack*]: One in a series of segmentally arranged lateral outpocketings of intestine.

intestine [G *Mitteldarm*]: Elongate section of digestive tract between stomach and rectum; may bear segmentally arranged lateral ceca.

jaw [G *Kiefer*]: Pair of large, sickle-shaped jaw structures at tip of everted proboscis; may also refer to entire complex of cuticular structures including maxillae and mandibles. (See also poison gland)

lamella [G *Borstenlappen, Lappen*]: Flattened, lobe-like portion of either branch of parapodium associated with setae. Portion of branch projecting anterior to point of insertion of setae may be termed notopodial or neuropodial presetal lamella, that posterior to point of insertion, notopodial or neuropodial postsetal lamella. (See also ligule)

lateral lappet (lateral lobe) [G *Seitenlappen*]: In sedentary polychaete, pair of lateral lobes on buccal segment or on one or more anterior segments.

lateral organ [G *Lateralorgan*]: Ciliated sensory structure located between notopodium and neuropodium of each parapodium.

ligule (ligula) [G *Lappen, Dorsallappen, Ventrallappen*]: Major process of either dorsal (notopodium) or ventral (neuropodium) branch of parapodium. (conical, leaf-like, rounded, strap-like, triangular) (See also lamella)

longitudinal muscle [G *Längsmuskel*]: Relatively thick inner muscle layer of body wall (cuticle, epidermis; circular, diagonal, longitudinal muscles; peritoneum); typically concentrated into two dorsolateral and two ventrolateral muscle bands.

macrognath [G *Makrognath*]: One of two large, denticulate teeth at distal end of everted proboscis; adjoined dorsally and ventrally by micrognaths.

mandible [G *Mandibel*]: Ventrally positioned jaw element in pharynx; adjoined dorsally by maxillae.

maxilla [G *Maxille*]: Dorsally positioned jaw element in pharynx; maxillary apparatus consists either of several longitudinal series of small pieces or of four to five pairs of toothed plates and posterior maxillary carrier. Posteriormost and largest pair of maxillae may be termed maxillae I or main fangs, forceps, or pincers. (distally dentate, distally falcate) (See also mandible)

maxillary carrier [G *Träger*]: Paired jaw element, occasionally with median unpaired piece, adjoining posteriormost maxilla (maxilla I) in pharynx. (long and slender, short and broad)

maxillary ring [G *maxillarer Ring*]: Distal division of proboscis (near jaws); may bear paragnaths or papillae. Separated from oral ring by transverse groove and may be further subdivided by longitudinal grooves into areas with roman numerals (I–IV).

metanephridium [G *Metanephridium*]: More advanced type of excretory organ consisting of nephrostome, nephridial tubule, and nephridiopore. May also function as gonoduct. (See also protonephridium)

metastomium (trunk) [G *Rumpf*]: Segmented division of body between head and pygidium; may be regionated into thorax and abdomen. Term has also been applied to part of head behind mouth (often fused to one or two trunk segments to form peristomium in the wider sense or buccal segment).

metatrochophore [G *Metatrochophora*]: Advanced trochophore larval stage characterized by first evidence of segmentation. (See also nectochaeta)

micrognath [G *Mikrognath*]: At tip of everted proboscis, one in a series of small, X- or Y-shaped teeth adjoining macrognaths in a ventral and dorsal arc.

mitraria [G *Mitraria-Larve*]: Type of polychaete larva characterized by numerous long flotation bristles.

mouth [G *Mund*]: Anterior opening of digestive tract; located ventrally between peristomium and overlying prostomium; opens into buccal cavity or pharynx. (See also proboscis)

nectochaeta [G *Nectochaeta*]: Larval stage in which first seta bundles and parapodia develop; ciliary girdle remains responsible for swimming.

nephridial papilla [G *Nephridialpapille*]: Lateral papilla bearing terminal nephridiopore; typically limited to anterior trunk segments of sedentary polychaete.

nephridial tubule [G *Nephridialkanal*]: Ciliated, tubular section of excretory system (protonephridium or metanephridium) between solenocytes or nephrostome and nephridiopore. Section of tubule anterior to body septum may be referred to as preseptal, that posterior to septum as postseptal. (simple, U-shaped, coiled)

nephridiopore [G *Nephroporus*]: Opening of excretory system (protonephridium or metanephridium) to exterior; typically located ventrally at base of each neuropodium, yet also restricted to more anterior segments (occasionally on nephridial papilla) or opening via single nephridiopore on head.

nephrostome [G *Nephrostom*]: Proximal opening of more advanced type of excretory system (metanephridium); typically funnel-shaped, ciliated, and located in body cavity of segment anterior to that bearing corresponding nephridiopore. May be associated with reproductive system.

neuropodium (neuropod) [G *Neuropodium, Neuropod*]: Ventral branch (ramus) of biramous parapodium; supported internally by aciculum and may bear ventral cirrus, ligules, and setae (neurosetae). (See also notopodium)

neuroseta [G *Borste des Neuropodiums*]: Seta arising from ventral branch (neuropodium) of parapodium. (See also notoseta)

notopodium (notopod) [G *Notopodium, Notopod*]: Dorsal branch (ramus) of biramous parapodium; supported internally by aciculum and may bear dorsal cirrus, ligules, and setae (notosetae). (See also neuropodium)

notoseta [G *Borste des Notopodiums*]: Seta arising from dorsal branch (notopodium) of parapodium. (See also neuroseta)

nuchal epaulette [G *Nuchalwulst*]: Modified, projecting type of nuchal organ extending from posterior margin of prostomium.

nuchal organ [G *Nuchalorgan*]: Sensory organ at posterior end of prostomium; typically in the form of a ciliated pit,

groove, or slit. (single, paired; retractile, nonretractile) (See also nuchal epaulette)

oblique muscle [G *Transversalmuskel*]: Segmentally arranged band of muscle spanning from midventral line to base of each parapodium.

ocular peduncle (ommatophore) [G *Ommatophor*]: Stalk-like projection of prostomium bearing eye. (fused, separated)

ommatophore: ocular peduncle

opercular stalk [G *Opercularstiel*]: Basal stalk-shaped section of operculum formed by modified radiole. (circular, oval, flattened; annulated, smooth, winged, with pinnules, with spines)

operculum [G *Operculum*]: Structure at anterior end of body serving to plug tube of retracting tubiculous polychaete; typically formed by modified radiole, yet occasionally also by fused body segments. (soft, calcareous; concave, conical, convex, flattened, funnel-shaped, globular, pear-shaped, spherical, with internal septa)

oral ring [G *oraler Ring*]: Proximal division of proboscis (near mouth); may bear paragnaths or papillae. Separated from maxillary ring by transverse groove and may be further subdivided by longitudinal grooves into areas with roman numerals (V–VIII).

ovary: gonad

palea [G *Palea*]: Strong, broad or flattened seta, typically associated with operculum or anterior part of body.

palp [G *Palpe*]: Paired projection of prostomium innervated by brain or circumesophageal connective; if jointed one may distinguish basal palpophore and distal palpostyle. With sensory function or used in feeding. (dorsal, ventral, frontal, occipital; simple, jointed, biarticulate; digitate, filiform, spherical)

palpode [G *Palpode*]: Tapered anterior projection of prostomium.

palpophore: Basal part of jointed palp. (See also palpostyle)

palpostyle [G *Styl*]: Distal part of jointed palp. (See also palpophore)

papilla [G *Papille*]: One of numerous short, soft projections in pharynx; come to lie on exterior when pharynx is everted as proboscis. (See also paragnath, proboscidial organ)

paragnath (denticle) [G *Paragnath*]: Minute, hard, tooth-like projection in pharynx; typically in groups or circlets which come to lie on exterior when pharynx is everted as proboscis. (conical, rod-shaped, pectinate bar, smooth bar) (See also papilla)

parapodium [G *Parapodium, Parapod*]: Lateral appendage of trunk segment; typically consists of two branches or lobes (rami)—a dorsal notopodium and ventral neuropodium—each supported by aciculum and bearing setae. May also bear gills (branchia) and cirri. (biramous, subbiramous, uniramous)

parathorax [G *Parathorax*]: In regionated thorax of tubiculous polychaete, posterior section consisting of three to four segments with biramous parapodia.

pedal ganglion [G *Podialganglion*]: Concentration of nerve tissue at base of each parapodium; joined to one another by longitudinal pedal nerve and to ventral cord by lateral nerves.

penis [G *Penis*]: ♂ copulatory organ. In echinoderm parasite, unpaired, papilla-shaped structure at base of third parapodium; in forms with internal fertilization, paired, setae-bearing organs on anterior segments.

peristomium (peristome) [G *Peristomium*]: Region of polychaete surrounding mouth laterally and ventrally; bears cirri and may be fused with trunk segments to form buccal segment as well as with prostomium to form head.

peritoneum [G *Peritoneum*]: Innermost layer of body wall (cuticle, epidermis; circular, diagonal, longitudinal muscles; peritoneum); lines body cavity and forms walls of septa between segments. Gonads develop on peritoneum.

pharyngeal ring: maxillary ring, oral ring

pharynx [G *Pharynx, Schlund*]: Cuticle-lined, muscular section of digestive tract between mouth or buccal cavity and esophagus; may bear jaws, teeth, paragnaths, or papillae which come to lie externally when anterior section of digestive tract is everted in the form of a proboscis.

pinnule [G *Pinnula*]: One in a series of regularly arranged side branches of radiole or branchia.

poison gland [G *Giftdrüse*]: Gland at base of each jaw element in proboscis; opens to exterior at tip of jaw by means of a duct.

proboscidial organ [G *proboscidiales Organ*]: One of numerous small, papilla-like structures covering surface of proboscis; contains central canal and terminal pore. May be uniformly distributed or arranged in longitudinal rows. (similar, dissimilar; conical, cordiform, cylindrical, digitiform, ringed, smooth, with distal flange) (See also papilla)

proboscis [G *Rüssel, Pharynx*]: Eversible anterior section of digestive tract, in essence corresponding to pharynx. (muscular, nonmuscular; armed, unarmed; bilobate, cylindrical, sac-like) (See also maxillary ring, oral ring)

prostomial horn [G *Hörner*]: Horn-shaped, laterally directed projection of T-shaped prostomium. (frontal, lateral)

prostomial peak [G *Prostomiumecke*]: In elytron-bearing polychaete, typically chitinized anterolateral projection of each side of bilobed prostomium.

prostomium [G *Prostomium, Prostom*]: Division of polychaete anterior to mouth (preoral lobe); consists of lobe-like dorsal projection containing brain and bearing numerous sensory structures (antennae, eyes, nuchal organs, palps). May be fused with peristomium and first trunk segments to form head. (annulated, bell-shaped, bifurcate, bilobed, cordiform, elongate, entire, notched, pointed, rounded, smooth, subquadrate, subtriangular, T-shaped, truncate)

protonephridium [G *Protonephridium*]: More primitive type of excretory system consisting of solenocytes, nephridial tubule, and nephridiopore. May also function as gonoduct. (See also metanephridium)

proventriculus [G *Proventrikel*]: Muscular, transversely striated section of digestive tract posterior to pharynx. May serve as pump.

pygidium [G *Pygid*]: Posteriormost of three divisions of body (prostomium, metastomium, pygidium) bearing anus; may also bear anal cirri, lappets, or lobes or be modified into anal funnel or anal plaque.

radiole [G *Radiolus*]: In tubiculous polychaete, one in a series of tentacle-like projections forming crown; typically bipinnate, with both pinnules and centrally located longitudinal groove bearing cilia. One radiole typically modified to form operculum. (free, united by web, with/without stylodes; dichotomously branching, bipinnate, simple)

ramus [G *Ast*]: One of two branches—either dorsal notopodium or ventral neuropodium—of parapodium.

receptaculum seminis (seminal receptacle) [G *Receptaculum seminis*]: In polychaete with internal fertilization, sperm storage organ in ♀.

rectum [G *Rektum, Enddarm*]: Posterior cuticle-lined section of digestive tract; opens to exterior via anus on pygidium.

retort organ: Large, eversible gland in head of pelagic polychaete; opens into buccal cavity.

rosette gland (rosette organ) [G *rosettenförmiges Organ*]: In pelagic polychaete, rosette-shaped gland at end of each branch (notopodium, neuropodium) of parapodium. (See also chromophil gland, hyaline gland)

rostrum (main fang) [G *Hauptzahn*]: Largest tooth or main fang at distal end of seta.

scale: elytron

scaphe [G *Scaphe*]: Modified posteriormost section of certain burrowing, tubiculous polychaetes; may or may not be distinctly set off from abdomen.

segment [G *Segment*]: One in a series of segments of trunk (metastomium); each segment typically bears pair of lateral parapodia externally and two nephridia internally. (apodous, asetigerous, setigerous, branchiferous)

seminal receptacle: receptaculum seminis

septum [G *Dissepiment*]: Transverse partition between two segments of trunk; formed by double layer of peritonium and penetrated by digestive tract, dorsal and ventral blood vessels, and nephridial canals. May be reduced. (See also diaphragm)

seta (chaeta) [G *Borste*]: One in a series of chitinous bristles projecting from parapodium; those of upper branch (notopodium) are termed notosetae, those of neuropodium, neurosetae. One may distinguish between simple, unjointed setae and compound or composite setae which are jointed and consist of shaft and distal article. (acicular, acuminate, alimbate, aristate, bifid, bilimbate, bipinnate, brush-tipped, camerated, capillary, crenulated, falcate, flanged, geniculate, harpoon-shaped, hirsute, hooded, hemigomph, heterogomph, homogomph, lancet-shaped, limbate, lyrate, multiarticulate, pectiniform, penicillate, pennoned, pseudopenicillate, punctate, sickle-shaped, spatulate, spirally serrulate, trumpet-shaped) (See also capillary, crochet, falciger, hook, palea, spiniger, uncinus)

setiger [G *Borstensegment*]: Setae-bearing segment.

sinus (gut sinus) [G *Sinus, Darmsinus*]: In circulatory system, variously developed network of blood vessels in wall of stomach or intestine.

solenocyte [G *Solenozyte*]: In more primitive type of excretory system (protonephridium), one of numerous cilium-bearing cells attached to nephridial tubule by thin canal.

spiniger [G *spinigere Borste*]: Seta, typically of the compound type, whose distal part tapers to a fine point. (See also falciger)

spinning gland [G *Spinndrüse*]: Between notopodium and neuropodium of elytron-bearing polychaete, gland producing fibers used in construction of tube.

spur gland: Gland accompanying chromophil gland in ventral branch (neuropodium) of parapodium in pelagic polychaete.

statocyst [G *Statozyste*]: Organ of equilibrium associated with brain or circumesophageal connective.

stolon [G *Stolon*]: Modified, sexually reproductive (epitokous) polychaete budding from posterior end of atokous individual.

stomach [G *Magen*]: Somewhat expanded section of digestive tract between esophagus and intestine; often absent.

stylode [G *Stylode*]: Small, finger-like projection, typically on parapodium yet also referring to projections on outer surface of radiole (remnant of web).

subpharyngeal ganglion [G *Unterschlundganglion*]: First ventral concentration of nerve tissue at point of merger of two circumesophageal connectives.

subuluncinus [G *Übergangsborste*]: Type of seta intermediate in form between uncinus and capillary; thick shaft narrows suddenly to a slender tip.

supernumerary seta: bacillary seta

tail [G *Schwanz*]: In regionated type of trunk (thorax, abdomen), section occasionally following abdomen; characterized by small segments lacking appendages.

tentacle crown: crown

tentacular cirrus [G *Tentakelzirrus*]: Elongate cirrus, representing parapodial remnant, projecting from body segment fused to peristomium.

testis: gonad

thorax [G *Thorax*]: In regionated type of trunk, anterior of two regions (thorax, abdomen). (See also parathorax, tail)

thread gland (fibrous gland) [G *Drüsensack*]: Fibrous gland between notopodium and neuropodium; gives rise to bacillary setae.

torus (uncigerous ridge) [G *Torus*]: On parapodium, modification of neuropodium into low glandular ridge from which setae arise.

trepan [G *Trepan*]: Ring of pharyngeal teeth.

trochophore [G *Trochophora*]: Early larval stage of polychaete; characterized by apical tuft of cilia and girdle of cilia around midregion. (See also metatrochophore, mitraria)

trunk: metastomium

tube [G *Rohr*]: Tube-like structure constructed by polychaete. May be calcareous, hyaline, membranous, or composed of foreign particles. Coiled calcareous tube may be either sinistral or dextral. (curved, straight; closed at one end, open on both ends; cross section: circular, triangular)

tubercle: dorsal tubercle

tubular gland (diffuse tubular gland) [G *Drüsenschlauch*]: Series of parallel tubular glands at distal end of parapodium of pelagic polychaete.

uncinus [G *Haken*]: Type of seta: modified hook with dentate tip, either S- or Z-shaped (avicular), gently curved (acicular), or plate-like with several teeth. (acicular, avicular, long-handled, short-handled, pectiniform, serpuliform)

ventral blood vessel [G *Bauch-gefäß*]: In circulatory system, longitudinal vessel lying ventral to intestine; gives rise to vessels extending into parapodia and carries blood in posterior direction. (See also dorsal blood vessel)

ventral cirrus [G *Ventralzirrus*]: Projection from lower side of ventral branch (neuropodium) of biramous parapodium.

ventral nerve cord [G *Bauchmark*]: Longitudinal nerve cord extending from subpharyngeal ganglion to posterior end of body; bears swellings and gives rise to lateral nerves (occasionally leading to pedal ganglia) in each segment. (fused, paired)

ventral sac [G *Ventralsack*]: Ventral glandular, sac-like structure formed by opposing membranes surrounding mouth at base of crown; serves in tube construction.

web [G *Membran*]: In crown, membrane uniting bases of radioles; may be reduced to stylodes.

ANNELIDA

OLIGOCHAETA

oligochaete [G *Oligochaet, Wenigborster*]

OLIGOCHAETA

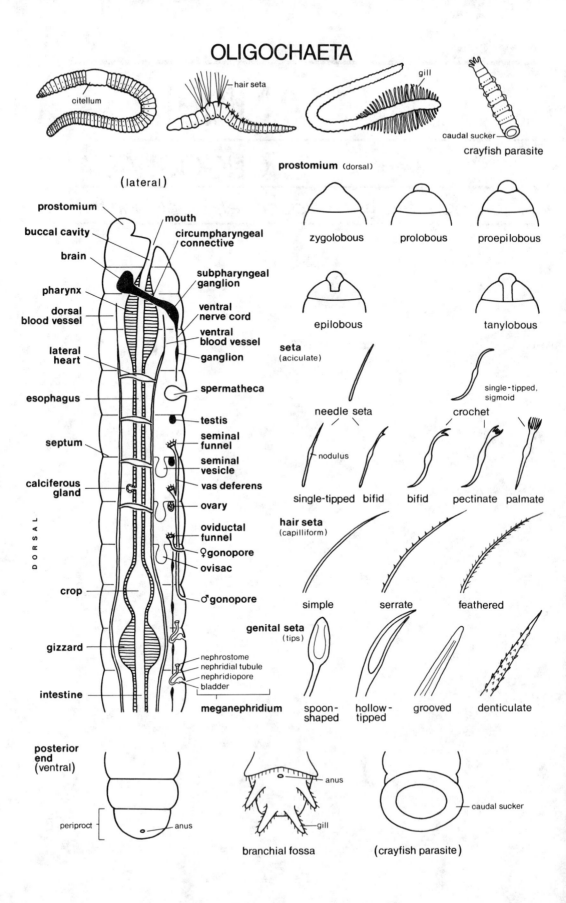

citellum

hair seta

gill

caudal sucker

crayfish parasite

prostomium (dorsal)

zygolobous

prolobous

proepilobous

epilobous

tanylobous

(lateral)

prostomium

buccal cavity

brain

pharynx

dorsal blood vessel

lateral heart

esophagus

septum

calciferous gland

crop

gizzard

intestine

mouth

circumpharyngeal connective

subpharyngeal ganglion

ventral nerve cord

ventral blood vessel

ganglion

spermatheca

testis

seminal funnel

seminal vesicle

vas deferens

ovary

oviductal funnel

♀gonopore

ovisac

♂gonopore

nephrostome

nephridial tubule

nephridiopore

bladder

meganephridium

D O R S A L

seta (aciculate)

needle seta

crochet

single-tipped, sigmoid

nodulus

single-tipped   bifid   bifid   pectinate   palmate

hair seta (capilliform)

simple   serrate   feathered

genital seta (tips)

spoon-shaped   hollow-tipped   grooved   denticulate

posterior end (ventral)

periproct   anus

branchial fossa

anus

gill

(crayfish parasite)

caudal sucker

ANNELIDA

OLIGOCHAETA

oligochaete [G *Oligochaet, Wenigborster*]

ampulla [G *Ampulle*]: In ♀ reproductive system, enlarged, sac-shaped distal section of spermatheca; opens to exterior via spermathecal duct and spermathecal pore.

annulus [G *Ringelfurche*]: Secondary furrow subdividing a segment externally. Two annuli, for example, one on each side of setae, result in triannulate condition of each segment. True segment can usually be distinguished from superficial ring by presence of setae.

anus [G *After*]: Posterior opening of digestive tract on last segment (periproct).

atrial gland [G *Atrialdrüse*]: Large, typically bilobed, glandulomuscular organ associated with terminal end of spermathecal duct.

atrium [G *Atrium*]: Muscular chamber at terminal end of ♂ reproductive system; either terminates in a copulatory apparatus (penis, pseudopenis) or opens to exterior as a simple pore. (single, paired; cylindrical, globular, horseshoe-shaped, ovoid, pear-shaped, spindle-shaped, tubular)

bladder (terminal bladder, nephridial vesicle) [G *Blase, Endblase, Sammelblase*]: In meganephridium, expanded terminal section of nephridial tubule; opens to exterior via nephridiopore.

brain (cerebral ganglion, suprapharyngeal ganglion) [G *Gehirn, Zerebralganglion, Oberschlundganglion*]: Main concentration of nerve tissue dorsal to pharynx; joined to subpharyngeal ganglion and longitudinal ventral nerve cord via pair of circumpharyngeal connectives. (See also subpharyngeal ganglion)

branchia: gill

branchial fossa [G *Korb, Atmungsapparat, Kiemenapparat, Analkieme*]: Shallow, cup-like structure at posterior end of certain aquatic oligochaetes; bears anus, gills, and in some cases palps.

buccal cavity [G *Mundhöhle*]: Short, cuticle-lined, muscular section of digestive tract between mouth and pharynx.

calciferous gland (gland of Morren, Morren's gland, chyle sac) [G *Kalkdrüse, Morren'sche Drüse, Chylustasche*]: Highly vascularized pair of series of glands associated with esophagus. Consisting of either inwardly directed folds (lamellae) of esophageal wall (intramural) or internally folded gland lying outside esophagus (extramural). May contain calcareous crystals and functions in calcium excretion. (See also tunnel)

capsule: cocoon

cerebral ganglion: brain

chaeta: seta

chloragogen tissue (chloragogenous cell, chloragogen cell) [G *Chloragoggewebe, Chloragog-Zelle*]: Metabolically active, often distinctly colored tissue developing from peritoneum surrounding intestine or blood vessels.

chromatophilic cell (chromatophile cell): pharyngeal gland

chyle sac: calciferous gland

cingulum: clitellum

circular muscle [G *Ringmuskel*]: Relatively thin layer of circular muscle between dermis and longitudinal muscles in body wall (cuticle, epidermis, dermis; circular, longitudinal muscles; peritoneum).

circumpharyngeal connective (circumpharyngeal commissure) [G *Schlundkonnektiv*]: One of two strands of nerve tissue surrounding pharynx and joining dorsal brain to ventral nerve cord; ventral point of merger is enlarged to form subpharyngeal ganglion.

clasper [G *Hafter*]: One of two lobelike or elongate, muscular copulatory structures; either retractile and emerging from pore on ventral surface or nonretractile and bearing penial setae and ♂ gonopores.

clitellum (cingulum, girdle, saddle) [G *Clitellum, Gürtel*]: Thickened glandular epidermis forming complete or incomplete ring around body; extends across several segments either at or posterior to gonopores. Secretes mucus for copulation and cocoon.

cocoon (capsule, egg capsule, egg case, oophore) [G *Kokon*]: Egg case secreted by clitellum and containing one to numerous eggs. (ovoid, spindle-shaped)

commissural vessel (lateral commissure) [G *Kommissurgefäß*]: In circulatory system, general term for paired, segmental blood vessels surrounding digestive tract and joining dorsal blood vessel to either ventral or subneural blood vessels. Anterior pairs may be contractile and termed lateral hearts.

copulatory seta: genital seta

crochet (crotchet): Type of hook-shaped seta.

crop [G *Kropf*]: Modified, expanded section of esophagus or intestine; relatively thin-walled and stores food before it passes into gizzard.

cuticle [G *Kutikula*]: Noncellular, multilayered outermost layer of body wall (cuticle, epidermis, dermis; circular, longitudinal muscles; peritoneum); lines buccal cavity and is secreted by epidermis.

dermis [G *Dermis*]: Layer of connective tissue between epidermis and circular muscles in body wall (cuticle, epidermis, dermis; circular, longitudinal muscles; peritoneum).

diagonal muscle [G *Diagonalmuskel*]: Thin layer of diagonal muscles between outer circular and inner longitudinal muscles of body wall.

diverticulum [G *Divertikel*]: In spermatheca, lateral outpocketing of spermathecal duct (occasionally of ampulla). (sessile, stalked; sacciform, tubular)

dorsal blood vessel [G *Rückengefäß*]: In circulatory system, often contractile longitudinal vessel lying dorsal to and closely associated with digestive tract; receives blood from commissural vessels and intestinal plexus and propels it anteriorly.

dorsal pore (middorsal pore) [G *Rückenporus*]: Small middorsal pore typical of terrestrial oligochaete; typically located in intersegmental furrows of all but most anterior segments. Equipped with sphincter muscle and serves as opening of body cavity to exterior.

egg capsule: cocoon

egg case: cocoon

epidermis [G *Epidermis*]: Outermost cellular layer of body wall (cuticle, epidermis, dermis; circular, longitudinal muscles; peritoneum); secretes cuticle and may be thickened to form clitellum.

epididymus [G *Epididymus*]: In ♂ reproductive system, coiled section of vas deferens; typically located in segment posterior to seminal funnel.

esophagus [G *Ösophagus*]: Narrow section of digestive tract between pharynx and intestine; may be modified to form stomach, crop, or gizzard and contains calciferous glands.

euprostate [G *Euprostata*]: Musculoglandular section of ♂ reproductive system; consists of looped terminal end of vas deferens. Functions as prostate gland.

eyespot: ocellus

fimbriated funnel: oviductal funnel, seminal funnel

funnel: nephrostome, oviductal funnel, seminal funnel

furrow (intersegmental furrow) [G *Furche, Intersegmentalfurche*]: One in a series of external circular grooves surrounding body and corresponding to internal septa; each segment may be further subdivided externally by superficial secondary furrows (annuli). Furrow number denoted by arabic numerals of segments to either side.

ganglion [G *Ganglion, Bauchmarkganglion*]: Concentration of nerve tissue, typically referring to segmentally arranged swellings of ventral nerve cord; each ganglion gives rise to pair of lateral nerves. (See also subpharyngeal ganglion)

genital seta (copulatory seta) [G *Genitalborste, Kopulationsborste*]: General term for modified ventral seta associated with ♂ gonopore (penial seta) or spermathecal pore (spermathecal seta).

gill (branchia) [G *Kieme*]: Highly vascularized projection of body wall; located in anal region or in one or more series at either anterior or posterior region of body. (digitiform, filamentous, finger-like, lamelliform, tuft-like)

girdle: clitellum

gizzard [G *Kaumagen, Muskelmagen*]: Modified, expanded section of esophagus, occasionally of intestine, posterior to crop; cuticle-lined, muscular, and typically extending across several segments. Serves to grind food.

gland of Morren: calciferous gland

gonopore [G *Geschlechtsöffnung, Gonoporus*]: Opening of ♂ or ♀ reproductive system to exterior. In ♂, either single or paired pores up to several segments posterior to testicular segment. In ♀, single or paired pores typically opening in segment posterior to ovarian segment. (anteclitellar, intraclitellar, postclitellar)

head pore: Pore in prostomium. Serves as opening of body cavity to exterior.

heart: lateral heart

holonephridium: meganephridium

intersegmental furrow: furrow

intestinal plexus (intestinal sinus, alimentary plexus) [G *Darmgefäßplexus, Darmblutsinus*]: Network or cavity of circulatory system located in wall of intestine. Blood from this plexus or sinus flows into dorsal blood vessel.

intestine [G *Mitteldarm, Intestinum*]: Elongate section of digestive tract between esophagus (or crop or gizzard) and rectum; bears longitudinal middorsal internal ridge (typhlosole) in terrestrial forms and may be covered by chloragogen tissue.

jaw [G *Kiefer*]: Toothed upper (dorsal) and lower (ventral) elements in buccal cavity of crayfish parasite. (teeth: equal, subequal)

lamella [G *Lamelle*]: One in a series of folds within calciferous glands of esophagus. (See also tunnel)

lateral commissure: commissural vessel

lateral heart (heart, pseudoheart) [G *Lateralherz, Herz*]: In circulatory system, enlarged, contractile anterior commissural vessels joining dorsal and ventral blood vessels.

lateral line [G *Seitenlinie*]: Narrow line extending along each side of body; consists of cell bodies (nuclei) of circular (and to a lesser extent, transverse) muscles in body wall.

lateral nerve (peripheral nerve, segmental nerve) [G *Segmentnerv*]: One of typically three or four pairs of nerves arising from ventral nerve cord in each segment.

lateral neural blood vessel (extraneural blood vessel) [G *Lateroneuralgefäß*]: Pair of narrow blood vessels, one along each side of ventral nerve cord.

lip [G *Lippe*]: One of typically two (dorsal and ventral) lobes surrounding mouth of crayfish parasite; functions as sucker and may bear digitiform or tentaculiform appendages.

longitudinal muscle [G *Längsmuskel*]: Relatively thick layer of longitudinal muscle between circular muscles and peritoneum of body wall (cuticle, epidermis, dermis; circular, longitudinal muscles; peritoneum); may be variable in structure. (intermediate, pennate, radiate, simple)

meganephridium (holonephridium) [G *Meganephridium*]: Typical type of large, paired metanephridium consisting of nephrostome, nephridial canal, bladder, and nephridiopore. (enteronephric, exonephric) (See also micronephridium, tufted nephridium)

meronephridium: micronephridium

metamere: segment

metanephridium [G *Metanephridium*]: Excretory organ; one may distinguish typically developed metanephridia (meganephridia) and modified micronephridia and tufted nephridia.

micronephridium (meronephridium) [G *Mikronephridium, Meronephridium*]: Modified type of metanephridium; typically relatively small and consisting of numerous tubular structures covering inner body wall and septa of each segment. Opens either into intestine (enteronephric) or to exterior via nephridiopore (exonephric). (See also metanephridium, tufted nephridium)

middorsal pore: dorsal pore

Morren's gland: calciferous gland

mouth [G *Mund, Mundöffnung*]: Anterior opening of digestive tract located beneath prostomium; opens into buccal cavity.

nephridial tubule [G *Nephridialkanal*]: Tubular section of metanephridium between nephrostome and nephridiopore; section of tubule anterior to body septum

may be referred to as preseptal, that posterior to septum as postseptal. Postseptal section may be coiled and differentiated into several regions, including bladder.

nephridial vesicle: bladder

nephridiopore [G *Nephridioporus, Nephroporus*]: One in a series of openings of metanephridia to exterior; typically located ventrolaterally on each side of body on intersegmental furrows.

nephridium: metanephridium

nephrostome [G *Nephrostom, Wimpertrichter*]: Proximal funnel-shaped and ciliated opening of meganephridium; located in body cavity of segment anterior to that bearing corresponding nephridiopore.

nodulus [G *Nodulus*]: Swelling on shaft of seta; serves as site of retractor muscle attachment.

ocellus (pigment-cup ocellus, eyespot) [G *Auge, Ocellus*]: Multicellular photosensitive organ on head of aquatic oligochaete. More primitive photosensitive organs are present in terrestrial forms.

oophore: cocoon

ovarian chamber [G *Eisack*]: In ♀ reproductive system, chamber formed by fusion of anterior and posterior septa of ovary-bearing (ovarian) segment; surrounds digestive tract, contains metanephridia, ovaries, and oviductal funnels, and gives rise posteriorly to ovisac.

ovary [G *Ovarium, Ovar*]: Section of ♀ reproductive system producing eggs; typically located at lower end of anterior septum of ovarian segment (ovarian segment follows testicular segment in anterior region of body). Eggs are released into body cavity or into ovisac. (bushy, discoidal, fan-shaped, finger-like, pear-shaped)

oviduct [G *Ovidukt, Eileiter*]: Ciliated, tubular section of ♀ reproductive system between oviductal funnel and ♀ gonopore. Relatively short, typically extending only to segment following ovarian segment.

oviductal funnel (female funnel, fimbriated funnel, ovarian funnel) [G *Eitrichter, Trichter*]: In ♀ reproductive system, ciliated, funnel-shaped proximal end of oviduct; located in same segment as ovary and typically associated with posterior septum. Takes up mature eggs and transports them to exterior.

ovisac (receptaculum ovorum) [G *Eisack, Eihälter*]: In ♀ reproductive system, pair of posteriorly directed outpocketings of posterior septum of ovary-bearing (ovarian) segment (occasionally of ovarian chamber); may extend back through several segments. Collects eggs before they are extruded to exterior via oviductal funnel and oviduct.

paratrium [G *Paratrium*]: Muscular outpocketing of atrium; may bear paratrial glands.

penial bulbus [G *Penialbulbus, Endapparat*]: Glandular, muscular opening of sperm duct.

penial seta [G *Penialborste*]: Modified, often enlarged seta associated with ♂ gonopore (prostatic pore). (shaft: straight, curved, undulating, smooth, ornamented, grooved; tip: bifid, claw-shaped, flattened, hollow-tipped, hooked, knobbed, lancet-shaped, sickle-shaped, spoon-shaped, spatulate, sharply pointed, bluntly pointed)

penial sheath [G *Penisscheide*]: Chitinous sheath surrounding penis. (conical, cylindrical, tub-shaped)

penis [G *Penis*]: Permanent ♂ copulatory organ within atrium; may be surrounded by penial sheath. (See also pseudopenis)

periproct (caudal segment, pygidium) [G *Aftersegment, Pygidium*]: Posteriormost projection of body; bears anus.

peristomium [G *Peristomium*]: First segment of body; surrounds mouth and bears prostomium dorsally. Lacks setae.

peritoneum [G *Peritoneum*]: Innermost layer of body wall (cuticle, epidermis, dermis; circular, longitudinal muscles; peritoneum); lines body cavity and forms walls of septa between segments. Gives rise to chloragogen tissue.

pharyngeal gland [G *Pharynxdrüse, Speicheldrüse*]: One of variously developed glandular masses associated with pharynx. (See also septal gland)

pharynx [G *Pharynx, Schlund*]: Expanded muscular section of digestive tract between buccal cavity and esophagus; may bear pharyngeal glands, septal glands, or diverticula, or be modified dorsally into muscular pad or plate.

porophore: Papilla-like elevation bearing ♂ gonopore terminally. (anteclitellar, intraclitellar)

proboscis (tentacle) [G *Rüssel, Tentakel*]: Elongate elaboration of prostomium. (retractile, nonretractile)

prostate gland [G *Prostata, Prostatadrüse*]: In ♂ reproductive system, paired gland associated with atrium and/or vas deferens or opening by separate duct into or near ♂ gonopore. (lobular, racemose, tubular; solid, diffuse)

prostomium [G *Prostomium*]: Lobe-like dorsal projection of first segment (peristomium). (epilobous, proepilobous, prolobous, tanylobous, zygolobous)

pseudoheart: lateral heart

pseudopenis [G *Pseudopenis*]: Papilla-shaped ♂ copulatory structure which is not a permanent (preformed) organ but rather represents eversion of terminal section of vas deferens or atrium. (See also penis)

pygidium: periproct

rectum [G *Rektum, Enddarm*]: Posterior section of digestive tract; opens to exterior via anus.

saddle: clitellum

segment (metamere, somite) [G *Segment*]: One in a series of segments of body; delimited by furrows corresponding to internal septa. Each primary segment may be further subdivided externally by superficial secondary furrows (annuli). Segment number is denoted by roman numerals.

segmental nerve: lateral nerve

seminal capsule: testis sac

seminal funnel (sperm funnel, male funnel) [G *Samentrichter*]: In ♂ reproductive system, ciliated, funnel-shaped proximal end of vas deferens; located in same segment as testis and typically associated with posterior septum. May be incorporated in testis sac. Takes up mature sperm from body cavity, seminal vesicle, or testis sac and transports them to exterior.

seminal groove [G *Samenrinne*]: Narrow groove extending posteriorly on outer surface of body wall from each ♂ gonopore to clitellum. Sperm travels along seminal groove during copulation.

seminal receptacle: spermatheca

seminal vesicle [G *Samenblase, Samensack*]: In ♂ reproductive system, typically paired outpocketings of posterior septum of testis-bearing (testicular) segment(s); serves to store sperm and may extend posteriorly through several segments.

septal gland [G *Septaldrüse*]: Modified pharyngeal gland located on septum adjacent to pharynx; opens dorsally into pharynx via long duct.

septum [G *Septum, Dissepiment*]: Transverse partition between two segments; corresponds externally to furrow. Formed by double layer of peritoneum and traversed by digestive tract, longitudinal blood vessels and nerve cords, and nephridial tubule.

seta (chaeta) [G *Borste*]: Variously shaped chitinous bristle projecting from body wall; produced in setal sac, movable due to special musculature, and typically emerging in pairs or bundles. One may distinguish lumbricine (octochaetine) or perichaetine distribution, with setae being designated by successive letters beginning from midventral line. (curved, sigmoid, straight; aciculate, bifid, biuncinate, blunt, capilliform, forked, pectinate, plumose, pointed, uncinate (See also crochet, genital seta, nodulus, penial seta)

setal follicle: setal sac

setal gland [G *Borstendrüse*]: One of various glands adjoining or associated with setal sacs and opening to exterior next to setae.

setal sac [G *Borstensack, Borstentasche*]: Tubular invagination of body wall containing setae; cuticle extends inward only as far as nodules. Base of sac contains glandular, seta-producing follicle and reserve or replacement follicle and is occasionally termed setal follicle.

somite: segment

spermatheca [G *Spermathek, Samentasche, Receptaculum seminis*]: In ♀ reproductive system, sac-shaped invagination of body wall for receiving sperm during copulation; typically paired, located in one or more anterior segments, and opening to exterior via spermathecal pore. May be differentiated into expanded ampulla and slender spermathecal duct.

spermathecal duct [G *Spermathekengang*]: In ♀ reproductive system, canal leading from enlarged section (ampulla) of spermatheca to exterior; may bear diverticulum.

spermathecal pore [G *Spermathekenporus*]: Opening of spermatheca to exterior in anterior region of body; typically located ventrolaterally in furrow. (paired, unpaired; middorsal, midventral, ventrolateral)

spermatophore [G *Spermatophor*]: Packet of sperm enclosed in membrane; produced at or near ♂ gonopore and placed on body of partner. May be regionated into distal capsule and proximal stalk. (club-shaped, leaf-shaped, oval, pear-shaped) (See also spermatozeugma)

spermatozeugma [G *Spermatozeugma, Spermiozeugma*]: Type of elongate spermatophore placed into spermatheca of partner; consists of central axial cavity around which sperm are regularly arranged.

sperm duct: vas deferens

sperm reservoir [G *Samenmagazin*]: In ♂ reproductive system, modified proximal end of vas deferens expanded to enclose testis and seminal vesicle.

stomach [G *Magen*]: Short, somewhat expanded section of digestive tract between esophagus and intestine; resembles intestine histologically.

subesophageal ganglion: subpharyngeal ganglion

subneural blood vessel [G *Subneuralgefäß*]: Longitudinal, noncontractile blood vessel below ventral nerve cord; typically restricted to intestinal region. Connected to dorsal blood vessel via paired commissural vessels in each segment. (See also ventral blood vessel)

subpharyngeal ganglion (subesophageal ganglion) [G *Unterschlundganglion*]: First ventral concentration of nerve tissue at point of merger of two circumpharyngeal connectives; joined to following segmental ganglia by ventral nerve cord.

sucker [G *Saugnapf, Endsaugnapf, Hintersaugnapf*]: Disc-like adhesive organ at posterior end of crayfish parasite. Occasionally termed caudal sucker to distinguish from sucker-like anterior structure formed by lips.

tentacle: proboscis

terminal bladder: bladder

testis [G *Hoden*]: Section of ♂ reproductive system producing sperm; typically paired organs on anterior septum of one or two (rarely three) (testicular) segments in anterior region of body. Sperm are released into body cavity or into testis sac/seminal vesicle. (free, in testis sac; digitate, lobed, pyriform)

testis sac (seminal capsule) [G *Testikelblase, Samenkapsel*]: In ♂ reproductive system, sac-like structure enclosing testes and typically also seminal funnels; adjoined by seminal vesicles.

tuberculum pubertatis (tubercle of puberty, ridge of puberty) [G *Pubertätshügel, Pubertätsleiste, Pubertätssaum*]: One of two glandular modifications of ventral surface in region of clitellum; consists of either ridge- or groove-like structure or papillose elevation.

tufted nephridium [G *Büschelnephridium*]: Modified, highly branched type of metanephridium. (See also meganephridium, micronephridium)

tumescence: Protruding, pad-like structure bearing glands and typically associated with gonopores or genital setae.

tunnel [G *Gang*]: In calciferous gland of esophagus, cavity formed between distally fused lamellae.

typhlosole [G *Typhlosolis*]: In terrestrial oligochaete, longitudinal ridge or fold of intestine wall; projects middorsally into lumen of intestine along its entire length and consists of simple or multiple folds.

valve [G *Ventilklappe, Klappe*]: In circulatory system, variously shaped valve or series of valves in lateral heart or dorsal blood vessel. (circular, double, single)

vas deferens (sperm duct) [G *Vas deferens, Samenleiter*]: Ciliated, tubular section of ♂ reproductive system between seminal funnel and ♂ gonopore. Relatively long, typically extending ventrally through several segments, and associated with prostate glands.

ventral blood vessel [G *Ventralgefäß*]: In circulatory system, longitudinal, noncontractile blood vessel lying ventral to digestive tract; carries blood in a posterior direction and is joined to dorsal blood vessel by lateral hearts.

ventral nerve cord [G *Bauchmark*]: Longitudinal nerve cord extending from subpharyngeal ganglion to posterior end of body; bears segmental ganglia and typically gives rise to three pairs of lateral nerves in each segment. (fused, paired)

HIRUDINEA

tapered — dorsoventrally flattened

elongate — pulsatile vesicle — cylindrical

regionated — trachelosome, gill, abdomen

with fin — lateral fin

vase-shaped — anterior sucker

(lateral)

head
- anterior sucker
- eye
- mouth
- brain
- buccal cavity
- circumpharyngeal connective
- subpharyngeal ganglion

preclitellar region
- pharynx
- ganglion
- penis sheath
- penis
- esophagus
- ♂ gonopore
- ejaculatory duct
- ♂

clitellar region
- stomach
- epididymis
- gastric ceca
- ♀ gonopore
- vagina
- oviduct
- ovary
- ovisac
- ♀

anterior trunk
- dorsal sinus
- testis
- vas efferens
- vas deferens
- DORSAL
- annulus
- nephridial bladder
- nephridiopore
- nephridial tubule
- nephrostome
- segment (triannulate)
- metanephridium
- ganglion
- ventral sinus
- intestine
- ventral nerve cord

posterior trunk
- rectal bladder
- rectum
- anal ganglion
- anus
- posterior sucker
- ocellus

with jaws
- anterior sucker
- jaws
- mouth
- salivary glands
- pharynx
- radial muscle

with proboscis
- anterior sucker
- proboscis
- proboscis sheath
- esophagus

no anterior sucker
with setae
- buccal cavity
- mouth
- pharynx
- setae

ANNELIDA

HIRUDINEA

leech [G *Egel*]

abdomen [G *Abdomen*]: In leech whose body is divided into two distinct regions, broad and flattened region posterior to clitellum; typically bears gills or pulsatile vesicles. (See also trachelosome)

albumen gland: oviducal gland

anal ganglion [G *Analganglion, Analknoten*]: Concentration of nerve tissue at posterior end of ventral nerve cord; consists of fused ganglia of posterior segments of body. Innervates posterior sucker.

annulus [G *Annulus, Ring*]: One of typically several external subdivisions of each segment. One annulus of each segment (considered to be either first or middle annulus) often surrounded by ring of sensilla and may be termed neural or sensory annulus.

anterior sucker (oral sucker) [G *Mundsaugnapf, vorderer Saugnapf*]: Relatively small, muscular anterior adhesive organ; surrounds mouth and may bear eyes dorsally. (continuous with body, set off; bell-shaped, discoid; with/without papillae) (See also lip, posterior sucker)

anus [G *After*]: Posterior opening of digestive tract, typically located dorsally at junction of body and posterior sucker.

atrial cornu (horn) [G *Horn des Atriums*]: Horn-shaped extension of atrium. (spirally coiled, not spirally coiled)

atrium (genital atrium) [G *Atrium*]: Muscular chamber at terminal end of ♂ reproductive system; may be surrounded by prostate glands. Either serves as site of spermatophore formation or contains copulatory apparatus (penis, penis sheath). (ellipsoidal, globular)

auricle [G *Aurikel*]: In terrestrial leech, membranous fold of posterior body wall at junction with posterior sucker; last pair of metanephridia open onto auricle.

bladder: nephridial bladder, rectal bladder

botryoidal tissue [G *Botryoidalgewebe, Botryoid-Zellen*]: In jawed leech, vascular tissue between body wall and digestive tract; consists of capillary channels and clusters of globular cells.

brain (supraesophageal ganglion) [G *Gehirn, Oberschlundganglion*]: Dorsal part of nerve ring surrounding pharynx; joined to subpharyngeal ganglion by pair of circumpharyngeal connectives.

branchia: gill

buccal cavity [G *Mundhöhle*]: Cuticle-lined section of digestive tract between mouth and pharynx; either bearing three jaws and separated from cavity of anterior sucker by velum or modified to accommodate proboscis.

buccal frill: Lobed margin of anterior sucker.

bursa [G *Bursa*]: Term occasionally applied to muscular, eversible section of atrium in ♂ reproductive system.

capsule (nephridial capsule) [G *Kapsel, Nephridialkapsel*]: Expanded, nonciliated section of metanephridium into which nephrostome opens; may or may not be directly connected with nephridial tubule.

cecum (caecum): esophageal cecum, gastric cecum, intestinal cecum

chaeta: seta

chlorogogenous cell [G *Chlorogog-Zelle*]: Metabolically active cell lining sinuses, associated with botryoidal tissue, or floating in body cavity.

ciliated organ: nephrostome

circular muscle [G *Ringmuskel*]: Outermost layer of muscle of body wall (cuticle, epidermis, dermis; circular, oblique, longitudinal muscles).

circumpharyngeal connective (peripharyngeal connective) [G *Schlundkonnektiv*]: One of two strands of nerve tissue surrounding pharynx and joining dorsal brain to large subpharyngeal ganglion.

clitellar gland [G *Klitellardrüse*]: One of numerous glands associated with clitellum; various types secrete outer wall of cocoon or fluid within cocoon.

clitellar region [G *Klitellarregion, Gürtelregion*]: Division of body (head, preclitellar and clitellar regions, anterior and posterior trunk) between preclitellar region and anterior trunk; consists of three segments, bears gonopores ventrally, and becomes modified into clitellum during reproductive period.

clitellum [G *Clitellum, Gürtel*]: Thickened glandular epidermis extending across three segments (clitellar region) between preclitellar region and anterior trunk; bears gonopores ventrally and contains clitellar glands. Conspicuous only during reproductive period.

cocoon [G *Kokon*]: Egg case secreted by clitellum; contains one to many eggs. One may distinguish outer and inner layers. (flat, rounded; outer surface: chitinoid, gelatinous, horny, spongy, tough and membranous)

conducting tissue: vector tissue

copulatory area [G *Area copulatrix, Begattungsfeld*]: Region of body onto or into which spermatophores are placed or injected. Sperm then migrates to ovisacs via underlying vector tissue.

copulatory gland: One of several glands opening ventrally, posterior to gonopores.

crop: stomach

cuticle [G *Kutikula*]: Noncellular, multilayered outermost layer of body wall (cuticle, epidermis, dermis; circular, oblique, longitudinal muscles); relatively thin and also lining proboscis, proboscis cavity, buccal cavity, and pharynx. Secreted by epidermis. May be produced into papillae or tubercles.

dermis [G *Dermis*]: Relatively thick layer of connective tissue between epidermis and musculature of body wall (cuticle, epidermis, dermis; circular, oblique, longitudinal muscles); may be penetrated by gland cells of epidermis.

diverticulum: cecum

dorsal blood vessel [G *Rückengefäß, dorsale Blutgefäß*]: In more primitive type of circulatory system, contractile, longitudinal vessel lying dorsal to digestive tract; propels blood anteriorly and may be equipped with valves. Connected to ventral blood vessel at anterior and posterior ends of body and gives rise to intestinal blood sinus. (See also dorsal sinus)

dorsal sinus [G *Dorsallakune, dorsaler Coelomkanal*]: In more advanced type of circulatory system, longitudinal channel dorsal to digestive tract; may contain dorsal blood vessel or replace vessel entirely. (See also lateral sinus, ventral sinus)

dorsoventral muscle [G *Dorsoventralmuskel*]: Band of muscle spanning between ventral and dorsal body walls and extending through other muscle layers (circular, oblique, longitudinal muscles); typically concentrated between ceca of digestive tract.

ejaculatory duct [G *Ductus ejaculatorius*]: Muscular section of ♂ reproductive system between epididymes (seminal vesicles) and common atrium.

epidermis [G *Epidermis*]: Outermost cellular layer of body wall (cuticle, epidermis, dermis; circular, oblique, longitudinal muscles); secretes cuticle and contains variously shaped unicellular glands. Thickened to form clitellum.

epididymis (seminal vesicle, sperm vesicle) [G *Samenblase*]: Enlarged, often coiled section of ♂ reproductive system at anterior end of each vas deferens; continues into ejaculatory duct.

esophageal cecum [G *Blindsack*]: One of two outpocketings of esophagus.

esophagus (oesophagus) [G *Ösophagus*]: Relatively short section of digestive tract between pharynx and intestine; in proboscis-bearing leech, between proboscis and intestine. May bear pair of esophageal ceca.

eye [G *Auge*]: One of typically several segmentally arranged, multicellular photosensitive organs located dorsally on anterior sucker or head. Number and position are of taxonomic importance. More primitive photosensitive organs (ocelli) may be present on posterior sucker. (crescent-shaped, rounded, fused)

Faivre's nerve [G *Faivre'scher Nerv*]: Small, longitudinal median nerve originating from subesophageal ganglion and accompanying ventral nerve cord.

fenestra [G *Fenestra*]: In fused type of gastric ceca, free spaces formed as result of incomplete fusion.

fin (lateral fin) [G *Flossensaum*]: Fin-like fold extending along each side of body; may be indented in region of clitellum.

folded organ [G *Faltenorgan*]: One of two highly folded lateral diverticula at border of intestine and rectal bladder.

funnel: nephrostome

ganglion [G *Ganglion*]: Concentration of nerve tissue, typically referring to segmentally arranged swellings of ventral nerve cord; each ganglion gives rise to lateral nerves. Concentrated anteriorly to form subpharyngeal ganglion, posteriorly to form anal ganglion.

gastric cecum [G *Magenblindsack, Blindsack*]: One of typically several pairs of highly distensible lateral outpocketings of stomach; posterior pair may be elongated and posteriorly directed. Serves to store ingested food. (bilobed, fused, simple) (See also fenestra)

gill (branchia) [G *Kieme*]: One in a series of highly vascularized lateral projections of body wall. (finger-shaped and branching, leaf-shaped and unbranched) (See also pulsatile vesicle)

gonopore [G *Geschlechtsöffnung, Gonoporus*]: Typically separate openings of ♂ and ♀ reproductive systems to exterior on ventral surface of clitellar region. ♂ gonopore always anterior to ♀ gonopore. Number of annuli between gonopores is of taxonomic importance.

head [G *Kopf, Kopfregion*]: Anteriormost of five divisions of body (head, preclitellar and clitellar regions, anterior and posterior trunk); consists of six segments and small prostomium. Bears dorsal eyes and ventral mouth surrounded by sucker.

horn: atrial cornu

intestinal blood sinus (intestinal plexus) [G *Darmgefäßplexus, Darmblutsinus*]: Expanded posterior region of dorsal blood vessel enveloping intestine and intestinal ceca.

intestinal cecum [G *Blindsack*]: One of typically two pairs of lateral outpocketings of intestine.

intestine [G *Hinterdarm*]: Relatively slender section of digestive tract between stomach (crop) and rectum; separated from stomach by pyloric sphincter. (tubular, with intestinal ceca)

jaw [G *Kiefer*]: One of three (one medio-dorsal, two ventrolateral) semicircular, muscular structures in buccal cavity. Each jaw contains ducts and openings of salivary glands, and cuticle coating is modified into numerous teeth.

lateral sinus [G *Seitenlakune*]: In more advanced type of circulatory system, one of two contractile, longitudinal channels along each side of body; gives rise to three branches per segment. (See also dorsal sinus, ventral sinus)

lip [G *Lippe*]: Term occasionally applied to rim of anterior sucker (i.e., anterior end of sucker is upper lip, posterior end is lower lip). (with/without median ventral groove)

longitudinal muscle [G *Längsmuskel*]: Relatively thick innermost muscle layer of body wall (cuticle, epidermis, dermis; circular, oblique, longitudinal muscles).

metanephridium [G *Metanephridium*]: One of 10–17 pairs of segmentally arranged excretory organs in midregion of body; consists of nephrostome, capsule, nephridial tubule, nephridial bladder, and nephridiopore.

mouth [G *Mund, Mundöffnung*]: Anterior opening of digestive tract; typically located somewhat ventrally and surrounded by anterior sucker. Opens into buccal cavity.

nephridial bladder (bladder) [G *Blase*]: Somewhat expanded terminal section of nephridial tubule prior to nephridiopore.

nephridial tubule [G *Nephridialkanal*]: Elongate and typically regionated section of metanephridium opening to exterior via nephridiopore. May be intracellular, continuous or not continuous with capsule and nephrostome, and interconnected with nephridial tubules of other segments.

nephridiopore [G *Nephridioporus, Nephroporus*]: Opening of metanephridium to exterior; paired and typically located ventrally in furrows between annuli, although first and last pairs in terrestrial leech may open next to suckers. (See also auricle)

nephridium: metanephridium

nephrostome [G *Nephrostom*]: Proximal end of metanephridium in body cavity or ventral sinus. Consists of either ciliated funnel or more complex ciliated organ. Adjoins testis and opens into capsule.

neural ring: sensory ring

oblique muscle [G *Diagonalmuskel*]: Layer of muscle between outer circular and inner longitudinal muscles of body wall (cuticle, epidermis, dermis; circular, oblique, longitudinal muscles).

ocellus (eyespot) [G *Ocellus, Augenfleck*]: One of numerous simple photosensitive pigment spots on posterior sucker. (punctiform, crescent-shaped) (See also eye)

oral sucker: anterior sucker

ovary [G *Ovarium, Ovar*]: Section of ♀ reproductive system producing eggs; consists of pair of elongate structures within coelomic sacs (ovisacs) in segment anterior to testes. Opens via oviduct, vagina, and ♀ gonopore.

oviducal gland (albumen gland): Aggregation of unicellular glands surrounding oviduct.

oviduct [G *Ovidukt, Eileiter*]: Short, relatively narrow section of ♀ reproductive system between each ovary and vagina. May be surrounded by oviducal gland.

ovisac [G *Ovarialsack, Ovarialschlauch*]: One of two tubes containing ovaries; may be located in ventral sinus. (elongate, looped)

parenchyma [G *Parenchym*]: Tissue between body wall and digestive tract. (See also botryoidal tissue)

penis [G *Penis*]: Eversible, muscular copulatory organ in atrium; enclosed in penis sheath.

penis sheath [G *Penisscheide*]: Muscular sheath enclosing penis in atrium.

peripharyngeal connective: circumpharyngeal connective

pharynx [G *Pharynx, Schlund*]: Cuticle-lined, muscular section of digestive tract between buccal cavity and esophagus; radial muscles of pharynx are interspersed with salivary glands.

posterior sucker (caudal sucker) [G *Hintersaugnapf, Endsaugnapf*]: Relatively large, muscular posterior adhesive organ; consists of seven fused segments and is typically positioned ventrally. Relative size (vs. trunk or anterior sucker) is of taxonomic importance. (discoid, deeply cupped; with/without ocelli, with/without retractile papillae, with/without subsidiary suckers)

preclitellar region [G *Präklitellarregion*]: Division of body (head, preclitellar and clitellar regions, anterior and posterior trunk) between head and clitellum; consists of four segments.

prepuce [G *Präputium*]: Fold of body wall of segment XII; projects over preceding two segments and covers ♀ and ♂ gonopores.

proboscis [G *Rüssel*]: Muscular tube located in proboscis sheath and forming anterior end of digestive tract; bears openings of salivary glands and is lined with cuticle both internally and externally. Projects from proboscis pore.

proboscis pore [G *Rüsselöffnung, "Mund"*]: Anterior opening of proboscis sheath through which proboscis is projected. Not strictly equivalent to mouth.

proboscis sheath [G *Rüsselscheide*]: Cuticle-lined sheath enclosing proboscis at anterior end of body. Opens to exterior via proboscis pore.

prostate gland [G *Prostata*]: In ♂ reproductive system, aggregation of unicellular glands surrounding atrium.

prostomium [G *Prostomium*]: Small, lobe-like projection preceding first segment; may form anterior rim of anterior sucker.

pulsatile vesicle [G *Seitenblase, Seitenbläschen, pulsierendes Bläschen*]: Hemispherical, contractile respiratory structure projecting from each side of segments of midbody; represents lateral extension of sinus system.

pyloric sphincter [G *Sphinkter*]: Sphincter muscle between stomach and intestine.

radial muscle [G *Radiärmuskel*]: Radial muscle spanning from pharynx to body wall or located in posterior sucker; those of pharynx are heavily interspersed with salivary glands.

rectal bladder [G *Rektalblase*]: Term applied to ciliated, distended section of rectum.

rectum [G *Enddarm*]: Relatively short, cuticle-lined section of digestive tract between intestine and anus; lacks ceca. May be preceded by ciliated rectal bladder.

salivary gland [G *Speicheldrüse*]: One of numerous unicellular glands surrounding pharynx and opening on proboscis or between teeth of jaws via long ducts.

scute [G *Schuppe*]: Small dorsal plate on dorsal surface of segment VIII.

segment (metamere, somite) [G *Segment, Metamer, Somit*]: One in a series of 34 segments of body; each primary segment (number denoted by roman numerals) is further subdivided externally into several annuli. Border of segment indicated externally by color pattern or sensory ring (considered to be either first or middle annulus), more accurately internally by innervation of corresponding ganglion.

seminal vesicle: epididymis

sensillum (sensory papilla) [G *Sinnesknospe*]: One in a series of small, often light-colored sense organs surrounding one annulus (considered to be either first or middle annulus) of each segment. (See also sensory ring)

sensory ring (neural ring) [G *Sinnesring*]: Ring of sensilla surrounding one annulus (sensory or neural annulus: considered to be either first or middle annulus) of each segment.

septum [G *Septum, Dissepiment*]: Transverse partition between two segments; present in embryo, rarely also in anterior-most segments of primitive adult leech.

seta (chaeta) [G *Borste*]: One in a series of bristles on first five segments of primitive leech.

sinus: dorsal sinus, lateral sinus, ventral sinus

somite: segment

spermatophore [G *Spermatophor*]: Packet of sperm enclosed by casing; formed in atrium and consists of two halves, each representing sperm bundle from one epididymis/ejaculatory duct. Placed on or partially injected into body wall of partner. (See also copulatory area)

stomach (crop) [G *Magen, Kropf, Magendarm*]: Large section of digestive tract between esophagus and intestine; separated from intestine by pyloric sphincter; typically with conspicuous gastric ceca. (tubular, with ceca)

subpharyngeal ganglion (sub-esophageal ganglion) [G *Unterschlundganglion*]: Large concentration of nerve tissue ventral to pharynx; consists of fused ganglia of several segments and is joined to dorsal brain by circumpharyngeal connectives.

subsidiary sucker [G *Nebensaugnapf, sekundärer Saugnapf*]: One of numerous stalked, small suckers on inner surface of posterior sucker.

supraesophageal ganglion: brain

sympathetic nerve ring (somatogastric nerve ring) [G *sympatischer Nervenring*]: Small nerve ring surrounding pharynx anterior to main nerve ring (brain, circumpharyngeal connectives, subpharyngeal ganglion); joined to network of nerves surrounding digestive tract (sympathetic nerve system).

testis [G *Hoden, Hodensäckchen, Hodenbläschen*]: In ♂ reproductive system, one of several to many pairs of segmentally arranged, spherical, sperm-producing structures; each testis is connected to common vas deferens via short vas efferens. Occasionally termed testis sac because it contains coelomic fluid and is considered to be derived from elongate body cavity. Typically positioned between gastric ceca.

testis sac: testis

tooth [G *Zähnchen*]: One in a series of small cuticular projections on each jaw; may be arranged in one or two rows. Salivary gland ducts may open between teeth.

trachelosome: In leech whose body is divided into two distinct regions, slender anterior region. (See also abdomen)

trunk [G *Rumpf*]: Posteriormost division of body (head, preclitellar and clitellar regions, anterior and posterior trunk). Divided into anterior region consisting of 15 segments and posterior region of 8 fused segments (posterior sucker).

tubercle [G *Warze*]: Type of ornamentation of body wall; forms series of symmetrically arranged projections on annuli.

vagina [G *Vagina*]: Muscular section of ♀ reproductive system between common oviduct and ♀ gonopore. (elongate, U-shaped; with/without ceca)

vas deferens [G *Vas deferens*]: In ♂ reproductive system, one of two longitudinal ducts into which testes empty via short vasa efferentia; modified anteriorly to form epididymis.

vas efferens [G *Vas efferens*]: Short, narrow section of ♂ reproductive system joining each testis to longitudinal vas deferens.

vector tissue (conducting tissue) [G *Leitgewebe*]: Tissue underlying copulatory area; sperm from injected spermatophore migrate to ovisac via vector tissue.

velum [G *Velum*]: Low fold of tissue separating buccal cavity from cavity of anterior sucker.

ventral blood vessel [G *Ventral-gefäß*]: In more primitive type of circulatory system, longitudinal vessel lying ventral to digestive tract; carries blood in posterior direction. (See also ventral sinus)

ventral nerve cord [G *Bauch-mark*]: Longitudinal pair of closely adjoining nerve cords extending from subpharyngeal ganglion to posterior end of body or anal ganglion; fused in each segment at ganglion.

ventral sinus [G *Ventrallakune*]: In more advanced type of circulatory system, longitudinal channel ventral to digestive tract; may contain ventral blood vessel (and ventral nerve cord, nephrostomes, ♀ reproductive organs) or replaces vessel entirely. (See also dorsal sinus, lateral sinus)

ONYCHOPHORA

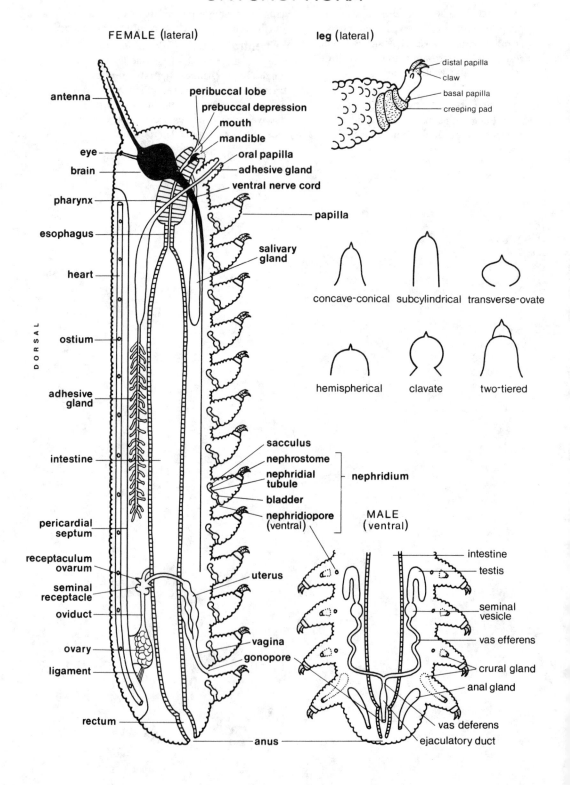

FEMALE (lateral)

leg (lateral)

- distal papilla
- claw
- basal papilla
- creeping pad

antenna

peribuccal lobe
prebuccal depression
mouth
mandible
oral papilla
adhesive gland
ventral nerve cord

eye
brain
pharynx
esophagus
heart

salivary gland

papilla

concave-conical subcylindrical transverse-ovate

DORSAL

ostium

adhesive gland

intestine

hemispherical clavate two-tiered

sacculus
nephrostome
nephridial tubule
bladder
nephridiopore (ventral)

nephridium

MALE (ventral)

pericardial septum
receptaculum ovarum
seminal receptacle
oviduct
ovary
ligament

uterus

vagina
gonopore

intestine
testis

seminal vesicle

vas efferens

crural gland
anal gland

vas deferens
ejaculatory duct

rectum

anus

ONYCHOPHORA

onchyophoran [G *Onychophore, Stummelfüßer*]

accessory tooth [G *akzessorischer Zahn*]: One in a series of smaller teeth adjoining primary tooth on inner or outer blade of each mandible. (See also denticle, diastema)

adhesive gland (slime gland) [G *Wehrdrüse, Schleimdrüse*]: Pair of elongate, anteriorly enlarged and tubular, posteriorly branched glands extending the length of the body; opens distally on oral papilla.

anal gland [G *Analdrüse, hintere akzessorische Genitaldrüse, Pygidialdrüse*]: Pair of accessory genital glands in pygidium of ♂; typically opens ventrally between gonopore and anus. (opening: dorsal/unpaired, ventral/paired or unpaired)

antenna [G *Antenne*]: Pair of elongate, annulated, movable appendages projecting from anterior end of onychophorans; innervated by brain and covered by various types of sensory organs.

anus [G *After*]: Posterior opening of digestive tract; located ventroterminally at posterior end of body.

apodeme [G *Apodem*]: Long, thin, sclerotized, posteriorly directed projection of mandibles; serves as base for muscle attachment.

archicerebrum: protocerebrum

basal papilla [G *Basalpapille*]: On leg, one of two papillae at junction of pads and foot. (See also distal papilla)

bladder [G *Harnblase*]: Expanded, contractile terminal section of nephridial canal; tapered section between bladder and nephridiopore is lined with cuticle.

brain [G *Gehirn, Oberschlundganglion*]: Large, bilobed main concentration of nerve tissue dorsal to pharynx; innervates eyes and antennae anteriorly and gives rise posteriorly to paired ventral nerve cords. Considered to consist of archicerebrum, deuterocerebrum, and tritocerebrum.

ciliated funnel: nephrostome

circular muscle [G *Ringmuskel*]: Outermost of three layers of muscle in body wall (cuticle, epidermis, dermis; circular, diagonal, longitudinal muscles); also overlying longitudinal muscles in intestine.

claw [G *Kralle*]: One of two movable, claw-shaped, sclerotized structures at end of each leg.

coxal vesicle [G *Coxalbläschen*]: Thin-walled, eversible vesicle in leg; opens to exterior ventrally, typically near nephridiopore, except in leg pairs four and five, where it lies between creeping pads.

crural gland [G *Cruraldrüse*]: Sac-shaped glandular invagination opening via crural papilla on leg near nephridiopore; typically found only in ♂. May be modified in first and last leg-bearing segments. (paired, unpaired)

crural papilla [G *Cruralpapille*]: In ♂, ventral cuticular projection on leg near nephridiopore; bears opening of crural gland distally. (retractile, nonretractile)

cuticle [G *Kutikula*]: Thin, flexible, noncellular layer covering outer surface of body and lining pharynx, esophagus, and rectum.

denticle (denticulus) [G *Zähnchen*]: One in a series of small teeth adjoining diastema on inner blade of mandible.

dermis [G *Dermis*]: Cell layer between musculature and epidermis in body wall (cuticle, epidermis, dermis; circular, diagonal, longitudinal muscles); contains collagen fibers.

deuterocerebrum [G *Deuterocerebrum*]: Middle and largest section of brain; innervates antennae. (See also protocerebrum, tritocerebrum)

diagonal muscle (oblique muscle) [G *Diagonalmuskel, Schrägmuskel*]: One of two layers of oblique muscles between inner longitudinal and outer circular muscles in body wall (cuticle, epidermis, dermis; circular, diagonal, longitudinal muscles).

diastema [G *Diastema*]: Gap between larger primary tooth and series of denticles on inner blade of mandible.

distal papilla [G *Fuß-Distalpapille*]: On leg, one of two to seven distal papillae surrounding claws. (See also basal papilla)

ejaculatory duct [G *Ductus ejaculatorius*]: Distal muscular end of vas deferens in ♂ reproductive system; opens to exterior via gonopore.

epidermis [G *Epidermis*]: Outermost cellular layer of body wall (cuticle, epidermis, dermis; circular, diagonal, longitudinal muscles); gives rise to cuticle. Papilla formation by epidermis determines ornamentation of body surface.

esophagus [G *Ösophagus*]: Short, relatively narrow, cuticle-lined section of digestive tract between pharynx and intestine; border between esophagus and intestine bears valvula cardiaca.

eye [G *Auge*]: Pair of small photoreceptive organs, one at base of each antenna; consists of cornea, lens, and retina.

foot [G *Fuß*]: Clearly delimited, claw-bearing distal section of leg.

foregut [G *Vorderdarm*]: Cuticle-lined anterior section of digestive tract consisting of pharynx and esophagus.

gonopore (genital pore) [G *Gonoporus, Geschlechtsöffnung*]: Unpaired opening of ♂ or ♀ reproductive system to exterior; location either between/behind last leg pair or between next-to-last (= penultimate) leg pair is of taxonomic importance. May be positioned on ovipositor in ♀.

heart [G *Herz*]: Elongate, tubular circulatory organ bearing series of paired lateral ostia; located dorsal to intestine within pericardial sinus and pumps blood anteriorly.

hindgut: rectum

intestine [G *Mitteldarm*]: Long, straight section of digestive tract between esophagus and rectum; bears inner longitudinal and outer circular muscle layers.

lateral sinus: ventrolateral sinus

leg (oncopodium) [G *Bein, Stummelbein*]: One of 13–43 pairs of large, conical, unjointed ventrolateral limbs; each leg bears two claws, basal (in a few species) and distal papillae, creeping pads, nephridiopore, and gland openings. Term occasionally used to distinguish proximal conical section of limb from clearly delimited, claw-bearing distal section.

longitudinal muscle [G *Längsmuskel*]: Relatively thick bands of longitudinal muscles underlying diagonal and circular muscles and forming innermost layer of body wall (cuticle, epidermis, dermis; circular, diagonal, longitudinal muscles). (dorsal, dorsolateral, ventral)

mandible (jaw) [G *Mandibel, Kieferklaue*]: Pair of movable, sclerotized jaws in prebuccal depression, one member on each side of mouth. Each consists of outer and inner sickle-shaped and toothed blades on a base with posteriorly directed apodeme.

mouth [G *Mund, Mundöffnung*]: Ventral opening of digestive tract at anterior end of body; located within prebuccal depression. (See also buccal cavity, mandible)

nephridial canal [G *Nephridialkanal*]: Narrow, tubular section of nephridium between nephrostome and nephridiopore; expanded distally, except in first leg pairs, to form bladder.

nephridiopore [G *Nephridialmündung*]: Opening of nephridium to exterior; typically located ventrally at base of each leg, in leg pairs four and five between creeping pads.

nephridium [G *Nephridium, Segmentalorgan*]: One in a series of segmentally arranged, paired excretory organs, each typically consisting of ventral nephrostome, nephridial tubule, and bladder; opens via nephridiopore at base or between creeping pads of leg. Anteriormost pair modified into salivary glands, posteriormost into anal glands.

nephrostome (ciliated funnel) [G *Nephrostom, Nephridialtrichter*]: Proximal ciliated section of each nephridium; closely associated with sacculus.

oblique muscle: diagonal muscle

oesophagus: esophagus

oncopodium: leg

oral papilla [G *Oralpapille*]: Pair of short, leg-like protuberances, one on each side of mouth; each bears opening of adhesive gland terminally.

ostium [G *Ostium*]: Series of paired lateral openings in heart, one in each segment.

ovary [G *Ovar*]: Proximal section of ♀ reproductive system located dorsal to intestine in posterior end of body; opens to exterior via paired oviducts, uteri, and vagina and gonopore. Either free or attached to pericardial septum directly or by ligament. (partially fused, totally fused)

oviduct [G *Ovidukt*]: Paired, relatively short section of ♀ reproductive system between ovary and expanded uterus; each oviduct typically bears seminal receptacle, in certain onychophorans also receptaculum ovorum.

ovipositor [G *Ovipositor*]: In oviparous and rarely in viviparous onychophorans, relatively large, unpaired midventral appendage at posterior end of ♀; bears gonopore distally.

pad (creeping pad, spiniferous pad, spinous pad, pedal ring) [G *Sohlenwulst, Sohlenring, Stachelplatte, Laufsohle*]: One of several (three to seven) papilla-bearing pads distally on ventral surface of each leg.

papilla [G *Papille*]: One of a great number of variously arranged cuticular protuberances covering outer surface of body. One may distinguish larger primary papillae with terminal sensory spine and smaller accessory papillae lacking spines. (clavate, concave-conical, concave-subcylindrical, convex-conical, hemispherical, ovate, transverse ovate)

penial papilla [G *Penispapille*]: In ♂ reproductive system of certain onychophorans, wart-covered, bill-shaped ventral projection closely associated with gonopore.

peribuccal lobe [G *Lippenpapille*]: Circle of fleshy lobes surrounding mouth and forming rim of prebuccal depression.

pericardial septum [G *Perikardialseptum*]: Horizontal septum dorsal to intestine; delimits pericardial sinus.

pericardial sinus [G *Perikardialsinus*]: Elongate body cavity dorsal to intestine, enclosing heart; separated from periintestinal sinus by pericardial septum.

periintestinal sinus [G *Periviszeralsinus*]: Median body cavity surrounding intestine, gonads, and adhesive glands; separated from ventrolateral sinuses by dorsoventral muscles, from pericardial sinus by pericardial septum.

pharynx [G *Pharynx, Schlund*]: Cuticle-lined, muscular section of digestive tract between mouth and esophagus.

placenta [G *Plazenta*]: In viviparous onychophorans, organ nourishing embryos.

Consists of stalk-like structure intimately associated with oviduct wall.

prebuccal depression [G *Mundvorraum, Mundhöhlenvorraum*]: Cavity formed around mouth by circle of peribuccal lobes; may be opened and closed and bears pair of mandibles internally.

primary tooth [G *Hauptzahn*]: First and largest tooth on inner or outer blade of each mandible; may be adjoined by series of accessory teeth. (See also diastema)

protocerebrum (archicerebrum) [G *Protocerebrum*]: Anterior section of brain innervating eye.

pygidium (postgenital terminal cone) [G *Pygidium*]: Term applied to conical, anus-bearing region of body posterior to last leg pair.

receptaculum ovorum [G *Receptaculum ovorum*]: In ♀ of certain onychophorans, pointed organ near seminal receptacle.

rectum (hindgut) [G *Enddarm*]: Short, cuticle-lined section of digestive tract posterior to intestine. Opens to exterior via anus.

sacculus [G *Sacculus*]: Sac-shaped body cavity located near base of each leg and into which proximal section of nephridium (nephrostome) opens.

salivary gland [G *Speicheldrüse*]: Pair of elongate, unbranched glands opening via common duct into foregut between prebuccal depression and pharynx.

scale [G *Schuppe*]: Minute, scale-shaped cuticular structure covering surface of dermal papillae in great numbers.

segmental organ: nephridium

seminal receptacle [G *Receptaculum seminis*]: Small, ciliated outpocketing of oviduct, one between ovary and each uterus; serves to store sperm.

seminal vesicle [G *Vesiculum seminalis, Samenblase*]: Short, expanded section of ♂ reproductive system between each testis and vas efferens.

slime gland: adhesive gland

spermatophore [G *Spermatophor*]: Packet of sperm formed in vas deferens and deposited on outer surface of ♀ or into ♀ gonopore.

spiracle [G *Atemöffnung, Trachealöffnung*]: Minute respiratory opening to exterior; not closable and continuing into cuticle-lined atrium and trachea. Distributed in great numbers over surface of body, especially dorsal side.

testis [G *Hoden*]: Paired, elongate section of ♂ reproductive system dorsal to intestine; each opens to exterior via separate seminal vesicles and vasa efferentia and common vas deferens, ejaculatory duct, and gonopore.

trachea [G *Trachee, Luftrohr*]: One of many long, thin respiratory tubules projecting into interior of body from each spiracle; cuticle-lined and typically unbranched.

transverse muscle (dorsoventral muscle) [G *Transversalmuskel, Dorsoventralmuskel*]: One of numerous dorsoventral muscles; divides body cavity into median periintestinal sinus and two ventrolateral sinuses.

tritocerebrum [G *Tritocerebrum*]: Small posterior section of brain innervating mouth and anterior section of digestive tract. (See also deuterocerebrum, protocerebrum)

uterus [G *Uterus*]: Expanded posterior section of each oviduct; opens to exterior via vagina and gonopore.

vagina [G *Vagina*]: Short, muscular section of ♀ reproductive system between uteri and gonopore.

valvula cardiaca [G *Ösophagealklappe*]: Valve-like structure at junction of esophagus and intestine; prevents reverse flow of food.

vas deferens [G *Vas deferens*]: Common sperm duct of ♂ reproductive system between junction of both vasa efferentia and gonopore. More narrow proximal section serves as site of spermatophore production, muscular distal section as ejaculatory duct.

vas efferens [G *Vas efferens*]: Long, narrow, coiled section of ♂ reproductive system between each seminal vesicle and common vas deferens.

ventral nerve cord [G *Bauchmark*]: Pair of longitudinal nerve cords extending from brain to posterior end of body, where they join above rectum. Connected via numerous commissures; innervates legs.

ventrolateral sinus [G *Lateralsinus*]: One of two longitudinal lateral body cavities; separated from main periintestinal sinus by dorsoventral muscle bands.

TARDIGRADA

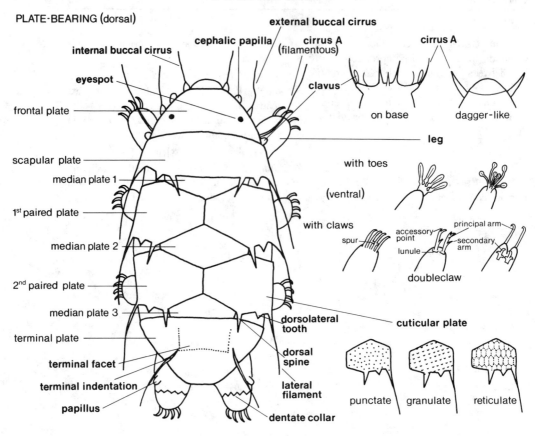

PLATE-BEARING (dorsal)

internal buccal cirrus
cephalic papilla
external buccal cirrus
cirrus A (filamentous)
eyespot
clavus
frontal plate
cirrus A
on base
dagger-like
scapular plate
leg
median plate 1
with toes
1ˢᵗ paired plate
(ventral)
median plate 2
with claws
accessory point
principal arm
secondary arm
spur
lunule
2ⁿᵈ paired plate
doubleclaw
median plate 3
dorsolateral tooth
terminal plate
cuticular plate
dorsal spine
terminal facet
terminal indentation
lateral filament
papillus
punctate   granulate   reticulate
dentate collar

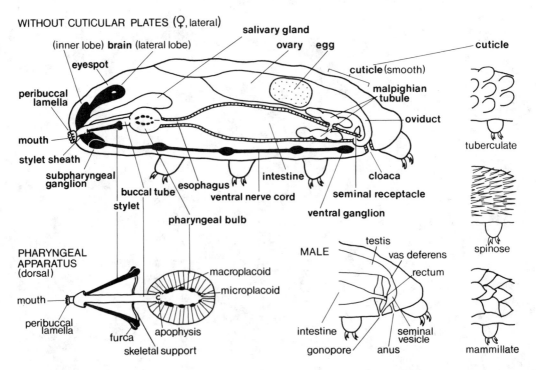

WITHOUT CUTICULAR PLATES (♀, lateral)

salivary gland
(inner lobe) brain (lateral lobe)
ovary   egg
cuticle
eyespot
cuticle (smooth)
peribuccal lamella
malpighian tubule
mouth
oviduct
stylet sheath
tuberculate
subpharyngeal ganglion
buccal tube
esophagus
intestine
cloaca
stylet
ventral nerve cord
seminal receptacle
pharyngeal bulb
ventral ganglion
spinose

PHARYNGEAL APPARATUS (dorsal)
macroplacoid
microplacoid
mouth
peribuccal lamella
furca
apophysis
skeletal support

MALE
testis
vas deferens
rectum
intestine
gonopore
seminal vesicle
anus
mammillate

TARDIGRADA

tardigrade, "water-bear" [G *Tardigrad, Bärtierchen*]

accessory point [G *Nebenspitze*]: Additional small point adjacent to tip of principal arm of doubleclaw.

adhesive disc [G *Haftlappen*]: Ventrally adhesive distal section of toe in certain tardigrades. (disc-shaped, finger-shaped)

apophysis [G *Apophyse*]: In digestive tract, cuticular thickening at border of buccal tube and pharyngeal bulb; functions as a valve.

brain (cerebral ganglion) [G *Gehirn*]: Main concentration of nerve tissue, composed of three median lobes and two lateral lobes lying over pharyngeal bulb and salivary gland; connected to subpharyngeal ganglion via two connectives surrounding pharynx.

buccal tube [G *Mundrohr*]: Section of digestive tract between mouth and pharyngeal bulb; pierced anteriorly by pair of stylets.

cephalic papilla [G *Papilla cephalica, Kopfpapille*]: In certain tardigrades, one of two sensory papillae around mouth at anterior end; typically located between internal and external buccal cirri.

cerebral ganglion: brain

chitinous rod [G *Schlundkopfleiste*]: One of three rod-shaped, chitinous structures within pharyngeal bulb; serves as skeletal support for pharynx musculature. (See also placoid)

cirrus A [G *Cirrus lateralis*]: In plate-bearing tardigrade, long, often anteriorly directed cirrus with clavus at its base; located laterally at junction of frontal and scapular plates.

clavus [G *Clava*]: Short, blunt cuticular projection adjacent to base of cirrus A. (thumb-shaped, club-shaped)

claw [G *Kralle*]: Variously shaped, claw-like structure at distal end of leg; size, shape, and arrangement are of taxonomic importance. (See also doubleclaw)

cloaca [G *Kloake*]: Term applied to posteriormost section of digestive tract in those tardigrades in which either oviduct or vasa deferentia open into rectum.

coelomocyte [G *Speicherzelle*]: Free storage cell in fluid-filled body cavity.

cuticle [G *Kutikula*]: Noncellular, multilayered, variously ornamented layer covering outer surface of body and extending into both foregut and rectum. (smooth, punctate, granulate, reticulate)

dentate collar [G *Dornfalte*]: Dentate, ventrally interrupted cuticular fold around last (fourth) pair of legs in certain tardigrades.

doubleclaw [G *Doppelhaken*]: Type of claw bifurcating into larger principal arm and smaller secondary arm; point of intersection of two arms is of taxonomic importance.

esophagus [G *Ösophagus*]: Tubular section of digestive tract between pharyngeal bulb and intestine.

external buccal cirrus [G *Cirrus medianus externus*]: Pair of sensory cirri, one to either side of mouth in certain tardigrades; typically located posterior to internal buccal cirri and cephalic papillae.

external claw [G *äußere Kralle*]: One of two outer claws on legs of certain plate-bearing tardigrades; differs in shape from adjacent internal claws.

eyespot [G *Auge, Pigmentbecherocelle, Ocellus*]: Pair of dorsal pigmented, light-sensitive cells on anterior end of body. (red, black)

foregut [G *Vorderdarm*]: Anterior cuticle-lined section of digestive tract between mouth and esophagus; comprises buccal tube and pharyngeal bulb.

frontal plate [G *Kopfplatte*]: In plate-bearing tardigrade, anteriormost plate covering greater part of head and bearing eyespots.

furca [G *Furca*]: Swollen, bifurcate posterior end of stylet; stylet supports are positioned in bifurcation.

gonopore [G *Gonoporus*]: Opening of ♂ or ♀ reproductive system; opens either at anus or as a separate pore between third and fourth leg pairs.

hindgut: rectum

internal buccal cirrus [G *Cirrus medialis internus*]: Pair of sensory cirri, one to either side of mouth in certain tardigrades; typically located anterior to external buccal cirri and cephalic papillae.

internal claw [G *innere Kralle*]: One of two median claws on legs of certain plate-bearing tardigrades; differs in shape from adjacent external claws.

intestine [G *Mitteldarm*]: Section of digestive tract between esophagus and rectum. Typically very prominent; may bear diverticula.

leg [G *Bein*]: One of four pairs of stubby ventral locomotory structures bearing terminal claws or toes.

lunule [G *Lunula*]: Half-moon-shaped cuticular thickening at base of claws in certain tardigrades. (smooth, dentate)

macroplacoid [G *Makroplacoide*]: In digestive tract, large, variously shaped cuticular thickening lining periphery of triradiate lumen of pharyngeal bulb; shape, number, and arrangement are of taxonomic importance. (rod-shaped, club-shaped) (See also microplacoid)

malpighian tubule [G *Malpighi'sches Gefäß*]: One of three glands, one located dorsally and two laterally, each consisting of three cells. Opens into digestive tract at junction of intestine and rectum.

median cirrus [G *Cirrus medianus*]: Terminal median cirrus at anterior end of head in certain tardigrades; may be flanked by internal and external buccal cirri as well as cephalic papillae.

median plates 1-3 [G *Schaltplatten 1-3*]: In plate-bearing tardigrade, small dorsomedian plates between scapular and first paired plates, between first and second paired plates, and between second and pseudosegmental or terminal plates, respectively.

microplacoid [G *Mikroplacoide, Komma*]: In digestive tract, small cuticular thickening posterior to macroplacoids within pharyngeal bulb; presence or absence is of taxonomic importance. (See also macroplacoid)

midgut: intestine

mouth [G *Mund*]: Anterior opening of digestive tract through which stylets are protruded during feeding; occasionally surrounded by peribuccal lamellae. (terminal, subterminal, ventral)

ovary [G *Ovar*]: Section of ♀ reproductive system located above intestine and connected to gonopore via single oviduct.

oviduct [G *Ovidukt*]: Section of ♀ reproductive system leading from dorsal ovary around right or left side of intestine to gonopore.

paired plates 1, 2 [G *paarige Rumpfplatten 1, 2*]: In plate-bearing tardigrade, large, often spine-bearing, dorsally divided plates extending down sides of body approximately at level of second and third leg pairs.

papillus [G *Papille*]: Small sensory projection on one or more leg pairs in certain tardigrades. (See also cephalic papilla)

pedal ganglion [G *Nebenganglion*]: Small ganglion in each leg; joined to ventral ganglion by nerve.

peribuccal lamellae [G *Mundlamellen*]: Ring of small cuticular thickenings surrounding mouth in certain tardigrades.

pharyngeal bulb (pharynx) [G *Pharynx, Schlundkopf*]: Muscular, bulbous section of digestive tract between buccal tube and esophagus, typically with triradiate lumen and placoids.

pharyngeal tube portion (of buccal tube) [G *präpharyngealer Teil der Mundröhre*]: Section of buccal tube between stylet supports and pharyngeal bulb.

pharynx: pharyngeal bulb

placoid [G *Placoide*]: General term for variously elaborated, larger (macroplacoid) or smaller (microplacoid) cuticular elements within pharyngeal bulb.

principal arm [G *Hauptast*]: In doubleclaw, larger of two arms. (See also secondary arm)

pseudosegmental plate [G *Pseudosegmentplatte, vierte Rumpfplatte*]: Additional large plate, with or without median division, between third median plate and terminal plate in certain tardigrades.

rectum (hindgut) [G *Rektum, Enddarm*]: Short, cuticle-lined section of digestive tract between intestine and anus. (See also cloaca)

salivary gland [G *Munddrüse, Speicheldrüse*]: Pair of large glands, one along each side of buccal tube and pharyngeal bulb; opens anteriorly into buccal tube.

scapular plate [G *Schulterplatte, erste Rumpfplatte*]: In plate-bearing tardigrade, large plate posterior to frontal plate;

extends down sides of body approximately at level of first leg pair.

secondary arm [G *Nebenast*]: In doubleclaw, smaller of two arms. (See also principal arm)

seminal receptacle [G *Receptaculum seminis*]: Small sac functioning for storage of sperm in ♀ of certain tardigrades; opens opposite oviduct into cloaca.

seminal vesicle [G *Samenblase*]: In ♂ reproductive system, expanded section of sperm duct (vas deferens).

septulum [G *Septulum*]: Cuticular thickening at border of pharyngeal bulb and esophagus; functions as a valve.

sperm duct: vas deferens

spine [G *Haar*]: Variously elongated (spine, filament) cuticular projection on leg and/or laterally, dorsolaterally, and dorsally on posterior margin of plates.

spur [G *Nebenhaken*]: In certain tardigrades, small, curved hook, typically at base of internal claw, occasionally also of external claw.

stylet [G *Stilett*]: Pair of lance-shaped cuticular structures, one on each side of buccal tube. Pointed anterior ends project into buccal tube; blunt posterior ends (furcae) joined to buccal tube by stylet supports.

stylet sheath [G *Stilettscheide*]: Pair of sheath-like structures, one on each side of anterior end of buccal tube; encloses anterior end of stylet.

stylet support [G *Stilettträger*]: Pair of thin, elastic rods joining blunt ends (furcae) of stylets to buccal tube.

subpharyngeal ganglion [G *Unterschlundganglion*]: First of five ventral ganglia; located below pharyngeal bulb and connected to brain by two connectives which surround pharyngeal apparatus.

terminal facet [G *Facette*]: Ridge or keel formed by strong downward angulation of terminal plate.

terminal indentation [G *Kleeblattkerbe*]: One of two indentations on posterior margin of terminal plate; imparts plate with cloverleaf-like appearance.

terminal plate [G *Endplatte*]: In plate-bearing tardigrade, posteriormost plate; extends down sides of body approximately at level of last leg pair.

testis [G *Hoden*]: Section of ♂ reproductive system located above intestine and connected to gonopore via two vasa deferentia.

throttling [G *Einschnürung*]: Median constriction in macroplacoids in pharyngeal bulb.

toe [G *Zehe*]: Term applied to finger-shaped distal projection of legs bearing termi-nal adhesive discs; number and arrangement are of taxonomic importance. (See also claw)

vas deferens (sperm duct) [G *Vas deferens, Samenleiter*]: Section of ♂ reproductive system between testis and gonopore; may be expanded to form seminal vesicle.

ventral ganglion [G *Bauchganglion*]: One of five ventral concentrations of nerve tissue (first = subpharyngeal ganglion) along longitudinal ventral nerve cords.

ventral nerve cord [G *Bauchkette*]: Pair of closely adjoining nerve cords extending from subpharyngeal ganglion to posterior end of body; bears series of ventral ganglia, one approximately at level of each leg pair.

ARTHROPODA

CRUSTACEA

Remipedia (Nectiopoda)

remipede [G *Remipedier*]

REMIPEDIA : Nectiopoda

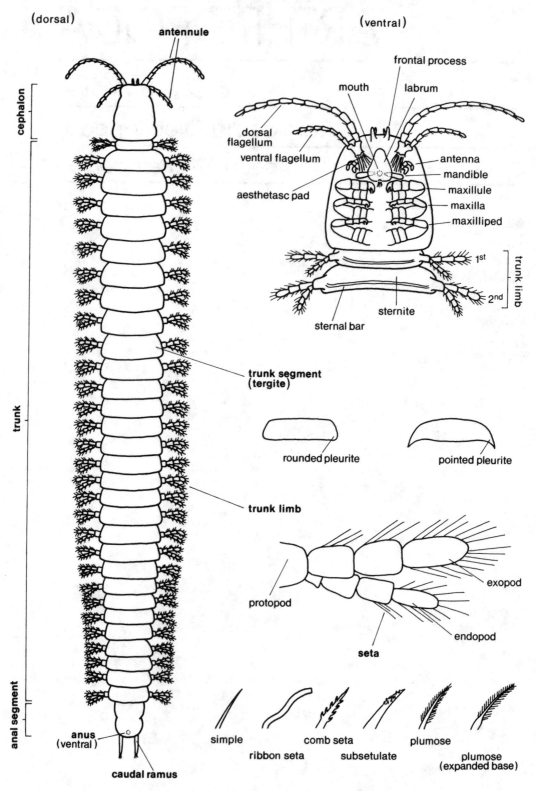

(dorsal)

antennule

cephalon

trunk

anal segment

anus
(ventral)

caudal ramus

trunk segment
(tergite)

(ventral)

frontal process

mouth

labrum

dorsal
flagellum

ventral flagellum

aesthetasc pad

antenna

mandible

maxillule

maxilla

maxilliped

1st

2nd

trunk limb

sternite

sternal bar

rounded pleurite

pointed pleurite

trunk limb

protopod

exopod

endopod

seta

simple

ribbon seta

comb seta

subsetulate

plumose

plumose
(expanded base)

ARTHROPODA

CRUSTACEA

Remipedia (Nectiopoda)

remipede [G *Remipedier*]

aesthetasc (esthetasc, esthete, olfactory hair) [G *Aesthetasc*]: One of numerous robust, long sensory projections aggregated in several dense rows at base of antennules (on pad on first article of peduncle); directed posteriorly and partially covering antennae.

anal flap [G *Anallappen*]: Small lobe covering anus at posterior end of anal segment.

anal segment [G *Analsegment*]: Posteriormost segment of body; partially fused to reduced last trunk segment and bearing both anus and caudal rami. Occasionally also referred to as telson. (length longer than, equal to, or shorter than width)

antenna (second antenna) [G *Antenne, zweite Antenne*]: Second and smaller, more paddle-shaped pair of antennae on underside of head (cephalon); biramous and consisting of protopod (with two articles) bearing scale-like exopod and three-segment endopod. Endopod articles arc around anterior aspect of exopod. Setae on antennae are plumose.

antennule (first antenna) [G *Antennula, erste Antenne*]: First and larger pair of antennae on underside of head (cephalon); biramous and consisting of basal peduncle (with two articles and bearing aesthetasc pad) and distal (dorsal and ventral) flagella. Setae on antennae are simple.

anus [G *After*]: Posterior opening of digestive tract; located terminally on underside of anal segment and may be covered by small anal flap.

article (joint, segment) [G *Glied*]: In antennule, one of 11 or 12 segments of larger dorsal flagellum or 3 or 8 segments of smaller ventral flagellum. Each article bears tuft of several setae at distoventral end as well as scattered setae along shaft. Last article bears terminal tuft of setae. Also synonymous with any appendage segment.

atrium oris [G *Atrium oris*]: Cavity around mouth formed by projecting posterior margin of labrum and by paragnaths; masticatory ends (incisor process, lacinia mobilis, molar process) of mandibles project into atrium oris.

brain (supraesophageal ganglion) [G *Gehirn, Oberschlundganglion*]: Main concentration of nerve tissue located above foregut in anterior region of head (cephalon); connected to ventral nerve cord by pair of circumesophageal commissures.

caudal ramus [G *Furca-Ast*]: One of two unbranched processes projecting posteriorly from anal segment; each ramus bears setae along median margin and terminally. (shorter than, longer than anal segment)

cephalic shield (head shield)
[G *Kopfschild*]: Shield-like cuticular structure dorsally covering and folding over front of head (cephalon) and extending somewhat laterally and posteriorly; tapers anteriorly (in one case with narrowest part in middle) and is marked by one or two transverse grooves. (subquadrangular, subrectangular, subtrapezoidal)

cephalon (head) [G *Kopf*]: Anterior division of body (cephalon, trunk) bearing frontal processes, antennules, antennae, labrum, mouth, mandibles, maxillules, maxillae, and maxillipeds; covered dorsally by cephalic shield. First trunk segment fused to cephalon.

circumesophageal connective
[G *Schlundkonnektiv*]: One of two strands of nerve tissue surrounding foregut and joining dorsal brain to ventral nerve cords.

claw (fang) [G *Kralle*]: Claw-like tip of last segment of maxillules, maxillae, and maxillipeds. One may distinguish the talon of the maxillules, which has a pore for the maxillulary gland, and the claws of the maxillae and maxillipeds, which usually have secretory? or sensory? pores. (arcuate, horseshoe-shaped, talon-like, trifid, 10-pronged)

cuticle [G *Kutikula*]: Noncellular, multilayered outer layer of body; secreted by epidermis and equipped with numerous pores and sensory organs. Also lines foregut and hindgut.

diverticulum (gut diverticulum)
[G *Darmtasche, Seitentasche*]: One of a pair of lateral outpocketings of midgut in each trunk segment; each diverticulum extends into pleurite of segment. Diverticula become smaller toward posterior end of body and almost completely disappear in posteriormost region.

endite [G *Endit*]: Inwardly (medially) directed lobe of first segments of maxillules, maxillae, and maxillipeds. Endites of maxillules may function as auxiliary mandibles. (arcuate, cone-like, digitiform, double-crested, lobe-like, subtriangular)

endopod (endopodite) [G *Endopod, Endopodit*]: Inner or more ventral branch (ramus) of biramous appendage (antennule, antenna, trunk limb).

esthetasc: aesthetasc

exopod (exopodite) [G *Exopod, Exopodit*]: Outer or more dorsal branch (ramus) of biramous appendage (antennule, antenna, trunk limb).

flagellum [G *Flagellum, Geißel, Antennengeißel*]: Distal section of antennule. One may distinguish a longer dorsal flagellum composed of 11 or 12 segments (articles) and a shorter ventral flagellum composed of 8 articles (in one case fused into a blade-like structure). Each article bears several setae.

foregut [G *Vorderdarm*]: Anteriormost of three divisions of digestive tract (foregut, midgut, hindgut); cuticle-lined, consists of muscular anterior portion and narrower tube-like posterior portion; extends from mouth to border of head (cephalon) and trunk.

frontal process (frontal filament, preantennular process) [G *präantennularer Fortsatz*]: One of two anteriorly projecting processes on underside of head (cephalon) near base of antennules; relatively small, each consisting of main shaft bearing stout spine. (club-like, "jointed," rod-like)

genital flap [G *Genitallappen*]: Small cuticular flap on base (protopod) of each limb of 14th trunk segment or at each end of sternal bar of 14th trunk segment; serves to cover ♂ gonopore or facilitate copulation.

gonopore (genital pore) [G *Gonoporus, Genitalporus, Geschlechtsöffnung*]: Opening of ♂ or ♀ reproductive system to exterior. Two gonopores are present, one on 7th trunk segment (♀), one on 14th trunk segment (♂), the latter borne on a genital papilla.

head: cephalon

head shield: cephalic shield

heart [G *Herz*]: Main longitudinal vessel of circulatory system; located middorsally above digestive tract. Muscular portion restricted to head (cephalon) region.

hindgut [G *Enddarm*]: Short posteriormost division of digestive tract (foregut, midgut, hindgut) located in anal segment. Opens to exterior via anus.

incisor process [G *Processus incivisus*]: Cutting process (vs. grinding = molar process) at tip of each mandible; incisor processes of left and right mandibles are generally asymmetrical, i.e., number (three or four) and shape of large denticles differ.

joint: article, segment

labrum [G *Labrum, Oberlippe*]: Relatively large, fleshy lobe anterior to mouth; divided by transverse furrow into subtriangular anterior part and bulbous posterior part, the latter with a fossa and dense array of ribbon setae on posterior margin. Projecting posterior margin forms atrium oris below and extending somewhat behind true mouth.

lacinia mobilis [G *Lacinia mobilis*]: Movable, toothed process between molar and incisor processes at tip of each mandible. Laciniae mobili of left and right mandibles are asymmetrical, i.e., of different shape and/or denticle number. (margin: concave and sickle-like, with three or six denticles)

limb (trunk limb, trunk appendage) [G *Rumpfextremität*]: One of a pair of appendages of each trunk segment. Each trunk limb is ventrolaterally directed and biramous, consisting of protopod with three-segmented exopod and four-segmented endopod. All limbs are homonomous, with the first being somewhat more slender and the posteriormost ones somewhat smaller. Bears different types of setae.

mandible [G *Mandibel*]: Anteriormost paired, asymmetrical mouthpart on underside of head (cephalon); lacks palp. Each mandible bears three masticatory processes (incisor process, lacinia mobilis, molar process) which lie in cavity (atrium oris) between labrum and mouth.

maxilla (second maxilla) [G *Maxille, zweite Maxille*]: Third paired mouthpart on underside of head (cephalon) posterior to maxillules. Uniramous, consisting of six or seven segments, and with main point of flexure between third and fourth segments. First two or three segments bear endites, the last bears terminal claw; all segments bear setae along inner margin. (prehensile, subchelate)

maxillary gland [G *Maxillennephridium*]: Pair of excretory organs, one located in each posterolateral corner of head (cephalon); open to exterior at base of each maxilla.

maxilliped (maxillipede) [G *Maxilliped, Kieferfuß*]: Paired appendage posterior to maxillae on underside of head (cephalon). Uniramous, consisting of seven or eight segments, and with main point of flexure between third and fourth segments. First one to three segments with endites, last with terminal claw, all with setae along inner margin. (prehensile, subchelate)

maxillule (first maxilla) [G *Maxilulla, erste Maxille*]: Second paired mouthpart on underside of head (cephalon) between mandibles and maxillae. Uniramous, consisting of seven segments, and with main point of flexure between fourth and fifth segments. First four segments with endites, seventh with terminal talon, all with setae along inner margin. (prehensile, subchelate)

midgut [G *Mitteldarm*]: Elongate section of digestive tract between foregut and hindgut; extends through trunk and bears pair of lateral diverticula in each segment.

molar process [G *Processus molaris*]: Grinding process (vs. cutting = incisor process) at tip of each mandible; located on pedestal and composed of broad, flat basin containing dense rows of spines.

mouth [G *Mund, Mundöffnung*]: Anterior opening of digestive tract on underside of head (cephalon); located in cavity (atrium oris) formed by projecting posterior margin of labrum and by paragnaths. Perioral area highly setose due to setae of labrum and paragnaths. Opens into foregut.

ovary [G *Ovar*]: Unpaired section of ♀ reproductive system in which eggs are produced; located dorsally in posterior region of head (cephalon) and opening to exterior on seventh trunk segment via oviducts.

oviduct [G *Ovidukt, Eileiter*]: Paired section of ♀ reproductive system between ovary and ♀ gonopore; extends dorsally above gut from head/trunk border to seventh trunk segment.

paragnath [G *Paragnath*]: One of a pair of lobes flanking opening (atrium oris) to mouth; margins of each paragnath bear dense cover of ribbon setae.

peduncle [G *Pedunculus, Stiel*]: Proximal part of antennule. Consists of two segments (articles), the proximal one bearing pad with numerous aesthetascs, the distal one being bifurcate and bearing two (dorsal and ventral) flagella. May also refer to base of antenna.

pleural lobe: pleurite

pleurite (pleural lobe) [G *Pleurit*]: Lateral extension of each adult trunk segment; gut diverticula extend into each pleurite. (posterolateral corner: rounded, pointed)

protopod [G *Protopod*]: Proximal part of biramous appendage. One-segmented, bearing endopod and exopod.

ramus [G *Ast*]: Branch of an appendage, typically referring to inner (endopod) and outer (exopod) branch of trunk appendage, antennule, or antenna. (See also caudal ramus)

segment [G *Segment*]: One in a series of divisions of body. Typically refers to one of approximately 30 trunk segments, the first being partially fused to cephalon, the last being reduced and partially fused to anal segment. Each trunk segment consists of dorsal tergite and ventral sternite, is produced later-ally into pleurite, and bears pair of biramous appendages. May also refer to segment of limb. (See also article)

seta [G *Borste*]: One of variously shaped, hair-like processes projecting from antennules, antennae, labrum, paragnaths, maxillules, maxillipeds, or trunk limbs. According to shape one may distinguish comb, plumose, ribbon, simple, and subsetulate setae.

sperm duct (vas deferens) [G *Samenleiter, Vas deferens*]: In ♂ reproductive system, pair of ducts extending from about 10th to 14th trunk segments. Joins testes with opening on genital papilla on base of 14th trunk limb.

standard measurements:
trunk segment number
body length
relative length of cephalon (cephalon length: total length)
length and width of cephalon, trunk segments 1, 2, and 14, anal segment, and caudal ramus

sternal bar (sternite bar) [G *Sternit*]: In trunk segment, transverse thickening sometimes reinforcing or even replacing sternite. Sternite bar of 14th trunk segment may bear lateral flap (genital flap) which covers gonopore. (bar-like, subtriangular; posterior margin: concave, convex)

sternite [G *Sternit*]: Ventral surface of body segment; sternites of trunk segment are sometimes plate-like, sometimes reinforced by or reduced to sternal bar. (See also tergite)

supraesophageal ganglion: brain

talon: claw

telson: anal segment

tergite [G *Tergit*]: Dorsal surface of body segment. Tergites of first and last trunk segments are reduced and/or partially fused to head (cephalon) and anal segment; typically produced laterally into pleurites. (posterolateral corner: rounded, pointed)

testis [G *Hoden*]: Paired section of ♂ reproductive system in which sperm are produced; located ventral and dorsal to midgut in posteriormost seventh to about tenth trunk segment.

trunk [G *Rumpf*]: Posterior division of body (cephalon, trunk). Consists of up to approximately 30 similar segments, the first (i.e., maxillipedal) being fused to head (cephalon), the second (= apparent first) being sometimes partially covered by head (cephalon), the last reduced and partially fused to anal segment; each segment is produced laterally into pleurites and bears pair of biramous appendages.

trunk appendage: limb

ventral nerve cord [G *ventraler Nervenstrang, Bauchmark*]: Pair of longitudinal nerve cords extending from circumesophageal connectives to posterior end of body; connected to each other via commissures and bearing only modestly developed trunk-segment ganglia.

CEPHALOCARIDA

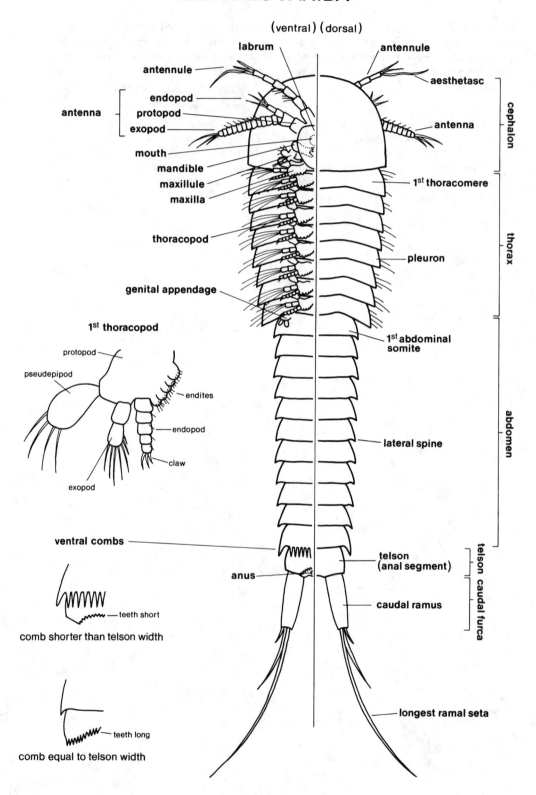

(ventral) (dorsal)

labrum

antennule

antennule

aesthetasc

antenna
- endopod
- protopod
- exopod

antenna

cephalon

mouth

mandible

maxillule

maxilla

1st thoracomere

thoracopod

thorax

pleuron

genital appendage

1st abdominal somite

1st thoracopod
- protopod
- pseudepipod
- endites
- endopod
- claw
- exopod

abdomen

lateral spine

ventral combs

telson (anal segment)

anus

caudal ramus

telson caudal furca

teeth short

comb shorter than telson width

teeth long

comb equal to telson width

longest ramal seta

ARTHROPODA

CRUSTACEA
Cephalocarida (Brachypoda)
cephalocarid, brachypodan
[G *Cephalocaride*]

abdomen [G *Abdomen*]: Posterior division (tagma) of body (cephalon, thorax, abdomen). Consists of 11 segments (somites) and bears terminal telson (anal somite) with caudal furca. With the exception of a reduced pair of appendages ("genital appendages"), abdomen lacks limbs. Thorax and abdomen together occasionally termed trunk.

abdominal somite [G *Abdominalsegment*]: One of 11 segments of abdomen. First abdominal somite (= 9th postcephalic segment) bears reduced pair of appendages ("genital appendages"). Last abdominal somite (19th postecephalic segment) bears telson.

aesthetasc (esthete) [G *Aesthetasc*]: Elongate sensory projection at tip of each antennule.

anal segment (anal somite): telson

antenna (second antenna) [G *Antenne, zweite Antenne*]: Second pair of antennae. Biramous, consisting of two-segmented protopod, relatively short, two-segmented endopod, and longer, multisegmented exopod. Serves in locomotion and feeding.

antennule (first antenna) [G *Antennula, erste Antenne*]: First pair of anten-nae. Uniramous, consisting of six segments; bears terminal aesthetasc. With locomotory and sensory functions.

anus [G *After*]: Posterior opening of digestive system to exterior; located terminally on telson (anal segment).

article [G *Glied*]: Segment of appendage.

atrium oris [G *Atrium oris*]: Large preoral cavity formed by labrum extending over mouth. Mandibles (and to a lesser extent the endites or gnathobases of maxillules) extend into atrium oris.

basis [G *Basis*]: protopod

brain [G *Gehirn*]: Main concentration of nerve tissue in cephalon. Connected with "eyes" by optic nerves and continues posteriorly to form subesophageal ganglion.

caudal furca: caudal ramus

caudal ramus [G *Furkalast, Furca-Ast*]: One of two posteriorly directed projections of last body segment (telson). Consists of single element with terminal setae (ramal setae, caudal setae). Both rami together are termed caudal furca. If telson is interpreted as representing anal somite, then caudal ramus may be termed uropod. (length relative to telson width: equal to or less than, less than twice, equal to twice)

cephalic hook: One in a series of minute, anteriorly directed hooks located on underside of head (cephalon); restricted to anterolateral margins.

cephalic shield (head shield) [G *Kopfschild*]: Broad, semicircular shield covering dorsal surface of head. Formed by fusion of dorsal surfaces (tergites) of first five segments.

cephalon (head) [G *Kopf*]: Anterior of three basic divisions (tagmata) of body (cephalon, thorax, abdomen). Relatively short and covered dorsally by cephalic shield. Consists of five fused segments (somites) bearing antennules, antennae, mandibles, maxillules, and maxillae.

claw (endopodal claw) [G *Endklaue, Endkralle, Krallendorn*]: One of up to several claw-like projections from last segment of inner branch (endopod) of maxillae or thoracopods. Inner border may bear series of fine denticles. Largest claw occasionally termed "dactyl."

comb: ventral comb

coxa [G *Coxa*]: protopod

dactyl: claw

diverticulum [G *Divertikel*]: One of two simple glandular outpocketings of digestive tract; located in head and opening into anterior region of midgut.

dorsoventral muscle [G *Dorsoventralmuskel*]: One of three pairs of muscles extending in a dorsoventral direction in each trunk segment.

endite [G *Endit*]: One of up to several inwardly (medially) directed, setose lobes of protopod. Termed gnathobase if considered to serve in feeding (e.g., maxillules, occasionally also maxillae and thoracopods).

endopod (endopodite) [G *Endopod, Endopodit*]: Inner branch of biramous appendage. Refers to two-segmented branch of antennae, lobate branch (in larvae only) of mandibles, annulate branch of maxillules, six-segmented branch of maxillae, and five- to six-segmented branch of thoracopods. (See also exopod)

epipod: pseudepipod

esophagus [G *Ösophagus*]: Anterior section of digestive tract.

exopod (exopodite) [G *Exopod, Exopodit*]: Outer branch of biramous appendage. Refers to annulate outer branch of antennae, six-segmented branch (in larvae only) of mandibles, one-segmented branch of maxillules, as well as two-segmented branches of maxillae and thoracopods. (See also endopod)

eye (compound eye) [G *Auge, Komplexauge*]: One of two sensory organs associated with anterior end of labrum on underside of head (cephalon). Located under surface of cuticle and connected to brain by optic nerve.

flagellum: aesthetasc

genital appendage [G *Genitalanhang*]: One of two reduced appendages on first abdominal somite (= ninth postcephalic segment). Ovisac of ♀ is attached to genital appendage.

gnathobase [G *Kauladen, Endit*]: One of several setose projections of base (protopod) of maxillule. Extends under labrum and serves in manipulating food. Endites of maxillae and thoracopods also occasionally termed gnathobases.

gonoduct [G *Gonodukt*]: Common duct formed by merger of vas deferens and oviduct on each side of sixth thoracic segment (thoracomere). Opens through gonopore.

gonopore (genital opening) [G *Gonoporus, Geschlechtsöffnung*]: Pair of common openings of ♂ and ♀ reproductive systems, one located on posterior face of protopod of each appendage (thoracopod) of sixth thoracomere.

head: cephalon

head shield: cephalic shield

heart [G *Herz*]: In circulatory system, muscular pumping organ extending through thorax; bears pair of ostia in each somite.

incisor process [G *Pars incisiva*]: Food-gripping process (vs. grinding = molar process) at end of each mandible. Extends into atrium oris under labrum. Variously armed with spines, setae, and denticles.

joint: article, segment

labrum (upper lip) [G *Labrum, Oberlippe*]: Large, bulbous, unpaired lobe covering mouth and extending over end (molar and incisor processes) of mandibles and to a lesser extend over maxillules. Preoral cavity formed under labrum is termed atrium oris. Anterior end of labrum may be associated with "eyes." (narrowly or broadly rounded anteriorly and posteriorly, somewhat pointed anteriorly and posteriorly; oval, obovate)

lateral spine [G *Epimer*]: Posteriorly directed, spine-shaped projection on each side of abdominal somites (9th–19th postcephalic segments). Also interpreted as representing pleura.

longitudinal muscle [G *Längsmuskel*]: Pair of relatively large dorsal and ventral muscles extending through entire body.

mandible [G *Mandibel*]: Third paired appendage of head (cephalon); located between antennae and maxillules and extends into atrium oris under labrum. Bears terminal molar and incisor processes.

maxilla (second maxilla) [G *Maxille, zweite Maxille*]: Fifth paired appendage of head (cephalon); located between maxillules and first thoracopods. Consists of protopod with five to six endites and large pseudepipod, six-segmented endopod, and two-segmented exopod. Basically biramous, although occasionally referred to as tri-, multi-, or polyramous due to numerous endites and pseudepipod.

maxillary gland [G *Maxillennephridium*]: One of two simple excretory organs in head (cephalon); each opens via pore on posterior surface of base of maxilla.

maxillule (first maxilla) [G *Maxillula, erste Maxille*]: Fourth paired appendage of head (cephalon); located between mandibles and maxillae. Biramous, consisting of protopod (with large endite), annulate endopod, and one-segmented exopod.

metanauplius [G *Metanauplius*]: Larval stage at which cephalocarids hatch. Includes a number of substages characterized by a gradual increase in number of segments and appendages.

midgut (mesenteron) [G *Mitteldarm*]: Greater part of digestive tract; extends posteriorly from esophagus through thorax and most of abdomen.

molar process [G *Pars molaris*]: Grinding process (vs. gripping = incisor process) at end of each mandible. Bears numerous small teeth and extends into atrium oris under labrum.

mouth [G *Mund, Mundöffnung*]: Anterior opening of digestive tract; located under labrum and associated with large preoral cavity (atrium oris).

mucous gland [G *Schleimdrüse*]: Gland located in proximal part (protopod) of each limb; empties through duct opening near endites.

ostium [G *Ostium*]: One of two openings of heart in each thoracic segment (thoracomere).

ovary [G *Ovar, Ovarium, Eierstock*]: One of two sections of ♀ reproductive system in which eggs are produced. Located in posterior end of head (cephalon) and opens to exterior via highly elongated oviduct and short common gonoduct.

oviduct [G *Ovidukt, Eileiter*]: Section of ♀ reproductive system between each ovary and gonopore. Highly elongated, extending posteriorly to 18th postcephalic segment and doubling back to 6th thoracomere. Eggs accumulate at point of flexure. Terminal section merges with vas deferens to form common duct (gonoduct).

ovisac [G *Eisack*]: Sac-like structure attached to rudimentary appendage ("genital appendage") of ninth postcephalic segment

(= first abdominal segment). Serves as brood chamber for egg. (single, paired)

pleuron (pleural lobe) [G *Epimer*]: Lateral part of integument (as opposed to sternite = ventral surface and tergite = dorsal surface); in first seven trunk somites (thoracomeres) developed as ventrolaterally directed lobes, in last (eighth) thoracomere small or reduced. (acute, rounded)

protopod (protopodite) [G *Protopod, Protopodit*]: Proximal part of appendage. Generally specified only when referring to biramous appendage (e.g., antennae, maxillules, maxillae, thoracopods). Consists basically of two segments (coxa, basis) and may bear endites and epipod (pseudepipod).

pseudepipod (pseudepipodite, epipod) [G *Pseudepipodit, Epipodit*]: Laterally directed lobe of proximal part (protopod) of maxillae and thoracopods. Relatively large, flattened, setose, and giving appendages a triramous or, with endites, a multi- or polyramous appearance. Represents an epipod, but is termed pseudepipod due to distal point of origin on protopod.

segment [G *Segment*]: One in a series of units of appendage (then also termed article or joint). May also refer to segment of body (then also termed somite). Shape and number of setae of appendage segments are of taxonomic importance.

seta [G *Borste*]: One of numerous bristle-shaped projections extending from appendage. Number and type are of taxonomic importance. According to shape one may distinguish, e.g., simple, bifid, brush, or plumose setae.

setule: One of numerous minute projections of a seta. May also refer to any small setae, e.g., those forming comb on 19th abdominal somite.

somite (segment) [G *Segment*]: One in a series of divisions of body. Head (cephalon) consists of 5 somites (fused dorsally to form cephalic shield), followed by 8 thoracic somites (thoracomeres) and 11 ab-

dominal somites. Last abdominal somite bears telson (anal somite).

standard measurements:
body length (excluding caudal setae)
ratio: cephalon + thorax/body length
cephalon width/body length
caudal ramus length/telson width
length of longest caudal seta relative to abdomen length
setal formula (number of setae per appendage segment)

sternite [G *Sternit*]: Ventral surface of body segment (somite). (See also tergite)

subesophageal ganglion [G *Unterschlundganglion*]: Concentration of nerve tissue below (posterior to) esophagus. Connected to brain by connectives surrounding esophagus. (See also ventral nerve cord)

tagma [G *Tagma*]: Major division of body (cephalon, thorax, abdomen). (See also trunk)

telson [G *Telson*]: Last segment of body (20th postcephalic segment) with ventral comb and pair of caudal rami. Bears terminal anus and therefore occasionally termed anal segment or anal somite.

tendon: Pair of dorsal and ventral supporting elements forming endoskeletal structures within body. Serves as site of muscle attachment.

tergite [G *Tergit*]: Dorsal surface of body segment. Tergites of first five somites are fused to form cephalic shield. (See also sternite)

testis [G *Hoden*]: One of two sausage-shaped sections of ♂ reproductive system in which sperm are produced. Located dorsal to digestive tract and extends from 12th to 7th postcephalic segments; anterior ends joined. Each testis opens to exterior on sixth thoracomere via vas deferens and gonoduct (common ♂ and ♀ duct).

thoracomere (thoracic segment) [G *Thoraxsegment, Thoracomer*]: One of eight segments (somites) of thorax, of which seven or eight bear biramous appendages (thoracopods).

thoracopod [G *Thoracopod, Thoraxbein*]: One of two appendages of each thoracic somite (thoracomere). Consists of protopod with four to six endites and pseudepipod, five- to six-segmented endopod, and two-segmented exopod. More posterior thoracopods may be somewhat modified, those of last (eighth) thoracomere may be reduced or absent. Basically biramous, although occasionally referred to as tri-, multi-, or polyramous due to large pseudepipod and numerous endites.

thorax [G *Thorax*]: Division (tagma) of body between head (cephalon) and abdomen. Consists of eight somites (thoracomeres), of which seven or eight bear biramous appendages (thoracopods). Thorax and abdomen together occasionally termed trunk.

trunk [G *Rumpf*]: Region of body posterior to head (cephalon). Consists of 19 segments (excluding telson), of which the first 8 represent the thorax, the last 11 the abdomen.

uropod [G *Uropod*]: Term applied to each caudal ramus if telson is interpreted as representing anal somite.

vas deferens [G *Vas deferens*]: Section of ♂ reproductive system between each testis and gonopore. Terminal section merges with oviduct to form common duct (gonoduct) opening on sixth thoracic segment (thoracomere).

ventral comb (comb row) [G *Kamm*]: Transverse series of ventral setae on telson (longer setae) and, in most cephalocarids, on posterior margin of 19th segment (shorter setae). (on telson: row shorter than or equal to telson width)

ventral nerve cord [G *ventraler Nervenstrang*]: Pair of longitudinal nerve cords extending from subesophageal ganglia to end of abdomen; bears pair of ganglia in each somite.

ANOSTRACA

FEMALE

(ventral) (dorsal)

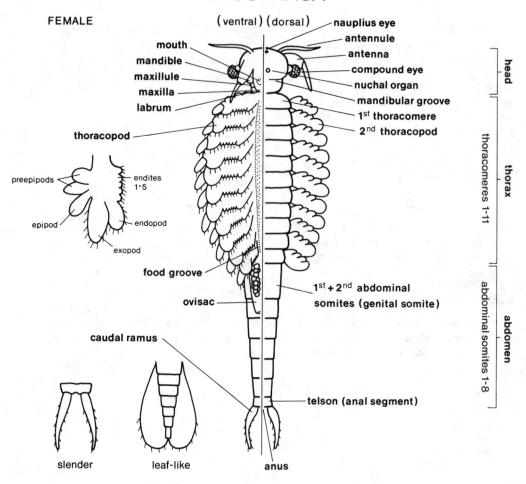

nauplius eye
antennule
antenna
compound eye
nuchal organ
mandibular groove
1st thoracomere
2nd thoracopod

mouth
mandible
maxillule
maxilla
labrum

thoracopod

preepipods

endites 1-5

epipod

endopod

exopod

food groove

ovisac

caudal ramus

head

thorax

thoracomeres 1-11

abdomen

abdominal somites 1-8

1st + 2nd abdominal
somites (genital somite)

telson (anal segment)

slender

leaf-like

anus

MALE (lateral)

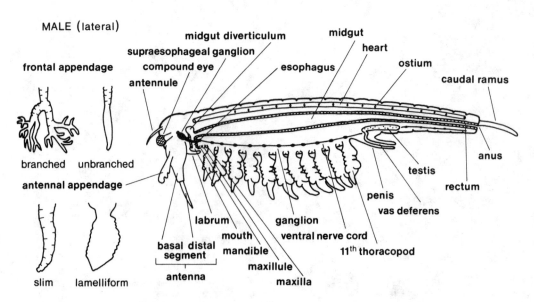

frontal appendage

branched   unbranched

antennal appendage

slim   lamelliform

midgut diverticulum
supraesophageal ganglion
compound eye
antennule

midgut
heart
esophagus
ostium
caudal ramus

anus

testis
rectum

penis
vas deferens
11th thoracopod

labrum
mouth
basal distal
segment
mandible
maxillule
maxilla

antenna

ganglion
ventral nerve cord

ARTHROPODA

CRUSTACEA

Anostraca

fairy shrimp [G *Kiemenfüßer, Kiemenfußkrebs*]

abdomen [G *Abdomen*]: Posterior division (tagma) of body (head, thorax, abdomen). Consists of eight segments (somites) and bears terminal telson with caudal rami. Lacks appendages except for anterior two somites, which are fused (genital somite) and bear reproductive structures.

abdominal spine [G *Stachel*]: Spine-shaped projection on each side of abdominal somite.

anal segment (anal somite): telson

antenna (second antenna) [G *Antenne, zweite Antenne*]: Second and larger of two pairs of antennae; uniramous, projecting ventrally from head. In ♀, typically unsegmented and reduced. In ♂, forms clasper generally consisting of expanded basal segment (with antennal appendage) and more slender distal segment. (fused at base, not fused at base; basal segment: with/without hand; with/without finger-like process on median surface near midlength; with prominent, dorsally directed, peg-like process on median surface just below midlength; with row of spines along median margin; with spinous pad, median protuberance, and many small distal spines; distal segment: laminate, not laminate, with/without calcar) (See also clypeus, hand)

antennal appendage (basal antennal appendage) [G *Anhang, Stirnlappen*]: Variously shaped process extending from basal segment of each antenna; more common in ♂. (conical, lamelliform, robust, slim; extending/not extending beyond basal segment of antenna; with small teeth, with large teeth; with teeth longer on one side, with teeth not greatly different in size; with/without conical processes)

antennule (first antenna) [G *Antennula, erste Antenne*]: First and smaller of two pairs of antennae; projecting from anterior end of head. Uniramous, either unsegmented or only superficially segmented. (with/without tuft of terminal setae)

anus [G *After*]: Posterior opening of digestive tract; located terminally on telson (between caudal rami).

brain [G *Gehirn*]: Main concentration of nerve tissue; consists of supraesophageal ganglion (protocerebrum, deutocerebrum) and, posterior to mouth, tritocerebrum (ganglion of antennal somite).

calcar: Spur-like process on distal segment of antenna. (one-half as long/one-eighth or less as long as basal segment of antenna)

caudal furca [G *Furca*]: caudal ramus

caudal ramus (furcal ramus, cercopod) [G *Furkalast, Furca-Ast*]: One of two posteriorly directed projections of last segment (telson). Unsegmented, typically fringed with setae. (freely movable/variously fused with last segment; thin and slender, broadly leaf-like or plate-like)

cecum: midgut diverticulum

cement gland [G *Schalendrüse*]: Pair of glands dorsal and proximal to eggs in ovisac; produces outer cover of eggs.

cephalon: head

cercopod: caudal ramus

clasper [G *Greiforgan*]: Term applied to well-developed antennae in ♂; used to clasp ♀ during copulation.

clypeus (frontal plate) [G *Frontalplatte*]: Plate-like structure formed by medial fusion of bases of antennae. (with/without digitiform process)

compound eye [G *Komplexauge*]: One of two relatively large, stalked photosensitive organs projecting from each side of head. (See also nauplius eye)

cuticle [G *Kutikula*]: Noncellular, multilayered structure forming outer layer of of body and lining esophagus and hindgut; relatively thin and not calcified. Secreted by epidermis.

deutocerebrum: brain

diverticulum: midgut diverticulum

egg sac: ovisac

endite [G *Endit*]: In thoracic appendage (thoracopod), one of basically five inwardly (medially) directed, setose lobes of protopod. Serves in feeding. (distinct, partially fused)

endopod (endopodite) [G *Endopod, Endopodit*]: Inner branch of thoracic appendage (thoracopod). Unsegmented, lobe-like, and bearing marginal setae. Not articulated with protopod and therefore occasionally termed sixth endite. Serves in feeding. (oval, rectangular) (See also exopod)

endoskeleton [G *Endoskelett*]: Term occasionally applied to various more rigid internal structures serving as sites of muscle attachment.

epipod (epipodite) [G *Epipod, Epipodit*]: In thoracic appendage (thoracopod), laterally (outwardly) directed lobe of protopod; located distal to preepipod(s). Serves in respiration.

esophageal connective [G *Schlundkonnektiv*]: One of two strands of nerve tissue extending around esophagus and connecting posterior part of supraesophageal ganglion with antennal ganglion (tritocerebrum).

esophagus [G *Ösophagus*]: Relatively short section of digestive tract between mouth and midgut. Directed dorsally, then posteriorly.

exopod (exopodite) [G *Exopod, Exopodit*]: Outer branch of thoracic appendage (thoracopod). Unsegmented, lobe-like, and bearing marginal setae; articulated with protopod. Serves in locomotion. (See also endopod)

eye: compound eye, nauplius eye

first maxilla: maxillule

food groove [G *Bauchrinne, Fanggasse*]: Elongate median groove between thoracopod bases on ventral side of thorax. Food is trapped by thoracopods, passed anteriorly along food groove, and transferred to mouth.

foregut: esophagus

frontal appendage [G *Frontalanhang*]: Unpaired median process projecting from head between antennae. (unbranched, branched; with two terminal branches, with three terminal branches; shorter/longer than antenna)

frontal organ (ventral frontal organ, paired ventral frontal organ) [G *Frontalorgan*]: Pair of sensory organs adjoining nauplius eye on anterior end of head; innervated by brain.

frontal plate: clypeus

furcal ramus: caudal namus

genital segment (genital somite) [G *Genitalsegment*]: Collective term for first and second abdominal somites. More or less fused and bearing penes in ♂ and ovisac in ♀.

gonopore: oviduct, penis

hand: Long, forked terminal outgrowth of basal segment of antenna in ♂ of certain anostracans.

head (cephalon) [G *Kopf*]: Anterior of three divisions (tagmata) of body (head, thorax, abdomen). Consists of five somites bearing antennules, antennae, mandibles, maxillules, and maxillae. Also bears pair of compound eyes laterally and nauplius eye middorsally; superficially subdivided dorsally by transverse mandibular groove.

heart [G *Herz*]: In circulatory system, elongate, muscular pumping organ extending through entire body above digestive tract. Blood from pericardial sinus enters heart through ostia and is pumped anteriorly.

hemocoel (hemocoele) [G *Haemocoel*]: Body cavity through which blood circulates; separated from pericardial sinus by pericardial septum.

hindgut [G *Enddarm*]: Short posteriormost section of digestive tract between midgut and anus.

labrum (upper lip) [G *Labrum, Oberlippe*]: Large, bulbous, unpaired lobe covering mouth and extending over end of mandibles. (with/without knob-like protuberance)

mandible [G *Mandibel*]: Third paired appendage of head; located between antennae and maxillules and extending under labrum. Relatively large, with undivided distal end. (without palp, with vestigial palp)

mandibular groove: Transverse groove across dorsal surface of head; superficially divides head into anterior and posterior regions.

maxilla (second maxilla) [G *Maxille, zweite Maxille*]: Fifth paired appendage of head; located between maxillules and first thoracopods. Relatively small, reduced to simple lobe.

maxillary gland [G *Maxillennephridium, Maxillardrüse*]: One of two excretory glands located in head and opening ventrally at level of (reduced) maxillae.

maxillule (first maxilla) [G *Maxillula, erste Maxille*]: Fourth paired appendage of head; located between mandibles and maxillae. Relatively small, uniramous, and typically bearing series of spines or setae along medial margin.

metanauplius [G *Metanauplius*]: Later larval stage between nauplius and adult. Characterized by gradual increase in number of segments and appendages.

midgut [G *Mitteldarm*]: Elongate section of digestive tract between esophagus and hindgut. Extends through thorax and most of abdomen and bears pair of midgut diverticula anteriorly.

midgut diverticulum (cecum) [G *Darmdivertikel, Divertikel*]: One of two short outpocketings of midgut. Located in head and opening into digestive tract posterior to esophagus/midgut border.

mouth [G *Mund, Mundöffnung*]: Anterior opening of digestive tract on underside of head; located under labrum and opening into esophagus.

nauplius [G *Nauplius*]: Early larval stage characterized by presence of only three pairs of appendages (antennules, antennae, mandibles) and an unsegmented trunk.

nauplius eye [G *Naupliusauge*]: Unpaired median photosensitive organ located somewhat posteriorly to antennules on anterior part of head; relatively small, consisting of three (two lateral, one ventral) cups. (See also compound eye)

nuchal organ (neck organ) [G *Nackensinnesorgan*]: Unpaired median sensory organ located on dorsal surface of head; visible beneath cuticle.

ostium [G *Ostium*]: One of 14 to 18 pairs of segmentally arranged openings of heart; blood from pericardial sinus enters heart through ostia.

ovary [G *Ovar, Ovarium*]: Paired section of ♀ reproductive system in which eggs are produced; extends anteriorly from fourth or fifth abdominal somite to eighth or seventh thoracic somite. Opens to exterior via oviduct originating near first or second abdominal somite (= genital somite).

oviduct [G *Ovidukt, Eileiter*]: Section of ♀ reproductive system between each ovary and single ovisac. Originates from ovary near first or second abdominal somite (= genital somite). Distal portion of oviduct (extending into ovisac) occasionally referred to as uterus, terminal end as vagina.

ovisac (egg sac) [G *Eisack, Brutsack*]: In ♀, unpaired, sac-like structure projecting from ventral surface of first and second abdominal somites (= genital somite). Contains cement glands and serves as brood chamber for eggs. (cylindrical, subglobular)

paragnath [G *Paragnath*]: On underside of head, structure posterior to mouth between mandibles and maxillules; considered to represent or at least function as lower lip.

penis [G *Penis*]: One of two eversible ♂ copulatory structures associated with fused first and second abdominal somites (= genital somite). Basically bipartite, consisting of expanded basal part and apical segment or copulatory part. Vas deferens opens at tip of each penis. (proceeding ventrally close to each other, proceeding ventrolaterally to laterally and widely separated from each other; basal part: rigid, soft and flexible; apical part: with/without spines, serrate-denticulate)

pericardial septum [G *Perikardialmembran*]: In circulatory system, ventral wall of pericardial sinus; separates pericardial sinus from underlying hemocoel. Perforated by series of holes to permit flow of blood from hemocoel to sinus.

pericardial sinus [G *Perikardialsinus*]: In circulatory system, elongate cavity surrounding heart. Blood collects within

sinus and enters heart through ostia; separated from underlying hemocoel by pericardial septum.

phyllopod: thoracopod

preepipod (preepipodite, proepipodite) [G *Exit*]: In thoracic appendage, laterally (outwardly) directed lobe projecting from base of protopod. (single, paired; serrated, unadorned) (See also epipod)

proepipodite: preepipod

protocerebrum: brain

protopod (protopodite) [G *Protopod, Protopodit*]: Proximal part of each thoracic appendage (thoracopod). Unsegmented, bearing five endites medially, one or more preepipods and epipod laterally, and distal endopod and exopod.

seminal vesicle [G *Vesicula seminalis*]: In ♂ reproductive system, expanded section of vas deferens at base of penis.

seta [G *Borste*]: One of numerous bristle-like projections extending from appendage. Those along thoracopod margins serve in feeding. Caudal rami typically fringed with plumose setae.

somite (segment) [G *Segment*]: One in a series of divisions of body. Head consists of 5 somites, followed by 11 (occasionally 17 or 19) thoracic somites (thoracomeres) and 8 abdominal somites. Last abdominal somite bears telson (anal somite).

spine: abdominal spine

sternite [G *Sternit*]: Ventral surface of body segment (somite). Sternite of genital segment considered to form ovisac.

supraesophageal ganglion [G *Oberschlundganglion*]: Main part of brain above (in front of) esophagus. Consists of protocerebrum (preantennular ganglion) and deutocerebrum (antennular ganglion); joined to tritocerebrum by connectives.

tagma [G *Tagma*]: Major division of body (head, thorax, abdomen). (See also trunk)

telson [G *Telson*]: Last segment of body; bears pair of caudal rami. Bears terminal anus and therefore occasionally termed anal somite.

testis [G *Hoden*]: Paired section of ♂ reproductive system in which sperm are produced; extends from third or fourth abdominal somite anteriorly to first and second abdominal somites (= genital somite). Each testis opens to exterior via vas deferens and penis.

thoracomere [G *Thoracomer, Thoraxsegment*]: One of typically 11 (in certain anostracans 17 or 19) somites of thorax; each bears pair of biramous appendages (thoracopods).

thoracopod (thoracic appendage) [G *Thoracopod, Thoraxbein*]: One of two appendages of each thoracic somite (thoracomere). Consists of proximal protopod (with endites, preepipod, and epipod), inner branch (endopod), and outer branch (exopod). Basically biramous, although occasionally referred to as multi- or polyramous due to large number of lobes; also termed phyllopod due to broad, flattened shape.

thorax [G *Thorax*]: Division (tagma) of body between head and abdomen. Typically consists of 11 somites (thoracomeres), each of which bears pair of biramous appendages (thoracopods). Ventral midline forms food groove. Thorax and abdomen together occasionally termed trunk.

tritocerebrum: brain

trunk [G *Rumpf*]: Region of body posterior to head. Typically consists of 19 postcephalic segments (excluding telson), of which the first 11 represent the thorax, the last 8 the abdomen.

uterus: oviduct

vagina: oviduct

vas deferens [G *Vas deferens, Samenleiter*]: Short, slender section of ♂ reproductive system between each testis and penis; expanded at base of penis to form seminal vesicle.

ventral frontal organ: frontal organ

ventral nerve cord [G *Bauchmark, ventraler Nervenstrang*]: Pair of widely separated, longitudinal nerve cords originating from posterior part (tritocerebrum) of brain and forming ladder-like chain (with series of ganglia) extending to second abdominal somite.

x-organ [G *Poren-X-Organ*]: Neurosecretory structure in dorsal region of head; innervated by brain.

NOTOSTRACA

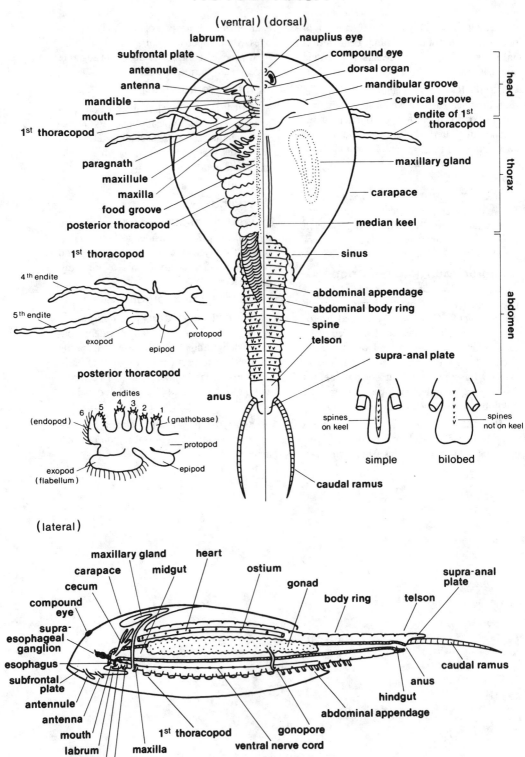

ARTHROPODA

CRUSTACEA

Notostraca

tadpole shrimp, notostracan [G *Notostrake*]

abdomen [G *Abdomen*]: Posterior division (tagma) of body (head, thorax, abdomen). Consists of large and variable number of segments (body rings), the anterior abdominal rings each bearing several appendages.

antenna (second antenna) [G *Antenne, zweite Antenne*]: Second and smaller pair of antennae; uniramous, extending from each side of labrum on underside of head. (very small, vestigial, absent)

antennule (first antenna) [G *Antennula, erste Antenne*]: First and larger pair of antennae; relatively short, uniramous, and slender. Extends from each side of labrum anterior to antennae. (simple, obscurely segmented)

anus [G *After*]: Posterior opening of digestive tract at end of telson; located between caudal rami and may be covered dorsally by supra-anal plate.

body ring (ring) [G *Körperring, Ring*]: One in a series of divisions of thorax and abdomen. Posterior body rings may bear more than one pair of appendages each and are therefore considered to consist of several fused segments. May be modified to form half rings or spiral rings. (with leg, apodous)

brain [G *Gehirn*]: Main concentration of nerve tissue in head; consists of supraesophageal ganglion (protocerebrum and deutocerebrum) and, posterior to mouth, tritocerebrum (ganglion of antennal somite). Continues posteriorly as ladder-like ventral nerve cords.

carapace (dorsal shield) [G *Carapax, Kopfschild, Ruckenschild*]: Large, broad shield covering dorsal surface of head, thorax, and anterior part of abdomen. Consists of single element (univalve) and is emarginated (notch, sinus) posteriorly. Bears median eye, compound eyes, and dorsal organ anteriorly and is sculptured by median keel and transverse mandibular and cervical grooves.

carina: median keel

caudal furca [G *Furca*]: caudal ramus

caudal ramus (furcal ramus) [G *Furkalast, Furca-Ast*]: One of two posteriorly directed projections of last segment (telson). Long and segmented (= annulate, multiarticulate). Both caudal rami together are termed caudal furca. (unarmed, armed with minute spinules and setae)

cecum (midgut diverticulum, digestive gland) [G *Divertikel, Hepatopankreas*]: One of two outpocketings of midgut. Elongate and extensively branched within carapace. Opens into digestive tract posterior to esophagus/midgut border.

cercopod: caudal ramus

cervical groove [G *Zervikalfurche, Cervicalfurche*]: Somewhat curved, transverse groove in anterior third of carapace; corresponds to border of head and thorax. (See also mandibular groove)

compound eye (lateral eye) [G *Komplexauge*]: One of two larger photosensitive organs located on sloping anterior part of carapace; sessile. (See also eye tubercle, nauplius eye)

dorsal organ (nuchal organ) [G *Dorsalorgan, Nackenorgan*]: Unpaired organ of unknown function located behind compound eyes on dorsal surface of carapace. (considerably behind eye tubercles, anterior part between eye tubercles; circular, somewhat triangular)

endite [G *Endit*]: One of five inwardly (medially) directed setose lobes of basal segment (protopod) of thoracic or abdominal appendage. Most distal (sixth) endite often termed endopod. Elongate endites of first thoracopod are termed flagella and may project beyond carapace margin. (lobelike, filiform)

endopod (endopodite) [G *Endopod, Endopodit*]: Inner lobe at distal end of thoracopod or abdominal appendage in certain tadpole shrimp. Most distal (sixth) endite also occasionally termed endopod.

epipod (epipodite) [G *Epipod, Epipodit*]: Dorsally directed lobe extending from lateral (outer) margin of thoracic appendage (thoracopod) or abdominal appendage. Serves in respiration.

esophagus [G *Ösophagus*]: Relatively short, narrow section of digestive tract between mouth and midgut. Directed dorsally, then posteriorly.

exopod (exopodite, flabellum) [G *Exopod, Exopodit*]: Large, setose outer lobe of thoracic appendage (thoracopod) or abdominal appendage. Extends dorsally and ventrally from narrow point of attachment; modified on last (11th) thoracopods in ♀ to form ovisac.

eye: compound eye, nauplius eye

eye tubercle: Slight elevation of carapace covering compound eye.

flabellum: exopod

food groove [G *Bauchrinne, Fangrinne*]: Elongate median groove between thoracopod bases on ventral side of thorax. Food is trapped by thoracopods, passed anteriorly along food groove, and transferred to mouth.

frontal organ [G *Frontalorgan*]: One of two sensory organs adjoining nauplius eye.

gonad [G *Gonade*]: Paired ♂ (testes) and/or ♀ (ovaries) reproductive system. Extends through thorax and part of abdomen and opens onto last (11th) thoracopods. In hermaphroditic tadpole shrimp, gonads consist of both ovarian and testicular tissue (ovotestis).

gonopore (genital pore) [G *Gonoporus, Geschlechtsöffnung*]: Opening of ♂ or ♀ reproductive system to exterior on last (11th) thoracic segment. In ♂, gonopores open on thoracopods; in ♀, oviducts open into ovisacs.

head [G *Kopf*]: Anterior of three divisions (tagmata) of body (head, thorax, abdomen). Consists of five somites bearing in part poorly developed antennules, antennae, mandibles, maxillules, and maxillae. Covered, along with thorax, by carapace; border between head and thorax indicated by cervical groove.

heart [G *Herz*]: In circulatory system, elongate, muscular pumping organ extending through 11 thoracic segments above digestive tract. Blood enters heart through 11 pairs of ostia and is pumped anteriorly.

hindgut (rectum) [G *Enddarm*]: Short posteriormost section of digestive tract between midgut and anus.

labrum (upper lip) [G *Labrum, Oberlippe*]: Large, unpaired lobe covering mouth and extending over end of mandibles.

leg (limb) [G *Bein, Blattbein*]: General term applied to appendage (thoracopod) of thorax or abdomen. Also termed phyllopod due to flattened construction.

mandible [G *Mandibel*]: Third paired appendage of head; located between small antennae and maxillules and extending under labrum. Relatively large, simple (without palp), and with toothed or spinous distal end.

mandibular groove: Transverse groove located between cervical groove and dorsal organ in anterior region of carapace.

maxilla (second maxilla) [G *Maxille, zweite Maxille*]: Fifth paired appendage of head; located between maxillules and first thoracopods. Associated with opening of maxillary gland. (with well-developed lobe, reduced, absent)

maxillary gland (shell gland) [G *Maxillardrüse, Maxillennephridium, Schalendrüse*]: One of two excretory organs located in head and opening ventrally on maxillae. Highly elongated, forming loops visible through carapace on each side of median keel.

maxillule (first maxilla) [G *Maxillula, erste Maxille*]: Fourth paired appendage of head; relatively small, located between mandibles and maxillae. Uniramous, simple, typically with spines along medial (inner) margin. (See also paragnath)

median keel (carina) [G *Kiel*]: Elevated, longitudinal keel along midline of carapace; extends anteriorly to transverse cervical groove.

metanauplius [G *Metanauplius*]: Later larval stage between nauplius and adult. Earlier metanauplii bear only first three pairs of appendages (as in nauplii), yet exhibit further segmentation of trunk. Most tadpole shrimp hatch as metanauplii.

midgut [G *Mitteldarm*]: Elongate section of digestive tract between esophagus and hindgut. Extends through thorax and most of abdomen and bears two ceca anteriorly.

midgut diverticulum: cecum

mouth [G *Mund, Mundöffnung*]: Anterior opening of digestive tract on underside of head; located under labrum and opening into esophagus.

nauplius [G *Nauplius*]: Early larval stage characterized by presence of only three pairs of appendages (antennules, antennae, mandibles) and unsegmented trunk. Certain tadpole shrimp hatch as nauplii. (See also metanauplius)

nauplius eye (median eye) [G *Naupliusauge*]: Small median photosensitive organ located in front of compound eyes on sloping anterior part of carapace; consists of four cups.

notch: sinus

nuchal organ: dorsal organ

ostium [G *Ostium*]: One of 11 pairs of segmentally arranged openings of heart in thorax; blood enters heart through ostia.

ovary: gonad

oviduct [G *Eileiter*]: gonad

ovisac [G *Eikapsel, Oothek, Ootheca*]: In ♀, brood chamber for eggs on each appendage of last (11th) thoracic segment; chamber formed by exopod, with epipod serving as lid.

ovotestis [G *Ovotestis*]: gonad

paragnath [G *Paragnath*]: One of a pair of lobe-like structures between mandibles and maxillules posterior to mouth. Considered to represent either true lower lip or additional lobe of each maxillule.

phyllopod [G *Phyllopod*]: leg

protopod (protopodite) [G *Protopod, Protopodit*]: Proximal part of thoracic appendage (thoracopod) or abdominal appendage. Bears series of five endites on inner margin, epipod on outer margin, distal endopod (sixth endite) and exopod (flabellum).

rectum: hindgut

ring: body ring

shell gland: maxillary gland

sinus (notch) [G *Einbuchtung*]: Posterior emargination of carapace; bears series of small denticles.

somite [G *Segment*]: One in a series of original divisions of body. Head consists of five somites, while somites of thorax and abdomen are variously fused to form units termed either segments or body rings.

spine [G *Stachel*]: One of numerous small, spine-like projections of body rings, telson, or supra-anal plate.

subfrontal plate: Broad, flattened, underturned anterior part of carapace; adjoins basal part of labrum on underside of head and partially covers antennules and antennae.

supra-anal plate [G *Supra-Analplatte*]: Posteriorly directed median extension of telson. (simple, bilobed; with/without median keel; median spines placed on/not placed on keel)

supraesophageal ganglion [G *Oberschlundganglion*]: Main part of brain above (in front of) esophagus. Consists of protocerebrum (preantennular ganglion) and deutocerebrum (antennular ganglion). (See also tritocerebrum)

tagma [G *Tagma*]: Major division of body (head, thorax, abdomen).

telson [G *Telson*]: Last segment of body; bears anus and pair of elongate caudal rami. (with/without supra-anal plate)

testis: gonad

thoracopod (thoracic appendage) [G *Thoracopod*]: One of two appendages of each thoracic segment (body ring). Consists of proximal protopod (with endites and epipod), endopod (= sixth endite), and exopod (flabellum). (See also leg)

thorax [G *Thorax*]: Division (tagma) of body between head and abdomen. Consists of 11 segments ("thoracic segments" or body rings), each bearing pair of polyramous appendages (thoracopods). Covered, along with head, by carapace. (See also trunk)

tritocerebrum: brain

trunk [G *Rumpf*]: Region of body posterior to head. Consists of thorax with 11 segments and abdomen with large and variable number of body rings; anterior region covered by carapace. Term is frequently applied due to poorly delimited border between thorax and abdomen.

vas deferens: gonad

ventral nerve cord [G *Bauchmark*]: Pair of widely separated longitudinal nerve cords extending posteriorly from brain; forms ladder-like chain (with series of ganglia).

ARTHROPODA

CRUSTACEA

Conchostraca

clam shrimp, conchostracan
[G *Muschelschaler, Conchostrake*]

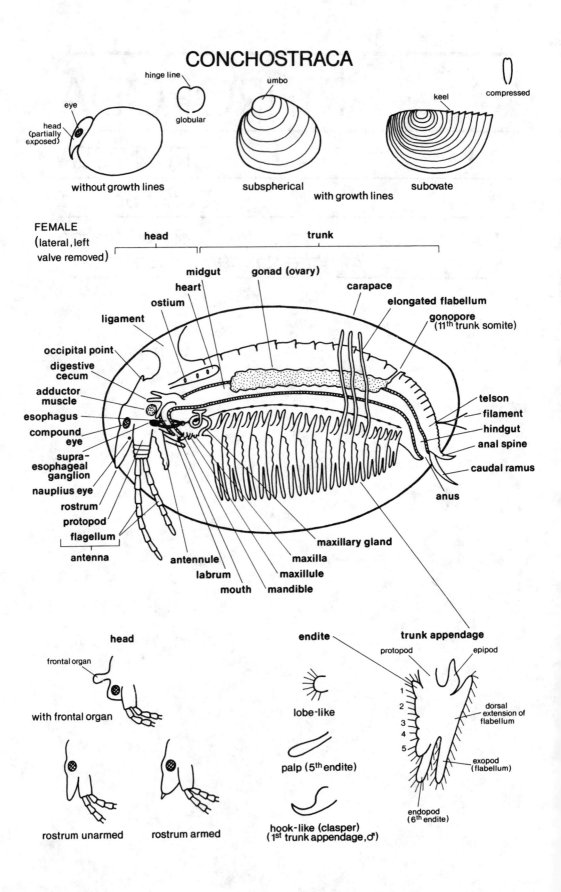

CONCHOSTRACA

ARTHROPODA

CRUSTACEA

Conchostraca

clam shrimp, conchostracan
[G *Muschelschaler, Conchostrake*]

abdomen [G *Abdomen*]: trunk

adductor muscle [G *Adduktormuskel, Schalenschließmuskel, Schließmuskel*]: Well-developed muscle attaching each side of head region to carapace. (See also ligament)

anal segment (anal somite): telson

anal spine [G *Abdominaldorn*]: One of two blade-like, serrate spines projecting from posterodorsal end of telson.

antenna (second antenna) [G *Antenne, zweite Antenne*]: Second and much larger pair of antennae on head; biramous, consisting of long, terminally annulate basal part (protopod, with basal lobe) and pair of segmented branches (flagella, rami).

antennule (first antenna) [G *Antennula, erste Antenne*]: First and much smaller pair of antennae; uniramous, typically with series of dorsal sensory papillae. (short, long; unsegmented = unjointed, with two segments, multijointed)

anus [G *After*]: Posterior opening of digestive tract at end of telson.

article [G *Glied*]: One of up to 25 segments of antennal branches (rami, flagella). May also refer to segment of other appendages.

atrium oris [G *Atrium oris*]: Large preoral cavity formed by upper lip (labrum) extending over mouth. Tips (molar processes) of mandibles extend into atrium oris.

attachment ligament: ligament

beak: rostrum

brain [G *Gehirn*]: Main concentration of nerve tissue in head. Consists basically of ganglion (supraesophageal ganglion = protocerebrum and deutocerebrum) anterior to esophagus and ganglion (tritocerebrum = circumesophageal ganglion = ganglion of antennal somite) to each side of or posterior to esophagus. Continues posteriorly as ladder-like ventral nerve cord.

carapace (shell) [G *Carapax*]: Bivalved, laterally compressed shield typically enclosing entire body; consists of two valves with dorsal hinge line, although occasionally interpreted as representing a single continuous element. Joined to body by ligament and pair of adductor muscles in head region. (broadly oval, compressed, globose = globular, oblong, ovate, subovate, subrectangular, subspherical, swollen, thick and globose, thin and pellucid; with/without growth lines, with/without umbones, with/without keel; sculpture: polygonal, punctate, reticulate, with dendritic or radial grooves, with striae)

caudal furca: caudal ramus

caudal ramus (cercopod, furcal claw) [G *Furkakralle, Endkralle*]: One of two relatively short, posteriorly directed projections of telson. Both caudal rami together are termed caudal furca. (claw-like, styliform) (See also uropod)

cecum: digestive cecum

cephalon: head

cercopod: caudal ramus

cervical groove [G *Zervikalfurche, Cervicalfurche*]: Transverse groove across dorsal surface of head; corresponds to level of mandibles.

circumesophageal   ganglion: brain

clasper [G *Greifer*]: Modified first and occasionally also second pair of trunk appendages in ♂; distal endites elongated and hook-like. Serves in grasping ♀ during copulation.

compound eye [G *Komplexauge*]: One of two large photosensitive organs located on anterior end of head; sessile, either closely adjoining or partially fused. Enclosed almost entirely within optic vesicle.

digestive   cecum   (caecum) [G *Leberschlauch, Hepatopankreas*]: One of two lateral outpocketings of midgut near border of esophagus and midgut. Highly branched, with lobe-shaped branches extending ventrally and dorsally into anterior part of head.

endite [G *Endit*]: One of five inwardly (medially) directed, setose lobes of basal part (protopod) of trunk appendage. Most distal (sixth) endite often termed endopod, while fifth may be modified as a palp (especially in ♂).

endopod (endopodite) [G *Endopod, Endopodit*]: Unsegmented, lobe-like inner branch of trunk appendage; often termed sixth endite.

epipod (epipodite) [G *Epipod, Epipodit*]: In trunk appendage, fleshy, lobe-like projection from base of protopod. Serves,

along with inner wall of carapace, in respiration. Dorsally directed part of flabellum may also represent an epipod.

epistome: rostrum

esophagus [G *Ösophagus*]: Relatively short and narrow anterior section of digestive tract. From mouth, esophagus curves forward, then upward and backward to more expanded midgut.

exopod (exopodite, flabellum) [G *Exopod, Exopodit, Flabellum*]: Outer branch of trunk appendage; unsegmented, prolonged dorsally and ventrally, and bearing marginal setae. Dorsal extension occasionally interpreted as representing additional epipod or flabellum, ventral extension as exopod proper.

eye: compound eye, nauplius eye

filament (forked filament) [G *Filament*]: Slender, basally forked projection on dorsal surface of telson. Relative position within series of spines on telson is of taxonomic importance.

flabellum [G *Flabellum*]: Term applied to entire outer branch (exopod) of trunk appendage or to dorsally directed part of exopod only. Modified in 9th through 11th appendages into elongate, slender structures for attachment of eggs.

flagellum [G *Flagellum, Ast*]: One of two multisegmented branches of antenna; each of the up to 25 segments (articles) bears a row of spine-like setae anteriorly and brush-like setae posteriorly. Number of segments is of taxonomic importance.

food groove [G *Bauchrinne*]: Elongate median groove between bases of trunk appendages. Food is trapped by trunk appendages, passed along food groove, and transferred to mouth.

frontal appendage: frontal organ

frontal organ (frontal appendage) [G *Frontalorgan, Scheitelorgan*]: In certain clam shrimp, unpaired, pyriform projection on middorsal surface of head.

furcal claw: caudal ramus

gonad [G *Gonade*]: Paired ♂ (testes) or ♀ (ovaries) reproductive system. Extends throughout trunk and opens onto 11th trunk somite via vasa deferentia (♂) or oviducts (♀).

gonopore (genital pore) [G *Gonoporus, Geschlechtsöffnung*]: One of two openings of ♂ or ♀ reproductive system to exterior on 11th trunk somite.

growth line [G *Anwachsstreifen, Zuwachsstreifen*]: One in a series of concentric lines on outer surface of each side (valve) of carapace; each growth line marks a successive molt. Number of lines is of taxonomic importance. (distinct, indistinct; widely spaced, crowded)

head (cephalon) [G *Kopf*]: Anterior of two divisions (tagmata) of body (head, trunk). Consists of five somites bearing antennules, antennae, mandibles, maxillules, and maxillae. Bears ventral rostrum, compound and nauplius eyes, and is typically entirely enclosed by carapace. (entirely/not entirely enclosed by carapace; with/without frontal appendage)

heart [G *Herz*]: In circulatory system, muscular pumping organ extending from head (from maxillule-bearing somite) into third trunk somite. Blood enters heart posteriorly and through typically four pairs of ostia.

hindgut [G *Enddarm*]: Short posteriormost section of digestive tract between midgut and anus.

hinge line: Longitudinal middorsal line of carapace. Considered either to mark connection of two valves of carapace or to represent thinner and most flexible region of continuous shield. (straight, in an arc)

keel [G *Kiel*]: Ridge-like elevation of posterior hinge line; produced into series of serrations by continued growth lines.

labrum (upper lip) [G *Labrum, Oberlippe*]: Relatively large, unpaired lobe covering mouth; free (posterior) margin bears prominent spine. Tips (molar pro-

cesses) of mandibles extend into preoral cavity (atrium oris) under labrum.

ligament (attachment ligament) [G *Ligament*]: Connective tissue spanning from top of head to umbonal region of carapace. Joins body to carapace (along with adductor muscles).

mandible [G *Mandibel*]: Third paired appendage of head; located between large antennae and maxillules. Simple (i.e., without palp), massive, with ridged or toothed terminal molar process extending into preoral cavity (atrium oris) under labrum.

maxilla (second maxilla) [G *Maxille, zweite Maxille*]: Fifth paired appendage of head; located between maxillules and first trunk appendages. If present, tiny, setose, and lobe-like. Associated with opening of maxillary gland. (vestigial, absent)

maxillary gland (shell gland) [G *Maxillardrüse, Maxillennephridium, Schalendrüse*]: One of two excretory glands located in head (and extending into carapace); opens ventrally at level of maxillae. Forms several loops.

maxillule (first maxilla) [G *Maxillula, erste Maxille*]: Fourth paired appendage of head; located between mandibles and maxillae. Relatively small, simple, setose lobe with small anterior accessory lobe.

metanauplius [G *Metanauplius*]: Late larval stage; carapace typically first produced by metanauplius. (See also nauplius)

midgut [G *Mitteldarm*]: Elongate, somewhat expanded section of digestive tract between esophagus and hindgut in trunk. Bears pair of digestive ceca anteriorly.

molar process [G *Pars molaris, Kaufläche*]: Terminal end of each mandible; extends into preoral cavity (atrium oris) under upper lip (labrum).

mouth [G *Mund, Mundöffnung*]: Anterior opening of digestive tract; located under labrum and associated with large preoral cavity (atrium oris). Opens into esophagus.

nauplius [G *Nauplius*]: Early larval stage; basically characterized by three pairs of appendages (antennules, antennae, mandibles), although antennules are typically reduced. Some nauplii bear broad dorsal shield. (See also metanauplius)

nauplius eye (median eye) [G *Naupliusauge*]: Small median photosensitive organ located ventrally on rostrum of head. (See also compound eye)

occipital notch: Notch-like depression in dorsal posterior region of head. May be preceded by occipital point. (shallow and inconspicuous, acute and pronounced, deeply cleft)

occipital point: On dorsal posterior region of head, protuberance in front of occipital notch.

optic vesicle: compound eye

ostium [G *Ostium*]: One of typically four pairs of lateral openings in heart (one pair in head region, three in first three trunk somites).

ovary: gonad

oviduct [G *Eileiter*]: gonad

palp (tactile process) [G *Palpus*]: Elongate process of fifth endite on trunk appendage.

phyllopod [G *Blattbein*]: trunk appendage

protopod (protopodite) [G *Protopod, Protopodit*]: Proximal, poorly delimited part of trunk appendage; bears series of endites along inner margin, epipod on outer margin, as well as distal endopod (= sixth endite) and exopod (flabellum). Also applied to proximal part of antenna.

ramus [G *Ast*]: Term applied to either branch (flagellum) of antenna.

rostrum (frontal process, rostral process, beak) [G *Rostrum*]: Prolongation of anterior end (underside) of head; located in front of upper lip (labrum) and therefore occasionally termed epistome. (armed, unarmed; acute, notched, pointed,

rounded, spatulate = spatuliform, subtriangular)

scape [G *Protopodit*]: Term occasionally applied to basal part of antenna.

seta [G *Borste*]: One of numerous bristle-shaped projections of appendage. Setae of trunk appendages serve in feeding. According to shape one may distinguish, e.g., simple and spine-like, or brush-like setae on flagellum of antenna.

shell: carapace

somite (segment) [G *Segment*]: One in a series of divisions of body. Head consists of five somites, trunk of 10–32 somites (excluding telson). Posteriormost somites drawn out into series of spine-like projections dorsally. Telson occasionally termed anal somite.

standard measurements:
carapace length, width, height
number of segments of antenna
number of trunk somites (and trunk appendages)
number of spines on telson

supraesophageal ganglion [G *Oberschlundganglion*]: brain

tagma [G *Tagma*]: Major division of body (head, trunk). Both tagmata are covered by carapace.

telson [G *Telson*]: Last segment of body; directed ventrally and bears pair of terminal anal spines as well as pair of caudal rami. Bears anus and is therefore occasionally termed anal segment or anal somite (in which case terminal lobe is interpreted as representing telson). Number of dorsal spines and position of forked filament are of taxonomic importance.

testis: gonad

thoracopod: trunk appendage

thorax: trunk

trunk (postcephalic body) [G *Rumpf*]: Posterior of two divisions (tagmata) of body (head, trunk); consists of 10–32 somites (excluding telson), each bearing pair of appendages. Considered to basically consist of thorax and abdomen, although the two are indistinguishable.

trunk appendage (thoracic appendage, thoracopod, trunk leg)
[G *Rumpfgliedmaß, Thoracopod*]: One of two relatively elongate appendages of each trunk somite. Basically biramous, consisting of poorly delimited, proximal protopod (with endites and epipod), endopod (= sixth endite), and exopod (flabellum). Decreases in size posteriorly. Unmodified trunk appendages are termed phyllopods due to flattened construction.

umbo [G *Umbo*]: Somewhat elevated, dorsalmost part of carapace; represents early growth stage of shell. Position relative to anterior end is of taxonomic importance.

upper lip: labrum

uropod [G *Uropod*]: Term applied to each caudal ramus if telson is interpreted as representing anal somite.

valve [G *Carapaxklappe*]: One of two shell elements comprising carapace. Covering left and right sides of body and joined together (or continuous) dorsally along hinge line. Consists of several layers and is variously ornamented on exterior. (See also carapace)

vas deferens [G *Samenleiter, Vas deferens*]: gonad

ventral nerve cord [G *Bauchmark, ventraler Nervenstran*]: Pair of widely separated, longitudinal nerve cords extending posteriorly from brain; forms ladder-like chain.

CLADOCERA

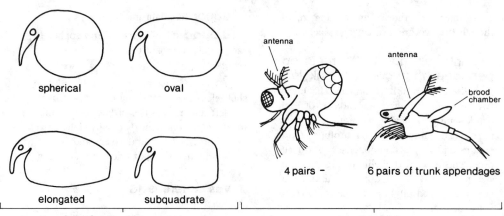

spherical   oval

elongated   subquadrate

fully developed carapace

antenna

antenna

brood chamber

4 pairs –   6 pairs of trunk appendages

reduced carapace

FEMALE (lateral)

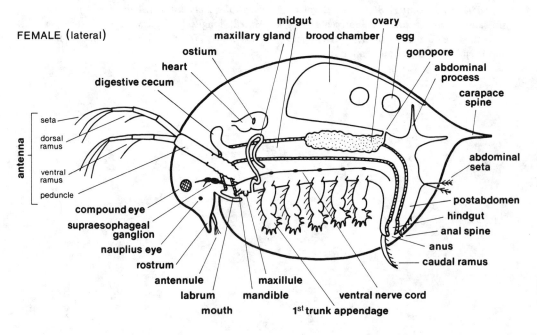

midgut

ovary

maxillary gland / brood chamber / egg

ostium   gonopore

heart   abdominal process

digestive cecum   carapace spine

seta

dorsal ramus

antenna

ventral ramus

peduncle

abdominal seta

compound eye

supraesophageal ganglion

nauplius eye

rostrum

postabdomen

hindgut

anal spine

anus

caudal ramus

antennule   maxillule

labrum   mandible

mouth   ventral nerve cord

1st trunk appendage

carapace sculpture

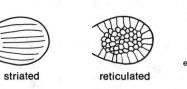

striated   reticulated

mucro

smooth   ventral margin with setae

trunk appendage

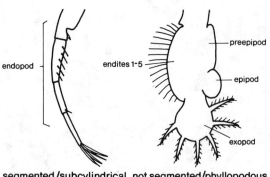

endopod

endites 1-5

preepipod

epipod

exopod

segmented/subcylindrical (predatory species)   not segmented/phyllopodous (filtering species)

ARTHROPODA

CRUSTACEA

Cladocera

water flea, cladoceran [G *Wasserfloh, Cladocere*]

abdomen: trunk

abdominal process (abdominal outgrowth) [G *Abdominalfortsatz, Rückenfortsatz, Zipfel*]: One of typically two dorsal processes projecting from border of trunk and postabdomen. Considered to serve in closing off brood chamber posteriorly.

abdominal seta (setae natatoriae) [G *Schwimmborste, Setae natatores*]: One of two elaborate setae located on small protuberance on dorsal surface of postabdomen.

abreptor: postabdomen

aesthetasc (esthete, olfactory seta) [G *Aesthetasc*]: One in a tuft of sensory projections at tip of each antennule.

anal spine (anal tooth) [G *kleine Kralle, Zahn*]: One in a series of spines along (morphological) dorsal surface of postabdomen.

antenna (second antenna) [G *Antenne, zweite Antenne*]: Second and much larger pair of antennae; located laterally near posterior margin of head. Biramous, consisting of relatively large basal part (peduncle) bearing two- to four-segmented dorsal and ventral branches (rami). Serves as principal locomotory organ and moved by relatively large muscles (antennal muscles).

(biramous and flattened, simple and cylindrical, branched, not branched)

antennal muscle [G *Antennenmuskel*]: One of several well-developed muscles spanning from antenna to dorsal part of head. According to position and function one may distinguish, e.g., antennal levator muscle and antennal adductor muscle.

antennal setae formula (setation formula) [G *Borstenformel*]: Notation specifying number of segments and number of setae per segment in dorsal and ventral branches (rami) of antenna. Example: 0-0-1-3/1-1-3.

antennule (first antenna) [G *Antennula, erste Antenne*]: First and typically much smaller pair of antennae; located ventrally near posterior margin of head. Uniramous, unsegmented, with tuft of terminal aesthetascs. (movable, immovable = fixed; fused/not fused with rostrum; inserted/not inserted at anterior end of ventral edge of head)

anus [G *After*]: Posterior opening of digestive tract at end of postabdomen. (terminal, on dorsal margin)

aorta [G *Aorta*]: heart

apical spine: carapace spine

beak: rostrum

brain [G *Gehirn*]: Main concentration of nerve tissue in head. Consists basically of ganglion (supraesophageal gnaglion = protocerebrum + deutocerebrum) anterior to esophagus and ganglion (tritocerebrum) to side of or posterior to esophagus. Continues posteriorly as ladder-like ventral nerve cord.

brood chamber (brood pouch) [G *Brutraum, Brutsack*]: In ♀, dorsal space between trunk and carapace; each ovary opens into brood chamber via oviduct. May be reduced to sac-like structure in predatory water fleas. Serves to brood various types of eggs.

bulbus arteriosus [G *Bulbus arteriosus*]: heart

carapace (shell) [G *Carapax, Schale*]: Single-piece (univalved) shield which, if fully developed, covers only posterior part of body (trunk). Laterally compressed, with ventral gape, and therefore occasionally described as bivalved (i.e., consisting of two "valves"). (angular, circular, elongate, oval, ovate, ovoid, spherical, subcircular, subquadrate; laterally compressed, thick-bodied; with/without carapace spine, with/without crest, narrowed and prolonged/not narrowed and prolonged posteriorly into short tube, anterior portion swollen/not swollen, with/without projection on anteroventral margin, ventral margin smooth/with fine spines, dorsal margin smooth/serrate; sculpture: reticulated, smooth; longitudinally, obliquely, transversely striated; with/without hairs, with polygonal or rhomboid markings)

carapace spine (apical spine, posterior spine, shell spine) [G *Carapaxstachel, Stachel, Spina*]: Large, posteriorly directed, spine-like structure formed by posterodorsal extension of carapace. Relative length may be of taxonomic importance.

caudal furca: caudal ramus

caudal ramus (claw, postabdominal claw, terminal claw) [G *Furcakralle, Endkralle*]: One of two relatively short, claw-shaped projections at end of postabdomen. May bear up to three series (pectens) of minute spines. Both caudal rami together are termed caudal furca. (denticulate, pectinate, with/without basal spine, with/without pecten)

cecum: digestive cecum

cephalon: head

cerebral ganglion: supraesophageal ganglion

cervical notch (cervical sinus) [G *Zervikalfurche, Cervicalfurche*]: Transverse groove between head and trunk. (with/without conspicuous folds)

cervical sinus: cervical notch

claw: caudal ramus

compound eye [G *Komplexauge*]: One of two large photosensitive organs on head; sessile, typically fused, and enclosed within optic vesicle. Movable by series of optic muscles. Relative position of compound eye on head (e.g., anterior, in middle near ventral margin) may be of taxonomic importance.

cyclomorph [G *Zyklomorph*]: Any one of variously modified individuals in a population. These modifications are seasonal and generally involve the head (development of helmet) or changes in eye size or carapace spine length.

denticle [G *Zähnchen, kleine Krallen, kleine Zähne*]: One in a series of small, tooth-like projections along postabdomen. Number, position, and relative size are of taxonomic importance. (equal, subequal; smooth, serrate; lateral, marginal)

deutocerebrum: supraesophageal ganglion

digestive cecum (cecum, hepatic cecum, midgut diverticulum) [G *Hepatopankreas, Hepatopankreasschlauch, Leberhörnchen*]: One of two lateral outpocketings of midgut; located in head posterior to border of esophagus and midgut. An additional unpaired cecum may open into posterior end of digestive tract.

endite [G *Endit*]: Inwardly (medially) directed, setose lobe of basal part (protopod) of trunk appendage. First endite typically well delimited, the following endites increasingly incorporated into endopod.

endopod (endopodite) [G *Endopod, Endopodit*]: Lobe-like inner branch (ramus) of trunk appendage; lacks articulation with protopodal part of appendage and is more or less continuous with distal endites. Forms entire distal part of limb in certain predatory water fleas.

ephippium [G *Ephippium*]: Egg case formed by walls of brood chamber. Typically separates from rest of carapace during molting. (knobby and with two eggs, reticulated and with single egg)

epipod (epipodite) [G *Epipod, Epipodit*]: In trunk appendage, laterally (outwardly) directed lobe projecting from base of protopod. Represents only lobe articulated with appendage and serves in respiration.

esophageal connective [G *Schlundring*]: One of two strands of nerve tissue surrounding esophagus and connecting posterior part of brain with ventral nerve cord.

esophagus [G *Ösophagus*]: Relatively short and narrow anterior section of digestive tract. From mouth, esophagus curves forward, then upward and backward to more expanded midgut. May be highly elongated in certain predatory water fleas.

exopod (exopodite) [G *Exopod, Exopodit*]: Lobe-like outer branch (ramus) of trunk appendage; lacks articulation with protopodal part of appendage. Reduced in certain predatory water fleas.

eye: compound eye, nauplius eye

food groove [G *Bauchrinne*]: In filter-feeding water fleas, elongate median groove between bases of trunk appendages. Food is trapped by trunk appendages, passed anteriorly along food groove, and transferred to mouth.

fornix [G *Fornix*]: Ridge-like extension of each side of head; may merge with rostrum to form ventrally directed, beak-like

structure. (covering/not covering most of antennules; simple, forming spinous processes)

frontal organ [G *Frontalorgan*]: General term for any one of several similar sensory organs in head. The different positions of these frontal organs has led to an array of terms for the individual structures, including nuchal or neck organ and, in the older literature, dorsal and ventral frontal organs. All are innervated by brain and considered to represent reduced photosensitive organs.

gonopore [G *Gonoporus, Geschlechtsöffnung*]: Opening of ♂ or ♀ reproductive system to exterior. In ♂, located either behind last pair of trunk appendages, on postabdomen, or terminally near caudal furca. In ♀, see vulva. (single, paired) (See also penis)

head (cephalon) [G *Kopf*]: Anterior of two divisions (tagmata) of body (head, trunk); consists of five somites bearing antennules, antennae, mandibles, maxillules, and (reduced) maxillae. Bears compound and naupliar eyes and may be covered by a head shield. Not enclosed by carapace. (clearly delimited/not clearly delimited; helmeted, rounded, with/without rostrum)

head pore: One of several minute pores in middorsal line of head shield.

head shield [G *Kopfschild*]: Variously developed, unpaired, shield-like structure covering head.

heart [G *Herz*]: In circulatory system, typically short, muscular pumping organ located anterior to brood chamber in trunk. Blood enters heart through single pair of ostia and is pumped anteriorly into a sinus, although certain predatory water fleas have been described as bearing anterior bulbus arteriosus and short anterior aorta. (elongate, oval, subovate)

helmet [G *Helm*]: Helmet-shaped elongation of anterior margin of head in certain individuals of a population. Appearance of helmeted individuals and shape of helmet is governed by environmental and genetic factors. (See also cyclomorph)

hepatic cecum: digestive cecum

hindgut (rectum) [G *Enddarm*]: Short posteriormost section of digestive tract between midgut and anus; may be associated with single ventral cecum.

intestine: midgut

labium (lower lip, paragnath) [G *Labium, Unterlippe*]: Unpaired, median lip-like structure posterior to mouth. (with/without median keel)

labrum (upper lip) [G *Labrum, Oberlippe*]: Relatively large, unpaired lobe covering mouth. Distal ends of mandibles extend under labrum. (ventral edge: smooth, toothed)

lower lip: labium

mandible [G *Mandibel*]: Third paired appendage of head; relatively small and simple (i.e., without palp). Typically with simple ridged or toothed distal end which extends under labrum.

mantle [G *Gallerthülle*]: Large, gelatinous envelope enclosing body in certain water fleas.

maxilla (second maxilla) [G *Maxille, zweite Maxille*]: Fifth and last paired appendage of head. Typically absent, yet if present, then reduced to tiny lobe. Associated with opening of maxillary gland. (absent, rudimentary)

maxillary gland (shell gland) [G *Maxillardrüse, Schalendrüse, Maxillennephridium*]: One of two excretory organs located in head and opening ventrally at level of maxillae. Forms several loops.

maxillule (first maxilla) [G *Maxillula, erste Maxille*]: Fourth paired appendage of head. Consists of very small, simple, setae-bearing lobe. (absent, minute)

median eye: nauplius eye

metanauplius [G *Metanauplius*]: Larval stage at which certain predatory water fleas hatch; otherwise, metanauplius is typically passed while still within egg.

midgut [G *Mitteldarm*]: Elongate section of digestive tract between esophagus and hindgut. Occasionally one may distinguish between a more expanded stomach region anteriorly and a more narrow posterior intestinal region. May bear pair of digestive ceca anteriorly. (straight, coiled = convoluted)

midgut diverticulum: digestive cecum

mouth [G *Mund, Mundöffnung*]: Anterior opening of digestive tract under labrum; opens into esophagus.

mucro [G *Mukro, Spitze*]: Sharp point at lower posterior end of carapace. (in immature forms: with minute dorsal notches, with minute ventral notches)

nauplius eye (median eye, ocellus) [G *Naupliusauge, Nebenauge*]: Small median photosensitive organ located ventrally on head between mouth and compound eye.

neck organ: nuchal organ

nuchal organ [G *Nackenorgan, Nackensinnesorgan*]: frontal organ

ocellus: nauplius eye

olfactory seta: aesthetasc

optic muscle [G *Augenmuskel*]: compound eye

optic nerve [G *Augenganglion, Ganglion opticum*]: Large nerve leading from supraesophageal ganglion to each (fused) compound eye.

optic vesicle [G *Augenkammer*]: Protuberance of outer wall of head whose cavity encloses compound eye. Optic vesicle does not communicate with exterior via pore. (touching/not touching margin of head)

ostium [G *Ostium*]: One of two lateral openings in heart. Blood enters heart through ostia and is pumped anteriorly.

ovary [G *Ovar, Ovarium, Eierstock*]: Paired section of ♀ reproductive system in which eggs are produced; typically extends through anterior part of trunk, one along each side of midgut. Opens dorsally via oviduct into brood chamber.

oviduct [G *Ovidukt, Eileiter*]: Short and narrow section of ♀ reproductive system between posterior part of each ovary and dorsal brood pouch.

paragnath: labium

pecten [G *Kamm*]: One of up to three (proximal, middle, distal) series of small teeth along each caudal ramus. (teeth: equal, not equal)

peduncle [G *Stammglied, Protopodit*]: Basal part of each antenna. Relatively long, bearing two- to four-segmented dorsal and ventral branches (rami).

penis [G *Penis*]: Typically unpaired ♂ copulatory structure at end of postabdomen. (single, paired)

postabdomen (abreptor, pygidium) [G *Postabdomen*]: Recurved (turned ventrally and forward) posteriormost region of trunk. Bears pair of abdominal setae proximally and caudal rami distally (telson indistinguishable), as well as series of spines (anal spines) and denticles. (cylindrical, quadrangular, subquadrate, bilobed/not bilobed, with/without spines; with marginal, with lateral, with marginal and lateral denticles)

postabdominal claw: caudal ramus

preepipod [G *Exit*]: On trunk appendage, laterally (outwardly) directed lobe projecting from base of protopod (proximal to epipod).

protocerebrum: supraesophageal ganglion

protopod (protopodite) [G *Protopod, Protopodit*]: Poorly delimited proximal part of trunk appendage; bears one or two distinguishable endites along inner margin and epipod on outer margin, as well as distal endopod and exopod.

pygidium: postabdomen

ramus [G *Ast*]: Branch of appendage. Refers either to dorsal and ventral branches of antennae or to inner (endopod) and outer (exopod) branches of trunk appendage.

rectum: hindgut

rostrum [G *Rostrum*]: Anterior (ventrally directed) extension of head. Relative length compared to antennules and antennae may be of taxonomic importance. (bent into hook, curved forward, recurved, straight; acute, blunt)

seminal receptacle [G *Receptaculum seminis*]: In ♀ reproductive system, term occasionally applied to terminal section of oviduct.

seta: abdominal seta, antennal setae formula, olfactory seta, trunk appendage

setae natatoriae: abdominal seta

setation formula: antennal setae formula

shell: carapace

shell gland: maxillary gland

shell spine: carapace spine

somite [G *Segment*]: One in a series of divisions of body. Head consists of five somites (bearing antennules, antennae, mandibles, maxillules, and reduced maxillae). Somites of trunk, especially of posterior region ("abdomen"), are indistinguishable, although trunk appendages reflect segmentation of anterior region ("thorax").

sperm duct: vas deferens

spine: anal spine, carapace spine

spinule [G *Dörnchen*]: Small spine along dorsal margin of carapace or along vertex.

standard measurements:
body length
body height
head length
number of trunk appendages
number of anal spines
antennal setae formula

stomach: midgut

supraesophageal ganglion [G *Oberschlundganglion*]: Main part of brain above (in front of) esophagus. Consists of protocerebrum (preantennular ganglion) and deutocerebrum (antennular ganglion).

swimming hair: antennal setae formula

tagma [G *Tagma*]: Major division of body. Typically a head and trunk are distinguished, although the body may be considered to consist of head, thorax, abdomen, and postabdomen. Fully developed carapace covers only trunk.

telson: postabdomen

terminal claw: caudal ramus

testis [G *Hoden*]: Paired section of ♂ reproductive system in which sperm are produced. Testes extend through anterior part of trunk, one to each side of midgut; opens to exterior via ventrally directed vas deferens.

thoracic appendage: trunk appendage

thoracopod: trunk appendage

thorax: trunk

tritocerebrum: brain

trunk [G *Rumpf*]: Posterior of two divisions (tagmata) of body (head, trunk). Trunk may be considered to consist of thorax, abdomen, and postabdomen, whereby thorax refers to anterior appendage-bearing region of trunk, abdomen to (reduced) region posterior to trunk appendages, and postabdomen to recurved posteriormost region. Somite borders indistinguishable in posterior region of trunk. (covered/not covered by valves)

trunk appendage (thoracic appendage, thoracopod) [G *Rumpfgliedmaß, Thoracopod*]: One of four to six pairs of appendages of anterior region ("thorax") of trunk. Basically biramous, consisting of indistinct proximal protopod (with endites and epipod) and more distal endopod and exopod. (flattened = foliaceous = phyllopodous, not clearly segmented; subcylindrical = prehensile, clearly segmented; similar, dissimilar)

upper lip: labrum

valve [G *Schalenklappe*]: Term applied to each half of carapace, one half (valve) covering each side of trunk.

vas deferens (sperm duct) [G *Vas deferens, Samenleiter*]: Section of ♂ reproductive system extending from posterior end of each testis to gonopore(s) or penis(es) on postabdomen.

ventral nerve cord [G *Bauchmark, ventraler Nervenstrang*]: Pair of widely separated, longitudinal nerve cords extending into trunk from posterior part of brain; forms ladder-like chain with relatively few, occasionally fused ganglia.

vertex [G *Vertex, Scheitel*]: Region of head between rostrum and compound eye (i.e., morphologically, part anterior to compound eye).

vulva [G *Vulva*]: In ♀ reproductive system, term occasionally applied to opening of oviduct into brood chamber.

ARTHROPODA

CRUSTACEA

Ostracoda

ostracod, ostracode, musselshrimp, seed shrimp [G *Ostracode, Ostrakode, Muschelkrebs*]

OSTRACODA

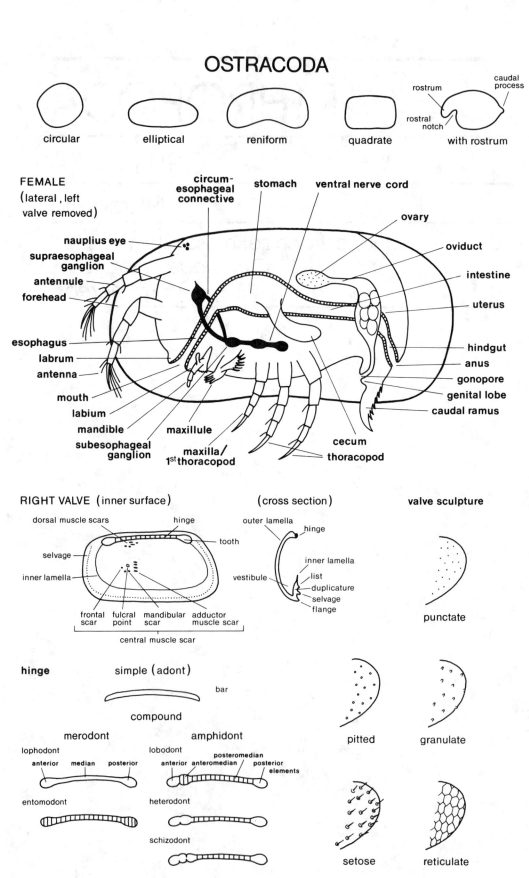

circular **elliptical** **reniform** **quadrate**

with rostrum
rostrum
caudal process
rostral notch

FEMALE (lateral, left valve removed)

circum-esophageal connective
stomach
ventral nerve cord
ovary
oviduct
intestine
uterus
nauplius eye
supraesophageal ganglion
antennule
forehead
esophagus
labrum
antenna
mouth
labium
mandible
subesophageal ganglion
maxillule
maxilla/1st thoracopod
cecum
thoracopod
hindgut
anus
gonopore
genital lobe
caudal ramus

RIGHT VALVE (inner surface)

dorsal muscle scars
hinge
tooth
selvage
inner lamella
frontal scar
fulcral point
mandibular scar
adductor muscle scar
central muscle scar

(cross section)

outer lamella
hinge
inner lamella
list
duplicature
selvage
flange
vestibule

valve sculpture

punctate

hinge

simple (adont)
bar

compound

merodont
lophodont
anterior median posterior
entomodont

amphidont
lobodont
anterior anteromedian
posteromedian
posterior elements
heterodont
schizodont

pitted
granulate
setose
reticulate

ARTHROPODA

CRUSTACEA

Ostracoda

ostracod, ostracode, musselshrimp, seed shrimp [G *Ostracode, Ostrakode, Muschelkrebs*]

abdomen [G *Abdomen, Hinterleib*]: Rudimentary region of body bearing caudal rami; indistinguishably fused to thorax.

accommodation groove [G *Ausweichfurche*]: Hinge structure; longitudinal groove above median hinge element. Receives dorsal edge of opposite valve.

adductor muscle [G *Adduktormuskel, Schließmuskel*]: Well-developed transverse muscle extending from body to inner surface of each valve. Forms characteristic muscle scars on valves and serves to close carapace.

adductor muscle scar [G *Schließmuskelansatz*]: muscle scar

aesthetasc [G *Aesthetasc*]: One in a series of small sensory projections along antennule or antenna; considered to be modified setae.

ala [G *flügelartige Erweiterung*]: Wing-like extension of valve, typically located ventrally and directed posteriorly. (simple, carinate)

antenna (second antenna) [G *Antenne, zweite Antenne*]: Second pair of antennae on head. Basically biramous, consisting of basal protopod and distal endopod and exopod (the latter often being reduced). Serves, along with antennules, as chief locomotory appendage. Carapace margin may be indented (rostral notch) for antenna. (biramous, uniramous)

antennal gland [G *Antennendrüse, Antennen-Nephridium*]: One of two excretory glands opening at base of antennae. (See also maxillary gland)

antennal muscle scar [G *antennaler Muskelansatz*]: muscle scar

antennal notch: rostral notch

antennule (first antenna) [G *Antennula, erste Antenne*]: First pair of antennae on head. Uniramous, typically consisting of five to eight segments. Bears setae and claws for swimming, digging, and copulation; bears one or more aesthetascs. Carapace margin may be indented (rostral notch) for antennules.

anterior element [G *vorderes Terminalelement*]: hinge

anus [G *After*]: Posterior opening of digestive tract at end of body (thorax/abdomen). May be located either dorsal or ventral to caudal rami.

aorta [G *Aorta, Kopfaorta*]: Unpaired blood vessel extending ventroposteriorly from heart. Separated from heart by anteroventral aortic valve and splits distally into several branches.

aortic valve [G *Klappe*]: Anteroventral opening in heart through which blood is pumped into aorta. (See also hepatic valve)

apophysis [G *Apophyse*]: Type of valve sculpture: extension of wall (murus) in reticulate pattern. May lead to pore.

area [G *Region*]: One in a number of somewhat arbitrarily delimited regions on lateral surface of valve. According to position one may distinguish anterior, anterodorsal, anteromedian, anteroventral, dorsal, dorsomedian, median, midanterior, middorsal, midposterior, posterior, posterodorsal, posteromedian, posteroventral, ventral, and ventromedian areas.

bar (hinge bar) [G *Schloßleiste*]: Hinge structure: ridge-like projection extending along and set off from dorsal margin of one valve. Fits into accommodation groove of opposite valve. (smooth, crenulate)

basis (basipodite) [G *Basis, Basipodit*]: Second segment of appendage; forms distal segment of protopod (coxa, basis). Often fused and indistinguishable.

beak: Downwardly directed (anteroventral) projection of carapace margin. Not equivalent to rostrum and not associated with rostral notch.

brain [G *Gehirn*]: supraesophageal ganglion

branchia: branchial plate, gill

branchial plate (vibratory plate, respiratory plate) [G *Atemplatte*]: Enlarged, flattened modification of outer branch (exopod) of either mandible, maxillule, maxilla, or first thoracic limb. Bears long setae along margin and maintains water circulation within carapace.

brush-shaped organ (brush-like structure) [G *bürstenförmiges Organ*]: Rudimentary third pair of thoracopods (ninth paired appendage of body). Consists of basal part (stem) bearing distal cluster of hairs. Typically restricted to ♂ and considered to be sensory.

bulb: Type of valve sculpture: relatively large, subspherical projection.

bulla: Type of valve sculpture: tubercle-like projection whose summit is elongated into a transverse or radial direction. May be considered to be type of verruca.

caperation: Type of valve sculpture: fine wrinkle.

carapace [G *Carapax*]: Shell-like outer covering; consists of pair of valves joined dorsally by hinge and entirely enclosing body. (calcareous, uncalcified; ornate, smooth; amplete, postplete, preplete)

carina [G *Carina*]: Type of valve sculpture: prominent ridge or keel. (disconnected, excavate, ponticulate = perforate, simple, undercut; dorsal, marginal, median, ventral)

caudal furca: caudal ramus

caudal process [G *caudaler Fortsatz*]: Posterior projection of valve margin. (posterodorsal, posteroventral and upwardly directed)

caudal ramus (furcal ramus) [G *Furkalast, Furca-Ast*]: One of two projections of posterior end (thorax/abdomen) of body; extends anteroventrally due to curvature of body. (paired, unpaired; symmetrical, asymmetrical; lamelliform, rod-shaped, claw-like)

caudal siphon: Posteroventral gape in valve margin; may be extended as tubular structure.

cecum (caecum, liver) [G *Divertikel, Leberschlauch*]: Pair of large outpocketings or one of numerous smaller outpocketings of anterior section (stomach) of midgut.

celation: Type of valve sculpture: elaboration (tegmen) of outer layer which overlaps and obscures underlying ornamentation.

cephalon: head

cephalothorax [G *Cephalothorax*]: Term occasionally applied to major part of body comprising fused head (cephalon) and thorax.

claw [G *Klaue*]: Any claw-shaped projection of limb or caudal ramus. (subterminal, terminal; smooth, toothed) (See also chela)

chela (pincer) [G *Zange*]: In maxilla of certain ♂, hook-like structure formed by segments of inner branch (endopod); used to hold ♀ during copulation. May also refer to pincer-like structure formed by middle segment of endopod in second (last) pair of thoracopods; used to clean food from maxillule.

circumesophageal connective [G *Schlundkonnektiv*]: One of two strands of nerve tissue surrounding esophagus and forming lateral part of circumesophageal nerve ring. Joins supraesophageal ganglion and complex formed by subesophageal ganglion and more or less concentrated ventral nerve cord.

circumesophageal nerve ring [G *Schlundring*]: Well-developed ring of nerve tissue surrounding esophagus; consists anteriorly of supraesophageal ganglion, laterally of circumesophageal connectives, and posteriorly of complex formed by subesophageal ganglion and more or less concentrated ventral nerve cord.

clava: Type of valve sculpture: tubercle-like projection elongated in longitudinal direction or parallel to margin. If arranged in a series, clavae may represent disconnected ridge (carina). (cruciform)

club-shaped organ: sense club

compound eye [G *Komplexauge, paariges Auge*]: One of two more complex anterodorsal organs of vision. Each is contained in ocular sinus beneath eye tubercle. (See also nauplius eye)

contact margin: margin

conulus [G *Conulus*]: Type of valve sculpture: conical spine.

costa [G *Rippe*]: Type of valve sculpture: prominent rib. Also used as synonym for carina.

coxa (coxopodite) [G *Coxa, Coxopodit*]: First, most proximal segment of appendage; typically forms protopod together with following segment (basis). Forms gnathobase in mandibles. (See also precoxa)

cuticle [G *Kutikula*]: Noncellular, multilayered, relatively thick layer forming valves, endoskeleton, and exoskeleton, and lining foregut and hindgut. Secreted by epidermis.

denticle [G *Zähnchen, Dentikel*]: Small, tooth-like projection on outer surface or hinge of valve.

dorsal muscle scar: muscle scar

dorsal process [G *Endbügel*]: Dorsal forked projection of basal segment of maxillule.

dorsum [G *Dorsum*]: In carapace, more or less flattened area to each side of hinge; delimited from lateral surface by angle in valve.

duplicature [G *verkalkter Teil der Innenlamelle*]: Along inner margin of valve, thickened, calcified section of inner lamella. Space between outer lamella and duplicature is termed vestibule. Medially directed surface of duplicature contacts opposite member when carapace is closed and may bear various ridges and grooves.

edge [G *Schalenrand*]: Outermost line of contact between two valves when carapace is closed (includes hinge region). Free edge refers to outermost line of contact of valves excluding hinge region.

ejaculatory duct [G *Ductus ejaculatorius*]: In ♂ reproductive system, modified distal end of each vas deferens. Typically associated with penis.

endite [G *Endit, Laden*]: Inwardly (medially) directed lobe of protopod, typically in maxillule, maxilla, and thoracopod.

endopod (endopodite) [G *Endopod, Endopodit, Innenast*]: Inner branch (ramus) of biramous appendage. Typically forms palp-like structure in mouthparts (e.g., mandibles, maxillules) or main component of thoracopods. (See also exopod)

endoskeleton [G *Endoskelett*]: Thin internal skeleton of body; forms complex framework supporting limbs and caudal rami.

epidermis [G *Epidermis*]: Outer cellular layer of body wall; secretes cuticle.

epipod (epipodite) [G *Epipodit*]: Laterally directed lobe of protopod. May also refer to modified exopod (branchial plate) of appendage.

esophagus [G *Ösophagus*]: Narrow, muscular section of digestive tract between mouth and midgut. Cuticle-lined and occasionally armed with ridges or modified into gastric mill. (See also foregut)

exopod (exopodite) [G *Exopod, Exopodit*]: Outer branch (ramus) of biramous appendage. Often developed as a branchial plate. (See also endopod)

eye: compound eye, nauplius eye

eye tubercle [G *Augenhöcker, Augenknoten*]: Protuberance on anterodorsal region of each valve; overlies ocular sinus and serves as lens for compound eye. (biconvex, concavo-convex, conical, hemispherical, tubular)

fenestra: Type of valve sculpture: one in a series of openings either in ridge (carina) or wall (murus) of reticulate pattern.

first antenna: antennule

first maxilla: maxillule

flange [G *Außenleiste*]: In closing apparatus of valve, ridge-like inner margin formed by projecting outer lamella. Fits into flange groove of opposite valve. (See also list, selvage)

flange groove: Groove-like depression along inner margin of valve. Separates major ridge (selvage) from flange on thickened part (duplicature) of inner lamella.

foregut [G *Vorderdarm*]: Anterior of three regions (foregut, midgut, hindgut) of

digestive tract. Cuticle-lined, corresponding to esophagus.

forehead [G *Stirn*]: Part of head anterior to mouth; bears antennules, antennae, and frontal organ.

fossa: Type of valve sculpture: single mesh in reticulate pattern (reticulum). Consists of central space (solum) and surrounding walls (muri).

foveola: Type of valve sculpture: minute pit (one order of magnitude smaller than punctum). (elliptical, irregular, rounded, stellate)

free edge: edge

free margin: margin

frill: Type of valve sculpture: wide, wing-like structure. Typically double-walled and extending beyond free edge of valves. (denticulate, nodulous, reticulate, septate, spinose, striate, undulating)

frontal filament: frontal organ

frontal muscle scar [G *Frontalnarbe*]: On inner wall of each valve, impression located anterior to adductor muscle scars. Formed by antennae.

frontal organ [G *Frontalorgan*]: Unpaired sensory structure located near eye on head; innervated by supraesophageal ganglion and typically developed as tentacle- or filament-like projection.

fulcral point [G *Mandibel-Stützfleck*]: On inner wall of each valve, impression formed at point at which (basal segment of) mandible pivots.

furrow: Type of valve sculpture: shallow groove.

gastric mill [G *Kauapparat*]: In certain ostracods, cuticular grinding apparatus in posterior part of esophagus.

gill (branchia) [G *Kieme, Branchie*]: In certain ostracods, term occasionally applied to lamellar respiratory process located dorsally on trunk.

genital lobe [G *Genitalhöcker*]: Lobe-like protuberance on posterior end of trunk in front of caudal rami. Bears gonopores.

gnathobase [G *Kaulade, Kaufortsatz*]: Basal part of mandible. Corresponds to coxa and consists of elongated projection whose toothed tip extends over mouth. Bears palp.

gonopore [G *Gonoporus, Geschlechtsöffnung*]: Opening of ♂ or ♀ reproductive system to exterior. In ♀, pore or larger opening (also termed uterine opening) of each oviduct on genital lobe. In ♂, paired or unpaired opening of vasa deferentia in front of caudal rami between or on penes. (See also vagina)

granule [G *Granula*]: Type of valve sculpture: minute protuberance. Often closely spaced.

hair [G *Härchen*]: Type of valve sculpture: filamentous projection whose diameter is one order of magnitude smaller than seta.

head (cephalon) [G *Kopf*]: Anterior of two major regions (head, thorax) of body; unsegmented and not clearly delimited from thorax. Bears five pairs of appendages (cephalic appendages), including antennules, antennae, mandibles, maxillules, and maxillae as well as eye. Entirely enclosed in carapace.

heart [G *Herz*]: In circulatory system of certain ostracods, relatively short, muscular pumping organ located anterodorsally behind eye. Receives blood from pericardium through pair of posterodorsal ostia and pumps it through anteroventral aortic valve and aorta as well as through posteroventral hepatic valves.

hepatic valve [G *Klappe*]: One of two posteroventral openings in heart through which blood is pumped to digestive tract. (See also aortic valve)

hindgut [G *Enddarm*]: Posterior of three regions (foregut, midgut, hindgut) of digestive tract. Relatively short, cuticlelined, and opening to exterior via anus.

hinge [G *Schloß*]: In carapace, structures along dorsal margin serving in articulation of two valves. If fully developed, consists of a series of teeth and ridges that fit into sockets and grooves of opposite valve; one may distinguish anterior, anteromedian, posteromedian, and posterior hinge elements. (adont, amphidont, archidont, desmodont, entomodont, heterodont, lobodont, merodont, schizodont, taxodont)

hinge margin [G *Schloßrand*]: In valve, dorsal margin adjoining hinge (as opposed to free = anterior, ventral, and posterior margins).

hinge selvage: In dorsal margin (hinge area) of valve, ridge-like elevation corresponding and occasionally continuous with selvage of anterior, ventral, and posterior margins (contact margin).

hingement: hinge

hypostome: labium

inner lamella [G *Innenlamelle, innere Lamelle*]: Relatively thin, lamella-like structure joined to outer wall (outer lamella) of valve. Secreted by epidermis; covers body except dorsally in hinge region. Junction with outer lamella along valve margin may be thickened to form duplicature. (membranous, calcareous)

intestine [G *Mitteldarmrohr*]: More narrow posterior section of midgut (stomach, intestine); opens into hindgut.

knee [G *Knie*]: Joint between segments of appendage at which limb forms an angle (e.g., between protopodite and endopodite of first thoracopod or between coxa and basis of antenna).

knob: Type of valve sculpture: prominent protuberance. Larger than node and differentiated from surrounding area by distinct angulation.

labium (lower lip, hypostome) [G *Unterlippe, Labium*]: Lip-like structure posterior to mouth on underside of head. May be bilobed, each lobe being termed paragnath.

labrum (upper lip) [G *Oberlippe, Labrum*]: Unpaired, lip-like structure in front of mouth. May contain salivary glands or glands which produce a luminescent secretion.

lamella: inner lamella, outer lamella

list [G *Innenleiste*]: In closing apparatus of valve, less prominent, ridge-like elevation along inner margin; located on thickened part (duplicature) of inner lamella, above major ridge (selvage). (See also flange, selvage groove)

liver: cecum

lobe [G *Lobus*]: Type of valve sculpture: rounded, major protuberance. According to position one may distinguish, e.g., anterior, median, and posterior lobes (designated L_1, L_2, and L_3). Typically separated by grooves (sulci). Also refers to lobes of appendages.

lower lip: labium

mandible [G *Mandibel*]: Third paired appendage of head. Mandibles flank mouth and are basically biramous, consisting of proximal gnathobase and distal palp.

mandibular scar [G *Mandibel-Narbe*]: On inner wall of each valve, one of typically two impressions formed by attachment of chitin rods supporting mandibles. Located anterior to adductor muscle scars.

margin (border) [G *Außenrand*]: Entire margin of carapace (or valve) when viewed from the side. Free margin refers to margin of carapace not held together by hinge (i.e., anteriorly, ventrally, posteriorly), while contact margin is restricted to that part of free margin of each valve that touches each other when carapace is closed. (See also edge)

maxilla (second maxilla, first thoracopod, fifth limb, maxilliped) [G *Maxille, zweite Maxille, erster Thoracopod, Kieferfuß*]: Fifth paired appendage of head. Serves, along with preceding mandibles and maxillules, as masticatory appendage or is modified for locomotion. Basically consists of a protopod (with endites) and variously developed endopod and exopod. (biramous, uniramous; maxilliform, pediform)

maxillary gland [G *Maxillendrüse, Maxillen-Nephridium*]: One of two excretory glands opening at base of maxillae. (See also antennal gland)

maxilliped: maxilla

maxillule (first maxilla, maxilla) [G *Maxillula, erste Maxille, Maxille*]: Fourth paired appendage of head; located posterior to mouth. Consists basically of proximal protopod with endites (masticatory processes) and variously developed distal branches (endopod, exopod). Serves, along with mandibles, as masticatory appendage or is modified for locomotion.

median element [G *medianer Schloßteil*]: hinge

midgut [G *Mitteldarm*]: Region of digestive tract between foregut and hindgut. Extends through most of body and may be divided into expanded stomach and narrower intestine. Bears ceca.

mouth [G *Mund, Mundöffnung*]: Anterior opening of digestive tract on underside of head. Bordered anteriorly by upper lip (labrum), posteriorly by lower lip (labium), and flanked by mandibles. Opens into esophagus.

murus: Type of valve sculpture: one element in a meshwork of walls forming a reticulate pattern. A single mesh in reticulum—consisting of central space (solum) and surrounding muri—is termed fossa. (caperate, excavate, foveolate, papillate, perforate, undercut)

muscle scar [G *Muskelnarbe*]: On inner wall of each valve, impression made by attachment of muscles. Largest and most conspicuous muscle scar (central or adductor muscle scar) is formed by muscles (closing or adductor muscles) spanning between valves. Additional muscle scars include those formed by antennules, antennae, mandibles, and other body muscles (dorsal muscle scars). Number and arrangement are of taxonomic importance. (See also frontal muscle scar, fulcral point)

nauplius [G *Nauplius*]: Early larval stage after hatching from egg. Bears three simple appendages (antennules, antennae, mandibles) and is covered by carapace.

nauplius eye [G *Naupliusauge*]: On dorsal surface of body near base of antennules, median unpaired photosensitive structure consisting of three ocelli. (See also compound eye)

node [G *Höcker, Knote*]: Type of valve sculpture: protuberance of intermediate size (between tubercle and knob). Node associated with attachment of different internal muscles is termed muscle scar node, subcentral tubercle, or submedian tubercle.

ocular sinus [G *Augensinus*]: Cavity in anterodorsal margin of each valve; contains compound eye and is overlaid by eye tubercle.

ornament: valve sculpture

ostium [G *Ostium*]: One of two posterodorsal openings in heart. Blood from pericardium enters heart through ostia and is pumped through anteroventral aortic valve and aorta as well as posteroventrally through hepatic valves.

outer lamella [G *Außenlamelle*]: Relatively thick outer wall of valve. Consists of a mineralized layer between two thinner, chitinous layers. (See also inner lamella)

ovary [G *Ovar, Ovarium, Eierstock*]: Section of ♀ reproductive system in which eggs are produced. Typically paired, located in posterior region of trunk, and opening to exterior via oviducts. (paired, unpaired)

oviduct [G *Ovidukt, Eileiter*]: Section of ♀ reproductive system leading from ovary to gonopore; may be developed distally into uterus. Seminal receptacle frequently opens into oviduct via canal (spiral canal). (paired, unpaired)

palp [G *Palpus, Taster*]: Distal part of mandible. Basically biramous (with endopod and exopod), although basis and endopod form its principal component.

papilla [G *Papille*]: Type of valve sculpture: small, steep-sided projection (one order of magnitude smaller than tubercle or spine).

paragnath [G *Paragnath*]: labium

penis [G *Penis*]: Complex, sclerotized copulatory structure in front of caudal rami on posterior end of trunk. Typically bears opening of vas deferens. (paired, unpaired)

pericardium [G *Perikard*]: In circulatory system, cavity surrounding heart. Divided into anterior and posterior sections (the former enclosing heart). Blood in pericardium enters heart through ostia.

pincer: chela

pit [G *Grube*]: Type of valve sculpture: relatively large, more or less circular depression. May also be applied as general term for fossae, puncta, etc.

plication: Type of valve sculpture: ridge or series of ridge-like folds involving entire valve wall.

podomere: segment

ponticulus: Type of valve sculpture: ridge (carina) modified in being perforated by a series of openings (fenestrae). (disconnected)

pore [G *Porus*]: Minute pore in outer surface of valve. Serves as opening of pore canal. (celate, funnel, intramural, rimmed)

pore canal [G *Porenkanal*]: One of variously developed, minute canals extending through valve; opens via pore. (false, normal, radial, sieve type)

posterior element [G *hinteres Terminalelement*]: hinge

precoxa [G *Präcoxa*]: Extra segment occasionally preceding coxa in protopod of appendage.

protopod (protopodite) [G *Protopodit*]: Proximal part of appendage; typically two-segmented (coxa, basis) and bearing variously developed inner branch (endopod) and outer branch (exopod). May bear endites.

pseudochaeta: Minute cuticular projection of body or limb. Lacks basal articulation, lumen, or innervation and therefore has only a mechanical function.

punctum: Type of valve sculpture: small, circular, pit-like depression (one order of magnitude larger than foveola).

pustula: Type of valve sculpture: small protuberance with pore at summit.

reticulum (reticulation) [G *Reticulum*]: Meshwork pattern consisting of numerous spaces or depressions (sola), each surrounded by walls (muri). A single mesh is termed fossa. (first order, second order; rounded, polygonal)

Rome's organ: In antennule, sensory organ on posterior border of second segment; derived from seta and may be bell-shaped.

rostral incisure [G *Rostralinzisur*]: rostral notch

rostral notch [G *Rostralinzisur*]: In anteroventral region of carapace, indentation of each valve below (posterior to) rostrum. Gaping, thus permitting protrusion of antennae when valves are closed (gape itself occasionally referred to as rostral incisure).

rostrum [G *Rostrum*]: Anterior gaping, beak-like extension of carapace. Serves, along with associated rostral notch, to permit protrusion of antennae when valves are closed. (See also beak)

second antenna: antenna

second maxilla: maxilla

segment (podomere) [G *Segment, Glied*]: One in a series of units of appendage (e.g., coxa, basis) or of branch (endopod, exopod) of appendage.

selvage [G *Saum*]: In closing apparatus of valve, ridge-like elevation along inner margin; fits into groove (selvage groove) of opposite valve. If margin is simple, selvage represents only ridge; if margin is complex (i.e., inner lamella with duplicature), selvage represents middle or principal ridge. (See also flange, hinge selvage, list)

selvage groove: In closing apparatus of valve, groove-like depression along inner margin. Separates major ridge (selvage) from minor ridge (list).

seminal receptacle [G *Receptaculum seminis*]: In ♀ reproductive system, one of two sperm-receiving organs. Each typically opens to exterior via vagina and is connected to end of oviduct by canal (spiral canal).

seminal vesicle [G *Vesicula seminalis, Samenblase*]: In ♂ reproductive system, enlargement of vas deferens for storing sperm.

sense club (club-shaped organ) [G *Riechborste*]: aesthetasc

seta [G *Borste*]: Innervated cuticular projection of body or appendage. Typically bipartite, being divided into proximal base and distal shaft by annulation. With both sensory and mechanical functions. (chelate, compound, pappose, plumed, plumose, serrate, simple) (See also hair, pseudochaeta)

setule: One in a series of minute secondary projections on seta. (flexible, suctorial, tooth-like)

shell: carapace

sieve plate [G *siebförmiger Porus*]: Minute, perforated plate covering pore in valve; may be associated with seta.

socket [G *Zahngrube*]: Hinge structure: pit-like (typically anterior or posterior) depression in hinge margin. Receives tooth of opposite valve. Major depression at end of hinge is termed cardinal socket; those with subdivided floors are termed compound sockets.

solum: Type of valve sculpture: in single mesh (fossa) or reticulate pattern, central space or depression surrounded by walls (muri).

sperm duct: vas deferens

spine [G *Stachel*]: Type of valve sculpture: elongate, larger, typically pointed projection. (clavellate, conjunctive, disjunctive, mamillate, marginal, perforate)

spinneret [G *Spinndrüse, Schalendrüse*]: Gland, located in forehead and opening to exterior via seta-like exopod of antenna. Secretes thread-like material used in locomotion.

spiral canal [G *Ausführungsgang*]: In ♀ reproductive system, very long, coiled tube connecting seminal receptacle to oviduct.

spur: Type of valve sculpture: flattened, spine-like projection. Represents modified, wing-like (velate) structure.

standard measurements:
length: greatest longitudinal dimension of valve or carapace, measured either parallel to hinge or between furthest anterior and posterior extremities
height: greatest dorsal to ventral dimension of valve or carapace, measured perpendicular to length
width: greatest dimension from outer surface of one valve to outer surface of opposing valve
thickness: distance from outer to inner surface of a valve
All values may be expressed in absolute and relative terms.

stomach [G *Magen*]: Expanded anterior section of midgut (stomach, intestine). May bear single pair of large ceca or numerous smaller ceca.

stria [G *Ritze*]: Type of valve sculpture: very fine groove.

subcentral tubercle (submedian tubercle): node

subesophageal ganglion [G *Unterschlundganglion*]: Concentration of nerve tissue posterior to esophagus. Fused to more or less concentrated ventral nerve cord to form posterior part of circumesophageal nerve ring.

sulcus [G *Sulcus*]: Type of valve sculpture: relatively prominent groove, typically running dorsoventrally. May be termed S_1, S_2, and S_3 and be separated by lobes. (deep, shallow; bisulcate, trisulcate, unisulcate; anterior, median, posterior, postocular)

supraesophageal ganglion (brain) [G *Oberschlundganglion, Gehirn*]: Concentration of nerve tissue anterior to esophagus and innervating, among others, compound eyes. Fused with circumesoph-

ageal connectives to form circumesophageal nerve ring.

tegmen: Elaboration of outer layer of valve which overlaps and obscures underlying ornamentation.

tegumental gland: One of numerous glands located within or below epidermis; opens to exterior via tegumental duct.

terminal element [G *Terminalelement*]: hinge

testis [G *Hoden*]: Section of ♂ reproductive system in which sperm are produced. Located in posterior region of trunk or between body and valve and opening to exterior via vasa deferentia. (paired, unpaired; coiled, elongate, four-lobed = quadripartite, lobed, simple, spheroid, tube-shaped)

thoracopod (thoracic appendage, trunk appendage) [G *Thoracopod, Thoraxbein*]: Sixth and seventh paired appendages of body; represents one of two pairs of limbs originating from (not clearly delimited) thoracic region. Basically consists of protopod, endopod, and exopod, although the latter may be modified as branchial plate or be reduced. (biramous, uniramous; lamelliform, multiarticulate, pediform, vermiform; locomotory, nonlocomotory) (See also brush-shaped organ)

thorax [G *Thorax*]: Posterior of two major regions (head, thorax) of body; unsegmented and not clearly delimited from head. Typically bears two pairs of appendages (thoracopods) and is indistinguishably fused with rudimentary abdomen. Entirely enclosed in carapace.

tooth (hinge tooth) [G *Zahn, Schloßzahn*]: Hinge structure: tooth-like (typically anterior or posterior) element projecting beyond hinge margin. Fits into socket of opposite valve. (bifid, conical, elongate, hemispherical, lobate, multilobate, reniform, simple, smooth, trilobate)

trunk [G *Rumpf*]: Region of body posterior to head; corresponds to thorax (and rudimentary abdomen). Unsegmented, typically bearing two pairs of appendages (thoracopods) and caudal rami. Entirely enclosed in carapace.

trunk appendage (postcephalic appendage) thoracopod

tubercle [G *Tuberkel, Buckel*]: Type of valve sculpture: fairly prominent, rounded projection. Larger than granule, smaller than knob. (simple, multifurcate) (See also eye tubercle, subcentral tubercle)

turret: Type of valve sculpture: large but short, spine-like projection whose summit bears series of smaller spines around its periphery. May be considered to be type of verruca.

upper lip: labrum

uterine opening: gonopore

uterus [G *Uterus*]: In ♀ reproductive system, expanded terminal section of oviduct in which eggs are stored. Opens via gonopore (uterine opening) to exterior and may be connected to seminal receptacle by canal.

vagina [G *Vagina, Begattungsporus, Begattungskanal*]: In ♀ reproductive system, large space leading to a seminal receptacle. Located next to gonopores (openings of oviduct = uterine openings) at posterior end of trunk.

valve: aortic valve, hepatic valve

valve [G *Carapaxklappe*]: One of two shell-like elements covering left and right sides of body (therefore left valve, right valve). Valves joined to each other dorsally by hinge and may bear complex system of ridges and grooves around margin that interfit when carapace is closed. Consists of inner and outer lamellae. (equal, subequal, unequal; amplete, postplete, preplete; lateral outline: circular, elliptical, elongate, ovate, quadrate, reniform; smooth, ornamented: e.g., alate, bilobate, bisulcate, bullate, caperate, carinate, celate, clavate, costate, foveolate, granuloreticulate, granulose, lobate, papillate, pitted, punctate, quadrilobate, reticulate, setose, spinose, striate, sulcate, trilobate, trisulcate, tuberculate, unilobate, unisulcate, verrucose) (See also valve sculpture)

valve sculpture (ornament)

[G *Skulptur*]: General term applied to any type of elevation or depression on exterior of valve; includes individual structures and continuous surface features.

1. individual structures
   depressions: fossa, foveola, pit, punctum
   elevation: bulb, bulla, clava, conulus, denticle, granule, knob, lobe, node, papilla, pustule, spine, tubercle, turret, verruca
2. continuous structures
   depressions: furrow, groove, stria, sulcus
   elevations: ala, caperation, carina, costa, murus, plication, velum

Size, shape, and occasionally position are important features distinguishing the various structures. Certain terms may be preferentially used for specific taxonomic groups.

vas deferens (sperm duct)

[G *Vas deferens, Samenleiter*]: Section of ♂ reproductive system leading from each testis to gonopore. May be expanded to form seminal vesicle, and distal end may be modified as an ejaculatory duct.

velum [G *Velum*]: Type of valve sculpture: wide, wing-shaped structure.

ventral nerve cord [G *Bauchmark*]: Paired, longitudinal nerve cord extending through trunk; bears pairs of ganglia near base of appendages. Typically more or less concentrated and forming, along with subesophageal ganglion, posterior part of circumesophageal nerve ring.

verruca: Type of valve sculpture: small, wart-like outgrowth. (See also bulla, turret)

vestibule [G *Vestibulum*]: Along inner margin of valve, space formed between outer lamella and thickened part (duplicature) of inner lamella.

vibratory plate: branchial plate

Zenker's organ [G *Zenker'sches Organ*]: In ♂ reproductive system, complex type of ejaculatory duct at end of vas deferens; serves as pump for elimination of giant sperm.

ARTHROPODA

CRUSTACEA

Mystacocarida

mystacocarid [G *Mystacocaride*]

MYSTACOCARIDA

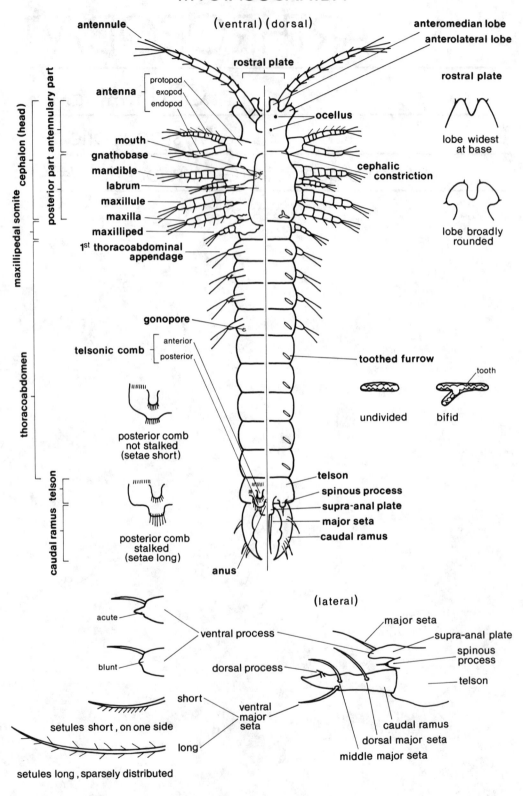

ARTHROPODA

CRUSTACEA

Mystacocarida

mystacocarid [G *Mystacocaride*]

abdomen [G *Abdomen*]: According to interpretation dividing body into cephalon, thorax, and abdomen, last five somites of body. Lacks appendages yet bears telson and caudal furca.

aesthetasc (aesthete) [G *Aesthetasc*]: Small sensory projection at tip of each antennule.

anal somite: telson

antenna (second antenna) [G *Antenne, zweite Antenne*]: Second and smaller pair of antennae. Biramous, consisting of basal protopod bearing nine-segmented exopod and four-segmented endopod. Serves in locomotion and feeding.

antennulary part (antennulary portion) [G *Kopfschild*]: Region of head (cephalon) anterior to cephalic constriction. Bears antennules, four ocelli, two anteromedian lobes (rostral plate), and two anterolateral (lateral) lobes. (posterolateral corner: obtuse, right-angled)

antennule (first antenna) [G *Antennula, erste Antenne*]: First and larger pair of antennae. Uniramous, consisting of eight setose segments and occasionally bearing terminal aesthetasc.

anterolateral lobe (lateral lobe): lobe

anteromedian lobe (anteromedial lobe): lobe

anus [G *After*]: Posterior opening of digestive tract; located on ventral surface of telson (anal somite).

basis: protopod

brain [G *Gehirn*]: Main concentration of nerve tissue in cephalon. Consists of anterior protocerebrum, median deutocerebrum (ganglion of antennular somite), and, at some distance posterior to deutocerebrum, a tritocerebrum (ganglion of antennal somite). Surrounds esophagus and gives rise to pair of ventral nerve cords.

caudal furca: caudal ramus

caudal ramus (furcal ramus, furcal claw) [G *Furkalast, Furca-Ast*]: One of two relatively large, posteriorly directed projections of last segment (telson). Claw-shaped, with serrate inner margin and bearing several major setae. Both rami together are termed caudal furca.

cephalic constriction: Transverse groove dividing head (cephalon) into anterior (antennulary) and posterior parts.

cephalon (head) [G *Kopf*]: Anterior of three basic divisions of body (according to interpretation, either cephalon, maxillipedal somite, thoracoabdomen, or cephalon, thorax, abdomen). Relatively large and divided into anterior (antennulary) and posterior parts by groove (cephalic constriction). Bears antennules, antennae, mandibles, maxillules, and maxillae.

comb: telsonic comb

coxa: protopod

cuticle [G *Kutikula*]: Noncellular, multilayered structure forming outer layer of body and lining foregut (stomodeum) and hindgut (proctodeum); relatively thin and pigment-free. Secreted by epidermis.

deutocerebrum: brain

endite [G *Endit*]: Inwardly (medially) directed lobe of protopod of maxillule, maxilla, and maxilliped. (See also gnathobase)

endopod (endopodite) [G *Endopod, Endopodite*]: Inner branch of biramous appendage. Refers to four-segmented branch of antenna and mandible or three-segmented branch of maxilliped. Single branch of uniramous appendage (e.g., maxilla) may also be interpreted as representing an endopod. (See also exopod)

esophageal connective [G *Schlundkonnektiv*]: One of two strands of nerve tissue extending from posterior part of brain (tritocerebrum) around esophagus.

esophagus [G *Ösophagus*]: Anterior section of foregut (stomodeum). From mouth, esophagus curves forward, then upward and backward. Surrounded laterally by posterior part of brain (tritocerebrum and esophageal connectives). May refer to entire foregut.

exopod (exopodite) [G *Exopod, Exopodit*]: Outer branch of biramous appendage. Refers to nine-segmented branch of antenna, seven-segmented branch of mandible, and one-segmented branch of maxilliped. (See also endopod)

eye: ocellus

first maxilla: maxillule

foregut: stomodeum

frontal organ [G *Frontalorgan*]: One of several sensory structures represented by

setae of anteromedian lobes (rostral plate) of cephalon.

furcal claw: caudal ramus

furrow: toothed furrow

gnathobase [G *Kauladen*]: Projection of base (protopod) of each mandible. Extends under labrum and serves in macerating food.

gonopore [G *Gonoporus, Geschlechtsöffnung*]: Single opening of ♂ or ♀ reproductive system; located on right side of ventral surface (sternite) of third thoracoabdominal somite.

head: cephalon

hindgut: proctodeum

labral ganglion (ganglion labrii) [G *Ganglion labrii*]: Large ganglion adjoining posterior end (tritocerebrum) of brain and innervating labrum.

labrum (upper lip) [G *Labrum, Oberlippe*]: Large, bulbous, unpaired lobe covering mouth. Contains large gland and labral ganglion and extends over gnathobases of mandibles. (distal = posterior end: gradually expanded, abruptly expanded)

lateral lobe: lobe

lobe [G *Lappen, Stirnlappen*]: One of four (two anteromedian and two anterolateral = lateral) extensions of anterior (antennulary) part of head (cephalon). Anteromedian lobes occasionally termed rostral plate. Configuration of lobes is of taxonomic importance.

mandible [G *Mandibel*]: Third paired appendage of cephalon; located between antennae and maxillules. Biramous, consisting of protopod, four-segmented endopod, and seven-segmented exopod. Projection (gnathobase) of protopod extends under labrum. Serves in locomotion and feeding.

maxilla (second maxilla) [G *Maxille, zweite Maxille*]: Fifth paired appendage of cephalon; located between maxillules and maxillipeds. Uniramous, consisting of protopod (with endites) and four-segmented shaft. Serves in locomotion and feeding.

maxilliped (maxillipede) [G *Maxilliped, Kieferfuß*]: Sixth paired appendage of body. Typically biramous, consisting of protopod (with endites) bearing three-segmented endopod and one-segmented exopod. Serves in feeding. (biramous, uniramous; with exopod, without exopod)

maxillipedal somite [G *maxillipedtragendes Segment*]: According to interpretation dividing body into cephalon, maxillipedal segment, and thoracoabdomen, one-segment tagma bearing maxillipeds. May also be considered as belonging to cephalon or thorax.

maxillule (first maxilla) [G *Maxillula, erste Maxille*]: Fourth paired appendage of cephalon; located between mandibles and maxillae. Uniramous, consisting of elongate protopod (with endites) and four-segmented shaft. Serves in locomotion and feeding.

metanauplius [G *Metanauplius*]: One of several substages in development between nauplius and adult. Characterized by gradual increase in number of segments and appendages.

midgut [G *Mitteldarm*]: Region of digestive tract between foregut (stomodeum) and hindgut (protocdeum). Extends throughout most of thoracoabdomen and lacks diverticula; with folded, relatively thick walls anteriorly, thinner walls posteriorly.

mouth [G *Mund, Mundöffnung*]: Anterior opening of digestive tract. Located under labrum; opens into esophagus.

nauplius eye: ocellus

nauplius larva [G *Nauplius-Larve*]: Larval stage at which most mystacocarids hatch. Characterized by presence of only three appendage pairs (antennules, antennae, mandibles). (See also metanauplius)

ocellus (eye, nauplius eye) [G *Ocellus, Naupliusauge*]: One of four photosensitive organs located dorsally on an-terior (antennulary) part of cephalon (head). (with/without lens)

ovary [G *Ovar, Ovarium, Eierstock*]: Unpaired section of ♀ reproductive system in which eggs are produced. Considered to consist of ovary proper and nutritive section ("yolk gland"). Extends throughout thoracoabdomen and opens to exterior via oviduct.

oviduct [G *Ovidukt, Eileiter*]: Unpaired, anteriorly directed section of ♀ reproductive system between ovary and gonopore.

precoxa: protopod

proctodeum (hindgut) [G *Proctodaeum, Enddarm*]: Posterior cuticle-lined region of digestive tract; extends through last three to four segments and opens to exterior via anus. (See also stomodeum)

protocerebrum: brain

protopod (protopodite) [G *Protopod, Protopodit*]: Proximal part of appendage. Generally specified only when referring to antennae, mandibles, maxillules, maxillae, and maxillipeds. Considered to consist of several indistinct (fused) segments (e.g., precoxa, coxa, basis). Bears gnathobase in mandibles and endites in maxillules, maxillae, and maxillipeds.

rostral plate [G *Rostralplatte*]: Term applied to two anteromedian lobes of anterior (antennulary) part of head (cephalon).

segment [G *Segment*]: One in a series of units of appendage. May also refer to segment of body (then also termed somite).

seta [G *Borste*]: One of numerous bristle-like projections extending from appendage or body. Number and type of setae are of taxonomic importance. Large setae on supra-anal plate and caudal rami are termed major setae. According to shape one may distinguish, e.g., simple and brush setae. (See also setule)

setule: One of numerous minute projections of a seta. (long, short; distribution: on one side only, sparse)

somite (segment) [G *Segment, Metamer*]: One in a series of divisions of body. Cephalon consists of five (fused) somites, followed by maxillipedal somite and nine thoracoabdominal somites. Last thoracoabdominal somite bears telson ("anal somite").

spermatophore [G *Spermatophor*]: Packet of sperm formed in vas deferens.

standard measurements:
body length (anterior tip of cephalon to base of supra-anal plate of telson)
length of labrum, antennule, caudal ramus
greatest width of antennulary part of cephalon
absolute and relative lengths of major setae
setal formula (number of setae per appendage segment)

sternite [G *Sternit*]: Ventral surface of body segment (somite). (See also tergite)

stomodeum (stomodaeum, foregut) [G *Stomodaeum, Vorderdarm*]: Anterior cuticle-lined region of digestive tract. Relatively long, extending through cephalon. Anterior section or entire region may be termed esophagus. (See also proctodeum)

supra-anal plate (supra-anal process) [G *Supra-Analklappe, Analoperculum*]: Unpaired median projection extending from posterodorsal end of last segment (telson). Posterior end characterized by large seta (major seta) and, ventral to it, variously shaped process with minute spines. Flanked on each side by spinous process.

tagma [G *Tagma*]: Major division of body. According to interpretation, one may distinguish either head (cephalon), maxillipedal somite, and thoracoabdomen or head, thorax, and abdomen.

telson [G *Telson*]: Last segment of body with telsonic combs ventrally, supra-anal plate posterodorsally, and caudal furca. Bears anus ventrally and therefore occasionally termed anal somite.

telsonic comb (comb, ventral comb) [G *Kamm*]: One of two pairs of comb-like cuticular structures on ventral surface of last segment (telson). Anterior and posterior pairs differ in configuration and tooth number. (stalked, not stalked)

tergite [G *Tergit*]: Dorsal surface of body segment (somite). Bears pair of toothed furrows laterally.

testis [G *Hoden*]: Unpaired section of ♂ reproductive system in which sperm are produced; extends throughout posterior region of thoracoabdomen. Bears several pairs of dorsal diverticula and opens to exterior through anteriorly directed vas deferens.

thoracoabdomen [G *Thoracoabdomen*]: According to interpretation dividing body into cephalon, maxillipedal somite, and thoracoabdomen, tagma including last nine somites as well as telson and caudal furca. First four thoracoabdominal somites bear appendages (thoracoabdominal appendages).

thoracomere [G *Thoracomer*]: According to interpretation dividing body into cephalon, thorax, and abdomen, one of five segments (somites) of thorax, the first bearing maxillipeds and the remaining bearing thoracopods.

thoracopod [G *Thoracopod, Thoraxbein*]: According to interpretation dividing body into cephalon, thorax, and abdomen, one of two appendages of each thoracic somite (thoracomere). Thus, first thoracopod represents maxilliped and the following four thoracopods are reduced to a single flap-like structure with two large terminal setae.

thorax [G *Thorax*]: According to interpretation dividing body into cephalon, thorax, and abdomen, five-segmented tagma including maxilliped-bearing (maxillipedal) somite and following four thoracopod-bearing somites (thoracomeres).

toothed furrow (furrow, lateral toothed furrow) [G *Streifenorgan*]: Pair of elongate, toothed depressions, one in dorsolateral surface of each thoracoabdominal somite, in maxillipedal somite (maxillipedal lateral toothed furrow), and in posterior region of cephalon (maxillary lateral toothed furrow). Number of teeth may be of taxonomic importance. (undivided, bifid)

tritocerebrum: brain

trunk [G *Rumpf*]: Entire region of body following head (cephalon); according to interpretation, may be subdivided into thorax and abdomen or maxillipedal somite and thoracoabdomen.

upper lip: labrum

uropod [G *Uropod*]: Term applied to each caudal ramus if supra-anal plate is interpreted as representing telson.

vas deferens [G *Vas deferens*]: Unpaired, anteriorly directed section of ♂ reproductive system between testis and gonopore. Produces spermatophores.

ventral nerve cord (ventral nerve chain) [G *ventraler Nervenstrang*]: Pair of longitudinal nerve cords extending from brain to posterior end of body; bears series of ganglia.

yolk gland: ovary

COPEPODA

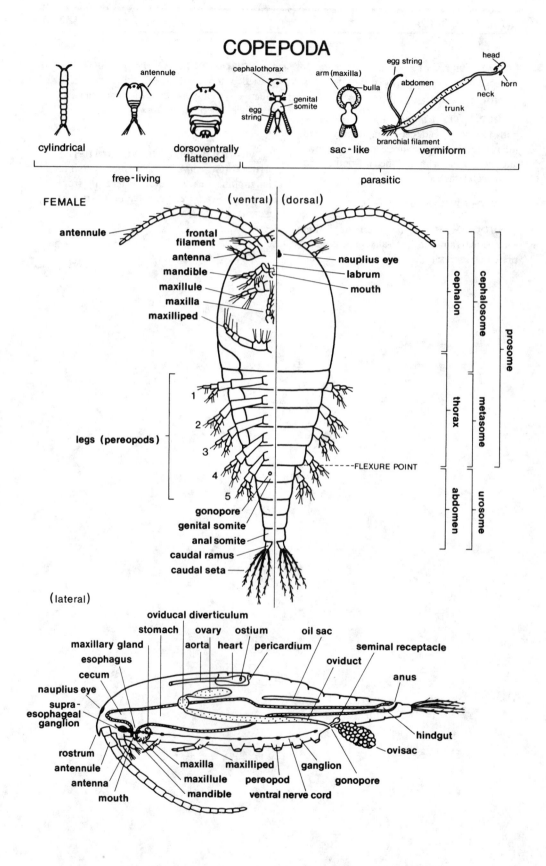

cylindrical

antennule

cephalothorax

genital somite

egg string

dorsoventrally flattened

arm (maxilla)

bulla

sac-like

egg string

abdomen

head

horn

neck

trunk

branchial filament

vermiform

free-living

parasitic

FEMALE

(ventral) (dorsal)

antennule

frontal filament

antenna

mandible

maxillule

maxilla

maxilliped

nauplius eye

labrum

mouth

legs (pereopods)

1

2

3

4

5

FLEXURE POINT

gonopore

genital somite

anal somite

caudal ramus

caudal seta

cephalon

cephalosome

thorax

metasome

abdomen

urosome

prosome

(lateral)

oviducal diverticulum

stomach

ovary

ostium

oil sac

maxillary gland

aorta

heart

pericardium

esophagus

seminal receptacle

cecum

oviduct

nauplius eye

anus

supra-esophageal ganglion

rostrum

antennule

maxilla

maxilliped

ganglion

hindgut

antenna

maxillule

pereopod

gonopore

ovisac

mouth

mandible

ventral nerve cord

ARTHROPODA

CRUSTACEA

Copepoda

copepod [G *Copepode, Ruderfußkrebs*]

abdomen (pleon) [G *Abdomen, Pleon, Hinterleib*]: Posterior and typically narrowest of three basic divisions of body (cephalon, thorax, abdomen). Consists of five somites (pleomeres), the first often being termed genital somite, the last termed anal somite (with posterior caudal rami). Typically lacks appendages. (symmetrical, asymmetrical; segmented, unsegmented)

aesthetasc (esthetasc, esthete) [G *Aesthetasc*]: One in a series of small sensory projections along antennule.

ala [G *Seitenlappen, Lappen*]: In parasitic copepod, one of two posterolateral extensions of cephalic shield.

anal operculum [G *Analdeckel, Analklappe*]: Lid-like structure covering anus. (smooth, toothed)

anal somite [G *Analsegment*]: Last (fifth) segment of abdomen; bears anus dorsally and pair of caudal rami posteriorly. Occasionally termed telson.

antenna (second antenna) [G *Antenne, zweite Antenne*]: Second and typically shorter pair of antennae on underside of head. Highly setose and serving in suspension feeding or, in certain parasitic copepods, with terminal claw and serving in attachment to host. If uniramous, outer branch (exopod) is missing. The term second

antenna is often preferentially applied. (biramous, uniramous; prehensile)

antennal gland [G *Antennendrüse, Antennennephridium*]: Excretory organ, one opening at base of each antenna. Characteristic for larval stages and certain parasitic copepods. (See also maxillary gland)

antennule (first antenna) [G *Antennula, erste Antenne*]: First and typically larger pair of antennae on underside of head. Uniramous, consisting of up to 26 articles. In ♂, one or both may be equipped with hinge joint for grasping ♀ during copulation. Bears aesthetascs. Number of articles is of systematic importance. The term first antenna is often preferentially applied. (geniculate)

antrum: atrium

anus [G *After*]: Posterior opening of digestive tract. Located terminally, on dorsal surface of last abdominal segment (anal somite); covered by anal operculum.

aorta (anterior aorta [G *Aorta, Kopfaorta, Aorta anterior*]: Unpaired dorsal blood vessel extending anteriorly from heart.

arm [G *Klammerarm*]: Term applied to elongate, arm-like maxilla of certain parasitic copepods. Fuses with opposite member to form distal attachment disc (bulla).

article (segment, joint) [G *Glied*]: One in a series of up to 26 segments of flagellum of antennule (fewer in antenna). Number of articles is of systematic importance. May also be used to refer to segments of other appendages.

articulation: body articulation

atrium (genital atrium, antrum [G *Atrium*]: In ♀ reproductive system, invagination of ventral surface (sternite) of first abdominal segment (genital somite). Receives common opening of both oviducts and openings of seminal receptacles.

basis (basipodite) [G *Basis, Basipodit*]: Second segment of appendage; forms distal segment of protopod (coxa, basis).

body articulation (body flexure, flexure point) [G *Körperhauptgelenk, Rumpfgelenk*]: Point at which body is capable of bending to greatest extent. Region of body anterior to flexure point is termed prosome, that posterior to it, urosome. According to group, it is located either between third/fourth, fourth/fifth, fifth/sixth thoracic somites or between sixth thoracomere and genital somite.

brain [G *Gehirn*]: supraesophageal ganglion

branchial filament [G *respiratorischer Anhang*]: In parasitic copepod, one element in tuft-like aggregations of slender processes of abdomen.

buccal chamber [G *Mundhöhle, präoraler Raum*]: In parasitic copepod, term applied to space formed around mouth by prolonged upper lip (labrum) and lower lip (labium). (See also mouth cone)

bulla (button) [G *Bulla, Haftscheibe*]: In parasitic copepod, attachment disc formed at fused end of both arm-like maxillae. (club-shaped, elliptical, spherical)

caudal ramus (furcal ramus) [G *Furkalast, Furca-Ast*]: One of two posteriorly directed projections of last abdominal segment (anal somite); typically bears a number of setae (caudal setae). (length greater than, equal to, or less than width; margins: hairy, smooth, spinulose)

caudal seta: seta

cecum (midgut cecum, anterior midgut cecum) [G *Caecum, Coecum, vorderes Mitteldarmdivertikel*]: Unpaired anterior extension of stomach. (See also lateral midgut diverticulum)

cephalic shield (head shield, dorsal shield) [G *Kopfschild*]: Dorsal cuticular shield covering head and thoracic somites (thoracomeres) fused to it. May project anteriorly to form rostrum, posteriorly to form pair of alae. (convex, flat, regionated)

cephalon (head) [G *Kopf, Cephalon*]: Anterior of three basic divisions of body (cephalon, thorax, abdomen). Consists of five (cephalic) somites and bears antennules, antennae, labrum, mouth, labium, mandibles, maxillules, and maxillae. Fused with one or more thoracic somites (thoracomeres) to form cephalosome.

cephalosome [G *Cephalosoma*]: One of three divisions (tagmata) of body (cephalosome, metasome, urosome); consists of head and one or two thoracic somites (thoracomeres) fused to it. Cephalosome and metasome together form prosome. (See also cephalothorax)

cephalothorax [G *Cephalothorax*]: Anterior region of body including head (cephalon) and thoracic somites (thoracomeres) fused to it. Typically applied only to parasitic copepod. (elongate, elongate oval, oval, pyriform, triangular)

circumesophageal connective (esophageal connective) [G *Schlundkonnektiv*]: One of two strands of nerve tissue surrounding esophagus and connecting supraesophageal ganglion to ventral nerve cord.

copepodid (copepodid stage) [G *Copepodit, Copepoditstadium*]: Second (and more advanced) of two basic stages in larval development (nauplius, copepodid). Typically involves five substages and is characterized by excretion through maxillary glands and a progressive increase in number of body segments and posterior appendages.

copulatory pore [G *Befruchtungsöffnung, Begattungsporus*]: In ♀ reproductive system, pore on ventral surface (sternite) of first abdominal segment (genital somite). Opens into seminal receptacle and thus represents, along with two gonopores, third opening of reproductive system.

coxa (coxopodite) [G *Coxa, Coxopodit*]: First, most proximal segment of appendage. Typically forms protopod together with following segment (basis). Generally specified only in maxillules (where it bears endites and exites) and pereopods (coxae of opposed limbs united).

cuticle [G *Kutikula*]: Noncellular, multilayered, relatively thick layer forming exoskeleton and lining both foregut and hindgut. Secreted by epidermis. Forms plate-like structures (cephalic shield, sternites, tergites).

diverticulum: lateral midgut diverticulum, oviducal diverticulum

egg sac: ovisac

egg string [G *Eischnur, Eischlauch*]: Elongate ovisac characteristic of certain parasitic copepods.

ejaculatory duct [G *Ductus ejaculatorius*]: In ♂ reproductive system, posteriormost section of vas deferens; opens to exterior via gonopore on first abdominal segment (genital somite).

endite [G *Endit, Laden*]: Inwardly (medially) directed lobe of proximal part (protopod) of maxillule. Used to clean filtering apparatus in suspension-feeding copepod. (See also exite)

endopod (endopodite) [G *Endopod, Endopodit, Innenast*]: Inner branch (ramus) of biramous appendage. Refers to inner, main, or only branch of antenna and maxilliped, as well as to inner branch of mandible, maxillule, and pereopod. (See also exopod)

endoskeleton [G *Innenskelett, Endoskelett*]: In head, weakly developed internal skeleton consisting of two endosternites.

endosternite [G *Endosternit*]: In head, one of two transverse skeletal elements arising from ventral surface (sternite) and forming internal skeleton (endoskeleton). Serves as site of attachment for muscles of antennae and mouthparts, one being located directly behind esophagus, the second in region of maxilla.

esophageal connective: circumesophageal connective

esophagus [G *Ösophagus*]: Narrow section of digestive tract between mouth and stomach; cuticle-lined, muscular, and basically corresponding to foregut.

exite [G *Exit*]: Outwardly (laterally) directed lobe of proximal part (protopod) of maxillule. (See also endite)

exopod (exopodite) [G *Exopod, Exopodit, Außenast*]: Outer branch (ramus) of biramous appendage. Refers to outer branch of antenna (occasionally reduced), mandible, maxillule, and pereopod. (See also endopod)

fertilization duct [G *Befruchtungsgang*]: In ♀ reproductive system, narrow duct connecting median seminal receptacle to end of each oviduct.

first antenna: antennule

first maxilla: maxillule

flagellum [G *Flagellum, Geißel*]: Multiarticulate distal section of antennule.

foregut [G *Vorderdarm, Stomodaeum*]: Anterior of three regions (foregut, midgut, hindgut) of digestive tract. Cuticle-lined, corresponding to esophagus (and occasionally including anterior part of stomach).

frontal filament [G *Frontalfaden, Rostralfilament*]: One of two slender projections extending from rostrum of free-living copepod. With sensory function.

frontal plate [G *Frontalplatte*]: In parasitic copepod, plate-like structure at anterior end of head; formed by fusion of frontal margin with bases of antennules. Bears pair of suckers (lunules).

furca: caudal ramus, sternal furca

genital somite (genital complex) [G *Genitalsegment, Genitalsomit*]: Segment of body bearing gonopores. Typically corresponds to first abdominal segment, yet also to fused last thoracic and first abdominal somites. May be enlarged in parasitic ♀.

gnathobase (blade) [G *Kauladen*]: Basal part of mandible. Basically consists of elongated projection whose toothed tip extends over mouth. Bears palp (with endopod and exopod).

gonad: ovary, testis

gonopore (genital pore) [G *Gonoporus, Geschlechtsöffnung, Genitalöffnung*]: Opening of reproductive system to exterior on ventral surface (sternite) of first abdominal segment (genital somite). In ♂, one or two openings, in ♀, either paired or in the form of a common atrium.

head: cephalon

head shield: cephalic shield

heart [G *Herz*]: In circulatory system of certain copepods, relatively short, muscular pumping organ located dorsally in anterior region of thorax. Receives blood from pericardium through three ostia and pumps it anteriorly via aorta.

hindgut (proctodeum) [G *Enddarm, Proctodaeum*]: Posterior of three regions (foregut, midgut, hindgut) of digestive tract. Relatively short, extending only through last two abdominal somites; opens to exterior via anus. Cuticle-lined and equipped with muscles.

hinge joint: In ♂, modified joint in midregion of geniculate type of antennule; by permitting distal half of antennule to fold down on basal half, antennule serves as grasping organ during copulation.

horn [G *Horn*]: In parasitic copepod, one of two to four horn-like projections of head. Used in attachment to host. (branched, unbranched)

hyaline lamella: Thin, membrane-like crest along certain segments of appendage (e.g., of antennule, pereopod). (broad,

narrow, notched, smooth, toothed, triangular)

intercoxal plate (median coupler plate) [G *Retinaculum*]: On underside of thorax, fold of sternite uniting basal segments (coxae) of each pereopod pair. Permits synchronous strokes during rapid swimming.

labium (lower lip) [G *Labium, Unterlippe*]: Lip-like structure posterior to mouth on underside of head; bilobed, each lobe being termed paragnath. May be drawn out, along with upper lip (labrum), into cone-like structure (mouth cone).

labrum (upper lip) [G *Labrum, Oberlippe*]: Unpaired, lip-like structure in front of mouth. May cover basal part (gnathobase) of mandible or be drawn out, along with lower lip (labium), into cone-like structure (mouth cone).

lateral midgut diverticulum [G *laterale Aussackung*]: One of up to several pairs of lateral extensions of midgut. (bifurcate, multilobed, simple)

lateral muscle [G *Längsmuskel*]: Prominent longitudinal muscle band, one along each side of body.

leg [G *Fuß*]: Preferred term for any appendage after cephalic appendages, i.e., for thoracopod or pereopod.

lunule [G *Lunula, Saugscheibe*]: In parasitic copepod, one of two anteroventral sucker-like structures on anterior margin (frontal plate) of head. (crescent-shaped, round)

mandible [G *Mandibel*]: Anteriormost paired mouthpart on underside of head. Basically a biramous, limb-like appendage consisting of proximal gnathobase and distal palp (the latter with endopod and exopod). (biramous, uniramous; falcate, stylet-like = styliform)

maxilla (second maxilla) [G *Maxille, zweite Maxille*]: Third paired mouthpart on underside of head; located between maxillules and maxillipeds. Uniramous, either setose and serving in suspension feeding, with terminal hooks/spines and serving as a raptorial organ, or arm-like and modified for attachment to host. Bears opening of maxillary gland basally. The term second maxilla is often preferentially applied.

maxillary gland [G *Maxillendrüse, Maxillennephridium, Schalendrüse*]: Excretory gland, one opening at base of each maxilla. (See also antennal gland)

maxillary hook [G *Maxillahaken, Kopfhaken*]: On ventral surface of head in parasitic copepod, one of two posteriorly directed, hook-like processes posterior to antennae.

maxilliped (maxillipede) [G *Maxilliped, Kieferfuß*]: Fourth and generally largest paired mouthpart on underside of head. Uniramous, located posterior to and often closely adjoining maxillae. Represents modified appendage (thoracopod) of first thoracic somite (thoracomere); second thoracopod may also be modified as maxilliped. (lamellate, prehensile, raptorial, setose, subchelate)

maxillule (first maxilla) [G *Maxillula, erste Maxille*]: Second paired mouthpart on underside of head; located between mandibles and maxillae. If fully developed, consists of proximal protopod with endites and exites as well as distal biramous palp with inner branch (endopod) and outer branch (exopod). The term first maxilla is often preferentially applied. (biramous, uniramous; setose, spinose, subchelate; prehensile)

metanauplius: nauplius

metasome [G *Metasoma*]: One of three divisions (tagmata) of body (cephalosome, metasome, urosome). Consists of free thoracic somites (thoracomeres) anterior to flexure point of body. Region comprising cephalosome and metasome may be termed prosome.

midgut [G *Mitteldarm*]: Region of digestive tract between foregut and hindgut. Extends through most of body and may be divided into expanded anterior stomach and more tube-like posterior section.

mouth [G *Mund, Mundöffnung*]: Anterior opening of digestive tract on underside of head. Bordered anteriorly by upper lip (labrum), posteriorly by lower lip (labium), and flanked by mandibles (and to a lesser extent by other mouthparts: maxillules, maxillae, maxillipeds). Opens into esophagus.

mouth cone (mouth tube siphon) [G *Mundkegel, Mundrohr, Sipho, Saugrohr*]: In parasitic copepod, cone- or tube-shaped structure around mouth. Formed by extension of upper lip (labrum) and lower lip (labium); open basally to accommodate mandibles.

mouthpart [G *Mundwerkzeug*]: Appendage closely associated with mouth. Includes paired mandibles, maxillules, maxillae, and maxillipeds as well as unpaired labrum and labium.

nauplius [G *Nauplius*]: First (earlier) of two basic stages in larval development (nauplius, copepodid). Characterized by presence of only three pairs of appendages (antennules, antennae, mandibles) and excretion by means of antennal glands. Typically involves five to six substages, the first two being referred to as orthonauplii, the remaining as metanauplii.

nauplius eye [G *Naupliusauge*]: On dorsal surface of cephalic shield, median photosensitive structure consisting of three ocelli.

neck [G *Hals*]: In parasitic copepod, relatively narrow region of body between head and trunk; composed of several thoacic somites (thoracomeres).

ocellus [G *Ocellus*]: One of three simple photosensitive structures forming nauplius eye. (closely adjoining, widely separated)

oil sac (oil gland) [G *Ölsack*]: Elongate, sac-like structure containing oil droplets; located dorsally along narrower section of midgut. Considered to serve in buoyancy and as a food reserve.

orthonauplius: nauplius

ostium [G *Ostium*]: One of three openings (two lateral, one posteroventral) in heart. Blood from pericardium enters heart through ostia and is pumped anteriorly via aorta.

ovary [G *Ovar, Ovarium, Eierstock*]: In free-living copepod unpaired, in parasitic copepod paired, expanded section of ♀ reproductive system in which eggs are produced. Located dorsally, above midgut, near border of head and thorax. Opens to exterior on first abdominal segment (genital somite) via oviduct. (paired, unpaired)

oviducal diverticulum [G *Ovardivertikel*]: In ♀ reproductive system, relatively large outpocketing of each oviduct near its point of origin at ovary.

oviduct [G *Ovidukt, Eileiter*]: Section of ♀ reproductive system leading from ovary to gonopore/atrium on first abdominal segment (genital somite). Paired, originating at anterior end of ovary and giving rise to oviducal diverticula. Secretes material forming ovisacs.

ovisac (egg sac) [G *Eisäckchen, Eiersack*]: In ♀, sac-like structure projecting from ventral surface of first abdominal segment (genital somite). Secreted by oviduct; serves as brood chamber for extruded, fertilized eggs. (single, paired)

palp [G *Palpus, Taster*]: Biramous, more distal part of fully developed mandible (gnathobase, palp) or maxillule (protopod, palp). Consists of inner branch (endopod) and outer branch (exopod).

paragnath [G *Paragnath*]: One of two lobes of lower lip (labium).

pereopod (pereiopod, peraeopod [G *Pereiopod, Peraeopod*]: One of four pairs of appendages used in locomotion. Pereopods represent appendages (thoracopods) of thorax segments (thoracomeres) two through five, those of first being modified as maxillipeds, those of sixth often as copulatory limbs. Biramous, consisting of proximal protopod with distal inner branch (endopod) and outer branch (exopod). The term leg is

often preferentially applied to all appendages following cephalic appendages.

pericardium (pericardial chamber, pericardial sac [G *Perikard, Perikardialsinus, Perikardialsack*]: Cavity surrounding heart; delimited ventrally from digestive tract by septum (pericardial floor). Blood in pericardium enters heart via ostia.

pleomere (abdominal somite) [G *Pleomer, Abdominalsegment*]: One of five segments (somites) of abdomen. First pleomere termed genital somite, last pleomere termed anal somite. Lacks appendages. (See also urosomal segment)

pleon: abdomen

prehensile antenna [G *Greifantenne*]: antenna

proctodeum: hindgut

prosome [G *Prosoma*]: In one interpretation of body segmentation, anterior of two divisions (tagmata) of body (prosome, urosome); consists of region anterior to flexure point of body (or anterior to genital complex in forms without body articulation) and comprises cephalosome (head and fused thoracic segments) and metasome (thoracic segments anterior to flexure point).

prosome-urosome articulation: body articulation

protopod (protopodite) [G *Protopodit*]: Proximal part of appendage. Typically two-segmented (coxa, basis) and bearing variously developed inner branch (endopod) and outer branch (exopod). May bear endites and exites. Generally specified only when referring to legs (i.e., biramous appendages).

ramus [G *Ast*]: Branch of an appendage, typically referring to inner branch (endopod) or outer branch (exopod) of biramous appendage (e.g., antenna, mandible, maxillule, pereopod).

rostrum [G *Rostrum*]: Anteriorly projecting part of cephalic shield; may fold over anterior margin. (paired, unpaired; pointed, long, short)

second antenna: antenna

second maxilla: maxilla

segment [G *Glied*]: One in a series of units of appendage (e.g., coxa, basis). May also refer to segment of body (then also termed somite). Segments of flagella of antennules are termed articles. (cylindrical, flattened; setose, smooth, spinose, spinulose)

seminal receptacle [G *Receptaculum seminis*]: In ♀ reproductive system, sperm-receiving structure in first abdominal segment (genital somite). Either paired, each receptacle opening into common atrium, or unpaired and joined to end of each oviduct via separate fertilization duct. (paired, unpaired)

seminal vesicle [G *Vesicula seminalis, Samenblase*]: In ♂ reproductive system, somewhat expanded section of vas deferens in which sperm are stored.

seta [G *Borste*]: One in a series of variously developed bristles projecting from margin or tip of appendage (e.g., caudal setae of caudal ramus). (simple, pinnate, plumose)

sinus [G *Sinus, Blutsinus*]: In circulatory system, space in which blood collects. Typically refers to system of ventral sinuses (supraneural sinus, sternal sinus).

siphon: mouth cone

somite (segment) [G *Segment*]: One in a series of divisions of body. Basically, head (cephalon) consists of five cephalic somites, thorax of five thoracic somites (thoracomeres), and abdomen (pleon) of five abdominal somites (pleomeres). Variously fused to form more or less distinct body divisions (cephalosome, metasome, urosome).

sperm duct: vas deferens

spermatophore [G *Spermatophor*]: Packet of sperm formed in modified region (spermatophoric sac) of vas deferens.

spermatophoric sac [G *Spermatophorenbildungssack*]: In ♂ reproductive system, somewhat expanded distal section of vas deferens in which spermatophores are produced.

sternal furca (sternal fork): On ventral surface of parasitic copepod, fork-shaped median projection between maxillipeds and first pereopods.

sternite [G *Sternit*]: Ventral surface of body segment (somite). Occasionally specified in connection with genital somite of ♀, with fold (intercoxal plate) of sternites which unites coxae of pereopod pairs, or with sternal furca.

stomach [G *Magen*]: Expanded anterior section of midgut. May bear median anterior outpocketing (cecum) and lateral diverticula.

supraesophageal ganglion (brain) [G *Oberschlundganglion, Gehirn, Supraösophagealganglion*]: Concentration of nerve tissue anterior (dorsal) to esophagus. Connected to ventral nerve cord by pair of circumesophageal connectives.

tagma [G *Tagma*]: Major division of body (cephalosome, metasome, and urosome, or prosome (cephalosome + metasome) and urosome).

telson: anal somite

tergite [G *Tergit*]: Dorsal surface of body segment (somite). More clearly delimited only in free thoracic somites and abdominal somites.

testis [G *Hoden*]: In free-living copepod typically unpaired, in parasitic copepod typically paired, expanded section of ♂ reproductive system in which sperm are produced. Located dorsally, above midgut, near border of head and thorax. Opens to exterior on first abdominal segment (genital somite) via regionated vas deferens. (paired, unpaired)

thoracomere (thoracic somite) [G *Thoracomer, Thoraxsegment*]: One of basically six segments (somites) of thorax, the first being incorporated into head; each thoracomere typically bears one pair of appendages (thoracopods).

thoracopod (thoracic appendage) [G *Thoracopod, Rumpfbein, Brustgliedmaß, Thoraxbein, Thoracalfuß)*]: One of two appendages of each thoracic somite (thoracomere). First (occasionally second) pair modified as mouthparts (maxillipeds), the remaining typically serving in locomotion (pereopods). Maxillipeds uniramous, the remaining biramous and consisting of basal protopod with distal inner branch (endopod) and outer branch (exopod). (See also leg)

thorax [G *Thorax*]: Second of three basic divisions of body (cephalon, thorax, abdomen). Consists of six somites (thoracomeres), of which at least the first is fused to head (to form cephalosome). Main flexure point of body may lie within thorax. (See also metasome, urosome)

trunk [G *Rumpf*]: Term applied to region of body posterior to head (i.e., thorax and abdomen). Refers also to more expanded part of thorax posterior to neck in parasitic copepod.

urosomal segment: urosome

urosome [G *Urosoma*]: Posterior division (tagma) of body (prosome, urosome, or cephalosome, metasome, urosome). Corresponds to region of body posterior to main flexure point (or posterior to genital complex in forms without body articulation) and includes both thoracic and abdominal somites.

vas deferens (sperm duct) [G *Vas deferens, Samenleiter*]: Section of ♂ reproductive system leading from testis to gonopore on first abdominal segment (genital somite). Originates at anterior end of testis and is typically regionated into seminal vesicle, spermatophoric sac, and ejaculatory duct. (paired, unpaired)

ventral nerve cord [G *Bauchmark*]: Unpaired, longitudinal nerve cord extending through thorax and joined to supraesophageal ganglion by pair of circumesophageal connectives. May be concentrated anteriorly in parasitic copepod, leading to prominent nerve ring around esophagus.

ARTHROPODA

CRUSTACEA

Branchiura

branchiuran, fish louse [G *Fischlaus, Branchiure*]

BRANCHIURA

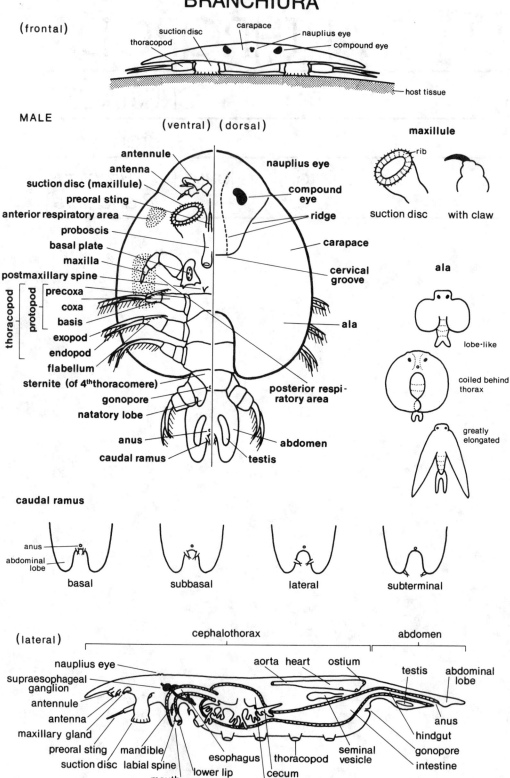

(frontal)

thoracopod · suction disc · carapace · nauplius eye · compound eye · host tissue

MALE

(ventral) (dorsal)

antennule
antenna
suction disc (maxillule)
preoral sting
anterior respiratory area
proboscis
basal plate
maxilla
postmaxillary spine
thoracopod — protopod — precoxa, coxa, basis
exopod
endopod
flabellum
sternite (of 4th thoracomere)
gonopore
natatory lobe
anus
caudal ramus

nauplius eye
compound eye
ridge
carapace
cervical groove
ala
posterior respiratory area
abdomen
testis

maxillule
rib
suction disc · with claw

ala
lobe-like
coiled behind thorax
greatly elongated

caudal ramus

anus
abdominal lobe
basal · subbasal · lateral · subterminal

(lateral)

cephalothorax · abdomen

nauplius eye
supraesophageal ganglion
antennule
antenna
maxillary gland
preoral sting
suction disc
mandible · labial spine
lower lip
mouth
esophagus
cecum
crop
thoracopod
seminal vesicle
aorta · heart · ostium
testis · abdominal lobe
anus
hindgut
gonopore
intestine

ARTHROPODA

CRUSTACEA

Branchiura

branchiuran, fish louse [G *Fischlaus, Branchiure*]

abdomen (pleon) [G *Abdomen, Pleon, Hinterleib*]: Unsegmented posterior division (tagma) of body (cephalothorax, abdomen, or cephalon, pereon, abdomen). Relatively small, dorsoventrally flattened, and drawn out posteriorly into two lobes; bears pair of caudal rami and anus in cleft (sinus) between two lobes. (acorn-shaped, broadly elliptical, cordate, obcordate, obovate, spindle-shaped; lobes: acutely pointed, bluntly pointed, rounded)

abdominal ganglion: Ganglion adjoining last thoracic ganglion in anterior region of thorax (forming last lobe of six-lobe ganglion mass). May bear smaller pair of lobes; gives rise to two pairs of nerves extending into abdomen.

ala [G *lateraler Flügel*]: One of two posterolateral extensions of carapace. (broad, broadly rounded, narrowly rounded, elongate and coiled behind thorax, greatly elongated; extending to second, third, or fourth thoracomere, to abdomen, beyond abdomen)

anal furca: caudal ramus

anal papilla: caudal ramus

antenna (second antenna) [G *Antenne, zweite Antenne*]: Second pair of appendages on underside of head. Relatively short, uniramous, and consisting of four to six segments. Armed basally with large spines/hooks, terminally with smaller spines. Serves in attachment to host.

antennule (first antenna) [G *Antennula, erste Antenne*]: First pair of appendages on underside of head. Relatively short and basically four-segmented. Bears palp and several large spines/hooks; anterior antennular spine/hook is associated with gland or sensory organ. Serves in attachment to host.

anus [G *After*]: Posterior opening of digestive tract. Located at posterior end between two lobes of abdomen; typically flanked on each side by minute caudal ramus.

aorta [G *Aorta, Aorta cephalica*]: Unpaired dorsal blood vessel extending anteriorly from heart.

bar: One in a series of chitinous skeletal elements supporting distal end of proboscis. According to position one may distinguish groove, lateral, lateral marginal, anterior and posterior lateral oblique, and inner and outer mandibular bars.

basal plate (maxillary plate) [G *Hakenplatte*]: Spine-studded, plate-like structure on basal segment of each maxilla. (spines: alike, unlike; close together, well-separated; blunt, curved, scale-like, sharp)

basis (basipodite) [G *Basis, Basipodit*]: In thoracopod, third of three segments (precoxa, coxa, basis) of protopod; may also be applied to segment of antennule.

blind capsule (blind gland) [G *Blindanhang*]: In ♂ reproductive system, pair of glands in anterior region of thorax. Duct from each blind capsule connects to vas deferens and leads to common ejaculatory duct and gonopore.

brain: supraesophageal ganglion

buccal cavity [G *Mundvorraum*]: In front of mouth, chamber formed by lip-like extensions of proboscis margins. Contains mandibles.

carapace [G *Carapax*]: Dorsoventrally flattened, shield-like structure covering head and first thoracic somite (thoracomere); characterized by pair of posterolateral wing-like extensions (alae). Bears pair of compound eyes dorsolaterally and set of median nauplius eyes. (circular, cordate, elliptical, trifoliate)

caudal ramus (anal furca, anal papilla) [G *Furca*]: Pair of minute projections between posterior lobes of abdomen; rami flank anus. (basal, lateral, subbasal, subterminal, terminal)

cecum [G *Mitteldarmdrüse, Darmblindsack*]: Highly branched diverticulum, one extending from each side of anterior midgut. Fills greater part of carapace after fish louse feeds on host tissues.

cephalon (head) [G *Kopf*]: Anterior part of cephalothorax (or anterior of three tagmata of body: cephalon, pereon, abdomen). Partially fused to first thoracic somite and delimited from last three thoracic somites (pereon) by cervical groove and ridges on carapace. Bears antennules, antennae, preoral sting, proboscis with mandibles, mouth and modified lips, maxillules, and maxillae.

cephalothorax [G *Cephalothorax*]: Anterior division (tagma) of body (cephalothorax, abdomen). Consists of head (cephalon)—fused with first thoracic somite—and

remaining three thoracomeres; anterior portion covered by carapace.

cervical groove (transverse groove) [G *Zervikalfurche, Cervicalfurche*]: On carapace, transverse groove delimiting cephalon (including first thoracic somite) from remaining three thoracic somites (pereon).

circumesophageal connective (paraesophageal connective) [G *Schlundkonnektiv*]: One of two short, relatively thick strands of nerve tissue surrounding esophagus and connecting dorsal supraesophageal ganglion with subesophageal ganglion. Innervates antennae.

circumgenital body cavity: gonocoel

compound eye [G *Komplexauge*]: One of two organs of vision suspended in a blood sinus on dorsal surface of carapace; each movable eye consists of numerous ommatidia, yet lacks true outer cornea.

coxa (coxopodite) [G *Coxa, Coxopodit*]: In thoracopod, second of three segments (precoxa, coxa, basis) of protopod; may also be applied to segment of antennule.

crop [G *Magendarm*]: midgut

ejaculatory duct [G *Ductus ejaculatorius*]: Unpaired terminal end of ♂ reproductive system; formed by merger of vasa deferentia.

endopod [G *Endopod*]: Setose inner branch (ramus) of thoracic appendage (thoracopod). (See also exopod)

end sac [G *Endsäckchen*]: Expanded proximal section of each maxillary gland. Opens to exterior via coiled duct.

epipod: flabellum

esophagus [G *Ösophagus, Vorderdarm*]: Section of digestive tract between mouth and midgut; partially contained within proboscis. (See also foregut)

exopod [G *Exopod*]: Setose outer branch (ramus) of thoracic appendage (thoracopod). Exopod of first two thoracopods may bear medially directed flabellum. (See also endopod)

eye: compound eye, nauplius eye

flabellum (epipod) [G *Flabellum*]: On first two pairs of thoracic appendages (thoracopods), medially (backwardly) directed projection from base of outer branch (exopod).

flagellum: palp

foregut [G *Vorderdarm*]: Anterior of three regions (foregut, midgut, hindgut) of digestive tract; corresponds to esophagus.

genital atrium: vagina

gonocoel (circumgenital body cavity) [G *Gonocoel*]: Cavity, separate from main body cavity (hemocoel), containing ovary.

gonopore (genital aperture) [G *Geschlechtsöffnung, Gonoporus*]: In ♂ reproductive system, unpaired opening on ventral surface of last thoracic somite (fourth thoracomere). In ♀, see vagina.

head: cephalon

heart [G *Herz*]: In circulatory system, muscular pumping organ located dorsally in last thoracic somite (fourth thoracomere). Receives blood through pair of ostia at posterior end and pumps it anteriorly via aorta, ventrally and posteriorly through additional openings.

hindgut (rectum) [G *Enddarm*]: Posterior of three regions (foregut, midgut, hindgut) of digestive tract; extends through abdomen and opens to exterior via anus.

intestine [G *Dünndarm*]: Posterior section of midgut; opens into hindgut at border of thorax and abdomen.

labial spine [G *Unterlippenstachel*]: One of two hollow spines in proboscis; each is located at tip of swelling and projects into anterior end of buccal cavity. Associated with gland.

labium: lower lip, paragnath

labrum (upper lip) [G *Oberlippe, Labrum*]: Lip-like structure in front of mouth; either consists of fleshy, papilla-shaped elevation or is reduced, along with paragnaths, to serrate flap enclosed in proboscis.

larva [G *Larve*]: One of up to nine stages of development after hatching from egg. One may distinguish larvae that swim by means of antennae and mandibles and those that swim with thoracopods.

lower lip [G *Unterlippe*]: Refers to either 1) paragnaths of labium surrounding mouth posteriorly or 2) lip-like lower margin (bearing labial spines and associated glands) of proboscis.

mandible [G *Mandibel*]: Paired, serrate mouthpart closely associated with mouth on underside of head. Typically enclosed, along with mouth and surrounding labrum and paragnaths, in proboscis. Lacks palp.

maxilla (second maxilla) [G *Maxille, zweite Maxille*]: Paired appendage on underside of head between maxillules (suction discs) and first thoracopods. Uniramous, consisting of several spinose segments and terminal claw. Bears basal plate and opening of maxillary gland basally.

maxillary gland [G *Maxillendrüse, Schalendrüse*]: One of two relatively large excretory organs located in head; each opens at base of maxilla and consists of end sac and duct.

maxilliped: maxilla

maxillule (first maxilla) [G *Maxillula, erste Maxille*]: Second paired mouthpart on underside of head (between mandibles and maxillae). Typically developed as prominent pair of suction discs, yet also as uniramous appendage with distal claws.

median eye: nauplius eye

midgut [G *Mitteldarm*]: Region of digestive tract in thorax between foregut (esophagus) and hindgut. One may distinguish an anterior crop and posterior intestine. Gives rise anteriorly to pair of highly branched ceca.

mouth [G *Mund*]: Anterior opening of digestive tract on underside of cephalothorax; typically enclosed, along with surrounding labrum, mandibles, and paragnaths, in proboscis. Opens into esophagus.

natatory lobe [G *Schwimmlappen*]: On last pair of thoracic appendages (fourth thoracopods), lobe-like projection of protopod. Serves in swimming.

nauplius eye (median eye, ocellus) [G *Naupliusauge*]: On dorsal midline of carapace, one to three simple photosensitive structures. (See also compound eye)

ocellus: nauplius eye

ommatidium [G *Ommatidium*]: One of up to 75 elongate, closely adjoining units forming each compound eye; lacks directly overlying true cornea.

ostium [G *Ostium*]: Opening at each posterolateral angle of heart. Blood enters heart through ostia and is pumped anteriorly through aorta, ventrally and posteriorly through additional openings.

ovary [G *Ovar, Ovarium, Eierstock*]: Large, unpaired section of ♀ reproductive system in which eggs are produced. Occupies greater part of thorax and opens to exterior via pair of short oviducts. Described as being located within separate body cavity (gonocoel).

oviduct [G *Ovidukt, Eileiter*]: One of two short ducts leading from ovary to gonopore (vagina) on last thoracic somite (fourth thoracomere). Considered to function alternately.

palp [G *Palpus, Taster*]: Slender medial projection of each antennule. Consists of two to three segments and may be tipped by setae. Formerly termed flagellum (along with thin epipods—flabella—of first two thoracopods). (uniramous, biramous)

paragnath [G *Paragnath*]: One of two lobes of lower lip (labium); either papilla-shaped elevations posterior to mouth or reduced, along with labrum, to serrate flap enclosed in proboscis.

peg: On last pair of thoracic appendages (fourth thoracopods) of ♂, peg-like process of protopod. Fits into socket of third thoracopods in a peg-and-socket arrangement and serves to hold both thoracopods together during copulation. (acute, bent, conical, cylindrical, finger-like, spherical, two-jointed)

pereon [G *Peraeon*]: Region of body comprising three free thoracic somites (the first thoracomere being fused to head).

pleon: abdomen

poison gland [G *Giftdrüse*]: Gland located at base of preoral sting; opens at tip of sting via duct.

postmaxillary spine: On underside of head, pair(s) of prominent, hook-shaped spines between or posterior to base of maxillae.

precoxa (precoxal ring) [G *Präcoxa*]: In thoracopod, ring-like proximal segment of protopod.

preoral sting (preoral spine) [G *präoraler Stachel, Stachel*]: On underside of cephalothorax, unpaired projection anterior to proboscis. Distal half of anteriorly directed preoral sting can be retracted into basal socket. Needle-like tip bears opening of poison gland. Used to puncture host.

proboscis [G *Rüssel*]: On underside of cephalothorax, unpaired median projection posterior to preoral sting. Retractile, enclosing mouth and surrounding labrum, mandibles, and paragnaths. Supported distally by series of chitinous bars. Used to feed on host tissues; lies in groove when at rest.

protopod (protopodite) [G *Protopodit*]: Proximal part of thoracic appendage (thoracopod); typically three-segmented, consisting of precoxa, coxa, and basis. Protopod of last (fourth) thoracopod may bear natatory lobe.

rectum: hindgut

respiratory area [G *Atemfeld*]: On underside of carapace, one of four areas characterized by thin cuticle and functioning as a respiratory surface. One may distinguish two smaller anterior respiratory areas (flanking maxillae) and two larger posterior areas flanking thoracopods. (anterior and posterior: equal, not equal; anterior: anterior, anteromesial, mesial, minute, narrow, prolonged laterally, triangular; posterior: deeply notched on inner margin, narrow, regular, symmetrical, very large, widened posteriorly)

rib: One in a parallel series of longitudinal supporting structures around rim of each suction disc. Consists of up to 50 segments. (segment: bead-like, cone-shaped, imbricated, doubly imbricated, quadrangular, rod-like)

ridge: Elevated ridge along carapace. Refers either to pair on longitudinal ridges delimiting cephalon laterally or to pair of median ribs between compound eyes. (simple, bifurcate anteriorly)

segment [G *Glied*]: One in a series of units of an appendage (e.g., coxa, basis). May also refer to segment of body (then also termed somite).

seminal papilla [G *Papille, Endpapille*]: In ♀ reproductive system, one of two papilla-like elevations flanking gonopore (vagina). Each seminal papilla is connected via duct to separate seminal receptacle. Fertilizes egg emerging from oviduct.

seminal receptacle (spermatheca) [G *Receptaculum seminis*]: In ♀ reproductive system, one of two sperm-receiving structures in anterior region of abdomen; each seminal receptacle is connected by duct to separate seminal papilla flanking gonopore (vagina). (circular, irregularly ovate)

seminal vesicle [G *Vesicula seminalis, Samenblase*]: In thorax, unpaired section of ♂ reproductive system in which sperm are stored. Connected to testes by paired vasa efferentia, to common ejaculatory duct and gonopore by paired vasa deferentia.

sinus [G *Sinus, Blutsinus, Schwanzlappensinus*]: Term referring either to cleft between posterior lobes of abdomen or to blood sinus such as that surrounding compound eyes.

socket: On third pair of thoracic appendages (thoracopods) of ♂, socket-like depression in protopod. Receives peg of fourth thoracopod in a peg-and-socket arrangement and serves to hold both thoracopods together during copulation.

somite (segment) [G *Somit, Segment*]: One in a series of divisions (segments) of body. Head (cephalon) consists of five somites plus first thoracic somite, pereon of three somites; abdomen is unsegmented.

spermatheca: seminal receptacle

spermatophore [G *Spermatophor*]: Packet of sperm produced in ♂ reproductive system.

sperm duct: vas deferens, vas efferens

sternite [G *Sternit, Sternum*]: Ventral surface of body segment (somite). More clearly delimited only on thorax and between maxillae. Postmaxillary spines located on sternite between maxillae, gonopore on sternite of last thoracic somite (fourth thoracomere).

subesophageal ganglion [G *Unterschlundganglion*]: Concentration of nerve tissue below (posterior to) esophagus. Connected to supraesophageal ganglion by pair of circumesophageal connectives and adjoined posteriorly by partially fused thoracic ganglia. Innervates mandibles, maxillules, and maxillae.

suction disc (sucking disc, suction cup) [G *Saugnapf*]: Modification of each maxillule into prominent, stalked, disc-shaped sucker; margins of sucker supported by series of ribs.

supraesophageal ganglion (brain) [G *Oberschlundganglion, Gehirn*]: Concentration of nerve tissue anterior (dorsal) to esophagus. A proto-, deutro-, and tritocerebrum may be distinguished. Connected to subesophageal ganglion by pair of circumesophageal connectives. Innervates antennules and eyes.

tagma [G *Tagma*]: Major division of body. According to interpretation, one may distinguish either cephalothorax and abdomen or head (cephalon), pereon, and abdomen (pleon).

tergite [G *Tergit*]: Dorsal surface of body segment (somite). More clearly delimited only in last three thoracic somites (pereon).

testis [G *Hoden*]: Paired, expanded section of ♂ reproductive system in abdomen. Each testis is connected to unpaired seminal vesicle in thorax by vas efferens. (elongate elliptical, elongate ovate, lunate)

thoracic ganglion [G *Thorax-ganglion*]: One in a series of four partially fused ganglia of thorax region. Innervates thoracic appendage (thoracopod). Adjoined anteriorly to subesophageal ganglion and posteriorly to abdominal ganglion to form six-lobe mass in anterior region of thorax.

thoracomere (thoracic somite) [G *Thoracomer, Thoraxsegment*]: One of four segments (somites) of thorax, the first being partially fused to head; each bears pair of appendages (thoracopods).

thoracopod [G *Thoracopod*]: One of two appendages of each thoracic segment (thoracomere). Biramous, consisting of three-segmented protopod bearing an endopod and exopod. (See also flabellum, natatory lobe, peg)

thorax [G *Thorax*]: Region of body between head (cephalon) and abdomen (pleon). First thoracic somite (thoracomere) is fused to head, the remaining three thoracomeres form pereon.

upper lip: labrum

vagina (genital atrium) [G *Vagina, Genitalatrium*]: ♀ gonopore on ventral surface of last thoracic somite (fourth thoracomere). Receives pair of oviducts, although only one oviduct is considered to be functional at any one time. Flanked by pair of seminal papillae.

vas deferens [G *Vas deferens*]: In ♂ reproductive system, pair of narrow sperm ducts leading from median seminal vesicle; each vas deferens connects to duct of blind capsule and leads to common ejaculatory duct.

vas efferens [G *Vas efferens*]: In ♂ reproductive system, narrow sperm duct leading from each testis to median seminal vesicle.

ARTHROPODA

CRUSTACEA

Cirripedia

cirriped [G *Rankenfüßer*]

CIRRIPEDIA

cypris larva

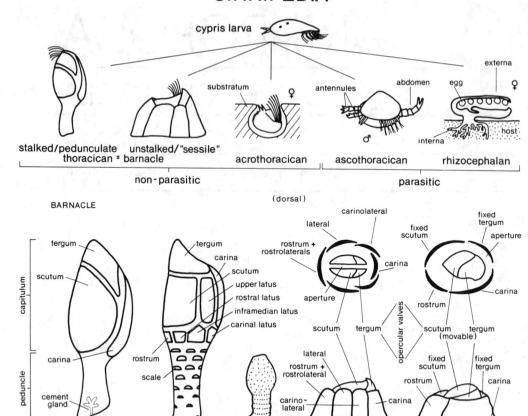

stalked/pedunculate
thoracican = barnacle

unstalked/"sessile"
barnacle

substratum

♀

acrothoracican

antennules

abdomen

ascothoracican

♂

externa

egg

♀

interna

host

rhizocephalan

non-parasitic

parasitic

BARNACLE

(dorsal)

capitulum

peduncle

tergum

scutum

carina

cement
gland

peduncle naked
lepadid

tergum

carina

scutum

upper latus

rostral latus

inframedian latus

carinal latus

rostrum

scale

peduncle with scales
scalpellid

unplated

stalked

carinolateral

lateral

rostrum +
rostrolaterals

aperture

scutum

tergum

lateral

rostrum +
rostrolateral

carino-
lateral

symmetrical
balanomorph

carina

carina

opercular valves

fixed
scutum

fixed
tergum

aperture

carina

rostrum

scutum
(movable)

tergum

fixed
scutum

fixed
tergum

rostrum

carina

asymmetrical
verrucomorph

unstalked

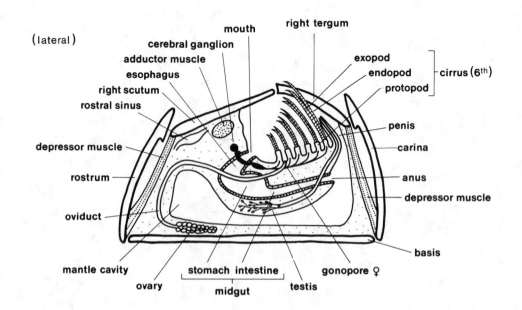

(lateral)

mouth

cerebral ganglion

adductor muscle

esophagus

right scutum

rostral sinus

depressor muscle

rostrum

oviduct

mantle cavity

ovary

right tergum

exopod

endopod

protopod

cirrus (6th)

penis

carina

anus

depressor muscle

basis

stomach intestine

midgut

testis

gonopore ♀

depressor muscle

ARTHROPODA

CRUSTACEA

Cirripedia

cirriped [G *Rankenfüßer*]

abdomen [G *Abdomen, Hinterleib*]: In ascothoracican, posteriormost division (tagma) of body (head, thorax, abdomen). Consists of four or five segments (somites), the last bearing caudal furca.

adductor muscle [G *Adduktormuskel, Schließmuskel*]: Conspicuous transverse muscle joining both scuta and serving to close aperture.

adductor muscle pit [G *Vertiefung für den Adduktor*]: Barnacle plate feature: on inner surface of scutum, depression marking attachment point of adductor muscle. In unstalked barnacle, located between adductor ridge and occludent margin.

adductor ridge [G *Leiste des Adduktors*]: Barnacle plate feature: on inner surface of scutum in unstalked form, linear elevation between adductor muscle pit and tergal margin.

adhesive gland: cement gland

ala [G *Ala*]: Barnacle plate feature: in unstalked form, triangular lateral part of compartmental plate. Delimited from median part (paries) and overlapped by lateral part (radius) of adjoining compartmental plate.

antenna (second antenna) [G *Antenne, zweite Antenne*]: Biramous second pair of antennae. Present, with the ex-

ception of certain ascothoracicans, only in early larval stages (nauplius).

antennule (first antenna) [G *Antennula, erste Antenne*]: First pair of antennae. present in larval (nauplius, cyprid) stages, yet typically reduced or absent in adults. Used in attachment by barnacle larvae and associated with cement glands. Well developed in adult acrothoracican and ascothoracican. (prehensile, raptorial, subchelate, uniramous)

anus [G *Anus, After*]: Posterior opening of digestive tract. Located near posteriormost thoracic appendages, i.e., between sixth pair of cirri; in ascothoracican at end of abdomen.

apertural hook: In acrothoracican, hook-like structure protecting aperture. (See also apertural spine)

apertural spine: In acrothoracican, spine-like structure protecting aperture. (See also apertural hook)

aperture (orifice) [G *Mantelspalt, Mantelöffnung*]: In barnacle, opening through which thoracic appendages (cirri) are protruded. In stalked barnacle, bordered on each side by scutum and tergum; bordered in unstalked barnacle by operculum (paired scuta and terga in symmetrical forms, single scutum and tergum in asymmetrical forms). (See also occludent margin)

apex [G *Apex*]: Barnacle plate feature; uppermost point (angle) of plate. Typically refers to scutum, tergum, or carina. (beaked)

apicoumbonal ridge: Barnacle plate feature; in scutum of stalked form, ridge extending from upper point (apex) to umbo.

article (joint, segment) [G *Glied*]: One in a series of segments of thoracic appendage (cirrus) or other appendage.

articular furrow: Barnacle plate feature: in unstalked form, groove on tergal margin of scutum or scutal margin of tergum. Furrow and ridge of scutum (or tergum) articulate with corresponding structures in tergum (or scutum).

articular ridge [G *Gelenksleiste, Crista articularis*]: Barnacle plate feature: in unstalked form, linear elevation on tergal margin of scutum or scutal margin of tergum. Ridge and furrow of scutum (or tergum) articulate with corresponding structures in tergum (or scutum).

attachment disc [G *Haftscheibe*]: In ♀ acrothoracican, anteroventral horny structure used in attachment to burrow. If somewhat elevated, referred to as attachment knob.

basal margin [G *Basalrand*]: Barnacle plate feature: lower edge of plate. Typically refers to lower edge of scutum or tergum.

basicarinal angle: Barnacle plate feature: corner of tergum formed by junction of basal and carinal margins.

basidorsal point [G *Horn*]: In certain unstalked barnacles, cone-shaped projection at base of penis.

basilateral angle: basitergal angle

basioccludent angle: Barnacle plate feature: corner of scutum formed by junction of basal and occludent margins.

basis [G *Basis*]: Basal element by which unstalked barnacle is attached to substratum. (calcareous = solid, membranous)

basiscutal angle: Barnacle plate feature; corner of tergum formed by junction of basal and scutal margins.

basitergal angle (basilateral angle): Barnacle plate feature; corner of scutum formed by junction of basal and tergal margins.

blood pump: rostral sinus

brain: supraesophageal ganglion

brood chamber (externa) [G *Brutsack, Externa, Sacculina externa*]: In ♀ rhizocephalan, pouch-like structure projecting from body of host; contains eggs and seminal receptacle and serves as site of attachment of ♂. May also refer to brood pouch of ascothoracican.

canal: longitudinal tube

capitulum [G *Capitulum*]: In stalked barnacle, one of two external divisions of body (capitulum, peduncle). Consists of mantle and associated plates enclosing body.

carapace: mantle

carina [G *Carina*]: One of five primary plates (carina, two scuta, two terga) on outer surface of barnacle: unpaired posterior plate. In stalked form, posterior plate on capitulum. In unstalked barnacle, compartmental plate between rostrum and fixed tergum (asymmetrical form) or plate opposite rostrum (symmetrical form); consists of median paries and lateral alae.

carinal latus (carinolatus) [G *Latus carinale*]: Barnacle plate: in stalked form, paired plates (latera), one on each side of carina. Homologous with carinolateral of unstalked form.

carinolateral (carinolatus) [G *Carino-lateral, Latus carinale*]: Barnacle shell plate: in unstalked form, paired posterolateral plate, one between carina and each lateral plate. Consists of median paries with radius on one (carinal) side, ala on other (lateral) side.

caudal furca: caudal ramus

caudal ramus [G *Furkalast, Furca-Ast*]: In ascothoracican, one of two posteriorly directed projections of last segment of abdomen. Both caudal rami together termed caudal furca.

cecum (caecum): digestive cecum

cement gland (adhesive gland) [G *Zementdrüse, Kittdrüse*]: Glandular structure, one associated with and opening at base of each antennule. Used by cyprid larva in attachment to substratum.

cerebral ganglion: supraesophageal ganglion

cirrus [G *Zirrus, Zirre, Ranke, Rankenfuß*]: One of up to six pairs of elongate appendages of thorax region (those of ascothoracican being termed thoracopods). Biramous, basically consisting of proximal protopod and distal endopod and exopod. Reduced and set-off pair of cirri in acrothoracican are termed mouth cirri. (uniramous, biramous; acanthopodous, ctenopodous, lasiopodous)

comb collar: In acrothoracican, collar-like structure surrounding upper part of aperture. Retractable, consisting of membranous fold with fringe of setae.

compartmental plate (compartment plate, mural plate) [G *Mauerplatte*]: In unstalked barnacle, one of up to eight overlapping/articulating plates forming wall which encircles body. May include from anterior to posterior end: single rostrum, paired rostrolaterals, laterals, carinolaterals, single carina, and in asymmetrical form, also fixed scutum and fixed tergum.

complemental male [G *Komplementärmännchen, Ersatzmännchen*]: In barnacle, term applied to relatively small ♂ attached to hermaphroditic individual. (See also dwarf male)

compound eye [G *Komplexauge*]: cypris larva

compound rostrum: rostrum

cyprid: cypris larva

cypris larva (cyprid, cyprid larva) [G *Cyprislarve, Cypris*]: Final larval stage, common to all cirripeds, characterized by bivalved carapace, compound eyes, prehensile antennules, and series of locomotory thoracic appendages (cirri). (See also nauplius)

depressor muscle [G *Depressor, Depressormuskel*]: In stalked barnacle, one of several paired muscles spanning from tergum and scutum to basis; serves in closing aperture.

depressor muscle crest: Barnacle plate feature: on inner surface of tergum in unstalked form, elevated denticles or ridges near basicarinal angle. Serves as site of attachment of depressor muscle.

digestive cecum (digestive gland, gut diverticulum, hepatic cecum) [G *Mitteldarm-Divertikel, Divertikel, Verdauungsdrüse*]: One of up to several pairs of outpocketings of anterior region of midgut.

digestive gland: digestive cecum

dwarf male [G *Zwergmännchen*]: In barnacle, term applied to relatively small ♂ attached to ♀ individual. (See also complemental male)

endopod (endopodite) [G *Endopod, Endopodit*]: Inner branch (ramus) of thoracic appendage (thoracopod in ascothoracican, cirrus in other cirripeds). (one- to three-jointed, multiarticulate) (See also exopod)

esophageal connective (circumesophageal connective) [G *Schlundkonnektiv*]: One of two strands of nerve tissue extending around esophagus and connecting supraesophageal and subesophageal ganglia.

esophagus [G *Ösophagus*]: Muscular, cuticle-lined, and relatively narrow section of digestive tract between mouth and midgut.

exopod (exopodite) [G *Exopod, Exopodit*]: Outer branch (ramus) of thoracic appendage (thoracopod in ascothoracican, cirrus in other cirripeds). (one- to three-jointed, multiarticulate) (See also endopod)

externa (reproductive sac)
[G *Externa, Sacculina externa*]: brood chamber

eye: compound eye, nauplius eye

first antenna: antennule

foregut [G *Vorderdarm*]: esophagus

furca: caudal furca

gnathobase [G *Kauladen*]: Inwardly (medially) directed lobe of proximal part (protopod) of mouthpart. Represents greater part of each mandible, maxillule, and maxilla.

gonad: ovary, testis

gonopore [G *Gonoporus, Geschlechtsöffnung*]: In ♀ reproductive system, opening of each oviduct to exterior; located on first thoracic somite at or near bases of first cirri. For ♂ see penis.

head [G *Kopf*]: Anterior of typically two basic divisions (tagmata) of body (head, thorax; in ascothoracican: head, thorax, abdomen). If fully developed consists of five fused segments (somites) bearing antennules, antennae, mandibles, maxillules, and maxillae.

heart: rostral sinus

hepatic cecum: digestive cecum

hindgut [G *Enddarm*]: Posteriormost, relatively narrow, cuticle-lined section of digestive tract between midgut and anus.

inframedian latus [G *Latus inframedium*]: Barnacle plate: paired plates (latera) in stalked form, one below each upper latus.

interlaminate figure: Barnacle plate feature: when plate of unstalked form is sectioned parallel to base, line running from outer to inner lamina through longitudinal septum. (simple, arborescent)

interna [G *Interna, Sacculina interna, Wurzelgeflecht*]: In ♀ rhizocephalan, rootlike section of body extending into host. (See also externa)

intestine: midgut

intraparies: Barnacle plate feature: in certain stalked forms, secondary lateral margins of carina.

intromittent organ: penis

kentrogon [G *Kentrogon*]: In rhizocephalan, dedifferentiated stage following cyprid larva. Injected into host through stylet.

labrum (upper lip) [G *Labrum, Oberlippe*]: Unpaired lobe covering mouth; typically bears pair of palps. Forms part of oral pyramid in ascothoracican. (bullate, not bullate; cutting edge: straight, concave, with median notch; with/without setae or spines)

lamella (lamina) [G *Lamelle*]: Barnacle plate feature: in unstalked form, outer and inner walls of paries and radius. Separated by longitudinal septa forming longitudinal tubes.

lateral: lateral plate

lateral bar [G *Lateralleiste*]: In acrothoracican, pair of chitinous thickenings reinforcing each side of mantle; extends from aperture halfway down body.

lateral plate [G *Laterale*]: Barnacle shell plate: in unstalked form, paired lateral plates, one located either between each rostrolateral and carinolateral or between carina and rostrolateral or rostrum. Consists of median paries with radius on one (carinal) side, ala on other (rostral) side.

latus [G *Latum*]: Barnacle plate: in stalked form, any plate on capitulum except scuta, terga, rostrum, subrostrum, carina, and subcarina. According to position one may distinguish carinal, inframedian, median, rostral, and upper paired latera.

longitudinal canal: longitudinal tube

longitudinal septum (parietal septum) [G *Septum*]: Barnacle shell feature: in unstalked form, one in a series of longitudinal walls between inner and outer lamellae of compartmental plate. Septa are arranged perpendicular to lamellae and form canals (longitudinal tubes). (primary, secondary)

longitudinal tube (longitudinal canal, parietal tube, parietal pore) [G *Kanälchen*]: Barnacle plate feature: in unstalked form, one in a series of canals between inner and outer lamellae of compartmental plate. Formed by series of longitudinal septa between the lamellae.

lower latus: Barnacle plate: in stalked form, general term for one of a number of smaller plates forming a series (whorl) below each upper latus.

mandible [G *Mandibel*]: First pair of mouthparts on head. Consists largely of gnathobase and, in acrothoracican, bears palps. Typically closely associated with labrum. (with/without palp; serrate, suctorial)

mantle [G *Mantel*]: In barnacle, soft outer part of body typically covered by a series of plates. In other cirripeds, outer cover of body. Often referred to as carapace.

mantle cavity [G *Mantlehöhle*]: In barnacle, space between main, appendage-bearing part of soft body and outer wall (mantle). Eggs may be stored in mantle cavity.

maxilla (second maxilla) [G *Maxille, zweite Maxille*]: Third paired mouthpart on head. Consists largely of gnathobase; associated with openings of maxillary glands. (fused, not fused) (See also oral pyramid)

maxillary gland [G *Maxillardrüse, Maxillennephridium*]: One of two excretory organs located in head and opening via ducts at level of maxillae.

maxilliped (maxillipede) [G *Maxilliped*]: One of up to three anterior pairs of thoracic appendages (cirri) modified as mouthparts. First maxillipeds of acrothoracican may also be termed mouth cirri.

maxillule (first maxilla) [G *Maxillula, erste Maxille*]: Second paired mouthpart on head; located between mandibles and maxillae. Consists largely of gnathobase. (with/without notch) (See also oral pyramid)

median latus: Barnacle plate: in stalked form with paired latera in one series (whorl), plate on each side of capitulum between rostral and carinal latera.

midgut [G *Mitteldarm*]: Elongate, somewhat expanded (in barnacle, U-shaped) section of digestive tract between esophagus and hindgut; bears digestive ceca. Expanded region of midgut may be termed stomach.

mouth [G *Mund, Mundöffnung*]: Anterior opening of digestive tract. Closely associated with labrum and mouthparts (trophi). Opens into esophagus.

mouth cirrus: cirrus

mouth cone: oral cone

mural plate: compartmental plate

muscle pit: adductor muscle pit

nauplius [G *Nauplius*]: Early larval stage; characterized by frontolateral horns, nauplius eye, and three pairs of appendages (antennules, antennae, and mandibles). (See also cypris larva)

nauplius eye [G *Naupliusauge*]: Unpaired median photosensitive organ in larvae (nauplius, cypris). May be retained under plates of adult barnacle.

occludent margin [G *Verschlußrand*]: Barnacle plate feature: margin of scutum or tergum facing aperture.

occludent tooth: Barnacle plate feature: one in a series of small projections along margin of scutum facing aperture (occludent margin). Formed by extensions of growth ridges; fits into corresponding structures of opposed scutum when aperture is closed.

opercular membrane [G *Gelenkhaut*]: Barnacle shell feature: in unstalked form, thin, flexible membrane attaching opercular valves (scuta, terga) to sheath.

opercular valve [G *Operkularplatte*]: Barnacle shell feature: in unstalked barnacle, calcareous plate forming part of closing apparatus (operculum). Consists in symmetrical forms of both scuta and terga, in asymmetrical forms of single (movable) scutum and single (movable) tergum. Attached to inner wall (sheath) of shell by opercular membrane.

operculum [G *Deckelapparat, Operculum*]: In unstalked barnacle, movable, lid-like structure used to close aperture. Consists of symmetrical forms of articulated scutum and tergum closing against opposite scutum and tergum, in asymmetrical forms of movable scutum closing against movable tergum.

oral cone: oral pyramid

oral pyramid (mouth cone, oral cone) [G *Mundkegel*]: Cone-like projection, especially well developed in ascothoracican, formed by close association of mouthparts (labrum, mandibles, maxillules, maxillae).

orifice: aperture

ovary [G *Ovar, Ovarium, Eierstock*]: Paired section of ♀ reproductive system located in peduncle (stalked barnacle) or anteriorly in basis and mantle wall (unstalked barnacle). Each ovary opens to exterior via oviduct.

oviduct [G *Ovidukt, Eileiter*]: Section of ♀ reproductive system between each ovary and gonopores at bases of first pair of cirri.

ovigerous frena: In certain stalked barnacles, one in a series of fleshy ridges to which egg masses (ovigerous lamellae) are attached.

ovigerous lamella: In certain stalked barnacles, one in a series of egg masses within mantle cavity. (See also ovigerous frena)

ovisac [G *Eisack*]: Region of mantle cavity in which eggs are brooded. (See also brood chamber)

palp [G *Palpus, Taster*]: One of two projections from lateral margins of labrum; in acrothoracican, one of two projections from each mandible.

paries [G *Paries*]: Barnacle plate feature: in unstalked form, median triangular part of each compartmental plate. Bears alae and/or radii laterally.

parietal canal: longitudinal tube

parietal plate: plate

parietal pore: longitudinal tube

parietal septum: longitudinal septum

parietal tube: longitudinal tube

pedicle: protopod

peduncle (stalk) [G *Stiel*]: In stalked barnacle, one of two external divisions of body (capitulum, peduncle). Attached to substratum by cement glands; contains ovaries. (naked, armed with scales)

penis (intromittent organ) [G *Penis*]: Male copulatory organ. In barnacle, located between bases of sixth (= last) pair of thoracic appendages (cirri); in ascothoracican, extends from first abdominal segment. (smooth, annulated; armed, unarmed; with/without basidorsal point)

pharynx: esophagus

plate [G *Platte*]: In barnacle, one of a number of calcareous elements on outer surface of body. Produced by mantle and with characteristic structure, shape, and sculpture. One generally distinguishes five primary plates (single carina and paired scuta and terga) and numerous secondary plates, including single rostrum, subrostrum, subcarina, as well as various lateral plates (lateralia or latera). (See also compartmental plate, operculum, parietal plate, scale)

primary plate [G *primäre Platte*]: One of five principal plates on outer surface of body. Includes single carina as well as paired scuta and terga (and occasionally single rostrum). (See also secondary plate)

primordial plate (primordial valve) [G *Primordialplatte*]: One of typically five more rigid, plate-like elements under shell of cypris larva.

protopod (pedicle) [G *Protopodit*]: Two-segmented proximal part of thoracic appendage (thoracopods in ascothoracican, cirri in other cirripeds). Bears distal endopod and exopod.

radius [G *Radius*]: Barnacle plate feature: in unstalked form, lateral part of compartmental plate. Delimited from median part (paries) by altered growth lines; overlaps sides (alae) of adjoining compartmental plates.

ramus [G *Ast*]: Branch of appendage, typically referring to inner (endopod) and outer (exopod) branches of thoracic appendage (thoracopods in ascothoracican, cirri in other cirripeds). (See also caudal ramus)

rostral angle: Barnacle plate feature: in stalked form, corner of scutum formed by junction of basal and occludent margins.

rostral latus [G *Latus rostrale*]: Barnacle plate: paired plates (latera) in stalked form, one on each side of rostrum or below scutum. Homologous with rostrolateral of unstalked form.

rostral plate: rostrum

rostral sinus (blood pump) [G *Rostralsinus, Sinus rostralis*]: In circulatory system, main pumping organ located in anterior region of body.

rostrolateral [G *Rostrolaterale*]: Barnacle shell plate: in symmetrical unstalked form, paired anterolateral plate, one between rostrum and each lateral plate. Consists of median paries and lateral radii. May be fused with rostrum or with each other (i.e., rostrum absent) to form compound rostrum.

rostrum [G *Rostrum*]: Barnacle shell plate: unpaired anterior plate. In stalked form, anteroventral plate on capitulum. In asymmetrical unstalked form, compartmental plate between carina and fixed scutum. In symmetrical unstalked form, anterior compartmental plate with median paries and lateral alae (simple or true rostrum) or with median paries and lateral radii (compound rostral plate: formed by fusion of rostrolaterals with rostrum or by fusion of rostrolaterals, with rostrum missing).

scale [G *Schuppe, Stielschuppe*]: One of numerous small calcareous elements on stalk (peduncle) of barnacle.

scutal margin [G *Scutalrand*]: Barnacle plate feature: edge of tergum or edge of any other plate articulating with or directly adjoining scutum.

scutum [G *Scutum*]: One of five primary plates (carina, two scuta, two terga) on outer surface of barnacle. Paired, one scutum located on each side of aperture. In unstalked barnacle, variously associated with tergum to form operculum. Features of scutum include angles (basioccludent, basitergal), apex, articular furrow, margins (basal, occludent, tergal, tergolateral), and ridges (apicobasal, apicoumbonal, articular). Bears markings (muscle pits) of muscle attachments on inner surface. (rhomboidal, triangular, split/not split into two parts)

second antenna: antenna

secondary plate [G *sekundäre Platte*]: Any one of numerous accessory calcareous elements on body (as opposed to primary plates). May include single rostrum, subrostrum, subcarina, and numerous lateral plates (latera, lateralia).

seminal receptacle [G *Receptaculum seminis*]: Within brood chamber of ♀ rhizocephalan, small cavity in which sperm are stored.

seminal vesicle [G *Vesicula seminalis*]: In ♂ reproductive system, expanded section of sperm duct (vas deferens).

seta [G *Borste*]: One of numerous bristle-like projections extending from appendage. Number, shape, and arrangement are of taxonomic importance. (lanceolate, plumose, simple)

sheath: Barnacle shell feature: in unstalked form, collar-like structure inside shell. Formed by thickened upper parts (parietes, alae) of compartmental plates; serves as site of attachment of opercular membrane.

shell [G *Skelett, Schale*]: In unstalked barnacle, general term for hard elements surrounding body; includes compartmental plates, basis (if calcareous), and opercular valves.

simple rostrum: rostrum

sinus [G *Sinus*]: In circulatory system, one of several blood-filled spaces. According to cirriped group and position one may distinguish epineural, mantle, prosomal, rostral, and thoracic sinuses. (See also rostral sinus)

somite (segment) [G *Segment*]: One in a series of divisions of body. Head consists of five somites, thorax of six, and in ascothoracican, abdomen of four to five somites.

sperm duct: vas deferens

spur [G *Sporn*]: Barnacle plate feature: in unstalked form, calcareous projection of basal margin of tergum.

spur furrow: Barnacle plate feature: in unstalked form, groove on outer surface of tergum, in line with spur and extending to upper angle (apex).

stomach [G *Magen*]: Enlarged anterior region of midgut.

stylet [G *Stachel, Kentron*]: In rhizocephalan, hollow tube used by kentrogon to pierce and enter host.

subcarina [G *Subcarina*]: Barnacle plate: in stalked form, relatively small, unpaired plate below carina.

subesophageal ganglion [G *Unterschlundganglion, Ganglion infraoesophageum*]: Concentration of nerve tissue below (posterior to) esophagus. Joined to supraesophageal ganglion by pair of esophageal connectives. Continuous posteriorly with ventral nerve cord or fused with ganglia of ventral nerve cord to form ventral nerve mass.

subrostrum [G *Subrostrum*]: Barnacle shell plate: in stalked form, relatively small, unpaired plate below rostrum.

supraesophageal ganglion (brain, cerebral ganglion) [G *Oberschlundganglion, Gehirn*]: Anterior concentration of nerve tissue. Joined to subesophageal ganglion by pair of esophageal connectives.

tagma [G *Tagma*]: Major division of body (head, thorax; in ascothoracican, head, thorax, abdomen).

tergal margin [G *Tergalrand*]: Barnacle plate feature: edge of scutum or edge of any other plate directly adjoining tergum.

tergolateral margin [G *Tergolateralrand*]: Barnacle plate feature: in stalked form bearing upper latera, angular edge of scutum adjoining tergum and upper latus.

tergum [G *Tergum*]: One of five primary plates (carina, two scuta, two terga) on outer surface of barnacle. Paired, one tergum located on each side of aperture. In unstalked barnacle, variously associated with scutum to form operculum. Features or regions of tergum include angles (basicarinal, basiscutal), apex, articular ridge, margins (basal, carinal, occludent, scutal), spur, and spur furrow. (fixed, movable)

testis [G *Hoden*]: Large, paired section of ♂ reproductive system in which sperm are produced. Located in anterior region of body and opening to exterior via pair of vasa deferentia and unpaired penis.

thoracopod [G *Thoracopod, Thoraxbein*]: One of up to six pairs of appendages of thorax. Typically refers to appendages of ascothoracican, those of other cirripeds being termed cirri. (uniramous, biramous) (See also maxilliped)

thorax [G *Thorax*]: Posterior of typically two basic divisions (tagmata) of body (head, thorax; in ascothoracican, head, thorax, abdomen). Consists of six segments (somites) bearing appendages (cirri).

trophi (pl.) [G *Trophi, Mundwerkzeuge*]: Collective term for mouthparts (mandibles, maxillules, maxillae). Closely associated with labrum with which, in ascothoracican, they form oral pyramid.

umbo [G *Umbo*]: Barnacle plate feature: in stalked barnacle, location on plate from which successive growth increments extend. (basal, subapical, subcentral)

upper latus [G *Latus superius*]: Barnacle plate: paired plates (latera) in stalked form, one on each side of capitulum between scutum and tergum or carina.

upper lip: labrum

valve: opercular valve

vas deferens (sperm duct)
[G *Vas deferens, Samenleiter*]: In ♂ reproductive system, duct extending from each testis; ducts merge at base of penis (to form ductus ejaculatorius).

vas efferens [G *Vas efferens*]: In ♂ ascothoracican, one of numerous small ducts leading from testicular lobes to vasa deferentia.

ventral nerve cord [G *Bauchmark, ventraler Nervenstrang*]: Pair of nerve cords extending posteriorly from subesophageal ganglion. Ganglia of ventral nerve cord may be fused with subesophageal ganglion to form large ventral nerve mass.

wall [G *Mauerkrone*]: In unstalked barnacle, calcareous wall surrounding body; closed dorsally by operculum. Consists of series of up to eight overlapping/articulating compartmental plates (from anterior to posterior: single rostrum, paired rostrolaterals, laterals, carinolaterals, and single carina) encircling body.

LEPTOSTRACA

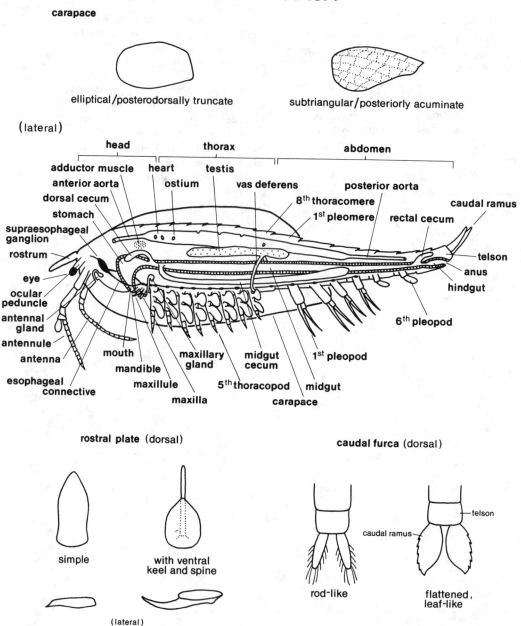

carapace

elliptical/posterodorsally truncate

subtriangular/posteriorly acuminate

(lateral)

head thorax abdomen

adductor muscle heart testis
anterior aorta ostium vas deferens
dorsal cecum
stomach 8th thoracomere
supraesophageal 1st pleomere
ganglion
rostrum

eye
ocular
peduncle
antennal
gland
antennule
antenna

posterior aorta caudal ramus
rectal cecum
telson
anus
hindgut
6th pleopod

mouth maxillary midgut 1st pleopod
mandible gland cecum
esophageal maxillule
connective maxilla 5th thoracopod midgut
carapace

rostral plate (dorsal)

simple with ventral
keel and spine

(lateral)

caudal furca (dorsal)

telson
caudal ramus

rod-like flattened,
leaf-like

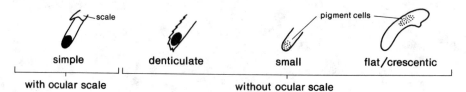

ocular peduncle

scale pigment cells

simple denticulate small flat/crescentic

with ocular scale without ocular scale

ARTHROPODA

CRUSTACEA

Leptostraca

leptostracan [G *Leptostrake*]

abdomen (pleon) [G *Abdomen, Pleon*]: Posterior division (tagma) of body (cephalon, thorax, abdomen). Consists of seven segments (somites), the first six each bearing pair of pleopods; bears telson and caudal furca posteriorly. Anterior region of abdomen covered, along with thorax and posterior part of head (cephalon), by carapace.

acicle: ocular scale

adductor muscle: carapace adductor muscle

aesthetasc [G *Aesthetasc, Sinnesschlauch, Riechschlauch*]: One in a series of sensory projections on antennules and occasionally also on antennae.

anal somite (anal segment): telson

antenna (second antenna) [G *Antenna, zweite Antenne*]: Second pair of antennae. Uniramous, relatively long, and consisting of three- or four-segmented peduncle and variously shaped endopod (flagellum). (peduncle: three-/four-segmented; flagellum: long with many small joints, small with few large joints)

antennal gland [G *Antennendrüse, Nephridium*]: Pair of excretory glands in head, one opening to exterior at base of each antenna. (See also maxillary gland)

antennule (first antenna) [G *Antennula, erste Antenne*]: First pair of antennae. Relatively long, basically biramous, typically consisting of four-segmented peduncle with flagelliform inner branch (endopod) and scale-like outer branch (exopod). (biramous, uniramous; fourth segment of peduncle: with/without process, with tuft of bristles)

anus [G *After*]: Posterior opening of digestive tract; located ventrally on telson between caudal rami.

aorta [G *Aorta*]: In circulatory system, major unpaired blood vessel extending from end of heart. One may distinguish an anterior aorta extending into head and a posterior aorta extending into abdomen; each is branched distally.

appendix interna [G *Appendix interna*]: In each of the four well-developed anterior pairs of pleopods, medially (inwardly) directed process at base of endopod. Serves to link pleopods together for swimming.

artery [G *Arterie*]: One of numerous blood vessels originating from heart or aortas. According to position or organ supplied, one may distinguish dorsal, labral, median, and segmental arteries, the latter with podial and visceral branches.

article (joint, segment) [G *Glied, Ringel*]: One in a series of segments of flagellum of antennule or antenna. May also refer to segment of other appendages.

brain: supraesophageal ganglion

brood chamber [G *Brutraum*]: In ♀, space between thoracopods and carapace in which eggs are brooded.

carapace [G *Carapax*]: Large shield covering thorax, posterior part of head, and anterior part of abdomen. Laterally compressed and therefore often described as being bivalved (although no dorsal hinge is present). Two halves of carapace can be drawn toward each other by carapace adductor muscle. Bears rostral plate anteriorly. (elliptical and posterodorsally truncate, subtriangular and posteriorly acuminate; smooth, with polygonal ornamentation)

carapace adductor muscle (adductor muscle) [G *Schalen-Schließmuskel, Schließmuskel des Karapax*]: Pair of transverse muscles in head region. Joined medially by horizontal ligament and attached on each side to inner surface of carapace. Serves to draw two halves of carapace together.

cardiac stomach [G *Kardiakalteil des Kaumagens*]: stomach

caudal furca [G *Furca, Furka*]: caudal ramus

caudal ramus (furcal ramus) [G *Furkalast, Furca-Ast*]: One of two posteriorly directed projections of last body segment (telson). Both rami together termed caudal furca. (dorsoventrally flattened and leaf-like, rod-like) (See also uropod)

cecum: midgut cecum, rectal cecum

cephalon (head) [G *Kopf*]: Anterior of three divisions (tagmata) of body (cephalon, thorax, abdomen). Consists of five somites bearing antennules, antennae, mandibles, maxillules, and maxillae. Covered by rostral plate and, posteriorly, by carapace.

compound eye: eye

coxa [G *Coxa, Koxa*]: Most proximal segment of appendage. Gonopores located on coxae of sixth (♀) or eighth (♂) thoracopods.

digestive gland: midgut cecum

diverticulum: midgut cecum

dorsal cecum: midgut cecum

endite [G *Endit*]: Inwardly (medially) directed lobe of proximal part (protopod) of appendage, e.g., of maxillules or maxillae.

endopod (endopodite) [G *Endopod, Endopodit*]: Inner branch (ramus) of biramous appendage. (simple, two-, three- to four-, five-segmented) (See also exopod)

epipod (epipodite) [G *Epipod, Epipodit*]: In thoracic appendage (thoracopod), laterally (outwardly) directed lobe of protopod. Serves in respiration.

esophageal connective [G *Schlundkonnektiv*]: One of two strands of nerve tissue surrounding esophagus and connecting brain (supraesophageal ganglion) with ventral nerve cord.

esophagus [G *Ösophagus*]: Short, cuticle-lined section of digestive tract between mouth and stomach. Esophagus and stomach form foregut.

exite [G *Exit*]: Outwardly (laterally) directed branch of proximal part (protopod) of appendage, e.g., of maxillae.

exopod (exopodite) [G *Exopod, Exopodit*]: Outer branch (ramus) of biramous appendage. (elongate and slender, flattened = lamellar = scale-like = plate-like)

eye (compound eye) [G *Auge, Komplexauge*]: One of two organs of vision located at tip of eyestalk (ocular peduncle).

eyestalk: ocular peduncle

first antenna: antennule

flagellum [G *Flagellum, Geißel*]: Distal division (peduncle, flagellum) of antennule. Represents inner branch (endopod). May also refer to endopod of antennule. (long with many small articles, small with few large articles)

foregut [G *Vorderdarm*]: Anterior region of digestive tract (foregut, midgut, hindgut); consists of esophagus and stomach.

gonad: ovary, testis

gonopore [G *Gonoporus, Geschlechtsöffnung*]: Opening of reproductive system to exterior. In ♂, opening at tip of papilla (genital papilla) on coxa of each eighth thoracopod; in ♀, opening on coxa of each sixth thoracopod.

head: cephalon

heart [G *Herz*]: In circulatory system, elongate, tubular pumping organ extending above digestive tract and gonads from posterior region of head to fourth pleomere. Bears 11 ostia and gives rise to numerous arteries as well as anterior and posterior aortas.

hindgut [G *Enddarm*]: Posterior region of digestive tract (foregut, midgut, hindgut); extends through one or more abdominal segments and opens to exterior via anus. If present, rectal cecum extends above hindgut.

incisor process [G *Pars incisiva*]: Gripping process (vs. grinding = molar process) of each mandible. (simple, with teeth)

joint: article

labium (lower lip) [G *Labium, Unterlippe*]: Lobe-like structure posterior to mouth; divided into two lobes (paragnaths).

labrum (upper lip) [G *Labrum, Oberlippe*]: Unpaired lobe-like structure anterior to and partially covering mouth.

lower lip: labium

manca stage (mancoid stage) [G *Manca-Stadium, Jungtier*]: Postlarval stage at which leptostracans hatch. Well developed, differing from adult in having rudimentary fourth pleopod pair.

mandible [G *Mandibel*]: Third paired appendage of head (cephalon); located between antennae and maxillules. Represents first pair of mouthparts and, if fully developed, consists of basal incisor and molar processes as well as distal palp.

maxilla (second maxilla) [G *Maxille, zweite Maxille*]: Fifth paired appendage of head (cephalon); located between maxillules and first pair of thoracopods. Represents third pair of mouthparts and, if fully developed, consists of proximal section with four endites as well as distal branches (endopod and exopod). Maxillary glands open at base of maxillae.

maxillary gland [G *Maxillendrüse, Nephridium*]: Pair of excretory glands located in head under carapace adductor muscle, one opening to exterior at base of each maxilla. (See also antennal gland)

maxillule (first maxilla) [G *Maxillula, erste Maxille*]: Fourth paired appendage of head (cephalon); located between mandibles and maxillae. Represents second pair of mouthparts and, if fully developed, consists of proximal section with two endites as well as distal palp.

midgut [G *Mitteldarm*]: Elongate section of digestive tract between foregut and hindgut. May bear one to several pairs of midgut ceca anteriorly and pair of rectal ceca posteriorly.

midgut cecum (digestive gland, diverticulum) [G *Mitteldarmdivertikel, Hepatopankreas, Divertikel*]: One of up to several pairs of variously developed digestive organs extending either anteriorly or posteriorly from anterior end of midgut. Relatively short, anteriorly directed pair of dorsal ceca are termed dorsal ceca. (sac-like, trilobed) (See also rectal cecum)

molar process [G *Pars molaris*]: Grinding process (vs. gripping = incisor process) of each mandible.

mouth [G *Mund, Mundöffnung*]: Anterior opening of digestive tract on underside of head. Bordered anteriorly by labrum, posteriorly by labium (paragnaths); opens into esophagus.

ocular peduncle (eyestalk) [G *Augenstiel*]: Movable, eye- or visual pigment-bearing projection on each side of head (cephalon). (simple, denticulate = serrate, flat and crescentic, scimitar-like; with/without ocular scale)

ocular scale (basal scale)
[G *Augenstachel*]: Scale- or spine-shaped structure projecting from base of each eyestalk (ocular peduncle); tip may cover ocular peduncle dorsally.

ostium [G *Ostium*]: One of 11 muscular openings in heart (three pairs of lateral ostia followed by three unpaired dorsal and one pair of lateral ostia). Blood in pericardium enters heart through ostia.

ovary [G *Ovar, Ovarium*]: In ♀ reproductive system, pair of relatively slender, tubular organs in which eggs are produced; located between digestive tract and heart and extending through most of body. Each ovary opens at base of sixth thoracopods via oviduct.

oviduct [G *Ovidukt, Eileiter*]: Relatively short, slender section of ♀ reproductive system between each ovary and gonopore.

palp [G *Palpus, Taster*]: Branch-like projection of mouthpart (mandible, maxillule, maxilla). May represent various parts of appendage (e.g., endopod, exite).

paragnath [G *Paragnath*]: One of two lobe-like structures forming lower lip (labium) posterior to mouth.

peduncle [G *Pedunculus, Stiel, Schaft*]: Proximal of two divisions (peduncle, flagellum) of antennule or antenna. (three-/four-segmented; fourth segment: with/without process, with tuft of bristles)

pericardium (pericardial sinus) [G *Perikard, Perikardialsack*]: Elongate cavity surrounding heart. Blood collected in pericardium enters heart through ostia.

pleomere (abdominal somite) [G *Pleomer*]: One in a series of seven segments (somites) of abdomen, the first four bearing well-developed appendages (pleopods), the fifth and sixth with reduced appendages. Bears telson posteriorly.

pleon: abdomen

pleopod (abdominal appendage) [G *Pleopod*]: One of six pairs of appendages of abdomen, the first four pairs being well developed, the last two pairs being reduced. Basically biramous, consisting of protopod with inner branch (endopod) and outer branch (exopod). (biramous, uniramous) (See also appendix interna)

protopod (protopodite) [G *Protopodit*]: Proximal part of appendage, e.g., of maxillae or pleopods.

pyloric stomach [G *Pylorikalteil des Magens*]: stomach

ramus [G *Ast*]: Branch of appendage; typically refers to inner (endopod) and outer (exopod) branches of biramous appendage. (lamellar = plate-like, flagelliform)

rectal cecum [G *Rektalcoecum*]: In digestive tract, unpaired dorsal cecum originating at end of midgut and extending over hindgut.

rostral plate (rostrum) [G *Rostralklappe, Rostrum*]: Plate-like structure extending over head. Articulated with carapace and movable by series of muscles. (simple, with ventral keel and spine)

rostrum: rostral plate

second antenna: antenna

segment [G *Glied*]: One in a series of divisions of an appendage. May also refer to segment of body (then also termed somite). Segments of flagella of antennule or antenna are often termed articles.

seta [G *Borste*]: One of numerous bristle-like projections extending from appendage or body.

somite [G *Segment*]: One in a series of divisions of body. Head (cephalon) consists of five fused somites, thorax of eight, and abdomen of seven somites (excluding posterior telson).

sperm duct: vas deferens

stomach [G *Magen, Kaumagen*]: Enlarged section of digestive tract between esophagus and midgut. If fully developed, consists of anterior cardiac region with grinding apparatus and posterior pyloric region with setose internal lobes. Forms foregut together with esophagus.

supraesophageal ganglion (brain) [G *Oberschlundganglion, Gehirn*]: Main concentration of nerve tissue in head (cephalon). Connected to ventral nerve cord by pair of esophageal connectives.

tagma [G *Tagma*]: Major division of body (cephalon, thorax, abdomen).

telson [G *Telson*]: Last segment of body; bears pair of caudal rami. Bears ventral anus and therefore also interpreted as representing anal somite.

testis [G *Hoden*]: In ♂ reproductive system, pair of slender, tubular organs in which sperm are produced. Located between digestive tract and heart and extending through most of body; each testis opens on coxa of eighth thoracopod via vas deferens.

thoracomere (thoracic somite) [G *Thoracomer, Thoraxsegment*]: One of eight segments (somites) of thorax, each bearing pair of appendages (thoracopods). All thoracomeres are covered by carapace.

thoracopod (thoracic appendage) [G *Thoracopod*]: One of eight pairs of appendages of thorax, one pair per thoracic somite (thoracomere). Biramous, typically flattened and consisting of proximal section (protopod) with inner branch (endopod) and outer branch (exopod). (foliaceous, phyllopodous; with/without epipods; projecting/not projecting beyond ventral margin of carapace)

thorax [G *Thorax*]: Division (tagma) of body between head (cephalon) and abdomen. Consists of eight somites (thoracomeres), each bearing pair of thoracopods. Covered, along with part of head and abdomen, by carapace.

upper lip: labrum

uropod [G *Uropod*]: Term applied to each caudal ramus if last body segment (telson) is interpreted as representing anal somite.

vas deferens (sperm duct) [G *Vas deferens, Samenleiter*]: Relatively short, slender section of ♂ reproductive system between each testis and gonopore.

ventral groove (food groove) [G *Bauchrinne*]: Elongate median line between thoracopod bases on ventral side of thorax. Food is trapped by thoracopods, passed anteriorly along ventral groove, and transferred to mouth.

ventral nerve cord [G *Bauchmark*]: Longitudinal nerve cord extending below digestive tract to sixth abdominal somite (next to last somite). Joined to brain (supraesophageal ganglion) by pair of esophageal connectives; bears series of segmentally arranged ganglia.

STOMATOPODA

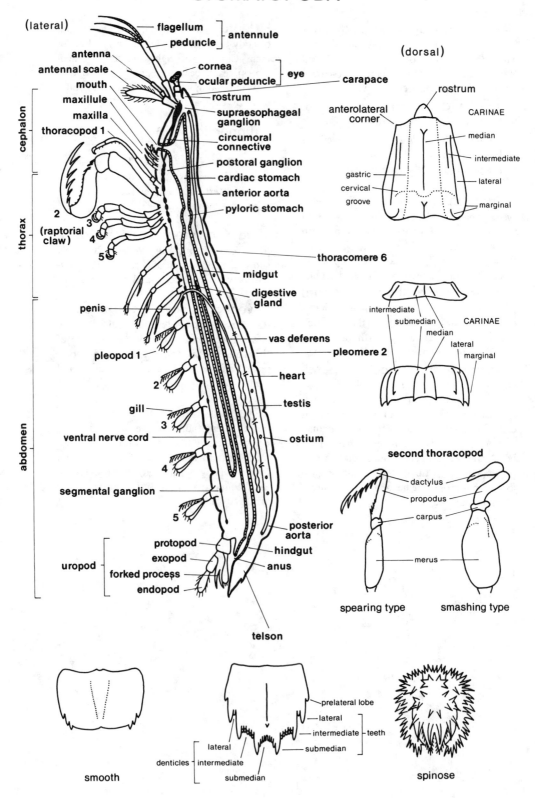

(lateral)

flagellum
peduncle — antennule

antenna
antennal scale
mouth
maxillule
maxilla
thoracopod 1

cornea
ocular peduncle — eye
rostrum
supraesophageal ganglion
circumoral connective
postoral ganglion
cardiac stomach
anterior aorta
pyloric stomach

cephalon

thorax

2
(raptorial claw)
3
4
5

thoracomere 6

midgut
digestive gland

penis

vas deferens

pleopod 1

pleomere 2

2

heart

gill

testis

3

ventral nerve cord

ostium

4

segmental ganglion

5

posterior aorta

protopod
exopod
forked process
endopod

hindgut
anus

uropod

abdomen

telson

(dorsal)

carapace

rostrum

anterolateral corner

CARINAE

median
intermediate
lateral
marginal

gastric
cervical
groove

intermediate
submedian
median

CARINAE

lateral
marginal

second thoracopod

dactylus
propodus
carpus

merus

spearing type   smashing type

smooth

prelateral lobe
lateral
intermediate — teeth
submedian

denticles
lateral
intermediate
submedian

spinose

ARTHROPODA

CRUSTACEA

Stomatopoda

stomatopod, mantis shrimp [G *Stomatopode, Maulfüßer, Fangschreckenkrebs*]

abdomen (pleon) [G *Abdomen, Hinterleib, Pleon*]: Posterior division (tagma) of body (cephalon, thorax, abdomen). Consists of six somites (pleomeres), each bearing pair of appendages (pleopods). Sixth pleomere (with uropods) bears or is fused with telson. (compressed, depressed)

abdominal appendage: pleopod

accessory gland [G *akzessorische Drüse*]: In ♂ reproductive system, pair of slender glands in thorax under heart; fused anteriorly and opening at tips of penes (via ducts running parallel to sperm ducts).

acron (ophthalmic somite) [G *Akron*]: Anteriormost segment of body (although not considered to be a true somite); divided into three plates—fastigial, ocular, and postocular—of which the median ocular plate bears the eyestalks (ocular peduncles). Not covered by carapace, although rostrum may partially or entirely extend across acron.

alima [G *Alimalarve*]: erichthus larva

ampulla (pars ampullaris) [G *Ampulle*]: In digestive tract, diverticulum at junction of posterior (pyloric) stomach, digestive ceca, and midgut.

antenna (second antenna) [G *Antenne, zweite Antenne*]: Second and smaller pair of antennae; originating from second somite of head (cephalon). Biramous, consisting of basal, two-segmented peduncle, two-segmented exopod (distal segment = antennal scale), and three-segmented endopod (with distal flagellum). (See also antennule)

antennal scale (scaphocerite) [G *Scaphocerit, Schuppe*]: On each (second) antenna, large, flattened terminal segment of exopod; bears setae.

antennular process: Tapering dorsolateral extension on each side of first head somite (antennular somite).

antennular somite (posterior dorsal plate) [G *Antennularsegment*]: First true somite of head (cephalon); located between acron and second (antenna-bearing) somite. Bears pair of anteriorly directed antennules as well as dorsolateral antennular processes. Not covered by carapace, although rostrum may partially or entirely extend across it.

antennule (first antenna) [G *Antennula, erste Antenne*]: First and larger pair of antennae; originating from first (antennular) somite of head. Consists of slender, three-segmented peduncle and three short flagella.

antizoea [G *Antizoea*]: Early larval stage of certain stomatopods; characterized by uniramous antennules, five pairs of biramous thoracopods, and no pleopods. (See also pseudozoea)

anus [G *After*]: Posterior opening of digestive tract; located anteroventrally on telson.

aorta [G *Aorta*]: Major blood vessel leading from heart. According to position one may distinguish an anterior aorta (giving rise to antennal, antennular, ophthalmic, and other cephalic arteries) and a posterior aorta.

appendix interna (stylamblys) [G *Appendix interna, Stylamblys*]: Projection of inner branch (endopod) of each pleopod. Bears hooks (retinacula) and, in first pleopod pair of ♂, hook-shaped and tubular processes (petasma).

artery [G *Arterie*]: One of numerous blood vessels originating from heart or aorta. According to position or appendages supplied, one may distinguish antennal, antennular, ophthalmic, lateral, and subneural arteries.

article [G *Glied*]: One in a series of numerous segments of flagellum at end of antennule or antenna.

basis (basipodite) [G *Basis, Basipodit*]: Third segment of thoracic appendage (thoracopod); located between coxa and merus. Also interpreted as being second segment (between coxa and ischium).

branchia: gill

carapace [G *Carapax*]: Shield-like cuticular structure covering cephalon (except for anterior acron and antennular somite) as well as first four thoracic somites (thoracomeres); overlaps but does not enclose bases of first five thoracopods. Broad, flattened, and typically sculptured with various carinae and grooves. Bears rostrum anteriorly. (with/ without carinae; anterolateral corners: armed, unarmed, rounded)

cardiac stomach [G *Kaumagen*]: Anterior and larger division of stomach occupying most of head (cephalon). Separated from smaller posterior pyloric stomach by filtering apparatus; contains only small pairs of plates (ossicles).

carina [G *Carina, Kiel*]: One in a series of longitudinal, keel-like elevations on dorsal surface of either carapace, thoracic somites (thoracomeres), abdominal somites (pleomeres), or telson. According to position (from midline outward) one may distinguish median, submedian, intermediate, lateral, marginal, and various accessory carinae or secondary ridges. Refers also to ventral elevations (e.g., postanal carina). (sharp, swollen/rounded; posteriorly spined, posteriorly rounded; straight, curved; median carina: bifurcate, not bifurcate)

carpus (carpopodite) [G *Carpus, Carpopodit*]: Fifth segment of thoracic appendage (thoracopod); located between merus and propodus.

cecum: digestive gland

cephalon (head) [G *Kopf*]: Anterior division (tagma) of body (cephalon, thorax, abdomen). Consists of an anterior acron and five somites and bears eyes, antennules, antennae, epistome, labrum, mouth, mandibles, maxillules, and maxillae. Covered, along with first four somites (thoracomeres) of thorax, by carapace.

cervical groove [G *Zervikalfurche, Cervicalfurche*]: Transverse groove across posterior third of carapace. (See also gastric groove)

circumoral connective (esophageal connective) [G *Schlundkonnektiv*]: One of two long strands of nerve tissue surrounding anterior section of digestive tract and joining supraesophageal ganglion to postoral ganglion.

cornea [G *Cornea*]: Outermost layer of compound eye; may be divided into upper and lower units. (bilobed, cylindrical, globular, subglobular, laterally expanded; at right angle to stalk, oblique on stalk)

coxa (coxopodite) [G *Coxa, Coxopodit*]: Second segment of thoracic appendage (thoracopod). Also interpreted as being first segment (forming fused precoxa and coxa).

dactyl (dactylus) [G *Dactylus, Dactylopodit*]: Seventh and terminal segment of thoracic appendage (thoracopod). In first five thoracopods, folded back on propodus to form subchela. Particularly well developed on second thoracopod to form raptorial claw. (inner margin: armed, unarmed; outer margin: curved, sinuate; base: inflated, not inflated)

denticle [G *Stachel*]: One of numerous smaller tooth-like projections between larger teeth on telson margin; according to position one may distinguish submedian, intermediate, and lateral denticles.

digestive gland (digestive cecum, cecum) [G *Mitteldarm-Drüsenschlauch, Hepatopankreas*]: Pair of large glands, one originating from each side of pyloric stomach and extending along midgut. Fills most of body, giving rise to pair of lateral pouches in posterior four thoracomeres, in all pleomeres, and branching into telson.

dorsal pit: median pore

endite [G *Endit*]: Inwardly (medially) directed lobe of first two segments of maxillules and maxillae. (bilobed, with terminal tooth)

endopod (endopodite) [G *Endopod, Endopodit*]: Inner branch (ramus) of biramous appendage. Refers to three-segmented, flagellum-bearing branch of antennae, appendix interna–bearing branch of pleopods, and one-segmented branch of uropods, while it is considered to represent outer branch of last three thoracopods. (See also exopod)

epipod (epipodite) [G *Epipod, Epipodit*]: Laterally directed branch of base of first five thoracic appendages (thoracopods); with respiratory function in thoracopods 3–5.

epistome [G *Epistom*]: Relatively large, plate-like structure in front of mouth on underside of head (cephalon); adjoined posteriorly by labrum.

erichthus larva [G *Erichthuslarve*]: Later larval stage of most stomatopods; similar stage in other stomatopods is termed alima.

esophageal connective: circumoral connective

exopod (exopodite) [G *Exopod, Exopodit*]: Outer branch (ramus) of biramous appendage. Refers to two-segmented branch of antennae, one-segmented, gill-bearing branch of pleopods, and two-segmented branch of uropods, while it is considered to represent inner branch of last three thoracopods. (See also endopod)

eye (compound eye) [G *Auge, Komplexauge*]: Large, paired organ of vision on head (cephalon); each eye is positioned on eyestalk (ocular peduncle) originating on acron. (See also cornea, nauplius eye)

eyestalk: ocular peduncle

fastigial plate: Anteriormost of three plates (fastigial, ocular, postocular) composing acron at tip of head.

filtering apparatus (posterior cardiac plate) [G *Zygokardialplatte + Rückenplatte = Cardiaplatte*]: Series of plates (median, upper lateral, lower lateral) separating larger anterior division of stomach (cardiac stomach) from smaller posterior division (pyloric stomach). Functions as sieve.

finger: dactyl

flagellum [G *Flagellum, Geißel, Antennengeißel*]: Distal section of antennule or antenna (peduncle, flagellum); slender, consisting of numerous segments (articles). Antennule bears three flagella (one dorsal and divided ventral flagella); antenna, one flagellum (on endopod).

forked process [G *Protopodit-Platte*]: Well-developed prolongation of basal part (protopod) of uropod. Flattened, extending between endopod and exopod. (with two terminal spines = bifurcate, with three terminal spines)

gastric groove [G *Gastrikalfurche, Gastricalfurche*]: Longitudinal groove in carapace, one on each side of median carina; extends from anterior end (rostrum) to posterior margin. (See also cervical groove)

genital papilla: penis

gill (branchia) [G *Kieme*]: Respiratory projections on appendages (pleopods) of first five abdominal appendages (pleomeres); located on exopod. Epipods of third through fifth thoracopods also with respiratory function.

gonopore [G *Gonoporus, Geschlechtsöffnung*]: Opening of ♂ or ♀ reproductive system to exterior. In ♀, single opening on underside of sixth thoracic segment (thoracomere); in ♂, pore at tip of penis on each appendage (thoracopod) of eighth thoracomere.

groove: cervical groove, gastric groove

head: cephalon

heart [G *Herz*]: Main longitudinal vessel of circulatory system. Located middorsally above digestive tract and gonads and extending through most of thorax and abdomen. Bears 13 pairs of ostia and gives rise to series of paired, segmentally arranged lateral arteries and to anterior and posterior aortas.

hindgut [G *Enddarm*]: Posterior section of digestive tract; follows midgut at level of fifth abdominal somite (pleomere). Opens to exterior via anus on telson.

incisor process (pars incisiva) [G *Processus incisivus*]: Well-developed food-gripping (vs. grinding = molar process) of each mandible; bears single row of teeth and lies over labrum.

ischiomerus: merus

ischium: merus

labium (lower lip) [G *Labium, Unterlippe*]: Lip-like structure posterior to mouth and mandibles on underside of head (cephalon); bulbous and bilobed, each fringed lobe being termed paragnath.

labrum (upper lip) [G *Labrum, Oberlippe*]: Relatively small, unpaired fleshy lobe in front of mouth; adjoined anteriorly and laterally by epistome.

lacuna [G *Lakune, Längslakune*]: In circulatory system, large, longitudinal space between digestive tract and ventral nerve cord in which blood collects before entering gills. (See also sinus)

lateral process: pleurite

lower lip: labium

mandible [G *Mandibel*]: Anteriormost paired mouthpart on underside of head (cephalon); borne on third cephalic somite. Well developed, bearing molar and incisor processes. (with/without palp)

maxilla (second maxilla) [G *Maxille, zweite Maxille*]: Third paired mouthpart on underside of head (cephalon); borne on fifth cephalic somite and located posterior and external to maxillules. Consists of four segments, the first being cylindrical, the others flattened; segments one and two with endites (that of second is bilobed). Bears opening of maxillary gland.

maxillary gland [G *Maxillennephridium, Maxillardrüse*]: Pair of excretory organs in head (cephalon), each opening to exterior via pore on basal segment of maxilla.

maxilliped (maxillipede) [G *Maxilliped, Kieferfuß*]: Term occasionally applied to first five appendages (thoracopods) of thorax.

maxillule (first maxilla) [G *Maxillula, erste Maxille*]: Second paired mouthpart on underside of head (cephalon); borne on fourth cephalic somite and located between mandibles and maxillae. Consists of two segments (coxa and basis, each with endite) and small endopod (palp).

median pore (dorsal pit): Dorsal pit in carapace; located along midline between cervical groove and anterior bifurcation of median carina.

merus [G *Merus*]: Fourth segment of thoracic appendage (thoracopod). Considered to represent fused ischium and merus (ischiomerus), yet also interpreted as being separate from and following ischium.

midgut [G *Darm*]: Narrow section of digestive tract between posterior division of stomach (pyloric stomach) and hindgut; extends through posterior somites of thorax and through abdomen. Gives rise anteriorly to pair of digestive glands.

molar process (pars molaris) [G *Processus molaris*]: Well-developed grinding process (vs. gripping = incisor process) of each mandible. Armed with two rows of teeth and extends beneath labrum into mouth.

mouth [G *Mund*]: Anterior opening of digestive tract on underside of head (cephalon); bordered anteriorly by labrum, posteriorly by labrum and mouthparts. Opens directly into stomach.

nauplius eye [G *Naupliusauge*]: Minute, unpaired light-sensitive organ on ocular plate between eyestalks (ocular peduncles).

ocular peduncle (eyestalk) [G *Augenstiel*]: One of two eye-bearing, movable projections of medial (ocular) plate of acron. (cylindrical, dilated, subtriangular)

ocular plate [G *Augenplatte*]: Second of three plates (fastigial, ocular, postocular composing acron at tip of head; bears eyestalks (ocular peduncles).

ocular scale: One of two small, dorsally directed projections of last plate (postocular plate) of acron.

ophthalmic somite: acron

ostium [G *Ostium*]: One of 13 pairs of openings of longitudinal heart (six pairs in thorax, seven in abdomen). Each ostium is equipped with a valve and serves to return blood into heart.

ovary [G *Ovar*]: Paired section of ♀ reproductive system in which eggs are produced. Elongate, extending along thorax and abdomen between heart and digestive glands. Fused posteriorly in telson and opening to exterior on sixth thoracic somite (thoracomere) via pair of oviducts, unpaired seminal receptacle, and gonopore.

oviduct [G *Ovidukt*]: Section of ♀ reproductive system between each ovary and unpaired seminal receptacle.

palp [G *Palpus*]: Relatively slender, typically three-segmented branch of mandible. Occasionally also refers to small endopod of maxillule. (one-, two-, three-segmented)

paragnath: labium

pars ampullaris: ampulla

peduncle [G *Pedunculus, Stiel*]: Proximal portion of antennule or antenna (two-segmented in the former, three-segmented in latter); bears flagella. (See also ocular peduncle)

penis (genital papilla) [G *Geschlechtspapille*]: Elongate ♂ copulatory structure on precoxa of each appendage (thoracopod) of eighth thoracic somite (thoracomere). Sperm ducts and ducts of accessory glands open at tips of penes.

pericardium (pericardial cavity) [G *Perikard, Perikardialsinus*]: Elongate cavity surrounding heart. Blood, coming from gills, collects in pericardium and enters heart via ostia.

peritrophic membrane gland [G *peritrophische Membran Drüse*]: Pair of small glands, one extending along each side of posterior (pyloric) stomach; opens into midgut and secretes membrane (peritrophic membrane) surrounding food material coming from digestive glands.

petasma [G *Petasma*]: In ♂ reproductive system, copulatory structure on each appendage (pleopod) of first abdominal somite (pleomere). Located on inner branch (endopod) of pleuropod; consists of hook-shaped and tubular processes.

pleomere [G *Pleomer*]: One of six segments (somites) of abdomen; each bears pair of appendages (pleopods). Last (sixth) somite bears uropods and is followed by telson. (See also pleotelson)

pleon: abdomen

pleopod [G *Pleopod*]: One of two appendages of each abdominal somite (pleomere). Biramous, flattened, and consisting of basal protopod and lamellar endopod and exopod, the former bearing an appendix interna, the latter gills. Appendages of last (sixth) pleomere are termed uropods.

pleotelson: telson

pleurite (pleuron, lateral process) [G *Pleurit*]: Lateral extension of body segment (somite).

pleuron: pleurite

posterior cardiac plate: filtering apparatus

postocular plate: Posteriormost of three plates (fastigial, ocular, postocular) composing acron at tip of head.

postoral ganglion (subesophageal ganglion) [G *Unterschlund-ganglion*]: Concentration of nerve tissue posterior to mouth; consists of fused ganglia of eight somites (mandibles to fifth thoracomeres). Joined to supraesophageal ganglion by pair of long circumoral connectives and continues posteriorly as ventral nerve cord.

precoxa [G *Präcoxa*]: First segment of thoracic appendage (thoracopod). Also interpreted as being fused to and therefore also termed coxa.

prelateral lobe: On each lateral margin of telson, lobe-like projection anterior to lateral tooth.

propodus [G *Propodus*]: Sixth segment of thoracic appendage (thoracopod); located between carpus and dactyl. In first five thoracopods, forms subchela together with dactyl (especially well developed in second thoracopod as raptorial claw). Bears tuft of setae for cleaning body on fifth thoracopod. (inner margin: smooth, pectinate, denticulate, grooved)

protopod (protopodite) [G *Protopod, Protopodit, Stamm*]: Proximal part of biramous fifth through eighth thoracopods or pleopods, three segments in the former, one (two) segment(s) in the latter. Bears endopod and exopod distally. May also refer to basal part (peduncle) of antennule or antenna.

pseudozoea [G *Pseudozoea*]: Early larval stage of certain stomatopods; characterized by biramous antennules, two unira-

mous thoracopods, and five pairs of pleopods. (See also antizoea, erichthus larva)

pyloric stomach [G *Pylorus, Filtermagen*]: Posterior and smaller division of stomach. Separated from large anterior cardiac stomach by filtering apparatus; opens into midgut via ampulla. Contains complex of plates (ossicles).

raptorial claw [G *Raubbein*]: Large grasping claw (subchela) at end of second thoracopod; formed by dactyl bearing down on propodus.

retinaculum [G *Retinaculum*]: One in a series of hooks on appendix interna of each abdominal appendage (pleopod); interlocks with retinacula of second member of pleopod pair and permits both pleopods to move in unison.

rostrum (rostral plate) [G *Rostrum*]: Unpaired, flattened anterior projection of carapace; relatively small, movable, and partially or entirely covering acron and antennular somite. (elongate triangular, subquadrate, subtriangular, triangular; with/without carina; deflexed)

scaphocerite: antennal scale

segment [G *Segment*]: One in a series of divisions either of body (also termed somites) or appendage. Segments of flagella at tip of antennules or antennae are often termed articles.

seminal receptacle [G *Receptaculum seminis*]: Unpaired expansion of ♀ reproductive system; located ventromedially in sixth thoracic somite (thoracomere), opens to exterior via gonopore.

sinus [G *Sinus, Blutsinus*]: In circulatory system, more clearly delimited space in which blood collects. Blood coming from gills passes through segmentally arranged, dorsally directed sinuses before entering large pericardial cavity surrounding heart. (See also lacuna)

somite (segment) [G *Segment*]: One in a series of divisions of body. Head (cephalon) consists of five somites (excluding acron), thorax of eight somites (thoracomeres), and abdomen of six somites (pleomeres) excluding telson. Each somite consists of dorsal tergite and ventral sternite and may be produced into lateral pleurites. (smooth, with carinae, denticulate) (See also antennular somite, ophthalmic somite)

sperm duct: vas deferens

stenopod [G *Stabbein*]: Slender, elongate appendage composed of rod-like segments. Refers to three pairs of slender appendages (thoracopods) on sixth through eighth thorax somites (thoracomeres).

sternite [G *Sternit*]: Ventral surface of body segment (somite); ♀ gonopore located on middle of sixth thoracic somite. (See also pleurite, tergite)

sternum [G *Sternum*]: Collective term for all sternites.

stomach [G *Magen*]: Expanded anterior section of digestive tract. Directly follows mouth and occupies most of head (cephalon). Consists of larger anterior cardiac stomach and smaller posterior pyloric stomach; opens posteriorly into midgut.

stylamblys: appendix interna

subchela (subcheliped, subchelipede) [G *Subcheliped*]: On thoracic appendages (thoracopods) 1–5, pincer-like structure formed by dactyl folded back and bearing down on propodus. Subchela of second thoracopod particularly well developed as a raptorial claw.

subesophageal ganglion: postoral ganglion

submedian tooth [G *submedianer Stachel*]: tooth

supraesophageal ganglion [G *Oberschlundganglion*]: Major concentration of nerve tissue in anterior region of head (cephalon); joined to ventral ganglia (postoral ganglia) and ventral nerve cord by pair of long circumoral connectives.

tagma [G *Tagma*]: Major division (cephalon, thorax, abdomen) of body.

tailfan [G *Schwanzfächer*]: Posterior fan-like structure formed by combination of uropods and telson.

telson [G *Telson*]: Well-developed posteriormost segment of body (although not considered to be a true somite); may be fused with sixth abdominal somite (pleomere) to form pleotelson. Dorsal surface typically sculptured with various carinae, margins bearing prelateral lobes and numerous teeth and denticles. (armed, smooth)

tergite [G *Tergit*]: Dorsal surface of body segment (somite). (See also sternite)

tergum [G *Tergum*]: Collective term for all tergites. (See also sternum)

testis [G *Hoden*]: Section of ♂ reproductive system in which sperm are produced. Elongate, paired, extending from third abdominal somite (pleomere) to telson between heart and digestive glands. Fused posteriorly in telson and opening to exterior on eighth thoracic somite (thoracomere) via vasa deferentia and penes.

thoracomere (thoracic segment) [G *Thoracomer, Thoraxsegment*]: One of eight segments (somites) of thorax, the first four being covered by carapace; each bears pair of appendages (thoracopods).

thoracopod (thoracic appendage) [G *Thoracopod*]: One of two appendages of each thoracic somite (thoracomere). First five pairs are uniramous (one through four subchelate, the second with raptorial claw), the last three (stenopods) biramous and used for walking. The former are composed of seven segments: precoxa, coxa, basis, merus, carpus, propodus, dactyl (also interpreted as coxa, basis, ischium, merus, propodus, dactyl). The latter are composed of three-segmented protopod with one-segmented outer ramus (considered to be endopod) and two-segmented inner branch (considered to be exopod).

thorax [G *Thorax*]: Division (tagma) of body between head (cephalon) and abdomen (pleon); consists of eight somites (thoracomeres), each bearing pair of appendages (thoracopods). First four somites are covered by carapace.

tooth [G *Dorn*]: One of typically six larger tooth-like projections along telson margin; according to position (from midline outward) one may distinguish submedian, intermediate, and lateral teeth. May also refer to teeth on distal segments (propodus, dactyl) of thoracopods.

upper lip: labrum

uropod [G *Uropod*]: One of two appendages of last (sixth) abdominal somite (pleomere). Well developed, flattened, consisting of one-segmented basal protopod (prolonged into forked process) bearing one-segmented endopod and two-segmented exopod. Forms tailfan together with telson.

vas deferens (sperm duct) [G *Vas deferens, Samenleiter*]: Section of ♂ reproductive system between each testis and each penis; elongate and slender, each extending from third abdominal somite (pleomere) anteriorly to penis on eighth thoracic appendage (thoracopod).

ventral nerve cord [G *ventraler Nervenstrang, Bauchmark*]: Longitudinal pair of fused nerve cords extending from postoral ganglion to posterior end of body; located midventrally and bearing ganglia (segmental ganglia) in each somite.

ARTHROPODA

CRUSTACEA

Syncarida (Anaspidacea, Stygocaridacea, Bathynellacea)

syncarid (anaspidacean, stygocaridacean, bathynellacean) [G *Syncaride: Anaspidacee, Stygocaridacee, Bathynellacee*]

SYNCARIDA

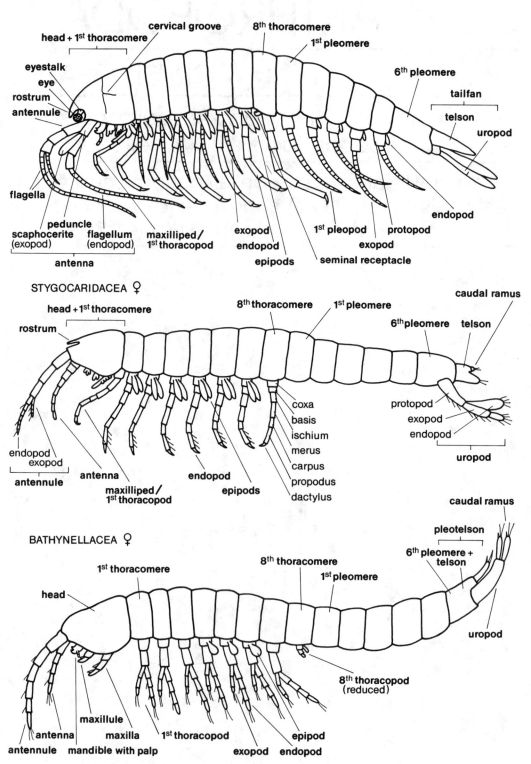

ARTHROPODA

CRUSTACEA

Syncarida (Anaspidacea, Stygocaridacea, Bathynellacea)

syncarid (anaspidacean, stygocaridacean, bathynellacean) [G *Syncaride: Anaspidacee, Stygocaridacee, Bathynellacee*]

abdomen (pleon) [G *Abdomen, Hinterleib, Pleon*]: Posterior of three divisions (tagmata) of body (head, thorax, abdomen). Consists basically of six somites (pleomeres) bearing posterior telson, although in bathynellacean, last pleomere is fused with telson to form pleotelson. Bears varying number of pleopods.

antenna (second antenna) [G *Antenne, zweite Antenne*]: Second pair of antennae. Consists of basal peduncle and either single flagellum (stygocaridacean), single or paired (flagellum and small exopod) branches (bathynellacean), or flagellum and scaphocerite (anaspidacean). (uniramous, biramous)

antennal scale: scaphocerite

antennule (first antenna) [G *Antennula, erste Antenne*]: First pair of antennae. Consists of basal peduncle and distal flagellum (or flagella). Biramous, with tiny lappet in bathynellacean interpreted as representing second branch. (with/without statocyst)

anterior aorta (anterior dorsal aorta) [G *Kopfaorta, Aorta dorsalis anterior*]: aorta

anus [G *After*]: Posterior opening of digestive tract. Located on telson. (terminal, ventral)

aorta [G *Aorta*]: In circulatory system of anaspidacean, major blood vessel extending from heart. According to position one may distinguish an anterior aorta extending into head, a posterior aorta extending into abdomen, and a descending aorta at level of eighth thoracic somite. (See also artery)

artery [G *Arterie*]: In anaspidacean, one of numerous blood vessels originating from heart or aorta. According to position one may distinguish seven pairs of lateral arteries and a subneural artery (arising from descending aorta).

basis (basipodite) [G *Basis, Basipodit*]: Second of two segments (coxa, basis) of proximal part (protopod) of appendage.

cardiac stomach: stomach

carpus (carpopodite) [G *Carpus, Carpopodit*]: Fourth of basically six segments (preischium, ischium, merus, carpus, propodus, dactylus) of inner branch (endopod) of thoracopod. Main bend (knee) of thoracopod typically occurs between carpus and propodus.

caudal ramus (furcal ramus) [G *Furkalast, Furca-Ast*]: In bathynellacean, one of two small protuberances at end of telson. Both caudal rami together termed caudal furca. May also be interpreted as representing lobes of telson.

cecum (digestive cecum) [G *Divertikel, Hepatopankreas, Mitteldarmdrüsenschlauch*]: In anaspidacean, one of numerous relatively thin tubules arising at anterior end of midgut and extending posteriorly to end of thorax. Also refers to short dorsal ceca of midgut, one at level of first abdominal somite, the other at level of fifth.

cephalon: head

cephalothorax [G *Cephalothorax*]: Term occasionally applied to part of body consisting of head and fused first thoracomere.

cervical groove (mandibular groove, mandibular sulcus) [G *Zervikalfurche*]: In anaspidacean, transverse groove across head.

coxa (coxopodite) [G *Coxa, Coxopodit*]: First of two segments (coxa, basis) of proximal part (protopod) of appendage. May bear lateral lobes (endites, epipods).

dactylus (dactyl) [G *Dactylus, Dactylopodit*]: Sixth, most distal segment (preischium, ischium, merus, carpus, propodus, dactylus) of inner branch (endopod) of thoracopod.

descending aorta [G *Aorta descendens*]: aorta

diaeresis [G *Diäresis, Quernaht, Sutur*]: In anaspidacean, transverse suture superficially subdividing outer branch (exopod) of uropod.

dorsal cecum: cecum

endite [G *Endit*]: One of two inwardly (medially) directed lobes of maxillule or two to four lobes of maxilla. Also refers to two lobes of first segment (coxa) of maxilliped. (See also gnathobase)

endopod (endopodite) [G *Endopod, Endopodit*]: Inner branch (ramus) of biramous appendage. Basically six-segmented in thoracopods, consisting of preischium, ischium, merus, carpus, propodus, and dactylus. (See also exopod, gonophysis)

epipod (epipodite) [G *Epipod, Epipodit*]: One of two projections of first segment (coxa) of thoracopod. Serves in respiration.

esophageal connective [G *Schlundkonnektiv*]: One of two strands of nerve tissue surrounding esophagus and connecting supraesophageal ganglion to ventral nerve cord.

esophagus [G *Ösophagus*]: In anaspidacean, relatively short, narrow section of digestive tract between mouth and stomach.

exite [G *Exit*]: Term applied to small lobe of maxillule in anaspidacean. Also variously interpreted as representing an endopod or exopod.

exopod (exopodite) [G *Exopod, Exopodit*]: Outer branch (ramus) of biramous appendage. In anaspidacean, represents scale-like branch (scaphocerite) of antenna (also occasionally in bathynellacean) and annulated ramus of thoracopods. (annulate, scale-shaped, spatulate, styliform)

eye [G *Auge, Komplexauge*]: If present (anaspidacean), one of two organs of vision. (sessile, stalked; anteriorly/laterally directed)

eyestalk [G *Augenstiel*]: In anaspidacean, one of two projections of head bearing eyes; may bear tubercle.

flagellum [G *Flagellum, Geißel*]: Distal division of antennule or antenna (peduncle, flagellum). (paired, unpaired) (See also scaphocerite)

four-celled sense organ: In anaspidacean, small pigmented area anterior to cervical groove on dorsal surface of head.

gnathobase [G *Gnathobasis*]: In anaspidacean, one of two spinose lobes (endites) of first segment (coxa) of maxilliped.

gonophysis [G *Gonophyse*]: In ♂, median process from base of first or second pleopods. Represents modified endopods and serves in copulation.

gonopore [G *Gonoporus, Geschlechtsöffnung*]: Opening of reproductive system to exterior. In ♂, single or paired opening on ventral surface (sternite) of last (eighth) thoracic somite. In ♀, pair of pores on first segments (coxae) of sixth thoracopods.

head (cephalon) [G *Kopf, Cephalon*]: Anterior of three major divisions (tagmata) of body (head, thorax, abdomen). Consists basically of five somites (bearing antennules, antennae, mandibles, maxillules, and maxillae), although in anaspidacean and stygocaridacean first thoracic somite (bearing maxillipeds) is fused to head. (See also cervical groove, rostrum)

heart [G *Herz*]: Muscular pumping organ of circulatory system. In anaspidacean, elongate, extending dorsal to digestive tract from first thoracic to fourth abdominal somite. Bears pair of ostia at level of third thoracic somite (thoracomere) and gives rise to aortas and arteries. In bathynellacean, relatively short, located in fourth thoracomere.

hindgut [G *Enddarm*]: In anaspidacean, relatively short section of digestive tract between midgut and anus; restricted to last (sixth) abdominal somite (pleomere).

incisor process [G *Pars incisiva*]: Distal gripping process (vs. proximal grinding = molar process) of each mandible.

ischium (ischiopod, ischiopodite) [G *Ischium, Ischiopodit*]: Second of basically six segments (preischium, ischium, merus, carpus, propodus, dactylus) of inner branch (endopod) of thoracopod.

knee [G *Knie*]: In thoracopod, main bend of inner branch (endopod). In anaspidacean, located between fourth and fifth segments (carpus and propodus).

labium: paragnath

labrum (upper lip) [G *Labrum, Oberlippe*]: Unpaired structure anterior to mouth. (See also paragnath)

lateral artery [G *Seitenarterie*]: artery

mandible [G *Mandibel*]: Third paired appendage of head; located between antennae and maxillules. Represents first pair of mouthparts. If fully developed, consists of molar and incisor processes as well as palp. (with/without palp) (See also penicilla)

mandibular groove: cervical groove

maxilla (second maxilla) [G *Maxille, zweite Maxille*]: Fifth paired appendage of head; located between maxillules and maxillipeds or between maxillules and first thoracopods. Represents third pair of mouthparts and bears two to four endites.

maxillary gland [G *Maxillardrüse, Maxillennephridium*]: One of two excretory glands located in posterior region of head. Consists of convoluted tubules opening on maxillae. In bathynellacean, loop of tubule extends posteriorly to level of fourth thoracic somite. (See also uropodal gland)

maxilliped (maxillipede) [G *Maxilliped, Kieferfuß*]: In anaspidacean and stygocaridacean, pair of mouthparts posterior to maxillae. Represents modified first pair of thoracopods.

maxillule (first maxilla) [G *Maxillula, erste Maxille*]: Fourth paired appendage of head; located between mandibles and maxillae. Represents second pair of mouthparts. Bears two endites and, in anaspidacean, a small lobe (exite, also variously interpreted as endopod or exopod).

merus (meropod, meropodite) [G *Merus, Meropodit*]: Third of basically six segments (preischium, ischium, merus, carpus, propodus, dactylus) of inner branch (endopod) of thoracopod.

midgut [G *Mitteldarm*]: Elongate section of digestive tract between stomach and hindgut. Extends through thorax and abdomen, giving rise anteriorly to series of ceca and posteriorly (at level of first and fifth pleomeres) to additional dorsal ceca.

molar process [G *Pars molaris*]: Proximal grinding process (vs. distal gripping = incisor process) of each mandible.

mouth [G *Mund, Mundöffnung*]: Anterior opening of digestive tract. Bordered anteriorly by upper lip (labrum), posteriorly by lower lip (paragraphs). Opens into esophagus.

ostium [G *Ostium*]: Pair of muscular openings in heart of anaspidacean. Located at level of third thoracic somite (thoracmere).

ovary [G *Ovar, Ovarium, Eierstock*]: Section of ♀ reproductive system in which eggs are produced. Paired, in anaspidacean, extending from anterior or posterior half of thorax to end of abdomen. Each ovary opens to exterior via oviduct. Restricted to abdomen in bathynellacean.

oviduct [G *Ovidukt, Eileiter*]: Section of ♀ reproductive system between each ovary and gonopores on sixth thoracopods.

palp [G *Palpus*]: In anaspidacean and bathynellacean, distal projection of mandibles. (one- to three-segmented; prehensile, not prehensile)

paragnath [G *Paragnath*]: One of two lobe-like structures forming lower lip (labium) posterior to mouth.

pars ampullaris [G *Pars ampullaris*]: In digestive tract of anaspidacean, diverticulum at junction of posterior (pyloric) stomach and ceca.

peduncle [G *Stamm, Pedunculus*]: Proximal division of antennule or antenna (peduncle, flagellum). First segment of peduncle in antennule of anaspidacean and bathynellacean bears statocyst. (two-, three-, four-segmented)

penicilla [G *Penicilla*]: In stygocaridacean, one in a series of dentate setae on mandible.

pereopod (peraeopod, pereiopod) [G *Pereopod, Peraeopod, Pereiopod*]: In anaspidacean and stygocaridacean, term occasionally applied to appendage of thoracic somite not fused with head, i.e., all appendages except those of first thoracomere, which is fused to head and bears maxillipeds.

petasma [G *Petasma*]: In ♂, copulatory structure formed by first two pleopods.

pleomere [G *Pleomer*]: One of basically six segments (somites) of abdomen. Anterior five may bear pleopods, with last (sixth) pleomere bearing uropods. In bathynellacean, last pleomere is fused with telson to form pleotelson. (anaspidacean: body flexed/not flexed at first pleomere)

pleon: abdomen

pleopod [G *Pleopod*]: One of two appendages of abdominal somite (pleomere). Full complement present only in anaspidacean, with stygocaridacean and bathynellacean bearing at most two anterior pairs of pleopods. First and second pairs modified for copulation in ♂. (uniramous, biramous)

posterior aorta (posterior dorsal aorta) [G *Schwanzaorta, Aorta dorsalis posterior*]: aorta

preischium [G *Präischium*]: First of basically six segments (preischium, ischium, merus, carpus, propodus, dactylus) of inner branch (endopod) of thoracopod. Located between basis and ischium and may be variously fused with basis.

propodus [G *Propodus*]: Fifth of basically six segments (preischium, ischium, merus, carpus, propodus, dactylus) of inner branch (endopod) of thoracopod. Main bend (knee) of thoracopod typically located between carpus and propodus.

protopod (protopodite) [G *Protopod, Protopodit*]: Proximal part of appendage. Consists basically of two segments (coxa, basis). May bear lateral lobes (endites, epipods) and distal branches (endopod, exopod).

pyloric stomach [G *Pylorus*]: stomach

retinaculum [G *Retinaculum*]: In ♂ anaspidacean, one in a series of small, hook-like structures along inner branch (endopod) of modified first and second pairs of pleopods. Serves to interlock left and right pleopods.

rostrum [G *Rostrum*]: Unpaired, unmovable anterior extension of head (cephalon). (deeply cleft, quadrate, strap-shaped, ventrally deflected)

scaphocerite (antennal scale) [G *Scaphocerit, Schuppe*]: In antenna of anaspidacean, scale-like outer branch (exopod) arising from second segment of peduncle. May also refer to small exopod in antenna of bathynellacean. (anaspidacean: longer/shorter than first two segments of endopod)

somite [G *Segment*]: One in a series of divisions of body. Head consists basically of five somites, thorax of eight somites (thoracomeres), and abdomen of six somites (pleomeres). First thoracomere may be fused with head.

spermatheca [G *Spermathek, Spermatheca*]: In ♀, modification of ventral surface (sternite) of last thoracic somite as sperm storage organ.

statocyst [G *Statozyste*]: In anaspidacean and stygocaridacean, organ of equilibrium located in first segment of each antennule.

sternite [G *Sternit*]: Ventral surface of body segment (somite). In anaspidacean, sternite of eighth thoracomere bears opening of reproductive system (♂) or is modified to form spermatheca (♀). (See also tergite)

stomach [G *Magen*]: In anaspidacean, expanded section of digestive tract between esophagus and midgut. Contains ridges internally and is divided into anterior cardiac stomach and posterior pyloric stomach.

supraesophageal ganglion (brain) [G *Oberschlundganglion, Gehirn*]: In anaspidacean, main concentration of nerve tissue above (anterior to) esophagus. Connected to ventral nerve cord via pair of esophageal connectives.

supraneural artery [G *Supraneuralarterie*]: artery

tagma [G *Tagma*]: One of three major divisions of body (head, thorax, abdomen).

tailfan [G *Schwanzfächer*]: In anaspidacean, posterior fan-like structure formed by combination of uropods and telson.

telson [G *Telson*]: Terminal segment of body. Forms tailfan together with uropods in anaspidacean, bears rudimentary caudal furca in stygocaridacean, and is fused with last (sixth) abdominal somite to form pleotelson in bathynellacean.

tergite [G *Tergit*]: Dorsal surface of body segment (somite). (See also sternite)

testis [G *Testis, Hoden*]: In anaspidacean, section of ♂ reproductive system in which sperm are produced. Paired, extending from fifth thoracic somite to posterior end of abdomen. Each testis opens to exterior via vas deferens. Restricted to abdomen in bathynellacean.

thoracomere [G *Thoracomer*]: One of basically eight segments (somites) between head and abdomen. In anaspidacean and stygocaridacean, first thoracomere (bearing maxillipeds) is fused with head. Each thoracomere bears pair of appendages (thoracopods), although in bathynellacean last pair may be absent. (anaspidacean: increasing/not increasing in length posteriorly)

thoracopod [G *Thoracopod*]: One of two appendages of each somite (thoracomere) of thorax. In anaspidacean and stygocaridacean, first pair is developed as maxillipeds. Basically biramous, consisting of two-segmented (coxa, basis) protopod bearing inner branch (endopod) and outer branch (exopod). (biramous, uniramous = with/without exopod; with/without epipods)

thorax [G *Thorax*]: Major division (tagma) of body between head and abdomen. Basically consists of eight somites (thoracomeres), although in anaspidacean and stygocaridacean, first thoracomere (bearing maxillipeds) is fused with head.

uropod [G *Uropod*]: Paired appendage of last somite (pleomere) of abdomen. Biramous, consisting of protopod bearing inner (endopod) and outer (exopod) branches. In anaspidacean, forms tailfan with telson. In bathynellacean, bears opening of uropodal gland. (spatulate, styliform)

uropodal gland [G *Uropodendrüse*]: In bathynellacean, relatively large accessory excretory gland located in last (sixth) abdominal somite (pleomere). Opens to exterior dorsally at base of uropod.

vas deferens [G *Vas deferens, Samenleiter*]: Section of ♂ reproductive system arising from anterior end of each testis and extending to gonopore(s) on ventral surface (sternite) of last (eighth) thoracic somite.

ventral nerve cord [G *Bauchmark*]: In anaspidacean, longitudinal nerve cord extending below digestive tract to end of abdomen. Band-shaped, bearing series of segmental ganglia. Connected anteriorly to supraesophageal ganglion by pair of esophageal connectives.

ARTHROPODA

CRUSTACEA

Mysidacea

mysidacean, mysid, opossum shrimp
[G *Schwebgarnele, Mysidacee*]

MYSIDACEA

carapace

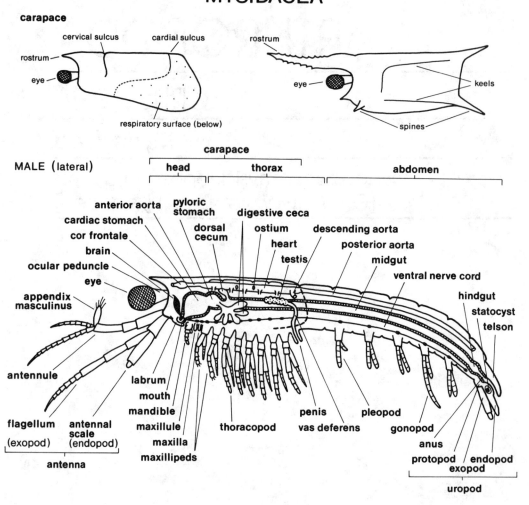

MALE (lateral)

thoracopod

telson

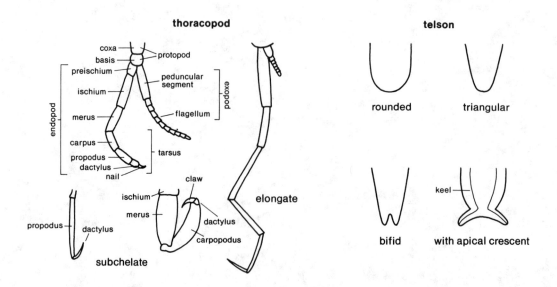

ARTHROPODA

CRUSTACEA

Mysidacea

mysidacean, mysid, opossum shrimp
[G *Schwebgarnele, Mysidacee*]

abdomen (pleon) [G *Abdomen, Hinterleib, Pleon*]: Posterior of three basic divisions (tagmata) of body (head, thorax, abdomen). Consists of six somites (pleomeres) bearing posterior telson.

abdominal aorta: posterior aorta

antenna (second antenna) [G *Antenne, zweite Antenne*]: Second pair of antennae. Biramous, consisting of peduncle bearing scale-like outer branch (antennal scale, squama, exopod) and elongate inner branch (flagellum).

antennal gland [G *Antennendrüse, Antennennephridium*]: Excretory gland located in head; consists of long, coiled excretory canal and bladder. Opens to exterior on basal segment of either antenna. (See also maxillary gland)

antennal scale (exopod, scaphocerite, squama) [G *Antennenschuppe, Scaphocerit, Exopodit*]: Scale-shaped outer branch (exopod) of antenna, frequently with terminal articulation. (scale-shaped, spine-shaped)

antennal spine: spine

antennular lamella: In antennule of certain mysidaceans, term occasionally applied to scale-like process projecting from last (third) peduncular segment.

antennule (first antenna) [G *Antennula, erste Antenne*]: First pair of antennae. Biramous, consisting of three-segmented peduncle and pair of flagella. Antennule of ♂ may bear processes masculinus and, rarely, an additional unsegmented accessory process.

anterior aorta [G *Kopfaorta, Aorta anterior, Aorta cephalica*]: Major unpaired blood vessel arising from anterior end of heart and extending over stomach into head. Bears accessory pumping organ (cor frontale).

anterior lappet: pleura

anus [G *After*]: Posterior opening of digestive tract; located on ventral surface of telson.

aorta [G *Aorta*]: In circulatory system, major blood vessel extending from heart. According to position one may distinguish an anterior aorta (bearing cor frontale), a posterior aorta, and a descending aorta.

artery [G *Arterie*]: One of numerous blood vessels originating from aorta or directly from heart. According to position and mysidacean group, one may distinguish hepatic, (anterior and posterior) lateral, sternal, subneural, and ventral arteries.

basis (basipodite) [G *Basis, Basipodit*]: Second of two segments (coxa, basis) of proximal part (protopod) of appendage.

branchia: gill

branchiostegal spine [G *Branchiostegaldorn*]: spine

brood pouch: marsupium

carapace [G *Carapax*]: Shield-like structure covering head and most of thorax. Typically fused to first three thoracic segments (thoracomeres) and typically bearing rostrum. (with/without rostrum; undecorated, ornamented: spinose, with keels; posterior margin: indented, produced into spines, with lobes) (See also cervical sulcus)

cardiac stomach [G *Cardia, Kardiakalteil des Magens*]: stomach

cardial sulcus [G *Cardialsulcus*]: Transverse groove across posterior region of carapace at level of heart. (See also cervical sulcus)

carpopropodus [G *Carpopropodus*]: On inner branch (endopod) of thoracopod, multisegmented unit formed by carpus and propodus.

carpus (carpopodite) [G *Carpus, Carpopodit*]: Fourth of basically six segments (preischium, ischium, merus, carpus, propodus, dactylus) of inner branch (endopod) of thoracopod. May form a multiarticulate unit with propodus (carpopropodus). (multiarticulate, one-segmented)

cecum: dorsal cecum, lateral cecum

cephalothorax [G *Cephalothorax*]: Anterior section of body consisting of head and, according to interpretation, either those anterior thoracomeres fused to head or entire thorax.

cervical sulcus (cervical groove) [G *Zervikalsulcus*]: Transverse groove across anterior region of carapace at level of mandibles.

cor frontale (anterior heart, frontal heart) [G *Cor frontale, Stirnherz*]: In circulatory system, muscular expansion of anterior aorta. Supports heart in pumping blood to anterior region of head.

coxa (coxopodite) [G *Coxa, Coxopodit*]: First of two segments (coxa, basis) of proximal part (protopod) of appendage.

dactylus (dactyl) [G *Dactylus, Dactylopodit*]: Sixth and most distal segment (preischium, ischium, merus, carpus, propodus, dactylus) of inner branch (endopod) of thoracopod. Usually minute, with claw. May form movable distal element in subchela.

descending aorta [G *Aorta descendens, Arteria descendens*]: Major unpaired blood vessel arising ventrally near posterior end of heart. Variously branched ventrally.

diaeresis [G *Sutur*]: Transverse suture superficially subdividing one or both branches of uropod, or antennal scale.

digestive cecum: dorsal cecum, lateral cecum

dorsal cecum (dorsal diverticulum) [G *dorsales Coecum, dorsaler Divertikel*]: Typically unpaired dorsal outpocketing of digestive tract at border of pyloric stomach and midgut. (paired, unpaired) (See also lateral cecum)

dorsal keel: keel

ejaculatory duct [G *Ductus ejaculatorius*]: In ♂ reproductive system, expanded terminal section of vas deferens. Opens to exterior on penis.

endite [G *Endit*]: Inwardly (medially) directed lobe of inner branch of biramous appendage.

endopod (endopodite) [G *Endopod, Endopodit*]: Inner branch (ramus) of biramous appendage. Represents palp of maxillae and maxillules as well as inner branch of maxillipeds, pereopods, and pleopods. Endopod of uropod may bear statocyst.

epipod (epipodite) [G *Epipodit*]: Dorsal projection from outer branch (exopod) of thoracopods. First thoracic epipod extends dorsally and posteriorly under carapace and serves to renew water supply for respiration. In certain mysidaceans, epipods of thoracopods 2–7 form ramified gills (branchiae).

esophageal connective [G *Schlundkonnektiv*]: One of two strands of nerve tissue surrounding esophagus and joining supraesophageal ganglion to subesophageal ganglion or directly to ventral nerve cord.

esophagus [G *Ösophagus*]: Relatively short section of digestive tract between mouth and stomach. May be lined internally with posteriorly directed spines.

exite [G *Exit*]: Outwardly (laterally) directed lobe of basal part of maxilla.

exopod (exopodite) [G *Exopod, Exopodit*]: Outer branch (ramus) of biramous appendage. Represents scale-like structure (antennal scale) of antenna and flagellum-shaped branch of maxillipeds, thoracopods, and pleopods. (annulate, elongate, flattened; natatory; reduced)

eye (compound eye) [G *Auge, Komplexauge*]: One of two organs of vision, each located distally on ocular peduncle. (divided, undivided, fused; plate-like, pyriform)

first antenna: antennule

flagellum [G *Flagellum, Geißel*]: In antennule, pair of elongate branches of peduncle. In antenna, elongate inner branch of peduncle consisting of three- to four-segmented basal section and multisegmented distal section. Also refers to annulate distal part of outer branch (exopod) of thoracopod.

frontal heart: cor frontale

genital papilla [G *Genitalpapille*]: penis

gill [G *Kieme*]: In certain mysidaceans, one in a series of branched respiratory structures arising dorsally from base of thoracopods. (See also epipod)

gnathopod [G *Gnathopod*]: Term applied to thoracic endopods when these are modified as mouthparts.

gonopod [G *Gonopod*]: In ♂, term applied to pleopod (usually third or fourth) modified for copulatory purposes.

gonopore [G *Gonoporus, Geschlechtsöffnung*]: Opening of reproductive system to exterior. In ♀, pair of pores, one located at base of each thoracopod of sixth thoracic somite. For ♂, see penis.

head [G *Kopf*]: Anterior of three basic divisions (tagmata) of body (head, thorax, abdomen). Fused with anterior thoracic segments (thoracomeres) to form cephalothorax; covered by carapace.

heart [G *Herz*]: In circulatory system, muscular pumping organ above digestive tract in thorax. Bears two or three pairs of ostia and gives rise to series of aortas and arteries.

hindgut [G *Enddarm*]: Relatively short, cuticle-lined section of digestive tract between midgut and anus. Restricted to last (sixth) abdominal somite (pleomere).

incisor process [G *Pars incisiva*]: Distal gripping process (vs. grinding = molar process) of each mandible.

ischium (ischiopod, ischiopodite) [G *Ischium, Ischiopodit*]: Second of basically six segments (preischium, ischium, merus, carpus, propodus, dactylus) of inner branch (endopod) of thoracopod.

keel [G *Kiel*]: In certain mysidaceans, one of several elevated ridges on carapace. According to position one may distinguish dorsal, midlateral, and lower lateral keels.

knee [G *Knie*]: In thoracopod, main bend of inner branch (endopod). Basically located between third and fourth segments (merus and carpus).

labium (lower lip) [G *Labium, Unterlippe*]: Lobe-like structure posterior to mouth. Divided into two well-developed lobes (paragnaths).

labrum (upper lip) [G *Labrum, Oberlippe*]: Unpaired, helmet-shaped structure anterior to mouth; extends posteriorly, covering mouth and part of mandibles. (bilobate, entire, with/without anterior spiniform projection)

lacinia mobilis [G *Lacinia mobilis*]: Articulated process between pars centralis (setal row) and incisor process on left mandible; frequently absent on right mandible.

lappet: One of two (anterior and posterior) lobe-like extensions of pleura.

lateral artery [G *Seitenarterie*]: artery

lateral cecum (lateral diverticulum, digestive cecum) [G *ventraler Divertikel, Hepatopankreas*]: One of up to five pairs of digestive glands opening ventrolaterally into digestive tract at border between pyloric stomach and midgut. (See also dorsal cecum)

lower lateral keel: keel

lower lip: labium

luminescent organ: photophore

mandible [G *Mandibel*]: Third paired appendage of head. Located between antennules and maxillules. Represents first pair of mouthparts. If fully developed, consists (proximally to distally) of molar process, pars centralis (spine row), lacinia mobilis, incisor process, and palp. Left and right mandibles differ considerably.

marsupium (brood pouch) [G *Marsupium, Brutbeutel*]: In ♀, chamber formed on ventral surface of thorax by two to seven pairs of oostegites. Serves to brood eggs and embryos.

maxilla (second maxilla) [G *Maxille, zweite Maxille*]: Fifth paired appendage of head; located between maxillules and maxillipeds. Represents third pair of mouthparts and consists of base bearing two to three endites, two-segmented endopod (palp), and exopod.

maxillary gland [G *Maxillardrüse, Maxillennephridium*]: In certain mysidaceans, one of two excretory glands in head, each opening to exterior at base of maxilla. (See also antennal gland)

maxilliped (maxillipede) [G *Maxilliped, Kieferfuß*]: Two pairs (rarely one, three, or four) of mouthparts posterior to maxillae. Represents modified endopod of thoracopod.

maxillule (first maxilla) [G *Maxillula, erste Maxille*]: Fourth paired appendage of head; located between mandibles and maxillae. Represents second pair of mouthparts. Consists of base bearing one or occasionally two endites, rarely also with endopodal palp.

merus (meropod, meropodite) [G *Merus, Meropodit*]: Third of basically six segments (preischium, ischium, merus, carpus, propodus, dactylus) of inner branch (endopod) of thoracopod.

midgut [G *Mitteldarm*]: Elongate section of digestive tract between stomach and hindgut. Extends through most of thorax and abdomen.

midlateral keel: keel

molar process [G *Pars molaris*]: Proximal grinding process (vs. gripping = incisor process) of each mandible.

mouth [G *Mund, Mundöffnung*]: Anterior opening of digestive tract. Bordered anteriorly by upper lip (labrum), posteriorly by lower lip (labium). Opens into esophagus.

nauplioid: First larval stage after hatching from egg membrane. With naupliar appendages only (antennae, antennules, mandibles).

ocular peduncle (eyestalk) [G *Augenstiel*]: One of two eye-bearing, movable projections of head. Consists of single segment and may bear papilla on inner or dorsal surface.

oostegite (brood plate) [G *Oostegit, Brutplatte*]: In ♀, one of two, three, or seven pairs of flattened plates projecting from bases of thoracopods. Overlap to form brood chamber (marsupium).

organ of Bellonci [G *Organ von Bellonci*]: Vesicular structure surrounded by sensory cells located directly under dorsal integument of eyestalk. Often connected to exterior by sensory pore. Frequently found at base or in ocular papilla. Presumably with neurosecretory and sensory functions.

ostium [G *Ostium*]: Muscular opening in heart, either two pairs at level of third and fourth thoracomeres or three pairs at level of second through fourth thoracomeres.

Blood from pericardial sinus enters heart via ostia.

ovary [G *Ovar, Ovarium, Eierstock*]: Paired section of ♀ reproductive system in which eggs are produced. Located dorsolateral to digestive tract in thorax. May be connected by bridge. Each ovary opens on base of sixth thoracopod via oviduct.

oviduct [G *Ovidukt, Eileiter*]: Section of ♀ reproductive system between each ovary and gonopores on bases (coxae) of sixth thoracopods.

palp [G *Palpus*]: Distal projection of mouthpart (mandible, maxilla, rarely maxillule). Three-segmented and well-developed in mandibles, two-segmented in maxillae. Represents endopod in posterior mouthparts.

papilla (ocular papilla) [G *Augenpapille*]: Tube- or finger-shaped projection from inner surface of each ocular peduncle.

paragnath: labium

pars centralis [G *Pars centralis*]: In mandible, central part between lacinia mobilis and molar process. Typically with spine row or with cutting or grinding structures.

peduncle [G *Pedunculus, Stiel, Schaft, Stamm*]: Three-segmented proximal division of antennule or antenna. Bears pair of flagella in former, flagellum and antennal scale in latter.

penis (genital papilla) [G *Penis, Genitalpapille*]: Pair of ♂ copulatory organs located on ventral surface (sternite) of last (eighth) thoracic somite.

pereomere (peraeomere) [G *Peraeomer*]: Term occasionally applied to thoracic segments (thoracomeres) not fused with head, i.e., somites of pereon.

pereon (peraeon) [G *Peraeon*]: Term occasionally applied to that part of thorax not fused with head, i.e., second (or third) through eighth thoracomeres.

pereopod (peraeopod, pereiopod) [G *Peraeopod, Pereiopod, Laufbein*]: Term occasionally applied to appendage of thoracic somites not fused with head, i.e.,

thoracic appendages not modified as mouthparts (maxillipeds).

pericardial sinus (pericardium) [G *Perikardialsinus, Perikard*]: Cavity surrounding heart. Blood collected in pericardial sinus enters heart via ostia.

photophore (luminescent organ) [G *Leuchtorgan, Photophor*]: In certain mysidaceans, bioluminescent gland emptying into reservoir which in turn opens near base of exopod of maxilla.

pleomere [G *Pleomer, Abdominalsegment*]: One of six segments (somites) of abdomen. Each pleomere may bear pair of appendages, those of first five being termed pleopods, those of sixth termed uropods.

pleon [G *Pleon*]: abdomen

pleopod [G *Pleopod*]: One of two appendages of abdominal somite (pleomere). Basically biramous, consisting of two-segmented protopod bearing multisegmented inner branch (endopod) and outer branch (exopod). Appendages of last pleomere termed uropods.

pleura [G *Pleura, Pleuralfalte*]: On each side of abdominal segment (pleomere) in certain mysidaceans, ventral extension of tergite. In ♀, pleura of first pleomere may be enlarged to support posterior part of brood chamber (marsupium). May be divided into anterior and posterior lappets.

pore [G *Porus*]: One of numerous cuticular openings on carapace and body; presumably with sensory function.

posterior aorta (abdominal aorta) [G *Aorta posterior, Schwanzaorta*]: Major blood vessel arising from posterior end of heart. Extends to end of abdomen and gives rise to series of segmental arteries.

posterior lappet: lappet

posterior lateral artery: artery

posterodorsal spine: spine

posterolateral spine: spine

postnauplioid: Second larval stage; characterized by all body segments and appendages of adult. Leaves brood pouch shortly before or after hatching to the juvenile stage. (See also nauplioid)

preischium [G *Präischium*]: In inner branch (endopod) of thoracopod, segment between basis and ischium.

processus masculinus [G *Processus masculinus*]: In antenna of ♂ in most mysidaceans, setose lobe projecting (along with two flagella) from last (third) peduncular segment.

propodus [G *Propodus*]: Fifth of basically six segments (preischium, ischium, merus, carpus, propodus, dactylus) of inner branch (endopod) of thoracopod. May form a multisegmented unit together with carpus (carpopropodus). (simple, multiarticulate) (See also subchela)

protopod (protopodite) [G *Protopodit*]: Proximal part of appendage. Consists basically of two segments (coxa, basis). May bear lateral lobes (endites, exites) and distal branches (endopod, exopod).

pseudobranchial lobe [G *Pseudobranchie*]: In ♂ of certain mysidaceans, posteriorly directed projection from basis of endopod of pleopods. (bifid, curved, rod-like, spirally coiled)

pyloric stomach [G *Pylorus*]: stomach

rib [G *Rippe*]: One in a series of cuticular reinforcements of oostegites; serves to maintain form and position of brood pouch.

rostrum [G *Rostrum*]: Unpaired, immovable anterior extension of carapace. (pointed, rounded, simple, serrate)

scaphocerite: antennal scale

segment [G *Segment*]: Segment of body or appendage, those of former often being termed somites. Segments of flagella of antennules, antennae, or thoracic exopods often termed articles.

seminal vesicle [G *Vesicula seminalis, Samenblase*]: Elongate section of ♂ reproductive system in which sperm are stored after leaving spermatic sacs.

somite (segment) [G *Segment*]: One in a series of divisions of body. Head

consists basically of five somites, thorax of eight somites (thoracomeres), and abdomen of six somites (pleomeres). Anterior thoracomeres (bearing maxillipeds) may be fused with head.

spermatic sac (spermatic pouch, follicle) [G *Säckchen, Bläschen*]: One in a series of sac-shaped structures into which testes are subdivided.

spine [G *Dorn*]: One of several thorn-like projections on carapace, body, or appendages. According to position, those on carapace may be termed antennal, branchiostegal, posterolateral, and posterodorsal spines (the latter unpaired).

spine row [G *Sägeborstenreihe*]: Row of spine-like projections in central position on cutting edge of either mandible.

squama: antennal scale

statocyst [G *Statozyste*]: Organ of equilibrium, one located at base of inner branch (endopod) of each uropod. Contains statolith and is innervated by last abdominal ganglion.

statolith [G *Statolith*]: Solid body within statocyst; supported by rows of sensory hairs.

sternal artery [G *Arteria sternalis*]: artery

sternal process [G *Sternalfortsatz*]: In certain mysidaceans, medial unpaired process on thoracic sternite.

sternite [G *Sternit*]: Ventral surface of body segment (somite). May be well developed in thoracic region of ♀ to accommodate brood pouch (marsupium). (See also tergite)

stomach [G *Magen, Vormagen*]: Expanded section of digestive tract between esophagus and midgut. Divided into larger anterior cardiac stomach and smaller posterior pyloric stomach. Characterized internally by numerous ridges and armed with spine- or tooth-like projections. Unpaired dorsal cecum and series of lateral ceca open into digestive tract at junction of stomach and midgut.

subchela [G *Subchela*]: Typically relatively small, pincer-like structure of thoracopod. Formed by terminal segment (dactylus) bearing down on propodus. Certain bottom-dwelling mysidaceans have powerful subchela on third thoracic endopod.

subesophageal ganglion [G *Unterschlundganglion*]: In certain mysidaceans, concentration of nerve tissue below (posterior to) esophagus. Joined to supraesophageal ganglion by pair of esophageal connectives and continues posteriorly as ventral nerve cord.

subneural artery [G *Arteria subneuralis, Subneuralarterie*]: artery

supraesophageal ganglion (brain) [G *Oberschlundganglion, Gehirn*]: Main concentration of nerve tissue above (anterior to) esophagus. Connected to ventral nerve cord (in certain mysidaceans to subesophageal ganglion) via pair of esophageal connectives.

tagma [G *Tagma*]: One of basically three major divisions of body (head, thorax, abdomen). (See also cephalothorax)

tailfan [G *Schwanzfächer*]: Posterior fan-like structure formed by combination of uropods and telson.

tarsus [G *Tarsus*]: Portion of thoracic endopod distal to knee. Consists of carpopropodus and dactylus.

telson [G *Telson*]: Terminal segment of body. Forms tailfan together with uropods and bears anus ventrally.

tergite [G *Tergit*]: Dorsal surface of body segment (somite). May extend ventrolaterally to form pleura in abdominal somites (pleomeres). (See also sternite)

testis [G *Hoden*]: Paired section of ♂ reproductive system in which sperm are produced. Located in thorax and differentiated into series of spermatic sacs. Opens to exterior via seminal vesicles, vasa deferentia, ejaculatory ducts, and penes.

thoracomere [G *Thoracomer, Thorakalsegment*]: One of basically eight segments (somites) between head and abdomen. Each thoracomere bears pair of appendages (thoracopods). (entirely/not entirely covered by carapace)

thoracopod [G *Thoracopod*]: One of two appendages of each somite (thoracomere) of thorax. First and second pairs typically developed as maxillipeds. Basically biramous, consisting of two-segmented base (coxa, basis) bearing inner branch (endopod) and flagellum-shaped outer branch (exopod). Those thoracopods not developed as maxillipeds occasionally termed pereopods. (with subchela = subchelate, with terminal claw; with/without exopod)

thorax [G *Thorax*]: Major division (tagma) of body between head and abdomen. Consists of eight somites (thoracomeres) and is covered by carapace. (See also cephalothorax)

transverse groove [G *Querfurche*]: Transverse suture superficially subdividing last (sixth) abdominal somite (pleomere).

upper lip: labrum

uropod [G *Uropod*]: Paired appendage of last somite (pleomere) of abdomen. Biramous and flattened, consisting of protopod bearing inner branch (endopod) and outer branch (exopod). Exopod and endopod may be subdivided by transverse suture (diaeresis); endopod usually bears a large statocyst.

vas deferens [G *Vas deferens, Samenleiter*]: Paired, slender, and elongate section of ♂ reproductive system between seminal vesicle and ejaculatory ducts.

ventral nerve cord [G *Bauchmark*]: Longitudinal nerve cord extending below digestive tract to last abdominal somite. Extends posteriorly from subesophageal ganglion or directly from supraesophageal ganglion via esophageal connectives. Bears series of segmental ganglia.

ventral artery [G *Arteria ventralis*]: artery

THERMOSBAENACEA

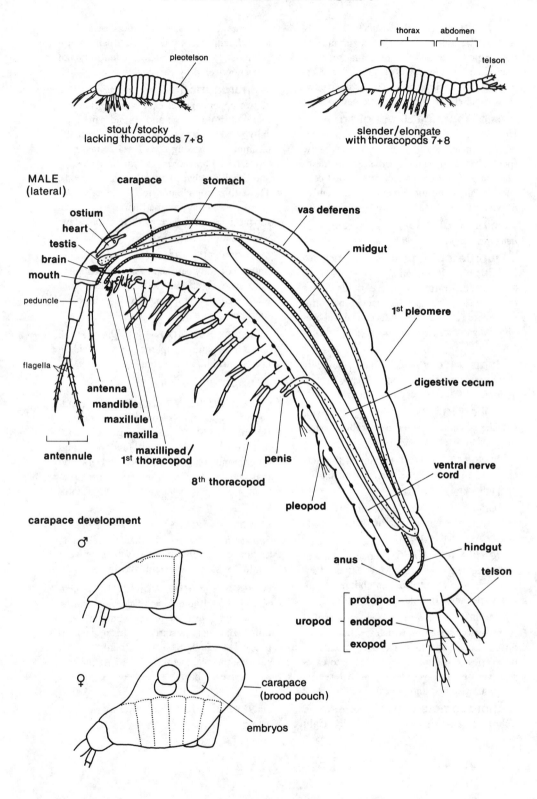

pleotelson

stout/stocky
lacking thoracopods 7+8

thorax · abdomen

telson

slender/elongate
with thoracopods 7+8

MALE
(lateral)

carapace · stomach

vas deferens

ostium

heart

testis

brain

mouth

midgut

1st pleomere

peduncle

digestive cecum

flagella

antenna

mandible

maxillule

maxilla

maxilliped/
1st thoracopod

ventral nerve
cord

antennule

penis

8th thoracopod

pleopod

hindgut

anus

telson

protopod

uropod endopod

exopod

carapace development

♂

♀

carapace
(brood pouch)

embryos

ARTHROPODA

CRUSTACEA

Thermosbaenacea

thermosbaenacean [G *Thermosbaenacee*]

---

abdomen (pleon) [G *Abdomen, Pleon, Hinterleib*]: Posterior of three divisions (tagmata) of body (cephalon, thorax, abdomen). Consists basically of six somites (pleomeres) bearing posterior telson. (See also pleotelson)

aesthetasc [G *Aesthetasc*]: One of several sensory projections on larger of two flagella of antennule.

antenna (second antenna) [G *Antenne, zweite Antenne*]: Second, relatively small pair of antennae. Uniramous, consisting of peduncle and flagellum.

antennule (first antenna) [G *Antennula, erste Antenne*]: First, relatively large pair of antennae. Biramous, consisting of three-segmented peduncle and two flagella.

anus [G *After*]: Posterior opening of digestive tract to exterior; apparently terminal.

aorta [G *Aorta, Aorta anterior, Aorta posterior*]: In circulatory system, blood vessel extending from end of heart. One may distinguish short anterior and posterior aortas.

article (joint, segment) [G *Glied, Segment*]: Any segment of appendage, although often applied to segment of flagella of antennules and antennae.

basis (basipodite) [G *Basis, Basipodit*]: Second of two segments (coxa, basis) of proximal part (protopod) of appendage. Frequently interpreted as being fused with following segment (ischium) in maxilliped endopod.

brain (supraesophageal ganglion) [G *Gehirn, Oberschlundganglion*]: Main concentration of nerve tissue above (anterior to) esophagus. Continues posterior to esophagus as ventral nerve cord.

brood pouch (marsupium) [G *Brutbeutel, Marsupium*]: In ♀, dorsal chamber between thorax and carapace. Serves to brood eggs.

carapace [G *Carapax*]: Shield-like structure fused to and covering head and first thoracic somite (thoracomere). Extends posteriorly and laterally to cover additional thoracomeres. Serves in respiration and forms brood pouch in ♀.

cardiac stomach [G *Cardia*]: stomach

carpus (carpopodite) [G *Carpus, Carpopodit*]: Third of basically five segments (ischium, merus, carpus, propodus, dactylus) of inner branch (endopod) of thoracopod.

cephalon (head) [G *Cephalon, Kopf*]: Anterior of three basic divisions (tagmata) of body (cephalon, thorax, abdomen). Fused with first thoracic somite, leading to alternate interpretation (cephalon, pereon,

---

409

abdomen). Covered by carapace. (See also cephalothorax)

cephalothorax [G *Cephalothorax*]: Term applied to anterior section of body consisting of cephalon and first thoracic somite (thoracomere). This unit may also be expressed by term cephalon.

coxa (coxopodite) [G *Coxa, Coxopodit*]: First of two segments (coxa, basis) of proximal part (protopod) of appendage.

dactylus (dactyl) [G *Dactylus, Dactylopodit*]: Fifth and most distal of basically five segments (ischium, merus, carpus, propodus, dactylus) of inner branch (endopod) of thoracopod.

digestive cecum (hepatopancreas) [G *Hepatopankreas, Blindschlauch*]: Pair of elongate digestive glands opening into digestive tract at border of stomach and midgut. Extends posteriorly to end of abdomen.

endite [G *Endit*]: Inwardly (medially) directed lobe of proximal section (protopod) of mouthpart (i.e., maxillule, maxilla, maxilliped).

endopod (endopodite) [G *Endopod, Endopodit*]: Inner branch (ramus) of biramous appendage. Represents only branch (flagellum) of antenna, palp of anterior mouthparts, and larger, basically five-segmented ramus (ischium, merus, carpus, propodus, dactylus) of thoracopods. (See also exopod)

epipod (epipodite) [G *Epipod, Epipodit*]: Projection of basal section (protopod) of each maxilliped. Extends posteriorly under carapace, serving in respiration and ventilation of brood pouch in ♀.

esophagus [G *Ösophagus*]: Relatively short, slender section of digestive tract between mouth and stomach.

exopod (exopodite) [G *Exopod, Exopodit*]: Outer branch (ramus) of biramous appendage. Present as small lobe on maxillae, palp-like structure of maxilliped, smaller, one- to two-segmented ramus of

thoracopods, and two-segmented ramus of uropod.

first antenna: antennule

first maxilla: maxillule

flagellum [G *Flagellum, Geißel*]: Distal division of antennule or antenna (peduncle, flagellum), paired in the former, unpaired in the latter. (See also aesthetasc)

gonopore [G *Gonoporus, Geschlechtsöffnung*]: Openings of reproductive system to exterior. In ♀, pair of pores on ventral surface (sternite) of sixth thoracic segment; in ♂, pair of pores on penes on last (eighth) thoracic segment.

guide setule: One in a series of fine setules on molar process of each mandible.

head: cephalon

heart [G *Herz*]: Muscular pumping organ of circulatory system. Short, located dorsally in first thoracic segment (thoracomere). Bears pair of ostia and gives rise to anterior and posterior aortas.

hepatopancreas: digestive cecum

incisor process [G *Pars incisiva*]: Distal gripping process (vs. grinding = molar process) of each mandible.

ischium (ischiopod, ischiopodite) [G *Ischium, Ischiopodit*]: First of basically five segments (ischium, merus, carpus, propodus, dactylus) of inner branch (endopod) of thoracopod. Frequently interpreted as being fused to second segment (basis) of protopod in maxilliped, to merus in the other thoracopods (pereopods).

joint: segment

labium (lower lip) [G *Labium, Unterlippe*]: Lobe-like structure posterior to mouth. Divided into two lobes (paragnaths). (See also labrum)

labrum (upper lip) [G *Labrum, Oberlippe*]: Unpaired, lobe-like structure anterior to mouth; relatively elongate, with terminal spines or setae. (See also labium)

lacinia mobilis [G *Lacinia mobilis*]: Articulated process between lifting spines and incisor process on each mandible.

lifting spine [G *Sägeborste*]: In mandible, row of spine-like projections between molar process and lacinia mobilis.

lower lip: labium

manca [G *Manca*]: In certain thermosbaenaceans, stage at which larva is released from brood pouch. Lacks last two pairs (seventh and eighth) of thoracopods and pleopods.

mandible [G *Mandibel*]: Third paired appendage of cephalon; located between antennae and maxillules. Represents first pair of mouthparts. Consists (proximally to distally) of molar process, lifting spines, lacinia mobilis, incisor process, and palp.

marsupium: brood pouch

maxilla (second maxilla) [G *Maxille, zweite Maxille*]: Fifth paired appendage of cephalon; located between maxillules and maxillipeds. Represents third pair of mouthparts and consists of protopod bearing three endites, palp (endopod), and small lobe (exopod).

maxilliped (maxillipede) [G *Maxilliped, Kieferfuß*]: Fourth and largest pair of mouthparts. Represents modified pair of appendages (thoracopods) of first thoracic segment (which is fused to cephalon). If fully developed, consists of protopod bearing two endites and an epipod as well as distal endopod and exopod. (sexually dimorphic, not sexually dimorphic)

maxillule (first maxilla) [G *Maxillula, erste Maxille*]: Fourth paired appendage of cephalon; located between mandibles and maxillae. Represents second pair of mouthparts. Consists of two-segmented protopod bearing two endites and small palp (endopod).

merus (meropod, meropodite) [G *Merus, Meropodit*]: Second of basically five segments (ischium, merus, carpus, propodus, dactylus) of inner branch (endopod) of thoracopod. Frequently interpreted as being fused to ischium.

midgut [G *Mitteldarm, Darm*]: Elongate section of digestive tract between stomach and anus. Extends through most of body. Digestive ceca open into digestive tract at border of stomach and midgut.

molar process [G *Pars molaris*]: Proximal grinding process (vs. gripping = incisor process) of each mandible. Bears series of fine setae (guide setules).

mouth [G *Mund, Mundöffnung*]: Anterior opening of digestive tract. Located on underside of cephalon and bordered anteriorly by upper lip (labrum), posteriorly by lower lip (labium). Opens into esophagus.

ostium [G *Ostium*]: One of two muscular openings of heart. Blood from carapace sinus enters heart via ostia.

ovary [G *Ovar, Ovarium, Eierstock*]: Section of ♀ reproductive system in which eggs are produced. Paired, elongate, extending through most of thorax. Each ovary opens to exterior on sixth thoracic segment via oviduct.

oviduct [G *Ovidukt, Eileiter*]: Relatively short, slender section of ♀ reproductive system between each ovary and gonopores on sixth thoracic segment.

palp [G *Palpus*]: Distal projection of mouthpart. Considered to represent inner branch (endopod) in mandible, maxillule, and maxilla, outer branch (exopod) in maxilliped.

paragnath: labium

peduncle [G *Stamm*]: Proximal division of antennule or antenna (peduncle, flagellum). Consists of three segments in the former; not as clearly delimited in the latter, interpreted as consisting of five segments.

penis [G *Penis*]: Paired, movable ♂ copulatory organs located on ventral surface (sternite) of last (eighth) thoracic somite.

pereon (peraeon, pereion) [G *Pereon, Peraeon*]: In one interpretation of main body regions, division between cephalon (head + first thoracic somite) and abdomen. Accordingly, pereon consists of seven somites (pereonites), their appendages being termed pereopods.

pereonite [G *Peraeomer*]: One of seven segments (somites) of pereon (i.e., all thoracomeres except first, which is fused to cephalon).

pereopod (peraeopod, pereiopod) [G *Peraeopod, Pereiopod*]: One of up to seven pairs of typically biramous appendages of pereon. Represents thoracopods 2–8 (i.e., excluding first thoracopods = maxillipeds).

pleomere (pleonite) [G *Pleomer*]: One of six segments (somites) of abdomen. Last pleomere may be fused to telson to form pleotelson. First two pleomeres bear pleopods, with last pleomere bearing uropods.

pleon: abdomen

pleonite: pleomere

pleopod [G *Pleopod*]: One of two pairs of appendages on first two somites (pleomeres) of abdomen. Small, uniramous, and consisting of single segment.

pleotelson [G *Pleotelson*]: In certain thermosbaenaceans, terminal segment of body formed by fusion of telson with last (sixth) abdominal somite (pleomere).

propodus [G *Propodus*]: Fourth of basically five segments (ischium, merus, carpus, propodus, dactylus) of inner branch (endopod) of thoracopod.

protopod (protopodite) [G *Protopodit*]: Proximal part of most appendages (with exception of antennules, antennae, mandibles, and pleopods). Consists basically of two segments (coxa, basis). May bear lateral lobes (endites; in maxillipeds also epipods) as well as distal branches (endopod, exopod). (See also peduncle)

pyloric stomach [G *Pylorus*]: stomach

second antenna: antenna

second maxilla: maxilla

segment [G *Segment*]: Segment of body or appendage, those of the former often termed somites. Segments of flagella of antennules or antennae often termed articles.

seminal vesicle [G *Samenblase, Vesicula seminalis*]: In ♂ reproductive system, expanded terminal section of each vas deferens. Serves to store sperm.

somite [G *Segment*]: One in a series of divisions of body. Cephalon consists basically of five somites, thorax of eight somites (thoracomeres), and abdomen of six somites (pleomeres). Because first thoracomere is fused to cephalon, an alternate interpretation distinguishes six cephalic somites, seven thoracomeres or pereonites, and six pleonites.

sternite [G *Sternit*]: Ventral surface of body segment (somite). In ♂, sternite of last (eighth) thoracomere bears pair of penes; in ♀, sternite of sixth thoracomere bears gonopores.

stomach [G *Magen, Kaumagen*]: Expanded anterior section of digestive tract between esophagus and midgut. May be divided into anterior cardiac and posterior pyloric sections. Digestive ceca open into digestive tract at border of stomach and midgut.

supraesophageal　　　ganglion: brain

tagma [G *Tagma*]: One of three major divisions of body (cephalon, thorax, abdomen).

telson [G *Telson*]: Posteriormost segment of body. May be fused to last (sixth) abdominal somite (pleomere) to form pleotelson.

tergite [G *Tergit*]: Dorsal surface of body segment (somite). Carapace fused to cephalon and to tergite of first thoracic segment (thoracomere).

testis [G *Hoden*]: Section of ♂ reproductive system in which sperm are produced. Paired, relatively short, located anteriorly in posterior cephalon/first thoracomere region. Each testis opens to exterior on last (eighth) thoracomere via elongate vas deferens.

thoracomere [G *Thoracomer*]: One of basically eight segments (somites) of thorax between cephalon proper and abdomen.

First thoracomere (bearing maxillipeds) is fused with head; accordingly, last seven thoracomeres may be interpreted as being pereonites.

thoracopod [G *Thoracopod*]: One of basically eight pairs of appendages of thorax. First pair developed as maxillipeds. Typically biramous, consisting of protopod bearing larger endopod and smaller exopod. (excluding maxillipeds: five pairs, seven pairs; equal/not equal in length; uniramous, biramous) (See also pereopod)

thorax [G *Thorax*]: Region (tagma) of body comprising eight somites between cephalon proper and abdomen. First thoracic somite (thoracomere) fused with cephalon. (See also pereon)

ungulus [G *Ungulus*]: In thoracopod, elongate, claw-shaped tip of last segment (dactylus) of endopod.

upper lip: labrum

uropod [G *Uropod*]: Paired appendage of last segment (pleomere) of abdomen. Biramous, consisting of protopod bearing one-segmented endopod and two-segmented exopod.

vas deferens [G *Vas deferens, Samenleiter*]: Elongate, slender section of ♂ reproductive system between each testis and penes on last (eighth) thoracomere. Extends through thorax and abdomen before turning anteriorly to gonopores. Expanded terminally to form seminal vesicle.

ventral nerve cord [G *Bauchmark*]: Longitudinal nerve cord extending posteriorly from brain. Located below digestive tract; bears segmentally arranged ganglion pairs.

CUMACEA

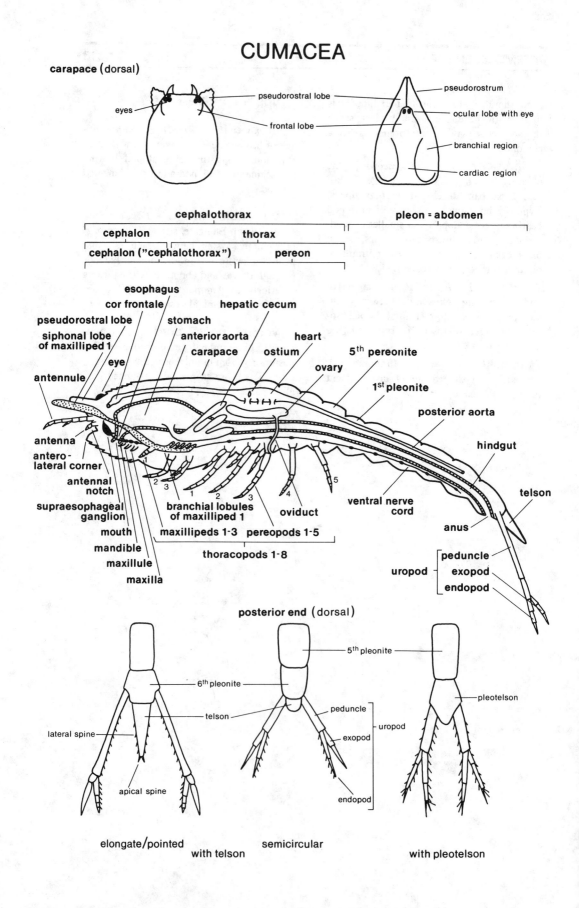

carapace (dorsal)

eyes — pseudorostral lobe
frontal lobe

pseudorostrum
ocular lobe with eye
branchial region
cardiac region

cephalothorax — pleon = abdomen

cephalon — thorax
cephalon ("cephalothorax") — pereon

esophagus
cor frontale
pseudorostral lobe
siphonal lobe of maxilliped 1
eye
antennule
antenna
antero-lateral corner
antennal notch
supraesophageal ganglion
mouth
mandible
maxillule
maxilla

stomach
anterior aorta
carapace

hepatic cecum
heart
ostium
ovary

5th pereonite
1st pleonite
posterior aorta
hindgut
telson

branchial lobules of maxilliped 1
maxillipeds 1-3
oviduct
pereopods 1-5
ventral nerve cord
anus

thoracopods 1-8

peduncle
uropod
exopod
endopod

posterior end (dorsal)

5th pleonite
6th pleonite
telson
peduncle
uropod
exopod
endopod
lateral spine
apical spine
pleotelson

elongate/pointed with telson

semicircular

with pleotelson

ARTHROPODA

CRUSTACEA
Cumacea
cumacean [G *Cumacee*]

abdomen: pleon

accessory flagellum [G *Nebengeißel*]: flagellum

accessory lobule [G *akzessorisches Kiemenelement*]: On posteriorly directed part of epipod of first maxilliped pair, single respiratory projection set apart from series of remaining branchial lobules.

aesthetasc [G *Aesthetasc*]: One in a series of sensory projections on main flagellum of antennule.

anal valve [G *Analklappe*]: anus

antenna (second antenna) [G *Antenne, zweite Antenne*]: Second pair of antennae; uniramous. If fully developed, consisting of proximal five-segmented peduncle and distal flagellum.

antennal notch (antennal sinus) [G *Antennenausschnitt, Subrostralschnitt*]: On each side of carapace, notch in anterior border below pseudorostrum. Antennules and occasionally antennae project through antennal notch. Lower extension of notch termed anterolateral angle or corner. (deep, shallow)

antennule (first antenna) [G *Antennula, erste Antenne*]: First pair of antennae. Basically biramous, consisting of proximal three-segmented peduncle bearing larger outer (main) and smaller inner (accessory) flagellum. May bear aesthetascs. (uniramous, biramous; geniculate)

anterior aorta [G *Aorta anterior*]: aorta

anterolateral angle (anterolateral corner): On each anterolateral margin of carapace, lower extension of antennal notch. (acute, rounded; armed, unarmed)

anus [G *After*]: Posterior opening of digestive tract. Located at end of last (sixth) pleonite or on ventral surface of telson (in the latter case occasionally dividing telson into preanal and postanal parts). Flanked by pair of anal valves.

aorta [G *Aorta*]: In circulatory system, major blood vessel extending from end of heart. One may distinguish 1) an unpaired anterior aorta extending into head and bearing a cor frontale and 2) paired posterior aortas extending to end of abdomen.

article (joint, segment) [G *Glied*]: Any segment of appendage, although often applied to segment of flagellum of antennule or antenna.

basis (basipodite) [G *Basis, Basipodit*]: Elongate second segment of thoracopod (maxillipeds and pereopods) or pleopods. May bear five-segmented endopod and smaller exopod. (See also epipod)

brain: supraesophageal ganglion

branchial apparatus [G *Kiemenapparat*]: Term applied to epipods of first maxilliped pair (or, in an alternate interpretation, to epipods and exopods). Modified for respiration, the posteriorly directed part extending into branchial cavity and bearing branchial lobules, the anteriorly directed part ("exopod") forming part of exhalent siphon.

branchial chamber [G *Kiemenhöhle, Atemhöhle*]: Space on each side of cephalothorax between body and inner wall of carapace. Contains posteriorly directed part of epipods of first maxillipeds. Closed posteriorly, with water entering anteriorly and exiting frontomedially via siphon.

branchial lobule [G *Kiemenschlauch, Kiemenelement*]: One in a series of respiratory projections along posteriorly directed part of epipods of first maxillipeds. Located in branchial chamber. (finger-like, lamelliform) (See also accessory lobule)

branchial region (branchial area) [G *Branchialregion*]: One of two relatively large lateral regions of carapace overlying branchial chamber.

brood chamber (brood pouch, marsupium) [G *Bruttasche, Brutbeutel, Marsupium*]: In ♀, chamber formed on ventral surface of pereon by oostegites on last (third) maxillipeds and on first three pereopod pairs.

carapace [G *Carapax*]: Large, shield-like structure covering head and three to six thoracic somites. Extends laterally to enclose branchial chamber. Regionated to form anterior pseudorostrum, frontal lobe, ocular lobe, as well as branchial and cardiac regions. (inflated, laterally compressed; dorsal outline: straight, arched, with slight undulations; with/without lateral horns; smooth, sculptured: hairy, rugose, with denticles, with median/lateral carina, with longitudinal depressions, with dorsal crest: serrate, smooth, with teeth)

cardiac region [G *Cardiacalregion*]: Unpaired median region in posterior half of carapace; overlies anterior region of heart. (See also branchial region)

carpus (carpopodite) [G *Carpus, Carpopodit*]: Third of five segments (ischium, merus, carpus, propodus, dactylus) of endopod of thoracopod (maxillipeds and pereopods).

cephalon (head) [G *Cephalon, Kopf*]: Anterior of three basic divisions (tagmata) of body (cephalon, thorax, abdomen = pleon). Fused with three thoracic somites to form cephalothorax; accordingly, body may be interpreted as being divided into cephalothorax and pleon or into cephalon, pereon, and pleon.

cephalothorax [G *Cephalothorax*]: Expanded anterior section of body consisting of head (cephalon) fused with first three thoracic somites. Covered by and fused dorsally and laterally with carapace.

cor frontale (cephalic heart) [G *Cor frontale, Stirnherz*]: In circulatory system, small, muscular expansion of anterior aorta. Located in head; supports heart in pumping blood anteriorly.

coxa (coxopodite) [G *Coxa, Coxopodit*]: First segment of thoracopod (maxillipeds and pereopods) or pleopod. May be more or less fused with ventral surface (sternites) of body. Coxae of second maxillipeds fused to one another.

dactylus (dactyl) [G *Dactylus, Dactylopodit*]: Fifth and most distal of five segments (ischium, merus, carpus, propodus, dactylus) of endopod of thoracopod (maxillipeds and pereopods).

endite [G *Endit*]: Inwardly (medially) directed lobe of proximal part of appendage, e.g., of maxillules and maxillae.

endopod (endopodite) [G *Endopod, Endopodit*]: Inner branch (ramus) of biramous appendage. Forms palp of maxillule and represents main five-segmented branch of thoracopod (maxillipeds and pereopods).

epimere [G *Pleura*]: Poorly developed lateral part/extension of somites in pereon and pleon region.

epipod (epipodite) [G *Epipod, Epipodit*]: In first pair of maxillipeds, enlarged lateral lobe of first segment (coxa). Consists of two parts, the first directed posteriorly into branchial chamber and bearing branchial lobules, the second (siphonal lobe) directed anteriorly to form part of siphon (the latter may alternately be interpreted as representing an exopod).

epistome [G *Epistom*]: On underside of head, pair of small plates between bases of antennules. Adjoined posteriorly by upper lip (labrum).

esophagus (oesophagus) [G *Ösophagus*]: Short, relatively narrow section of digestive tract between mouth and stomach.

exopod (exopodite) [G *Exopod, Exopodit*]: Outer branch (ramus) of biramous appendage. Represents smaller branch of pereopods; occasionally interpreted as representing flabellum of maxilla as well as siphonal lobe of first maxillipeds.

eye [G *Auge, Komplexauge*]: Photosensitive organ on dorsal surface of carapace; typically unpaired and positioned on ocular lobe. (paired, unpaired)

eye lobe: ocular lobe

first antenna: antennule

first maxilla: maxillule

flabellum [G *Flabellum*]: Flattened lateral branch of maxilla. Considered to represent an exopod.

flagellum [G *Flagellum, Geißel*]: Distal division of antennule or antenna (peduncle, flagellum). In antennule, one may distinguish a larger outer (main) and smaller inner (accessory) flagellum. Flagellum of antennule bears aesthetascs. (multiarticulate, uniarticulate)

frontal lobe [G *Frontallobus*]: On dorsal surface of carapace, bell-shaped prominence marking anterior margin of head. Bears ocular lobes anteriorly and is adjoined laterally and anteriorly by pseudorostrum.

gill: branchial lobule

gill plate [G *Kiemenplatte, Kiementräger*]: In first pair of maxillipeds, plate-like structure on posteriorly directed part of epipod. Located in branchial chamber; bears series of branchial lobules.

gonad: ovary, testis

gonopore [G *Gonoporus, Geschlechtsöffnung*]: Opening of reproductive system to exterior. In ♀, pair of pores on first segments (coxae) of third pereopods; in ♂, pair of pores on ventral surface (sternite) of last (fifth) pereon segment.

head: cephalon

heart [G *Herz*]: Muscular pumping organ of circulatory system. Relatively short, located dorsally in more anterior pereon segments (pereonites). Bears one pair of ostia and gives rise to anterior and posterior aortas as well as series of lateral arteries.

hepatic cecum (digestive caecum) [G *Caecum, Leberblindschlauch*]: One of up to four pairs of relatively short, lateral digestive ceca opening into digestive tract at stomach/midgut border.

hindgut [G *Enddarm*]: Elongate section of digestive tract between midgut and anus. Extends through most of pereon and abdomen and may be coiled in the former.

incisor process (pars incisiva) [G *Pars incisiva*]: Distal gripping process (vs. grinding = molar process) of each mandible.

ischium (ischiopod, ischiopodite) [G *Ischium, Ischiopodit*]: First of five segments (ischium, merus, carpus, propodus, dactylus) of endopod of thoracopod (maxillipeds and pereopods).

joint: article, segment

labium (lower lip) [G *Labium, Unterlippe*]: Lobe-like structure posterior to mouth. Divided into two lobes. (See also labrum)

labrum (upper lip) [G *Labrum, Oberlippe*]: Unpaired, lobe-like structure anterior to mouth; adjoins bases of antennules and antennae. (See also labium)

lacinia mobilis [G *Lacinia mobilis*]: Small, articulated process between setal row and incisor process on mandible. Fully developed only on left mandible.

lower lip: labium

main flagellum [G *Hauptgeißel*]: flagellum

manca [G *Manca, Mancastadium*]: First larval stage after release from brood pouch (marsupium). Resembles adult, yet lacks last (fifth) pair of pereopods.

mandible [G *Mandibel*]: Third paired appendage of head (cephalon); located between antennae and maxillules. Represents first pair of mouthparts and, if fully developed, consists (proximally to distally) of molar process, spine row, lacinia mobilis, and incisor process. (base: pointed, truncate)

marsupium: brood chamber

maxilla (second maxilla) [G *Maxille, zweite Maxille*]: Fifth paired appendage of head (cephalon); located between maxillules and first maxillipeds. Represents third pair of mouthparts and consists of base (protopod) with two endites as well as a lateral flabellum (exopod).

maxillary gland [G *Maxillendrüse, Maxillennephridium*]: Pair of excretory glands located in head, one opening to exterior at base of each maxilla.

maxilliped (maxillipede) [G *Maxilliped, Kieferfuß*]: One of first three pairs of thoracopods serving as mouthparts and belonging to three thoracic somites fused to head. Basically consists of basal section (coxa, basis) bearing five-segmented palp (endopod: ischium, merus, carpus, propodus, dactylus) as well as an exopod. (See also epipod, retinaculum)

maxillule (first maxilla) [G *Maxillula, erste Maxille*]: Fourth paired appendage of head (cephalon); located between mandibles and maxillae. Represents second pair of mouthparts and consists of base (protopod) with two endites as well as distal palp (endopod).

merus (meropod, meropodite) [G *Merus, Meropodit*]: Second of five segments (ischium, merus, carpus, propodus, dactylus) of endopod of thoracopod (maxillipeds and pereopods).

midgut [G *Mitteldarm*]: Relatively short section of digestive tract between stomach and hindgut. Wall of midgut distinguished from that of stomach by larger (syncytial) cell structure.

molar process (pars molaris) [G *Pars molaris*]: Proximal grinding process (vs. gripping = incisor process) of each mandible. (columnar = cylindrical = truncate, styliform)

mouth [G *Mund, Mundöffnung*]: Anterior opening of digestive tract. Bordered anteriorly by upper lip (labrum), posteriorly by lower lip (labium); opens into esophagus.

ocular lobe (eye lobe) [G *Ozellarlobus*]: On anterior section of carapace, small, unpaired median projection of frontal lobe; typically bears eyes.

oostegite (brood plate) [G *Oostegit, Brutplatte*]: In ♀, one in a series of flattened plates projecting from basal segments (coxae) of last maxilliped pair and first three pereopod pairs. Overlap to form brood chamber. (See also rudimentary oostegite)

ostium [G *Ostium*]: Pair of muscular openings in posterior region of heart. Blood in pericardial sinus enters heart through ostia.

ovary [G *Ovar, Ovarium, Eierstock*]: Section of ♀ reproductive system in which eggs are produced. Paired, tube-like, located on each side in person. Each ovary opens to exterior on third pereopods via oviduct.

oviduct [G *Ovidukt, Eileiter*]: Section of ♀ reproductive system between each ovary and gonopores on first segments (coxae) of third pereopods.

palp [G *Palpus*]: Posteriorly directed distal part of maxillule. Elongate and typically bearing filaments extending into branchial chamber.

pars incisiva: incisor process

pars molaris: molar process

peduncle [G *Stamm*]: Proximal division of antennule or antenna (peduncle, flagellum). If fully developed, consists of three segments in the former, five in the latter. May also refer to proximal part of pleopod or uropod.

pereon (peraeon, pereion) [G *Peraeon*]: Division (tagma) of body between head (cephalon) and abdomen (pleon) or, more accurately, between cephalothorax and abdomen. Consists of five somites (pereonites), each bearing appendages (pereopods). Morphologically not equivalent to thorax (with a total of eight somites) because first three somites are fused to cephalon (i.e., cephalothorax).

pereonite [G *Peraeomer*]: One of five segments (somites) of pereon (i.e., all thoracomeres except first three which are fused to cephalon).

pereopod (peraeopod, pereiopod) [G *Peraeopod, Pereiopod*]: One of five pairs of appendages of pereon. Basically biramous, consisting of basal section (coxa, basis) bearing five-segmented endopod (ischium, merus, carpus, propodus, dactylus) as well as smaller exopod. Number of pereopods bearing exopods is of taxonomic importance. (with/without exopod)

pericardial sinus (pericardium) [G *Perikardialsinus, Perikard*]: Cavity surrounding heart. Blood collected in pericardial sinus enters heart via ostia.

pleon (abdomen) [G *Pleon, Abdomen, Hinterleib*]: Posterior division (tagma) of body (either cephalon, pereon, pleon, or cephalothorax, pleon). Narrow, consisting of six somites (pleonites) bearing posterior telson.

pleonite [G *Pleomer*]: One of six segments (somites) of abdomen (pleon). Cylindrical, anterior pleonites occasionally bearing appendages (pleopods). Posterior pleonite bears uropods and may be fused to telson to form pleotelson.

pleopod [G *Pleopod*]: One of a variable number of paired appendages of abdomen (pleon); typically present only in ♂. Consists of two-segmented basal part (peduncle) bearing two-segmented endopods and one-segmented exopods. Number of pleopods in ♂ is of taxonomic importance.

pleotelson [G *Pleotelson*]: Structure formed by fusion of last (sixth) abdominal somite (pleonite) with telson.

posterior aorta [G *Aorta posterior*]: aorta

propodus [G *Propodus*]: Fourth of five segments (ischium, merus, carpus, propodus, dactylus) of endopod of thoracopod (maxillipeds and pereopods).

protopod (protopodite) [G *Protopodit*]: Term occasionally applied to proximal part of certain appendages (e.g., maxillules, maxillae). (See also peduncle)

pseudorostral lobe [G *Pseudorostrallobus*]: One of two anterior projections of carapace. May be separated, yet typically closely adjoining to form pseudorostrum.

pseudorostrum [G *Pseudorostrum*]: Anterior projection of carapace formed by two adjoining pseudorostral lobes. (horizontal, upturned, reflexed; acute, blunt; simple, with long setae)

retinaculum (coupling hook) [G *Retinaculum*]: In first pair of maxillipeds, one of several small, hook-like structures along inner margin of basis. Serves to interlock left and right maxillipeds.

rudimentary oostegite [G *rudimentärer Oostegit*]: In second maxillipeds of ♀, flattened plate extending from first segment (coxa). Posterior margin bears series of elongate setae extending into brood chamber.

second antenna: antenna

segment (article, joint) [G *Segment*]: One in a series of divisions of an appendage. May also refer to segment of body (then also termed somite). Segments of flagella of antennule or antenna often termed articles.

setal row (spine row) [G *Sägebor-stenreihe*]: In mandible, row of spine-like projections between molar process and lacinia mobilis (or between molar and incisor processes when lacinia mobilis is absent).

siphon [G *Sipho, Atemsipho*]: Tubular, exhalent respiratory structure consisting dorsally of pseudorostral lobes (pseudorostrum) of carapace and laterally of anteriorly directed lobes (according to interpretation: siphonal lobe of epipod or exopod) of first maxilliped pair. Water from branchial chamber exists via siphon extended in front of head.

siphonal lobe [G *Siphonalast*]: epipod

somite [G *Segment*]: One in a series of divisions of body. Head consists basically of five somites, thorax of eight somites (thoracomeres), and abdomen of six somites (pleomeres).

sperm duct: vas deferens

sternite [G *Sternit*]: Ventral surface of body segment (somite). In ♂, sternite of last (fifth) pereonite bears gonopores. First segments (coxae) of thoracopods may be more or less fused with sternites. (See also tergite)

stomach [G *Magen, Kaumagen*]: Expanded section of digestive tract between esophagus and midgut. Characterized by various lateral folds, channels, and ridges and may be differentiated into anterior cardiac and posterior pyloric regions. One to four pairs of hepatic ceca open into stomach at stomach/midgut border.

supraesophageal   ganglion (brain) [G *Oberschlundganglion, Gehirn*]: Main concentration of nerve tissue above (anterior to) esophagus. Continues posterior to esophagus as ventral nerve cord.

telson [G *Telson*]: Posteriormost segment of body. May be fused to last (sixth) abdominal somite (pleonite) to form pleotelson. May bear anus and thus be divided into preanal and postanal parts. (free, fused; elongate, short; pointed, semicircular, truncate; with/without apical spines)

tergite [G *Tergit*]: Dorsal surface of body segment (somite). (See also sternite)

testis [G *Hoden*]: Section of ♂ reproductive system in which sperm are produced. Paired, tube-like, located on each side of pereon. Each testis opens to exterior on last (fifth) pereon segment via vas deferens.

thoracomere (thoracic somite) [G *Thoracomer, Thoraxsegment*]: One of basically eight segments (somites) between head (cephalon) and abdomen (pleon). First three thoracomeres are fused with cephalon, with remaining thoracomeres (forming pereon) therefore also termed pereonites.

thoracopod [G *Thoracopod*]: One of basically eight pairs of appendages of thorax. First three pairs of thoracopods (those of thoracic somites fused to head) are developed as maxillipeds, with the remaining five pairs being termed pereopods.

thorax [G *Thorax*]: Region of body comprising eight somites between head and abdomen. First three thoracic somites fused with head to form cephalothorax, with remaining thoracic somites collectively forming pereon.

upper lip: labrum

uropod [G *Uropod*]: Long, thin, paired appendage of last segment (pleonite) of abdomen (pleon). Biramous, consisting of one-segmented basal part (peduncle) bearing one- to three-segmented endopod and two-segmented exopod. Number of endopod segments is of taxonomic importance. (styliform)

vas   deferens   (sperm   duct) [G *Vas deferens, Samenleiter*]: Section of ♂ reproductive system between each testis and gonopores on ventral surface of last (fifth) pereon segment.

ventral nerve cord [G *Bauchmark*]: Longitudinal nerve cord extending posteriorly from brain (supraesophageal ganglion). Located below digestive tract; bears 16 pairs of segmentally arranged ganglia.

ARTHROPODA

CRUSTACEA

Tanaidacea (Anisopoda)

tanaidacean [G *Tanaidacee, Scherenassel*]

TANAIDACEA

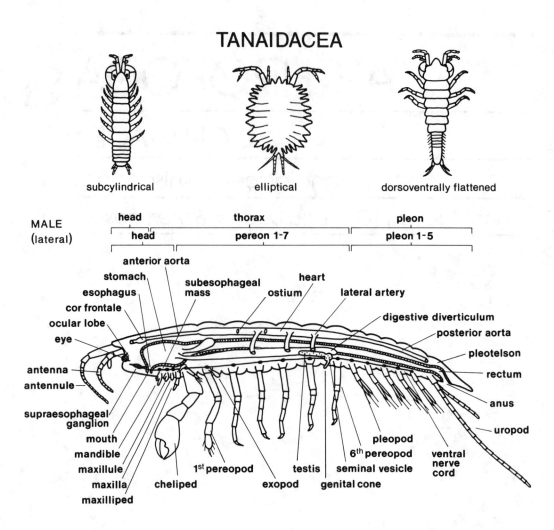

subcylindrical   elliptical   dorsoventrally flattened

MALE
(lateral)

head | thorax | pleon

head | pereon 1-7 | pleon 1-5

anterior aorta
stomach
esophagus
cor frontale
ocular lobe
eye
antenna
antennule
supraesophageal
ganglion
mouth
mandible
maxillule
maxilla
maxilliped

subesophageal
mass

heart
ostium
lateral artery

digestive diverticulum
posterior aorta
pleotelson
rectum
anus
uropod

1st pereopod
cheliped
exopod

testis
seminal vesicle
genital cone

pleopod
6th pereopod
ventral
nerve
cord

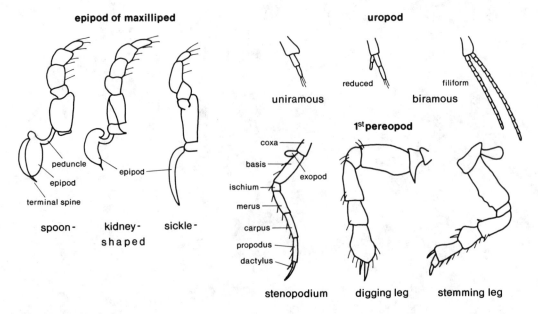

epipod of maxilliped   uropod

peduncle
epipod
terminal spine

epipod

spoon-   kidney-   sickle-
shaped

uniramous   reduced   filiform
biramous

1st pereopod

coxa
basis
ischium
merus
carpus
propodus
dactylus

exopod

stenopodium   digging leg   stemming leg

ARTHROPODA

CRUSTACEA

Tanaidacea (Anisopoda)

tanaidacean [G *Tanaidacee, Scherenassel*]

abdomen: pleon

aesthetasc [G *Aesthetasc*]: One in a series of sensory projections on flagella of antennules.

antenna (second antenna) [G *Antenna, zweite Antenne*]: Second, relatively small pair of antennae. If fully developed, consists of two-segmented proximal section (variously termed peduncle or protopod) bearing distal flagellum and antennal scale (exopod). (uniramous, biramous)

antennal scale [G *Schuppe, Squama, Exopodit*]: Scale-like structure projecting from second segment of peduncle of antenna. Considered to represent outer branch (exopod).

antennule (first antenna) [G *Antennula, erste Antenne*]: First, relatively large pair of antennae. If fully developed, biramous, consisting of proximal four-segmented peduncle and two flagella. (uniramous, biramous)

anterior aorta (cephalic aorta) [G *Kopfaorta, Aorta anterior, Aorta cephalica*]: aorta

anus [G *After*]: Posterior opening of digestive tract to exterior. Located on ventral surface of telson.

aorta [G *Aorta*]: In circulatory system, major blood vessel extending from end of heart. One may distinguish an anterior aorta extending into head and paired posterior aortas. Anterior aorta bears cor frontale. (See also lateral artery)

basis (basipodite) [G *Basis, Basipodit*]: Second of seven segments (coxa, basis, ischium, merus, carpus, propodus, dactylus) of thoracopod. Bears endite in maxillae and maxillipeds.

branchial chamber (branchial cavity) [G *Atemhöhle, Carapaxhöhle, Carapaxfalte*]: Space on each side of head between body and inner wall of carapace. Epipods of maxillipeds extend into branchial chamber. Water enters chamber primarily at posteroventral margins of carapace and exits ventrally at bases of maxillipeds.

brood pouch: marsupium

carapace [G *Carapax*]: Shield-like structure covering head and first two thoracic segments (thoracomeres). Extends laterally to enclose branchial chamber. (See also cephalothorax)

carpus (carpopodite) [G *Carpus, Carpopodit*]: Fifth of basically seven segments (coxa, basis, ischium, merus, carpus, propodus, dactylus) of thoracopod.

cephalothorax [G *Cephalothorax*]: Term applied to anterior section of body consisting of head and first two thoracic segments (thoracomeres); covered dorsally by carapace.

chela [G *Chela*]: Pincer-like structure at end of cheliped (first pereopod). Formed by movable finger (dactylus) opposed by immovable distal extension of propodus. Highly variable according to sex and stage of maturity. (See also subchela)

cheliped (chelipede) [G *Cheliped, Scherenfuß, Gnathopod*]: Modified second thoracopod (= first pereopod). Typically consists of only five segments (basis, merus, carpus, propodus, dactylus), with coxa and ischium being absent. (chelate, subchelate; symmetrical, asymmetrical)

clypeus [G *Clypeus*]: On underside of head, plate-like structure bearing upper lip (labrum). Separated from labrum by suture.

cor frontale (cephalic heart) [G *Cor frontale, Stirnherz*]: In circulatory system, small, muscular expansion of anterior aorta. Located in head; supports heart in pumping blood anteriorly.

coxa (coxopodite) [G *Coxa, Coxopodit*]: First of basically seven segments (coxa, basis, ischium, merus, carpus, propodus, dactylus) of thoracopod. Bears well-developed epipod in maxillipeds (= first thoracopods) and may be indistinct or absent in chelipeds (= second thoracopods). (See also precoxa)

dactylus (dactyl) [G *Dactylus, Dactylopodit*]: Seventh, most distal of basically seven segments (coxa, basis, ischium, merus, carpus, propodus, dactylus) of thoracopod. In second pair of thoracopods (= chelipeds or first pereopods) forms movable distal element in chela or subchela.

digestive diverticulum [G *Mitteldarmdivertikel, Mitteldarmdrüse*]: One or two pairs of digestive glands opening into digestive tract at border of stomach and midgut. ›

endite [G *Endit*]: Inwardly (medially) directed lobe of mouthpart (i.e., maxillule, maxilla). Basis of maxilliped also bears endite. (simple, deeply cleft)

endopod (endopodite) [G *Endopod, Endopodit*]: Inner branch (ramus) of binamous appendage. Represents only branch or main, typically five-segmented branch (ischium, merus, carpus, propodus, dactylus) of thoracopod. (See also exopod).

epipod (epipodite) [G *Epipod, Epipodit, Epignath, Epipodialapparat, Kiemenanhang, Kiemenlamelle*]: Well-developed projection of first segment (coxa) of each maxilliped. If fully developed, consists of stalk-like proximal section (muscular peduncle) and distal expanded section. The latter may further bear small lobe basally and finger-like process (terminal spine) terminally. Projects posteriorly into branchial chamber and serves in respiration. (kidney-shaped, sickle-shaped, spoon-shaped)

esophageal connective [G *Schlundkonnektiv*]: One of two strands of nerve tissue surrounding esophagus and connecting supraesophageal ganglion with subesophageal mass.

esophagus [G *Ösophagus*]: Relatively short section of digestive tract between mouth and stomach.

exopod (exopodite) [G *Exopod, Exopodit*]: Outer branch (ramus) of biramous appendage. May be present as small ramus on chelipeds and anterior pereopods. Scale-like projection (antennal scale) on antenna considered to represent exopod.

eye [G *Auge*]: One of two photosensitive organs on carapace; positioned anterolaterally on short projections (ocular lobes).

finger [G *Finger*]: chela

first antenna: antennule

first maxilla: maxillule

flagellum [G *Flagellum, Geißel*]: Distal division of antennule or antenna (peduncle, flagellum). Antennule may bear two flagella (with distal segments bearing aesthetascs).

genital cone [G *Zapfen*]: In ♂ reproductive system, cone-shaped process on ventral surface (sternite) of last thoracic segment (= last or seventh pereonite). Bears ♂ gonopore(s). (paired, unpaired)

gonopore [G *Gonoporus, Geschlechtsöffnung*]: Opening of ♂ or ♀ reproductive system to exterior. In ♂, pair of pores on genital cone(s) on last (seventh) pereonite; in ♀, pair of pores ventrally on fifth pereonite.

head [G *Kopf*]: Anterior of three basic divisions (tagmata) of body (head, thorax, abdomen). Consists of five somites with five pairs of appendages (antennules, antennae, mandibles, maxillules, maxillae); fused with first thoracic somite (bearing maxillipeds), this entire complex also being termed head.

heart [G *Herz*]: Elongate, muscular pumping organ of circulatory system. Located within pericardial sinus above digestive tract; typically extends through entire pereon. Bears two pairs of ostia (in pereonites 2 and 3) and gives rise to single anterior aorta, pair of posterior aortas, and up to several pairs of lateral arteries.

hypopharynx: labium

incisor process [G *Pars incisiva*]: Distal gripping process (vs. grinding = molar process) of each mandible.

ischium (ischiopod, ischiopodite) [G *Ischium, Ischiopodit*]: Third of basically seven segments (coxa, basis, ischium, merus, carpus, propodus, dactylus) of thoracopod. Relatively short; may be fused with basis.

labium (lower lip, hypopharynx [G *Labium, Unterlippe, Hypopharynx, Hypostom*]: Lobe-like structure posterior to mouth. Divided into two main lobes (paragnaths) which may bear hairs, spines, or comb-like rows of short setae. (with/without palp)

labrum (upper lip) [G *Labrum, Oberlippe*]: Unpaired, lobe-like structure anterior to mouth. Adjoined anteriorly by clypeus.

lacinia mobilis [G *Lacinia mobilis*]: Small, articulated process between setal row

and incisor process on mandibles. May be present only on left mandible. (present/absent on right mandible)

lateral artery [G *Lateralarterie, Seitenarterie*]: One of up to several pairs of blood vessels originating from sides of heart. Terminal pair often termed posterior aortas.

manca [G *Manca*]: Larval stage at which tanaidaceans are released from brood pouch (marsupium). Consists of several substages.

mandible [G *Mandibel*]: Third paired appendage of head; located between antennae and maxillules. Represents first pair of mouthparts and, if fully developed, consists (proximally to distally) of molar process, spine row, lacinia mobilis, incisor process, as well as a palp. Mouthparts may be absent in ♂. (monocondylic, dicondylic)

marsupium (brood pouch) [G *Marsupium, Bruttasche*]: In ♀, chamber formed on ventral surface of pereon by oostegites. (See also ovisac)

maxilla (second maxilla) [G *Maxille, zweite Maxille*]: Fifth paired appendage of head; located between maxillules and maxillipeds. Represents third pair of mouthparts and, if fully developed, consists of protopod bearing two endites. May be reduced to simple lobe. Mouthparts may be absent in ♂.

maxillary gland [G *Maxillendrüse, Maxillennephridium*]: Pair of excretory organs in head, one opening to exterior at base of each maxilla.

maxilliped (maxillipede) [G *Maxilliped, Kieferfuß*]: Fourth and largest pair of mouthparts. Represents modified pair of appendages (thoracopods) of first thoracic segment (which is fused to head). Basically consists of proximal section (coxa, basis) bearing segmented palp. Coxa bears well-developed epipod; basis bears endite. Variously fused to one another basally.

maxillule (first maxilla) [G *Maxillula, erste Maxille*]: Fourth paired appendage of head. Located between mandibles and maxillae; represents second pair of mouth-parts. If fully developed, consists of protopod bearing two endites as well as a palp (endo-pod). Mouthparts may be absent in ♂.

merus (meropod, meropodite) [G *Merus, Meropodit*]: Fourth of basically seven segments (coxa, basis, ischium, merus, carpus, propodus, dactylus) of thoracopod.

midgut [G *Mitteldarm*]: Elongate section of digestive tract between stomach and rectum. Digestive diverticula open into digestive tract at border of stomach and midgut.

molar process [G *Pars molaris*]: Proximal grinding process (vs. gripping = incisor process) of each mandible.

mouth [G *Mund, Mundöffnung*]: Anterior opening of digestive tract. Bordered anteriorly by upper lip (labrum), posteriorly by lower lip (labium); opens into esophagus.

neutrum [G *Neutrum*]: Sexually neutral juvenile. Depending on sex of accompanying adults, neutrum develops into ♂ or ♀.

ocular lobe [G *Augenlobus, Augenhügel*]: Pair of relatively short, immovable projections, one located anterolaterally on each side of carapace; bears eyes.

oostegite (brood plate) [G *Oostegit, Brutplatte*]: In ♀, one in a series of flattened plates projecting from basal segments (coxae) of second through fifth pereopods. Overlap to form marsupium or ovisac.

ostium [G *Ostium*]: One of typically two pairs of muscular openings in anterior region of heart. Blood from pericardial sinus enters heart through ostia.

ovary [G *Ovar, Ovarium, Eierstock*]: Section of ♀ reproductive system in which eggs are produced. Paired, tube-like, extending through most of pereon. Each ovary opens to exterior via oviduct.

oviduct [G *Ovidukt, Eileiter*]: Section of ♀ reproductive system between each ovary and gonopores on sixth thoracic segment (= fifth pereonite).

ovisac [G *Ovisack*]: In ♀ of certain tanaidaceans, term applied to brood chamber

formed by single pair of oostegites of fifth pereopod pair. (See also marsupium)

palp [G *Palp, Taster*]: Distal projection of mouthpart (i.e., of mandible, maxillule). If fully developed, consists of three segments in the former, two in the latter. Also refers to distal section (endopod) of maxilliped or to terminal process on each lobe (paragnath) of lower lip (labium).

peduncle [G *Pedunculus, Stamm*]: Proximal division of antennule or antenna (peduncle, flagellum). If fully developed, consists of four segments in the former, two in the latter (the second bearing an antennal scale). Proximal peduncular segment is large.

pereon (peraeon, pereion) [G *Peraeon*]: Division (tagma) of body between head and abdomen (pleon). Morphologically not equivalent to thorax (with a total of eight somites) because first thoracomere (bearing maxillipeds) is fused to head, the second being covered by and fused to carapace; accordingly, pereon can be considered to consist of seven or six segments (pereonites).

pereonal segment: pereonite

pereonite [G *Peraeomer*]: One of seven segments (somites) of pereon (i.e., all thoracomeres except first, which is fused to head).

pereopod (peraeopod, pereiopod) [G *Peraeopod, Pereiopod*]: One of seven pairs of appendages (thoracopods 2–8) of pereon. First pair developed as chelipeds (leading to alternate nomenclature in which only six pairs of pereopods are recognized). (uniramous, biramous = with/without exopod; with/without ischium)

pericardial sinus (pericardium) [G *Perikardialsinus, Perikard*]: Cavity surrounding heart. Blood collected in pericardial sinus enters heart via ostia.

pleomere (pleonite) [G *Pleomer, Pleonit*]: One of six segments (somites) of abdomen (pleon). Last (sixth) or last two (fifth and sixth) pleomeres fused to telson to form pleotelson. First five pleomeres typically bear pleopods; last pleomere bears uropods.

pleon (abdomen) [G *Pleon, Abdomen, Hinterleib*]: Posterior of three divisions (tagmata) of body (either head, thorax, pleon, or head, pereon, pleon). Consists basically of six somites (pleomeres) bearing posterior telson. (See also pleotelson)

pleopod [G *Pleopod*]: Paired appendage of first five somites (pleomeres) of abdomen (pleon). Consists basically of two-segmented protopod (coxa, basis) bearing one-segmented endopod and one- to two-segmented exopod. Typically absent in ♀. (uniramous, biramous; slender and elongate, flattened and short; simple, setose) (See also uropod)

pleotelson [G *Pleotelson*]: Terminal segment of body formed by fusion of telson with last (sixth) or fifth and sixth abdominal somites (pleomeres).

posterior aorta [G *Aorta posterior*]: aorta

precoxa [G *Präcoxa*]: In certain tanaidaceans, apparent segment of pereopod proximal to coxa. According to this interpretation, basal part (protopod) of pereopod consists of three segments (precoxa, coxa, basis).

propodus [G *Propodus*]: Sixth of basically seven segments (coxa, basis, ischium, merus, carpus, propodus, dactylus) of thoracopod. In second pair of thoracopods (cheliped or first pereopod) forms chela or subchela together with dactylus.

protopod (protopodite) [G *Protopod, Protopodit*]: Proximal part of certain appendages. Typically specified only when referring to maxillules and maxillae, yet basically applicable to proximal section (coxa, basis) of other appendages. (See also peduncle)

rectum [G *Rektum, Enddarm*]: In pleotelson, relatively short posterior section of digestive tract between midgut and anus.

rostrum [G *Rostrum*]: Relatively short anteromedian extension of carapace.

seminal vesicle [G *Samenblase*]: Unpaired median expansion of ♂ reproductive system into which both testes open. Located in last thoracic somite (= eighth thoracomere or seventh pereonite). Opens to exterior via two gonopores. (See also genital cone)

somite [G *Segment*]: One in a series of divisions of body. Head consists of five somites (plus first thoracic somite), pereon of seven somites (pereonites), and abdomen (pleon) of six somites (pleomeres).

spine row [G *Sägeborstenreihe*]: In mandible, row of spine-like projections between molar process and lacinia mobilis (or between molar and incisor processes when lacinia mobilis is absent).

spinning gland [G *Spinndrüse*]: One in a series of paired glands located in first pereonites. Opens at tip (dactylus) of corresponding pereopod. Serves in tube construction.

squama: antennal scale

sternite [G *Sternit*]: Ventral surface of body segment (somite). Gonopores of ♂ located on sternite of last (eighth) thoracic somite (thoracomere). (See also tergite)

stomach [G *Magen, Kaumagen*]: Expanded section of digestive tract between esophagus and midgut. If fully developed, divided into anterior cardiac and posterior pyloric sections. Digestive diverticula open into digestive tract at border of stomach and midgut.

subchela [G *Subchela*]: Pincer-like structure at end of cheliped (first pereopod). Formed by terminal segment (dactylus) bearing down on next to last segment (propodus).

subesophageal mass [G *Unterschlundmasse*]: Concentration of nerve tissue below (posterior to) esophagus. Connected to supraesophageal ganglion by pair of esophageal connectives. Consists of ganglia of mandibles, maxillules, maxillae, maxillipeds, and chelipeds. Continues posteriorly as ventral nerve cord.

supraesophageal ganglion [G *Oberschlundganglion*]: Main concentration of nerve tissue above (anterior to) esophagus. Connected to subesophageal mass by esophageal connectives.

tagma [G *Tagma*]: One of three major divisions of body (according to interpretation: head, thorax, pleon = abdomen; head, pereon, pleon; cephalothorax, pereon, pleon).

telson [G *Telson*]: Posteriormost segment of body. May be fused to last (sixth) or to fifth and sixth abdominal somites (pleomeres) to form pleotelson.

tergite [G *Tergit*]: Dorsal surface of body segment (somite). (See also sternite)

testis [G *Hoden*]: Paired, relatively short section of ♂ reproductive system in which sperm are produced. Both testes open into common seminal vesicle.

thoracomere [G *Thoracomer, Thorakalsegment*]: One of basically eight segments (somites) of thorax between head proper and abdomen (pleon). First thoracomere fused with head, with first and second being fused with and covered by carapace.

Accordingly, last seven or six thoracomeres are termed pereonites.

thoracopod [G *Thoracopod*]: One of eight pairs of appendages of thorax. First pair developed as maxillipeds, second pair as chelipeds (with either second through eighth or third through eighth pairs thus being termed pereopods). Basically consists of seven segments (coxa, basis, ischium, merus, carpus, propodus, dactylus).

thorax [G *Thorax*]: Region of body comprising eight somites between head proper and abdomen (pleon). First thoracic somite (thoracomere) fused with head, with first and second thoracomeres being covered by and fused with carapace. (See also pereon)

uropod [G *Uropod*]: Paired appendage of last segment (pleomere) of abdomen (pleon). Basically consists of one-segmented (basis) proximal section, multisegmented endopod, and few-segmented exopod. (uniramous, biramous; short, flagelliform)

ventral nerve cord [G *Bauchmark*]: Longitudinal nerve cord extending posteriorly from subesophageal mass. Located below digestive tract; bears segmentally arranged ganglia.

ARTHROPODA

CRUSTACEA

Isopoda

isopod [G *Isopode, Assel*]

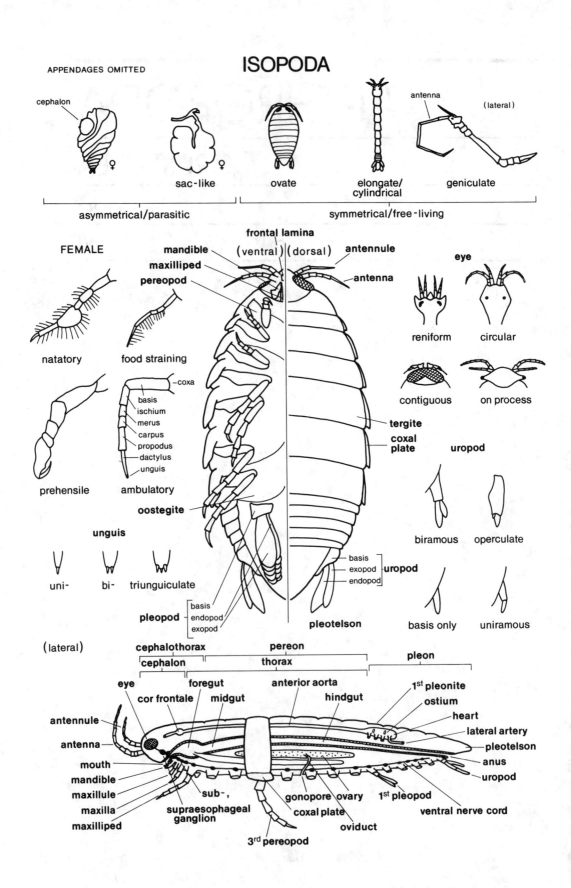

ISOPODA

APPENDAGES OMITTED

cephalon

♀

sac-like ♀

ovate

elongate/
cylindrical

antenna (lateral)

geniculate

asymmetrical/parasitic

symmetrical/free-living

FEMALE

frontal lamina
(ventral)(dorsal)

mandible
maxilliped
pereopod

antennule
antenna

eye

reniform   circular

contiguous   on process

natatory

food straining

—coxa
basis
ischium
merus
carpus
propodus
dactylus
unguis

tergite
coxal
plate

uropod

biramous   operculate

prehensile

ambulatory

oostegite

basis
exopod   uropod
endopod

unguis

uni-   bi-   triunguiculate

pleopod

basis
endopod
exopod

pleotelson

basis only   uniramous

(lateral)

cephalothorax
cephalon

pereon
thorax

pleon

eye
cor frontale

foregut
midgut

anterior aorta
hindgut

1st pleonite
ostium
heart

antennule
antenna

lateral artery
pleotelson
anus
uropod

mouth
mandible
maxillule
maxilla
maxilliped

sub-,
supraesophageal
ganglion

gonopore ovary
coxal plate
oviduct

1st pleopod

1st pleopod
ventral nerve cord

3rd pereopod

ARTHROPODA

CRUSTACEA
Isopoda
isopod [G *Isopode, Assel*]

abdomen [G *Abdomen, Hinterleib*]: pleon

accessory flagellum [G *Nebengeißel*]: In antenna of certain isopods, small, many-segmented flagellum projecting from fourth segment of peduncle.

aesthetasc [G *Aesthetasc, Ästhetask*]: One in a series of sensory projections on antennules. (uniramous, biramous, triramous)

ampulla: One of two spinose or setose internal ridges on each side of foregut. Serves in crushing food.

antenna [G *Antenne*]: Second, usually longer appendage of cephalon. Uniramous, consisting of basal peduncle and distal flagellum. (pediform) (See also accessory flagellum, antennal scale)

antennal gland [G *Antennendrüse, Antennennephridium*]: Pair of poorly developed excretory organs in head of certain isopods, one opening at base of each antenna.

antennal scale (squama) [G *Schuppe, Exopodit*]: Projection extending from base (peduncle) of antenna; considered to represent reduced outer branch (exopod).

antennule (first antenna) [G *Antennula, erste Antenne*]: First, usually shorter appendage of cephalon; typically uniramous, consisting of basal peduncle and distal flagellum. (uniramous, biramous)

anus [G *After*]: Posterior opening of digestive tract to exterior; located ventrally at base of telson.

aorta (anterior aorta) [G *Aorta, Aorta cephalica, Kopfaorta*]: In circulatory system, major unpaired blood vessel extending from anterior end of heart. Extends into head and bears small accessory heart (cor frontale).

appendix masculina (male stylet) [G *Appendix masculina*]: Rod-shaped median process on endopod of second pleopod in ♂; serves in sperm transfer.

artery [G *Arterie*]: One of numerous blood vessels originating from heart or aorta. According to position one may distinguish anterior and posterior arteries as well as a subneural artery.

article (joint, segment) [G *Glied, Segment*]: Any segment of an appendage, although often applied to segments of flagellum of antennae or antennules.

basis (basipodite) [G *Basis, Basipodit*]: Second segment of pereopod; located between coxa and ischium.

brain: supraesophageal ganglion

branchia [G *Kiemen*]: Respiratory structure on or modification (thin cuticle, folding, etc.) of pleopod for gas exchange.

branchial chamber [G *Atemhöhle*]: Area on ventral surface of abdomen (pleon) containing branchiae; may be covered by expanded pleopods or uropods.

brood pouch (brood chamber, marsupium) [G *Bruttasche, Marsupium*]: Chamber formed on ventral surface of pereon in ♀ by three to five pairs of oostegites.

carpus (carpopodite) [G *Carpus, Carpopodit*]: Fifth segment of pereopod; located between merus and propodus.

cephalon (head) [G *Kopf*]: Anterior division (tagma) of body bearing eyes, mouth, two antennules, two antennae, and three pairs of mouthparts; fused with first thoracic segment, in certain cases also with second, and therefore occasionally referred to as a cephalothorax.

cephalothorax [G *Cephalothorax*]: Term applied to anterior section of body consisting of head (cephalon) fused with first (and occasionally second) thoracic segment.

clypeus [G *Clypeus*]: Part of head (cephalon) bearing upper lip (labrum).

cor frontale (accessory heart) [G *Cor frontale, Stirnherz*]: In circulatory system, small, muscular expansion of anterior aorta. Located at border of cephalon and pereon; supports heart in pumping blood to anterior region of head.

coupling seta (coupling hook) [G *Hakenborste*]: One in a series of cuticular projections on inner branch (endopod) of maxilliped or on medial margin of pleopods; serves to hook maxilliped or pleopod to its opposite member.

coxa (coxapodite) [G *Coxa, Coxopodit, Laufbeinhüfte*]: First segment of pereopod; often expanded to form projecting coxal plate.

coxal plate (epimere) [G *Epimer*]: Lateral extension of proximal segment (coxa) of pereopod; may be fused to body wall.

cryptoniscus [G *Cryptoniscium*]: Third of three larval stages (epicaridium, microniscus, cryptoniscus) in certain parasitic isopods. Free-living, more elongate, and characterized by seven pairs of grasping pereopods. Swims by means of antennae and pleopods and attaches to final host.

dactylus (dactyl, dactylopodite) [G *Dactylus, Dactylopodit*]: Seventh and terminal segment of perepod. In gnathopod, may form movable distal element in subchela. (biunguiculate, triunguiculate)

endite [G *Endit, Laden*]: Inwardly (medially) directed lobe of proximal part of posterior mouthparts (maxillules, maxillae, maxillipeds). If endites of first two pairs of mouthparts are paired, one may distinguish a proximal = inner and a distal = outer endite/plate/lobe. (immovable, movably articulated; rounded, cleft)

endopod (endopodite) [G *Endopod, Endopodit*]: Inner branch (ramus) of biramous appendage.

epicaridium [G *Epicaridium*]: First of three larval stages (epicaridium, microniscus, cryptoniscus) of certain parasitic isopods. Free-living, characterized by stout, segmented body lacking last (= seventh) pair of pereopods. Swims by means of antennae and pleopods and attaches to intermediate host.

epimere (epimeron) [G *Epimer, Epimeron*]: Lateral extension of body, either of the body wall proper or, in the pereon, more frequently of coxae of pereopods (more accurately termed coxal plate).

esophageal connective [G *Schlundkonnektiv*]: One of two strands of nerve tissue surrounding foregut and connecting brain (supraesophageal ganglion) with subesophageal ganglion.

esophagus: foregut

exopod (exopodite) [G *Exopod, Exopodit, Außenast*]: Outer branch (ramus) of biramous appendage (e.g., pleopod).

eye (compound eye) [G *Auge, Komplexauge*]: Organ of vision on cephalon; when present, usually small and sessile. (sessile, stalked; pear-shaped, reniform, oval, contiguous)

flagellum [G *Flagellum, Geißel, Antennengeißel*]: Distal division of antennule or antenna (peduncle, flagellum); composed of a varying number of articles.

foregut [G *Vorderdarm*]: Anterior of two basic divisions (foregut, hindgut) of digestive tract. Characterized by various internal folds (lamellae), chambers, and ridges.

frontal lamina [G *Lamina frontalis*]: Plate-like structure anterior to mouthparts, usually situated between antennae. (pentagonal, quadrate, spiniform, triangular)

gill [G *Kieme*]: branchia

gnathopod [G *Gnathopod, Greifbein*]: Term applied to more anterior pereopod(s) if modified as prehensile (grasping) organ. Last segments typically form subchela.

gonopod [G *Gonopod*]: Second pleopod of ♂; inner branch (endopod) may be modified into a rod-shaped structure (appendix masculina) for reproductive purposes.

gonopore [G *Gonoporus, Geschlechtsöffnung*]: In ♀ reproductive system, opening of each oviduct to exterior; located on ventral surface (sternite) of fifth pereon segment (pereonite). (See also penis)

head [G *Kopf*]: cephalon

heart [G *Herz*]: Muscular pumping organ of circulatory system; relatively short, located above hindgut in abdomen (pleon). Bears one to two pairs of ostia and gives rise to anterior aorta and series of lateral arteries.

hepatic cecum: midgut cecum

hindgut [G *Enddarm*]: Posterior of two basic divisions (foregut, hindgut) of digestive tract; elongate, typically extending through part of pereon and entire abdomen (pleon). Opens posteriorly via anus at base of telson.

hypopharynx: labium

incisor process [G *Pars incisiva, Processus incisivus*]: Distal, often toothed gripping process (vs. grinding = molar process) of each mandible.

ischium (ischiopod, ischiopodite) [G *Ischium, Ischiopodit*]: Third segment of pereopod; located between basis and merus.

labium (lower lip, hypopharynx, posterior lip) [G *Labium, Hypopharynx, Unterlippe*]: Lobe-like structure posterior to mouth. Typically divided into two lobes (paragnaths).

labrum (upper lip, frontal lip) [G *Labrum, Oberlippe*]: Unpaired, lobe-like structure anterior to and partially covering mouth.

lacinia mobilis [G *Lacinia mobilis*]: Toothed, movably articulated process between setal row and incisor process on left (and occasionally also on right) mandible.

lower lip: labium

manca [G *Manca, Mancastadium*]: First free-living larval stage after release from brood pouch (marsupium). Resembles adult, yet characterized by lack of last (= seventh) pair of pereopods.

mandible [G *Mandibel*]: Third paired appendage of head (cephalon); located between antennae and maxillules. Represents first pair of mouthparts and, if fully developed, consists (proximally to distally) of molar process, spine row, lacinia mobilis, incisor process, and palp. (biting/chewing, piercing/sucking = styliform)

marsupium [G *Marsupium*]: brood pouch

maxilla (second maxilla) [G *Maxille, zweite Maxille*]: Fifth paired appendage of head (cephalon); located between maxillules and maxillipeds. Represents third pair of mouthparts. Relatively small, consisting of base and two or three endites.

maxillary gland [G *Maxillardrüse, Maxillennephridium*]: Pair of excretory organs in head (cephalon), one opening at base of each maxilla. (See also antennal gland)

maxilliped (maxillipede) [G *Maxilliped, Kieferfuß*]: Fourth and largest pair of mouthparts. Represents highly modified pair of appendages (thoracopods) of first thoracic segment (which is fused to head). Basically consists of basal section (coxa, basis) bearing five-segmented (ischium, merus, carpus, propodus, dactylus) palp. Segments may bear variously developed endites. Maxillipeds may be coupled to one another by retinacula.

maxillule (first maxilla) [G *Maxillula, erste Maxille*]: Fourth paired appendage of head (cephalon); located between mandibles and maxillae. Represents second pair of mouthparts and typically consists of base with two differently developed endites.

merus (meropod, meropodite) [G *Merus, Meropodit*]: Fourth segment of pereopod; located between ischium and carpus.

microniscus [G *Microniscus, Microniscium*]: Second of three larval stages (epicaridium, microniscus, cryptoniscus) in certain parasitic isopods. Parasitic on intermediate host. Characterized by loss of natatory function of antennae and pleopods as well as by development of last (= seventh) pereopod pair.

midgut [G *Mitteldarm*]: Poorly defined section of digestive tract between foregut and hindgut. Term either applied to relatively short section of digestive tract near origin of midgut ceca or considered to be represented by midgut ceca themselves.

midgut cecum (digestive cecum, hepatic cecum) [G *Mitteldarm-Divertikel, Hepatopankreas*]: One of up to three pairs of elongate outpocketings of digestive tract. Originates in anterior region of pereon at border of foregut and hindgut and extends through most of pereon.

molar process [G *Pars molaris, Processus molaris*]: Proximal grinding process (vs. gripping = incisor process) of mandible. (immovable, movably articulated; blade-like, blunt-pointed, broad and truncate, pointed

with apical setae, spiniform, truncate with teeth)

mouth [G *Mund, Mundöffnung*]: Anterior opening of digestive tract on underside of head (cephalon). Bordered anteriorly by upper lip (labrum), posteriorly by lower lip (labium).

oostegite (oöstegite, brood plate) [G *Oostegit, Brutplatte*]: On ♀, one in a series of flattened plates typically projecting from basal segments (coxae) of second through fifth or sixth pereopods. Overlap to form brood chamber (marsupium).

operculum [G *Operculum*]: Protective cover on ventral surface of abdomen (pleon); usually derived from anterior pleopod pair(s) or uropods. (See also valve).

ostium [G *Ostium*]: One of one or two pairs of muscular openings of heart. Blood from pericardial sinus enters heart through ostia.

ovary [G *Ovar, Ovarium, Eierstock*]: Paired section of ♀ reproductive system in which eggs are produced. Located dorsolateral to digestive tract in pereon, each ovary opening to exterior on ventral surface (sternite) of fifth pereon segment (pereonite) via oviduct.

oviduct [G *Ovidukt, Eileiter*]: Section of ♀ reproductive system between each ovary and gonopores on ventral surface (sternite) of fifth pereon segment (pereonite).

palp [G *Palpus, Taster*]: Distal projection of mandible or maxilliped. Represents endopod and basically consists of three segments in mandible, five segments in maxilliped.

paragnath [G *Paragnath*]: One of two lobes of labium.

peduncle [G *Pedunculus, Stiel*]: Proximal division of antennule or antenna (peduncle, flagellum). Typically consists of three segments in antennule, five segments in antenna.

penis (pl. penes) [G *Penis, Geschlechtspapille*]: In ♂, paired ventral copulatory papillae (occasionally a single cone) on ventral surface (sternite) of last (seventh) pereon segment (pereonite); transfers sperm to gonopods.

pereon (peraeon, pereion) [G *Peraeon, Pereion*]: Second and largest division (tagma) of body between head (cephalon) and abdomen (pleon) or more accurately between cephalothorax and pleon. Typically consists of seven segments (pereonites); morphologically not equivalent to thorax because first (and occasionally second) thoracic segment (bearing maxillipeds) is fused to cephalon.

pereonite (pereiomere) [G *Peraeomer*]: One of typically seven segments of pereon (i.e., all thoracomeres except first and occasionally second, which are fused to cephalon). Appendages of pereonites are termed pereopods.

pereopod (peraeopod, pereiopod) [G *Peraeopod, Pereiopod*]: One of up to seven pairs of appendages of pereon. Uniramous, consisting of coxa, basis, ischium, merus, carpus, propodus, and dactylus. Anterior one or two pairs may be differentiated as gnathopods.

pericardial sinus (pericardium) [G *Perikardialsinus, Perikard*]: Cavity surrounding heart. Blood collected in pericardial sinus enters heart via ostia.

pleomere: pleonite

pleon (abdomen) [G *Pleon, Abdomen, Hinterleib*]: Posterior division (tagma) of body (cephalon, pereon, pleon) consisting basically of six segments (pleonites). (See also pleotelson)

pleonite (pleomere) [G *Pleomer*]: One of basically six segments of pleon. Appendages of pleonite are termed pleopods; often considerably narrower than pereonites. Last (= sixth) pleonite typically fused to telson to form pleotelson.

pleopod [G *Pleopod*]: One of two appendages of each of the first five abdominal somites (those of sixth somite being termed uropods). Biramous, flattened, and typically consisting of basal segment and two branches (rami). Usually with respiratory function. Second pair in ♂ modified as gonopods.

pleotelson [G *Pleotelson*]: Terminal structure of body formed by fusion of one or more abdominal segments (pleonites) with telson.

praniza [G *Praniza, Pranizalarve*]: Parasitic larval stage of certain isopods. Characterized by highly expanded posterior region of pereon.

precoxa [G *Präcoxa*]: Segment of appendage occasionally distinguished proximal to coxa.

propodus [G *Propodit*]: Sixth or next to last segment of thoracopod; located between carpus and dactylus.

pseudotrachea [G *Tracheenlunge*]: In certain terrestrial isopods, modification of outer branch (exopod) of pleopod for respiratory purposes.

pylopod [G *Pylopod*]: Modified appendage of first pereon segment (pereonite) in certain isopods; functions as maxilliped.

ramus [G *Ast*]: Branch of appendage; typically refers to inner (endopod) and outer (exopod) branches of biramous appendage (e.g., pleopods, uropods).

retinaculum (coupling seta, coupling hook) [G *Retinaculum*]: In maxilliped, one in a series of small, hook-like structures along margin of endite of second segment (basis). Serves to interlock left and right maxillipeds.

rostrum [G *Rostrum*]: Anterior median extension of head (cephalon).

sclerite: Term occasionally applied to any one of several, difficult to identify, proximal segments of mouthparts.

setal row [G *Sägeborsten, Borstenreihe*]: Row of setae between lacinia mobilis and molar process on mandible.

somite [G *Somit*]: One in a series of divisions of body. Cephalon consists of five somites (cephalothorax of these five plus first thoracic somite), pereon of six or seven somites (pereonites), and pleon of basically six somites (pleomeres).

spiracle [G *Stigma*]: Opening to pseudotracheae on pleopods of certain terrestrial isopods.

statocyst [G *Statozyste*]: Organ of equilibrium located in telson of certain isopods; contains statolith.

sternite (sternal plate) [G *Sternit*]: Ventral surface of body segment (somite). Sternite of fifth (♀) or seventh (♂) pereon segment bears gonopores.

subchela [G *Subchela*]: In modified anterior pereopod(s), pincer-like structure formed by dactylus bearing down on propodus or propodus down on carpus.

subesophageal ganglion [G *Unterschlundganglion*]: Concentration of nerve tissue below (posterior to) foregut. Consists of fused ganglia of mouthparts (mandibles, maxillules, maxillae, maxillipeds). Continues posteriorly as ventral nerve cord.

supraesophageal ganglion (brain) [G *Oberschlundganglion, Gehirn*]: Concentration of nerve tissue above (anterior to) foregut. Connected to subesophageal ganglion by pair of esophageal connectives.

tagma [G *Tagma*]: One of three major divisions of body (cephalon, pereon, pleon). (See also cephalothorax)

telson [G *Telson*]: Terminal segment of body, typically fused with one or more abdominal segments (pleonites) to form pleotelson. (lanceolate, ovate, pointed, with subparallel margins; posterior margin: concave, convex)

tergite (tergal plate) [G *Tergit*]: Dorsal surface of body segment (somite); often expanded laterally.

testis [G *Hoden*]: Paired section of ♂ reproductive system in which sperm are produced. Located dorsolateral to digestive tract in pereon, each testis opening to exterior on last (seventh) pereon segment via vas deferens.

thoracomere [G *Thoracomer, Thorakalsegment*]: One of basically eight segments (somites) between head (cephalon) and abdomen (pleon). First (and occasionally second) thoracomere is fused to cephalon, and remaining thoracomeres (forming pereon) therefore also termed pereonites.

thoracopod [G *Thoracopod*]: One of two appendages of each somite (thoracomere) of thorax. First pair developed as maxillipeds; second through eighth pairs may be termed pereopods, second and third differentiated as gnathopods. (See also pylopod)

thorax [G *Thorax*]: Region of body basically comprising eight somites between head (cephalon) and abdomen (pleon). First thoracic somite (and occasionally second) is fused with cephalon to form cephalothorax; remaining somites of thorax collectively form pereon.

unguis [G *Unguis*]: Terminal claw-like portion of dactylus. (uniunguiculate, biunguiculate, triunguiculate)

upper lip: labrum

uropod [G *Uropod*]: Paired appendage on next to last segment of abdomen (pleon); rami always a single segment. (biramous, uniramous, simple uniramous; spatulate, styliform)

valve [G *Klappe*]: Term applied to uropods covering branchial chamber in certain isopods.

vas deferens (sperm duct) [G *Vas deferens, Samenleiter*]: Section of ♂ reproductive system between each testis and ♂ gonopore(s).

ventral nerve cord [G *Bauchmark, ventraler Nervenstrang*]: Longitudinal nerve cord extending below digestive tract from subesophageal ganglion into abdomen. Bears series of segmentally arranged ganglia (anteriorly displaced in pleon).

ARTHROPODA

CRUSTACEA

Amphipoda

amphipod [G *Amphipode, Flohkrebs*]

AMPHIPODA

inflated

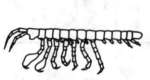

cylindrical

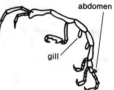

abdomen

gill

elongate

gill

dorsoventrally
flattened
(parasitic)

FEMALE (lateral)

cephalon — pereon (pereonites 1-7)

pleon (pleomeres 1-6)

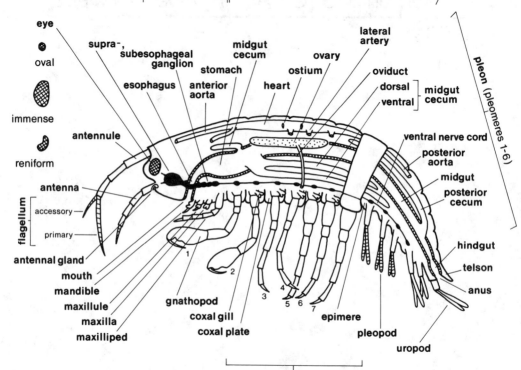

eye

oval

immense

reniform

supra-,
subesophageal
ganglion

esophagus

antennule

antenna

accessory

primary

flagellum

antennal gland

mouth

mandible

maxillule

maxilla

maxilliped

stomach

anterior
aorta

midgut
cecum

heart

ostium

ovary

lateral
artery

oviduct

dorsal
ventral

midgut
cecum

ventral nerve cord

posterior
aorta

midgut

posterior
cecum

hindgut

telson

anus

gnathopod

coxal gill

coxal plate

epimere

pleopod

uropod

pereopods 1-7
(right)

pereopod

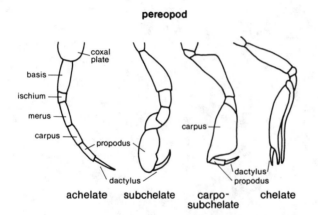

coxal
plate

basis

ischium

merus

carpus

propodus

dactylus

achelate subchelate

carpus

dactylus
propodus

carpo-
subchelate

chelate

telson

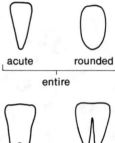

acute rounded

entire

emarginate cleft

ARTHROPODA

CRUSTACEA

Amphipoda

amphipod [G *Amphipode, Flohkrebs*]

abdomen: pleon

accessory flagellum [G *Nebengeißel*]: Small inner flagellum of antennule (as opposed to larger primary flagellum). Number of articles is of taxonomic importance.

accessory gill: One in a series of small respiratory structures projecting from bases of coxal gills.

aesthetasc [G *Aesthetasc, Ästhetask*]: One in a series of thin-walled sensory projections on antennule. (See also calceolus)

ampulla: One in a series of spinose and setose ridges surrounding border between esophagus and stomach.

antenna (second antenna) [G *Antenna, zweite Antenne*]: Second pair of antennae. Uniramous, consisting of proximal five-segmented peduncle and distal flagellum. May bear sensory structures (aesthetascs, calceoli). (short, long; slender, robust; naked, setose, spinose; similar, sexually dimorphic; pediform)

antennal gland [G *Antennendrüse, Antennenephridium*]: Pair of excretory glands in head, one opening to exterior at base (peduncle) of each antenna. (See also gland cone)

antennule (first antenna) [G *Antennula, erste Antenne*]: First pair of antennae. Basically biramous, consisting of proximal three-segmented peduncle bearing large outer (primary) and smaller inner (accessory) flagella. (short, long; slender, robust; naked, setose, spinose; geniculate; with/without accessory flagellum = biramous/uniramous)

anterior aorta [G *Kopfaorta, Aorta cephalica*]: aorta

anus [G *After*]: Posterior opening of digestive tract to exterior; located ventrally at base of telson.

anterior head lobe: lateral lobe

aorta [G *Aorta*]: In circulatory system, major blood vessel extending from end of heart. One may distinguish an anterior aorta extending into head and a paired or unpaired posterior aorta extending to end of abdomen.

artery [G *Arterie*]: One of numerous blood vessels originating from heart or aortas. According to position one may distinguish anterior lateral or facial artery as well as three pairs of lateral arteries.

article (joint, segment) [G *Glied, Segment*]: Any segment of an appendage, although often applied to segments of flagella of antennules or antennae.

basis (basipodite) [G *Basis, Basipodit*]: Second of seven segments of thoracopod (coxa, basis, ischium, merus, carpus, propodus, dactylus); represents first movable (free) segment.

brain: supraesophageal ganglion

branchia: coxal gill

brood plate: oostegite

brood pouch (marsupium) [G *Bruttasche, Brutraum, Marsupium*]: In ♀, chamber formed on ventral surface of pereon by oostegites on second to fifth pereopods.

calceolus [G *Calceolus*]: One in a series of club-shaped sensory projections on antennules and antennae. More common in ♂. (See also aesthetasc)

carpus (carpopodite) [G *Carpus, Carpopodit*]: Fifth of basically seven segments (coxa, basis, ischium, merus, carpus, propodus, dactylus) of thoracopod; represents fourth movable (free) segment. In certain amphipods, may form part of subchela (i.e., dactyl and propodus bearing down on carpus)

cecum: midgut cecum

cephalon (head) [G *Cephalon, Kopf*]: Anterior of three major divisions (tagmata) of body (cephalon, pereon, pleon). Typically laterally compressed and fused with first (occasionally also second) thoracic somite to form cephalothorax. The term head is often used synonymously with cephalothorax. (with/without rostrum, anteriorly truncate; laterally compressed, dorsoventrally flattened) (See also lateral lobe, postantennal sinus)

cephalothorax [G *Cephalothorax*]: Term applied to anterior section of body consisting of head (cephalon) fused with first (and occasionally with second) thoracic somite.

chela [G *Chela*]: Pincer-like structure of first two pereopod pairs (gnathopods). Formed by movable finger (dactylus) opposed by immovable distal extension of propodus. (See also subchela)

coxa (coxopodite) [G *Coxa, Coxopodit*]: First of basically seven segments (coxa, basis, ischium, merus, carpus, propodus, dactylus) of thoracopod; typically developed as large, immovable coxal plate.

coxal gill (branchia) [G *Kieme*]: Respiratory structure projecting from basal segment (coxa) of pereon appendage (pereopod); consists of modified epipod. Six pairs of coxal gills (on pereon segments 2–8) represent full complement. (digitiform, foliate, sac-like; walls: simple = smooth, convoluted, pleated) (See also accessory gill, sternal gill)

coxal gland [G *Coxaldrüse*]: Segmentally arranged series of small excretory structures, one pair in each somite (pereonite) of pereon as well as in first four abdominal somites.

coxal plate [G *Coxalplatte*]: Modification of proximal segment (coxa) of appendage (e.g., pereopod) into large, immovable, plate-like structure. Serves to protect gills and accentuates laterally compressed shape of body. (contiguous, discontiguous; bilobed, rectangular, rounded, spiniform; margin: setose, serrate, smooth; ornamentation: setose, with ridges, spines, tubercles)

cuticle [G *Kutikula*]: Noncellular, multilayered layer forming exoskeleton and lining foregut and hindgut. (smooth, spinose, with plate-like projections)

dactylus (dactyl) [G *Dactylus, Dactylopodit*]: Seventh and most distal of basically seven segments (coxa, basis, ischium, merus, carpus, propodus, dactylus) of thoracopod. In gnathopod, may form movable distal element in chela or subchela. (oriented anteriorly/posteriorly)

endite [G *Endit*]: Inwardly (medially) directed lobe of proximal part of posterior mouthparts (e.g., maxillules, maxillae, maxillipeds). First (proximal) endite may be termed inner plate, second termed outer plate. Shape, degree of development, and armature are of taxonomic importance. (equal, subequal, rudimentary, absent; with spines/ setae)

endopod (endopodite) [G *Endopod, Endopodit*]: Inner branch (ramus) of biramous appendage. Forms palp in mouthparts (mandibles, maxillules, maxillae, maxillipeds) and represents main five-segmented part of pereopods. (See also exopod)

epimere (epimeral plate) [G *Epimer, Epimerialplatte*]: Lateral projection of segments of anterior abdomen (pleomere). (posterodistal angle: acute, rounded, quadrate, prolonged into large tooth; margins: smooth, serrate, setose)

epipod (epipodite) [G *Epipod, Epipodit*]: In appendage (pereopod) of pereon, inwardly directed lobe of protopod. Serves in respiration and therefore frequently termed coxal gill or branchia.

epistome [G *Epistom*]: On underside of head, plate-like structure in front of upper lip (labrum). (pointed, rounded)

esophageal connective [G *Schlundkonnektiv*]: One of two strands of nerve tissue surrounding esophagus and connecting brain (supraesophageal ganglion) with subesophageal ganglion.

esophagus [G *Ösophagus*]: Relatively short, cuticle-lined section of digestive tract between mouth and stomach. Esophagus and stomach together form foregut.

exopod (exopodite) [G *Exopod, Exopodit*]: Outer branch (ramus) of biramous appendage (e.g., pleopod, uropod)

eye (compound eye) [G *Auge, Komplexauge*]: One of two unstalked (sessile) organs of vision on each side of head. (contiguous, convex = globular, immense, oval, reniform; blind, oculate)

flagellum [G *Flagellum, Geißel*]: Distal division of antennule or antenna (peduncle, flagellum). May bear sensory structures (aesthetascs on antennules, calceoli on antennules and antennae). If two flagella are present on antennule, one distinguishes a larger (primary) and smaller (accessory) flagellum.

foregut [G *Vorderdarm*]: Anterior of three basic divisions (foregut, midgut, hindgut) of digestive tract. Typically consists of esophagus and stomach.

ganglion [G *Ganglion*]: ventral nerve cord

gill: accessory gill, coxal gill, sternal gill

gland cone [G *Drüsenkegel*]: In each antenna, conical projection on second segment of base (peduncle). Bears opening of antennal gland apically.

gnathopod [G *Gnathopod, Gnathopodium*]: Modified second and third thoracopod pairs (= first and second pereopods). Uniramous, consisting of modified, immovable coxa (coxal plate) and six free segments (basis, ischium, merus, carpus, propodus, dactylus). Last segments typically form subchela. (sexually/not sexually dimorphic, polymorphic; carpo-subchelate, chelate, subchelate)

gonopore (genital pore) [G *Gonoporus, Geschlechtsöffnung*]: In ♀ reproductive system, opening of each oviduct to exterior. Located on ventral surface (sternite) of fifth pereon segment (pereonite). (See also penis)

head: cephalon

heart [G *Herz*]: Elongate, muscular pumping organ of circulatory system; located within pericardial sinus above digestive tract and extending through most of pereon (first to sixth segments). Bears three pairs of ostia and gives rise to anterior and posterior aortas as well as series of lateral arteries.

hindgut (proctodeum, rectum) [G *Enddarm*]: Posterior of three basic divisions (foregut, midgut, hindgut) of digestive tract. Cuticle-lined and opening to exterior via anus.

incisor process [G *Pars incisiva*]: Gripping process (vs. grinding = molar process) of each mandible.

inferior antennal sinus: postantennal sinus

inner lobe: labium

inner plate [G *Innenlade*]: endite

ischium (ischiopod, ischiopodite) [G *Ischium, Ischiopodit*]: Third of basically seven segments (coxa, basis, ischium, merus, carpus, propodus, dactylus) of thoracopod; represents second movable (free) segment.

labium (lower lip) [G *Labium, Unterlippe*]: Lobe-like structure posterior to mouth. Typically divided into two lobes (paragnaths), each of which may be further subdivided into a large section (outer lobe), a median inner lobe, and a posterolateral extension (mandibular lobe).

labrum (upper lip) [G *Labrum, Oberlippe*]: Unpaired, lobe-like structure anterior to and partially covering mouth. (rounded, medially incised = apically notched = bilobed; lobes: symmetrical/asymmetrical)

lacinia mobilis [G *Lacinia mobilis*]: Small, articulated accessory process between setal row and incisor process on each mandible. Differs in shape on left and right mandibles. (serrate, smooth)

lateral lobe (anterior head lobe): Extension of each anterolateral margin of head; may accommodate eyes. (angular, rounded; prolonged)

lower lip: labium

mandible [G *Mandibel*]: Third paired appendage of head (cephalon); located between antennae and maxillules. Represents first pair of mouthparts and, if fully developed, consists (proximally to distally) of molar process, spine row, lacinia mobilis, incisor process, and palp. (with/without palp)

mandibular lobe (mandibular process): In lower lip (labium), posterolateral extension of each paragnath.

marsupium: brood pouch

maxilla (second maxilla) [G *Maxille, zweite Maxille*]: Fifth paired appendage of head (cephalon); located between maxillules and maxillipeds. Represents third pair of mouthparts. Relatively small, consisting of base and two endites.

maxilliped (maxillipede) [G *Maxilliped, Kieferfuß*]: Fourth and largest pair of mouthparts. Represents highly modified pair of appendages (thoracopods) of first thoracic segment (which is fused to head). Basically consists of basal section (coxa, basis) bearing a five-segmented (ischium, merus, carpus, propodus, dactylus) palp. Basis and ischium typically with endites. Maxilliped pair variously fused. (with/without palp; maxilliped pair: with fused coxae, with fused coxae and endites, reduced to single unpaired plate)

maxillule (first maxilla) [G *Maxillula, erste Maxille*]: Fourth paired appendage of head (cephalon); located between mandibles and maxillae. Represents second pair of mouthparts and, if fully developed, consists of base with two endites as well as distal palp.

mesosome: pereon

merus (meropod, meropodite) [G *Merus, Meropodit*]: Fourth of basically seven segments (coxa, basis, ischium, merus, carpus propodus, dactylus) of thoracopod; represents third movable (free) segment.

metasome: pleon

midgut [G *Mitteldarm*]: Elongate section of digestive tract between foregut (esophagus and stomach) and hindgut. Gives rise to various ceca anteriorly (at stomach/midgut border) and posteriorly.

midgut cecum (cecum) [G *Caecum, Coecum, Darmblindsack, Hepatopankreas*]: One of several tubular outpocketings of digestive tract. According to position, one may distinguish 1) two pairs of posteriorly directed lateral ceca (two dorsal and two ventral) arising at stomach/midgut border and extending through most of body, 2) single or paired, anteriorly directed ceca arising middorsally at stomach/midgut border, and 3) single or paired, anteriorly directed ceca arising from posterior end of midgut. In certain amphipods may also refer to large, unpaired pouch into which esophagus opens.

molar process [G *Pars molaris*]: Proximal grinding process (vs. gripping =

incisor process) of each mandible. (laminar, ridged, rounded, setulose, smoothly truncated, absent)

mouth [G *Mund, Mundöffnung*]: Anterior opening of digestive tract on underside of head. Bordered anteriorly by upper lip (labrum), posteriorly by lower lip (labium); opens into esophagus.

ocular lobe (eye lobe): One of two small, articulated lobes on head of certain eyeless amphipods. Function unknown.

oostegite (oöstegite, brood plate) [G *Oostegit, Brutplatte*]: In ♀, one in a series of flattened plates projecting from basal segments (coxae) of second through fifth pereopod pairs. Overlapping or interlocking to form brood chamber (marsupium). (pentagonal, rectangular, subrectangular)

ostium [G *Ostium*]: One of basically three pairs of muscular openings in heart at level of second, third, and fourth segments of pereon. Blood in pericardial sinus enters heart through ostia.

outer lobe: labium

outer plate [G *Außenlade*]: endite

ovary [G *Ovar, Ovarium, Eierstock*]: Paired section of ♀ reproductive system in which eggs are produced. Elongate, cylindrical, extending between heart and digestive tract in pereon. Each ovary opens to exterior on fifth pereon segment (pereonite) via oviduct and gonopore.

oviduct [G *Ovidukt, Eileiter*]: Section of ♀ reproductive system between each ovary and gonopores on ventral surface of fifth pereon segment (pereonite).

palm [G *Palma, Hand*]: In subchela of first two pairs of pereopods (gnathopods), posterior margin of propodus against which dactylus bears down.

palp [G *Palpus, Taster*]: Distal projection of mouthpart (mandible, maxillule, maxilla, maxilliped). Represents endopod and, if fully developed, consists of five segments (ischium, merus, carpus, propodus, dactylus). (one- to five-segmented/articulate)

paragnath: labium

peduncle [G *Pendunculus, Stiel, Schaft*]: Proximal division of antennule or antenna (peduncle, flagellum), although also applied to base of other appendages (e.g., maxilliped, pleopod, uropod). Typically consists in antennule of three segments, in antenna of five; second peduncular segment of latter bears opening of antennal gland.

penis (penis papilla) [G *Penis, Penispapille*]: Pair of relatively small ♂ copulatory structures on ventral surface (sternite) of last (seventh) pereon segment (pereonite), one associated with each vas deferens.

pereon (peraeon, pereion, mesosome) [G *Peraeon*]: Division (tagma) of body between head (cephalon) and abdomen (pleon) or more accurately between cephalothorax and abdomen. Consists of four to seven segments (pereonites), each bearing appendages (pereopods). Morphologically not equivalent to "thorax" because first (and occasionally second) thoracic segment (bearing maxilliped) is fused to cephalon.

pereonite [G *Peraeomer*]: One of up to seven segments (somites) of pereon (i.e., all thoracomeres except first and occasionally second, which are fused to cephalon).

pereopod (peraeopod, pereiopod, walking leg) [G *Peraeopod, Pereiopod*]: One of up to seven pairs of appendages of pereon. Uniramous, consisting of modified (immovable) coxa (coxal plate) and six free segments (basis, ischium, merus, carpus, propodus, dactylus). Anterior two pairs may be differentiated as gnathopods. (simple, chelate, subchelate; ambulatory, raptorial)

pericardial sinus (pericardium) [G *Perikardialsinus, Perikard*]: Cavity surrounding heart. Elongate, often extending beyond heart as far posteriorly as telson. Blood collected in pericardial sinus enters heart via ostia.

pleomere [G *Pleomer*]: One of six segments (somites) of abdomen (pleon). Each pleomere bears appendages (pleopods). If abdomen is subdivided into anterior pleosome and posterior urosome, then posterior three somites may be termed urosomites.

pleon (abdomen) [G *Pleon, Abdomen, Hinterleib*]: Posterior division (tagma) of body (cephalon, pereon, pleon) consisting basically of six somites (pleomeres). May be further subdivided into anterior pleosome and posterior urosome. Reduced or absent in certain amphipods.

pleopod [G *Pleopod*]: Paired appendage of each somite of anterior subdivision (pleosome) of abdomen (pleon). Basically biramous, consisting of basal segment (peduncle) and two branches (rami). Each pair may be coupled basally by hooks for swimming. First two uropod pairs represent modified pleopods.

pleosome [G *Metasom*]: Anterior of two subdivisions (pleosome, urosome) of abdomen (pleon). Consists of three somites (pleomeres), each bearing pair of appendages (pleopods).

podopericardial channel [G *Podoperikardialkanal*]: One in a series of channels through which blood flows from ventral sinus to gills and appendages.

postantennal sinus (inferior antennal sinus): On each side of head, typically excavate margin below or behind lateral lobe, adjacent to origin of antennae.

posterior aorta [G *Schwanzaorta, Aorta posterior*]: aorta

primary flagellum [G *Hauptgeißel*]: flagellum

proctodeum: hindgut

propodus [G *Propodus*]: Sixth of basically seven segments (coxa, basis, ischium, merus, carpus, propodus, dactylus) of thoracopod; represents fifth movable (free) segment. In gnathopod, may be greatly enlarged to form subchela or chela together with dactylus. (See also palm)

protopod (protopodite) [G *Protopodit*]: Proximal part of appendage, basically comprising coxa and basis. Only

rarely applied in amphipods (e.g., maxillules) due to modification of coxae as coxal plates.

ramus [G *Ast*]: Branch of appendage. Typically refers to inner (endopod) and outer (exopod) branches of biramous appendage (e.g., pleopods, uropods). (one-articulate, multiarticulate)

rectum: hindgut

rostral hood: In certain amphipods, broad anterodorsal extension of head; covers basal segments of antennae.

rostrum [G *Rostrum*]: Unpaired, immovable anterior extension of head between bases of antennules. (deflected, spine-like, triangular)

setal row (spine row) [G *Sägeborstenreihe*]: Row of spine-like projections between molar process and lacinia mobilis on each mandible.

somite [G *Segment*]: One in a series of divisions of body. Cephalon consists of five somites (cephalothorax of these five plus first thoracic somite), pereon of four to seven somites (pereonites), and pleon of six somites (pleomeres). (See also urosomite)

spine row: setal row

statocyst [G *Statozyste*]: One of two small organs of equilibrium located dorsally in head; each statocyst contains one to five statoliths.

sternal gill: One in a series of small structures projecting from ventral surfaces (sternites) of pereon segments. Considered to have respiratory or osmoregulatory function.

sternite [G *Sternit*]: Ventral surface of body segment (somite). Sternites of anterior division (pleosome) of abdomen may bear sternal gills; in ♂, sternite of last (seventh) pereon segment bears penes.

stomach [G *Magen, Kaumagen*]: Posterior expanded section of foregut (esophagus, stomach). Characterized by various internal folds, channels, and ridges and may be differentiated into cardiac and pyloric regions. Opens into midgut and may give rise to series of ceca at stomach/midgut border.

subchela [G *Subchela*]: Pincer-like structure of first two pereopod pairs (gnathopods). Formed by terminal segment (dactylus) bearing down on posterior margin (palm) of enlarged next to last segment (propodus); occasionally also formed by dactylus and propodus bearing down on carpus.

subesophageal ganglion [G *Unterschlundganglion*]: Concentration of nerve tissue below (posterior to) esophagus. Consists of fused ganglia of mouthparts (mandibles, maxillules, maxillae, maxillipeds) and is connected to brain (supraesophageal ganglion) by pair of esophageal connectives. Continues posteriorly as ventral nerve cord.

supraesophageal ganglion (brain) [G *Oberschlundganglion, Gehirn*]: Concentration of nerve tissue above (anterior to) esophagus in head. Connected to subesophageal ganglion by pair of esophageal connectives.

tagma [G *Tagma*]: One of three major divisions of body (cephalon = head, pereon, abdomen = pleon). (See also cephalothorax)

telson [G *Telson*]: Posteriormost, relatively small segment of body. Shape is of taxonomic importance. (fleshy, laminar; entire, emarginate = bilobed = cleft; acute, rounded; naked, setose, spinose)

testis [G *Hoden*]: Paired section of ♂ reproductive system in which sperm are produced. Elongate, cylindrical, extending between heart and digestive tract from third to sixth pereon segments (pereonites); each testis opens to exterior on last (seventh) pereonite via vas deferens and penis.

thoracomere [G *Thoracomer, Thorakalsegment*]: One of basically eight segments (somites) between head (cephalon) and abdomen (pleon). First (and occasionally second) thoracomere is fused with cephalon, and remaining thoracomeres (forming pereon) therefore also termed pereonites.

thoracopod [G *Thoracopod*]: One of two appendages of each somite (thoracomere) of thorax. First pair developed as maxillipeds; second through eighth pairs may be termed pereopods, second and third differentiated as gnathopods. (chelate, subchelate; ambulatory, raptorial)

thorax [G *Thorax*]: Region of body basically comprising eight somites between cephalon and pleon. First thoracic somite (and occasionally second) is fused with cephalon to form cephalothorax; remaining somites of thorax collectively form pereon.

upper lip: labrum

uropod [G *Uropod, Sprunggriffel*]: Paired appendage of each segment (urosomite) of posterior subdivision (urosome) of abdomen. Basically biramous, consisting of basal segment (peduncle) and pair of rami. (lamellate, styliform)

urosome [G *Urosom*]: Posterior of two subdivisions (pleosome, urosome) of abdomen (pleon). Consists basically of three somites (urosomites), each bearing pair of appendages (uropods).

urosomite: If abdomen (pleon) is subdivided into anterior pleosome and posterior urosome, one of three segments (somites) of urosome.

vas deferens (sperm duct) [G *Vas deferens, Samenleiter*]: Section of ♂ reproductive system between each testis and each penis. Characterized by glandular walls and terminal modifications.

ventral nerve cord [G *Bauchmark*]: Longitudinal nerve cord extending below digestive tract from subesophageal ganglion to end of abdomen. Bears series of segmentally arranged (posteriorly occasionally fused) ganglia.

ventral sinus (sternal sinus) [G *Ventralsinus*]: In circulatory system, large ventral space below digestive tract in which blood collects before entering appendages and gills.

walking leg: pereopod

EUPHAUSIACEA

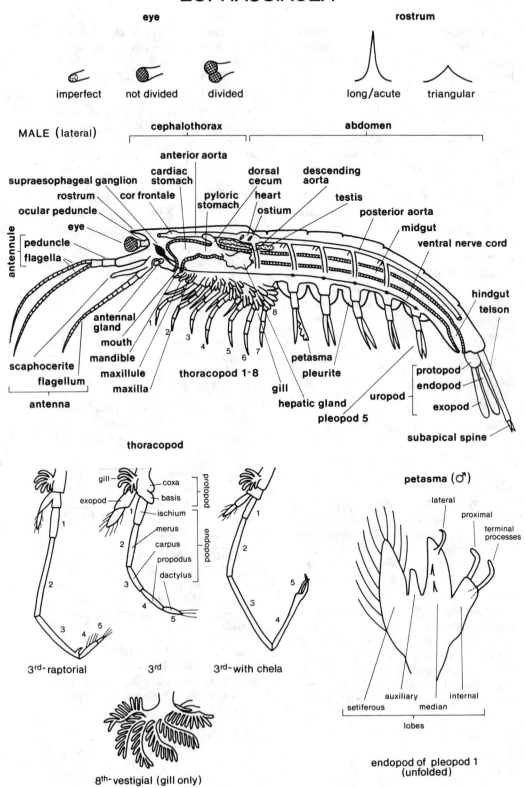

eye

imperfect   not divided   divided

rostrum

long/acute   triangular

MALE (lateral)

cephalothorax   abdomen

anterior aorta

cardiac stomach

supraesophageal ganglion   dorsal cecum   descending aorta

rostrum   cor frontale   pyloric stomach   heart   testis

ocular peduncle   ostium   posterior aorta

eye   midgut

antennule   peduncle   ventral nerve cord

flagella

antennal gland   hindgut   telson

mouth

mandible   protopod

maxillule   endopod

scaphocerite   maxilla   exopod

flagellum   thoracopod 1-8   petasma   pleurite

antenna   gill   uropod   subapical spine

hepatic gland   pleopod 5

thoracopod

gill — coxa   protopod

exopod — basis

ischium

merus   endopod

carpus

propodus

dactylus

3rd-raptorial   3rd   3rd-with chela

8th-vestigial (gill only)

petasma (♂)

lateral

proximal

terminal processes

auxiliary   internal

setiferous   median

lobes

endopod of pleopod 1 (unfolded)

ARTHROPODA

CRUSTACEA

Euphausiacea

euphausiacean, krill [G *Leuchtkrebs, Krill*]

abdomen [G *Abdomen, Hinterleib*]: Posterior division (tagma) of body (cephalothorax, abdomen); consists of six somites (pleomeres), the first first five bearing pleopods, the sixth bearing uropods. Bears posterior telson.

aesthetasc [G *Aesthetasc*]: One in a series of small sensory projections on each antennule.

antenna (second antenna) [G *Antenne, zweite Antenne*]: Second pair of antennae. Biramous, consisting of proximal single-segmented peduncle, scale-shaped outer branch (exopod = scaphocerite), and inner branch (endopod) consisting of three larger segments and multiarticulate flagellum. Peduncle bears opening of antennal gland.

antennal gland [G *Antennendrüse, Antennennephridium*]: Pair of excretory glands in head, one opening to exterior at base (peduncle) of each antenna.

antennal scale: scaphocerite

antennular lappet (leaflet) [G *Läppchen*]: On each antennule, small lobe extending dorsally from end of first peduncular segment. (bifid, laminar, obtuse, triangular)

antennule (first antenna) [G *Antennula, erste Antenne*]: First pair of antennae. Biramous, consisting of proximal three-segmented peduncle and two distal flagella. Antennules joined to each other by setae on first peduncular segment. (with/without antennular lappet)

anterior aorta [G *Kopfaorta, Aorta cephalica*]: Major unpaired blood vessel arising from anterior end of heart and extending over stomach into head. Bears accessory pumping organ (cor frontale) and gives rise to series of smaller arteries supplying brain and eyes.

anus [G *After*]: Posterior opening of digestive tract; located ventrally at posterior margin of last (sixth) abdominal somite (pleomere).

aorta: anterior aorta, descending aorta, posterior aorta

appendix interna [G *Appendix interna*]: Medially (inwardly) directed process of each pleopod. Serves to link pleopods of each somite together for swimming.

artery [G *Arterie*]: One of numerous blood vessels originating from aorta or directly from heart. According to position or body region/appendages supplied, one may distinguish, among others, abdominal, antennal, antennular, anterior lateral, mandibular-maxillary, ophthalmic, posterior lateral, and sternal arteries.

basis (basipodite) [G *Basis, Basipodit*]: Second segment of appendage; forms distal segment of protopod (coxa, basis) and bears variously developed branches (endopod, exopod). Typically specified only when referring to thoracopods.

brain: supraesophageal ganglion

calyptopis [G *Calyptopis, Calyptopis-Stadium*]: Third of several larval stages (nauplius, metanauplius, calyptopis, furcilia). Characterized by development of carapace, elongation of abdomen, differentiation of somites, and development of eyes under carapace. (See also cyrtopia)

carapace [G *Carapax*]: Shield-like, relatively thin structure fused to thoracic somites and covering cephalothorax. Does not enclose gill-bearing bases of thoracopods. Typically bears cervical groove. (with/without rostrum)

cardiac stomach [G *Kardiakalteil des Kaumagens*]: Anterior and larger division of stomach; separated from smaller posterior pyloric stomach by cardiopyloric valve. Serves to grind food. (with/without denticles)

cardiopyloric valve [G *Kardiopylorikalklappe*]: In digestive tract, valve-like structure between anterior cardiac stomach and posterior pyloric stomach.

carpus (carpopodite) [G *Carpus, Carpopodit*]: Fifth segment of thoracopod (third segment of five-segmented endopod); positioned between merus and propodus.

caudal fan: tailfan

caudal furca: subapical spine

cephalon (head) [G *Cephalon, Kopf*]: Anterior part of cephalothorax. Consists of five cephalic somites bearing antennules, antennae, mandibles, maxillules, and maxillae. Covered, along with thorax, by carapace. (See also protocephalon)

cephalothorax [G *Cephalothorax*]: Anterior division (tagma) of body (cephalothorax, abdomen). Consists of head (cephalon) fused with thorax. Covered dorsally (incompletely on sides) by carapace.

cervical groove [G *Zervikalfurche, Cervicalfurche*]: Transverse groove across anterior region of carapace.

chela [G *Chela, Schere*]: Pincer-like structure at end of third thoracopod in certain euphausiaceans. Formed by movable dactylus bearing down on rigid spine(s) projecting distally from propodus.

compound eye [G *Komplexauge*]: Pair of larger organs of vision, each located distally on eyestalk (ocular peduncle). (divided = bilobed = multilobed, undivided = single-lobed = spherical, imperfect)

cor frontale [G *Cor frontale, Stirnherz*]: In circulatory system, muscular expansion of anterior aorta. Supports heart in pumping blood to anterior region of head.

coxa (coxopodite) [G *Coxa, Coxopodit*]: First, most proximal segment of appendage; forms protopod together with following segment (basis). Typically specified only when referring to thoracopods. Coxae of maxillae bear endites, those of thoracopods bear gills (epipodites), with coxae of second and seventh thoracopods (occasionally more) containing photophores.

cyrtopia [G *Cyrtopia-Stadium*]: furcilia

dactylus (dactyl) [G *Dactylus, Dactylopodit*]: Seventh segment of thoracopod (fifth or last segment of endopod); may serve as movable element in terminal pincer (chela).

descending aorta [G *Aorta descendens, Arteria descendens*]: Major blood vessel arising from ventral surface of heart. Extends around gonads and branches above ganglia of thorax into an anteriorly and posteriorly directed artery (sternal and abdominal arteries). (paired, unpaired)

diaeresis [G *Diäresis, Pseudogelenk, Quernaht*]: Transverse groove across outer branch (exopod) of uropod in certain euphausiaceans.

dorsal cecum [G *dorsales Coecum*]: One of two small dorsal outpocketings of digestive tract at border of pyloric stomach and midgut. (See also hepatic gland)

dorsal spine: spine

endite [G *Endit*]: One of two inwardly (medially) directed, setose lobes of base (protopod) of maxillules and maxillae. (simple, bilobed, trilobed)

endopod (endopodite) [G *Endopod, Endopodit*]: Inner branch (ramus) of biramous appendage. Represents flagellum-bearing branch of antenna, palp-like branch of maxillules and maxillae, five-segmented main branch of thoracopods, and flattened inner branch of pleopods and uropods. (See also exopod)

epipod (epipodite) [G *Epipod, Epipodit*]: Laterally (outwardly) directed lobe of basal segment (coxa) of thoracic appendage (thoracopod). Those of second through eighth thoracopods serve in respiration and are therefore termed gills or podobranchs.

esophageal connective [G *Schlundkonnektiv*]: One of two strands of nerve tissue surrounding esophagus and connecting brain (supraesophageal ganglion) with ventral nerve cord.

esophagus [G *Ösophagus*]: Short section of digestive tract between mouth and stomach.

exite [G *Exit*]: In maxillule, outwardly (laterally) directed, setose, lobe-like branch of base (protopod). (plate-like, rudimentary, absent) (See also endite)

exopod (exopodite) [G *Exopod, Exopodit*]: Outer branch (ramus) of biramous appendage. Represents scale (scaphocerite) of antenna, one-segmented lobe of maxilla in certain euphausiaceans, relatively small, two-segmented outer branch of thoracopods, and flattened outer branch of pleopods and uropods. (See also endopod)

eye: compound eye, nauplius eye

eyestalk: ocular peduncle

flagellum [G *Flagellum, Geißel*]: Elongate, multiarticulate distal end of antennule or antenna. Paired in the former, un-

paired and representing inner branch (endopod) in the latter.

frontal plate: Somewhat extended anterior margin of carapace. (short and triangular, truncate)

furcilia [G *Furcilia, Furcilia-Stadium*]: Fourth of several larval stages (nauplius, metanauplius, calyptopis, furcilia). Characterized by emergence of movable eyes from under carapace and progressive development of thoracic and abdominal somites. Last substage, formerly termed cyrtopia, involves final differentiation of second antennae, mandibles (palps), first thoracopods, and telson and development of pleopods for swimming.

gill (podobranch, podobranchial gill, epipod) [G *Kieme, Schlauchkieme, Epipodit, Podobranchie*]: Respiratory structure associated with second through eighth thoracopods. Projects from basal segment (coxa) of appendages, thus representing an epipod and often being termed podobranch. Not covered by carapace. Gill forms greater part of reduced eighth (and occasionally seventh) thoracopods.

gonad: ovary, testis

gonopore [G *Gonoporus, Geschlechtsöffnung*]: Opening of reproductive system to exterior. In ♀, opening of oviduct on coxae of sixth thoracopods; in ♂, openings of both vasa deferentia on last (eighth) thoracomere.

head: cephalon

heart [G *Herz*]: In circulatory system, relatively short, dorsoventrally flattened pumping organ; located within pericardial sinus in posterodorsal region of cephalothorax. Bears two pairs of ostia and gives rise to numerous anterior, posterior, and lateral blood vessels (aortas, arteries).

hepatic cecum: hepatic gland

hepatic gland (digestive cecum, hepatic cecum, hepatopancreas) [G *Mitteldarmdrüse*]: One of two large, branching digestive organs arising ventrally at border of pyloric stomach and midgut. Extends through most of thorax.

hindgut [G *Enddarm*]: Posteriormost section of digestive tract between midgut and anus; may give rise to short blind diverticulum.

incisor process [G *Pars incisiva*]: Gripping process (vs. grinding = molar process) of each mandible.

interior spine: spine

ischium (ischiopod, ischiopodite) [G *Ischium, Ischiopodit*]: Third segment of thoracopod (first segment of endopod); positioned between basis and merus.

labium (lower lip) [G *Labium, Unterlippe*]: Lobe-like structure posterior to mouth; divided into two lobes (paragnaths).

labrum (upper lip) [G *Labrum, Oberlippe*]: Unpaired, lobe-like structure anterior to and partially covering mouth.

light organ: photophore

luminescent organ: photophore

mandible [G *Mandibel*]: Third paired appendage of head (cephalon); located between antennae and maxillules. Represents first pair of mouthparts and typically consists of basal molar and incisor processes as well as a distal palp. (with/without palp)

maxilla (second maxilla) [G *Maxille, zweite Maxille*]: Fifth paired appendage of head (cephalon); located between maxillules and first thoracopods. Represents third pair of mouthparts and, if fully developed, consists of three-segmented base (protopod) with two endites as well as distal endopod (palp) and exopod. (foliaceous)

maxillule (first maxilla) [G *Maxillula, erste Maxille*]: Fourth paired appendage of head (cephalon); located between mandibles and maxillae. Represents second pair of mouthparts and, if fully developed, consists of three-segmented base with two endites and one exite as well as distal endopod (palp). (foliaceous)

merus (meropod, meropodite) [G *Merus, Meropodit*]: Fourth segment of thoracopod (second segment of five-segmented endopod); positioned between ischium and carpus.

metanauplius [G *Metanauplius*]: Second of several larval stages (nauplius, metanauplius, calyptopis, furcilia). Characterized by development of labrum and rudiments of maxillules, maxillae, and first thoracopods. (See also cyrtopia)

midgut [G *Mitteldarm*]: Elongate section of digestive tract between stomach and hindgut. Gives rise to pair of small dorsal ceca and pair of large ventral hepatic glands.

midgut cecum: dorsal cecum, hepatic gland

molar process [G *Pars molaris*]: Grinding process (vs. gripping = incisor process) of each mandible.

mouth [G *Mund, Mundöffnung*]: Anterior opening of digestive tract on underside of head. Bordered anteriorly by upper lip (labrum), posteriorly by lower lip (labium); opens into esophagus.

nauplius [G *Nauplius*]: First of several larval stages (nauplius, metanauplius, calyptopis, furcilia). Characterized by three pairs of appendages (antennules, antennae, mandibles). (See also cyrtopia)

nauplius eye [G *Naupliusauge*]: Small median photosensitive organ located between eyestalks (ocular peduncles) under rostrum. (See also compound eye)

ocular peduncle (eyestalk) [G *Augenstiel*]: One of two compound eye-bearing, movable projections of head. Consists of two indistinctly separated segments and bears photophore.

ostium [G *Ostium*]: One of two pairs of muscular openings in heart. Blood from pericardial sinus enters heart through ostia.

ovary [G *Ovar, Ovarium, Eierstock*]: Section of ♀ reproductive system in which eggs are produced. Paired (connected anteriorly), located between heart and digestive tract, and extending through most of digestive tract. Each ovary opens to exterior at base of sixth thoracopod via oviduct.

oviduct [G *Ovidukt, Eileiter*]: Slender section of ♀ reproductive system between each ovary and corresponding ♀ gonopore on coxae of sixth thoracopods.

ovisac [G *Brutsack*]: In certain ♀ euphausiaceans, sac-like structure associated with posterior thoracopods and serving to brood eggs. (single, double, fused)

palp [G *Palpus*]: Distal projection of mouthpart (mandible, maxillule, maxilla). Generally considered to be endopodal in nature. (unsegmented, two-/three-segmented)

paragnath [G *Paragnath*]: One of two lobe-like structures forming lower lip (labium) posterior to mouth.

peduncle [G *Pedunculus, Schaft, Stamm, Stiel*]: Proximal division of antennule or antenna. Consists in the former of three segments, the first bearing antennular lappet; in the latter, consists of a single, superficially divided segment which bears opening of antennal gland.

pereopod: thoracopod

pericardial sinus (pericardium) [G *Perikardialsinus, Perikard*]: Cavity surrounding heart. Blood collected in pericardial sinus enters heart via ostia.

petasma [G *Petasma*]: In ♂, one of first two pairs of pleopods modified as copulatory structure. Inner branch (endopod) basically consists of four lobes (auxiliary = additional, internal, median, setiferous lobes), each bearing a variety of projections (blades, heel, processes, setae, spines). Serves to transfer spermatophores into thelycum of ♀.

photophore (light organ, luminescent organ) [G *Leuchtorgan, Photophor*]: One of numerous light-producing organs. Located at upper end of eyestalks (ocular peduncles), in coxae of second and eighth (often also third through sixth) thoracopods, and in middle of ventral surface (sternite) of first four abdominal somites.

pleomere [G *Pleomer*]: One of six segments (somites) of abdomen. Each pleomere bears pair of appendages, those of first five being termed pleopods, those of sixth being termed uropods. Last pleomere bears posterior telson. (smooth, with dorsal spine, with dorsal keel)

pleopod [G *Pleopod*]: One of two appendages of each of the first five abdominal somites (those of sixth somite being termed uropods). Biramous, consisting of two-segmented base as well as one-segmented endopod and exopod. First (and to a lesser extent second) pair of pleopods in ♂ modified to form petasma. (See also appendix interna)

pleurite (pleuron, epimere) [G *Pleura, Epimer*]: Lateral part of integument of somite; forms prominent, ventrally directed extensions in each abdominal somite (pleomere). (See also sternite, tergite)

podobranch: gill

posterior aorta [G *Schwanzaorta, Arteria posterior*]: Major blood vessel arising from posterior end of heart. Extends to posterior end of abdomen and gives rise to series of segmental arteries supplying appendages (pleopods) and photophores. (single, paired)

posterior spine: spine

preanal spine: spine

propodus [G *Propodus*]: Sixth segment of thoracopod (fourth and next to last segment of five-segmented endopod); positioned between carpus and dactylus.

protocephalon [G *Protocephalon*]: Anteriormost region of head (cephalon); consists of eyes and first two head segments (antennular and antennal somites) and is movably articulated with rest of head.

protopod (protopodite) [G *Protopod, Protopodit*]: Proximal part of biramous appendage. Typically specified only when referring to thoracopod, where protopod consists of two segments (coxa, basis), although occasionally also applied to three-segmented (precoxa, coxa, basis) base of maxillules and maxillae or to two-segmented base of pleopods.

pyloric stomach [G *Pylorikalteil des Magens*]: Posterior and smaller division of stomach; separated from larger anterior cardiac stomach by cardiopyloric valve. Opens into midgut.

ramus [G *Ast*]: Branch of appendage; typically refers to inner (endopod) or outer (exopod) branch of biramous appendage.

rostrum [G *Rostrum*]: Unpaired, unmovable anterior extension of carapace; projects between eyestalks (ocular peduncles). (absent, obsolete, weakly produced, reaching/not reaching beyond eyes; acute, triangular)

scaphocerite (antennal scale, squama) [G *Scaphocerit, Schuppe, Squama*]: Scale-shaped outer branch (exopod) of each antenna.

seminal vesicle [G *Vesicula seminis*]: In ♂ reproductive system, expanded section of each vas deferens. Serves to store sperm.

somite (segment) [G *Segment*]: One in a series of divisions of body. Head (cephalon) consists of five somites, thorax of eight somites (thoracomeres), and abdomen of six somites (pleomeres). Each somite basically consists of dorsal tergite, ventral sternite, and lateral pleurites.

spermatophore [G *Spermatophor*]: Packet of sperm formed in distal end of each vas deferens. Emerges from gonopore and is transferred by modified pleopods (petasma) to ♀ thelycum.

spine [G *Stachel, Dorn*]: One of various spine-like projections of body. According to position one may distinguish median preanal spine on ventral surface of last abdominal somite (pleomere), dorsal or middorsal spine on dorsal midline of one or more pleomeres, as well as interior or posterior spine associated with lateral margin of carapace. (See also subapical spine)

sternite [G *Sternit*]: Ventral surface of body segment (somite). Sternites of first four abdominal somites (pleomeres) bear median photophores. (See also pleurite, tergite)

stomach: cardiac stomach, pyloric stomach

subapical spine (caudal furca) [G *Subapikalanhang*]: One of two movable spines at end of telson.

supraesophageal ganglion (brain) [G *Oberschlundganglion, Gehirn*]: Concentration of nerve tissue in anterior region of cephalothorax; located above (anterior to) esophagus and somewhat anteroventral to stomach. Joined to chain of postoral ganglia by pair of esophageal connectives.

tagma [G *Tagma*]: One of two major divisions of body (cephalothorax, abdomen).

tailfan (caudal fan) [G *Schwanzfächer*]: Posterior fan-like structure formed by combination of uropods and telson.

telson [G *Telson*]: Narrow and pointed posteriormost segment of body. Forms tailfan together with appendages (uropods) of sixth (last) abdominal somite. Typically bears pair of posterior subapical spines.

tergite [G *Tergit*]: Dorsal surface of body segment (somite). Tergite of first abdominal somite (pleomere) has clearly delimited anterior section which may give the impression of a separate body segment. (See also pleurite, sternite)

testis [G *Hoden*]: Section of ♂ reproductive system in which sperm are produced; located between heart and digestive tract in thorax. Each testis (or each side of unpaired testis) opens to exterior on last (eighth) thoracomere via vas deferens. (single, paired)

thelycum [G *Thelycum*]: In ♀ reproductive system, pouch-like accessory copulatory structure on underside of thorax; associated with ♀ gonopores and formed by sternite and coxal plates of posterior thoracic somite. Receives spermatophore of ♂.

thoracomere [G *Thoracomer, Thoraxsegment*]: One of eight segments (somites) of thorax, each bearing pair of appendages (thoracopods, the most posterior pairs occasionally reduced or vestigial).

thoracopod (pereopod) [G *Thoracopod*]: One of two appendages of each thoracic somite (thoracomere). Basically biramous, consisting of proximal part (protopod) bearing gill (epipod) as well as distal five-segmented endopod and two-segmented exopod. Posterior pair(s) may be reduced except for gill. (chelate, prehensile, rudimentary, vestigial)

thorax [G *Thorax*]: Posterior part of cephalothorax. Consists of eight somites (thoracomeres), most bearing well-developed appendages (thoracopods). Fused with and largely covered by carapace.

uropod [G *Uropod*]: One of two appendages of last (sixth) abdominal somite (pleomere). Flattened and biramous, consisting of short, one-segmented base and two more elongate, one-segmented branches (endopod and exopod). Forms tailfan together with telson.

vas deferens (sperm duct) [G *Vas deferens, Samenleiter*]: Section of ♂ reproductive system between each testis (or from each side of unpaired testis) and gonopores on eighth thoracomere. May be expanded distally to form seminal vesicle. (coiled, straight)

ventral nerve cord [G *Bauchmark, ventraler Nervenstrang*]: Longitudinal nerve cord extending below digestive tract to last (sixth) abdominal somite. Joined to brain (supraesophageal ganglion) by pair of esophageal connectives; bears series of segmentally arranged ganglia.

DECAPODA (crab-like/brachyuran)

carapace

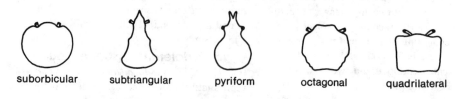

suborbicular   subtriangular   pyriform   octagonal   quadrilateral

(ventral) (dorsal)

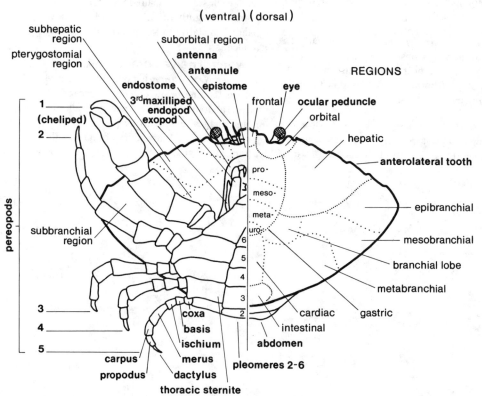

REGIONS

subhepatic region
pterygostomial region
suborbital region
antenna
antennule
endostome
epistome
3rd maxilliped
endopod
exopod
eye
frontal
ocular peduncle
orbital
hepatic
anterolateral tooth

1 ___ (cheliped)
2 ___
pro-
meso-
meta-
uro-
epibranchial

subbranchial region
mesobranchial
branchial lobe
metabranchial

pereopods

3 ___
4 ___
5 ___

coxa
basis
ischium
merus
dactylus
carpus
propodus
thoracic sternite

6
5
4
3
2

cardiac
intestinal
abdomen
pleomeres 2-6

gastric

FEMALE (lateral)

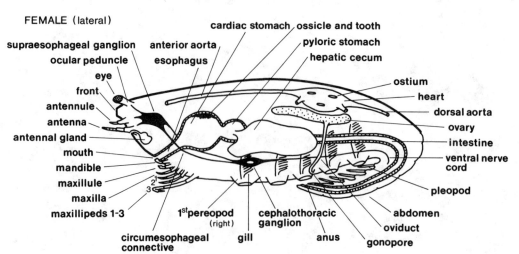

cardiac stomach   ossicle and tooth
pyloric stomach
hepatic cecum

supraesophageal ganglion
ocular peduncle
eye
front
antennule
antenna
antennal gland
mouth
mandible
maxillule
maxilla
maxillipeds 1-3

anterior aorta
esophagus

ostium
heart
dorsal aorta
ovary
intestine
ventral nerve cord
pleopod

circumesophageal connective
1st pereopod (right)
gill
cephalothoracic ganglion
anus
abdomen
oviduct
gonopore

DECAPODA (shrimp-like/natantian)

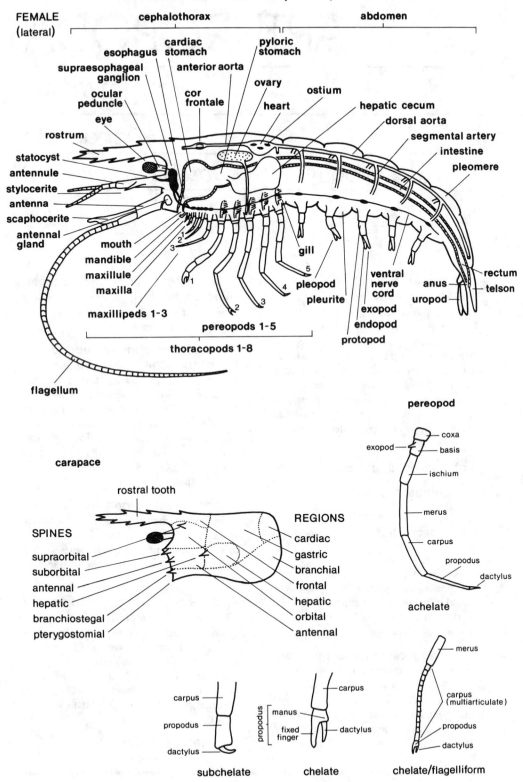

FEMALE (lateral)

cephalothorax — abdomen

esophagus, cardiac stomach, pyloric stomach, supraesophageal ganglion, anterior aorta, ocular peduncle, cor frontale, ovary, eye, heart, ostium, rostrum, hepatic cecum, statocyst, dorsal aorta, antennule, segmental artery, stylocerite, intestine, antenna, pleomere, scaphocerite, antennal gland, mouth, mandible, maxillule, maxilla, gill, maxillipeds 1-3, pereopods 1-5, thoracopods 1-8, pleopod, pleurite, ventral nerve cord, exopod, endopod, protopod, rectum, anus, telson, uropod, flagellum

carapace

rostral tooth

SPINES
supraorbital
suborbital
antennal
hepatic
branchiostegal
pterygostomial

REGIONS
cardiac
gastric
branchial
frontal
hepatic
orbital
antennal

pereopod
coxa
exopod — basis
ischium
merus
carpus
propodus
dactylus
achelate

subchelate
carpus
propodus
dactylus

chelate
propodus — manus, fixed finger
carpus
dactylus

chelate/flagelliform
merus
carpus (multiarticulate)
propodus
dactylus

ARTHROPODA

CRUSTACEA

Decapoda

decapod [G *Decapode, Zehnfußkrebs*]

abdomen (pleon) [G *Abdomen, Hinterleib, Pleon*]: Posterior division (tagma) of body (cephalothorax, abdomen). Consists of six somites (pleomeres), each bearing pair of appendages (pleopods). Sixth pleomere (with uropods) bears posterior telson. (extended, bent upon itself, folded under cephalothorax, spirally coiled; symmetrical, asymmetrical; compressed, depressed; acutely triangular, broadly triangular, broadly oval, subcircular, T-shaped; smooth, sculptured)

abdominal somite: pleomere

acanthosoma [G *Acanthosoma*]: zoea

accessory heart [G *Nebenherz*]: cor frontale

acicle [G *Aciculum*]: Spine-shaped outer branch (exopod) of antenna; represents reduced antennal scale (scaphocerite). May also refer to spine-like ophthalmic scale on eyestalk. (unarmed, spinose)

acron (ophthalmic somite, pre-segmental region) [G *Acron, Ocular-segment*]: Anterior segment of body (although not considered to be a true somite); bears eyestalks (ocular peduncles). Acron and first cephalic somite occasionally interpreted as representing protocephalon.

aesthetasc (esthetasc, esthete, olfactory hair) [G *Aesthetasc*]: One in a series of small sensory projections on outer flagellum of antennule.

afferent respiratory channel (afferent branchial channel) [G *Einströmungskanal*]: Opening through which water enters branchial chamber under carapace. Typically applied to more restricted opening in crab-like (brachyuran) decapod; located anteroventrally on each side of body, either behind pterygostomial regions or in front of chelipeds. (See also efferent respiratory channel)

androgenic gland [G *androgene Drüse*]: In ♂, small gland associated with terminal end of each vas deferens. Secretes hormones stimulating spermatogenesis and development of secondary sex characteristics.

antenna (second antenna) [G *Antenne, zweite Antenne*]: Second and typically larger pair of antennae. Originates from second (antennal) somite of head. Biramous, consisting of basal peduncle (typically composed of protopod, endopod, and scaphocerite) and distal flagellum. Bears pore of antennal gland. (elongate, spatulate, squamiform)

antennal acicle: acicle

antennal artery (antennary artery) [G *Arteria lateralis cephalica*]: artery

antennal carina (postannular crest) [G *Crista antennalis*]: On each side of carapace, narrow, longitudinal ridge extending posteriorly from antennal spine.

antennal gland (green gland, excretory organ) [G *Antennendrüse, grüne Drüse, Antennennephridium, Exkretionsorgan*]: One of two glands located in anterior region of head and consisting basically of end sac (divided into saccule and labyrinth), bladder, and excretory duct. Opens via excretory pore on basal segment (coxa) of each antenna. Functions in excretion and controls internal fluid pressure as well as ion concentration.

antennal groove [G *Antennalfurche*]: On each side of carapace, longitudinal groove extending posteriorly from vicinity of antennal spine.

antennal region [G *Antennalregion*]: In shrimp-like (natantian) decapod, one of two relatively small regions along anterolateral margin of carapace; corresponds to position of antenna. Adjoining regions include orbital, hepatic, and pterygostomial regions.

antennal scale: scaphocerite

antennal spine [G *Antennalstachel*]: Anteriorly directed, spine-like projection on anterior region on each side of carapace; located below orbit, adjacent to base of antenna, and may extend beyond carapace margin.

antennular artery [G *Arteria antennularis*]: artery

antennular fossette [G *Antennularhöhle*]: On each side of carapace, anteroventral depression containing basal portion of antennule. (See also orbitoantennulary pit)

antennular scale: stylocerite

antennular spine: stylocerite

antennule (first antenna) [G *Antennula, erste Antenne*]: First and typically smaller pair of antennae. Originates from first (antennular) somite of head, each consisting of basal peduncle and one or more distal flagella. Peduncle bear statocysts and stylocerites, flagella bear aesthetascs. In crab-like (brachyuran) decapod, may be folded into orbitoantennulary pit of carapace. (biramous, triramous)

anterior aorta (ophthalmic artery) [G *Kopfaorta, vordere Aorta, Aorta anterior*]: Major unpaired blood vessel arising from anterior end of heart and extending into head. May bear accessory heart (cor frontale); gives rise to series of smaller arteries supplying brain and eyestalks.

anterolateral region [G *Anterolateralregion*]: In crab-like (brachyuran) decapod, relatively small region along each anterolateral margin of carapace. Adjoining regions include hepatic or subhepatic regions. (See also posterolateral region)

anterolateral tooth [G *Anterolateralstachel, Anterolateralzahn*]: On carapace of crab-like (brachyuran) decapod, one (or one in a series of) projection(s) along each anterolateral margin. (acute, conical, obtuse, quadrangular, spiniform, subtriangular; ciliated, denticulate, spiny; tip: acuminate, rounded)

anus [G *After*]: Posterior opening of digestive tract; located on ventral surface of telson.

aorta [G *Aorta*]: anterior aorta, dorsal aorta

apodeme [G *Apodem*]: Any one of numerous infolds of exoskeleton serving for attachment of muscles (e.g., head apodeme). May be highly developed as endopleurites and endosternites to form endoskeleton.

appendix interna (stylamblys) [G *Appendix interna, Stylamblys*]: Projection of inner branch (endopod) of each pleopod; serves to hook pleopod to its opposite member (for swimming).

appendix masculina [G *Appendix masculina*]: Median process of inner branch (endopod) of second pleopod in ♂; serves in copulation or spermatophore transfer.

areola [G *Areola*]: On dorsal surface of carapace, area between two longitudinal branchiocardiac grooves and posterior to transverse cervical groove.

artery [G *Arterie*]: One of numerous blood vessels originating from aorta or directly from heart. According to position or body region/appendages supplied, one may distinguish, among others, antennal, antennular, hepatic, lateral cephalic, ophthalmic, optic, pleopodal, rostral, segmental, sternal, and subneural arteries.

arthrobranch (arthrobranchia) [G *Arthrobranchie*]: Type of gill attached to point of articulation (articular membrane) between thoracopod and body wall. If full complement is developed, may be found associated with first through seventh thoracopods (i.e., all except last pereopods). (See also pleurobranch, podobranch)

article (joint, segment) [G *Glied*]: One in a series of numerous segments of flagellum at end of antenna or antennule. May also refer to segment of pereopod.

attractor epimeralis muscle (tergoepimeral muscle) [G *Musculus attractor epimeralis*]: In each side of cephalothorax, prominent muscle extending from epimeral fold to carapace, where it inserts along branchiocardiac groove (or lateral gastrocardiac marking).

basicerite [G *Basicerit*]: In antenna, term applied to second of two segments (coxicerite, basicerite) of peduncle; may bear antennal scale (scaphocerite). (See also carpocerite, ischiocerite, merocerite)

basi-ischium: ischiobasis

basiophthalmite [G *Basiophthalmit, Basalglied*]: Proximal of typically two segments (basiophthalmite, podophthalmite) of eyestalk.

basis (basipodite) [G *Basis, Basipodit*]: Second segment of appendage (thoracopod); forms distal segment of protopod (coxa, basis) and bears variously developed branches (endopod, exopod). May be

fused to following segment (ischium) to form ischiobasis.

bladder [G *Sammelblase, Harnblase*]: Expanded section of antennal gland between end sac and excretory pore. Consists of simple vesicle or is subdivided into a number of lobes or elongate diverticula extending into different regions of body.

brain: supraesophageal ganglion

branchia: gill

branchial carina [G *branchiale Carina*]: On each side of carapace in shrimp-like (natantian) decapod, narrow, longitudinal ridge extending posteriorly from orbit over branchial region.

branchial chamber (gill chamber) [G *Kiemenkammer, Kiemenhöhle*]: Lateral or dorsolateral space on each side of cephalothorax between body wall and carapace; encloses gills. In shrimp-like (natantian) decapod, water enters branchial chamber along posterior or ventral edges of carapace; in crab-like (brachyuran) decapod, it typically enters via afferent respiratory channels near bases of chelipeds. (See also epibranchial space, hypobranchial space)

branchial region [G *Branchialregion, Kiemenregion*]: One of two relatively large, lateral regions of carapace overlying gills (branchiae). May be subdivided (from anterior to posterior) into epibranchial, mesobranchial, and metabranchial regions/lobes/areas.

branchiocardiac carina [G *Crista branchiocardiaca*]: On each side of carapace, narrow ridge separating branchial region from cardiac region.

branchiocardiac groove [G *Branchiocardiacalfurche*]: On each side of carapace, oblique groove separating branchial and cardiac regions. Considered to represent former transverse border between two cephalic somites. May also refer to pair of grooves connecting cervical and postcervical grooves or extending posteriorly from postcervical groove.

branchiostegal spine [G *Branchiostegaldorn*]: Anteriorly directed, spine-like projection on branchial region on each side of carapace in shrimp-like (natantian) decapod; located at anterior margin, below branchiostegal groove and between antennal and pterygostomial spines.

branchiostegite [G *Branchiostegit*]: Lateral or dorsolateral part of carapace extending over gills and forming outer wall of branchial chamber.

buccal cavity [G *Bukkalhöhle, Mundfeld*]: On underside of head, cavity in which mouthparts (labrum, mandibles, maxillules, maxillae, maxillipeds, labium) are located.

buccal frame (buccal cavern) [G *Mundfeld*]: In crab-like (brachyuran) decapod, well-defined, frame-like depression on underside of head; contains mouthparts (labrum, mandibles, maxillules, maxillae, maxillipeds, labium) and may be covered by expanded (operculiform) maxillipeds. (quadrate, triangular) (See also endostome)

buccal groove [G *Bukkalfurche*]: On each side of carapace, transverse groove behind antennal spine; connects gastroorbital and antennal grooves.

carapace [G *Carapax*]: Large, shield-like cuticular structure covering cephalothorax; bears rostrum anteriorly and is typically regionated and sculptured with various carinae, grooves, spines, and teeth. Encloses gill-bearing bases of thoracopods and thus forms branchial chambers. (general form: cylindrical, compressed, depressed; shape: convex, circular, globular, hemispherical, hexagonal, octagonal, orbicular, oval, ovate triangular, pentagonal, polygonal, pyriform, quadrilateral, rectangular, subcircular, subcylindrical, subglobose, subhexagonal, suborbicular, subovoid, subpentagonal, subquadrate, subquadrilateral, subtriangular, transversely oval, tumid; surface: areolated, carinate, eroded, glabrous, granulate, nodular, nodulose, plicate, polished, pubescent, punctuate, with rugae, smooth, squamose, tuberculate, wrinkled)

carapace region (carapace area) [G *Carapaxregion, Region*]: One of numerous, more or less clearly delimited and elevated subdivisions of carapace. According to position or underlying internal organs, as well as according to group, one may distinguish antennal, anterolateral, branchial, cardiac, epibranchial, epigastric, frontal, gastric, hepatic, intestinal, mesobranchial, mesogastric metabranchial, metagastric, orbital, posterolateral, protogastric, pterygostomial, subhepatic, and urogastric regions.

cardiac notch (cardiac incision): In shrimp-like (natantian) decapod, notch on posterior margin of carapace.

cardiac region [G *Cardiacalregion, Herzregion*]: Relatively large, unpaired median region in posterior half of carapace; overlies heart. Adjoined anteriorly by gastric, posteriorly by intestinal, and laterally by branchial regions.

cardiac stomach [G *Kaumagen, Cardia*]: Anterior and larger cuticle-lined division of stomach; separated from smaller posterior pyloric stomach by cardiopyloric valve and characterized by variously developed grinding apparatus (gastric mill).

cardiac tooth [G *Cardiacalzahn*]: On midline of carapace, tooth-like projection in cardiac region just posterior to cervical groove.

cardiopyloric valve [G *Cardiopyloricalklappe*]: In digestive tract, valve-like structure between anterior cardiac stomach and posterior pyloric stomach.

caridean lobe [G *Eucyphidenanhang*]: In first maxilliped, setose broadening or lobe at base of outer branch (exopod).

carina (ridge, keel) [G *Carina, Crista, Kiel*]: One of a number of narrow, elevated ridges along carapace or abdomen. According to position and according to group, those on carapace may be termed antennal, branchial, branchiocardiac, gastroorbital, lateral, orbital, posterior, postorbital, postrostral, rostral, subhepatic, submedian, and supraorbital carinae.

carpocerite [G *Carpocerit*]: In antenna, term applied to last of three segments (ischiocerite, merocerite, carpocerite) of endopod (alternate interpretation: fifth segment of peduncle).

carpus (carpopod, carpopodite, wrist) [G *Carpus, Carpopodit*]: Fifth segment of thoracopod; located between merus and propodus. (simple, multiarticulated = subdivided)

caudal fan: tailfan

cecum: hepatic cecum

cephalic flexure: On underside of cephalothorax, forward, or occasionally upward, deflection of anterior sternum.

cephalic somite (cephalomere) [G *Kopfsegment*]: Any one of five segments (somites) of head (not including anteriormost acron). According to appendages borne, one may distinguish an antennular, antennal, mandibular, maxillular, and maxillar somite. Fused to each other and to thoracic somites to form cephalothorax.

cephalomere: cephalic somite

cephalon (head) [G *Kopf*]: Anterior part of cephalothorax. Consists of anterior acron and five cephalic somites and bears eyes, antennules, antennae, labrum, mouth, mandibles, maxillules, maxillae, and labium. First three thoracic appendages (maxillipeds) closely associated with head. Covered, along with thorax, by carapace.

cephalothoracic ganglion (postesophageal ganglion) [G *Thoraxganglion, Brustganglion, ventrale Thoracalganglienmasse*]: In crab-like (brachyuran) decapod, large concentration of nerve tissue posterior to esophagus. Consists of fused ganglia of thoracic somites (thoracomeres).

cephalothorax [G *Cephalothorax*]: Anterior division (tagma) of body (cephalothorax, abdomen). Consists of head (cephalon) fused with thorax and is largely or entirely covered by carapace.

cervical groove (cervical suture, cervical furrow) [G *Cervicalfurche, Sul-cus cervicalis, Nackenfurche*]: On carapace, well-defined, transverse groove separating gastric and cardiac regions. Considered to represent former border between two cephalic somites.

cervical notch (cervical incision): On dorsal surface of carapace, strong indentation at level of cervical groove.

cervical suture: cervical groove

chela (pincer, claw) [G *Chela, Schere*]: Pincer-like structure at end of appendage. Formed by movable finger (dactylus) opposed by immovable distal extension (fixed finger) of expanded propodus (manus). (heterochelous, crushing, tearing) (See also subchela)

cheliped [G *Cheliped, Scherenfuß*]: Any chela-bearing thoracopod; typically refers to first pair(s) of pereopods. (equal, unequal: major, minor)

chromatophore [G *Chromatophor*]: Pigment-filled, sac-shaped cell; contraction and expansion of chromatophores result in color changes. (bichromatic = dichromatic, monochromatic, polychromatic)

circumesophageal connective (esophageal connective) [G *Schlundkonnektiv*]: One of two strands of nerve tissue surrounding esophagus and connecting dorsal supraesophageal ganglia with posterior subesophageal ganglia. May bear ganglia (paraesophageal ganglia).

claw: chela

cor frontale (accessory heart) [G *Cor frontale, Stirnherz*]: In circulatory system, muscular expansion of anterior aorta. Supports heart in pumping blood to anterior region of head.

cornea [G *Cornea*]: Transparent cuticle covering ommatidia of each eye.

coxa (coxopodite) [G *Coxa, Coxopodit*]: First, most proximal segment of appendage; forms protopod together with following segment (basis). Depending on appendage, coxa may bear opening of antennal gland, epipodite, gill (podobranch), or gonopore.

coxicerite [G *Coxocerit*]: In antenna, term applied to first of two segments (coxicerite, basicerite) of peduncle.

crista dentata [G *Kauleiste, Crista dentata*]: In third maxilliped, toothed median margin of third segment (ischium).

cuticle [G *Kutikula*]: Noncellular, multilayered, relatively thick layer forming exoskeleton and endoskeleton and lining foregut and hindgut. Secreted by epidermis.

dactylus (dactyl, dactylopodite) [G *Dactylus, Dactylopodit*]: Seventh and terminal segment of appendage (thoracopod). Follows propodus and may serve as distal element in subchela or as movable finger in chela. (simple, many jointed; bifurcate, curved, falcate, falciform, obtuse, rounded, sharply pointed, spatulate, subacute, straight; ciliate, furred, grooved, hairy, smooth, spinose)

decapodid stage [G *Decapodid-Stadium, Decapodit*]: postlarva

dendrobranch (dendrobranchia) [G *Dendrobranchie*]: Type of gill whose axis bears series of highly subdivided branches. (See also phyllobranch, trichobranch)

dermal gland [G *Tegumentaldrüse*]: One of numerous glands in epidermis underlying cuticle; opens to exterior through long ducts traversing cuticle.

deutocerebrum: supraesophageal ganglion

diaeresis [G *Diäresis, Pseudogelenk, Quernaht*]: Transverse groove across outer branch (exopod) of uropod; may divide exopod into two movable parts.

digestive gland: hepatic cecum

dorsal abdominal artery: dorsal aorta

dorsal aorta (dorsal abdominal artery, posterior aorta, superior abdominal aorta) [G *Aorta posterior, Arteria dorsalis pleica*]: Major unpaired blood vessel arising from posterior end of heart. Extends to posterior end of abdomen and gives rise to series of segmental arteries in abdominal somites (pleomeres).

dorsal plate: Spindle-shaped division of carapace intercalated with median suture.

dorsomedian groove: On carapace, longitudinal dorsomedian groove extending from tip of rostrum to posterior margin.

dorsoventralis posterior muscle [G *Musculus dorsoventralis posterior*]: In cephalothorax, prominent muscle extending from head apodemes to carapace, where it inserts posterior to cervical groove.

efferent respiratory channel (efferent branchial channel) [G *Ausströmungskanal*]: Opening through which water exits from branchial chamber under carapace. Typically applied to more restricted opening in crab-like (brachyuran) decapod; located anteriorly on each side of endostome.

ejaculatory duct [G *Ductus ejaculatorius*]: In ♂ reproductive system, muscular terminal end of vas deferens.

elaphocaris [G *Elaphocaris*]: protozoea

endite [G *Endit, Lade*]: Inwardly (medially) directed lobe of protopod, typically on maxillules, maxillae, and maxillipeds. (bilobed, entire) (See also exite)

endognath [G *Endognath*]: Inner or principal branch (endopod) of maxilliped. (See also exognath)

endophragm (arthrophragm) [G *Arthrophragma*]: In cephalothorax, septum of endoskeleton formed at border of each somite by inward projections (endopleurites, endosternites) of exoskeleton.

endophragmal skeleton: endoskeleton

endopleurite [G *Endopleurit*]: In endoskeleton, one in a series of rigid structures projecting into body from lateral part (pleurite, epimeral fold) of exoskeleton. Serves also as site of muscle attachment (apodeme) and may be branched. (See also endosternite)

endopod (endopodite) [G *Endopod, Endopodit, Innenast*]: Inner branch (ramus) of biramous appendage. Refers to three-segmented part of peduncle of antenna, palp-shaped or more elongate inner branch of mouthparts, inner branch of pleopods and uropods; considered to represent main, basically five-segmented part of pereopod. (paddle-shaped, palp-shaped, pediform)

endoskeleton (endophragmal skeleton, endophragmal system) [G *Endoskelett, Innenskelett, Endophragmalsystem*]: Internal skeletal structure of head and thorax (cephalothorax). When fully developed, consists of complex system of fused projections (endopleurites, endosternites). (See also exoskeleton)

endosternite [G *Endosternit*]: In endoskeleton, one in a series of rigid structures projecting into body from ventral part (sternites) of exoskeleton. Serves also as site of muscle attachment (apodeme) and may be branched.

endostome (palate) [G *Endostom*]: In crab-like (brachyuran) decapod, posterior part of epistome; forms roof of mouth (buccal frame) and may be separated from epistome by transverse ridge.

end sac [G *Endsäckchen*]: Proximal section of antennal gland. Typically divided into saccule and labyrinth; opens into bladder (occasionally via excretory tubule).

epibranchial region [G *Epibranchialregion*]: branchial region

epibranchial space [G *Epibranchialraum*]: In each branchial chamber in cephalothorax, space above (exterior to) gills, i.e., between gills and lateral wall of carapace. (See also hypobranchial space)

epidermis [G *Epidermis*]: Outer cellular layer of body wall; secretes cuticle. Formerly termed hypodermis.

epigastric region [G *Epigastricalregion*]: gastric region

epimeral fold [G *Epimeralfalte*]: Steep fold of certain lateral elements (endopleurites) of exoskeleton. Lies parallel to and

attaches to side walls (branchiostegites) of carapace, thus forming branchial chamber enclosing gills.

epimere (epimeron): pleurite

epipodite (epipod) [G *Epipodit*]: Inwardly (medially) directed lobe of first two segments of thoracopods (maxillipeds and pereopods). Typically with respiratory function. (simple, lobate) (See also mastigobranch)

episternum [G *Episternum*]: On ventral surface of cephalothorax, one in a series of posterolateral projections of successive sternites; serves as ventral support for articulation of pereopod.

epistome [G *Epistom*]: Relatively large, plate-like structure in front of mouth on underside of head. May be divided into more narrow anterior part extending between antennae and broader posterior portion (endostome). Adjoined posteriorly by labrum. Also considered to be sternum of second cephalic segment (antennal somite). (See also metapon)

esophageal connective: circumesophageal connective

esophagus [G *Ösophagus*]: Relatively short, narrow, cuticle-lined section of digestive tract between mouth and cardiac stomach.

excretory duct [G *Harnleiter, Ureter, Ausführgang der Blase*]: In antennal gland, short duct leading from bladder to excretory pore on basal segment (coxa) of antenna. (See also excretory tubule)

excretory organ: antennal gland

excretory pore (nephropore) [G *Exkretionsporus, Nephroporus*]: Opening of each antennal gland to exterior; located on basal segment (coxa) of antenna. May bear operculum.

excretory tubule [G *Nephridialkanal*]: Narrow section of antennal gland occasionally joining second part of end sac (labyrinth) to bladder. (See also excretory duct)

exhalent passage [G *Ausführkanal, Präbranchialkammer, präbranchialer Kanal*]: On each side of cephalothorax, narrow chamber or canal leading from branchial chamber to exterior at anterior end. Water current driven by scaphognathite.

exite [G *Exit*]: Outwardly (laterally) directed, lobe-like branch of protopod.

exognath [G *Exognath*]: Outer branch (exopod) of maxilliped. (See also endognath)

exopod (exopodite) [G *Exopod, Exopodit, Außenast*]: Outer branch (ramus) of biramous appendage (endopod, exopod). Refers to antennal scale (scaphocerite) of antenna, scaphognathite of maxillula, flagelliform branch of maxillipeds, and variously developed outer branch of pereopods, pleopods, and uropods. (annulate, flagelliform, scale-like, spine-like)

exorbital process (external orbital tooth) [G *Exorbitalzahn*]: On carapace of crab-like (brachyuran) decapod, tooth-like projection on margin of orbit.

exoskeleton [G *Exoskelett, Panzer*]: Chitinous or calcified outer integument of body (pleurites, sternites, tergites) and appendages. (See also endoskeleton)

eye [G *Auge*]: One of two organs of vision, each consisting of numerous ommatidia covered by cornea; typically positioned at tip of eyestalk. (sessile, stalked; club-shaped, reniform, spherical)

eyestalk: ocular peduncle

finger [G *Finger*]: One of two finger-like elements of chela. One may distinguish a movable finger (dactyl) bearing down on a fixed finger (thumb, pollex), which is an extension of expanded propodus (manus, palm). (straight, curved, agape, closely fitting; cutting edge: corneous, entire, pectinate, serrate, sharp, sinuous, tuberculate, toothed; tips: acute, bidentate, corneous, incurved, spoon-shaped, subacute)

first antenna: antennule

first maxilla: maxillule

fixed finger (immovable finger, pollex, thumb) [G *unbeweglicher Finger*]: finger

flagellum [G *Flagellum, Geißel*]: Distal of two divisions (peduncle, flagellum) of antennule or antenna; typically paired in the former, unpaired in the latter. Slender, consisting of numerous segments (articles). May also refer to branch of mouthpart. (straight, bent; filiform, fusiform, plate-like; cross section: circular, dorsoventrally flattened, U-shaped; hairy, smooth, spinose)

foregut (stomodeum, stomodaeum) [G *Vorderdarm, Stomatodaeum*]: Anterior cuticle-lined region of digestive tract comprising esophagus, cardiac stomach, and anterior half of pyloric stomach. (See also hindgut, midgut)

front (frontal margin) [G *Stirnrand*]: In crab-like (brachyuran) decapod, median anterior margin of carapace between orbits. (defected, deflexed, sinuous, triangular, tridentate, trilobate, truncate)

frontal plate [G *ventraler Stirnfortsatz*]: In carapace of crab-like (brachyuran) decapod, modified rostrum which bears downwardly directed process united with ventral epistome.

frontal region [G *Frontalregion, Stirnregion*]: Anteromedian region of carapace. May bear rostrum and is adjoined posteriorly by gastric region, laterally by orbital region.

frontal tooth: rostral tooth

gastric mill [G *Magenmühle, Kauapparat*]: Grinding apparatus in posterior region of cardiac stomach. If fully developed, consists of muscular stomach walls bearing plates (ossicles), projecting teeth, and various pads and folds.

gastric region [G *Gastricalregion, Magenregion*]: Relatively large, unpaired region of carapace overlying stomach. May be variously subdivided (from anterior to posterior) into paired epigastric and protogastric as well as unpaired mesogastric, metagastric, and urogastric regions or areas. Adjoined anteriorly by frontal region, posteriorly by cardiac region.

gastric tooth [G *Gastricalzahn*]: On midline of carapace, tooth-like projection on gastric region just anterior to cervical groove.

gastrolith [G *Gastrolith*]: Disc-shaped, calcareous nodule in anterior region of cardiac stomach.

gastroorbital carina (supraorbital carina) [G *Crista gastroorbitalis*]: On each side of carapace, narrow, longitudinal ridge extending posteriorly from supraorbital spine.

gastroorbital groove [G *Sulcus gastroorbitalis*]: On each side of carapace, short, longitudinal groove extending from cervical groove toward orbit.

genital region: urogastric region

gill (branchia) [G *Kieme, Branchie*]: One of several respiratory projections that may be associated with each thoracic appendage (thoracopod). According to position one may distinguish arthrobranchs, pleurobranchs, or podobranchs; according to structure, dendrobranchs, phyllobranchs, or trichobranchs. Typically contained in branchial chamber on each side of cephalothorax. (See also gill formula)

gill bailer: scaphognathite

gill chamber: branchial chamber

gill formula (branchial formula) [G *Kiemenformel*]: Notation indicating type and number of gills associated with each thoracopod. Full complement would include one podobranch, two arthrobranchs, and one pleurobranch per thoracopod.

glaucothöe [G *Glaucothöe*]: postlarva

gonad: ovary, testis

gonopod [G *Gonopod*]: Modified pleopod serving in reproduction. Typically refers to first or second pleopod of ♂. (See also appendix masculina, petasma)

gonopore [G *Gonoporus, Geschlechtsöffnung*]: One of two openings of reproductive system to exterior. In ♂, pore on

basal segment (coxa) of last (eighth) thoracopods; in ♀, pore on coxa of sixth thoracopods (third pereopods).

green gland: antennal gland

grimothea [G *Grimothea*]: postlarva

groove [G *Furche, Sulcus*]: One of numerous furrows extending along carapace. According to position and according to group, one may distinguish antennal, branchiocardiac, buccal, cervical, dorsomedian, gastroorbital, hepatic, inferior, intercervical, intestinal, marginal, parabranchial, postcervical, seller, submedian, and urogastric grooves.

hair: seta

hand: manus

head: cephalon

head apodeme: apodeme

heart [G *Herz*]: In circulatory system, relatively short, muscular pumping organ located dorsally in posterior region of cephalothorax. Receives blood from pericardium through three pairs of ostia and pumps it anteriorly, posteriorly, and ventrally via numerous blood vessels (aortas, arteries).

hepatic artery [G *Arteria hepatica*]: artery

hepatic cecum (hepatopancreas, digestive gland) [G *Hepatopankreas, Mitteldarmdrüse*]: One of typically two large digestive glands occupying greater part of cephalothorax, each opening into junction of pyloric stomach and intestine. (paired, unpaired)

hepatic groove [G *Hepaticalfurche*]: On each side of carapace, short, longitudinal groove connecting cervical with postcervical and branchiocardiac grooves; more or less continuous with antennal groove.

hepatic region [G *Hepaticalregion, Leberregion*]: One of two lateral regions of carapace more or less corresponding to underlying hepatic ceca. Adjoining regions include antennal, branchial, cardiac, and pterygostomial regions.

hepatic spine [G *Hepaticaldorn*]: In shrimp-like (natantian) decapod, anteriorly directed, spine-like projection on hepatic region on each side of carapace.

hepatopancreas: hepatic cecum

hindgut (proctodeum, proctodaeum) [G *Enddarm, Proctodaeum*]: Posteriormost cuticle-lined region of digestive tract (foregut, midgut, hindgut). Opens via rectum and anus to exterior.

hindgut gland (rectal gland) [G *Enddarmdrüse*]: Relatively small, unpaired gland opening into end of digestive tract.

hypobranchial space [G *Hypobranchialraum*]: In each branchial chamber in cephalothorax, space below gills. (See also epibranchial space)

hypodermis: epidermis

hypopharynx: labium

hypostome (hypostoma): labium

immovable finger: finger

incisor process (pars incisiva) [G *Pars incisiva*]: Food-gripping process (vs. grinding = molar process) of each mandible. May be fused with molar process.

inferior abdominal artery: subneural artery

inferior groove: On each side of carapace, transverse groove extending from junction of hepatic and cervical grooves toward lateral margin; more or less continuous with cervical groove.

interantennular septum [G *medianes Septum*]: On anterior margin of carapace, median, ventrally directed, cuticular outgrowth extending from front to proepistome; separates left and right antennules as well as cavities into which each antennule may be retracted.

intercervical groove [G *Intercervicalfurche*]: On each side of carapace, oblique groove connecting postcervical and cervical grooves.

interpleurite: endopleurite, endosternite

intestinal groove [G *Intestinalfurche*]: On posterior part of carapace, short, transverse groove across dorsomedian surface; interrupted by tubercle.

intestinal region (posterior cardiac lobe) [G *Intestinalregion*]: Unpaired region at posterior end of carapace. Adjoined anteriorly by cardiac region.

intestine [G *Mitteldarmrohr*]: Elongate section of digestive tract extending through abdomen from pyloric stomach to rectum.

ischiobasis (basi-ischium) [G *Basiischium*]: Segment of appendage (thoracopod) formed by fusion of basis and ischium.

ischiocerite [G *Ischiocerit*]: In antenna, term applied to first of three segments (ischiocerite, merocerite, carpocerite) of inner branch (endopod). Alternate interpretation: third segment of peduncle.

ischiomerus [G *Ischiomerus*]: Segment of appendage (thoracopod) formed by fusion of ischium and merus.

ischiopod (ischiopodite): ischium

ischium (ischiopod, ischiopodite) [G *Ischium, Ischiopodit*]: Third segment of appendage (thoracopod), positioned between basis (last segment of protopod) and merus and thus representing first segment of endopod. May be fused with basis to form ischiobasis or with merus to form ischiomerus.

joint: article, segment

jugal region: pterygostomial region

labium (metastome, hypostoma, hypostome, hypopharynx, paragnath) [G *Labium, Unterlippe, Metastoma*]: Lip-like structure posterior to mouth on underside of head; symmetrically bilobed, each lobe being termed paragnath. (paragnath: oval, distally attenuated, sinuous)

labrum (upper lip) [G *Labrum, Oberlippe*]: Relatively large, unpaired, fleshy lobe in front of and partially covering mouth; adjoined anteriorly by epistome.

labyrinth [G *Labyrinth*]: In antennal gland, more distal of two divisions (saccule, labyrinth) of end sac. Characterized by complexly folded walls and opening either directly or indirectly (via excretory tubule) into bladder.

lateral artery (lateral cephalic artery) [G *Arteria lateralis cephalicus*]: artery

lateral carina [G *lateraler Carina*]: Narrow, longitudinal ridge extending along each side (lateral margin) of carapace.

lateral gastrocardiac marking [G *Muskeleindruck des Musculus attractor epimeralis*]: In crab-like (brachyuran) decapod, marking on each side of carapace formed by insertion of attractor epimeralis muscle (if normal point of insertion—branchiocardiac groove—is absent).

lateral margin [G *Seitenkante*]: In crab-like (brachyuran) decapod, outer edge of body formed by sharp bend in carapace.

lateral tooth [G *Lateralzahn*]: On carapace, one in a series of tooth-like projections along each side. According to position relative to cervical groove, one may distinguish anterolateral, mediolateral, and posterolateral teeth. (See also anterolateral tooth)

leg: swimming leg, walking leg

linea anomurica [G *Linea anomurica*]: On carapace of certain decapods, longitudinal groove or uncalcified line.

linea branchiostegalis [G *Linea branchiostegalis*]: In shrimp-like (natantian) decapod, longitudinal groove or uncalcified line extending from anterior margin of carapace (slightly above branchiostegal spine) to or slightly beyond hepatic spine.

linea dromica (linea dromiidica) [G *Linea dromica*]: On each side of carapace in certain decapods, longitudinal groove or uncalcified line—comparable to linea thalassinica—extending from anterior to posterior margin.

linea homolica [G *Linea homolica*]: On each side of carapace in certain decapods, longitudinal groove or uncalcified line—comparable to linea thalassinica—extending from anterior to posterior margin.

linea lateralis [G *Linea lateralis, Längsfurche*]: In shrimp-like (natantian) decapod, longitudinal groove or uncalcified line extending from frontal margin of carapace (below orbit) as far as posterior extremity of carapace.

linea thalassinica [G *Linea thalassinica*]: On each side of carapace in certain decapods, longitudinal groove or uncalcified line extending from anterior region below antennal spine to posterior margin.

lower lip: labium

mandible [G *Mandibel*]: Anteriormost paired mouthpart on underside of head; borne by third cephalic segment (mandibular somite). If fully developed, bears terminal molar and incisor processes as well as palp.

mandibular artery [G *Arteria mandibularis*]: artery

manus (palm, hand) [G *Palma, Hand, Manus*]: In pincer (chela), broad proximal part of next to last segment (propodus); movable finger (dactylus) closes on distal extension (fixed finger) of propodus.

marginal groove (lateromarginal groove, posteromarginal groove) [G *Randfurche*]: On each side of carapace, groove close to and parallel with margin; according to position one may distinguish lateromarginal and posteromarginal grooves.

mastigobranch [G *Mastigobranchie*]: In basal segment (coxa) of thoracic appendage (thoracopod), that part of epipodite which is not modified as gill (podobranch). Projects between gills and serves as both supporting and respiratory structure.

mastigopus [G *Mastigopus-Stadium*]: postlarva

maxilla (second maxilla) [G *Maxille, zweite Maxille*]: Third paired mouthpart on underside of head; borne on fifth cephalic segment (maxillar somite) and located between maxillule and first maxilliped. Typically consists of protopod, two endites, palp, and scaphognathite.

maxilliped (maxillipede) [G *Maxilliped, Maxillarfuß, Kieferfuß*]: One of three pairs of mouthparts posterior to maxillae on underside of head. Maxillipeds represent highly modified and anteriorly displaced first three pairs of thoracic appendages (thoracopods); basically consist of protopod (bearing endite and epipodite), endopod, and exopod. Posterior pair(s) increasingly resemble fourth through eighth thoracopods (pereopods) and may bear gills (podobranchs). (See also crista dentata)

maxillule (first maxilla) [G *Maxillula, erste Maxille*]: Second paired mouthpart on underside of head; borne on second cephalic segment (maxillular somite) and located between mandibles and maxillae. If fully developed, consists of protopod (with two endites), endopodal palp, and exite.

medulla terminalis X-organ [G *Medulla-X-Organ*]: X-organ

megalopa [G *Megalopa, Megalopa-Stadium*]: postlarva

merocerite [G *Merocerit*]: In antenna, term applied to second of three segments (ischiocerite, merocerite, carpocerite) of outer branch (endopod); alternate interpretation: fourth segment of peduncle.

merus (meropod, meropodite) [G *Merus, Meropodit*]: Fourth segment of appendage (thoracopod); positioned between ischium and carpus. May be fused to ischium to form ischiomerus.

mesenteron: midgut

mesobranchial region [G *Mesobranchial region*]: branchial region

mesogastric region [G *Mesogastricalregion*]: gastric region

mesosternum [G *Mesosternum*]: On underside of cephalothorax in crab-like (brachyuran) decapod, median plate of sternum; may give rise to inwardly directed endosternite.

metabranchial region [G *Metabranchialregion*]: branchial region

metagastric region [G *Metagastricalregion*]: gastric region

metanauplius [G *Metanauplius*]: Second of basically five larval stages (nauplius, metanauplius, protozoea, zoea, postlarva). Resembles nauplius, but bears additional appendages. Typically still contained within egg; if free-swimming, with antennal locomotion.

metapon [G *Metapon*]: Entire preoral area on underside of head; includes interannular septum, epistome, endostome, and part of mandibular somite.

metastome: labium

metazoea [G *Metazoea*]: In crab-like (brachyuran) decapod, last zoeal substage before metamorphosis to postlarva.

midgut (mesenteron) [G *Mitteldarm, Mesenteron*]: Region of digestive tract between foregut and hindgut. Either restricted to posterior section of pyloric stomach or corresponding to greater part of intestine. Not lined by cuticle.

molar process (pars molaris) [G *Pars molaris*]: Grinding process (vs. gripping = incisor process) of each mandible. May be fused with incisor process.

mouth [G *Mund, Mundöffnung*]: Anterior opening of digestive tract on underside of head; bordered anteriorly by upper lip (labrum), posteriorly by lower lip (labium), and flanked by mandibles, maxillules, and maxillae (and to various degrees by maxillipeds). Leads into cardiac stomach via esophagus.

movable finger (dactyl) [G *beweglicher Finger, Dactyl*]: finger

mysis (mysis stage, schizopod larva) [G *Mysis, Mysis-Stadium*]: zoea

nauplius [G *Nauplius*]: First of basically five larval stages (nauplius, metanauplius, protozoea, zoea, postlarva). Characterized by three pairs of appendages (antennules, antennae, and mandibles). Hatching from egg may take place at nauplius stage (certain shrimp-like decapods) or later.

nephropore: excretory pore

nisto [G *Nisto*]: postlarva

notum [G *Notum*]: Term occasionally applied to posterior part of dorsal region of carapace in shrimp-like (natantian) decapod or to dorsal plate of carapace in crab-like (brachyuran) decapod.

ocular acicle: acicle

ocular peduncle (eyestalk) [G *Augenstiel*]: One of two eye-bearing, movable projections of head; typically consists of two segments (proximal basiophthalmite and distal cornea-bearing podophthalmite). Segment of head (ophthalmic segment) bearing ocular peduncles is termed acron and is not considered to be true somite. (retractile, nonretractile; two-segmented, three-segmented; club-shaped, conical, dilated, elongate, oval, triangular; pubescent, setiferous)

ommatidium [G *Ommatidium*]: One of numerous elongate, closely adjoining units forming eye; covered by cornea.

operculum [G *Operculum*]: In crab-like (brachyuran) decapod, small, lid-like structure serving to close opening (excretory pore) of antennal gland.

ophthalmic artery: anterior aorta

ophthalmic scale (eye scale) [G *Augenschuppe, Augenplättchen*]: Scale-like projection on proximal segment of eyestalk (ocular peduncle). (unarmed, acuminate, serrate, with spines) (See also ocular acicle)

ophthalmic somite [G *Ocularsegment*]: acron

optic artery [G *Arteria optica*]: artery

orbit [G *Orbita, Augenhöhle*]: On anterior margin of carapace, one of two openings enclosing eyestalks (ocular peduncles). (circular, tubular)

orbital carina: On carapace, narrow ridge on margin of each orbit.

orbital region [G *Orbitalregion*]: One of two relatively narrow regions along anterior margin of carapace; corresponds to to point of projection of eyestalks. Adjoining regions include frontal, antennal or hepatic, and gastric regions.

orbital tooth: exorbital process

orbitoantennulary pit [G *orbitoantennulare Grube*]: On anterior margin of carapace, one of two cavities into which eyestalk and antennule may be retracted. (See also antennular fossette)

ossicle [G *Ossiculum, Os*]: One in a complex of rigid, plate-like structures associated with wall of cardiac stomach and, to a lesser extent, with pyloric stomach. Serves as site of attachment of muscles operating gastric mill or gives rise to tooth-like projections.

ostium [G *Ostium*]: One of typically three pairs of muscular openings in heart. Blood from pericardial sinus enters heart through ostia. According to position one may distinguish dorsal, lateral, and ventral pairs of ostia.

ovary [G *Ovar, Eierstock*]: Paired, expanded section of ♀ reproductive system typically located in posterodorsal region of cephalothorax. Connected by bridge, each ovary opening to exterior on basal segment (coxa) of third pereopod via oviduct.

oviduct [G *Ovidukt, Eileiter*]: Narrow section of ♀ reproductive system extending from each dorsal ovary to gonopore on basal segment (coxa) of sixth thoracopod (= third pereopod). May be modified terminally to form seminal receptacle and vagina.

palate: endostome

palm: manus

palp [G *Palpus, Taster*]: Branch-like projection of mouthpart (mandible, maxillule, maxilla, or first maxilliped); consists of distal portion of mouthpart and is typically endopodal in nature. (one-, two-, three-segmented/jointed; club-shaped, flattened, recurved)

parabranchial groove: On each side of carapace, groove below, behind, and almost parallel with branchiocardiac and postcervical grooves; joins latter in lower part.

paraesophageal ganglion [G *Ganglion connectivale*]: Swelling of each circumesophageal connective between supraesophageal and subesophageal ganglia.

paragnath [G *Paragnath*]: labium

pars incisiva: incisor process

pars molaris: molar process

parva (parva stage) [G *Parva-Stadium*]: postlarva

peduncle [G *Pedunculus, Stiel, Schaft*]: Proximal of two divisions (peduncle, flagellum) of antennule or antenna. In antennule, typically consists of three segments (also termed protopod) and bears stylocerite. In antenna, typically consists of two-segmented protopod and two- or three-segmented endopod and bears scaphocerite. (See also ocular peduncle)

penis (sexual tube) [G *Penis, Genitalpapille*]: Tubular ♂ copulatory structure associated with first segment (coxa) of last pair of pereopods; consists of extensible terminal part of vas deferens and may be variously enclosed in cuticular sheath. Functions in conjunction with modified anterior pleopod pairs. (single, paired)

pereopod (pereiopod, peraeopod) [G *Pereiopod, Peraeopod, Schreitbein*]: One of two appendages on each side of last five thoracic somites (appendage pairs of first three being modified as maxillipeds). Basically biramous, consisting of protopod, endopod, and exopod; exopod is often reduced, resulting in seven-segmented limb (coxa, basis, ischium, merus, carpus, propodus, dactylus). (ambulatory, natatory; biramous, uniramous; achelate, chelate, subchelate; equal, subequal, unequal; elongate, filiform, flagelliform, flattened, paddle-shaped, slender, stout; denticulate, granulate, hairy, pubescent, smooth, spinose, spinulose, tuberculate, unarmed)

pericardial sinus (pericardial sac, pericardium) [G *Perikardialsinus, Perikardialsack, Perikard*]: Cavity surrounding heart. Blood, coming from gills, collects in pericardial sinus and enters heart via ostia.

peritrophic membrane [G *peritrophische Membran*]: Chitinous membrane, secreted in anterior region of midgut and surrounding feces; considered to protect midgut lining from damage during passage of undigestible material.

petasma [G *Petasma*]: In reproductive system of ♂ in shrimp-like (natantian) decapod, expanded structure formed by modified inner branches (endopods) of first pair of pleopods; may bear series of distal lobes (e.g., distolateral, distoventral, distomedian lobes) and other processes. Functions in transferring spermatophores into thelycum of ♀.

photophore [G *Leuchtorgan, Photosphaere*]: Luminous organ; typically consisting of lens and reflector and variously distributed in and on body.

phyllobranch (phyllobranchia) [G *Phyllobranchie*]: Type of gill whose axis bears series of broad and flattened, leaf-like branches. (See also dendrobranch, trichobranch)

phyllosoma [G *Phyllosoma, Phyllosoma-Larve*]: zoea

pleomere (pleonite, abdominal somite) [G *Pleomer*]: One of six segments (somites) of abdomen (pleon); each bears pair of appendages (pleopods). Last pleomere bears uropods and is followed by telson.

pleon: abdomen

pleopod (swimmeret) [G *Pleopod, Schwimmfuß*]: One of two appendages of each abdominal somite (pleomere); typically consists of base (protopod) and two branches (endopod, exopod). Serves in swimming or variously modified as copulatory structures (e.g., gonopod, petasma) in ♂, egg-brooding structures in ♀. (biramous, uniramous; symmetrical, asymmetrical) (See also uropod)

pleopodal artery [G *Arteria pedis spurii*]: artery

pleura: pleurite

pleural suture [G *Pleuralnaht*]: suture

pleurite (epimere, epimeron, pleura, pleural lobe, pleurepimere, pleuron, tergal fold) [G *Pleura, Epimer*]: Lateral part of integument of somite (as opposed to sternite = ventral surface and tergite = dorsal surface). Most clearly visible in abdomen of shrimp-like (natantian) decapod, where they may form prominent lateral extensions (those of second pleomere occasionally being largest and overlapping preceding and following pleurites). The multiplicity of terms stems from differing early interpretations of cephalothorax region, i.e., lateral side of primary somites and lateral side of overlying carapace.

pleurobranch (pleurobranchia) [G *Pleurobranchie*]: Type of gill attached directly to body wall above base of thoracopods 2–8 (i.e., if full complement is present, above all thoracopods except first maxilliped). (See also arthrobranch, podobranch)

pleuron: pleurite

podobranch (podobranchia) [G *Podobranchie*]: Type of gill attached to first segment (coxa) or on epipod of coxa of thoracopods 2–7 (i.e., if full complement is present, on all thoracopods except first maxilliped and last pereopod). (See also arthrobranch, pleurobranch)

podomere: segment

podophthalmite [G *Podophthalmit, Endglied*]: Distal of typically two segments (basiophthalmite, podophthalmite) of eyestalk (ocular peduncle); bears cornea.

pollex: finger

postcervical groove [G *Postcervicalfurche, Sulcus postcervicalis*]: On carapace, groove posterior to and parallel with cervical groove; bisects cardiac region.

posterior aorta: dorsal aorta

posterior cardiac lobe: intestinal region

posterior carina: On carapace, narrow, transverse ridge in front of marginal groove.

posterior gastric pit [G *Muskeleindruck der Musculi gastrici posteriores*]: On carapace, one of two small depressions along midline marking insertion of stomach muscle.

posterior tooth (posteromedian tooth): On midline of carapace, tooth-like projection between marginal groove and posterior margin.

posterolateral region [G *Posterolateralregion*]: In crab-like (brachyuran) decapod, relatively small region along each posterolateral margin of carapace. (See also anterolateral region)

postesophageal ganglion: cephalothoracic ganglion

postlarva [G *Postlarva*]: Last of basically five larval stages (nauplius, metanauplius, protozoea, zoea, postlarva). Characterized by initial appearance of adult characters and, in shrimp-like (natantian) decapod, by pleopodal locomotion. According to group, this stage may be termed glaucothöe, grimothea, mastigopus, megalopa, nisto, parva, or pseudibaccus.

postorbital carina: On each side of carapace, narrow ridge slightly posterior and parallel to margin of orbit.

postorbital groove [G *Postorbitalfurche*]: On each side of carapace, groove close to and parallel with margin of orbit.

postorbital spine [G *Postorbitaldorn*]: In shrimp-like (natantian) decapod, anteriorly directed, spine-like projection somewhat behind orbit on each side of carapace.

postrostral carina [G *Carina dorsalis, Crista dorsalis*]: In shrimp-like (natantian) decapod, narrow, longitudinal ridge along dorsal midline of carapace; originates behind rostrum and typically extends almost to posterior end of carapace. Usually with teeth.

postrostral spine: In shrimp-like (natantian) decapod, anteriorly directed, spine-like projection on carapace directly behind rostrum.

pregastric tooth [G *Prägastricalzahn*]: On midline of carapace, tooth-like projection between gastric tooth and rostrum. (broadly rounded, bilobed and incised)

proctodeum (proctodaeum): hindgut

propodus [G *Propodus*]: Sixth segment of appendage, between carpus and dactylus. May serve as proximal element of subchela or be divided into proximal manus and distal (fixed) finger of chela.

prosartema (dorsal eye brush) [G *Prosartema, innere Schuppe*]: In shrimp-like (natantian) decapod, long, lobe- or scale-like projection from inner (medial) margin of first antennular segment.

protocephalon [G *Protocephalon, Vorderkopf*]: Term applied to anteriormost region of head (cephalon), especially when this is free of carapace. Consists of eyes and eyestalks, antennules, antennae, and labrum (i.e., of acron and first cephalic somite).

protocerebrum: supraesophageal ganglion

protogastric region [G *Protogastricalregion*]: gastric region

protopod (protopodite, sympod, sympodite) [G *Protopodit*]: Proximal part of biramous appendage. Typically two-segmented (coxa, basis) and bearing variously developed inner (endopod) and outer (exopod) branches.

protozoea [G *Protozoea*]: Third of basically five larval stages (nauplius, metanauplius, protozoea, zoea, postlarva). Applied to shrimp-like (natantian) decapod and characterized by antennal locomotion. May be subdivided into several substages. Protozoea of certain forms termed elaphocaris.

pseudibaccus [G *Pseudibaccus*]: postlarva

pseudorostrum [G *Pseudorostrum*]: In certain crab-like (brachyuran) decapods, anterior extension of carapace formed by fusion of two projecting teeth of frontal margin.

pterygostome [G *Pterygostom*]: ptergostomial region

ptergostomial region (jugal region) [G *Pterygostomialregion*]: On ventral side of carapace, one of two anterolateral regions flanking buccal cavity.

pterygostomial spine [G *Pterygostomialdorn*]: In shrimp-like (natantian) decapod, anteriorly directed, spine-like projection located below branchiostegal spine on anterolateral corner or carapace.

puerulus [G *Puerulus*]: zoea

pyloric stomach [G *Pylorus*]: Posterior and smaller division of stomach; separated from large anterior cardiac stomach by cardiopyloric valve. Characterized by highly folded, setose walls (filtering apparatus), plates, and valve-like structures. Opens into intestine; pair of ducts from hepatic cecum enter posterior end of pyloric stomach.

ramus [G *Ast*]: Branch of an appendage, typically referring to inner (endopod) and outer (exopod) branches of biramous appendage.

rectal gland: hindgut gland

rectum [G *Rektum*]: Posteriormost muscular and cuticle-lined section of digestive tract. May be somewhat expanded; opens to exterior through anus. Typically considered to represent only most terminal region of hindgut.

ridge: carina

rostral carina: On carapace, narrow, longitudinal ridge continuous with lateral margin of rostrum; may join submedian carina.

rostral tooth [G *Rostralzahn*]: One in a series of tooth- or spine-like projections along anterior extension (rostrum) of carapace. According to position one may distinguish upper, lower, and lateral teeth.

rostrum [G *Rostrum*]: Unpaired anterior extension of carapace. Projects between eyestalks (ocular peduncles) and represents extension of frontal region. (compressed, depressed; acute, arched, circular, deflexed, obsolete, obtuse, sinuous, straight, triangular; tip: attenuate, bifid, bifurcate, styliform, subtruncate, truncate, upturned; unarmed, armed: ciliated, spinuliferous, toothed, tridentate, trifid)

saccule [G *Sacculus*]: In antennal gland, proximalmost of two divisions (saccule, labyrinth) of end sac. Consists of simple vesicle or is partitioned internally.

scaphocerite (antennal scale, squama) [G *Scaphocerit, Schuppe, Squama*]: In antenna, variously shaped outer branch (exopod) projecting from peduncle. (broad, flat, lanceolate, ovoid, rounded, slender; serrate, setose)

scaphognathite (gill bailer) [G *Scaphognathit, Atemplatte*]: Relatively large outer branch (exopod) of each maxilla. Typically consists of two branches or lobes with setose margins, one of which may extend under carapace to generate respiratory current.

schizopod larva: mysis

second antenna: antenna

second maxilla: maxilla

segment [G *Glied*]: One in a series of units of an appendage, e.g., coxa, basis, ischium, merus, carpus, propodus, and dactylus of pereopod. May also refer to segment of body (then also termed somite). Segments of flagella of antennule or antenna are often termed articles. (segment of pereopod: not subdivided, subdivided = many-jointed = multiarticulate; compressed, cylindrical, flattened, obcordate, prismatic, subcubical, subcylindrical, subquadrate; bispinose, ciliate, granulate, rugose, serratogranulate, smooth, spinous, spinulose, with crest)

segmental artery [G *Arteria lateralis plica*]: artery

sella turcica [G *Sella turcica, Türkensattel*]: In endoskeleton of crab-like (brachyuran) decapod, fused and anteriorly extended endosternite on posterior border of last thoracic somite.

seller groove: On carapace, short, transverse groove extending across dorsomedian surface in front of cervical groove.

seminal receptacle [G *Receptaculum seminis*]: In ♀ reproductive system, sperm-receiving structure consisting of either expanded terminal section of each oviduct or single median pouch (thelycum).

sensory pore X-organ: X-organ

seta (bristle, hair) [G *Borste, Haar*]: Small, bristle-like projection articulating with or extending through cuticle. According to shape one may distinguish brush, cuspidate, feathered, hamate, nonplumose, pappose, plumodenticulate, plumose, serrate, setose, simple, triserrate, and triserrulate setae.

shield (anterior shield) [G *verkalkter mittlerer Teil des Carapax*]: Anterior, better calcified section of cephalothorax in front of cervical groove in hermit crabs. (subcordate, subquadrate; convex, flattened; with granules, with tufts of hair)

sinus [G *Blutsinus, Sinus*]: In circulatory system, one of several spaces in which deoxygenated blood collects before entering gill and returning to pericardium and heart. According to position one may distinguish branchial, dorsal, infrabranchial, and sternal sinuses.

sinus gland [G *Sinusdrüse*]: Small structure located in eyestalk (ocular peduncle) and serving to store and release hormones produced by X-organ.

somite (segment) [G *Segment*]: One in a series of divisions of body. Head (cephalon) consists of five somites (antennular, antennal, mandibular, maxillular, maxillar somites) excluding acron, thorax of eight somites (thoracomeres), and abdomen of six somites (pleomeres) excluding telson. Each somite basically consists of dorsal tergite, ventral sternite, and lateral pleurites.

spermatophore [G *Spermatophor*]: Packet of sperm formed in vas deferens, emerging from gonopore, and transferred to ♀ with the aid of modified first pleopod pair(s) (gonopods, petasma, appendix masculina). (stalked, unstalked; pear-shaped, rod-shaped, lobed, bilobed)

sperm duct: vas deferens

spine [G *Stachel, Dorn*]: One of numerous sharp, spine-like projections on carapace. Typically refers to anterior projections in shrimp-like (natantian) decapod. According to position and group one may distinguish, e.g., antennal, branchiostegal, hepatic, postorbital, postrostral, pterygostomial, suborbital, and supraorbital spines.

squama: scaphocerite

standard measurements:
carapace length and width
relative dimensions of appendage segments
relative length of appendages
(See also gill formula)

statocyst [G *Statozyste*]: Organ of equilibrium, one located in first segment of peduncle of each antennule. Innervated by branch of antennular nerve coming from brain and may contain statolith supported by sensory hairs. (open, closed; with/without statolith)

statolith [G *Statolith*]: Solid body within statocyst; supported by rows of sensory hairs. May be composed of sand grains.

stenopod (stenopodium) [G *Stabbein*]: Any slender, elongate appendage composed of rod-like segments. May refer to inner branch (endopod) of third maxillipeds or main part (endopod) of pereopods.

sternal artery [G *Arteria descendens*]: artery

sternal groove [G *Sternalfurche*]: On ventral surface of cephalothorax in certain crab-like (brachyuran) decapods, one of two oblique grooves along sternum.

sternal plastron: sternal plate

sternal plate (sternalplastron) [G *Sternalplastron*]: Plate-like ventral surface of cephalothorax formed by fusion of sternites. May be depressed in decapods whose abdomen is folded up against cephalothorax.

sternite [G *Sternit*]: Ventral surface of body segment (somite). May be fused with other sternites to form sternal plate in cephalothorax. (See also episternum, pleurite, tergite)

sternum [G *Sternum*]: Collective term for all sternites. Occasionally used as synonym for individual sternite or for fused sternites (sternal plate) of cephalothorax region.

sternum canal [G *Sternalkanal*]: In endoskeleton of crab-like (brachyuran) decapod, tube-like structure enclosing nerve cord; formed by fused inward projections (endosternites) of sternites.

stomach: cardiac stomach, pyloric stomach

stomodeum (stomodaeum): foregut

stylamblys: appendix interna

stylocerite (antennular scale, antennular spine) [G *Stylocerit*]: In antennule of shrimp-like (natantian) decapod, process projecting from outer (lateral) part of first segment of peduncle. Extending over and considered to protect statocyst. (scale-like, spiniform; tip: lanceolate, pointed, rounded, truncate; margins: convex, fringed with hairs, linear, sinuous, straight)

subchela [G *Subchela*]: Pincer-like structure formed by terminal segment of appendage (dactylus) bearing down on next to last segment (propodus). (See also chela)

subesophageal ganglion [G *Subösophagealganglion, Unterschlundganglion*]: Concentration of nerve tissue below (posterior to) esophagus. Consists of fused ganglia of somites bearing mouthparts (mandibles to third maxillipeds) and is connected to supraesophageal ganglion by pair of circumesophageal connectives. Continues posteriorly as ventral nerve cord. (See also postesophageal ganglion)

subhepatic carina [G *Crista hepatica*]: On each side of carapace, narrow, longitudinal ridge extending posteriorly from branchiostegal spine.

subhepatic region [G *Subhepaticalregion*]: On ventral side of carapace, region below each hepatic region. Adjoining regions include pterygostomial and suborbital regions.

submedian carina: On carapace, narrow, longitudinal ridge extending along each side of median postrostral carina; may join rostral carina.

submedian groove: On each side of carapace, submedian longitudinal groove to each side of postrostral carina.

subneural artery (posterior subneural artery, inferior abdominal artery, ventral abdominal artery) [G *Arteria subneuralis*]: artery

suborbital region [G *Suborbitalregion*]: On anteroventral surface of carapace, region below each orbit.

suborbital spine [G *Suborbitaldorn*]: Anteriorly directed, spine-like projection on orbital region on each side of carapace; located at anterior edge, on lower rim of orbit, between supraorbital and antennal spines.

sulcus: groove

superior abdominal aorta: dorsal aorta

supraesophageal ganglion (brain) [G *Supraösophagealganglion, Oberschlundganglion, Gehirn*]: Main concentration of nerve tissue lying above esophagus

at anterior end of cephalothorax; consists basically of protocerebrum, deutocerebrum, and tritocerebrum. Gives rise to numerous nerves extending to anterior appendages and is connected to subesophageal ganglion and ventral nerve cord by pair of circumesophageal connectives.

supraorbital carina: gastroorbital carina

supraorbital spine [G *Supraorbitaldorn*]: Anteriorly directed, spine-like projection on orbital region on each side of carapace; located at anterior edge, above and behind orbit.

suture [G *Naht*]: Weakly calcified lines along which exoskeleton splits during molting (e.g., pleural suture).

swimmeret: pleopod

swimming leg [G *Schwimmbein*]: Term referring either to pereopod modified (flattened) for swimming or to pleopods used in swimming.

tagma [G *Tagma*]: Major division of body (cephalothorax, abdomen).

tailfan (caudal fan) [G *Schwanzfächer*]: Posterior fan-like structure formed by combination of uropods and telson. (asymmetrical, symmetrical)

telson [G *Telson*]: Posteriormost segment of body (although not considered to be a true somite). May form tailfan together with uropods and bears anus ventrally. (single element, several elements; rectangular, rounded, spiniform, styliform, subquadrangular, subrectangular, subtriangular, subtruncate, triangular; tip: acuminate, rounded, subacute, truncate; armed: laterally serrate, with carina, with median groove, with movable spines, with spines, with spinules)

tergal fold: pleurite

tergite [G *Tergit*]: Dorsal surface of body segment (somite). Tergites of thoracic somites (thoracomeres) typically replaced by carapace. (See also pleurite, sternite)

tergum [G *Tergum*]: Collective term for all tergites; occasionally used as synonym for tergite.

testis [G *Hoden*]: Typically paired, expanded section of ♂ reproductive system in posterodorsal region of cephalothorax. Connected by bridge, each testis opening to exterior on basal segment (coxa) of last pereopod via vas deferens. (paired, fused; multilobed, tubular)

thelycum [G *Thelycum*]: In ♀ reproductive system of certain shrimp-like (natantian) decapods, pouch-like accessory copulatory structure formed by sternites of last and next to last thoracic somites. Serves as seminal receptacle and may be composed of several lobes or plates.

thoracomere (thoracic somite) [G *Thoracomer, Thoraxsegment*]: One of eight segments (somites) of thorax, the first three being incorporated into head; each bears pair of appendages (thoracopods).

thoracopod (thoracopodite) [G *Thoracopod*]: One of two appendages of each thoracic somite (thoracomere). First three pairs are modified as mouthparts (maxillipeds), the remaining five pairs serve in locomotion (pereopods).

thorax [G *Thorax*]: Posterior part of cephalothorax. Consists of eight somites (thoracomeres), each bearing pair of appendages (thoracopods). Covered, along with cephalon, by carapace.

thumb: finger

tooth [G *Zahn*]: Tooth-like structure, either in gastric mill of cardiac stomach or referring to blunt, relatively broad projection on outer surface of carapace. In cardiac stomach, one may distinguish one median and two lateral teeth. On carapace, one may distinguish, according to position and according to group, cardiac, gastric, lateral, orbital, pregastric, and rostral teeth.

trichobranch (trichobranchia) [G *Trichobranchie*]: Type of gill whose axis bears series of undivided, filament-shaped branches. (See also dendrobranch, phyllobranch)

tritocerebrum: supraesophageal ganglion

upper lip: labrum

ureter: excretory duct

urogastric groove [G *Urogastricalfurche*]: On carapace, short, transverse groove posterior to and occasionally joined to postcervical groove.

urogastric region (genital region) [G *Urogastricalregion*]: gastric region

uropod [G *Uropod*]: One of two appendages of last (sixth) abdominal somite (pleomere); typically flattened and consisting of basal protopod and two branches (endopod, exopod). May form tailfan together with telson. (elongate, falciform, ovate; with spine, with tooth)

vagina [G *Vagina*]: In ♀ reproductive system, terminal modification of oviduct to accommodate penis of ♂.

vas deferens (sperm duct) [G *Vas deferens, Samenleiter*]: Narrow section of ♂ reproductive system extending from each dorsal testis to gonopore on basal segment (coxa) of last thoracopod. May be modified into various regions for spermatophore production; forms terminal ejaculatory duct.

ventral ganglion [G *Bauchganglion*]: One in a series of ganglion pairs of ventral nerve cord in thorax (thoracic ganglion) or abdomen (abdominal ganglion). Segmentally arranged, each giving rise to pair of nerves. Ganglion pairs may shift anteriorly and be fused with one another to form cephalothoracic ganglion.

ventral nerve cord [G *Bauchmark*]: Longitudinal pair of more or less fused nerve cords extending from subesophageal ganglion to telson. Basically bears pair of (fused) ganglia in each somite.

walking leg [G *Schreitbein*]: pereopod

wrist: carpus

X-organ [G *X-organ*]: Neurosecretory structure in eyestalk (ocular peduncle). Secretes hormones that inhibit molting and are involved in gonad development and metabolism. In shrimp-like (natantian) decapod one may distinguish medulla terminalis X-organ and sensory pore X-organ.

Y-organ [G *Y-Organ, Y-Drüse, Carapaxdrüse*]: Neurosecretory structure in maxillae or base of antennae; secretes hormones promoting molting.

zoea [G *Zoea*]: Fourth of basically five larval stages (nauplius, metanauplius, protozoea, zoea, postlarva). If hatched from egg and free-swimming, then with thoracopodal locomotion. According to group this stage may be termed acanthosoma, mysis, phyllosoma, or puerulus.

ARTHROPODA

CHELICERATA

Xiphosura

horseshoe crab, king crab
[G *Pfeilschwanzkrebs, Schwertschwanzkrebs, Molukkenkrebs, Schwertschwanz, Pfeilschwanz*]

XIPHOSURA

(ventral) (dorsal)

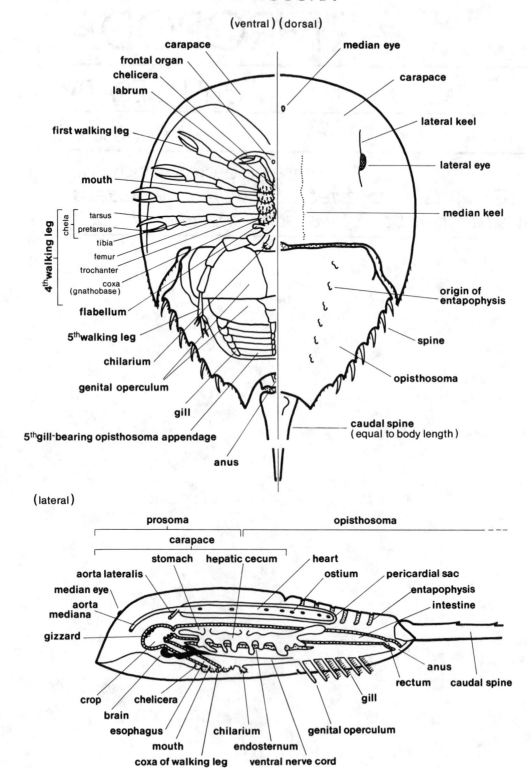

carapace
frontal organ
chelicera
labrum
first walking leg
mouth
tarsus
pretarsus
tibia
femur
trochanter
coxa (gnathobase)
chela
4th walking leg
flabellum
5th walking leg
chilarium
genital operculum
gill
5th gill-bearing opisthosoma appendage
anus

median eye
carapace
lateral keel
lateral eye
median keel
origin of entapophysis
spine
opisthosoma
caudal spine (equal to body length)

(lateral)

prosoma
opisthosoma
carapace
stomach   hepatic cecum
aorta lateralis
median eye
aorta mediana
gizzard
crop
brain
esophagus
mouth
coxa of walking leg
chelicera
chilarium
endosternum
ventral nerve cord
heart
ostium
pericardial sac
entapophysis
intestine
anus
rectum   caudal spine
gill
genital operculum

ARTHROPODA

CHELICERATA

Xiphosura

horseshoe crab, king crab
[G *Pfeilschwanzkrebs, Schwertschwanzkrebs, Molukkenkrebs, Schwertschwanz, Pfeilschwanz*]

abdomen: opisthosoma

acron [G *Acron*]: Anteriormost part of body; indistinguishably fused to prosoma.

anus [G *After*]: Posterior opening of digestive tract; located on underside of opisthosoma in front of base of caudal spine.

aorta [G *Aorta*]: One of three major blood vessels arising from anterior end of heart. Some or all of these blood vessels are also occasionally termed arteries. (See also aorta lateralis, aorta mediana)

aorta lateralis [G *Aorta lateralis, laterale Aorta, Aortenbogen*]: Pair of major blood vessels, one arising from each side of anterior end of heart. Extends anteriorly for a short distance, then proceeds ventrally around digestive tract to supply ventral region of prosoma and appendages.

aorta mediana [G *Aorta mediana, mediane Kopfaorta, Arteria frontalis, Frontalarterie*]: Single median blood vessel arising from anterior end of heart; proceeds dorsally over crop and gizzard to anterior margin of prosoma, where it bifurcates and continues along each lateral margin as an arteria marginalis. Also supplies hepatic ceca and gonads.

arteria collateralis [G *Arteria collateralis, Kollateralarterie*]: Large longitudinal blood vessel, one along each side of heart; each is considered to be a branch of the second lateral artery and joins all four lateral arteries of one side near their bases. Also gives rise to branches near each additional ostium pair. Joined to opposite member behind heart.

arteria hepatica [G *Arteria hepatica, Arterie der Mitteldarmdivertikel*]: In circulatory system, one of two large blood vessels branching from second pair of lateral arteries (arteriae laterales). Each proceeds anteriorly and supplies hepatic ceca.

arteria lateralis (lateral artery) [G *Arteria lateralis, Seitenarterie*]: One of eight blood vessels arising laterally from heart, one pair adjacent to each of the first four pairs of ostia. All four lateral arteries on each side are connected to one another near their bases by arteria collateralis. Each artery of second (= largest) pair gives rise to arteria collateralis, arteria hepatica, and arteria marginalis.

arteria marginalis [G *Arteria marginalis, Randarterie*]: In circulatory system, one of two large blood vessels branching from second pair of lateral arteries (arteriae laterales). Each proceeds along margin of prosoma in an anterior direction where it is continuous with branch of aorta frontalis.

artery: arteria collateralis, arteria lateralis, arteria marginalis

article [G *Glied*]: Segment of appendage (e.g., coxa, trochanter, femur, tibia, tarsus, pretarsus).

bladder [G *Blase*]: In excretory system, expanded, posteriorly directed section of each excretory tubule; opens to exterior via excretory pore.

book gill [G *Buchkieme*]: Respiratory structure formed by series of gill lamellae along each side of third to seventh opisthosoma appendages. Term is also applied to entire gill-bearing opisthosoma appendage itself; each flap-like appendage originally represents pair of limbs and is regionated into membranous median section and more rigid lateral sections (bearing gills).

brain [G *Gehirn*]: Main concentration of nerve tissue forming a ring around esophagus in prosoma. Upper (anterior) part of ring represents protocerebrum (with various substructures being distinguished, e.g., corporae pedunculi = globuli) and is occasionally termed supraesophageal ganglion. Lateral and lower (posterior) parts represent tritocerebrum fused with ganglia of first seven appendage-bearing segments and are occasionally termed subesophageal ganglion. Gives rise posteriorly to ventral nerve cord.

carapace [G *Prosomaduplikatur, Duplikatur*]: Large, horseshoe-shaped cuticular structure produced by and covering prosoma. Somewhat regionated by longitudinal keels and bears pair of median and lateral eyes. Dorsal joint between carapace and "opisthosoma" does not correspond exactly to true prosoma/opisthosoma border.

caudal spine (tail spine, telson) [G *Schwanzstachel, Telson*]: Elongate, movable terminal projection of opisthosoma.

cephalothorax: prosoma

chela (pincer) [G *Chela, Schere*]: Pincer-like structure at end of chelicerae and first four pairs of walking legs (reduced in first pair in ♂). In walking legs consists of movable sixth segment (pretarsus) bearing down on fifth segment (tarsus).

chelicera [G *Chelicere*]: First, relatively small, paired appendage of prosoma; located anterior to mouth. Consists of three segments, the distal two forming a pincer (chela).

chilarium [G *Chilarium*]: Paired appendage borne by first segment of opisthosoma (fused to prosoma). Located between last (fifth) pair of walking legs and consists of single flattened segment bearing spines and hairs.

compound eye: lateral eye

corpora pedunculata: brain

coxa [G *Coxa, Hüfte*]: First, most proximal segment of walking leg. Modified as gnathobase in first four pairs of walking legs; coxae of fifth pair bear flabella.

coxal gland [G *Coxaldrüse*]: One in a series of four excretory organs along each side of gizzard in prosoma; successive glands correspond in position with second to fifth appendage-bearing body segments. Glands on each side empty via common sacculus, excretory tubule, and bladder to exterior.

crop [G *Kaumagen*]: In digestive tract, anterior portion of expanded, cuticle-lined crop/gizzard complex between esophagus and stomach. Corresponds to anteriormost part (bend) of digestive tract and serves to store food before it passes into gizzard.

cuticle [G *Kutikula*]: Noncellular, multilayered structure forming exoskeleton and endoskeleton and lining foregut and rectum. Secreted by epidermis and is not calcified.

endochondrite [G *Endochondrit*]: One in a series of six small, plate-like elements of internal skeleton (endoskeleton). Located below ventral nerve cord in opisthosoma.

endoparietal eye [G *Endoparietalauge*]: Pair of reduced median eyes associated with conspicuous median eyes, yet not extending to carapace surface.

endoskeleton [G *Endoskelett*]: Internal skeletal structures of body consisting mainly of endosternum and endochondrites. (See also exoskeleton)

endosternum (endosternite) [G *Endosternum, Endosternit*]: Large, plate-like element of internal skeleton (endoskeleton) located between ventral nerve cord and digestive tract in prosoma. Irregularly shaped and bearing numerous lateral processes for muscle attachment.

endostome [G *Endostom, Endostoma*]: Term applied to more rigid sternal element between mouth and chilaria.

entapophysis [G *Entapophyse*]: One of six pairs of inwardly directed, cuticular projections from dorsal surface of opisthosoma; indicated on opisthosoma as six pairs of pit-like depressions. Serves as attachment site for muscles of opisthosomal appendages.

esophagus [G *Ösophagus*]: Narrow, anteriorly directed, cuticle-lined section of digestive tract between mouth and expanded crop/gizzard complex.

excretory pore [G *Exkretionsporus*]: One of two openings of excretory system to exterior; located at base of last pair of walking legs.

excretory tubule [G *Exkretionskanal*]: Elongate section of excretory system between sacculus and excretory pore; initially narrow, anteriorly directed, and coiled, then expanded (as bladder) and posteriorly directed.

exoskeleton [G *Exoskelett*]: Chitinous outer integument of body and appendages. (See also endoskeleton)

eye: endoparietal eye, lateral eye, median eye, ventral eye

femur [G *Femur*]: Rod-shaped third segment of walking leg (coxa, trochanter, femur, tibia, tarsus, pretarsus).

flabellum [G *Flabellum*]: Short projection of basal segment (coxa) of last (fifth) walking legs. Serves to clean gills.

foregut [G *Vorderdarm*]: Anterior cuticle-lined region of digestive tract comprising esophagus and crop/gizzard complex. (See also midgut)

frontal organ [G *Frontalorgan*]: Unpaired, wart-like sense organ anterior to mouth on underside of prosoma. Considered to be chemosensory in nature. Innervated by brain.

genital operculum (operculum) [G *Genitaloperculum, Operculum*]: Appendage of genital segment (eighth appendage-bearing segment = second opisthosoma segment); consists of single regionated, flap-like structure (originally representing paired limb). Bears gonopores basally and covers gill-bearing appendages.

genital segment [G *Geschlechtssegment*]: Eight appendage-bearing segment of body (second of opisthosoma); bears genital operculum. (See also pregenital segment)

gill [G *Kieme*]: Respiratory structure formed by series of approximately 150 lamellae on each side of last 5 appendages (3rd to 7th opisthosoma appendages). (See also book gill)

gill book: book gill

gill lamella (lamella) [G *Kiemenblatt, Lamelle*]: One in a series of approximately 150 closely spaced, leaf-like respiratory structures along each side of last 5 appendages (3rd to 7th opisthosoma appendages). (See also book gill)

gizzard [G *Kaumagen*]: In digestive tract, posterior portion of expanded crop/gizzard complex following esophagus. Equipped with well-developed muscle layers and lined internally with longitudinal cuticular folds bearing denticles. Opens into stomach via cone-shaped papilla.

gnathobase [G *Lade, Kaulade*]: Modified basal segment (coxa) of first four pairs of walking legs. Flattened, with margin bearing a series of bristles and hairs. Used in macerating and passing food into mouth.

gonad: ovary, testis

gonopore (genital pore) [G *Gonoporus, Geschlechtsöffnung*]: One of two openings of ♂ or ♀ reproductive system to exterior on underside of opisthosoma. Located basally on rear side of genital operculum.

heart [G *Herz*]: In circulatory system, elongate, tubular pumping organ extending throughout most of the body dorsally above midgut; posterior end is blind. Receives blood from pericardium through eight pairs of ostia and pumps it anteriorly through three aortas and laterally through four pairs of lateral arteries (arteriae laterales).

hepatic cecum [G *Mitteldarmdrüse*]: One of two pairs of highly branched digestive glands opening into stomach. Extensive, filling large part of prosoma.

intestine [G *Mitteldarmrohr*]: In digestive tract, narrower posterior section of midgut (stomach, intestine); opens into rectum.

labrum (upper lip) [G *Labrum, Oberlippe*]: More rigid skeletal element in front of mouth; associated with bases of chelicerae and with frontal organ.

lateral artery: arteria lateralis

lateral eye (compound eye) [G *Seitenauge, Komplexauge, Fazettenauge*]: Pair of large photosensitive organs, one below each lateral keel on upper surface of carapace. Each consists of numerous units (ommatidia) which are not closely adjoining. (See also endoparietal eye, median eye, ventral eye)

median eye [G *Medianauge*]: Pair of relatively small photosensitive organs, one to each side of median keel on upper surface (anterior end) of carapace. (See also endoparietal eye, lateral eye, ventral eye)

mesosoma [G *Mesosoma*]: Anterior subdivision of opisthosoma (mesosoma, metasoma). According to this interpretation, mesosoma represents greater, appendage-

bearing part of opisthosoma and thus consists of first 7 of originally 10 segments.

metasoma [G *Metasoma*]: Posterior subdivision of opisthosoma (mesosoma, metasoma). According to this interpretation, metasoma represents smaller, appendage-free part of opisthosoma and thus consists of last 3 of originally 10 segments.

midgut [G *Mitteldarm*]: Section of digestive tract between gizzard and rectum. Consists of more expanded anterior stomach and narrower intestine and is not lined by cuticle. (See also foregut)

mouth [G *Mund*]: Elongate anterior opening of digestive tract; located on ventral surface of prosoma between modified coxae (gnathobases) of walking legs. Opens into esophagus.

ommatidium [G *Ommatidium*]: One of numerous elongate units forming each lateral eye. Each basically consists of cornea, lens, and a number of retinular cells associated with rhabdome. Not closely adjoining and thus round rather than prismatic in cross section.

operculum: genital operculum

opisthosoma (abdomen) [G *Opisthosoma*]: Posterior division of body (prosoma, opisthosoma). Consists of seven appendage-bearing segments: pregenital segment (fused to prosoma and bearing chilaria), genital segment (with genital operculum), and five gill-bearing segments. Characterized by series of six spines along each side and a large terminal spine (caudal spine). (See also mesosoma, metasoma)

ostium [G *Ostium*]: One of eight pairs of openings in heart, two in prosoma region and six in opisthosoma. Blood from pericardium enters heart through ostia.

ovary [G *Ovar, Ovarium, Eierstock*]: Originally paired section of ♀ reproductive system in which eggs are produced. Forms network in prosoma (and to a lesser extent in opisthosoma) and opens to exterior via two short oviducts.

oviduct [G *Ovidukt, Eileiter*]: In ♀ reproductive system, one of two ducts joining net-like ovary to gonopores.

patella: tibia

pedipalp [G *Pedipalp*]: Appendage considered to be represented by first pair of walking legs (= second appendages).

pericardium (pericardial sinus) [G *Perikard, Perikardialsinus*]: Elongate cavity surrounding heart. Blood, coming from gills, collects in pericardium and enters heart via eight pairs of ostia.

pregenital segment [G *Prägenitalsegment*]: Seventh appendage-bearing segment of body (= first segment of opisthosoma). Bears chilaria and, although belonging to opisthosoma, is fused to prosoma.

pretarsus (praetarsus) [G *Prätarsus, Telotarsus*]: Sixth segment of walking leg (coxa, trochanter, femur, tibia, tarsus, pretarsus). Forms movable finger in pincer (chela) of first four pairs of walking legs.

prosoma [G *Prosoma*]: Anterior division of body (prosoma, opisthosoma). Consists of acron and first six appendage-bearing segments and is covered by carapace. First segment (pregenital segment with chilaria) of opisthosoma is fused to prosoma. Occasionally termed cephalothorax.

protocerebrum: brain

rectum [G *Rektum, Enddarm, Proctodaeum*]: Relatively short, cuticle-lined posterior section of digestive tract between intestine and anus,

sacculus [G *Sacculus*]: In excretory system, sac-like cavity into which four coxal glands on each side empty; opens into excretory tubule.

sclerite [G *Sklerit*]: More rigid element supporting membranous median section of each gill-bearing appendage (last five appendages).

segment [G *Segment*]: One of basically 13 recognizable (i.e., appendage-bearing) segments of body, the first 6 appendage-bearing segments comprising the

prosoma, the remaining 7 comprising the opisthosoma. May also refer to unit (e.g., coxa, trochanter, femur, tibia, tarsus, pretarsus) of walking leg.

sinus [G *Sinus, Blutsinus*]: In circulatory system, one of several longitudinal spaces in which deoxygenated blood collects before entering gills and returning to pericardium and heart.

sperm duct [G *Samenleiter*]: In ♂ reproductive system, network of ducts through which sperm are transported from testes to gonopores.

spine [G *Stachel*]: One of six short, movable spines projecting from each posterolateral edge of opisthosoma.

sternite [G *Sternit*]: Ventral surface of body segment. Highly reduced, the sternites of the original body segments being fused into a narrow strip (sternum) between bases of appendages. (See also endosternum, tergite)

sternum [G *Sternum*]: Collective term for all sternites.

stomach [G *Magen*]: In digestive tract, more expanded anterior section of midgut (stomach, intestine) following gizzard; border between stomach an intestine is indistinct. Receives pair of ducts of hepatic ceca on each side.

subesophageal ganglion [G *Unterschlundganglion*]: brain

supraesophageal ganglion [G *Oberschlundganglion*]: brain

tail spine: caudal spine

tarsus [G *Tarsus, Basitarsus*]: Fifth segment of walking leg (coxa, trochanter, femur, tibia, tarsus, pretarsus). Forms proximal part of pincer (chela) in first four pairs of walking legs.

telopodite [G *Telopodit*]: One of two relatively slender, segmented branches in midregion of gill-bearing appendages (third to seventh opisthosoma appendages = last five appendages). Considered to represent inner branches (endopods) of originally paired limbs.

telson: caudal spine

tergite [G *Tergit*]: Dorsal surface of body segment. Obscured in prosoma region by carapace and indistinguishably fused in opisthosoma. (See also sternite)

testis [G *Hoden*]: In ♂ reproductive system, one of numerous small vesicles in which sperm are produced. Located for the most part in prosoma and attached to a network of sperm ducts.

tibia [G *Tibia*]: Rod-shaped fourth segment of walking leg (coxa, trochanter, femur, tibia, tarsus, pretarsus). May be interpreted as consisting of fused patella and tibia.

trilobite larva [G *Trilobitenlarve*]: First free-living larval stage. Characterized by presence of all segments, yet only first 9 (of 13) appendages, and by indistinct caudal spine.

tritocerebrum: brain

trochanter [G *Trochanter*]: Second, relatively flattened segment of walking leg (coxa, trochanter, femur, tibia, tarsus, pretarsus).

ventral eye [G *Ventralauge*]: Small pair of eyes anterior to mouth on underside of prosoma. Well developed in larvae and juveniles, reduced and incorporated into frontal organ in adults.

ventral nerve cord [G *Bauchmark, Bauchstrang*]: Longitudinal pair of closely adjoining nerve cords extending from lower (posterior) part of brain into opisthosoma. Fused at position of ganglia.

walking leg [G *Laufbein*]: One of five pairs of uniramous appendages following chelicerae on prosoma. Consists of six to seven segments which may basically be considered to represent coxa, trochanter, femur, tibia, tarsus, and pretarsus. (chelate, nonchelate)

ARTHROPODA

CHELICERATA

Pycnogonida (Pantopoda)

sea spider, pantopod, pycnogonid
[G *Asselspinne, Pantopod, Pycnogonide*]

PYCNOGONIDA

(dorsal)

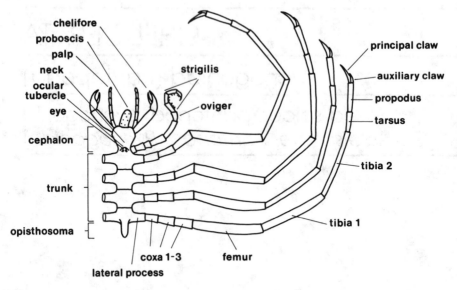

chelifore
proboscis
palp
neck
ocular tubercle
eye
cephalon
trunk
opisthosoma

strigilis
oviger

principal claw
auxiliary claw
propodus
tarsus
tibia 2
tibia 1

coxa 1-3
lateral process
femur

chela

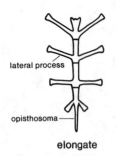

tooth
hand
immovable finger
movable finger

globular

curved fingers/toothless

elongate with teeth

body shape

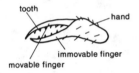

lateral process

opisthosoma

elongate

4- 5- 6-
segmented

compact

oval/unsegmented

leg

tibia 2
tibia 1
femur

with comb rows

recticulate

with long setae

with fringed setae

ARTHROPODA

CHELICERATA

Pycnogonida (Pantopoda)

sea spider, pantopod, pycnogonid
[G *Asselspinne, Pantopod, Pycnogonide*]

abdomen: opisthosoma

antimere [G *Antimer, Längsteil*]: One of three longitudinal elements (one dorsal, two ventrolateral) forming proboscis and imparting it with a triradial symmetry. Each antimere ends in chitinous lip.

anus [G *After*]: Posterior opening of digestive tract at end of opisthosoma.

auxiliary claw [G *Nebenkralle, Nebenklaue*]: Paired, claw-shaped structures associated with principal claw of propodus; typically smaller but occasionally larger than principal claw.

brain (supraesophageal ganglion) [G *Gehirn, Oberschlundganglion*]: Main concentration of nerve tissue beneath ocular tubercle in cephalon innervating eyes and chelifores; connected to subesophageal ganglion by two circumesophageal commissures.

cecum (caecum) [G *Darmschenkel, Blindsack*]: One in a series of blind extensions of intestine into legs and occasionally into ovigers, palps, chelifores, and proboscis.

cement gland [G *Kittdrüse*]: Multicellular glands, typically with a common opening, on dorsal surface of femur of legs in ♂.

cephalon (head, cephalic somite) [G *Kopf*]: Anteriormost division of body (cephalon, trunk, opisthosoma) bearing proboscis, chelifores, and palps; decreasing in diameter posteriorly to form neck.

chela [G *Schere*]: Variously elaborated, pincer-like structure at distal end of each chelifore; consists of a fixed and movable segment (finger). (elongate, globular, imperfect, reduced and knob-like)

chelifore (chelicera) [G *Chelicere*]: Anteriormost pair of appendages of cephalon, each consisting of a one- or two-segmented shaft (scape) and chela. (chelate, reduced and knob-like)

circumesophageal commissure [G *Schlundkonnektiv*]: One of two strands of nerve tissue surrounding esophagus and joining brain to subesophageal ganglion.

claw [G *Kralle, Endklaue*]: General term applied to main or principal claw.

coxa (1–3) [G *Coxa, Hüftglied*]: Term applied to three proximal, typically relatively short segments of leg; second coxa of one or more legs bears gonopores.

cruriger: lateral process

cuticula [G *Kutikula*]: Noncellular, multilayered, variously ornamented layer covering outer surface of body and extending into pharynx, esophagus, and rectum.

487

diverticulum: cecum

dorsal median vessel: heart

esophagus (oesophagus) [G *Ösophagus*]: Short section of digestive tract between pharynx and intestine; located in narrow neck region of cephalon and separated from intestine by valvula.

eye [G *Auge*]: One of typically four lens-bearing organs of vision on distal end of ocular tubercle on cephalon.

femur [G *Femur*]: Fourth, typically relatively long segment of leg between coxa 3 and tibia 1; bearing cement glands in ♂ and serving as storage site for mature eggs in ♀.

finger [G *Finger*]: One of two finger-like elements of pincer (chela) at end of chelifore. One may distinguish a movable finger bearing down on a fixed finger. (with/without teeth)

gonad [G *Gonade*]: ovary, testis

gonopore (genital pore) [G *Gonoporus, Geschlechtsöffnung*]: Opening of ♂ or ♀ reproductive system on ventral surface of coxa 2 of one or more (usually posterior) leg pairs.

head: cephalon

heart (dorsal median vessel) [G *Herz*]: Elongate, contractile circulatory organ bearing two or three pairs of ostia and extending along dorsal surface of trunk; pumps blood anteriorly into proboscis and chelifores.

hindgut: rectum

intestine (midgut) [G *Mitteldarm*]: Largest section of digestive tract between esophagus and rectum; extends through entire trunk and sends ceca into legs and occasionally into anterior appendages.

lateral process (cruriger) [G *Seitenfortsatz*]: Pair of lateral extensions of each segment of trunk. Bears legs. (smooth, with setae, with spines)

lateral sense organ [G *seitliches Sinnesorgan*]: Sense organ on each side of ocular tubercle. Considered to be chemosensory or thermosensory.

leg [G *Bein*]: One of typically four to six pairs of variously elaborated locomotory appendages consisting of eight segments (coxae 1–3, femur, tibia 1–2, tarsus, propodus, and terminal claw). (with four, five, or six pairs of legs; ambulatory, natatory; naked, ornamented: with setae/spines/tubercles, reticulate)

lip [G *Lippe*]: Term applied to one of three movable, cuticular structures surrounding mouth; each lip consists of inner tooth and outer brush-shaped structure.

main claw: principal claw

mouth [G *Mund*]: Anterior triangular opening of digestive tract at distal end of proboscis; typically reinforced by three cuticular lips.

neck [G *Hals*]: Narrowed posterior section of cephalon; typically bearing ocular tubercle dorsally and corresponding internally to esophagus.

ocular tubercle (eye tubercle) [G *Augenhügel*]: Variously elaborated prominence bearing eyes; typically located on neck, but occasionally also anteriorly between chelifores or posteriorly at border of neck and trunk. (elongate, rounded)

opisthosoma (abdomen, anal process) [G *Opisthosoma, Abdomen, Schwanzabschnitt*]: Small, unsegmented, and upwardly angled posteriormost division of body (cephalon, trunk, opisthosoma); bears rectum and terminal anus.

ostium [G *Ostium*]: One of two pairs of openings through which blood enters heart; located at level of second and third lateral processes of trunk. Unpaired posterior ostium may also be present.

ovary [G *Ovar, Ovarium*]: U-shaped ♀ reproductive organ above digestive tract in trunk; typically extending into legs. (See also gonopore)

oviger (ovigerous leg) [G *Oviger, Eierträger*]: Third pair of appendages, typically positioned on cephalon just anterior to first pair of walking legs; less well developed than legs and articulating ventrally. Consists of a maximum of 10 segments and serves to carry egg mass in ♂, often reduced in ♀. Number and configuration of segments are of taxonomic importance.

ovigeral spine [G *Dorn des Brustbeins*]: One in a series of broad, denticulate spines on modified distal segments (strigilis) of oviger.

ovigerous leg: oviger

palp [G *Palp*]: Second pair of appendages, between chelifores and ovigers on cephalon; consists of a maximum of 11 segments and bears sensory hairs.

pharyngeal plate [G *Chitinplatte*]: One of three articulating cuticular plates forming wall of pharynx and bearing bristle-shaped structures posteriorly; serves to maintain suction and in maceration of food.

pharynx [G *Pharynx*]: Muscular, cuticle-lined section of digestive tract between mouth and esophagus; extends the length of proboscis and is triradiate in cross section due to pharyngeal plates.

principal claw (main claw) [G *Hauptkralle, Kralle, Klaue*]: Movable, claw-shaped structure on distal end of each leg and often accompanied by pair of auxiliary claws; relative length of principal and auxiliary claws is of taxonomic importance.

proboscis [G *Proboscis, Rüssel*]: Muscular, tubular organ at anterior end of cephalon bearing terminal mouth and containing pharynx; may be directed anteriorly or folded under. (downcurved, straight, upcurved; clavate, oval, spindle-shaped; with/without dilation)

propodus [G *Propodus*]: Eighth and terminal segment of leg; bearing principal and occasionally pair of auxiliary claws. (with/without ventral spines, proximal spines, or lamina)

prosoma [G *Prosoma*]: Term applied to region of body encompassing cephalon and trunk.

rectum (hindgut) [G *Enddarm*]: Short, cuticle-lined section of digestive tract between intestine and anus; located in opisthosoma.

scape [G *Schaft*]: Basal section of cheliphore. (one-, two-segmented)

segment: somite

somite (segment) [G *Segment*]: One of basically four segments of body, each with pair of lateral processes bearing legs.

standard measurements:
proboscis: length, greatest diameter
trunk: total length, length of individual segments
abdomen: length
third leg: length of each segment

strigilis [G *Ringbürste*]: Modified terminal end of third pair of appendages (ovigers); distal four segments curved into sickle-shaped configuration, each segment bearing spines (ovigeral spines). Used to clean other limbs.

subesophageal ganglion [G *Unterschlundganglion*]: First ventral concentration of nerve tissue innervating palps and ovigers; located below esophagus and connected to brain via two circumesophageal commissures.

tarsus [G *Tarsus*]: Seventh and next to last segment of leg between tibia 2 and propodus.

testis [G *Hoden*]: U-shaped ♂ reproductive organ above digestive tract in trunk; typically extending into legs. (See also gonopore)

tibia (1, 2) [G *Tibia 1, 2*]: Fifth and sixth, typically relatively elongated segments of leg between femur and tarsus.

trunk [G *Rumpf*]: Second and largest division of body (cephalon, trunk, opisthosoma); consists of four to six segments, each bearing a pair of lateral extensions (lateral processes) on which legs articulate. (four-, five-, six-segmented, polymerous; ornamentation: with columnar processes, distal spurs, spines, spires, tubercles, arborescent tubercles)

valvula [G *Valvula*]: Cuticular, tripartite structure functioning as valve between esophagus and intestine.

ventral nerve cord [G *Bauchmark*]: Longitudinal pair of nerve cords extending posteriorly from subesophageal ganglion. Located midventrally and bearing ganglia for ovigers and each pair of legs.

PENTASTOMIDA

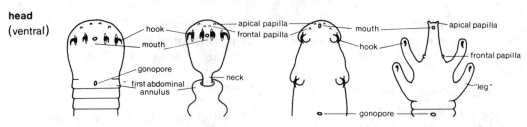

head
(ventral)

hook

mouth

gonopore

first abdominal annulus

neck

apical papilla

frontal papilla

mouth

hook

gonopore

apical papilla

frontal papilla

"leg"

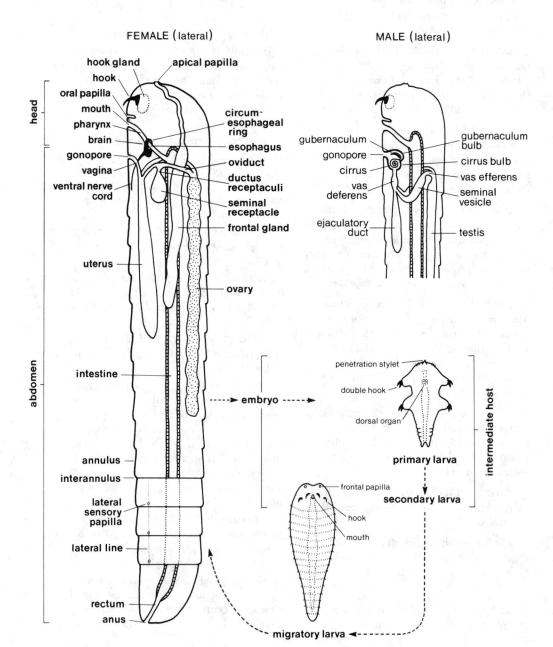

FEMALE (lateral)

MALE (lateral)

head

hook gland

apical papilla

hook

oral papilla

mouth

pharynx

brain

gonopore

vagina

ventral nerve cord

circum-esophageal ring

esophagus

oviduct

ductus receptaculi

seminal receptacle

frontal gland

uterus

ovary

abdomen

intestine

annulus

interannulus

lateral sensory papilla

lateral line

rectum

anus

gubernaculum

gonopore

cirrus

vas deferens

ejaculatory duct

gubernaculum bulb

cirrus bulb

vas efferens

seminal vesicle

testis

embryo

penetration stylet

double hook

dorsal organ

primary larva

intermediate host

frontal papilla

hook

mouth

secondary larva

migratory larva

PENTASTOMIDA (LINGUATULIDA)

pentastomid, linguatulid, tongue worm [G *Zungenwurm*]

abdomen [G *Rumpf, Hinterkörper, Abdomen*]: Posterior and larger of two divisions of body (head, abdomen); typically annulated with terminal anus and ventral gonopore on either anterior or posterior annulus.

annulus [G *Ring, Körperring, Annulus, Segment*]: One in a series of elevated, cuticular rings on abdomen; corresponds internally to segmental arrangement of abdominal muscles and may bear series of sensory papillae or pore plates laterally. (See also interannulus)

anus [G *After*]: Posterior opening of digestive tract; located terminally on last abdominal annulus and occasionally flanked by caudal papillae. (present, absent) (See also cloaca)

apical papilla [G *Apikalpapille*]: One of two small sensory projections anterior to mouth at tip of head; innervated by brain. (See also frontal papilla)

brain [G *Gehirn*]: Main concentration of nerve tissue in head, in form of either three separate, closely adjoining pairs of ventral subesophageal ganglia (lobi cephales, lobi pharyngeales, mandibular ganglia) or single fused subesophageal ganglion. Gives rise posteriorly to ventral nerve cord. (separate, fused)

buccal cavity [G *Mundhöhle*]: Cuticle-lined cavity following mouth cavity and bearing muscular oral papilla.

cardia [G *Cardia, Valvula cardiaca*]: Funnel-shaped valve between esophagus and intestine; prevents regurgitation of food.

caudal papilla [G *Terminalanhang, Appendix caudalis*]: Soft, flexible, lobelike projection, one to each side of anus.

circular muscle [G *Ringmuskel*]: Relatively thin, circular outer layer of muscle fibers in body wall. (See also longitudinal muscle)

circumesophageal ring [G *Schlundkommissur*]: In brain, anterior band of nerve tissue passing dorsally over esophagus.

cirrus [G *Zirrus*]: In cirrus bulb of ♂ reproductive system, highly elongated, thin, spirally coiled tube; receives sperm from vas deferens and is protruded, along with gubernaculum, outside ♂ gonopore during copulation.

cirrus sac [G *Zirrustasche*]: In ♂ reproductive system, expanded section of genital bulb enclosing coiled cirrus.

claw: hook

cloaca [G *Kloake*]: In certain pentastomids, common opening of ♀ reproductive system and digestive tract on terminal abdominal annulus.

copulatory spicule: gubernaculum

cuticle [G *Kutikula*]: Multilayered, noncellular surface covering body and lining buccal cavity, pharnyx, esophagus, rectum, and certain terminal sections of reproductive system. Outer layer may bear various projections for locomotion.

dorsal organ [G *Dorsalorgan*]: Large, multicellular gland on dorsal surface of embryo or primary larva; opens to exterior via median pore and secretes outer muous coating.

dorsoventral muscle [G *Dorsoventralmuskel*]: One in a series of transverse muscle bands spanning from dorsal to ventral surface on side of each annulus; may form lateral chamber in body cavity.

double hook [G *Doppelhaken, Doppelkralle*]: Bipartite hook apparatus consisting of hook and adjoining hook-shaped sheath into which the former can be retracted; typically restricted to various larval stages.

ductus receptaculi [G *Ductus receptaculi*]: In ♀ reproductive system, short, thin, curved duct joining each seminal receptacle to oviduct or to junction of oviduct and uterus.

egg [G *Ei*]: ♀ reproductive body; contains well-developed embryo when ripe and may fill uterus in large numbers.

ejaculatory duct [G *Ductus ejaculatorius, Ejakulationsschlauch*]: Section of ♂ reproductive system consisting of paired, elongate, muscular, blind tubes; opens at junction of seminal vesicle and vas deferens. Receives sperm from seminal vesicle and pumps them through vas deferens.

embryo [G *Embryo*]: Series of developmental stages of pentastomid while still enclosed in egg; early stages characterized by four pairs of extremities, stylet, and dorsal organ.

epidermis [G *Epidermis*]: Cellular layer of body wall between circular muscle and cuticle.

esophagus [G *Ösophagus*]: Section of digestive tract between pharynx and considerably wider intestine; less strongly cuticularized and more rounded in cross section than pharynx.

frontal gland [G *Frontaldrüse*]: Pair of glandular organs, either small and located between head and abdomen or large and extending posteriorly along each side of intestine; open separately on head adjacent to or on apical papillae.

frontal papilla [G *Frontalpapille*]: Pair of small sensory projections posterior to apical papillae on ventral surface of head; innervated by brain.

fulcrum [G *Fulcrum*]: Posteriorly directed, cuticular extension of hook; serves as site of attachment for hook muscles.

genital bulb [G *Bulbus genitalis*]: Paired, expanded, cuticle-lined section of ♂ reproductive system ventral to intestine between vasa deferentia and gonopore; either bipartite, consisting of cirrus bulb and gubernaculum bulb, or single and bearing only gubernaculum.

genital papilla [G *Genitalpapille, Geschlechtspapille*]: Pair of ventral sensory papillae, one to each side of ♂ gonopore on first abdominal annulus.

gonopore (genital pore) [G *Genitalporus, Geschlechtsöffnung*]: Unpaired opening of reproductive system to exterior. In ♂, median ventral opening on first abdominal annulus; may be flanked on both sides by genital papillae. In ♀, located either midventrally on first abdominal annulus or on or near last annulus. (See also cloaca)

gubernaculum [G *Gubernaculum, Kopulationszapfen*]: In gubernaculum bulb of ♂ reproductive system, slightly curved, cone-shaped, cuticular structure; protrudes, along with cirrus (if present), outside ♂ gonopore during copulation.

gubernaculum bulb [G *Dilatator-sack*]: Expanded, muscular, cuticle-lined section of genital bulb bearing gubernaculum; may be adjoined by cirrus bulb.

gut: intestine

head [G *Kopf, Vorderkörper*]: Anterior and smaller of two divisions of body (head, abdomen); formed by fusion of several segments. Bears mouth, two pairs of hooks, and apical and frontal papillae. (flat, pointed)

hook (claw) [G *Haken, Hakenkralle, Kralle*]: One of four movable, claw-shaped, cuticular structures embedded in ventral surface of head or positioned terminally on fleshy projections (legs); may be variously modified in larvae and serve in clinging to host. (See also double hook, fulcrum)

hook gland [G *Hakendrüse*]: Glandular organ in head; similar to parietal and frontal glands in structure and opening to exterior at base of each hook or in hook pouch.

hook lobe [G *Hakenlappen*]: Fleshy lobe adjoining hook dorsally and covering it when retracted.

hook sac [G *Hakentasche*]: Cuticle-lined depression into which hook may be retracted.

interannulus [G *Interannulus*]: Any rings of thinner cuticle between annuli of abdomen.

intestine [G *Darm*]: Straight, non-cuticularized, longest (and widest) section of digestive tract between esophagus and rectum; lumen may bear longitudinal folds.

larva: migratory larva, nymph, primary larva, secondary larva

lateral line [G *Seitenlinie*]: Longitudinal line, one along each side of abdomen, formed by series of pore plates.

lateral nerve [G *Lateralnerv*]: One in a series of paired lateral branches arising from brain or ventral nerve cord. Anterior lateral nerves supply hooks; posterior lateral nerves, lateral sensory papillae or pore plates.

lateral sensory papilla [G *laterale Sinnespapille*]: One in a series of sensory papillae along each side of head and abdomen; in the latter, one per annulus.

leg [G *Stummelfuß*]: Term occasionally applied to one of four unjointed, fleshy processes bearing terminal hook.

lobus cerebralis (cerebral lobe) [G *Lobus cerebralis*]: In type of brain characterized by three closely adjoining pairs of ganglia, flat lateral pair of ganglia bearing circumesophageal commissure. (See also lobus pharyngealis, mandibular ganglion)

lobus parapodialis [G *Lobus parapodialis*]: Fleshy lobe to each side of hook.

lobus pharyngealis [G *Lobus pharyngealis*]: In type of brain characterized by three closely adjoining pairs of ganglia, anteroventral ganglion pair; bears subesophageal commissure and innervates frontal papillae. (See also lobus cerebralis, mandibular ganglion)

longitudinal muscle [G *Längsmuskel*]: Relatively thick, metamerically arranged bands of longitudinal muscle underlying circular muscle in body wall.

mandibular ganglion [G *Mandibularganglion*]: In type of brain characterized by three closely adjoining pairs of ganglia, large posteroventral ganglion pair. Innervates first pair of hooks; continues posteriorly in form of ventral nerve cord bearing several paired, ganglionated masses.

maxillary ganglion [G *Maxillenganglion*]: In pentastomid characterized by brain with three closely adjoining pairs of ganglia, first paired, ganglionated mass of ventral nerve cord; innervates second pair of hooks.

migratory larva [G *Wanderlarve, Terminallarve, Stachellarve*]: Final larval stage in definitive host; characterized by more elongate abdomen and occasionally by transverse rows of small cuticular projections on annuli. Wanders to final location where it molts into adult.

mouth [G *Mund*]: Anterior opening of digestive tract located ventrally on flat head or on tip of pointed head; typically surrounded by cuticular buccal cavity and closed by oral papilla.

neck [G *Hals*]: Narrow constriction between head and abdomen.

nymph [G *Nymphe*]: Juvenile stage after molt of embryo or primary larva; typically characterized by loss of penetrating stylets and dorsal gland. May also be synonymous with primary larva.

oral papilla [G *Oralpapille, Oberlippe*]: Muscular structure anterior to mouth opening; serves, together with pumping activity of pharynx, in food uptake.

ovary [G *Ovar, Ovarium*]: Elongate, unpaired section of ♀ reproductive system extending posteriorly above intestine; attached to dorsal body wall by mesentery. Joins uterus via paired or unpaired oviduct.

oviduct [G *Ovidukt, Eileiter*]: Tubular section of ♀ reproductive system passing around intestine and joining ovary to uterus. (paired, unpaired)

parietal gland [G *Parietaldrüse*]: One of numerous multicellular clusters of gland cells in body wall; opens to exterior via thin, cuticular ductule.

penetration stylet [G *Bohrstachel*]: One to several anterior cuticular projections on head of early larval stages; may also refer only to median and largest projection. Used to penetrate intestine of intermediate host.

pharynx [G *Pharynx*]: Cuticularized, muscular section of digestive tract between buccal cavity and esophagus; sickle-shaped in cross section.

pore plate [G *Porenplatte*]: One in a series of sensory structures, on both sides of annuli, forming lateral line along each side of abdomen; consists of round, cuticular plate with peripheral pores and underlying sensory cells.

primary larva [G *Primärlarve, Erstlarve*]: First juvenile stage after hatching from egg; may refer to freshly emerged "embryo" or, after first molt, to more advanced stage. Typically characterized by two pairs of extremities, penetration stylet, dorsal gland, and short abdomen. (See also nymph)

rectum [G *Enddarm*]: Short, cuticle-lined posteriormost section of digestive tract between intestine and anus.

secondary larva [G *Sekundärlarve, Zweitlarve*]: Second major juvenile stage, often encapsulated in intermediate host, following primary larva or nymph; typically characterized by loss of paired anterior extremities (legs) and penetration stylets.

seminal receptacle (spermatheca) [G *Receptaculum seminis, Spermathek*]: Expanded section of ♀ reproductive system ventral to intestine; opens into oviduct or junction of oviduct and uterus. Serves to store sperm after copulation.

seminal vesicle [G *Vesicula seminalis, Samenblase*]: Typically posteriorly directed and looped section of ♂ reproductive system between vas efferens and junction of vas deferens and ejaculatory duct; proximally unpaired, distally bifurcating into each ejaculatory duct.

spermatheca: seminal receptacle

sperm sieve [G *Spermiensieb*]: Section of ♂ reproductive system at junction of testis and seminal vesicle; consists of thickened tissue surrounding vas efferens.

stigmal gland [G *Stigmadrüse*]: One of numerous typically serially arranged, multicellular, flask-shaped glands opening to exterior via small pore.

testis [G *Hoden*]: Sperm-producing section of ♂ reproductive system extending posteriorly above intestine; attached to dorsal body wall by mesentery. Opens into seminal vesicle through short, thin vasa efferentia or through intervening sperm sieve.

uterus [G *Uterus*]: In adult ♀ pentastomid, large, unpaired section of reproductive system ventral to intestine; joined to ovary by oviduct and often filled with eggs. Opens to exterior either anteriorly or posteriorly via short vagina. (saccate, tubular)

vagina [G *Vagina*]: In adult ♀ pentastomid, short, muscular section of reproductive system between uterus and gonopore.

valvula cardiaca: cardia

vas deferens [G *Vas deferens*]: Section of ♂ reproductive system at junction of seminal vesicle and ejaculatory duct; typically short, enclosed by thick layer of gland cells, and lined by thin cuticle. Sperm passes through vas deferens from ejaculatory duct to genital bulb.

vas efferens [G *Vas efferens*]: Short, thin section of ♂ reproductive system between testis and seminal vesicle.

ventral nerve cord [G *Bauchmark, ventraler Nervenstrang*]: Paired nerve cords extending below digestive tract from brain to posterior end of abdomen; bears metamerically arranged lateral branches and, in more primitive pentastomid, several paired ganglionated masses anteriorly.

BRYOZOA

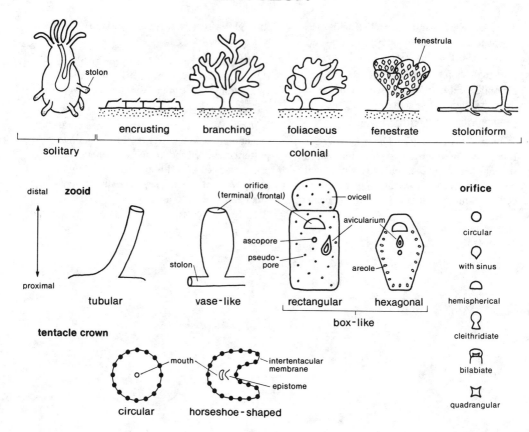

solitary

encrusting branching foliaceous fenestrate stoloniform

fenestrula

colonial

distal   **zooid**

proximal

tubular   vase-like   rectangular   hexagonal

box-like

orifice (terminal) (frontal)   ovicell

ascopore

pseudo-pore

avicularium

areole

stolon

orifice

circular

with sinus

hemispherical

cleithridiate

bilabiate

quadrangular

tentacle crown

mouth

intertentacular membrane

epistome

circular   horseshoe-shaped

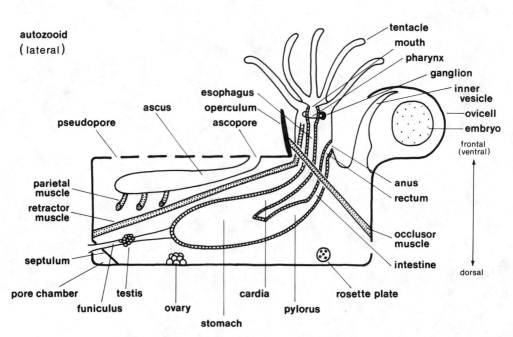

autozooid
(lateral)

tentacle
mouth
pharynx
ganglion
inner vesicle
ovicell
embryo
frontal (ventral)
anus
rectum
occlusor muscle
intestine
dorsal
rosette plate
pylorus
stomach
cardia
ovary
testis
funiculus
pore chamber
septulum
retractor muscle
parietal muscle
pseudopore
ascus
esophagus
operculum
ascopore

LOPHOPHORATA (TENTACULATA)

BRYOZOA (ECTOPROCTA)

bryozoan, ectoproct, moss animal [G Bryozoe, Bryozoon, Moostierchen]

alveole [G *Alveolus*]: In certain encrusting bryozoans, one in a series of extrazooidal cavities formed by vertical, calcareous partitions between zooids. If fully developed, roofed over by calcareous layer.

ancestrula [G *Ancestrula*]: First bryozoan individual (zooid) of a colony; typically differs in size and shape from subsequently formed zooids. Arises from metamorphosis of free-swimming larva or, in freshwater forms, from statoblast. (See also proancestrula)

annulus (pneumatic ring) [G *Schwimmring*]: In reproductive bud (statocyst), flattened ring of air-filled cells surrounding central capsule. Acts as float. (with/without spines)

anter [G *vorderer Deckelabschnitt*]: Distal main part of lid (operculum) covering orifice. May be hinged to proximal part (poster).

anus [G *After*]: Posterior opening of digestive tract. Located outside of tentacle crown on tentacle sheath.

apertural bar: Skeletal (zooecium) structure: first (fused) pair of spines (costae) proximal to orifice.

aperture: orifice

areola (areole, lateral punctuation) [G *Areola*]: Skeletal (zooecium) structure: one in a series of spaces between buttress-like secondary thickenings. Located above areolar pore on calcified frontal surface.

areolar pore: Skeletal (zooecium) structure: one in a series of marginal pseudopores along margin of calcified frontal surface.

ascopore [G *Ascoporus*]: Skeletal (zooecium) structure: separate opening of compensation sac (ascus) on calcified frontal wall.

ascus (compensation sac, compensatrix) [G *Ascus*]: Water-filled, sac-like organ located below frontal wall. Opens to exterior at orifice or via separate pore (ascopore). Serves as hydrostatic organ in bryozoan with calcified frontal wall to compensate for pressure differences caused by protraction and retraction of lophophore.

atrium (vestibule) [G *Atrium, Vestibulum*]: In zooid with retracted lophophore, space between orifice and sheath (tentacle sheath) surrounding tentacles, i.e., at level of diaphragm.

autozooid [G *Autozoid*]: In colony (zoarium), individual whose function is to capture and ingest food. (caducous) (See also gonozooid, heterozooid, interzooid, kenozooid, nanozooid, zooid)

avicellarium [G *Avicellarium*]: In type of zooid (avicularium) modified to protect colony, small, fixed chamber containing mandibular muscles.

avicularium [G *Avicularium*]: In colony (zoarium), modified individual (zooid) with reduced internal organs and in which orifice region and operculum are well developed (the latter into mandible). Serves to protect colony. (adventitious = dependent, interzooidal, vicarious; single, paired; sessile, pedunculate; elliptical, oval, setiform, spatulate, triangular) (See also onychocellarium)

brain: ganglion

brown body [G *brauner Körper*]: In individual (zooid) in which polypide has degenerated, spherical, colored structure representing encapsulated remnants of internal organs.

capitulum [G *Capitulum*]: Colony (zoarium) structure: apical group of feeding individuals (autozooids) in stalked species.

cardella: condyle

cardia [G *Cardialteil des Magens*]: In digestive tract, tubular anterior section of stomach.

cauda [G *Cauda*]: Skeletal (zooecium) structure: thread-like proximal portion of club-shaped feeding zooid (autozooid).

cecum (caecum) [G *Blindsack, Magen-Blindsack, Caecum, Coecum*]: In digestive tract, outpocketing of stomach. Attached to zooid wall by funiculus.

circumoral ring [G *oraler Nervenring*]: In freshwater bryozoan, ring of nerve tissue surrounding anterior part of digestive tract and extending into median tentacles in convex (outer = ventral) part of horseshoe-shaped lophophore. (See also ganglionic horn)

coelocyst [G *Coelozyste*]: Skeletal (zooecium) structure: type of wall in which living tissues surrounding zooidal wall continue on outer side and give rise to extrazooidal tissues. Has double epithelium. (See also stictocyst)

coelom [G *Coelom*]: Peritoneum-lined body cavity. One may distinguish cavity of epistome (protocoel), ring coelom (mesocoel) at base of lophophore, and general or trunk coelom (metacoel).

coelomopore (supraneural pore) [G *Coelomoporus, Zölomoporus, Supraneuralporus*]: Opening of body cavity (coelom) to exterior. Located at base of tentacles; serves as opening for emerging eggs. (See also intertentacular organ)

collar (collare, pleated collar, setigerous collar) [G *Colare, Collare, Kragen*]: Circular, variously stiffened membrane borne by diaphragm. Serves to seal atrium (vestibule) when tentacles are retracted.

colony: zoarium

columella [G *Columella*]: In type of zooid (avicularium) modified to protect colony, small column on midpoint of bar (pivotal bar) on which mandible articulates.

communication pore: Skeletal (zooecium) structure: opening in wall between two zooids.

compensation sac: ascus

compensatrix: ascus

condyle (cardella) [G *Gelenk*]: Skeletal (zooecium) structure: one of two lateral teeth projecting from rim of orifice. Serves as pivot for lid (operculum) or modified operculum (mandible).

costa (costula) [G *Costa, Randborste*]: Skeletal (zooecium) structure: one in a series of spines originating along margin and extending over frontal membrane. Typically paired, fused medially with opposite member and fused at intermittent intervals with adjoining costae to form costate shield or frontal shield.

costate shield (costal shield):
Skeletal (zooecium) structure; shield-like frontal wall formed by series of fused spines (costae).

costula: costa

cryptocyst [G *Kryptozyste, Cryptocyste*]: Skeletal (zooecium) structure: horizontal calcareous shelf parallel to and below frontal membrane. Formed by ingrowth of vertical walls of zooecium and penetrated by pores (opesiules). (smooth, finely granular)

cyphonautes (cyphonautes larva) [G *Cyphonautes*]: Free-swimming, bivalved larva of certain bryozoans. Bears fully developed and functional digestive tract.

cystid [G *Cystid*]: One of two major sections (cystid, polypide) of single bryozoan individual (zooid). Consists of skeletal outer structure (zooecium) and attached layers of body wall.

cystigenic valve (dorsal valve):
Dorsal of two valves (cystigenic, deutoplasmic) forming capsule of statoblast. Delimited by equatorial suture.

deutoplasmic valve (ventral valve): Ventral of two valves (cystigenic, deutoplasmic) forming capsule of statoblast. Delimited by equatorial suture.

diaphragma [G *Diaphragma*]:
Transverse constriction, operated by muscles, near distal end of zooid; marks lower border of atrium (vestibule) when tentacles are retracted. Also refers to membranous or skeletal partition that extends transversely across entire zooidal chamber.

dietella (pore chamber) [G *Porenkammer*]: Skeletal (zooecium) structure: in distal region of bryozoan individual (zooid), small chamber at base of vertical wall. Bears pores (rosette plate) on one side and communicates with adjacent zooid by pore on other side.

divaricator muscle [G *Divaricator, Öffnermuskel*]: Pair of muscles spanning from frontal membrane just behind hinge line

of orifice lid (operculum) to zooidal wall. Opens operculum.

ectocyst [G *Ektozyste*]: Term applied either to cuticular layer of body wall or to epidermis, cuticle, and skeleton.

ectooecium: Outer, typically calcified layer of ooecial wall. (non/partially/wholly calcified) (See also endooecium).

endooecium: Inner, typically membranous layer of ooecial wall. (membranous, calcified) (See also ectooecium)

entosaccal cavity [G *intrasakaler Raum*]: Innermost of two cavities (entosaccal, exosaccal) formed by subdivision of main body cavity (metacoel) by membranous sac.

epistegal space: epistege

epistege (epistegal space):
In bryozoan bearing frontal shield, space between frontal membrane and overlying shield.

epistome [G *Epistom*]: In freshwater bryozoan, small, hollow, movable lobe overhanging mouth. (See also coelom)

epistomial ring [G *epistomialer Nervenring*]: In freshwater bryozoan, ring of nerve tissue surrounding epistome and extending into tentacles of concave (dorsal = inner) part of horseshoe-shaped lophophore. (See also circumoral ring)

esophagus (oesophagus) [G *Ösophagus*]: Nonciliated section of digestive tract between pharynx and stomach. Foregut considered to consist of pharynx and esophagus.

exosaccal cavity [G *extrasakaler Raum*]: Outermost of two cavities (entosaccal, exosaccal) formed by subdivision of main body cavity (metacoel) by membranous sac.

fenestra: Skeletal (zooecium) structure: in outer wall (ectooecium) of ooecium, uncalcified area through which endooecium is visible.

fenestrula/fenestrule [G *Fenestrula*]: Colony (zoarium) structure: one in a series of open spaces between branches in reticulate (fenestrate) colony. (See also trabecula)

floatoblast [G *Flottoblast*]: Type of reproductive bud (statoblast) which floats. Annulus well developed, without marginal hooks or spines. (See also leptoblast, sessoblast, spinoblast)

forked canal [G *Gabelkanal*]: In freshwater bryozoan, one of two canals originating from lophophoral coelom and surrounding epistomial coelom.

frontal membrane [G *Frontalmembran*]: Skeletal (zooecium) structure: relatively thin, membranous part of frontal (ventral) wall. Flexible, allowing for compensation of hydrostatic changes due to lophophore protrusion and retraction. (exposed, covered by spines/frontal shield)

frontal shield [G *Frontalschild*]: Skeletal (zooecium) structure: variously derived, calcified frontal surface.

frontal wall [G *Frontalwand*]: Skeletal (zooecium) structure: rigid, calcified frontal surface (covering ascus).

funicular strand (funicular cord) [G *Funiculus*]: One of various strands of mesenchymal tissue traversing body cavity and joining internal organs to zooid wall, where it may serve in interzooidal communication by being associated with pore plates or with cord in stolon. Conspicuous cord spanning from stomach cecum to posterior wall is termed funiculus.

ganglion (brain) [G *Ganglion*]: Main concentration of nerve tissue in coelom on dorsal side of pharynx. Gives rise to nerve ring encircling anterior part of digestive tract and to series of additional nerves. (See also circumoral ring, epistomial ring, ganglionic horn)

ganglionic horn [G *Ganglion-Horn*]: In nerve system of freshwater bryozoan, tract of nerve tissue extending from each side of brain (ganglion) into lophophoral arm of that side.

gizzard [G *Kaumagen*]: In digestive tract of certain bryozoans, modified section of cardia. Muscular, lined with rigid plates or teeth.

gonad: ovary, testis

gonozooid [G *Gonozoid*]: In colony (zoarium), individual modified and enlarged to serve as brood chamber for embryos.

gymnocyst [G *Gymnozyste, Gymnocyste*]: Skeletal (zooecium) structure: calcified part of frontal wall between frontal membrane and vertical walls. May give rise to series of spines extending over membrane.

hemiphragm [G *Hemiphragma*]: Skeletal (zooecium) structure: one in an alternating series of inwardly directed, shelf-like projections in tube-shaped zooid.

hemiseptum [G *Hemiseptum*]: Skeletal (zooecium) structure; one of one or two inwardly directed, shelf-like projections near aperture in tube-shaped zooid.

heterozooid [G *Heterozoid*]: In colony (zoarium), modified individual lacking functional feeding apparatus (lophophore). Includes avicularia, gonozooids, and vibracula.

hibernaculum [G *Hibernaculum*]: Reproductive bud. Consists of capsule containing yolk-like material and partly developed feeding and digestive organs. Serves as resistent overwintering or dispersive body. (fusiform, irregular)

hypostegal coelom [G *Hypostegalcoelom*]: Extension of main body cavity (coelom) overlying various types of frontal calcifications.

hypostege: In certain bryozoans lacking compensation sac (ascus), cavity between double wall consisting of ectocyst and cryptocyst. Serves as hydrostatic organ.

inner vesicle: ooecial vesicle

internode (segment) [G *Internodium*]: In certain branching colonies (zoaria), relatively thick, zooid-bearing segment joined to other internodes by narrow articulation (node). (cylindrical, flattened)

intertentacular membrane (calyx) [G *Intertentakularmembran*]: In freshwater bryozoan, thin, transparent membrane connecting bases of tentacles.

intertentacular organ [G *Intertentakularorgan*]: Ciliated tube bearing opening (coelomopore) of body cavity to exterior. Located on outer side of tentacle crown and formed by fusion of two most dorsal tentacle bases.

interzooid (interzooidal polymorph): In colony (zoarium), relatively small individual (zooid) positioned between and serving as communication between two normally sized zooids. Considered to include dietellae.

intestine [G *Intestinum*]: Elongate section of digestive tract between posterior part (pylorus) of stomach and rectum. Occasionally termed rectum.

introvert: tentacle sheath

joint: node

kenozooid [G *Kenozoid*]: In colony (zoarium), highly reduced individual (zooid) typically lacking orifice and most internal organs. May serve to support or interconnect colony. (See also rhizoid)

labellum: lip

lacuna [G *Lakuna*]: Skeletal (zooecium) structure: one in a series of performations (true pores) between adjoining spines (costae) making up frontal shield. (See also lumen)

leptoblast [G *Leptoblast*]: Reproductive bud (statoblast). Represents type of floatoblast which bypasses diapause and may complete polypide development while still within parental colony. Germinates almost immediately after release.

lip (labellum) [G *Lippe*]: Skeletal (zooecium) structure: calcified, tongue-shaped extension of ovicell extending into orifice. May also refer to one of two lip-like structures closing orifice in certain bryozoans.

lophophore [G *Lophophor, Tentakelträger*]: Anterior fold of body wall bearing numerous hollow, ciliated tentacles. Encloses mouth. (circular, horseshoe-shaped; bell-shaped, campylonemidan, truncate)

lucida [G *Lucida*]: Skeletal (zooecium) structure: cavity between two layers of lid (operculum) covering orifice.

lumen: Skeletal (zooecium) structure: one in a row of small pores on midpoint of spine (costa) making up frontal shield.

lyrula (lyrule) [G *Lyrula*]: Skeletal (zooecium) structure: median tooth projecting from proximal rim of orifice. (anvil-shaped)

mandible [G *Mandibel*]: In type of zooid (avicularium) serving to protect colony, movable structure representing modified operculum. (beak-shaped, pointed, setiform, spatulate, triangular, vibraculoid)

maternal zooid [G *Mutterzoid*]: Proximal reproductive zooid which extrudes eggs into brood chamber (ooecium).

membranous sac [G *Membransack*]: Longitudinal, cylindrical membrane subdividing main body cavity (metacoel) into entosaccal and exosaccal cavities.

mesocoel [G *Mesocoel*]: coelom

metacoel [G *Metacoel*]: coelom

mouth [G *Mund, Mundöffnung*]: Round or oval anterior opening of digestive tract; surrounded by tentacles of lophophore. Opens into pharynx.

mucro [G *Mucro*]: Skeletal (zooecium) structure: well-developed elevation of proximal rim or orifice. (blunt, spinous)

mural rim: Skeletal (zooecium) structure: on frontal surface, raised edge at border of gymnocyst and frontal membrane.

mural spine: Skeletal (zooecium) structure: one in a series of inwardly directed spines in tube-shaped zooid.

muscle: divaricator muscle, occlusor muscle, parietal muscle, retractor muscle

muscle dot [G *Muskelansatz*]: Skeletal (zooecium) structure: one of two small depressions in lid (operculum) covering orifice.

myozooid [G *Myozoid*]: In colony (zoarium), special kind of kenozooid without polypide but with well-developed muscle fibers; forms stalk (peduncle) of colony.

nanozooid [G *Nanozoid*]: In colony (zoarium), minute individual (zooid) with reduced internal organs.

node [G *Knoten*]: In certain branching colonies (zoaria), relatively narrow articulation between two zooid-bearing segments (internodes).

occlusor muscle [G *Deckelschließmuskel*]: Pair of muscles spanning from orifice lid (operculum) to zooid wall. Closes operculum.

olocyst [G *Olozyste*]: Skeletal (zooecium) structure; innermost of three layers (olocyst, pleurocyst, tremocyst) in calcified front of zooid with ascus. Smooth.

onychocellarium [G *Onychocellarium*]: In certain bryozoans, type of protective zooid (avicularium) with mandible bearing membranous expansion.

ooecial vesicle (inner vesicle) [G *innere Brutkammer, Embryosack, inneres Ovicell, inneres Ooecium*]: Inner membrane of ooecium.

ooeciopore [G *Ooeciopore*]: Skeletal (zooecium) structure: aperture of bryozoan individual (gonozooid) modified as brood chamber.

ooeciostome [G *Ooeciostom*]: Skeletal (zooecium) structure: rim-like structure surrounding aperture of bryozoan individual (gonozooid) modified as brood chamber. (fan-shaped, funnel-like, tubular; with/without hood)

ooecium [G *Ooecium*]: Skeletal (zooecium) structure: ovicell or brood chamber, excluding inner (ooecial) vesicle. (acleithral, cleithral; dependent, independent; endotoichal, endozooidal; funnel-shaped, tubular)

opercular gland: vestibular gland

operculum (lid) [G *Operculum, Deckel*]: Skeletal (zooecium) structure: lid-like structure closing orifice when lophophore is retracted. (crescentic, horseshoe-shaped, semicircular) (See also mandible)

opesia [G *Opesium*]: Skeletal (zooecium) structure: opening below frontal membrane which remains after development of cryptocyst. (oval, triangular, trifoliate)

opesiule [G *Opesiula*]: Skeletal (zooecium) structure: small notch or pore in calcareous plate (cryptocyst) below frontal membrane. Permits passage of depressor muscle from body to frontal membrane.

oral gland: vestibular gland

orifice (primary orifice) [G *Orificium, Apertur*]: Skeletal (zooecium) structure: opening in zooid wall through which lophophore is protruded and retracted. In certain bryozoans having a calcified frontal wall and peristome, the resulting secondary opening is termed aperture. (with/without operculum; frontal, subterminal, terminal; bilabiate, circular, cleithridiate, orbicular, sinuate, subcircular, tridentate)

orifical collar: peristome

ovary [G *Ovar, Ovarium, Eierstock*]: ♀ reproductive organ consisting of one or two clusters of ovocytes enclosed in thin peritoneum; attached to body wall, digestive tract, or funiculus.

ovicell [G *Ovicelle*]: Skeletal (zooecium) structure: globular structure enclosing brood chamber. Represents modified individual (gonozooid) in certain bryozoans. (dependent, independent; perforate, imperforate; immersed = endozooidal, subimmersed, prominent; carinate, mucronate, trilobate)

palate: rostrum

parietal muscle [G *Parietalmuskel*]: One in a series of muscles spanning from floor of compensation sac (ascus) to zooid wall. Functions in hydrostatic system.

peduncle [G *Stiel*]: Stalk-like part of pedunculate avicularium.

pericyst: Skeletal (zooecium) structure: calcified frontal wall not covered by ectocyst. Generally consists of fused marginal spines.

peripheral lamina: Colony (zoarium) structure: marginal extension of basal gymnocystal wall in certain bryozoans with tubular zooids.

peristome (orificial collar) [G *Peristom*]: Skeletal (zooecium) structure: variously elevated, collar-like extension of orifice rim. (coalescent, independent; biserial, irregular; deciduous, not deciduous; pointed, not pointed, bicuspidate, tricuspidate)

pharynx [G *Pharynx*]: Ciliated, muscular section of digestive tract into which mouth opens. Foregut may be divided into pharynx and esophagus.

piptoblast: Type of reproductive bud (statoblast) lacking marginal hooks or annulus. Retained within narrow, tubular zooecium rather than being cemented to substrate.

pivotal bar: In type of zooid (avicularium) modified to protect colony, calcareous rim or bar on which mandible is articulated.

pleurocyst [G *Pleurozyste*]: Skeletal (zooecium) structure: second of three layers (olocyst, pleurocyst, tremocyst) in calcified front of zooid with ascus. Granular.

pneumatic ring: annulus

polymorph: In colony (zoarium), general term for individual (zooid) modified for reproductive, defensive, cleaning, supportive, or connective purposes.

polypide [G *Polypid*]: One of two sections (cystid, polypide) of single bryozoan individual (zooid). Consists of internal organs including lophophore and digestive tract, as well as muscular and nervous systems.

pore chamber: dietella

porta: Skeletal (zooecium) structure: anterior part of orifice.

poster [G *hinterer Deckelabschnitt*]: Skeletal (zooecium) structure: small, tab-like extension of lid (operculum) fitting into proximal slit or notch (sinus) of orifice. May be hinged to main part (anter) of operculum.

primary zooid [G *Primärzoid*]: ancestrula

proancestrula [G *Proancestrula*]: In certain bryozoans, hemispherical cell mass derived from metamorphosis of free-swimming larva. Gives rise to first individual (ancestrula) and to basal part of future colony.

protocoel [G *Protocoel*]: coelom

pseudopore [G *Pseudoporus*]: Skeletal (zooecium) structure: pore-shaped space in calcified layer of frontal wall. Filled with tissue and not penetrating entire frontal wall.

pseudosinus [G *Pseudosinus*]: Skeletal (zooecium) structure: notch or hole in elevated rim (orificial collar, peristome) surrounding orifice.

pseudospiramen [G *Pseudospiramen*]: Skeletal (zooecium) structure: asymmetrical sinus on proximal part of extended rim (peristome) of orifice.

pylorus [G *Pylorus*]: In digestive tract, relatively narrow, ciliated posterior section of stomach. Opens into intestine.

rectum [G *Enddarm, Rektum*]: In digestive tract, term referring either to intestine or to terminal part of intestine near anus.

retractor muscle [G *Retraktor, Retraktormuskel, Rückziehmuskel*]: One in a series of muscles spanning from base of lophophore and other internal organs to zooid wall. Retracts body (polypide).

rhizoid (radicular fiber) [G *Rhizoid, Wurzelfaden*]: In certain erect colonies, root-like anchoring structure composed of modified tubular zooids (kenozooids).

rimula (rimule): Skeletal (zooecium) structure: notch, representing ascus opening (sinus), in extended rim (orificial collar, peristome) of orifice.

rosette plate [G *Rosettenplatte*]: Skeletal (zooecium) structure: in walls between two individuals (zooids), small, thin, circular area bearing numerous minute pores.

rostrum [G *Rostrum*]: In type of zooid (avicularium) modified to protect colony, thickened, rigid structure against which movable mandible closes.

sclerite [G *Sklerit*]: Skeletal (zooecium) structure: ridge-like thickening of 1) lid (operculum) covering orifice, 2) mandible, or 3) frontal membrane.

scutum [G *Scutum, Fornix*]: Skeletal (zooecium) structure: spine projecting from margin of zooecium over frontal membrane. Typically flattened, branched, and forming protective shield.

secondary orifice [G *sekundäre Apertur*]: Skeletal (zooecium) structure: orifice at terminal end of extended orificial collar (peristome).

segment: internode

septulum: Perforation in wall between two zooids; associated with several cell types functioning in interzooidal communication. (multiporous, uniporous)

septum [G *Septum*]: Skeletal (zooecium) structure: interior wall or partition not associated with cuticle, for example, transverse vertical wall.

sessoblast [G *sessiler Statoblast, sitzender Statoblast, Sessoblast*]: Type of reproductive bud (statoblast) which sinks to or is cemented to substrate; annulus typically rudimentary, lacking marginal hooks or spines. (See also floatoblast, spinoblast)

seta (flagellum) [G *Borste, Geißel, Mandibel*]: In vibraculum, highly elongated, freely movable bristle representing modified operculum. (flexible, stiff; distal end: smooth, toothed on one/on both sides)

setiferous organ [G *Fühlknopf*]: In type of zooid (avicularium) modified to protect colony, reduced internal body (polypide). Consists of protuberance bearing tuft of long, stiff bristles and may serve as sense organ.

sinus [G *Sinus*]: Skeletal (zooecium) structure: slit or notch in proximal margin of orifice. Serves as opening for compensation sac (ascus). (broad, shallow; pointed, rounded, V-shaped)

spinoblast [G *Spinoblast*]: Type of reproductive bud (statoblast) which floats. Annulus well developed, with marginal hooks or spines. (See also floatoblast, sessoblast)

spinula: Skeletal (zooecium) structure: one in a series of simple projections ("false spines") along margin of cryptocyst. Lacks inner canal.

spiramen [G *Spiramen*]: Skeletal (zooecium) structure: hole in extended rim (orificial collar, peristome) of orifice. Formed by closure of notch (rimula).

statoblast [G *Statoblast*]: Asexually produced reproductive bud in freshwater bryozoan. Consists of disc-shaped capsule and peripheral ring (annulus). (See also floatoblast, sessoblast, spinoblast)

sternum: Skeletal (zooecium) structure: central separation between spines (costae) converging from each side of zooid.

stictocyst: Skeletal (zooecium) structure: type of wall consisting of peritoneum, epidermis, calcareous layer, and cuticle (the latter pierced by numerous pseudopores). (See also coelocyst)

stolon [G *Stolo*]: Colony (zoarium) structure: slender tube of modified individuals (kenozooids) bearing feeding individuals (autozooids) along its length.

stomach [G *Magen*]: Section of digestive tract between esophagus and intestine. Occupies turn in U-shaped tract and variously subdivided into cardia, stomach sac, and pylorus. Bears cecum.

sulcus: Skeletal (zooecium) structure: groove delineating boundary between adjacent individuals (zooids).

supraneural pore: coelomopore

tata ancestrula (tatiform) [G *Tata-Form*]: Type of ancestrula with circle of spines around orifice or frontal membrane.

tentacle [G *Tentakel*]: One in a series of hollow, ciliated projections arising from anterior end of lophophore.

tentacle sheath [G *Tentakelscheide*]: Free (unattached to skeleton) anterior part of body wall. Introverts to enclose tentacles in their retracted position.

termen: Skeletal (zooecium) structure: calcareous edge surrounding frontal area. Continuous with, yet differing in structure from gymnocyst. Typically bearing marginal spines.

terminal membrane [G *Terminalmembran*]: In certain bryozoans with cylindrical zooids, uncalcified distal portion of zooid which bears orifice. Considered to be homologous to frontal membrane.

testis [G *Hoden*]: ♂ reproductive organ consisting of one to several cell aggregations attached to body wall, digestive tract, or funiculus.

trabecula [G *Trabekel*]: Colony (zoarium) structure: zooid-bearing branch separating open spaces (fenestrulae) in reticulate colony.

tremocyst [G *Tremozyste*]: Skeletal (zooecium) structure: outermost of three layers (olocyst, pleurocyst, tremocyst) in calcified front of zooid with ascus. Bears pseudopores.

umbo [G *Umbo*]: Skeletal (zooecium) structure: blunt prominence on frontal wall or ovicell.

vanna: Skeletal (zooecium) structure: posterior margin of orifice where compensation sac (ascus) opens.

vestibular gland (opercular gland, oral gland): Pair of glands opening into vestibule close to diaphragm. (flask-shaped, globular, multilobular, tubular)

vestibule: atrium

vibraculum [G *Vibracularium, Vibrakular*]: In colony (zoarium), modified individual (zooid) in which operculum is developed as highly elongated, freely movable bristle (seta).

zoarium (colony) [G *Zoarium, Bryozoenstock, Kolonie*]: Group of adjoining individual zooids that intercommunicate with each other and stem from a common zooid (ancestrula) by asexual reproduction. (calcified, not calcified: fleshy, gelatinous; adnate, creeping, encrusting, erect, frondose, unilaminar; biserial, multiserial; jointed, unjointed; adeoniform, celleporiform, idmoneiform = idmidroneiform, lunulitiform, membraniporiform, petraliiform, pustuliporiform, reteporiform, setoselliniform, stomatoporiform, tubuliporiform, vinculariiform)

zooecium (zoecium) [G *Zoecium*]: Skeletal outer structure encasing single bryozoan individual (zooid).

zooid [G *Zoid, Zooid*]: Single bryozoan individual, encased in zooecium and typically adjoined by and connected with other zooids to form colony (zoarium). Generally described as consisting of outer section or cystid (zooecium and underlying body wall layers) and internal section or polypide (lophophore, digestive tract, nervous and muscular systems). (bottle-shaped, box-like, cylindrical, hexagonal, oval, polygonal, rectangular, tubular, vase-like) (See also autozooid, gonozooid, heterozooid, kenozooid, maternal zooid, nanozooid)

PHORONIDA

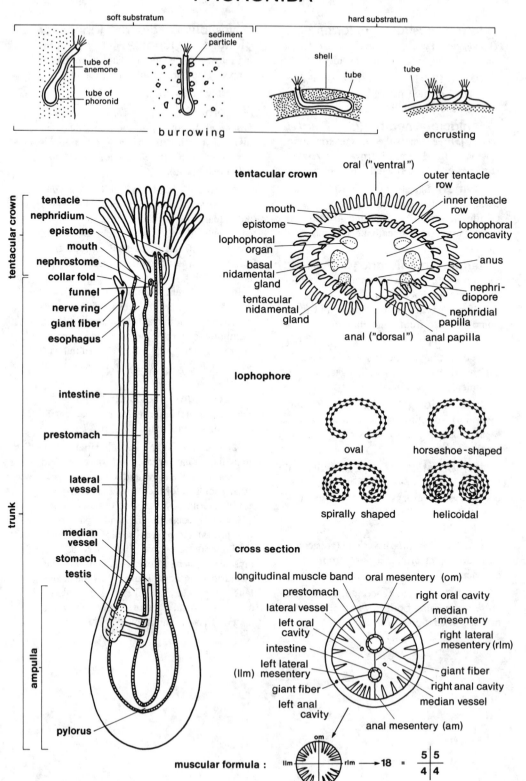

soft substratum

tube of anemone
tube of phoronid

sediment particle

hard substratum

shell
tube

tube

b u r r o w i n g

encrusting

tentacular crown

oral ("ventral")
outer tentacle row
inner tentacle row
mouth
epistome
lophophoral organ
basal nidamental gland
lophophoral concavity
anus
nephridiopore
nephridial papilla
tentacular nidamental gland
anal ("dorsal")   anal papilla

tentacular crown

tentacle
nephridium
epistome
mouth
nephrostome
collar fold
funnel
nerve ring
giant fiber
esophagus

intestine

prestomach

lateral vessel

median vessel
stomach
testis

trunk

ampulla

pylorus

lophophore

oval   horseshoe-shaped

spirally shaped   helicoidal

cross section

longitudinal muscle band
prestomach
lateral vessel
left oral cavity
intestine
left lateral (llm) mesentery
giant fiber
left anal cavity

oral mesentery (om)
right oral cavity
median mesentery
right lateral mesentery (rlm)
giant fiber
right anal cavity
median vessel
anal mesentery (am)

muscular formula :   llm —— rlm → 18 = $\dfrac{5}{4}\Big|\dfrac{5}{4}$

LOPHOPHORATA (TENTACULATA)

PHORONIDA

phoronid [G *Phoronide*]

accessory mesentery [G *Neben-mesenterium, akzessorisches Mesenterium*]: One of two longitudinal septa, additional to oral and right lateral mesenteries, extending from prestomach to body wall in trunk.

accessory reproductive gland [G *Fortpflanzungsnebendrüse, akzessorische Fortpflanzungsdrüse*]: General term applied to all structures aiding in reproduction, including lophophoral organ and both basal and tentacular nidamental glands.

accessory spermatophoral organ: lophophoral organ

afferent lophophoral ring vessel [G *aufsteigendes Lophophorringgefäß, afferentes Lophophorringgefäß, zuführendes Lophophorringgefäß*]: Section of lophophoral vessel receiving blood from median vessel and sending a single vessel into each tentacle. (See also efferent lophophoral ring vessel)

ampulla (end bulb) [G *Ampulle*]: Enlarged posterior region of trunk containing stomach and initial expanded section of intestine.

anal funnel [G *Analtrichter*]: In phoronid with two openings (nephrostomes) of nephridium into metacoel, expanded, funnel-shaped section of descending nephridial

branch opening closer to anus. (See also oral funnel)

anal mesentery [G *Analmesenterium*]: Longidutinal septum extending from intestine to body wall in trunk opposite mouth.

anal papilla [G *Analpapille, Afterpapille*]: Elevated, more or less well-developed posteriormost section of digestive tract consisting of tall cells and located opposite mouth on tentacular crown; forms rectum and bears terminal anus.

anus [G *After*]: Posterior opening of U-shaped digestive tract opposite mouth on tentacle crown; positioned in certain phoronids on anal papilla.

ascending nephridial branch [G *aufsteigender Nephridienkanal*]: Long, thick, ascending section of U-shaped nephridium opening to exterior on tentacle crown via nephridiopore. (See also descending nephridial branch)

basal nidamental gland [G *basale Nidamentaldrüse, Basalbrutdrüse*]: Pair of accessory reproductive glands, one on each side of lophophoral concavity between tentacular nidamental gland and lophophoral organ.

brain: nerve ring

buccal groove [G *Bukkalrinne*]: Ciliated groove extending between inner and outer tentacle rows from both sides of mouth.

buccal tube [G *Mundhöhle, Bukkalgrube*]: Term applied to short, ciliated anterior section of digestive tract—bounded on one side by epistome—between mouth and esophagus.

capillary cecum [G *Gefäßcoecum, Kapillarcoecum, Coecalgefäß*]: One of numerous short, simple, occasionally branched, highly contractile blind tubes projecting into metacoel from lateral blood vessel; more numerous posteriorly in ampulla region.

ciliated strip [G *Wimperrinne*]: Narrow band of strongly ciliated cells in prestomach (corresponding to line of attachment of median mesentery = "anal" side) and extending into stomach.

coelom [G *Coelom*]: mesocoel, metacoel, protocoel

collar fold [G *Ringfalte*]: In certain phoronids, an anteriorly directed, collar-like fold of outer body wall marking border between tentacular crown and trunk; corresponds internally with diaphragma.

descending nephridial branch [G *absteigender Nephridienkanal*]: Short, thin section of U-shaped nephridium descending posteriorly into trunk and terminating in one or two nephrostomes or funnels. (See also ascending nephridial branch)

diaphragm [G *Diaphragma*]: Oblique septum perpendicular to longitudinal axis, separating mesocoel from metacoel; traversed by esophagus, descending nephridial branch, and both lateral and median blood vessels. Corresponds externally to collar fold at tentacular crown/trunk border.

dorsal vessel: median vessel

efferent lophophoral ring vessel [G *absteigendes Lophophorringgefäß, efferentes Lophophorringgefäß, ableitendes Lophophorringgefäß*]: Section of lophophoral vessel conveying blood collected from tentacles to posterior end of body via lateral vessel. (See also lophophoral ring vessel)

end bulb: ampulla

epistome [G *Epistom, Mundklappe*]: Median crescentic flap or "upper lip" between mouth and inner tentacle row and projecting over the former. Cavity within epistome is considered to represent protocoel.

epithelial nervous plexus [G *epitheliales Nervennetz, epithelialer Nervenplexus*]: Fine nerve net continuous with nerve ring and extending throughout body wall.

esophagus [G *Ösophagus*]: Thick-walled, ciliated, tubular descending section of U-shaped digestive tract between mouth and prestomach.

funnel [G *Trichter*]: Funnel-shaped expansion surrounding opening(s) (nephrostome) of nephridium into metacoel. (See also anal funnel, oral funnel)

giant fiber [G *Kolossalfaser*]: left giant fiber, right giant fiber

gonad: ovary, testis

inner tentacle row (internal tentacle row) [G *innere Tentakelreihe*]: Tentacles originating from inner ridge of tentacular crown; continuous with and parallel to outer tentacle row and, relative to mouth, convex in configuration.

intestine [G *Intestinum, Dünndarm*]: Initially somewhat expanded, then tubular ascending arm of U-shaped digestive tract between stomach and rectum; constricted junction of stomach and intestine (pylorus) located at posterior end of trunk.

lateral vessel (ventral vessel) [G *Lateralgefäß*]: Longitudinal blood vessel in left oral cavity conveying blood from efferent lophophoral vessel posteriorly to stomach region, where it is connected with median vessel; bears capillary ceca. If two lateral vessels are present, one distinguishes a left and right lateral vessel.

left anal cavity [G *linke Analhöhle*]: Longitudinal compartment in trunk delimited by left lateral mesentery, intestine, and anal mesentery.

left giant fiber (left lateral nerve)
[G *linker Lateralnerv*]: Giant nerve fiber originating on right side of preoral nervous field in nerve ring and running along left side of trunk at junction of left lateral mesentery and body wall.

left lateral mesentery [G *linkes Lateralmesenterium*]: Longitudinal septum
extending from intestine to left body wall in trunk; line of attachment to body wall corresponds with position of left giant fiber.

left oral cavity [G *linke Oralhöhle, linke Oralkammer*]: Longitudinal compart-
ment in trunk delimited by oral mesentery, prestomach, median mesentery, intestine, and left lateral mesentery; contains lateral vessel.

longitudinal muscle band/fold/ridge [G *Längsmuskelband*]: Prominent
aggregation of muscle into bundle projecting into trunk. Number and distribution are of taxonomic importance. (See also muscular formula)

lophophoral concavity [G *Lophophorkonkavität, Lophophorbucht*]: Sur-
face of tentacle-bearing anterior end enclosed by inner tentacle row and also bordered in certain phoronids by anal and nephridial papillae; may bear accessory reproductive glands laterally.

lophophoral organ (accessory spermatophoral organ) [G *Lopho-
phororgan*]: Pair of glandular depressions or thickenings, one on each side of lophophoral concavity; each is connected to corresponding nephridiopore via ciliated groove.

lophophoral vessel [G *Lophophoralgefäß, Lophophorgefäß, Gefäßring*]:
Complex, horseshoe-shaped "ring" blood vessel in mesosome. (See also afferent lophophoral ring vessel, efferent lophophoral ring vessel)

lophophore [G *Lophophor*]: Ante-
rior fold of body wall forming two parallel, continuous ridges enclosing mouth and bearing numerous hollow, ciliated tentacles (inner and outer tentacle rows); may also refer only to ridges from which tentacles originate. (crescent-shaped, helicoidal, horseshoe-shaped, oval, spirally shaped)

lophophore coelom: mesocoel

median mesentery [G *Median-
mesenterium*]: Longitudinal septum extending between prestomach and intestine in trunk.

median vessel (dorsal vessel, afferent vessel) [G *Mediangefäß, af-
ferentes Gefäß, zuführendes Gefäß, Dorsalgefäß*]: Longitudinal blood vessel between limbs of digestive tract (prestomach, intestine) in right anal cavity; connected to lateral vessel in stomach region by network of vessels and conveying blood anteriorly into tentacle crown.

mesentery [G *Mesenterium*]: One of
a number of thin, longitudinal membranes (anal, median, oral, left lateral, right lateral mesenteries) extending between limbs of U-shaped digestive tract or joining digestive tract to body wall in trunk; divides metacoel into compartments.

mesocoel (mesocoelom, lophophor coelom) [G *Mesocoel, Lophophor-
coelom*]: Second largest of three body cavities (protocoel, mesocoel, metacoel) consisting of cavities in lophophore and canals within tentacles; in communication with protocoel in epistome.

mesosome [G *Mesosom, Meso-
soma*]: Region of body corresponding with and enclosing mesocoel.

metacoel (metacoelom, trunk coelom) [G *Metacoel, Rumpfcoelom*]:
Posteriormost and largest of three body cavities (protocoel, mesocoel, metacoel) consisting of cavity within trunk; typically divided into a number of longitudinal compartments by mesenteries between digestive tract and body wall (left oral, right oral, left anal, right anal cavities).

metasome [G *Metasom, Meta-
soma*]: Region of body corresponding with and enclosing metacoel. Equivalent to trunk.

mouth [G *Mund*]: Cresent-shaped an-
terior opening of digestive tract between

inner and outer tentacle rows; overhung by epistome.

muscular formula [G *Muskelformel*]: Formula indicating number of longitudinal muscle bands/folds in each of the four major cavities (left oral, right oral, left anal, right anal) in trunk. Numbers range from 12 to more than 200.

nephridial papilla [G *Nephridienpapille, Nephridialpapille*]: In certain phoronids, one of two papillae on each side of anal papilla bearing nephridiopore.

nephridiopore [G *Nephroporus*]: Opening of ascending nephridial branch of nephridium to exterior; located on each side of anus, in certain phoronids on separate nephridial papillae.

nephridium [G *Nephridium*]: One of two ciliated, U-shaped tubes with descending nephridial branch opening into metacoel via funnel and with ascending nephridial branch opening to exterior via nephridiopore.

nephrostome [G *Nephrostom*]: Opening of descending nephridial branch of nephridium into metacoel; typically expanded to form funnel.

nerve ring (ring nerve, brain) [G *Ringnerv, Nervenring*]: Main concentration of nerve tissue forming a ring in epidermis of body wall around a mouth at border of tentacle crown and trunk; gives rise to nerve for each tentacle and to left and right giant fibers (lateral nerves).

nidamental gland: basal nidamental gland, tentacular nidamental gland

oral funnel [G *Oraltrichter*]: In phoronid with two openings of nephridium into metacoel, expanded, funnel-shaped section of descending nephridial branch opening into metacoel closer to mouth. (See also anal funnel)

oral mesentery [G *Oralmesenterium*]: Longitudinal septum extending from prestomach to body wall opposite anus in metacoel.

outer tentacle row [G *äußere Tentakelreihe*]: Tentacles originating from ridge forming outer perimeter of tentacle crown; continuous with and parallel to inner tentacle row.

ovary [G *Ovar, Ovarium*]: Mass of ♀ reproductive cells surrounding capillary ceca along one side of lateral vessel; testis located on other side.

preoral nervous field [G *Zerebralganglion*]: Term applied to thickest part of nerve ring in lophophoral concavity between mouth and anus.

prestomach [G *Vormagen*]: Long, weakly ciliated section of U-shaped digestive tract between esophagus and stomach; forms greater part of posteriorly directed limb of U.

protocoel (protocoelom, epistome coelom) [G *Prosoma-Coelom, Protocoel, Protocoelom, Epistomcoelom*]: Smallest and anteriormost of three body cavities (protocoel, mesocoel, metacoel) consisting of cavity within epistome; in communication with mesocoel.

protsome [G *Prosom, Prosoma*]: Part of body corresponding to and enclosing protocoel. Equivalent to epistome.

proventriculus: prestomach

pylorus [G *Ringklappe, Pylorusklappe*]: Constricted section of digestive tract between stomach and intestine in posterior end of trunk.

rectum [G *Enddarm, Rektum*]: Short terminal section of digestive tract between intestine and anus.

right anal cavity [G *rechte Analhöhle, rechte Analkammer*]: Longitudinal compartment in trunk delimited by right lateral mesentery, prestomach, median mesentery, intestine, and anal mesentery; contains median vessel.

right giant fiber (right lateral nerve) [G *rechter Lateralnerv*]: Longitudinal nerve fiber, typically smaller than left giant fiber, originating on left side of preoral nervous field and running along right side of trunk at junction of right lateral mesentery and body wall.

right lateral mesentery [G *rechtes Lateralmesenterium*]: Longitudinal septum extending from prestomach to right body wall in trunk; line of attachment to body wall corresponds with position of right giant fiber.

right oral cavity [G *rechte Oralhöhle, rechte Oralkammer*]: Longitudinal compartment in trunk delimited by oral mesentery, prestomach, and right lateral mesentery.

stomach [G *Magen*]: Enlarged section of U-shaped digestive tract between prestomach and intestine in posterior end of trunk.

tentacle [G *Tentakel*]: One in a series of hollow, basally fused, innervated and ciliated projections arising from a double ridge (inner and outer tentacle rows) on anterior end. Number and configuration in adult phoronid are of taxonomic importance.

tentacular cilium [G *Tentakelzilien*]: One in a series of cilia on tentacles; longest and most dense on surface facing mouth.

tentacular crown [G *Tentakelkrone*]: Term applied to tentacles and ridges from which tentacles originate (see lophophore). May also refer to entire tentacle-bearing anterior end comprising protsome and mesosome, as opposed to trunk.

tentacular nidamental gland [G *Tentakelbrutdrüse, Nidamentaldrüse*]: Accessory reproductive glands on inner surface of certain tentacles opposite nidamental glands.

testis [G *Hoden*]: Mass of ♂ reproductive cells surrounding capillary ceca along one side of lateral vessel; ovary located on other side.

trunk (metasome) [G *Rumpf, Metasoma*]: Posterior and largest of two externally discernible divisions (tentacular crown, trunk) of body; corresponds to metacoel. Divided internally into several longitudinal cavities by mesenteries.

tube [G *Wohnröhre*]: Translucent, chitinous tube inhabited by all phoronids; often incorporates sediment particles in soft substrata.

BRACHIOPODA

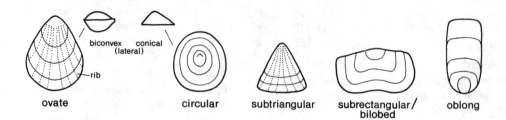

biconvex conical
(lateral)

rib

ovate

circular

subtriangular

subrectangular/
bilobed

oblong

lophophore arm
(cross section)

(lateral)

tentacle

brachial gutter

brachial fold

food groove

brachial loop

small large

lophophore canal
(coelom)

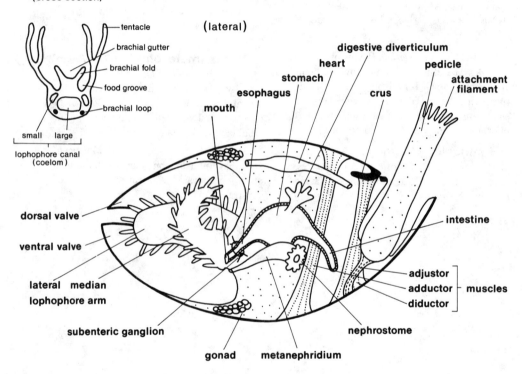

digestive diverticulum

heart

pedicle

stomach

attachment
filament

esophagus

crus

mouth

dorsal valve

intestine

ventral valve

**lateral median
lophophore arm**

adjustor
adductor muscles
diductor

subenteric ganglion

nephrostome

gonad metanephridium

ARTICULATE SHELL
(inner side)

dorsal = brachial valve

ventral = pedicle valve

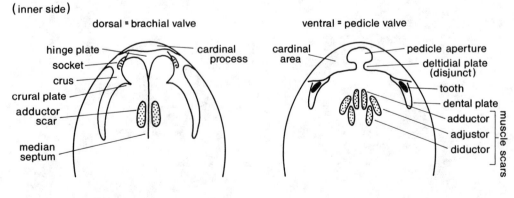

hinge plate

cardinal
process

cardinal
area

pedicle aperture

socket

deltidial plate
(disjunct)

crus

tooth

crural plate

dental plate

adductor
scar

adductor

adjustor

median
septum

diductor

muscle scars

LOPHOPHORATA (TENTACULATA)

BRACHIOPODA

brachiopod, lamp shell [G *Brachiopode, Armfüßer*]

acinus [G *Acinus*]: In digestive tract, one of numerous blind sacs of digestive diverticulum.

adductor muscle (occlusor muscle) [G *Schalenschließer*]: General term for one of several muscles spanning from dorsal to ventral valves in posterior region of shell. Serves to close valves.

adjustor muscle (pedicle adjustor muscle) [G *Stielmuskel*]: One of two pairs (dorsal and ventral) of muscles spanning between valves and stalk (pedicle). Serves to move shell relative to substratum.

anus [G *After*]: Posterior opening of digestive tract. (median; dorsal, on right side)

apex [G *Apex*]: Shell structure: first-formed part of valve, from which growth lines emanate. (central, posterior, subcentral)

arm: brachium

beak [G *Schnabel*]: Shell structure: curved posterior section of ventral valve extending beyond dorsal valve. May bear hole for pedicle.

beak ridge [G *Schnabelkante*]: Shell structure: linear ridge, one extending from each side of umbo so as to delimit all or most of cardinal area.

body cavity: coelom

brachial cavity (mantle cavity) [G *Mantelkammer*]: Cavity between two valves occupied by lophophore.

brachial fold (brachial lip): Fold running along arms of lophophore and forming one wall of food groove.

brachial groove: food groove

brachial gutter: Along median or lateral arms of plectophorous type of lophophore, broad median depression between two brachial folds.

brachial loop [G *Skelettschleife*]: Delicate, shelly, loop-like support for lophophore. Extends from crura on dorsal valve and may be further supported by median septum. (complete, incomplete)

brachial valve: dorsal valve

brachidium (brachial support) [G *Armskelett, Armgerüst*]: Entire internal skeleton supporting lophophore.

brachium [G *Lophophorenarm*]: One of up to three (two lateral, one median) arms of lophophore.

cardinal area [G *Cardinalarea*]: Shell structure: posterior section of valve exclusive of delthyrium.

cardinalia [G *Cardinalia*]: Collective term for shell structures associated with posterior margin of dorsal valve. Includes structures serving in articulation (hinge plates), for muscle attachment (cardinal process), and lophophore support (crura).

cardinal process [G *Cardinalplatte*]: Shell structure: on inner side of dorsal valve, posteromedian elevation between sockets. Fits between teeth when these are locked into sockets. Associated with diductor muscles.

cecum (caecum) [G *Kragenpapille, Epidermispapille*]: One of numerous outgrowths of mantle tissue into blind canals (endopunctae) in valves.

circumenteric connective (esophageal nerve ring) [G *Ringkommissur*]: Ring of nerve tissue surrounding esophagus and joining subenteric and supraenteric ganglia.

cirrus: tentacle

coelom [G *Coelom*]: Peritoneum-lined body cavity. In tripartite division of coelom one may distinguish a protocoel and mesocoel associated with epistome, lophophore, and esophagus region, as well as a metacoel forming the main body cavity. The latter encloses the remainder of the digestive tract, valve muscles, gonads, and metanephridia.

commissure [G *Kommissur*]: Variously developed line of contact between edges or margins of valves. (rectimarginate)

crural plate [G *Kruralfortsatz, Fortsatz des Crus*]: Shell structure: in dorsal valve, one of two plates extending medially from crura. May attach crura to inner surface of valve.

crus [G *Crus*]: Shell structure: on inner side of dorsal valve, one of two shelly processes extending from region of sockets. Projects anteriorly into base of lohophore. (See also brachial loop)

delthyrium [G *Delthyrium*]: Shell structure: median aperture in posterior margin of ventral valve. Serves wholly or par-

tially as opening for pedicle. (See also foramen, pedicle aperture)

deltidial plate [G *Deltidialplatte*]: Shell structure: on inner side of ventral valve, pair of plates projecting medially from margin of delthyrium; constricts aperture for pedicle. (conjunct, disjunct)

dental plate [G *Stützplatte, Zahnplatte*]: Shell structure: on inner surface of ventral valve, one of two plate-like structures bearing teeth (hinge teeth).

dental socket: socket

diductor muscle (divaricator) [G *Schalenöffner*]: General term for one of several muscles spanning from dorsal to ventral valve in posterior region of shell. Serves to open valves.

digestive diverticulum (digestive gland) [G *Mitteldarmdrüse, Magendrüse*]: In digestive tract, one of several, often paired digestive glands opening into stomach via one or more large ducts. (grapelike, lobate, tubular)

dorsal valve (brachial valve) [G *Rückenschild, dorsale Schalenklappe*]: valve

endopuncta [G *Porenkanal, Kragenporus, Endopuncta*]: Shell structure: one of numerous minute canals extending from inner valve surface almost to exterior. Canals accommodate mantle tissue (ceca).

enteric ganglion: subenteric ganglion, supraenteric ganglion

epistome [G *Epistom*]: Anterior body region; considered to be represented by median part of brachial fold above mouth.

esophageal nerve ring: circumenteric connective

esophagus [G *Ösophagus*]: Relatively narrow section of digestive tract between mouth and stomach.

filament: tentacle

food groove (brachial groove) [G *Nahrungsrinne*]: Groove running along inner side of lophophore. Food is transported along food groove into mouth.

foramen (pedicle foramen)
[G *Stielloch*]: Hole through posterior end of dorsal valve for pedicle. (amphithyridid, epithyridid, hypothyridid, mesothyridid, permesothyridid, submesothyridid; circular, subcircular)

gastroparietal band [G *Gastroparietalband*]: Pair of lateral bands spanning across body cavity from stomach to body wall. Supports openings (nephrostomes) of anterior pair of excretory organs (metanephridia) if two pairs are present. (See also ileoparietal band)

heart (contractile vesicle)
[G *Herz*]: In circulatory system, one or more poorly developed contractile vesicles along a longitudinal blood vessel dorsal to stomach. (See also main dorsal channel)

hinge axis [G *Schloßachse*]: Straight line about which valves open and close; located posteriorly, passing between teeth (hinge teeth).

hinge plate (socket plate)
[G *Schloßplatte*]: Shell structure: general term for supporting ridges on inner side of dorsal valve. Located to either side of sockets (dental sockets). (See also socket ridge)

ileoparietal band [G *Ileoparietalband*]: Pair of lateral mesenteries spanning across body cavity from digestive tract to body wall. Supports openings (nephrostomes) of excretory organs (metanephridia) and may bear gonads. (See also gastroparietal band)

intestine [G *Darm, Dünndarm*]: Relatively narrow section of digestive tract following stomach. Ends blindly or opens to exterior via rectum and anus. (with/without expansion)

liver: digestive diverticulum

loop: brachial loop

lophophore [G *Lophophor, Tentakelträger*]: Anterior fold of body wall. Variously shaped, bearing series of tentacles and serving as feeding organ. (plectolophous, ptycholophous, schizolophous, spirolophous, trocholophous, zugolophous = zygolophous)

main dorsal channel (middorsal blood channel) [G *Rückengefäß*]: In circulatory system, middorsal blood channel bearing heart(s) and extending along digestive tract. Splits into left and right branches both anteriorly (at level of esophagus) and posteriorly.

mantle [G *Mantel*]: Fleshy fold of body wall which lines brachial cavity and secretes valves.

mantle canal (mantle sinus)
[G *Coelomkanal*]: One in a series of flattened, tube-like canals extending from main body cavity (metacoel) into mantle. May leave impressions on inner surface of valves. Branches of canals termed vascula.

mantle cavity: brachial cavity

median septum [G *medianes Septum, Medianseptum*]: Shell structure: longitudinally oriented, vertical plate along inner surface of either dorsal or ventral valve. May support brachial loop.

mesocoel [G *Mesocoel, Hydrocoel*]: coelom

metacoel [G *Metacoel, Somatocoel*]: coelom

metanephridium [G *Nephridium*]: Paired excretory organ consisting of nephrostome, tubular section, and opening to exterior (nephridiopore). (one pair, two pairs)

mosaic [G *Mosaik*]: Type of shell sculpture: minute mosaic pattern on inner surface of valve. Formed by structure of inner valve layer.

mouth [G *Mund, Mundöffnung*]: Anterior, transversely elongated opening of digestive tract. Located at base of lophophore and leads into esophagus.

nephridiopore [G *Nephroporus*]: Small opening of each excretory organ (metanephridium) to exterior. Located on anterior body wall, close to midline and ventral to mouth.

nephridium: metanephridium

nephrostome [G *Nephrostom*]: Broad, funnel-shaped opening of each excretory organ (metanephridium) into body cavity (coelom); associated with lateral mesenteries (ileoparietal bands).

occlusor muscle: adductor muscle

ovary [G *Ovar, Ovarium, Eierstock*]: ♀ reproductive organ. Develops on wall of body cavity, on mesenteries (e.g., ileoparietal bands), or in mantle canals. Eggs are discharged through metanephridia.

palintrope [G *Palintrop*]: Shell structure: recurved, anteriorly directed part of ventral valve between beak and hinge line.

pedicle (peduncle, stalk) [G *Stiel*]: Fleshy, cuticle-covered stalk by which brachiopod attaches to or anchors itself in substratum. Emerges through ventral valve via pedicle aperture or foramen.

pedicle adjustor muscle: adjustor muscle

pedicle aperture [G *Stielloch*]: Shell structure: opening, typically in ventral valve, through which stalk (pedicle) extends.

pedicle foramen: foramen

pedicle valve [G *Stielklappe*]: ventral valve

peduncle: pedicle

periostracum [G *Periostracum*]: Relatively thin, organic outermost layer of each valve.

protegulum [G *Protegulum*]: First minute valves secreted by juvenile brachiopod. Differs from subsequently produced shell by absence of sculpture.

pustule [G *Pustel*]: Type of shell sculpture: small, nodose protuberance on outer surface of valve.

rectum [G *Enddarm, Rectum*]: Terminal section of digestive tract between intestine and anus.

rib [G *Rippe*]: Type of shell sculpture: one in a series of more prominent radial ribs along outer surface of valve. (See also stria)

shell [G *Schale*]: Entire calcareous structure, consisting of dorsal and ventral valves enclosing soft body.

socket (dental socket) [G *Gelenkgrube*]: Shell structure: one of two deep indentations located posteriorly on inner side of dorsal valve. Receives hinge tooth of ventral valve.

socket ridge [G *Gelenkplatte*]: Shell structure: on inner side of dorsal valve, one of two ridges bordering inner side of sockets (dental sockets).

spicule [G *Spiculum, Spikel, Kalksklerit*]: One of numerous variously shaped, calcareous plates helping to support lophophore tissue.

spine [G *Stachel*]: Type of shell sculpture: more elongate projection on outer surface of valve. May be formed at intersection of growth ridges and radiating ridges.

statocyst [G *Statozyste*]: In certain brachiopods, pair of equilibrium organs on mesenteries (gastroparietal bands) near anterior adductor muscles. Consists of sac containing statoliths.

statolith [G *Statolith*]: One of approximately 30 minute, solid bodies in each statocyst.

stomach [G *Magen*]: Expanded section of digestive tract between esophagus and intestine. Digestive diverticula open into stomach.

stria [G *Stria, Streife*]: Type of shell sculpture: one in a series of very fine, radial ridges along outer surface of valve.

subenteric ganglion [G *Unterschlundganglion*]: In nervous system, larger ganglion located on ventral side of esophagus. Joined to smaller supraenteric ganglion by circumenteric connectives.

sulcus [G *Sulcus, Furche*]: Shell structure: major depression of valve surface. May lead to sinuous anterior commissure.

supraenteric ganglion [G *Supraösophagealganglion*]: In nervous system, smaller ganglion located on dorsal side of esophagus. Joined to larger subenteric ganglion by circumenteric connectives.

tentacle (cirrus, filament) [G *Cirrus, Tentakel*]: One in a series of up to several hundred hollow, ciliated projections arising from arms (brachia) of lophophore.

testis [G *Hoden*]: ♂ reproductive organ. Develops on wall of body cavity, on mesenteries (e.g., ileoparietal bands), or in mantle canals. Sperm are discharged through metanephridia.

tooth (hinge tooth) [G *Zahn, Gelenkkopf*]: Shell structure: one of two toothlike projections located posteriorly on inner side of ventral valve. Fits into socket of dorsal valve.

tubercle [G *Tuberkel*]: Type of shell sculpture: large, nodose protuberance on outer surface of valve. Typically associated with ribs. (See also pustule)

umbo [G *Umbo*]: Shell structure: median, apical posterior region of either valve.

valve [G *Klappe, Schalenklappe*]: One of two shell elements secreted by mantle and covering dorsal and ventral sides of soft body (therefore dorsal and ventral valves). Valves joined together posteriorly either by hinge and muscles or by muscles alone. (outline: elongate, oblong, ovate, subcircular, subquadrate, subrectangular, subtriangular, transversely oval, triangular; biconvex, circular, conical, globose; impunctate, endopunctate; chitinophosphatic, calcareous; sculpture: ribbed, smooth, striated, tuberculate, with pustules)

vasculum [G *Vasculum*]: Any one of several branches of mantle canal system.

ventral valve (pedicle valve) [G *Bauchklappe, Bauchschild, Stielklappe*]: valve

CHAETOGNATHA

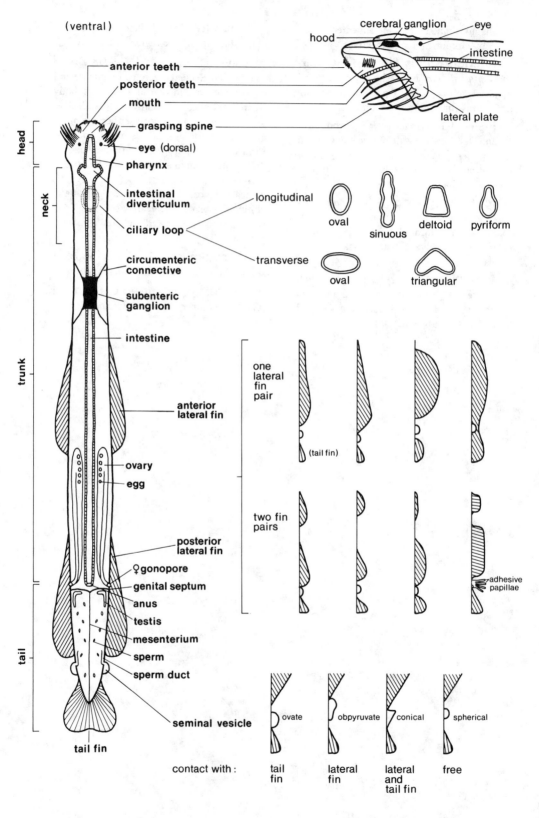

HEAD (lateral)

(ventral)

cerebral ganglion — eye

hood

intestine

anterior teeth

posterior teeth

mouth

grasping spine

lateral plate

head

eye (dorsal)

pharynx

neck

intestinal diverticulum

ciliary loop

longitudinal

oval   sinuous   deltoid   pyriform

transverse

oval   triangular

circumenteric connective

subenteric ganglion

intestine

trunk

one lateral fin pair

anterior lateral fin

(tail fin)

ovary

egg

two fin pairs

posterior lateral fin

♀ gonopore

genital septum

anus

testis

adhesive papillae

mesenterium

tail

sperm

sperm duct

seminal vesicle

ovate   obpyruvate   conical   spherical

tail fin

contact with :   tail fin   lateral fin   lateral and tail fin   free

CHAETOGNATHA (HOMALOPTERYGIA)

chaetognath, arrowworm [G *Chaetognath, Pfeilwurm*]

accessory fertilization cell [G *akzessorische Zelle, Zweizellenapparat, Aufhängeapparat*]: Term applied to two cells from inner wall of oviduct which allow fertilization by forming hollow connection between egg in ovary and seminal receptacle.

adhesive papilla [G *Anheftungs-papille*]: In certain chaetognaths, one in a series of adhesive structures on ventral surface, at beginning of tail; in part on finger-like projections behind posterior lateral fins. Also refers to frontal glandular structures in juveniles of these chaetognaths; used to adhere to substratum.

anterior lateral fin [G *Vorderflosse, vordere Seitenflosse*]: In chaetognath bearing two pairs of lateral fins, anterior fins located on each side of trunk.

anterior teeth [G *vordere Zahn-reihe*]: One of two rows of short teeth, typically 3 to 10 in number, on each side of anterior tip of head. (See also posterior teeth)

anus [G *After*]: Posterior opening of digestive tract located midventrally at trunk/tail border (immediately anterior to genital septum) between posterior lateral fins.

base [G *Basis*]: Expanded proximal section of grasping spines serving as attachment point for muscles. (See also shaft, tip)

brain: cerebral ganglion

bristle receptor (sensory bristle) [G *Tastorgan, Tasthügel, Tasthaarbündel*]: Mechanoreceptive sensory structure consisting of two cell types: cilia-bearing cells (neuromasts), whose cilia are fused to form bristle, and supporting cells forming prominence; typically arranged in longitudinal rows along trunk. Also refers to cells (neurocristae) in which cilia are arranged in fan-like configuration; scattered on trunk.

caudal fin: tail fin

caudal septum: genital septum

cement gland [G *Kittdrüse*]: In certain chaetognaths, glands surrounding vagina and gonopore of ♀; secretes adhesive coat and stalk for each egg laid.

cerebral ganglion (brain) [G *Zerebralganglion, Gehirn*]: Main concentration of nerve tissue dorsally in anterior region of head. Innervates eyes and ciliary loop; connected to vestibular and subenteric ganglia.

ciliary loop (corona, corona ciliata) [G *Corona, Wimperschleife*]: Variously elaborated, typically longitudinally oriented, closed tract of cilia extending dorsally from retrocerebral pore on head over various distances along neck or trunk.

ciliary pit [G *Wimpergrube*]: Sensory organs distributed over entire body of certain chaetognaths; consists of elongated cells embedded in epidermis and bearing numerous independent, undulating hairs.

circumenteric connective (circumesophageal connective) [G *Schlundkonnektiv*]: One of two strands of nerve tissue running laterally and posteriorly from cerebral ganglion to subenteric ganglion.

collarette [G *Collarette, Halsleiste*]: Multilayer, thickened epidermis posterior to head on both sides of neck; may extend along trunk as far as lateral fins or tail fin.

corona: ciliary loop

coronal nerve [G *Coronanerv*]: One of two strands of nerve tissue running posteriorly from cerebral ganglion to ciliary loop.

cupola [G *Cupola*]: Term applied to each of typically five fused pigment cups constituting eye.

cuticle [G *Kutikula*]: Thin, noncellular layer covering ventral areas of head and extending into vestibule.

epidermis [G *Epidermis*]: Two- to three-layer body cover; differentiated on both sides of neck into a multilayer stratum (collarette).

esophagus: pharynx

eye [G *Auge*]: One of two photoreceptive organs beneath epidermis on posterior region of head; each typically consists of five partially fused pigment-cup ocelli (cupolae).

fin ray [G *Flossenstrahl*]: Delicate, ray-like supports of lateral fin and tail fin.

frontal connective [G *Frontalnerv*]: One of two strands of nerve tissue running anteriorly from cerebral ganglion to vestibular ganglion.

genital septum (caudal septum, trunk septum, tail septum) [G *Genitalseptum*]: Mesodermal septum protruding lateroventrally into body cavity (coelom) and thus separating trunk coelom from tail coelom without interrupting longitudinal muscle.

gonopore [G *Geschlechtsöffnung*]: Opening of reproductive system to exterior; in ♀, a pair of lateral openings, one on each side of anus at trunk/tail border. For ♂, see seminal vesicle.

grasping spine (spine, prehensile spine, seizing jaws) [G *Greifhaken*]: One in a series of 4 to 14 large, curved, movable, chitinous spines forming two rows, one on each side of head; each spine consists of expanded base, curved shaft, and distinct tip.

head [G *Kopf*]: Anteriormost division of body (head, trunk, tail) bearing cerebral ganglion, eyes, mouth, and grasping spines.

head septum [G *Kopfseptum*]: Mesodermal septum separating body cavity (coelom) and musculature of head and trunk.

hood [G *Kapuze, Kopfkappe*]: Fold in body wall encircling head; attached ventrally at head/trunk border, extending dorsally almost to anterior tip. Covers grasping spines when pulled forward.

intestinal diverticulum [G *Darmdivertikel, Darmblindsack*]: One of two lateral diverticula of intestine extending anteriorly from pharynx/intestine border to head/trunk border.

intestine [G *Darm, Mitteldarm*]: Long, straight section of digestive tract between pharynx and anus; may send pair of intestinal diverticula forward to head/trunk border.

lateral fin [G *Seitenflosse*]: One of up to two pairs of thin, horizontal epidermal extensions supported by fin rays; first pair (anterior lateral fins) typically on trunk, second pair (posterior lateral fins) typically extending across trunk/tail border. Number, position, and shape are of taxonomic importance.

lateral plate [G *Lateral spange, Seitenplatte*]: One of two posteriorly widening lateral plates between cuticle and epidermis in head; extends from tip of head to head/trunk border and serves as support for teeth and grasping spines as well as for muscle attachment.

longitudinal muscle [G *Längsmuskel*]: One of four thick, longitudinal muscle bands (two dorsolateral, two ventrolateral) spanning from head/trunk border to tail fin in body wall.

mesenterium (longitudinal septum) [G *Längsseptum*]: Longitudinal, double-walled septum formed by medial walls of body cavities (coeloms). Extends above and below gut (e.g., in trunk region) and is uninterrupted in tail region.

mouth [G *Mund*]: Anterior opening of digestive tract on underside of head at base of vestibule.

neck [G *Hals*]: Narrowed section of body immediately posterior to head and often bearing lateral collarette.

ovary [G *Ovar*]: Elongate, tubular, egg-bearing section of ♀ reproductive system, one tube on each side of intestine in posterior region of trunk.

oviduct [G *Ovidukt*]: Elongate, flattened section of ♀ reproductive system running along ovary between ovary and body wall; contains a second tube constituting seminal receptacle.

pharyngeal bulb [G *Pharynx*]: Expanded, bulbous posterior section of pharynx.

pharynx (esophagus) [G *Pharynx, Ösophagus*]: Muscular section of digestive tract between mouth and intestine; expanded posteriorly to form pharyngeal bulb.

posterior lateral fin [G *hintere Seitenflosse*]: lateral fin

posterior teeth [G *hintere Zahnreihe*]: One of two curved rows of teeth, occasionally exceeding 30 in number, on each side of vestibule. (See also anterior teeth)

rectum [G *Rektum*]: Term applied to histologically differentiated posteriormost section of intestine.

retrocerebral organ [G *Retrocerebralorgan, retrocerebrale Drüse*]: Pair of sacs embedded in but separate from posterior part of cerebral ganglion; opens to exterior through retrocerebral pore.

retrocerebral pore [G *Mündung des Retrocerebralorgans*]: Common dorsal opening of two sacs (retrocerebral organ) to exterior behind cerebral ganglion.

seminal receptacle [G *Receptaculum seminis*]: Tube-shaped, posteriorly expanded section of ♀ reproductive system within oviduct; opens to exterior via short vagina and gonopore.

seminal vesicle [G *Samenblase, Vesicula seminalis*]: Variously elaborated pair of lateral projections of body wall, one on each side of tail between lateral fin and tail fin; section of ♂ reproductive system into which mature sperm are transferred via sperm duct and formed into spermatophore. Shape is of taxonomic importance.

sensory bristle: bristle receptor

septum: genital septum, head septum, mesenterium

shaft [G *Schaft*]: Curved section of grasping spine between base and tip; typically wedge-shaped in cross section.

sperm duct [G *Vas deferens*]: Section of ♂ reproductive system between testes and seminal vesicle; extends midlaterally in body wall on each side of tail and also opens into body cavity of tail where sperm undergo maturation.

spine: grasping spine

subenteric ganglion (ventral ganglion) [G *Ventralganglion, Bauchganglion*]: Conspicuous midventral concentration of nerve tissue in anterior region of trunk; connected to cerebral ganglion by circumenteric connectives. Gives rise to numerous nerves extending posteriorly to trunk and tail.

tail [G *Schwanz*]: Posteriormost division of body (head, trunk, tail) posterior to anus and bearing tail fin and seminal vesicles; separated from trunk by internal transverse genital septum.

tail fin (caudal fin) [G *Schwanzflosse*]: Thin, unpaired horizontal extension of epidermis on tail; supported by fin rays. (See also lateral fin)

tentacle [G *Tentakel*]: Pair of tentacle-like projections of hood, one on each side of head, in one species.

testes [G *Hoden*]: Band-shaped section of ♂ reproductive system, running along each side of body cavity in anterior region of tail; joined to seminal vesicle by sperm duct.

tip [G *Endspitze*]: Short, pointed distal section of grasping spine; set into end of shaft. (See also base)

tooth: anterior teeth, posterior teeth

trunk [G *Rumpf*]: Second and largest division of body (head, trunk, tail) bearing one or two pairs of lateral fins and separated from both head and tail by internal septum.

vagina [G *Vagina*]: Short section of ♀ reproductive system between gonopore and seminal receptacle.

ventral ganglion: subenteric ganglion

ventral plate [G *Ventralspange*]: One of two triangular plates located ventrally between cuticle and epidermis in head; serves for muscle attachment.

vestibular ganglion [G *Vestibularganglion*]: Concentration of nerve tissue, one on each side of mouth; connected to cerebral ganglion by pair of frontal connectives.

vestibular organ [G *Vestibularorgan*]: Paired chemoreceptive sense organ of transversely arranged papillae behind posterior rows of teeth.

vestibular pit [G *Vestibulargrube*]: Round or oval, histologically differentiated, glandular concavity in vestibulum, one on each side of mouth behind posterior teeth.

vestibule [G *Vestibulum*]: Large depression on underside of head leading to mouth.

ECHINODERMATA

CRINOIDEA

crinoid, sea lily, feather star [G *Crinoide, Seelilie, Haarstern*]

CRINOIDEA

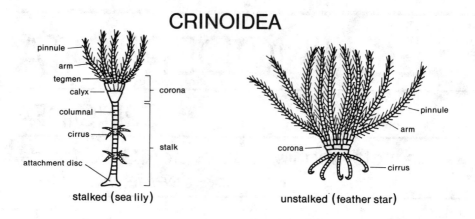

pinnule
arm
tegmen
calyx — corona
columnal
cirrus — stalk
attachment disc

stalked (sea lily)

pinnule
arm
corona
cirrus

unstalked (feather star)

(lateral)

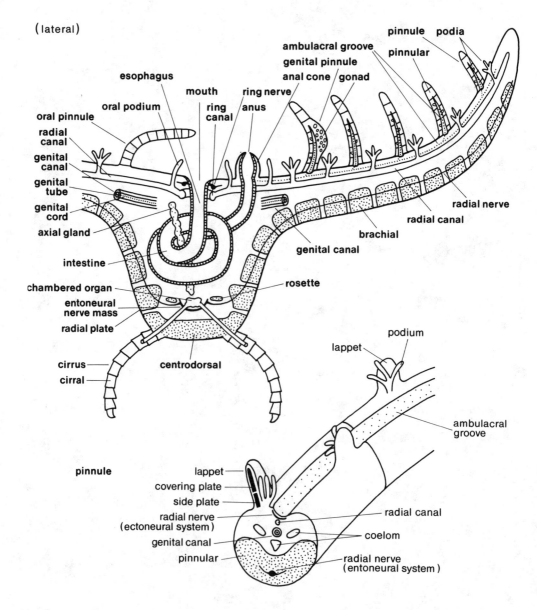

esophagus
oral pinnule
radial canal
genital canal
genital tube
genital cord
axial gland
intestine
chambered organ
entoneural nerve mass
radial plate
cirrus
cirral

oral podium
mouth
ring canal
ring nerve
anus

ambulacral groove
genital pinnule
anal cone
gonad
pinnule podia
pinnular

radial nerve
radial canal
brachial
genital canal
rosette
centrodorsal

podium
lappet
ambulacral groove

pinnule
lappet
covering plate
side plate
radial nerve (ectoneural system)
genital canal
pinnular

radial canal
coelom
radial nerve (entoneural system)

ECHINODERMATA

CRINOIDEA

crinoid, sea lily, feather star [G *Crinoide, Seelilie, Haarstern*]

aboral coelom (perivisceral coelom) [G *aborales Coelom, aborales Somatocoel*]: One of two main divisions of body cavity (coelom). Spongy, surrounding intestine, and communicating with exterior via hydropores. Gives rise to aboral coelomic canals extending into arms and pinnules. (See also ciliated pit)

aboral nerve system: entoneural nerve system

adoral coelom (subambulacral coelom) [G *Oralcoelom, adorales Coelom*]: One of two main divisions of body cavity (coelom). Consists of central space, oral extension (axial sinus), and five adoral coelomic canals (subtentacular coelom) extending into arms and pinnules.

adoral nerve system: ectoneural nerve system, hyponeural nerve system

ambulacral groove (food groove) [G *Ambulakralfurche, Ambulakralrinne, Wimperfurche, Nahrungsrinne*]: Furrow-like depression along oral surface of arms and pinnules. Margins bear series of tube feet (podia) and lappets. Grooves cross tegmen and converge at mouth.

anal cone (anal tube) [G *Afterhügel, Afterröhre, Analtubus*]: Cone- or tube-like projection on margin of membrane (tegmen) covering crown. Encloses rectum and bears anus at tip. May be supported by endoskeletal elements.

anus [G *After*]: Posterior opening of digestive tract. Located at tip of anal cone on oral surface of crown.

arm: brachium

attachment disc [G *Anheftungsscheibe*]: At base of stalk, disc-shaped expansion serving as holdfast. Represents modified endoskeletal element (columnal) of stalk. (See also radix)

axial canal [G *Axialkanal*]: Canal extending through stalk and cirri; contains extension of body cavity (chambered organ) and of entoneural nerve system.

axial gland (dorsal organ, genital stolon, heart, ovoid gland) [G *Axialdrüse*]: In main body (crown), gland extending along central axis of body from region of mouth to chambered organ. Enclosed within axial sinus.

axial sinus [G *Axialsinus*]: Part of adoral body cavity (coelom) extending along central axis of body. Previously considered to be separate cavity.

axillary [G *Axillar, Axillarglied*]: Endoskeletal structure: in arm (brachium), type of brachial element preceding fork in arm. In arm with several forks, one may distinguish (proximally to distally) a primaxil, secundaxil, and tertaxil.

basal plate (basal) [G *Basal-platte*]: Endoskeleton structure: one in a circlet of five plates of calyx. Located interradially and adjoined orally by circlet of radial plates, aborally occasionally by circlet of infrabasal plates. May be modified and displaced internally to form rosette. (fused, not fused)

basal ray [G *Kalkstäbchen*]: Endoskeleton structure: in calyx, one of five rod-like elements on oral surface of centrodorsal plate. Oriented interradially and adjoined orally by radial plates.

bivium [G *Bivium*]: In certain crinoids, section of body including two smaller arms. (See also trivium)

blood-lacunar system: hemal system

brachial (brach) [G *Brachiale, Armskelettplatte, Armglied, Armwirbel*]: Endoskeleton structure: one in a series of elements supporting each arm (brachium). Variously joined together by muscles and ligaments. According to position one may distinguish axillaries, primibrachs, secundibrachs, tertibrachs = palmars, postpalmars, and fixed brachials. (biserial, uniserial; cylindrical, disc-shaped, wedge-shaped; movable, immovable)

brachial water canal: radial canal

brachitaxis [G *Brachitaxis*]: In forked type of arm (brachium), section of arm between two successive forks, i.e., comprising series of elements (brachials) extending from 1) radial plate of calyx up to and including first axillary or 2) from one axillary up to and including next axillary.

brachium (arm) [G *Arm*]: One in a circlet of basically 10 elongate, movable major extensions of main body (crown). Supported by series of endoskeletal elements (brachials) bearing two alternating rows of pinnules and accommodating ambulacral groove on oral surface.

brood pouch: marsupium

calyx (dorsal cup) [G *Calyx, Kelch*]: Endoskeletal structure surrounding main part (crown) of body. Consists basally of centrodorsal plate and circlets of more lateral plates, e.g., basal, radial, and infraradial plates. (dicyclic, cryptodicyclic, monocyclic)

centrodorsal (centrodorsal plate) [G *Centrodorsale*]: Endoskeleton structure: top element (columnal) of stalk. Forms large aboral plate in unstalked crinoid (sea feather). (cirriferous, noncirriferous; columnar, conical, discoidal, hemispherical, subconical)

chambered organ [G *gekammertes Organ*]: In main body, organ consisting of circlet of five body cavities (coeloms). Located between endoskeletal elements (centrodorsal and rosette) and enclosed by entoneural nerve mass. Gives rise to canals extending into cirri and stalk.

ciliated funnel: hydropore

ciliated pit [G *Wimpersäckchen*]: One in a series of small, cilia-bearing depressions in body cavity (aboral coelom) which extends through pinnules (and to a lesser extent in arms). Occurs in groups.

cirral [G *Rankenglied, Cirrale*]: Endoskeleton structure: one in a series of elements (ossicles) supporting cirrus. Traversed by axial canal. (with/without aboral crests, spines, tubercles)

cirrus [G *Cirrus, Ranke*]: One in a series of appendages of stalk or of aboral surface of crown. Supported by series of endoskeletal elements (cirrals). Typically occurs in multiples of five. Number of segments and length relative to arms are of taxonomic importance. (discontinuously arranged, evenly distributed)

cirrus socket (cirrus facet) [G *Cirrusansatz, Ansatz des Cirrus*]: Depression in plate (centrodorsal) at aboral end of crown or in certain ossicles (nodia) of stalk. Marks point of attachment of cirrus. (arrangement: in columns, in transverse, alternating rows, irregularly)

coelom [G *Coelom*]: Epithelium-lined body cavity. Consists basically of adoral = subambulacral coelom and aboral =

periviceral coelom. One may also distinguish chambered organ and hyponeural sinus. (See also axial sinus)

column: stalk

columnal [G *Columnale, Stielglied*]: Endoskeleton structure; one in a series of elements (ossicles) of stalk. Traversed by axial canal and entoneural nerve system. (elliptical, oval, pentagonal, rounded; articulation: bifacial, synarthrial, synostosial)

comb: terminal comb

corona: crown

covering plate [G *Deckplatte, Deckplättchen*]: Endoskeleton structure: one in a series of supporting elements of lappets along ambulcral grooves of pinnules and arms. Represents ambulacral plate and may close, along with underlying side plate, over groove.

crown (corona) [G *Krone, Corona*]: Main part of body or, in stalked crinoid, one of two divisions (stalk, crown) of body. Consists of cup-like calyx and oral membrane (tegmen); bears arms (brachia) and encloses internal organs. (discoidal, hemispherical, oval, rounded)

cystidean stage: Term occasionally applied to attached, early larval stage characterized by stalk and absence of arms. (See also doliolaria, pentacrinoid stage)

deltoid plate (deltoid) [G *Deltoid, Oralplatte*]: Endoskeleton structure: on crown, one of five triangular plates associated with tegmen. Positioned interradially; surrounds mouth.

disc: tegmen

distal pinnule [G *distale Pinnula*]: One of three types of lateral projections (oral, genital, distal pinnules) along arms (brachia). Includes all pinnules distal to genital papillae. Always with ambulacral grooves and podia.

diverticulum [G *Divertikel, Blindsack*]: In digestive tract of crinoid with endocyclic mouth, one or more outpocketings of intestine. (simple, branching)

doliolaria [G *Doliolaria*]: Early, free-swimming larval stage. Barrel-shaped, characterized by bands of cilia.

dorsal cup: calyx

ectoneural nerve system (oral nerve system, superficial nerve system, adoral nerve system) [G *ektoneurales Nervensystem*]: One of three main components (ectoneural, entoneural, hyponeural systems) of nerve system. Consists of 1) nerve sheath around esophagus, 2) five bands radiating across tegmen, 3) branches extending under ambulacral grooves in arms and pinnules. Innervates surface of ambulacral grooves and inner sides of podia.

entoneural nerve system (aboral nerve system) [G *entoneurales Nervensystem*]: One of three main components (ectoneural, entoneural, hyponeural systems) of nerve system. Consists of 1) cup-shaped mass at aboral end of crown, 2) basically five radial nerves, 3) pentagonal commissure, 4) five main brachial nerves extending and further branching into arms. Main motor nerve system, innervating cirri, stalk, and muscles in arms and pinnules.

epizygal [G *Epizygal, epizygiales Armglied*]: Endoskeleton structure: in arm (brachium), type of supporting element (brachial). Represents distal of two brachials (hypozygal, epizygal) in syzygial articulation. (See also syzygy)

esophagus [G *Ösophagus, Oesophagus, Schlund*]: Section of digestive tract between mouth and intestine.

fixed brachial (fixed brach): Endoskeleton structure: one of several proximal supporting elements (brachials) of arm (brachium) which are associated with membrane (tegmen) covering main body (crown).

fossa [G *Vertiefung, Höhlung*]: Endoskeleton structure: any one of numerous depressions in plate or ossicle for attachment of ligaments or muscles. (See also cirrus socket)

genital canal [G *Genitalkanal*]: In reproductive system, slender body cavity (coelom) extending through arms and pinnules. Contains genital tube and genital cord; becomes expanded to form gonad. Considered to be part of perihemal system.

genital cord [G *Genitalstrang*]: In reproductive system, strand of cells located within genital tube, which is itself enclosed within genital canal extending through arms and pinnules. Consists of series of sex cells which apparently multiply and make their way to expanded genital canal (gonad).

genital pinnule [G *Genitalpinnula*]: One of three types of lateral projections (oral, genital, distal pinnules) along arms (brachia). Located between oral and distal pinnules. Contains gonads and therefore appears swollen on sexually mature individuals. (with/without ambulacral grooves, with/without podia)

genital tube [G *Genitalschlauch*]: Tube-like canal enclosed within genital canals extending through arms and pinnules. Contains genital cord. Considered to be part of hemal (blood-lacunar) system.

gonad [G *Gonade*]: Term applied to mass of ♂ or ♀ sex cells in expanded genital canals in arms or pinnules (genital pinnules).

hemal system (haemal system, blood-lacunar system) [G *Blutgefäßsystem, Haemalsystem*]: System of fluid-filled canals consisting of periesophageal plexus, subtegminal plexus, spongy organ, and genital tube.

hydropore (ciliated funnel) [G *Hydroporus, Kelchporus, Porenkanal*]: In water-vascular system, one of several hundred to more than 1,000 minute pores in membrane (tegmen) covering crown. Opens into body cavity (coelom).

hyponeural nerve system (deeper oral nerve system) [G *hyponeurales Nervensystem*]: One of three main components (ectoneural, entoneural, hyponeural systems) of nerve system. Consists of 1) pentagonal circumesophageal nerve ring and 2) 10 main branches extending into arms. Innervates labial podia, internal organs, anal cone, radial canals, and external sides of podia.

hyponeural sinus [G *hyponeuraler Sinus*]: Narrow body cavity (coelom) extending along arms. Located between radial canals and ambulacral grooves. Considered to represent part of perihemal system.

hypozygal [G *Hypozygal, hypozygiales Armglied*]: Endoskeleton structure: in arm (brachium), type of supporting element (brachial). Represents proximal of two brachials (hypozygal, epizygal) in syzygial articulation. (See also syzygy)

infrabasal plate (infrabasal) [G *Infrabasalplatte, Infrabasale*]: Endoskeleton structure: one in a circlet of five plates of calyx. Located aboral to basal plates. (fused, not fused)

interambulacral plate (interambulacral) [G *Interambulakralplatte, Interambulacrale*]: Endoskeleton structure: type of plate (perisomatic plate) associated with membrane (tegmen) covering crown. Located between arm bases.

internode [G *Internodium*]: Endoskeleton structure: type of ossicle (columnar) of stalk which does not bear cirri. Smaller than cirrus-bearing nodes.

interradial plate (interradial) [G *Interradialplatte, Interradiale*]: Endoskeleton structure: type of plate (perisomatic plate) associated with membrane (tegmen) covering crown. Located between radial plates of calyx.

intestine (midgut) [G *Mitteldarm*]: Elongate section of digestive tract between esophagus and rectum. Characterized by one coil (crinoid with central = endocyclic mouth) to four coils (crinoid with peripheral = exocyclic mouth). (with/without diverticula)

joint: segment

labial plexus: spongy organ

labial podium: oral podium

lappet [G *Saumlamelle, Saumläppchen*]: One in a series of flattened extensions of ambulacral groove margins. Lappets adjoin podia and are variously supported internally (covering plates, side plates). May be folded over ambulacral groove.

ligament [G *Ligament, Band*]: Bundle of elastic fibers spanning between ossicles (columnals, cirrals) of stalk and cirri.

marsupium (brood pouch) [G *Marsupium, Bruttasche*]: In reproductive system, expanded, sac-like structure in which eggs develop. Located adjacent to gonads in pinnules (genital pinnules).

mouth [G *Mund, Mundöffnung*]: Anterior opening of digestive tract. Located in membrane (tegmen) on top of crown; opens into esophagus. Five ambulacral grooves lead from arms into mouth. Defines oral (= ventral) surface of crinoid.

nerve system: ectoneural nerve system, entoneural nerve system, hyponeural nerve system

node [G *Nodium*]: Endoskeleton structure: type of ossicle (columnal) of stalk to which cirri attach. Bears depressions (cirrus sockets) and is typically larger than internode.

noditaxis: Section of stalk consisting of one node and adjoining series of proximal internodes down to next node. Stalk consists of several noditaxes.

oral nerve system: ectoneural nerve system

oral pinnule [G *Oralpinnula*]: One of three types of lateral projections (oral, genital, distal pinnules) along arms (brachia). Located proximally at base of arms and often folded over main part of body (crown). Elongated, lacking ambulacral grooves and podia. (with/without terminal comb)

oral podium (labial podium) [G *Labialtentakel, Mundtentakel*]: Part of water-vascular system; individual (i.e., not in groups of three) tube feet (podia) around mouth. Arranged along sections of ambulacral grooves on membrane (tegmen) covering crown.

ossicle [G *Skelettelement, Kalkkörper, Kalkstück*]: Endoskeleton structure: one of numerous skeletal elements supporting crinoid body. Includes, for example, brachials, cirrals, columnals, pinnulars, and plates of crown. (See also plate)

ovary: gonad

palmar: tertibrach

pentacrinoid stage [G *Pentacrinoid-Phase*]: Attached, late larval stage characterized by stalk (even in unstalked crinoid), five radial plates, and developing arms. (See also cystidean stage, doliolaria)

periesophageal plexus [G *periösophagealer Plexus*]: Part of hemal (bloodlacunar) system; irregular network of channels surrounding esophagus. Certain branches of plexus are closely associated with intestine, while others lead to subtegminal plexus and spongy organ.

perihemal system [G *Perihaemalsystem*]: Part of body cavity (coelom) system; considered to consist of hyponeural sinus and sheaths surrounding certain organs.

perisomatic skeleton (perisomic skeleton, adoral skeleton) [G *perisomales Skelett, perisomatisches Skelett*]: Endoskeleton structure: general category of plate associated with membrane (tegmen) covering crown. Includes interbrachials, interradials, side and covering plates.

peristome [G *Peristom, Mundhaut*]: Five-sided, membranous, and flexible region of tegmen surrounding mouth.

perivisceral coelom: aboral coelom

pinnular [G *Pinnulare*]: Endoskeleton structure: one in a series of elements supporting each pinnule. Joined together by muscles and ligaments.

pinnule [G *Pinnula*]: One in a series of lateral projections along each side of arm (brachium). Supported by series of endoskeletal elements (pinnulars) and bearing ambulacral groove on oral surface. According to position, one may distinguish oral, genital, and distal pinnules. (smooth, spiny)

plate [G *Platte, Skelettplatte*]: Endoskeleton structure: general term for flattened supporting elements (ossicles) of body. Includes, for example, basals, infrabasals, interambulacrals, interradials, radials, radianals, covering plates, and side plates. (dicyclic, monocyclic)

podium (tube foot, tentacle) [G *Tentakel, Füßchen*]: Part of water-vascular system; one in a series of papilla-like extensions along arms and pinnules. Arranged in groups of three, each group adjoining a lappet of ambulacral groove. (See also oral podium)

postpalmar [G *Postpalmare*]: Endoskeleton structure: in forked type of arm, any supporting element (brachial) beyond third section (brachitaxis) of arm, i.e., distal to tertibrachs. (See also palmar = tertibrach)

primaxil [G *Primaxil*]: Endoskeleton structure: in arm (brachium), brachial element (axillary) at which arm forks for first time. (See also secundaxil, tertaxil)

primibrach [G *Primibrachiale, Primibrach*]: Endoskeleton structure: in forked type of arm, any supporting element (brachial) of first section (first brachitaxis) of arm. (See also secundibrach, tertibrach)

radial canal (brachial water canal) [G *Radiärkanal, Ambulakralkanal, Mesocoelkanal*]: Part of water-vascular system; one of five canals originating on ring canal and extending (and branching) into arms (brachia), pinnules, and tube feet (podia). Located under ambulacral grooves.

radial pentagon: Endoskeleton structure: ring-like structure representing circlet of radial plates. Forms major part of calyx in crinoid lacking basal plates.

radial plate (radial) [G *Radialplatte*]: Endoskeleton structure: one in a circlet of five plates of calyx. Located orally and adjoined aborally by circlet of basal plates. (fused, not fused)

radianal plate (radianal) [G *Radianale*]: Endoskeleton structure: on crown, type of plate (perisomatic plate) associated with tegmen. Adjoins anal cone.

radicle (radicular cirrus) [G *Wurzelcirrus*]: At base of stalk, one in a network of branches forming holdfast (radix).

radicular cirrus: radicle

radix [G *Wurzel*]: At base of stalk, root-like network serving as holdfast. Individual branches of radix termed radicles. (See also attachment disc)

rectum [G *Enddarm*]: Section of digestive tract between intestine and anus. Located within anal cone.

ring canal (water ring) [G *Ringkanal, Ringkanal des Hydrocoels*]: Part of water-vascular system; circular canal surrounding esophagus. Bears series of stone canals, sends branches into oral podia, and gives rise to five main radial canals extending (and branching) into arms.

root: radix

rosette [G *Rosette*]: Endoskeleton structure: small, 10-sided element representing modified and displaced basal plates. Located within calyx above centrodorsal plate. Bears central hole.

saccule [G *Sacculus*]: One in a series of small, spherical bodies within tissue along ambulacral grooves in arms and pinnules. Alternating in position with podia.

secundaxil [G *Secundaxil*]: Endoskeleton structure; in arm (brachium), brachial element (axillary) at which arm forks for second time. (See also primaxil, tertaxil)

secundibrach [G *Sekundibrachiale, Secundibrach*]: Endoskeleton structure: in forked type of arm, any supporting element (brachial) of second section (second brachitaxis) of arm. (See also primibrach, tertibrach)

segment (joint) [G *Glied*]: Term applied to any externally visible unit of stalk, arm, cirrus, or pinnule. Corresponds internally with endoskeletal elements of above structures.

side plate (adambulacral plate, adambulacral) [G *Randplättchen, Seitenplatte*]: Endoskeleton structure: one in a series of small elements of lappets along ambulacral grooves of pinnules and arms. Represents adambulacral plate and may close, along with overlying covering plate, over groove.

spongy organ [G *schwammiges Gewebe, spongiöses Gewebe*]: Part of hemal (blood-lacunar) system; consists of spongy masses of thick-walled channels (lacunae) associated with rounded cells. Joined to periesophageal plexus and closely adjoining axial gland.

stalk (column, stem) [G *Stiel, Säule, Stamm*]: In stalked crinoid (sea lily), one of two divisions (stalk, crown) of body. Forms an elongate, slender attachment structure supported by series of endoskeletal elements (columnals). Bears variously developed holdfast basally. (with/without cirri; cross section: cylindrical, polygonal)

stem: stalk

stone canal [G *Steinkanal*]: Part of water-vascular system; one of numerous minute canals extending from ring canal around esophagus. Opens into body cavity (coelom).

subambulacral coelom: adoral coelom

subtegminal plexus [G *subtegminales Lakunengeflecht*]: Part of hemal (blood-lacunar) system; network of channels under membrane (tegmen) covering crown. Gives rise to five radial hemal canals extending into arms.

subtentacular coelom: adoral coelom

superficial nerve system: ectoneural nerve system

synarthry [G *Synarthrie*]: In arm (brachium), type of articulation between two supporting endoskeletal elements (brachials): ligamentary ("immovable"), with two brachials close together and joined by two large bundles of ligaments. Also refers to similar articulation between elements (columnals) of stalk. (See also syzygy)

syzygy [G *Syzygie*]: In arm (brachium), type of articulation between two supporting endoskeletal structures (brachials): ligamentary ("immovable"), with two brachials close together and joined by many very short ligaments. Visible externally as wavy line or row of dots. (See also epizygal, hypozygal)

tegmen (disk, vault) [G *Tegmen, Kelchdecke, Scheibe*]: In main body (crown), oral membrane between arm bases on top of calyx. Divided by position of ambulacral grooves into ambulacral and interambulacral areas. (naked = not calcified, with plates = calcified; delicate, leathery) (See also perisomatic skeleton)

tentacle: podium

terminal comb (comb) [G *Kamm*]: On oral pinnules, comb-like structure formed by series of triangular projections on distal endoskeletal elements (pinnulars).

tertaxil [G *Tertaxil*]: Endoskeleton structure: in arm (brachium), brachial element (axillary) at which arm forks for third time. (See also primaxil, secundaxil)

tertibrach (palmar) [G *Tertibrachiale, Tertibrach, Palmare*]: Endoskeleton structure: in forked type of arm, any supporting element (brachial) of third section (third brachitaxis) of arm. (See also primibrach, secundibrach)

testis: gonad

theca [G *Theca*]: General term for skeletal structure enclosing main body. Reduced in recent crinoids to plates of calyx.

trivium [G *Trivium*]: In certain crinoids, section of body including three larger arms. (See also bivium)

tube foot: podium

vault: tegmen

water ring: ring canal

water-vascular system [G *Wassergefäßsystem, Ambulakralsystem, Ambulakralkanalsystem*]: System of canals consisting of hydropores, ring canal bearing series of stone canals, radial canals extending into each arm, as well as smaller branches into tube feet (podia).

ECHINODERMATA

ASTEROIDEA

sea star, starfish [G *Seestern*]

ASTEROIDEA

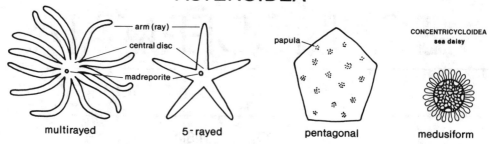

multirayed   5-rayed   pentagonal   medusiform

arm (ray)
central disc
madreporite
papula

CONCENTRICYCLOIDEA
sea daisy

(lateral)

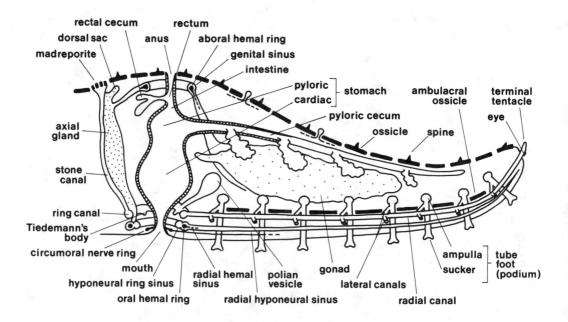

PHANEROZONIC SEA STAR

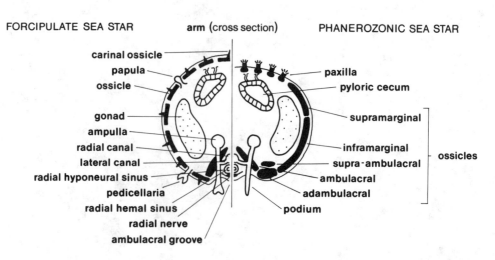

FORCIPULATE SEA STAR   arm (cross section)   PHANEROZONIC SEA STAR

ECHINODERMATA

ASTEROIDEA

sea star, starfish [G *Seestern*]

aboral hemal ring (aboral haemal ring) [G *aboraler Lakunenring, aboraler Haemalring*]: In blood-vascular (hemal) system, ring-like canal within genital sinus around rectum. Connected with oral hemal ring via axial gland.

aboral sinus: genital sinus

adambulacral ossicle [G *Adambulakralplatte*]: Endoskeleton structure: one in a row of elements (ossicles) along oral surface of arm. Located lateral to ambulacral ossicles; forms edges of ambulacral groove. Bears movable spines.

adambulacral peristome: peristome

adambulacral spine (furrow spine) [G *Adambulakralstachel, Furchenstachel*]: One in a series of spines on adambulacral ossicles. May protect ambulacral groove and retracted tube feet (podia).

ambulacral groove [G *Ambulakralfurche, Ambulakralrinne*]: Furrow-like depression along oral surface of each arm from which tube feet (podia) project. Formed by inverted V-shaped arrangement of skeletal elements (ambulacral ossicles).

ambulacral ossicle [G *Ambulakralplatte*]: Endoskeleton structure: one in a double row of elements (ossicles) along aboral surface of arm. Positioned in an inverted V-shaped arrangement to form ambulacral groove; tube feet (podia) pass between ambulacral ossicles.

ambulacral peristome: peristome

ambulacral pore [G *Ambulakralporus*]: Endoskeleton structure: one in a series of pores between ambulacral ossicles. Formed by concavity or half pore in each adjoining ossicle; serves as opening of tube foot (podium) to exterior. (biserial, quadriserial)

ambulacral ridge: Endoskeleton structure: skeletal ridge along aboral part of ambulacral groove. Formed at apex of inverted V-shaped groove by abutting pairs of ambulacral ossicles.

ambulacral system: water-vascular system

ampulla [G *Ampulle, Füßchenampulle*]: Part of water-vascular (ambulacral) system; expanded, bulb-like section of each tube foot (podium). Attached to radial canal of arm by lateral canal. May also refer to space (ampulla of stone canal) beneath madreporite into which pore canals open. (simple = single, bilobed = double)

anus [G *After*]: Opening of digestive system to exterior. If present, located on aboral surface of central disc in interradius BC.

arm (ray) [G *Arm*]: One of typically five ray-like extensions of main body (central disc), each bearing internal organs. Oral surface of arm bears open groove (ambulacral groove) with tube feet (podia) serving in locomotion. (cross section: cylindrical, flattened; elongate, petaloid)

axial gland (brown gland, dorsal organ, heart, ovoid gland, septal organ) [G *Axialdrüse*]: In central disc, dark, spongy, aboral-orally directed gland extending along stone canal within axial sinus. Attached orally to hyponeural ring sinus and elaborated aborally into terminal process within dorsal sac.

axial sinus [G *Axialsinus, Axialcoelom, linker Protocoelschlauch*]: In central disc, tubular, aboral–orally directed body cavity (coelom) enclosing axial gland and stone canal. Closely associated with interbrachial septum of interradius CD. Communicates aborally with genital sinus and ampulla (of stone canal), orally with hyponeural ring sinus.

basal ossicle [G *Basalsklerit, basales Skelettstück, Basalstück*]: Basalmost of three skeletal elements (single basal ossicle, two jaws or valves) in stalked (pedunculate) pedicellaria.

bipinnaria [G *Bipinnaria*]: Earlier of two larval stages (bipinnaria, brachiolaria); characterized by ciliary band and presence of arm-like projections.

bivium [G *Bivium*]: Section of sea star body including two arms (C and D) on either side of madreporite. (See also trivium)

blood-vascular system: hemal system

brachial cecum: pyloric cecum

brachiolar arm [G *Brachiolarfortsatz*]: One of three relatively long, contractile arms of brachiolaria larva; used, along with sucker between arms, in attachment to substratum.

brachiolaria [G *Brachiolaria*]: Later of two larval stages (bipinnaria, brachiolaria); characterized by development of three adhesive arms (brachiolar arms) around sucker-like structure.

brood pouch [G *Bruttasche*]: In sea daisy, one of five pairs of pouches within body cavity; contains embryos and juvenile stages.

cardiac stomach [G *Cardia*]: stomach

carinal plate (carinal) [G *Carinalplatte, sekundäre Radialplatte, Kielplatte*]: Endoskeleton structure: one in a series of well-developed elements (ossicles) along medial aboral line or arm. Followed laterally (from aboral to oral side) by series of dorsolaterals, supramarginals, and inframarginals.

central disc (disk) [G *Scheibe, Zentralscheibe, Körperscheibe*]: Central part of body excluding arms (rays).

central plate [G *Centrodorsalplatte, Centrodorsale, Centrale*]: Endoskeleton structure: central element (ossicle) on aboral surface of disc; surrounded by five interradial plates. Typically unrecognizable in adult sea star.

circumoral nerve ring [G *ectoneuraler Ringnerv*]: ectoneural nerve system

coelom [G *Coelom*]: Epithelium-lined body cavity. Considered to be subdivided into 1) protocoel surrounding axial gland, 2) mesocoel forming water-vascular system, and 3) metacoel forming main body cavity and complex system of canals (sinuses) of so-called perihemal system.

comet [G *Kometenstadium*]: In certain sea stars, term applied to regeneration stage in which single cast-off arm regenerates central disc and complete complement of (still small) arms.

cribriform organ [G *cribriformes Organ, Siebrinnenorgan*]: In certain sea stars, series of thin, vertical plates or vertical rows of papillae on margin of central disc between arms. May also be present between marginal ossicles along entire arm. Serves in respiration, in the latter configuration also in feeding. (between all/not between all marginal plates)

dermal branchia: papula

digestive gland: pyloric cecum

disk: central disc

dorsal sac (madreporic vesicle, terminal sac) [G *Dorsalsack, Dorsal-*

blase, rechtes Protocoel]: In central disc, aboral body cavity (coelom) enclosing terminal process of axial gland. Located under madreporite.

dorsolateral plate (dorsolateral) [G *Dorsolateralplatte*]: Endoskeleton structure: one in a series of elements (ossicles) along arm. Bordered aborally by series of carinals, orally by supramarginals and inframarginals.

ectoneural nerve system (ectoneural system, oral nerve system) [G *ektoneurales Nervensystem, Ektoneuralsystem*]: One of two major components (ectoneural and hyponeural systems) of nerve system. Consists of circumoral nerve ring, radial nerves in arms, and subepithelial plexus.

endoskeleton [G *Endoskelett, Skelett*]: Internal skeleton consisting of numerous calcareous elements (ossicles). Divisible into main, deeper supporting skeleton and more superficial skeletal elements.

epiproctal cone [G *zentrale Papille*]: On aboral surface of central disc in certain sea stars, elevation of central region into conical structure.

esophageal pouch [G *Schlundtasche*]: In digestive tract, one of 10 greatly folded outpocketings of esophagus.

esophagus [G *Ösophagus, Oesophagus, Schlund*]: Relatively short and wide section of digestive tract between mouth and stomach. (with/without esophageal pouches)

eye spot (eye) [G *Auge*]: Light-sensitive organ consisting of numerous photoreceptors (pigment-cup ocelli in advanced forms) within optic cushion at tip of each arm.

fan [G *Fächer*]: One in a series of transverse, fan-like structures along ambulacral groove on oral surface of arm. Formed by membrane uniting spines of adambulacral ossicles; continuous with lateral membrane.

fasciole: vibratile spine

furrow spine: adambulacral spine

gastric hemal tuft [G *Darmstrang*]: In blood-vascular (hemal) system, one to four tufts of hemal strands extending from wall of stomach to axial gland.

gastric ligament [G *Magenmesenterium*]: One of 10 pairs of triangular mesenteries spanning between cardiac stomach and ambulacral ridges. (See also nodule)

genital sinus (aboral sinus) [G *Genitalsinus, Genitalkanal, aboraler Metacoelkanal*]: Aboral body cavity (coelom) in central disc. Forms pentagonal canal and gives off extensions enclosing gonads in each arm. (See also perihemal system)

gill: papula

gonad [G *Gonade*]: One of typically 10 ♂ (testes) or ♀ (ovaries) reproductive organs, one located along each side of arm; bush-shaped, branched, and attached at base of arm near interbrachial septa. Each gonad opens to exterior via proximal gonopore or, if serial, by series of gonopores. (simple, serial)

gonopore [G *Geschlechtsöffnung, Gonoporus*]: Opening of gonad to exterior. Typically located orally on base of arm or, in the case of serial gonads, in a series of openings along arm.

granule [G *Körnchen, Granula*]: Endoskeleton structure: small calcareous protuberance resting upon but not fused to underlying skeletal element (ossicle). (See also spine, tubercle)

hemal system (haemal system, blood-vascular system) [G *Blutgefäßsystem, Haemalsystem*]: System of narrow, fluid-filled canals contained within so-called perihemal system; consists of oral hemal ring (within hyponeural ring sinus), radial hemal sinus (within radial hyponeural sinus), and aboral hemal ring (within genital sinus).

hepatic cecum: pyloric cecum

hyponeural nerve system (hyponeural system) [G *hyponeurales Nervensystem, hyponeurales System*]: One of two major components (ectoneural and hyponeural systems) of nerve system. Located in walls of perihemal canal system (hyponeural ring sinus, radial hyponeural sinus); has primarily motor function.

hyponeural ring sinus (oral ring sinus, peribuccal sinus) [G *Hyponeuralkanal, Perihämalkanal, Perihaemalring, oraler Kanal des Metacoels*]: In central disc, oral body cavity (coelom) forming canal around mouth and enclosing oral hemal ring. Divided by septum into small inner and larger outer rings. Gives rise to radial hyponeural sinus into each arm. (See also perihemal system)

inframarginal plate (inframarginal) [G *Inframarginalplatte*]: Endoskeleton structure: one in a series of elements (ossicles) along arm. Located along adambulacrals and followed aborally by supramarginals, dorsolaterals, and carinals.

interbrachial septum (interradial septum) [G *Interradialseptum*]: In central disc, continuation of body wall inward between arms. (membranous, supported by ossicles)

interradial plate [G *Interradialplatte*]: Endoskeleton structure: on aboral surface of five-arm sea star, one of five elements (ossicles) surrounding central plate and located in interradial position. One interradial plate incorporates madreporite.

interradial septum: interbrachial septum

interradius [G *Interradius*]: In five-arm sea star, one of five sections of central disc between arms. (See also radius)

intestinal cecum: rectal cecum

intestine [G *Enddarm, Darmkanal*]: Very short, narrow section of digestive tract between pyloric stomach and anus. Bears rectal cecum or ceca. (present, absent)

jaw (valve) [G *Zangenbacken, Zangenstück, Greifarm, Klappe*]: One of typically two movable calcareous elements in pedicellaria. Associated with basal ossicle or directly with underlying skeletal element(s).

lateral canal (lateral branch, podial branch, connecting canal) [G *Füßchenkanal, Seitenzweig*]: Part of water-vascular system; one in a series of short lateral canals extending from radial canal in each arm. Connects ampullae of tube feet (podia) to radial canal. Bears valve.

lateral membrane [G *Flossensaum, Randsaum, Haut*]: On oral surface, membranous extension of fan-like structure (fan) formed by spines of adambulacral ossicles. Supported by series of long, parallel, transverse spines.

madreporic vesicle: dorsal sac

madreporite [G *Madreporenplatte*]: Opening of water-vascular (ambulacral) system to exterior. Consists of circular calcareous plate, perforated by pore canals, located in and defining interradius CD on aboral surface of central disc. (single, multiple; naked, with paxillae)

marginal plate [G *Marginalplatte*]: inframarginal plate, supramarginal plate

marginal sinus [G *marginaler Hyponeuralkanal*]: Narrow, tube-like body cavity (coelom), one extending along margin of each arm. Joined to central hyponeural radial sinus by series of channels.

marginal spine: In sea daisy, one in a circlet of excavate spines surrounding periphery of body.

mesocoel [G *Mesocoel, Hydrocoel, Wassergefäßsystem, Ambulakralsystem*]: coelom

metacoel [G *Metacoel, Somatocoel*]: coelom

mouth [G *Mund, Mundöffnung*]: Anterior opening of digestive tract. Defines oral surface of sea star. Surrounded by variously developed peristome; opens into esophagus.

mouth frame (peristomial skeletal ring) [G *Mund-Skelettring, Mundskelett*]: Endoskeleton structure: one in a series of elements (ossicles) surrounding mouth. Composed of alternating ambulacral and adambulcral ossicles. (adambulacral, ambulacral) (See also peristome)

nerve system: ectoneural nerve system, hyponeural nerve system

nidamental chamber [G *Brutkammer, Brutraum, Atemraum*]: In sea star in which supradorsal membrane forms aboral surface, space between membrane and underlying body wall. Serves as brood chamber and opens to exterior via osculum.

nodule: Tough connective tissue at junction of gastric ligament and cardiac stomach.

ocellus (pigment-cup ocellus) [G *Becherauge*]: eye spot

optic cushion [G *Augenwulst, Polster*]: At tip of each arm, cushion-like expansion at base of terminal tentacle. Bears numerous photoreceptors and serves as eye.

oral hemal ring [G *oraler Haemalring, oraler Lakunenring*]: In blood-vascular (hemal) system, ring-like canal within hyponeural ring sinus. Gives rise to radial hemal sinus extending into each arm.

oral ring sinus: hyponeural ring sinus

osculum [G *Osculum*]: In sea star with supradorsal membrane and underlying nidamental chamber, large, central aperture on aboral surface serving as excurrent opening of chamber. Supported by spines. (See also spiracle)

ossicle [G *Skelettplatte, Platte*]: Endoskeleton structure: one of numerous calcareous elements variously joined to one another to form internal skeleton. Includes adambulacrals, ambulacrals, carinals, dorsolaterals, inframarginals, paxillae, and supramarginals. (elongated, four-lobed, lobulated, paxilliform, rod-shaped, rounded) (See also ring ossicle)

ovary [G *Ovar, Ovarium, Eierstock*]: gonad

papula [G *Papula, Kieme*]: One of numerous small, retractile projections emerging between skeletal elements (ossicles). Represents hollow evagination of body wall and serves in respiration. (branched, bush-shaped, conical, simple, tubular; restricted to aboral surface, on aboral and oral surfaces)

papular area: papularium

papularium (papular area) [G *Papularium, Kiemenfeld*]: On aboral surface, one of several delimited areas bearing respiratory structures (papulae). (on arm, on arm base; irregularly scattered, in distinct rows along sides of arm)

paxilla [G *Paxille*]: Endoskeleton structure: type of element (ossicle) consisting of basal part and erect column with expanded top covered with small, movable projections (spinelets, tubercles).

pedicellaria [G *Pedicellaria*]: On outer surface of body, any one of numerous variously shaped, pincer-like organs. Serves as protective and food-gathering structure. (alveolar: bivalved, excavate, spatulate; pedunculate: crossed, furcate, straight; sessile: fasciculate, pectinate, spiniform, subvalvulate, valvulate = valvate)

peribuccal sinus: hyponeural ring sinus

perihemal system (perihaemal system) [G *Perihaemalsystem*]: Term applied to canal-like system of body cavities (coelom) enclosing blood-vascular (hemal) system. Consists of genital sinus (enclosing aboral hemal ring), hyponeural ring sinus (enclosing oral hemal ring), and radial hyponeural sinus (enclosing radial hemal sinus).

peristome [G *Peristom, Mundhaut*]: Membranous region around mouth opening. Supported by ring of skeletal elements (ossicles) to form mouth frame. Ring in which adambulacral ossicles are more conspicuous is termed adambulacral peristome or mouth frame, that in which ambulacral ossicles are larger is termed ambulacral peristome or mouth frame. (adambulcaral, ambulacral)

podial branch: lateral canal

podium (tube foot) [G *Ambulakralfüßchen, Füßchen*]: Part of water-vascular (ambulacral) system; one in a series of tube-like extensions projecting from ambulacral groove along oral side of arms. May be subdivided into terminal sucker, tube-like section, and expanded ampulla, the latter being connected to radial canal of arm by lateral canal. In sea daisy, arranged in a single series around outer ring canal. (pointed, with sucker; in two series, in four series)

polian vesicle (Polian vesicle) [G *Polische Blase, Poli'sche Blase*]: Part of water-vascular (ambulacral) system; larger type of muscular outpocketing of ring canal. Up to several polian vesicles are located in each interradius. (See also Tiedemann's body)

pore canal [G *Porenkanälchen*]: Part of water-vascular (ambulacral) system; one of numerous minute canals leading from surface of madreporite into stone canal.

protocoel [G *Protocoel, Axocoel*]: coelom

pseudopaxilla [G *Pseudopaxille*]: Endoskeleton structure: type of element (ossicle) resembling paxilla in that it consists of erect column tipped with spinelets.

pyloric cecum (hepatic cecum, digestive gland, brachial cecum) [G *Pylorusdivertikel, Pylorusblindsack, Mitteldarmdrüse*]: In digestive tract, one of 10 elongate digestive glands opening into pyloric stomach. Two pyloric ceca, consisting of series of lateral lobules and suspended by pair of mesenteries, extend through each arm.

pyloric stomach [G *Pylorus*]: stomach

radial canal [G *Radiärkanal, Radialkanal, Radiärkanal des Mesocoels, Ambulakralkanal*]: Part of water-vascular (ambulacral) system; canal originating on outer side of ring canal and extending to tip of each arm. Gives rise to series of lateral canals supplying tube feet (podia).

radial hemal sinus (radial haemal channel) [G *Radiärlakune*]: In blood-vascular (hemal) system, tube-like canal extending into each arm from oral hemal ring; gives rise to branches extending

into tube feet (podia). Enclosed by radial hyponeural sinus.

radial hyponeural sinus [G *Perihaemalkanal, Radiärkanal des oralen Metacoels*]: Tube-like oral body cavity (coelom) extending from hyponeural ring sinus into each arm. Gives rise to series of smaller channels to marginal sinuses and tube feet (podia). Encloses radial hemal sinus. (See also perihemal system)

radial nerve [G *ectoneuraler Radiärnerv*]: ectoneural nerve system

radius [G *Radius*]: In five-ray/arm sea star, one of five sections of central disc corresponding to axes of arms. (See also interradius)

ray: arm

rectal cecum (intestinal cecum, rectal sac) [G *Rectaldivertikel*]: In digestive tract, one of up to several outpocketings of intestine. (paired, unpaired; with/without long ducts; two-/three-branched)

rectum [G *Enddarm*]: In digestive tract, term occasionally applied to section of intestine aboral to intestinal (rectal) ceca.

ring canal (water ring) [G *oraler Ambulakralring, Mesocoelring*]: Part of water-vascular (ambulacral) system; circular canal surrounding mouth and lying parallel to and above (aboral to) hyponeural ring sinus. Gives rise to radial canal in each arm and bears polian vesicles and Tiedemann's bodies. Connected to madreporite by stone canal. In sea daisy, one of two circular canals (inner = circumoral ring canal and outer = "radial" ring canal) around margin of body; joined interradially.

ring ossicle: Endoskeleton structure: in sea daisy, one in a circlet of elements near margin of body. Outer ring canal passes around outer side of ring ossicles.

skeletal arch [G *Bogen*]: Endoskeleton structure: continuous skeletal structure formed by arrangement of individual elements (ossicles) into rows aligned transverse to arm axis.

spine [G *Stachel*]: Endoskeleton structure: elongate calcareous projection resting upon but not fused to underlying skeletal element (ossicle); movable by muscles. (See also granule, tubercle, marginal spine, vibratile spine)

spinelet (spinule) [G *Stachelchen*]: In endoskeleton, one in a series of small, spine-like projections on marginal plates, pseudopaxillae, or crown of paxillae. Movable by muscles in the latter.

spiracle [G *Spiraculum*]: In sea star with supradorsal membrane and underlying nidamental chamber, one of several contractile pores serving as incurrent opening of chamber. (See also osculum)

standard measurements:
R and r, or relation of R to r, whereby
R = center of disc to arm tip
r = disc radius (center of disc to edge of disc midway between arms
Orientation: Arm opposite madreporite = A. Remaining arms follow in alphabetic order in clockwise direction (B,C,D,E). In five-arm sea star, madreporite positioned in interradius CD.

stomach [G *Magen*]: Expanded section of digestive tract in central disc. May be divided by horizontal constriction into larger oral cardiac stomach and smaller aboral pyloric stomach, the latter giving rise to pyloric ceca.

stone canal [G *Steinkanal*]: Part of water-vascular (ambulacral) system; vertical canal between madreporite on aboral surface and ring canal surrounding mouth. Walls strengthened by calcareous deposits. May be variously subdivided internally by ridges.

sucker (terminal disk, suctorial disc) [G *Saugscheibe*]: Part of water-vascular (ambulacral) system; disc-shaped terminal expansion of tube foot (podium).

suctorial disc: sucker

superambulacral ossicle: supra-ambulacral ossicle

supra-ambulacral ossicle (super-ambulacral ossicle) [G *Supraam-bulakralplatte*]: Endoskeleton structure: in arm, one in a row of small elements (ossicles) above each ambulacral ossicle.

supradorsal membrane [G *Supradorsalmembran, Deckmembran, Rückenhaut*]: Membrane uniting crowns of certain endoskeletal elements (paxillae) and forming aboral surface. (See also nidamental chamber)

supramarginal plate (supramarginal) [G *Supramarginalplatte*]: Endoskeleton structure: one in a series of elements (ossicles) along arm. Bordered aborally by dorsolaterals and carinals, orally by inframarginals.

tentacle: terminal tentacle

terminal plate [G *Terminalplatte*]: Endoskeleton structure: on aboral surface of juvenile sea star, one of five skeletal elements (ossicles) surrounding interradial plates and located in radial position. With growth, terminal plates shift to tips of arms. (unarmed, with spines)

terminal sac: dorsal sac

terminal tentacle (tentacle) [G *Terminaltentakel*]: Unpaired, hollow, tentacle-like projection at tip of each arm. Represents modified podium at distal end of each radial canal of water-vascular system. Base of tentacle bears eye spot.

testis [G *Hoden*]: gonad

Tiedemann's body [G *Tiedemannsches Körperchen, Tiedemann'sches Körperchen*]: Part of water-vascular (ambulacral) system; one of typically five pairs of small outpocketings of ring canal. Located interradially on inner side of canal. (See also polian vesicle)

Tiedemann's pouch [G *Tiedemannsche Tasche, Tiedemann'sche Tasche*]: In digestive tract, one of typically five outpocketings of stomach. Extends below (oral to) pyloric ceca in arms.

trivium [G *Trivium*]: Section of sea star body comprising those three arms not included in bivium (i.e., A,B,E).

tube foot: podium

tubercle [G *Tuberkel, Höcker*]: Endoskeleton structure: moderate calcareous elevation resting upon but not fused to underlying skeletal element (ossicle). (See also granule, spine)

valve [G *Ventil*]: In water-vascular (ambulacral) system, valve-like structure in lateral canal leading to each tube foot (podium). May also refer to each jaw of certain types of pedicellaria.

velum [G *Velum*]: In sea daisy, thin membrane stretching across lower surface of body. Considered to be derived from former stomach tissue.

vibratile spine [G *Wimperstachel*]: One in a series of minute, heavily flagellated spines in vertical grooves (fascioles) between marginal plates. Considered to produce respiratory current.

water ring: ring canal

water-vascular system (ambulacral system) [G *Wassergefäßsystem, Ambulakralsystem, Ambulacral-Kanalsystem, Mesocoel*]: System of canals consisting of opening to exterior (madreporite), stone canal, ring canal bearing polian vesicles and Tiedemann's bodies, as well as radial canals connected to tube feet (podia) via series of lateral canals. In sea daisy, consists of two concentric ring canals joined by five interradial canals; tube feet (podia) arise from outer ring canal.

ECHINODERMATA

OPHIUROIDEA

brittle star [G *Schlangestern*]

OPHIUROIDEA

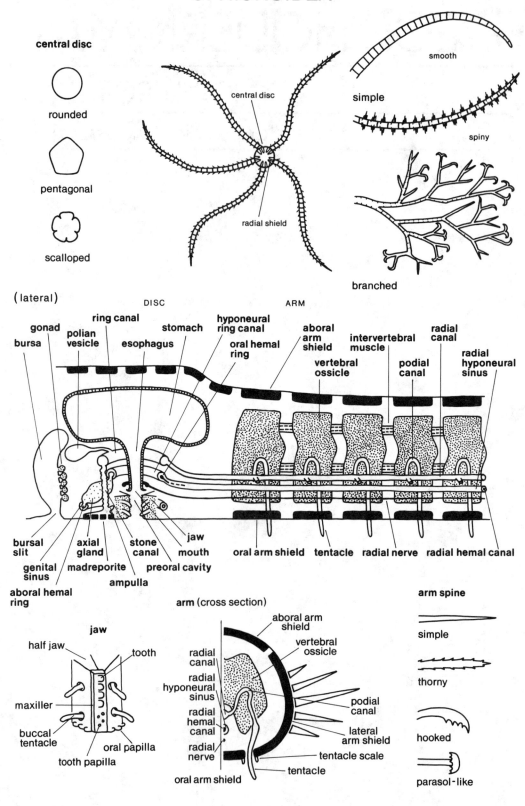

central disc

rounded

pentagonal

scalloped

central disc

radial shield

smooth

simple

spiny

branched

(lateral)

DISC

ARM

gonad
bursa
polian vesicle
ring canal
esophagus
stomach
hyponeural ring canal
oral hemal ring
aboral arm shield
vertebral ossicle
intervertebral muscle
podial canal
radial canal
radial hyponeural sinus

bursal slit
genital sinus
aboral hemal ring
axial gland
madreporite
stone canal
ampulla
jaw
mouth
preoral cavity
oral arm shield
tentacle
radial nerve
radial hemal canal

jaw

half jaw
tooth
maxiller
buccal tentacle
tooth papilla
oral papilla

arm (cross section)

aboral arm shield
vertebral ossicle
radial canal
radial hyponeural sinus
radial hemal canal
radial nerve
oral arm shield
podial canal
lateral arm shield
tentacle scale
tentacle

arm spine

simple

thorny

hooked

parasol-like

ECHINODERMATA

OPHIUROIDEA

brittle star [G *Schlangestern*]

aboral arm shield [G *aborale Platte*]: Endoskeleton structure: one in a series of plate-like elements along aboral surface of arm, one forming outer dorsal cover of each arm segment. (elliptical, fan-shaped, hexagonal, polygonal, rhombic) (See also lateral arm shield, oral arm shield)

aboral hemal ring (aboral haemal sinus) [G *aboraler Lakunenring, aboraler Haemalring*]: In blood-vascular (hemal) system, ring-like canal within genital sinus. Connected with oral hemal ring via axial gland. Encloses genital rachis.

aboral sinus: genital sinus

adambulacral ossicle [G *Adambulakralplatte*]: Category of skeletal elements modified in brittle star to form smaller piece of half jaw (= first pair), adoral shields (= second pair), and lateral arm shields (= third and successive pairs). (See also ambulacral ossicle)

adoral shield [G *adorale Platte*]: Endoskeleton structure: one in a series of 10 plates on oral surface of central disc, one pair flanking each buccal shield.

ambulacral ossicle [G *Ambulakralplatte*]: Category of skeletal elements modified in brittle star to form peristomial plate (= first pair), larger piece of half jaw (= second pair), and vertebral ossicles (= third and successive pairs). (See also adambulacral ossicle)

ambulacral system: water-vascular system

ampulla [G *Ampulle*]: Part of water-vascular system; in central disc, pouch-like expansion of stone canal adjacent to opening (madreporite, hydropore) to exterior. Axial sinus also opens into ampulla.

arm (ray) [G *Arm*]: One of typically five ray-like extensions of main body (central disc). Consists of series of arm segments and contains extensions of water-vascular system, nerves, and hemal system. (branched, simple; smooth, spiny, carinate dorsally)

arm comb [G *Kamm*]: At junction of arm and disk, comb-like series of papillae surrounding arm laterally and aborally. (continuous/not continuous dorsally)

arm plate [G *Armplatte, Armschild*]: Endoskeletal structure: one of typically four elements forming outer cover of each arm segment. Includes single oral and aboral arm shields as well as pair of lateral arm shields.

arm segment (arm joint, joint) [G *Armsegment*]: One in a series of units forming arm. The skeletal elements of each arm segment consist of internal element (vertebral ossicle) surrounded by single oral and aboral arm shields as well as pair of lateral arm shields.

arm spine [G *Armstachel, Stachel*]: One in a series of spine-like projections along sides of arm. Both lateral arm shields of each arm segment give rise to transverse row of arm spines. Number and shape are of taxonomic importance. (appressed, erect; united/ not united into fan; shape: blunt, pointed, slender, stout, smooth, thorny, flattened, rounded, club-shaped, conical, parasol-shaped = umbrella-shaped, spatulate, with squared tips)

axial gland [G *Axialdrüse*]: In central disc, gland extending along stone canal. Connects oral and aboral hemal rings and consists of darker and lighter sections. Enclosed, along with stone canal, by axial sinus.

axial sinus [G *Axialsinus, Protocoel*]: In central disc, body cavity (coelom) enclosing stone canal and axial gland, the latter dividing sinus into left and right parts. Communicates with ampulla of stone canal and genital sinus on one end, with hyponeural ring sinus on other end.

blood-vascular system: hemal system

brood chamber [G *Brutraum*]: In brooding brittle star species, term applied to bursa when filled with embryos.

buccal podium: buccal tentacle

buccal shield (oral shield) [G *Bukkalplatte, Oralplatte*]: Endoskeleton structure: one in a series of five plates on oral surface of central disc. Located interradially, flanked by adoral shields, and more or less covering jaws. One or more buccal shields may serve as madreporite.

buccal tentacle [G *Mundfüß-chen*]: Part of water-vascular system; refers to first and second pairs of podia (tentacles) at base of arms, i.e., those located on jaws within preoral cavity.

bursa [G *Bursa*]: One of typically 10 pouch-like invaginations of body wall on periphery of central disc. Located to each side of arm base and opening to exterior via bursal slits. Associated with gonads; also serves in respiration. (number: 20, 10, partially fused = 5, entirely fused = 1).

bursal slit (genital slit) [G *Bursa-spalt, Bursalspalt, Bursalspalte*]: Opening(s) of each bursa to exterior. Located on oral

surface of central disc, typically one pair along margins of each arm base.

central disc (disk) [G *Scheibe, Zentralscheibe, Körperscheibe*]: Central part of body excluding arms (rays). (aboral surface: granular, leathery, naked, tuberculate)

central plate [G *Centrodorsalplatte, Centrodorsale, Centrale*]: Endoskeleton structure: central element on aboral surface of disc; surrounded by several concentric circles of plates. Typically unrecognizable in adult.

circumoral nerve ring (nerve ring) [G *Ringnerv*]: In nerve system, ring of nerve tissue surrounding esophagus. Consists of tissue from ectoneural and hyponeural systems. Innervates jaws, buccal tentacles, esophagus, and stomach and sends radial nerves into arms.

coelom [G *Coelom*]: Epithelium-lined body cavity. Considered to be subdivided into 1) protocoel surrounding axial gland, 2) mesocoel forming water-vascular system, and 3) metacoel forming complex system of canals of so-called perihemal system as well as body coelom.

dental papilla: tooth papilla

disk: central disc

dorsal plate: aboral arm shield

ectoneural nerve system [G *ektoneurales Nervensystem, Ektoneuralsystem*]: nerve system

endoskeleton [G *Endoskelett, Skelett*]: Internal skeleton consisting of numerous calcareous plates or shields as well as vertebral ossicles in arms.

epineural sinus [G *Epineuralsinus*]: One of five body cavities (not coelomic) between nerve system and body wall.

esophagus [G *Ösophagus, Oesophagus*]: Short section of digestive system between mouth opening and stomach.

fan [G *Fächer*]: In certain brittle stars, one in a series of fan-shaped structures along each side of arm. Formed by interconnection of arm spines of one arm segment by web-like membrane.

genital papilla [G *Genitalpapille, Papille*]: On oral surface of central disc, one in a series of small projections lining edges of opening (bursal slit) of each bursa.

genital plate: genital shield

genital rachis (genital stolon) [G *Genitalstrang*]: Cylindrical strand of reproductive tissue within aboral hemal ring (in turn enclosed within genital sinus). Gives rise to gonads.

genital scale [G *Genitalschuppe*]: Endoskeleton structure: on oral surface of central disc, one in a series of small elements between each genital shield and buccal shield.

genital shield (genital plate) [G *Genitalplatte*]: Endoskeleton structure: on oral surface of central disc, elongate plate bordering each side of opening (bursal slit) of each bursa.

genital sinus (aboral sinus, aboral coelomic sinus) [G *Genitalsinus, aboraler Ringkanal des Metacoels, aboraler Perihaemalring*]: Part of perihemal system; tube-like body cavity (coelom) enclosing aboral hemal ring in central disc. Located ventrally, on substrate side of jaws; represents displaced aboral sinus of sea star. Forms five-angle system of loops adjoining bursae and gives rise to sac around each gonad.

genital slit: bursal slit

genital stolon: genital rachis

gonad [G *Gonade*]: ♂ (testis) or ♀ (ovary) reproductive organ, typically attached to walls of bursae in central disc. Eggs and sperm are discharged via bursae and bursal slits to exterior. (in central disc, in arms)

granule [G *Körnchen, Granula*]: Endoskeleton structure: one of numerous minute, rounded skeletal elements on outer surface of shields.

half jaw [G *Kieferhälfte*]: One of two main elements fused to form each jaw. Each half jaw consists of two pieces.

hemal system (haemal system, blood-vascular system) [G *Blutgefäßsystem, Haemalsystem*]: System of narrow, fluid-filled canals contained within so-called perihemal system. Consists of oral hemal ring (within hyponeural ring sinus), radial hemal canal (within radial hyponeural sinus), and aboral hemal ring (within genital = aboral sinus).

hydropore [G *Hydroporus*]: Opening of water-vascular system to exterior on oral surface of central disc. Single or numerous, either on madreporite or not associated with skeletal elements.

hyponeural nerve system [G *hyponeurales Nervensystem, hyponeurales System*]: nerve system

hyponeural radial canal: radial hyponeural sinus

hyponeural ring sinus (hyponeural ring canal, oral perihemal ring) [G *oraler Perihaemalring, Hyponeuralkanal, Perihaemalkanal*]: Part of perihemal system; in central disc, tube-like body cavity forming ring at level of esophagus between jaws and stomach. Encloses oral hemal ring and gives rise to radial hyponeural sinus extending into each arm.

interradius [G *Interradius*]: In five-ray/arm brittle star, one of five sections of central disc between arms. (See also radius)

intervertebral muscle [G *Intervertebralmuskel*]: In each arm, one of four muscle bands spanning between successive vertebral ossicles. One may distinguish pair of oral muscles and pair of aboral intervertebral muscles.

jaw [G *Kiefer*]: One of five triangular, interradial structures surrounding preoral cavity and forming mouth frame. Movable, each jaw consisting of two half jaws. Variously armed with oral papillae, tooth papillae, and teeth.

jaw plate: maxiller

joint: arm segment

lateral arm shield (lateral shield) [G *Seitenplatte, seitliches Armschild, Seiten-*

armplatte]: Endoskeleton structure; one in a series of elements along sides of arms, one pair forming outer lateral cover of each arm segment. Bears arm spines. (See also aboral arm shield, oral arm shield)

madreporite [G *Madreporenplatte*]: Plate-like opening of water-vascular system to exterior. Located on oral surface of central disc; consists of modified oral shield(s) perforated by pore canals. Opens into ampulla and stone canal. (number: one, in interradius CD; five, in each interradius) (See also hydropore)

maxiller (jaw plate, torus angularis) [G *Kieferleiste*]: At tip of each jaw, slender, vertical element bearing teeth and tooth papillae. Perforated for passage of muscles operating teeth.

mesocoel [G *Mesocoel, Hydrocoel, Wassergefäßsystem*]: coelom

metacoel [G *Metacoel, Somatocoel*]: coelom

mouth [G *Mund, Mundöffnung*]: Anterior and only opening of digestive system located above jaws. Defines oral surface of brittle star, leading to distinction between oral (= ventral) and aboral (= dorsal) surfaces. Surrounded by peristome; opens into esophagus.

mouth frame [G *Mund-Skelettring, Mundskelett*]: On oral side of central disc, frame-like structure enclosing cavity (preoral cavity) leading to mouth. Formed by five jaws and thus five-angled.

nerve ring: circumoral nerve ring

nerve system [G *Nervensystem*]: System of nerve tissue consisting of two closely adjoining, parallel ectoneural (= sensory and motor) and hyponeural (= motor) systems. Composed basically of circumoral nerve rings, radial nerves into each arm, and series of lateral nerves in each arm segment.

ophiopluteus [G *Ophiopluteus*]: Planktonic larval stage bearing four pairs of arms (in order of appearance: posterolateral, anterolateral, postoral, posterodorsal) supported by skeletal rods.

oral arm shield [G *Oralplatte, Epineuralplatte*]: Endoskeleton structure: one in a series of elements along oral surface of arm, one shield forming outer ventral cover of each arm segment. (See also aboral arm shield, lateral arm shield)

oral hemal ring [G *oraler Haemalring*]: In blood-vascular (hemal) system, ring-like canal at level of esophagus between jaws and stomach. Enclosed by hyponeural ring canal; gives rise to radial hemal sinus extending into each arm.

oral papilla (mouth papilla) [G *Mundpapille*]: On each jaw, one in a series of calcareous projections fringing margin closest to substrate (i.e., ventral = oral margin). (conical, flattened, papilliform, scale-shaped, serrate, spine-shaped = spiniform, simple)

oral perihemal ring: hyponeural ring sinus

oral shield: buccal shield

ovary [G *Ovar, Ovarium, Eierstock*]: gonad

perihemal system (perihaemal system) [G *Perihaemalsystem*]: Term applied to canal-like system of body cavities (coelom) enclosing blood-vascular (hemal) system. Consists of 1) genital sinus (= aboral sinus), 2) hyponeural ring canal (= oral perihemal ring) enclosing oral hemal ring, and 3) radial hyponeural canal enclosing radial hemal sinus.

peristome (peristomial membrane) [G *Peristom, Mundhaut*]: Membranous region around mouth; attached to upper (dorsal) side of jaw apparatus.

peristomial plate [G *Peristomplatte*]: Endoskeleton structure: one of five pairs of internal plates located on aboral (dorsal) side of jaws in central disc.

plate (shield) [G *Platte, Schild*]: Endoskeleton structure: one of numerous plate-like elements covering outer surface of body. Includes, for example, primary and secondary plates of aboral disc surface, adoral, buccal, and genital shields of oral disc surface, as well as series of aboral, oral, and lateral arm shields along each arm.

podial canal [G *Tentakelkanal, Seitenkanal*]: Part of water-vascular system; in each arm segment, paired lateral canals extending from radial canals to podia (tentacles). (straight, curved)

podium: tentacle

polian vesicle (Polian vesicle) [G *Polische Blase, Poli'sche Blase*]: Part of water-vascular system; muscular outpocketing of ring canal, typically one in each interradius except CD (= interradius bearing stone canal).

pore canal [G *Porenkanälchen*]: Part of water-vascular system; single canal leading from hydropore to stone canal or one of numerous minute canals leading from surface of madreporite into stone canal.

prebuccal cavity: preoral cavity

preoral cavity (prebuccal cavity) [G *Mundvorraum*]: On oral side of central disc, cavity between mouth frame formed by jaws. Leads via mouth into esophagus.

primary plate [G *Primärplatte*]: Endoskeleton structure: on aboral surface of disc, one of numerous first-formed elements. Includes central plate and several series (with five plates each) of concentric plates (e.g., terminal plates). (See also secondary plate)

protocoel [G *Protocoel, Axocoel*]: coelom

radial canal [G *Radiärkanal, Radialkanal, Radiärkanal des Mesocoels*]: Part of water-vascular system; one of five canals originating on ring canal and extending to tip of each arm. Gives rise in each arm segment to pair of lateral podial canals leading to podia (tentacles).

radial hemal sinus (radial haemal channel) [G *Radiärlakune*]: In blood-vascular (hemal) system, tube-like canal extending into each arm from oral hemal ring; gives rise to branches into podia (tentacles). Enclosed by radial hyponeural sinus.

radial hyponeural sinus (hyponeural radial canal) [G *Hyponeuralkanal, Radiärkanal des oralen Metacoels*]: Part of perihemal system; tube-like body cavity extending from hyponeural ring sinus into each arm. Encloses radial hemal sinus.

radial nerve [G *Radiärnerv*]: In nerve system, one of five nerves arising from circumoral nerve ring and extending through arms. Consists of ectoneural and hyponeural components, the former innervating tentacles and arm spines, the latter innervating muscles between vertebral ossicles.

radial shield [G *Adradialplatte, Radialplatte*]: Endoskeleton structure: one of 10 relatively large elements on aboral surface of central disc. Present in pairs on disc margin, one radial shield along either side of each arm where it joins disc.

radius [G *Radius*]: In five-ray/arm brittle star, one of five sections of central disc corresponding to axes of arms. (See also interradius)

ray: arm

ring canal (water ring) [G *Ambulakralring, Mesocoelring*]: Part of water-vascular system; circular canal surrounding mouth. Gives rise to radial canal in each arm and bears polian vesicles. Connected to exterior via stone canal, ampulla, and madreporite or hydropore.

rosette [G *Rosette*]: On aboral surface of disc, series of primary plates arranged in a distinct circle around central plate.

scale (disc scale) [G *Schuppe*]: On central disc, one of numerous plates modified to resemble small scales. (smooth, spiny, thorny; imbricating, separated by furrows)

secondary plate [G *Sekundärplatte*]: Endoskeleton structure: on aboral surface of disc, one of numerous elements which, with growth of the brittle star, become interpolated between first-formed (primary) plates.

shield: plate

Simroth's appendage [G *Simroth'scher Anhang*]: Part of water-vascular system in certain brittle stars; one of numerous long, slender, tubular appendages of ring canal.

skin [G *Haut*]: Variously developed outer layer of body; when well developed, often concealing underlying shields. (naked, granulate, tuberculate)

spine [G *Stachel*]: One of numerous small spines (spinelets) on central disc. (simple, bifid, trifid; cross section: triangular, not prismatic) (See also arm spine)

stomach [G *Magen*]: Expanded section of digestive system in central disc. Folded into stomach pouches peripherally, yet almost without exception not sending ceca into arms.

stomach pouch [G *Magentasche*]: One in a series of 10 peripheral folds of stomach in central disc.

stone canal [G *Steinkanal*]: Part of water-vascular system; canal between madreporite or hydropore on oral surface of central disc and ring canal. Oral end is developed into ampulla. Enclosed, along with axial gland, in axial sinus. (number: one, at interradius CD; five, at each interradius)

supplementary upper arm plate [G *akzessorische aborale Armskelettplatte*]: Endoskeleton structure: one in a series of small plates between aboral and lateral arm shields along arm.

tentacle (podium, tube foot) [G *Tentakel, Füßchen, Ambulakralfüßchen*]: In each arm segment, pair of movable, papillae-like extensions of water-vascular system. Joined to radial canals by lateral podial canals. Projects from pores (tentacle pores) on oral surface of arm between oral and lateral arm shields. (See also buccal tentacle, tentacle scale)

tentacle pore [G *Tentakelporus, Porus*]: On oral surface of each arm, pair of pores between oral and lateral arm shields from which podia (tentacles) emerge. Variously protected by one or more tentacle scales.

tentacle scale [G *Tentakelschuppe, Schuppe*]: On oral surface of each arm segment, one or more skeletal elements at base of each podium (tentacle). May close over tentacle pore when tentacle is retracted.

terminal plate [G *Terminalplatte*]: Endoskeleton structure: on aboral surface of juvenile brittle star, one of five skeletal elements on outer margin of disc. With growth, terminal plates shift to tips of arms.

terminal tentacle (terminal podium) [G *Terminaltentakel*]: Unpaired, hollow, papilla-like projection at tip of each arm. Represents podium (tentacle) at distal end of each radial canal of water-vascular system.

testis [G *Hoden*]: gonad

tooth [G *Zahn*]: At tip of each jaw, one in a series of tooth-like projections. Typically refers only to larger elements either in a single row or on more ventral (= oral) part of cutting edge. (See also oral papilla, tooth papilla)

tooth papilla (dental papilla) [G *Zahnpapille*]: At tip of each jaw, one in a group of smaller projections between oral papillae and teeth. (scale-like, tooth-like)

tube foot: tentacle

ventral plate: oral arm shield

vertebra: vertebral ossicle

vertebral ossicle (vertebra) [G *Armwirbel, Wirbel*]: Endoskeleton structure: one in a series of internal elements within each arm, a single vertebral ossicle in each arm segment. Typically surrounded by four plates (aboral, oral, and lateral arm shields). (articulation: streptospondylous = streptospondyline, zygospondylous = zygospondyline = zygophiuroid)

water ring: ring canal

water-vascular system (ambulacral system) [G *Wassergefäßsystem, Ambulakralsystem*]: System of canals consisting of opening to exterior (madreporite or hydropore), stone canal with ampulla, ring canal, radial canals into arms, and series of lateral podial canals to podia (tentacles).

ECHINODERMATA

ECHINOIDEA

sea urchin, echinoid [G *Seeigel, Echinoide*]

ECHINOIDEA

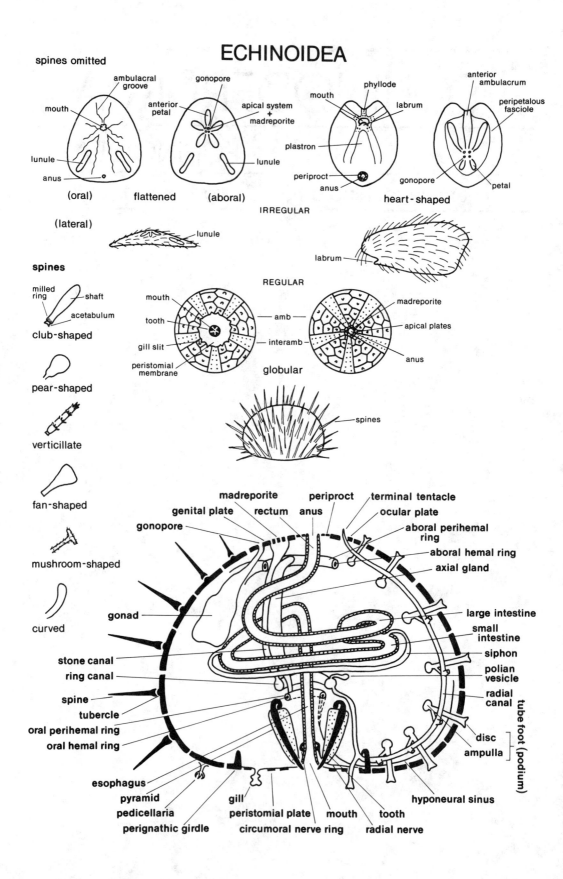

spines omitted

ambulacral groove

mouth

anterior petal

gonopore

apical system + madreporite

mouth

phyllode

labrum

anterior ambulacrum

peripetalous fasciole

lunule

anus

(oral)   flattened   (aboral)

lunule

plastron

periproct

anus

gonopore

petal

heart-shaped

(lateral)

IRREGULAR

lunule

labrum

spines

milled ring

shaft

acetabulum

club-shaped

REGULAR

mouth

tooth

gill slit

peristomial membrane

amb

interamb

globular

madreporite

apical plates

anus

pear-shaped

verticillate

fan-shaped

spines

mushroom-shaped

curved

madreporite

genital plate

gonopore

rectum

periproct

anus

terminal tentacle

ocular plate

aboral perihemal ring

aboral hemal ring

axial gland

gonad

large intestine

small intestine

siphon

polian vesicle

radial canal

stone canal

ring canal

spine

tubercle

oral perihemal ring

oral hemal ring

disc

ampulla

esophagus

pyramid

pedicellaria

perignathic girdle

gill

peristomial plate

circumoral nerve ring

mouth

tooth

radial nerve

hyponeural sinus

tube foot (podium)

ECHINODERMATA

ECHINOIDEA

sea urchin, echinoid [G *Seeigel, Echinoide*]

aboral hemal ring (aboral haemal ring) [G *aboraler Haemalring*]: Ring-like canal of hemal system at top (apex) of shell (test). Joined to oral hemal ring by network of vessels associated with axial gland.

aboral sinus (genital sinus) [G *aboraler Metacölkanal*]: In regular sea urchin, equivalent to genital sinus; in irregular sea urchin, body cavity corresponding with fused perianal, periproctal, and genital sinuses.

acetabulum (socket) [G *Gelenkpfanne*]: Endoskeleton structure: concave basal socket of spine. Articulates with projection (tubercle) of shell (test).

adapical suture [G *Adapikalnaht*]: Endoskeleton structure: between individual ambulacral or interambulacral plates, transverse = horizontal line of contact (suture) closer to anus. (See also adoral suture)

adoral suture [G *Adoralnaht*]: Endoskeleton structure: between individual ambulacral or interambulacral plates, line of contact closer to mouth. (See also adapical suture)

adradial suture [G *Adradialnaht*]: Endoskeleton structure: line of contact (suture) between ambulacrum and interambulacrum. (See also suture)

ambital plate [G *Ambitalplatte*]: Endoskeleton structure: one in a series of plates around greatest circumference (ambitus) of shell (test).

ambitus [G *Ambitus*]: Greatest horizontal circumference of shell (test). (outline from above: circular, pentagonal, oval, heart-shaped)

ambulacral furrow (food groove) [G *Ambulakralfurche*]: Endoskeleton structure: in irregular sea urchin, one in a series of narrow grooves radiating from mouth. Located in ambulacra and typically restricted to underside of test. Bears specialized tube feet.

ambulacral plate [G *Ambulakralplatte, Ambulakralschild*]: Endoskeleton structure: in shell (test), one in a series of plates along each ambulacrum. Typically paired (i.e., forming two columns) and perforated by pores for passage of tube feet of underlying water-vascular system. (bigeminate, oligoporous, polyporous, trigeminate) (See also compound plate, primary plate)

ambulacrum (amb) [G *Ambulacrum, Ambulakrum, Ambulakralfeld*]: Endoskeleton structure: one of five narrower radial sections of shell (test) basically extending from mouth to apex and alternating with interambulacra. Consists of double row of plates (ambulacral plates) bearing pores for tube feet (podia). Corresponds internally with radial canals of water-vascular system. (petaloid, nonpetaloid)

ampulla [G *Ampulle*]: Part of water-vascular system; expanded, bulb-like structure of each tube foot (podium). Arranged in a series along inner side of ambulacra; connected with podia via pore pairs in ambulacral plates, with radial canal via short branch. May also refer to space ("dorsal sac") beneath madreporite into which pore canals open into stone canal.

anal fasciole [G *anale Fasciole*]: In irregular sea urchin, fasciole partially surrounding anal region (periproct). Termed lateroanal fasciole if connecting with lateral fascioles.

anus [G *After*]: Posterior opening of digestive tract located either centrally at top (apex) of shell (test) or posteriorly on underside of test. Located within periproct and surrounded by membrane (periproctal membrane) bearing plates (periproctal plates). (aboral, oral and posterior)

apex [G *Apex, Scheitel*]: Highest point of shell (test). Bears circlets of apical plates and, in regular sea urchin, anus. (See also ambitus)

apical plate [G *Apikalplatte, Scheitelplatte*]: Endoskeleton structure: one of basically four types of plates (apical, coronal, periproctal, peristomal). Consists of series of plates (ocular, genital) at top (apex) of test. Collectively termed apical system. (dicyclic, monocyclic; apical system: ethmolytic = ethmolysian, ethmophract)

apophysis [G *Apophyse*]: Endoskeleton structure: on inner side of shell (test), one in a series of processes of interambulacral plates around mouth (peristome). Serves as site of muscle attachment for chewing apparatus (Aristotle's lantern).

arbacioid plate (arbacioid compound plate) [G *arbacioide Platte*]: Endoskeleton structure: in shell (test), type of compound ambulacral plate basically composed of three elements—a median normal (primary) plate and a reduced plate (demiplate) above = adapical and below = adoral to it.

areole [G *Areole, Hof, Warzenhof*]: Endoskeleton structure; on shell (test) plate, depression around base (boss) of tubercle. Serves as attachment site of spine muscles. (See also scrobicule)

Aristotle's lantern (lantern of Aristotle) [G *Laterne des Aristoteles, Kaugerüst, Kiefergerüst*]: Endoskeleton structure: complex chewing apparatus consisting of numerous calcareous elements (40 if fully developed) and muscles. Held in place within shell (test) by muscles spanning to perignathic girdle. (aulodont, camerodont, stirodont) (See also compass, demipyramid, epiphysis, foramen magnum, pyramid, rotule, tooth)

auricle [G *Aurikel, Auricula*]: Endoskeleton structure: on inner side of shell (test), one in a series of processes of ambulacral plates around mouth (peristome). Serves as site of muscle attachment for chewing apparatus (Aristotle's lantern). (See also apophysis)

axial gland (axial organ, brown gland, heart, kidney, ovoid gland) [G *Axialdrüse, Axialorgan*]: Dark, aboral-orally directed organ extending along stone canal. Encloses axocoel. (See also ampulla)

axocoel [G *Axocoel*]: coelom

base [G *Basis*]: Endoskeleton structure: section of spine below milled ring. Hollowed proximally to form socket (acetabulum) articulating with tubercle on plate of shell (test).

bivium [G *Bivium*]: In irregular sea urchin, posterior two ambulacra. Because of anteriorly displaced mouth, these posterior ambulacra are longer than anterior three ambulacra of trivium.

boss: Endoskeleton structure: on shell (test) plate, basal, cone-shaped part of tubercle; flat-topped, bearing distal rounded, spine-bearing mamelon.

bourrelet: Endoskeleton structure: on underside of irregular sea urchin shell (test), swollen and protruding section of each interambulacrum around mouth. Bourrelets and adjoining five phyllodes form floscelle.

buccal cavity [G *Mundhöhle*]: In sea urchin with chewing apparatus (Aristotle's lantern), short section of digestive tract between mouth and pharynx. Surrounded by circumoral nerve ring; encloses tips of teeth.

buccal membrane [G *Peristomialmembran, Peristom-Membran, Mundhaut*]: Circular membrane surrounding mouth and covering peristome. Bears plates (peristomial plates).

buccal podium [G *Mundfüßchen, Bukkalfüßchen, Peristomialfüßchen*]: Part of water-vascular system; one of two modified tube feet (podia) in each radius. Located on membrane (buccal membrane) surrounding mouth.

cecum (caecum) [G *Blinddarm*]: Small, blind outpocketing of digestive tract between esophagus and intestine.

circumoral nerve ring (peripharyngeal nerve ring) [G *oraler Nervenring, ectoneuraler Nervenring, ectoneuraler Mundring*]: In main = ectoneural nerve system, ring of nerve tissue around anterior section of digestive tract. Located within chewing apparatus (Aristotle's lantern); gives rise to five radial nerves.

clavule [G *Clavula*]: Endoskeleton structure: in irregular sea urchin, one in a series of small, basally ciliated and distally expanded spines arranged in fascioles.

coelom [G *Coelom*]: Epithelium-lined body cavity. Divided into protocoel (axocoel), mesocoel (hydrocoel) forming water-vascular system, and metacoel (somatocoel) forming main body cavity. The latter is further subdivided into various small cavities (e.g., aboral = genital sinus, perianal sinus, peripharyngeal cavity, periproctal sinus).

collar (collerette) [G *Kragen*]: Endoskeleton structure: at base of spine, smooth, tapering section between milled ring and neck.

collateral sinus [G *Collateralsinus*]: Part of hemal system in regular sea urchin;

ring-like channel between chewing apparatus (Aristotle's lantern) and intestine. Connected to parallel outer marginal sinus by series of canals.

compass [G *Compas, Gabelstück, Bügelstück*]: Endoskeleton structure: one of five elements at top of chewing apparatus (Aristotle's lantern). Located in ambulacral position above rotules and oriented radially.

compound plate (compound ambulacral plate) [G *zusammengesetzte Platte, Großplatte*]: Endoskeleton structure: in shell (test), plate formed by fusion of several ambulacral plates. Bears more than one pore pair. Individual elements bound together by tubercle. According to composition of normal (primary) and/or reduced ambulacral, one may distinguish arbacioid, diadematoid, echinoid, and echinothurioid plates. (oligoporous, polyporous)

corona [G *Corona*]: Major part of shell (test) consisting of all ambulacral and interambulacral plates (i.e., minus plates around mouth = peristomial plates, around apex = apical plates, around anus = periproctal plates).

coronal plate [G *Coronalplatte*]: Endoskeleton structure: one of basically four types of plates (apical, coronal, periproctal, peristomal). Consists of ambulacral and interambulacral plates and forms main part (corona) of shell (test). Collectively termed coronal system.

cortex [G *Cortex, Rindenschicht*]: Endoskeleton structure: outer wall of spine. Variously ornamented and may bear hairs. May be restricted to cap-like structure at tip of spine. (See also medulla)

demiplate [G *Halbplatte*]: Endoskeleton structure: in shell (test), type of modified (reduced) ambulacral plate which touches ambulacrum/interambulacrum border but not midline of ambulacrum (i.e., adradial but not perradial suture).

demipyramid (half pyramid) [G *Halbpyramide, Kieferpaar, Zahnstück*]: Endoskeleton structure: in chewing apparatus (Aristotle's lantern), one half of each pyramid. Two halves joined by longitudinal suture. (top surface: pitted, smooth)

dental sac [G *Zahnblase*]: Body cavity (coelom) enclosing soft, curled upper end of each tooth. Represents oupocketing of peripharyngeal sinus.

diadematoid plate (diadematoid compound plate [G *diadematoide Platte*]: Endoskeleton structure: in shell (test), type of compound ambulacral plate composed of three normal (primary) plates. Middle plate is largest.

disc (disk, sucker, sucking disc) [G *Saugscheibe, Saugscheibchen*]: Disc-shaped terminal expansion of tube foot (podium). May be supported by various minute calcareous elements (frame, rosette, spicule). (circular, oval, reniform)

echinoid plate (echinoid compound plate) [G *echinoide Platte*]: Endoskeleton structure: in shell (test), type of compound ambulacral plate basically composed of three elements—a reduced plate (demiplate) and normal (primary) plates above = adapical and below = adoral to it. Lower plate is largest.

echinopluteus (pluteus) [G *Echinopluteus, Pluteus*]: Planktonic larval stage bearing projections (anterodorsal, anterolateral, posterodorsal, posterolateral, postoral, preoral arm pairs) supported by skeletal rods.

echinothurioid plate (echinothurioid compound plate, echinothuriid plate) [G *echinothurioide Platte*]: Endoskeleton structure: in shell (test), type of compound ambulacral plate basically composed of three elements—a normal (primary) plate and two reduced plates (included plates) below = adoral to it.

ectoneural nerve system [G *ektoneurales Nervensystem*]: Largest of three components (ectoneural, entoneural, hyponeural systems) of nerve system. Consists of circumoral nerve ring, radial nerves along ambulacra, podial nerves, and subepidermal plexus.

endoskeleton [G *Endoskelett*]: Internal skeleton consisting of numerous calcareous plates of test as well as elements of chewing apparatus (Aristotle's lantern) and spicules.

entoneural nerve system (aboral nerve system) [G *entoneurales Nervensystem, aborales Nervensystem*]: In certain regular sea urchins, one of three components (ectoneural, entoneural, hyponeural systems) of nerve system. Consists of nerve ring around anal region (periproct) at top (apex) of test. Nerve ring gives rise to branches innervating gonads.

epineural sinus [G *Epineuralkanal*]: One of five radial, fluid-filled channels lying between radial nerves and test wall.

epiphysis [G *Epiphyse, Knochen-Bogen, Ergänzungsstück*]: Endoskeleton structure: in chewing apparatus (Aristotle's lantern), bar-shaped element joined to top of each demipyramid. Epiphysis pairs of each pyramid may be fused (thus defining upper border of foramen magnum). (separate, sutured)

episternal plate [G *Episternalplatte*]: Endoskeleton structure: on underside of irregular sea urchin shell (test), second pair of plates posterior to labrum.

esophagus [G *Ösophagus, Oesophagus*]: In sea urchin with chewing apparatus (Aristotle's lantern), section of digestive tract between pharynx and first loop of intestine. Descends in direction of mouth from top of lantern. In sea urchin without lantern, section of digestive tract between mouth and intestine.

eye: terminal tentacle

fasciole [G *Fasciole, Semita*]: On shell (test) of irregular sea urchin, narrow band of minute, densely spaced tubercles bearing modified spines (clavules). According to position, one may distinguish anal, internal, lateral, lateromarginal, marginal, peripetalous, and subanal fascioles.

floscelle [G *Floscelle*]: Endoskeleton structure: on underside of irregular sea urchin shell (test), flower-like structure around mouth. Formed by ambulacra (phyllodes) alternating with modified interambulacra (bourrelets).

food groove: ambulacral furrow

foramen magnum (external foramen) [G *Foramen magnum, Foramen externum*]: Endoskeleton structure: in chewing apparatus (Aristotle's lantern), space between upper ends of demipyramid pairs forming each pyramid. May be bordered above by epiphyses.

frame [G *Skelettspange*]: Endoskeleton structure: calcareous structure supporting expanded terminal end (disc) of tube foot (podium). (See also rosette)

genital papilla [G *Genitalpapille, Geschlechtspapille*]: In reproductive system, papilla-like projection bearing terminal gonopore; lacks muscles.

genital plate [G *Genitalplatte, Basalplatte*]: Endoskeleton structure: one of five larger plates which, together with smaller ocular plates, form apical system at top of test. Perforated by one or more gonopores. Located in (and define) five interradia of test. (See also madreporite)

genital pore: gonopore

genital sinus (aboral sinus) [G *aboraler Metacoelkanal, genitaler Ringsinus, apicaler Ringsinus*]: In regular sea urchin, ring-shaped aboral body cavity (coelom) under apical plates at top of shell (test). Associated with gonads.

gill (external gill, branchia) [G *Kieme, äußere Kieme*]: On underside of regular sea urchin, one of 10 respiratory structures extending through gill slits at margin of peristomial membrane.

gill cut: gill slit

gill slit (gill cut, branchial slit) [G *Kiemenbucht, Kiemeneinschnitt*]: Endoskeleton structure: on underside of shell (test), one in a series of smooth grooves at borders of interambulacra and peristome. One slit located on each side of each interambulacrum. Accommodates gill.

gonad [G *Gonade*]: ♂ (testis) or ♀ (ovary) reproductive organ. If fully developed, consists of five organs extending along interambulacra. Each gonad opens to exterior via gonoduct and gonopore on genital plate.

gonoduct [G *Ausfuhrgang, Ausfuhrkanal*]: Short, narrow section of reproductive system between each gonad and gonopores on genital plates.

gonopore (genital pore) [G *Genitalporus*]: Opening of each gonad to exterior. Located at top (apex) of shell (test), one pore in each genital plate.

granule [G *Granula*]: Endoskeleton structure: on outer surface of shell (test) plate, one of numerous minute elevations lacking appendages. (See also tubercle)

hair [G *Wimperhärchen*]: One element in a dense coat of fine projections of outer wall (cortex) of spine. (short and thin, long and fluffy, anastomosing, branching; tips: doubly hooked, minutely tuberculate, plain, singly hooked)

handle [G *Bügel*]: Endoskeleton structure: in ophiocephalous pedicellaria, one of three arched elements at base of jaws. Serves to lock grip when jaws close.

head [G *Kopf, Pedicellarienköpfchen*]: Distal of three sections (stem, neck, head) of pedicellaria. Consists of three movable jaws supported by endoskeletal elements (valves).

hemal ring: aboral hemal ring, oral hemal ring

hemal system (haemal system) [G *Haemalsystem, Blutgefäßsystem*]: System of fluid-filled canals. Consists of oral hemal ring, five radial hemal sinuses, inner = ventral marginal sinus, outer = dorsal marginal sinus, collateral sinus, and aboral hemal ring.

hydropore [G *Hydroporus*]: Opening of water-vascular system to exterior. One of up to several pores in plate(s) (madreporite) within apical system of plates at top of shell (test).

hyponeural nerve system (deeper oral nerve system) [G *hyponeurales Nervensystem*]: One of three components (ectoneural, entoneural, hyponeural systems) of nerve system. Consists of five radially positioned plaques of nerve tissue below circumoral nerve ring of ectoneural system.

hyponeural sinus (hyponeural canal, pseudohemal canal, perihemal canal) [G *Hyponeuralkanal, Perihaemalkanal*]: One of five radial, fluid-filled channels (radial metacoel) paralleling radial nerve and lying between nerve and radial hemal sinus. May represent perihemal system.

included plate [G *Halbplatte*]: Endoskeleton structure: in shell (test), modified (reduced) ambulacral plate which touches neither ambulacrum/interambulacrum border nor midline of ambulacrum (i.e., neither adradial nor perradial suture). Together with primary plate, two included plates form echinothurioid plate.

inner marginal sinus (ventral marginal sinus) [G *inneres Darm-Längsgefäß*]: Part of hemal system; larger channel arising from oral hemal ring and running parallel to inner side of digestive tract (small and large intestines). Supplies digestive tract with dense network of branches. (See also outer = dorsal marginal sinus)

interambulacral plate [G *Interambulakralplatte*]: Endoskeleton structure: in shell (test), one in a series of plates along each interambulacrum. Typically paired, i.e., forming two columns of alternating plates. Variously ornamented with tubercles which bear spines.

interambulacrum (interamb) [G *Interambulakrum, interambulakrales Feld*]: Endoskeleton structure: one of five interradial sections of shell (test) basically extending from mouth to anus and alternating with ambulacra. Consists of double row of plates (interambulacral plates).

internal fasciole [G *interne Fasciole*]: In irregular sea urchin, fasciole enclosing apex of shell (test) and crossing petals.

interradial suture [G *interradiale Naht*]: Endoskeleton structure: line of contact (suture) between two rows (columns) of interambulacral plates forming each interambulacrum.

interradius [G *Interradius*]: interambulacrum

intestine [G *Darm*]: Elongate section of digestive tract between pharynx or esophagus and rectum. Typically forms two loops within shell (test), the first being termed stomach/small intestine/inferior intestine, the second termed intestine/large intestine/ superior intestine.

jaw [G *Zangenbacke, Klappzange*]: One of typically three movable units at distal end (head) of pedicellaria. Each jaw is supported internally by calcareous element (valve).

labrum [G *Labrum, Lippe*]: Endoskeleton structure: on underside of irregular sea urchin shell (test), lip-like protrusion bordering posterior margin of mouth area (peristome). Formed by enlarged plate of interambulacrum.

lantern of Aristotle: Aristotle's lantern

lateral fasciole [G *laterale Fasciole*]: In irregular sea urchin, pair of fascioles, one extending posteriorly from each side of peripetalous fasciole. Termed lateroanal fasciole if connecting with anal fasciole.

lateroanal fasciole: anal fasciole, lateral fasciole

lip [G *Lippenring*]: Ring-shaped bulge of membrane (peristomial membrane) around mouth.

lunule [G *Lunula*]: Endoskeleton structure: one of two or more holes through shell (test) of irregular sea urchin. Corresponds in position with (perradial or interradial) sutures between plates. (round, elongate)

madreporic plate: madreporite

madreporite (madreporic plate)

[G *Madreporenplatte*]: Endoskeleton structure: one or more plates in apical system of plates at top of shell (test). Bears openings (madrepores) of water-vascular system to exterior and is connected to ring canal by stone canal. Represents modified genital plate. (ethmophract, ethmolytic)

mamelon

[G *Gelenkkopf*]: Endoskeleton structure: rounded distal part of tubercule on shell (test) plate. Represents site of spine articulation. (imperforate, perforate)

marginal fasciole

[G *marginale Fasciole*]: In irregular sea urchin, fasciole encircling greatest circumference (ambitus) of shell (test).

medulla (central core, central mesh)

[G *zentraler Markschicht*]: Endoskeleton structure: central core of spine. Consists of skeletal meshwork. (See also cortex)

mesocoel

[G *Mesocoel*]: coelom

metacoel

[G *Metacoel*]: coelom

miliary spine (miliary, tertiary spine)

[G *Tertiärstachel*]: One of numerous very small spines projecting from shell (test) plates. Articulates with miliary tubercles.

miliary tubercle

[G *Tertiärtuberkel*]: Endoskeleton structure: on outer surface of shell (test) plate, one in a series of numerous very small tubercles bearing miliary spines or pedicellariae. (See also tubercle)

milled ring

[G *Annulus*]: Endoskeleton structure: flared proximal section of spine between base and collar. Grooved, serving as site of muscle attachment.

mouth

[G *Mund, Mundöffnung*]: Anterior opening of digestive tract on underside of shell (test). Located within peristome and surrounded by buccal membrane (bearing peristomial plates). Teeth of chewing apparatus (Aristotle's lantern) typically project from mouth. Opens into buccal cavity. (anterior, central)

muscle

[G *Muskel*]: One of numerous typically smooth muscles operating chewing apparatus (Aristotle's lantern), spines, pedicellaria, tube feet (podia), and even regulating shape of flexible tests.

neck

[G *Hals*]: Endoskeleton structure: smooth, cylindrical section at base of spine between shaft and collar. May also refer to distal flexible section (lacking internal skeletal element) of pedicellaria.

nerve system:

ectoneural nerve system, entoneural nerve system, hyponeural nerve system

neuropore:

Endoskeleton structure: in each ambulacral plate of shell (test), single pore adjoining pore pair. Accommodates nerve supplying tube foot.

ocular plate (terminal plate)

[G *Okularplatte, Terminalplatte, Ocellarplatte, Radialplatte*]: Endoskeleton structure: one of five smaller plates which, together with five larger genital plates, form apical system at top of test. Each ocular plate is perforated by pore (ocular pore) for last tube foot (terminal tentacle). Located in (and define) five radii of test. (exsert, insert)

ocular pore

[G *Terminalporus, Okularporus*]: Endoskeleton structure: pore in each ocular plate at top (apex) of shell (test). Serves for passage of last tube foot (terminal tentacle) of water-vascular system.

oral hemal ring

[G *oraler Haemalring*]: Ring-like canal of hemal system around anterior part of digestive tract. Gives rise to five interradial canals supplying polian vesicles and five radial hemal sinuses. Joined to aboral hemal ring via axial gland.

outer marginal sinus (dorsal marginal sinus)

[G *äußerer Darm-Sinus*]: Part of hemal system; smaller channel running parallel to outer side of first loop of digestive tract (stomach = small intestine = inferior intestine). Supplies digestive tract with dense network of branches. (See also inner = ventral marginal sinus)

ovary

[G *Ovar, Ovarium*]: gonad

pedicellaria

[G *Pedicellaria*]: One in a series of appendages articulating with small tubercles on shell (test) plates. Each pedicellaria consists of stem (supported by spicule) and head with movable jaws (valves). (bidentate, biphyllous, claviform, dactylous, gemmiform = glandular = globiferous, ophiocephalous, rostrate, tridactyle = tridentate, trifoliate = triphyllous)

perianal sinus [G *Perianalsinus*]: In regular sea urchin, ring-shaped body cavity (coelom) around anus.

perignathic girdle [G *perignathischer Gürtel*]: Endoskeleton structure: on inner side of shell (test), ring of processes around mouth area (peristome). Consists of apophyses and auricles and serves as site of muscle attachment for chewing apparatus (Aristotle's lantern). (continuous, discontinuous)

perihemal system: hyponeural sinus

peripetalous fasciole [G *peripetale Fasciole*]: In irregular sea urchin, fasciole enclosing most or all petals and apical plate system.

peripharyngeal cavity [G *Peripharyngealsinus*]: Body cavity (coelom) enclosing chewing apparatus (Aristotle's lantern).

peripharyngeal nerve ring: circumoral nerve ring

periproct [G *Periproct, Afterfeld, Analfeld*]: Opening in shell (test) for anus. Covered by membrane (periproctal membrane) bearing plates (periproctal plates).

periproctal membrane [G *Periproctmembran*]: Circular membrane surrounding anus and covering periproct. Bears plates (periproctal plates).

periproctal plate [G *Periproctplatte, Apikalplatte*]: Endoskeleton structure: one of basically four types of plates (apical, coronal, periproctal, peristomial). Consists of series of plates associated with membrane (periproctal membrane) around anus. Collectively termed periproctal system. (arrangement: random, spiral, in irregular circlets; dissociated, overlapping)

periproctal sinus [G *Periproctalsinus*]: In regular sea urchin, ring-shaped body cavity (coelom) at top of shell (test);

formed by membrane connecting rectum to test.

peristome (actinosome) [G *Peristom, Mundfeld, Bukkalfeld*]: Opening on underside of shell (test) for mouth. Covered by buccal membrane bearing plates (peristomial plates). (crescentic, D-shaped, oval, pentagonal)

peristomial plate [G *Peristomplatte, Peristomalplatte*]: Endoskeleton structure: one of basically four types of plates (apical, coronal, periproctal, peristomial). Consists of series of plates associated with membrane (buccal membrane) surrounding mouth.

perradial suture [G *Mediannaht, Mediansutur, Perradialnaht*]: Endoskeleton structure: line of contact (suture) between two rows (columns) of ambulacral plates forming each ambulacrum.

petal (petaloid) [G *Petalodium, petaloider Ambulakrum, Ambulakralblatt*]: Endoskeleton structure: on upper side of irregular sea urchin shell (test), expanded section of each ambulacrum. Five petals are arranged in flower-like configuration and bear tube feet (podia) modified for respiration. (open, closed, subpetaloid)

pharynx [G *Schlund*]: In sea urchin with chewing apparatus (Aristotle's lantern), section of digestive tract between buccal cavity and esophagus. Ascends vertically through lantern. (See also esophagus)

phyllode [G *Phyllodium*]: Endoskeleton structure: on underside of irregular sea urchin shell (test), section of each ambulacrum around mouth. Expanded into petal-like shape; bears large pores for modified tube feet (podia). (See also bourrelet)

plastron (sternum) [G *Plastron*]: Endoskeleton structure: on underside of irregular sea urchin shell (test) posterior to labrum, area between two elongate ambulacra (i.e., posterior ambulacra of bivium). Represents widened interambulacrum and may bear modified spines. (See also episternal plate)

plate [G *Platte*]: Endoskeleton structure: one of numerous plate-like calcareous elements forming shell (test). One may distinguish plates forming greater part of test (coronal system) as well as those around mouth (peristomial system), around anus (periproctal system), and aboral end (apical system). (oligoporous, polyporous)

pluteus [G *Pluteus*]: echinopluteus

podium (tube foot) [G *Füßchen, Ambulakralfüßchen*]: Movable extension of water-vascular system outside of shell (test). Located in a series along ambulacra; connected to ampullae via podial canals through pores in ambulacral plates. According to position or function one may, for example, distinguish branchial, buccal, and subanal podia as well as terminal tentacles. (blade-like, papillate, penicillate, suckered)

poison bag [G *Giftbeutel*]: At tip of certain spines, small, sac-like structure enclosed in muscle and connective tissue and containing toxic fluid.

poison sac [G *Giftsack*]: In jaws of certain pedicellaria (globiferous pedicellaria), sac-like structure within each jaw. Lined with glands; contains toxic fluid.

polian vesicle [G *Polische Blase, Poli'sche Blase*]: Part of water-vascular system; small outpocketing of ring canal, one aligned with top end of tooth in each interradius.

pore pair [G *Porenpaar*]: Endoskeleton structure: in each ambulacral plate of shell (test), pair of pores accommodating single tube foot (podium) of underlying water-vascular system. (similar, dissimilar; conjugate, nonconjugate; elongated, oval, pyriform, round)

primary plate (primary ambulacral plate) [G *Primärplatte*]: Endoskeleton structure: in shell (test), basic type of ambulacral plate. Bears one pore pair and extends from midline of ambulacrum to ambulacrum/interambulacrum border (i.e., from perradial to adradial suture). (See also compound plate)

primary spine (radiole) [G *Primärstachel*]: Endoskeleton structure: one of numerous large spines projecting from coronal plate of shell (test). Articulates with primary tubercles and represents first-formed spines. (See also miliary spine, secondary spine)

primary tubercle [G *Primärtuberkel, primäre Stachelwarze*]: Endoskeleton structure: on outer surface of coronal plate, large, more or less central tubercle bearing primary spine. (See also miliary tubercle, secondary tubercle)

pyramid [G *Pyramid, Kieferstück, Hauptstück*]: Endoskeleton structure: one of five large, interradial elements of chewing apparatus (Aristotle's lantern). Encloses most of tooth and consists of two demipyramids.

radial canal (radial water canal) [G *Radiärkanal, Radiärkanal des Ambulakralsystems, radiärer Ambulakralkanal, Radiärkanal des Mesocoels*]: Part of water-vascular system; one of five canals originating on ring canal and extending along ambulacra of shell (test). Gives rise to series of branches to various tube feet (podia).

radial hemal sinus (radial hemal vessel) [G *radiärer Haemalkanal, Radiärkanal der Blutlakune, radiäre Lakune des Blutgefäßsystems*]: One of five radial canals of hemal system. Extends along ambulacra parallel to radial water canal.

radial nerve [G *Radiärnerv, ectoneuraler Radiärstrang*]: In main = ectoneural nerve system, one of five nerves arising from circumoral nerve ring and running along midline of ambulacra; gives rise to podial nerve into each tube foot.

radiole: primary spine

radius [G *Radius*]: ambulacrum

rectum [G *Rektum, Enddarm*]: In digestive tract, terminal part of intestine. Opens to exterior via anus (i.e., ascending to apex of test in regular sea urchin or extending posteriorly in irregular forms).

ring canal (water ring) [G *Ambulakralring, Mesocoelring*]: Part of water-vascular system; circular canal surrounding anterior part of digestive tract. Located above chewing apparatus (Aristotle's lantern) when present. Gives rise to radial canal at each ambulacrum, polian vessel at each interambulacrum. Connected to exterior via stone canal and madreporite.

rosette (calcareous rosette) [G *Rosette, Kalkrosette*]: Endoskeleton structure: minute calcareous structure supporting expanded terminal end (disc) of tube foot (podium). Located distal to frame. May also refer to flower-like arrangement of ambulacra (petals) on upper side of irregular sea urchin.

rotule (rotula, brace) [G *Rotula, Schaltstück*]: Endoskeleton structure: one of five larger elements interlocking with epiphyses at top of chewing apparatus (Aristotle's lantern). Located in ambulacral position and oriented radially.

scrobicular ring [G *Warzenring*]: On shell (test) plate, ring of small tubercles (scrobicular tubercles) bordering smooth area (scrobicule) around primary tubercle.

scrobicular tubercle [G *Tuberkel des Warzenrings*]: Endoskeleton structure: on outer surface of shell (test) plate (interambulacral plate), one in a series of secondary tubercles of scrobicular ring around primary tubercle.

scrobicule [G *Höfchen*]: Endoskeleton structure: on shell (test) plate, smooth, ring-like depression around base (boss) of tubercle. Surrounded by scrobicular ring; serves as attachment site for spine muscles. (confluent, contiguous) (See also areole)

secondary spine [G *Sekundärstachel, Nebenstachel, Stachel zweiter Ordnung*]: Endoskeleton structure: one of numerous smaller spines projecting from shell (test) plate. Develops later than larger primary spines and articulates with secondary tubercles.

secondary tubercle [G *Sekundärtuberkel, sekundärer Tuberkel*]: Endoskeleton structure: on outer surface of shell (test) plate, smaller tubercle bearing secondary

spine. (See also military tubercle, primary tubercle, scrobicular tubercle)

septum [G *Septum*]: Endoskeleton structure: in spine, one in a series of longitudinal partitions radiating from central medulla to outer wall (cortex).

shaft [G *Schaft*]: Endoskeleton structure: distal and longest section of spine.

shell: test

siphon [G *Siphon, Nebendarm*]: In digestive tract, slender tube running parallel to first loop of intestine (= stomach = inferior intestine = small intestine). Funnels water directly from first to second loop of intestine.

sphaeridium [G *Sphäridium*]: Endoskeleton structure: one in a series of minute, movable spines articulating on small tubercles. Restricted to ambulacral section of shell (test), typically near mouth. (oval, spherical; free, in grooves, in pits, in closed chambers)

spicule [G *Sklerit, Spicula*]: Endoskeleton structure: one of variously shaped, minute calcareous elements located, for example, in tube feet, around mouth, and in wall of digestive tract. (biacerate = bihamate, dumbbell-shaped, forked = H-shaped, triradiate)

spine [G *Stachel*]: Endoskeleton structure: one in a series of movable calcareous structures articulating with projection (tubercle) of shell (test) plate. According to size, one may distinguish primary, secondary, and miliary spines. If fully developed, consists of socket (acetabulum), base, milled ring, collar, neck, and shaft. (club-shaped, fan-shaped, flat-topped, hair-like, mushroom-shaped, oar-like, paddle-shaped; curved, straight; cylindrical, flattened; banded, fluted, smooth, tessellate, with secondary spines or thorns) (See also cortex, medulla, septum)

spongy body [G *Tiedemann'sches Bläschen*]: Outpocketing of oral hemal ring in association with ring canal.

standard measurements: ambitus diameter

length of mouth-anus axis
peristome diameter

stem (stalk) [G *Stiel, Pedicellarien-stiel*]: Elongate, stalk-like basal part of pedicellaria. Supported internally by rod-like endoskeletal element.

stereom [G *Stereom*]: Endoskeleton structure: structural meshwork of which all endoskeleton structures (except teeth and spicules) are composed. Permeated by living tissue.

sternum: plastron

Stewart's organ [G *Gabelblase, Stewart'sches Organ*]: One of five large body cavities (coeloms) extending radially from periesophageal sinuses.

stomach: intestine

stone canal [G *Steinkanal*]: Part of water-vascular system; vertical canal between madreporite at apex of shell (test) and ring canal surrounding digestive tract. Walls strengthened by calcareous deposits. Accompanied by axial gland.

subanal fasciole [G *subanale Fasciole*]: In irregular sea urchin, elliptical fasciole below anal region (periproct) at posterior end of shell (test).

sucker: disc

suranal plate (primary anal plate) [G *Zentralplatte*]: Endoskeleton structure: largest and first-formed element on plate system (periproctal plates) around anus.

suture [G *Sutur, Naht*]: Endoskeleton structure: line of contact between plates forming shell (test). According to position, one may distinguish adapical, adoral, adradial, interradial, and perradial sutures. May also refer to line of contact between the two demipyramids of each pyramid of chewing apparatus (Aristotle's lantern).

terminal plate: ocular plate

terminal tentacle (ocular tentacle) [G *Terminaltentakel, Endtentakel*]: Part of water-vascular system; one of five modified tube feet (podia) at apex of shell (test). Represents end of each radial canal and extends through pore (ocular pore) of ocular plate.

tertiary spine: miliary spine

test (shell) [G *Schale*]: Endoskeleton structure: rigid shell determining shape of body and enclosing internal organs. Consists of series of fused plates. (elongate, flattened, globular, hemispherical, ovoid, nearly cylindrical; rigid, somewhat flexible, very flexible)

testis: gonad

tooth [G *Zahn*]: Endoskeleton structure: one of five elongate elements in chewing apparatus (Aristotle's lantern); each tooth is surrounded by pyramid. Located in interambulacral position, with tips projecting from mouth. Also refers to tooth-like tip of valve in pedicellaria. (equal, unequal)

trivium [G *Trivium*]: In irregular sea urchin, anterior three ambulacra. Ambulacra of trivium extend to anteriorly displaced mouth and are thus shorter than posterior ambulacra of bivium.

tube foot: podium

tubercle [G *Tuberkel, Warze*]: On outer surface of shell (test) plate, knob-like elevation bearing spine. According to position and size one may distinguish primary, secondary, and miliary tubercles. (perforate, imperforate; crenulate, noncrenulate) (See also boss, mamelon, scrobicular tubercle)

valve [G *Greifzange, Klappe*]: Endoskeletal element within each jaw of pedicellaria. May bear distal tooth and be closely associated with poison gland.

water ring: ring canal

water-vascular system (ambulacral system) [G *Wassergefäßsystem, Ambulakralsystem*]: System of canals consisting of opening to exterior (hydropores in madreporite), stone canal, ring canal with polian bodies, and radial canals connected to tube feet (podia).

zygopore: pore pair

HOLOTHUROIDEA

papillate podium

locomotory podium

papilla

creeping sole

sail

tentacle

web

vermiform / cylindrical | flattened

creeping

swimming

(lateral)

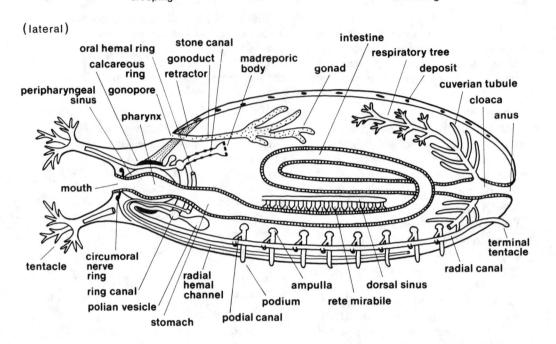

oral hemal ring
calcareous ring
peripharyngeal sinus
gonopore
pharynx
stone canal
gonoduct
retractor
madreporic body
intestine
respiratory tree
gonad
deposit
cuverian tubule
cloaca
anus

mouth

tentacle
circumoral nerve ring
ring canal
polian vesicle
stomach
radial hemal channel
podial canal
podium
ampulla
rete mirabile
dorsal sinus
terminal tentacle
radial canal

tentacle

branched | pinnate | peltate | digitate

deposit (ossicle, sclerite)

button
hole

miliary granule

anchor plate
hole
socket

rods

simple

rosette

basket

anchor
vertex
fluke
shank
stock

thorny

C-rod | increased branching
hole

table
spire

wheel
spoke

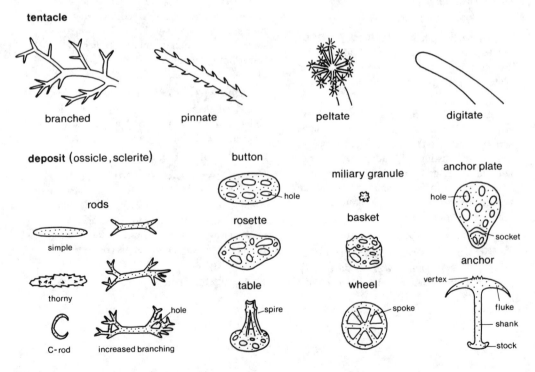

ECHINODERMATA

HOLOTHUROIDEA

holothurian, sea cucumber [G *Holothurie, Seegurke, Seewalze*]

ambulacrum [G *Ambulacrum*]: One of five radial tracts running longitudinally along body; consists of radial canal of water-vascular system and row of tube feet (podia). (See also interambulcrum)

ampulla [G *Ampulle*]: Part of water-vascular system; expanded section of canal at base of tube foot (podium). Joined to radial canal via pidial canals. Ampullae of tentacles often termed tentacular ampullae. (single, branched)

anal papilla (anal tooth) [G *Analpapille*]: Part of water-vascular system; one of five projections surrounding anus. Represents modified tube foot (podium) at end of radial canals. May be calcified and tooth-like.

anchor [G *Anker*]: Endoskeleton structure: anchor-shaped deposit consisting of shaft (shank) bearing pair of arms (flukes) on one end and terminal bar (stock) at other end. Central area between flukes is termed vertex. Anchor is attached at an angle to anchor plate. (stock: rugose, smooth; branching, unbranched; flukes: smooth, laterally toothed = serrate; vertex: smooth, with minute knobs)

anchor plate [G *Ankerplatte*]: Endoskeleton structure: perforated, plate-like deposit. Bears arched transverse bar on its often narrowed posterior end; bar is site of attachment between anchor plate and an-chor. (irregular, oval, pear-shaped, rectangular, rounded, subrectangular; transverse bar: smooth, toothed)

antimesenterial sinus: ventral sinus

anus [G *After*]: Posterior opening of digestive tract. Typically located at posterior end of body and may be surrounded by anal papillae, plates, or teeth. (dorsal, terminal, ventral)

aquapharyngeal bulb [G *peripharyngealer Unterhaut-Ringwulst*]: Term applied to complex structure consisting of pharynx and sac-like surrounding tissue; outer wall of latter bears calcareous ring and section of radial canals. Outer wall of bulb separated from pharynx by cavity (peripharyngeal sinus). (See also pharynx suspensor)

auricularia [G *Auricularia, Auricularia-Larve*]: Early, free-swimming larval stage. Characterized by continuous curving, flagellated band. (See also doliolaria, pentactula)

axial gland [G *Axialdrüse*]: Term variously applied to spongy tissue associated with and extending posteriorly from water ring.

basket [G *Gitterhalbkugel*]: Endoskeleton structure: type of deposit consisting of concave, perforated plate. (rim: smooth, toothed)

bivium [G *Bivium*]: Section of sea cucumber body including two of five radii (C and D) as well as interradii BC, CD, and DE. May form dorsal surface of creeping species. (See also trivium)

bridge: table

buccal membrane (peristome) [G *Bukkalmembran, Peristom, Mundhaut*]: Circular membrane surrounding mouth and bordered by circlet of tentacles. Associated with circumoral nerve ring and peribuccal sinus.

buccal podium [G *Bukkal-Füßchen*]: tentacle

button [G *Schnalle*]: Endoskeleton structure: flat type of deposit perforated with four, six, or more similarly sized holes in two rows. (circular, elongate, oval; knobbed, smooth) (See also fenestrated ellipsoid, pseudobutton, rosette)

calcareous ring [G *Kalkring*]: Ring of basically 10 calcareous elements surrounding pharynx (associated with outer wall of aquapharyngeal bulb). One may distinguish five larger radial and five smaller interradial plates. Serves as site of attachment of longitudinal muscles and retractors, with radial elements accommodating radial canals and radial nerves. Plate shape is of taxonomic importance.

cartilaginous ring: Ring of dense connective tissue directly behind calcareous ring.

caudal appendage [G *Schwanzanhang*]: In certain sea cucumbers, abruptly narrowed, tail-like posterior section of body. Bears terminal anus. Serves as respiratory tube. (See also postanal tail)

cecum [G *Caecum, Blindsack*]: In certain sea cucumbers, anteriorly directed blind sac arising from cloaca.

ciliated funnel: ciliated urn

ciliated urn (ciliated funnel) [G *Wimperurne, Urne*]: In certain sea cucumbers, one of numerous minute, concave structures projecting into body cavity. Attached to mesenteries or body wall by

stalks. Considered to serve in excretion. (See also tree of urns)

circumoral nerve ring (nerve ring) [G *oraler Nervenring, Ringnerv*]: In nerve system, ring of nerve tissue associated with buccal membrane surrounding mouth. Innervates tentacles and pharynx and sends radial nerves through calcareous ring into posterior end of body.

cloaca [G *Kloake*]: Expanded, relatively short posterior section of digestive tract between intestine and anus. Attached to inner body wall by cloacal suspensors. Respiratory trees originate from cloaca.

cloacal suspensor [G *Aufhängebänder der Kloake*]: Radiating series of muscles and connective tissue spanning from posterior section (cloaca) of digestive tract to inner body wall.

coelom [G *Coelom*]: Epithelium-lined body cavity. Divided into mesocoel forming water-vascular system and metacoel forming main body cavity. The latter is further subdivided into various smaller cavities (e.g., hyponeural, perianal, peribuccal, and peripharyngeal sinuses).

collecting vessel [G *Sammelsinus*]: In blood-lacunar (hemal) system, longitudinal channel running between rete mirabile and small intestine. Connected to both by series of branches.

creeping sole: sole

cruciform body [G *kreuzförmiger Körper, Kreuzstab*]: Endoskeleton structure: type of deposit resembling a cross.

cup [G *Napf*]: Endoskeleton structure: cup-shaped, fenestrated type of deposit.

cuvierian tubule (Cuvierian organ, tubule of Cuvier) [G *Cuvierscher Schlauch, Cuvier'scher Schlauch*]: One of numerous hollow, tube-like organs arising from base of one or both respiratory trees. Sea cucumbers may emit cuvierian tubules from anus when disturbed.

deposit (ossicle, spicule) [G *Sklerit*]: Endoskeleton structure: one of numerous small calcareous elements embedded in outer body wall or located in tentacles and podia. Highly variable in shape, including anchors, anchor plates, baskets, buttons, cups, military granules, plates, pseudobuttons, rods, rosettes, sigmoid bodies, tables, and wheels.

doliolaria (pupa) [G *Doliolaria*]: Free-swimming larval stage following auricularia. Characterized by series of flagellated rings around body. (See also pentactula)

dorsal mesentery [G *dorsales Mesenterium, oberes Mesenterium*]: Thin, longitudinal membrane joining anterior part of digestive tract (esophagus, stomach, anterior intestine) to body wall. Attaches to body wall along middorsal interradius (CD). Gonad associated with dorsal mesentery. (See also left mesentery, ventral mesentery)

dorsal papilla (papillate podium, wart) [G *Dorsalpapille, Papille*]: Part of water-vascular system; in creeping sea cucumber species, modified, nonlocomotory type of tube foot (podium) on dorsal side of body.

dorsal sinus (mesenterial sinus) [G *dorsaler Sinus, mesenteriale Darm-Längslakune, dorsaes Darmgefäß*]: In blood-lacunar (hemal) system, large, longitudinal channel extending along that side of intestine attached to body wall by dorsal mesentery. Communicates with intestine via numerous branches and rete mirabilis.

ectoneural nerve system [G *ektoneurales Nervensystem*]: radial nerve

ellipsoid: fenestrated ellipsoid

end plate [G *Endscheibchen, Sklerit*]: Endoskeleton structure: small, perforated calcareous deposit supporting sucker of tube foot (podium).

epineural canal (epineural sinus) [G *Epineuralkanal*]: One of five longitudinal (radial) body cavities running between radial nerves and radial water canals. (See also hyponeural canal)

esophageal ossicle: calcareous ring

esophagus [G *Ösophagus*]: Slender, relatively short section of digestive tract between pharynx and stomach.

eye spot [G *Augenfleck*]: In certain sea cucumbers, pair of photosensitive organs at base of each tentacle.

fenestrated ellipsoid [G *gefenstertes Ellipsoid*]: Endoskeleton structure: type of deposit representing modified button. Hollow, fenestrated, three-dimensional structure formed by interconnecting knobs.

fenestrated sphere [G *gefensterte Kugel*]: Endoskeleton structure: type of deposit representing modified table. Hollow, rounded, fenestrated structure formed by interconnection between spines of spire and margin of plate.

fluke [G *Ankerbogen*]: anchor

genital papilla [G *Genitalpapille*]: Papilla-like elevation of body wall bearing opening (gonopore) of reproductive system.

gonad [G *Gonade*]: ♂ (testis) or ♀ (ovary) reproductive organ located in anterior part of body cavity. Consists of numerous tubes (gonadial tubules) often forming a tuft. Associated with dorsal mesentery and opening to exterior via gonoduct.

gonadial tubule [G *Gonadenröhrchen*]: One of typically numerous tubes forming gonad. United basally into tuft-like structure. (simple, branched)

gonoduct [G *Gonodukt*]: Section of reproductive system between tubules of ovary or testis and gonopore. Associated with dorsal mesentery.

gonopore [G *Gonoporus, Genitalporus, Geschlechtsöffnung*]: Opening(s) of reproductive system to exterior. Located anteriorly along middorsal line (interradius CD) between or somewhat posterior to tentacles. May be located on genital papilla.

hemal ring: oral hemal ring

hydropore [G *Hydroporus*]: Opening of water-vascular system to exterior. General term for one or numerous pores on madreporite or madreporic body.

hyponeural canal: radial hyponeural sinus

hyponeural nerve system [G *hyponeurales Nervensystem*]: radial nerve

interambulacrum [G *Interambulacrum*]: Longitudinal section of body between ambulacra.

interradial plate [G *interradiärer Sklerit*]: calcareous ring

interradius [G *Interradius*]: One of five longitudinal sections of body between radii. Interradius CD is defined by location of madreporite and gonopore. (See also bivium, trivium)

intestine [G *Darm*]: Highly elongated section of digestive tract between stomach and cloaca. Forms loops within body cavity and may be variously subdivided into narrower small intestine (with descending and ascending parts) and more expanded large intestine. Connected to body wall by mesenteries (dorsal, left, ventral mesenteries).

introvert [G *Introvert*]: In certain sea cucumbers, smooth-skinned, collar-like region of body wall posterior to tentacles. May be retracted into body. (See also tentacular collar)

large intestine: intestine

left mesentery [G *linkes Mesenterium*]: Thin, longitudinal membrane joining part of small intestine to body wall. Attaches to body wall along left dorsal interradius (DE). (See also dorsal mesentery, ventral mesentery)

lenticulate plate (scale) [G *linsenförmiges Plättchen*]: Endoskeleton structure: type of thick deposit resembling double-convex lens.

lithocyte [G *Statolith, Lithozyte*]: Type of cell within equilibrium sense organ (statocyst). Contains inorganic material and functions as statolith.

longitudinal muscle [G *Längsmuskel, radiärer Längsmuskel*]: One of five concentrations of muscle running along inner body wall from anterior to posterior end.

Located radially and adjoining radial canals and radial nerves.

madrepore [G *Madreporus*]: One or more openings of madreporite.

madreporic body [G *Madreporenköpfchen*]: Part of water-vascular system; in sea cucumbers in which stone canal does not open to exterior, expanded distal end of stone canal within body cavity. Bears one or more canals and pores and is supported by calcareous elements.

madreporite [G *Madreporenplatte*]: Opening of water-vascular system to exterior. Consists of calcareous plates perforated by pore canals. Attached to ring canal via stone canal. (See also madreporic body)

mesenterial sinus: dorsal sinus

mesocoel [G *Mesocoel*]: coelom

metacoel [G *Metacoel*]: coelom

miliary granule [G *Körnchen*]: Endoskeleton structure: small, solid type of deposit. (rounded, slightly elongate)

mouth [G *Mund, Mundöffnung*]: Anterior opening of digestive tract. Surrounded by buccal membrane and circlet of tentacles. Opens into pharynx.

nerve ring: circumoral nerve ring

oral disc: buccal membrane

oral hemal ring (haemal ring) [G *oraler Haemalring, Ringsinus, Ringkanal*]: In blood-lacunar (hemal) system, ring-like canal surrounding pharynx. Gives rise to five radial hemal channels extending to posterior end of body.

ossicle: deposit

ovary: gonad

papilla: anal papilla, dorsal papilla, genital papilla

papillate podium: dorsal papilla

pedicel [G *Lokomotionsfüßchen*]: Part of water-vascular system; term applied to those tube feet (podia) modified for locomotion. (with/without sucker)

pentactula [G *Pentactula*]: Late larval stage. Characterized by five primary tentacles and one or two podia.

perforated plate: plate

perianal sinus [G *Perianalsinus*]: Body cavity (coelom) surrounding anus.

peribuccal sinus [G *Peribukkalsinus*]: Body cavity (coelom) associated with peribuccal membrane and located anterior to peripharyngeal sinus.

peripharyngeal sinus [G *Peripharyngealsinus*]: Body cavity (coelom) between pharynx and outer wall of aquapharyngeal bulb. Traversed by pharynx suspensors.

pharynx [G *Pharynx, Schlund*]: Expanded, muscular, cuticle-lined section of digestive tract between mouth and esophagus or stomach. Located within aquapharyngeal bulb complex and passing through calcareous ring and ring canal.

pharynx suspensor [G *Pharynxaufhängung*]: One of numerous strands connecting pharynx with outer wall of aquapharyngeal bulb (where they attach to calcareous ring). Passes through peripharyngeal sinus and consists of connective tissue and muscle fibers. (arrangement: regular, irregular)

phosphatic body: Endoskeleton structure: type of deposit coated with or consisting entirely of iron phosphate. Represents modified anchor or rosette.

plate [G *Platte*]: Endoskeleton structure: refers to radial and interradial plates of calcareous ring or to a wide variety of flattened, plate-like deposits in body wall (anchor, end, lenticulate, perforated plates).

podial canal [G *Füßchenkanal*]: Part of water-vascular system; one in a series of lateral branches of radial canals. Extends to both ampulla and to tube foot (podium).

podium (tube foot) [G *Ambulakralfüßchen, Füßchen*]: Part of water-vascular system; general term for several types of projections of body wall to exterior. Associated with ampullae and supplied by channels (podial canals) branching from radial canals. Locomotory podia are termed pedicels; nonlocomotory dorsal types are termed papillate podia, papillae, or warts, while enlarged anterior podia are termed buccal podia or tentacles. (retractable, unretractable; with/without sucker; evenly/not evenly distributed)

polian vesicle (Polian vesicle) [G *Polische Blase, Poli'sche Blase*]: Part of water-vascular system; one of up to several or numerous muscular outpocketings of ring canal. (elongated, ovoid, rounded)

postanal tail [G *Rückenfortsatz*]: In certain sea cucumbers, tail-like extension of body posterior to anus. (See also caudal appendage)

pseudobutton [G *irreguläre Schnalle*]: Endoskeleton structure: type of deposit representing incomplete or imperfect button. Usually twisted and bearing only single row of holes.

pupa: doliolaria

radial canal [G *Radiärkanal, Radiärkanal des Mesocoels*]: Part of water-vascular system; one of five canals originating on ring canal and extending anteriorly into tentacles (as tentacular canals), posteriorly along calcareous ring and body wall into tube feet (podia).

radial hemal channel [G *Radiärlakune, radiäre Haemallakune*]: In blood-lacunar (hemal) system, one of five longitudinal channels arising from oral hemal ring. Extends radially to posterior end of body along with radial water canals.

radial hyponeural sinus (hyponeural canal) [G *Hyponeuralkanal, oraler Radiärkanal des Metacoels*]: Body cavity (coelom) running longitudinally along outer side of each radial nerve.

radial nerve [G *Radiärnerv*]: In nerve system, one of five nerves arising from circumoral nerve ring, passing through radial plates of calcareous ring, and extending radially along body wall to posterior end of body. Consists of closely adjoining, parallel ectoneural (sensory and motor) and hyponeural (motor) components.

radial plate [G *radiärer Sklerit*]: calcareous ring

radius [G *Radius*]: One of five longitudinal sections of body corresponding internally to radial canals of water-vascular, hemal, and nerve systems, externally to rows of tube feet (podia). Radius opposite gonopore and madrepodite is designated as A, remaining radii follow in alphabetic order in clockwise direction (B,C,D,E). (See also bivium, interradius, trivium)

rectum [G *Enddarm*]: Term occasionally applied to somewhat distended posterior section (= large intestine) of intestine or to cloaca.

respiratory tree [G *Wasserlunge*]: One of two (left and right) anteriorly directed, tube-like organs arising from posterior section (cloaca) of digestive tract. Highly branched, with branches ending in thin-walled vesicles. (equal, unequal)

rete mirabile [G *Wundernetz*]: In blood-lacunar (hemal) system, dense network of channels between dorsal sinus and small intestine. Consists of a series of tuft-like structures. (See also collecting vessel)

retractor muscle [G *Retraktor*]: One in a series of five radial muscle bands spanning from calcareous ring of aquapharyngeal bulb to body wall. Serves to retract anterior end.

ring canal (water ring) [G *Ambulakralring, Ringkanal des Mesocoels*]: Part of water-vascular system; circular canal surrounding pharynx. Bears series of polian vesicles and stone canal(s) and gives rise to radial canals supplying tentacles and various other types of tube feet (podia).

rod [G *Stab*]: Endoskeleton structure: type of deposit that, in its basic form, is elongate and bar-like. (smooth, spinose, warty, with/without branches, with/without perforated ends)

rosette [G *Rosette, Gitterplatte*]: Endoskeleton structure: type of deposit representing modified rod. Highly branched, with branch ends variously fused to form perforations. Differs from button in that perforations vary in size.

sail [G *Schultermembran*]: In certain sea cucumbers, fan-shaped, membranous extension of body wall. Located anteriorly on dorsal side of body. Serves in swimming.

scale [G *Panzerplatte*]: Endoskeleton structure: in certain sea cucumbers, one of numerous large, perforated plates covering outer surface of body. (rhomboidal, rounded; smooth, warty; imbricating, not imbricating) (See also lenticulate plate)

sensory cup [G *Sinnesgrube*]: One in a series of minute, cup-shaped sensory organs attached to base of tentacles by stalks.

shank: anchor

sigmoid body [G *sigmoider Körper*]: Endoskeleton structure: type of smooth, curved rod with curve at one end being pointed, the other being blunt. Three-dimensional, two halves curving in planes at right angles to each other.

small intestine: intestine

sphere: fenestrated sphere

spicule: deposit

spire: table

sole (creeping sole) [G *Kriechsohle*]: In creeping sea cucumber, modified ventral surface bearing locomotory feet (podia). Corresponds to trivium and consists of radii E,A,B as well as interradii AB and AE.

statocyst [G *Statozyste, Statocyste*]: One in a varying number of organs of equilibrium associated with circumoral nerve ring or with radial nerves at anterior end of body. Contains lithocytes.

stock: anchor

stomach [G *Magen*]: Expanded section of digestive tract between esophagus and intestine.

stone canal [G *Steinkanal*]: Part of water-vascular system; one or more canals arising from ring canal and strengthened by calcareous deposits. Opens either to exterior via pore(s) (madrepores, hydropores) or into body cavity via madreporic body.

sucker (terminal disk, suctorial disc) [G *Saugscheibe*]: Part of water-vascular system; disc-shaped terminal expansion of tube foot (podium). Often supported by endoskeletal element (end plate).

table [G *Stühlchen, Turm*]: Endoskeleton structure: type of deposit consisting of perforated, plate-like part and erect spire. Spire ends in several short spines and consists of two to four vertical elements linked by transverse bars (bridges). (See also fenestrated sphere)

tail: postanal tail

tentacle (buccal podium) [G *Tentakel, Bukkalfüßchen, Mundfüßchen*]: Part of water-vascular system; variously modified anterior tube feet (podia). Highly retractile, supplied by radial canals or directly by ring canal. (dendritic, digitate, peltate, peltato-digitate, pinnate)

tentacle ampulla [G *Tentakelampulle, Bukkalampulle*]: ampulla

tentacular canal [G *Tentakelkanal*]: Part of water-vascular system; narrow canal extending through each tentacle. Originates either from radial canal or directly from ring canal.

tentacular collar [G *Kragen*]: In certain sea cucumbers, modified body wall at anterior end of body. Forms projecting, often fringed collar when tentacles are retracted.

terminal disk: sucker

terminal tentacle [G *Terminaltentakel, Terminalfüßchen*]: In water-vascular system, term applied to last tube foot (podium) arising from each radial canal. Located at posterior end of body near anus.

testis [G *Hoden*]: gonad

tree of urns [G *Wimperbäumchen*]: Dense aggregation of ciliated urns on fold of mesentery.

trivium [G *Trivium*]: Section of sea cucumber body including three of five radii (E,A,B) as well as interradii AB and AE. May form ventral creeping sole.

tube foot: podium

tubule of Cuvier: cuvierian tubule

urn: ciliated urn

ventral mesentery [G *unteres Mesenterium*]: Thin longitudinal membrane joining large intestine to body wall. Attaches to body wall ventrally along interradius AB. (See also dorsal mesentery, left mesentery)

ventral sinus (antimesenterial sinus) [G *ventraler Darmsinus, antimesenteriale Darm-Längslakune*]: In blood-vascular (hemal) system, large, longitudinal channel extending along side of intestine not attached to body wall by dorsal mesentery. Communicates with intestine via numerous branches. (See also dorsal sinus)

vertex: anchor

wart [G *Warze*]: papilla

water lung: respiratory tree

water ring: ring canal

water-vascular system (ambulacral system) [G *Wassergefäßsystem, Ambulakralsystem, Mesocoel*]: System of canals consisting of ring canal bearing polian vesicles and stone canal(s) (with or without opening to exterior) as well as series of radial canals extending into tentacles and various other types of tube feet (podia) via podial canals.

wheel [G *Rädchen*]: Endoskeleton structure: type of deposit resembling a wheel. Consists of six or more spokes leading from central hub to peripheral rim.

PTEROBRANCHIA

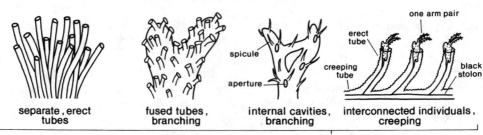

separate, erect tubes | fused tubes, branching | internal cavities, branching | interconnected individuals, creeping

spicule | aperture

erect tube | creeping tube | one arm pair | black stolon

with coenecium

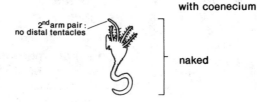

2nd arm pair: no distal tentacles

naked

protosome = proboscis (buccal shield)   mesosome = collar

collar ganglion
stomochord
heart vesicle
central sinus
ventral shield sinus
glomerulus
mouth
buccal tube
oral lamella
pharynx
gill slit
esophagus
ventral trunk nerve
ventral trunk sinus

knob
arm
tentacle
gonopore
anus
rectum
gonad
dorsal sinus
dorsal recess
intestine
stomach

stalk

bud

metasome = trunk

PTEROBRANCHIA

pterobranch [G *Pterobranchier*]

anus [G *After*]: Posterior opening of digestive tract. Located on metasome = trunk at varying distances from metasome/mesosome border; may be somewhat elevated on a protuberance.

arm [G *Arm*]: One of a single pair or several pairs of hollow projections on dorsal side of mesosome (= collar). Typically bears ventral groove and series of small tentacles along each side. Serves in respiration and feeding. (with/without knob)

arm nerve [G *Armnerv*]: One in a series of nerves originating from collar ganglion and extending into each arm; runs dorsally along arm.

black stolon [G *schwarzer Stolo*]: In creeping type of colony with erect tubes, narrow, tubular cord in floor of creeping part. Extends short distance into each erect tube and connects all individuals of colony.

branchial sac (dorsolateral pharyngeal pocket, gill pouch) [G *Kiemensack, Kiementasche*]: One of two sac-like pouches associated with pharynx and into which gill slits open. Connected to exterior via gill pore.

buccal diverticulum: stomochord

buccal shield [G *Mundschild*]: protosome

buccal tube [G *Mundhöhle*]: Short anterior section of digestive tract in mesosome (= collar) between mouth and pharynx. Gives rise to anteriorly directed buccal diverticulum (= stomochord). (See also dorsal recess of pharynx)

bud [G *Knospe*]: In asexual reproduction, young individual developing from base of stalk.

central sinus [G *Zentralsinus, Herz*]: Anterior part of circulatory system in front of buccal diverticulum (= stomochord) in protosome. Located under and largely enclosed by heart vesicle. Receives blood from dorsal sinus and pumps it posteriorly via ventral shield sinus, glomerulus, peribuccal channels, and ventral trunk sinus.

cephalic shield [G *Kopfschild*]: protosome

circumenteric connective (circumbuccal connective) [G *Nervenring*]: Ring of nerve tissue surrounding digestive tract and connecting ventral and dorsal trunk nerves; located at mesosome/metasome (= collar/trunk) border.

coelom [G *Coelom*]: Epithelium-lined body cavity. Basically divided into protocoel of protosome, mesocoel of mesosome, and metacoel of metasome.

coenecium [G *Coenecium, Cönöcium, Zoecium, Gehäuse*]: Rigid outer structure encasing a colony of individuals (zooids). May incorporate foreign particles. (with separate, erect tubes; branching, with fused tubes; branching, without tubes; spongy, without tubes; creeping, with erect tubes)

collar [G *Kragen*]: mesosome

collar canal [G *Kragenpforte, Coelomodukt*]: One of two dorsolateral canals leading from collar cavity (mesocoel) to exterior via collar pores.

collar ganglion [G *Kragengang-lion, Kragenmark, dorsales Ganglion*]: Main middorsal concentration of nerve tissue in mesosome (= collar). Gives rise to series of arm nerves, posteriorly to dorsal trunk nerve, and anteriorly to plexus in protosome.

collar pore [G *Kragenporus*]: External pore of each canal (collar canal) leading from collar cavity (mesocoel). Located anterior to gill pores.

dorsal recess of pharynx [G *Pharynxblindtasche*]: Short, broad outpocketing of digestive tract originating at buccal tube/ pharynx border.

dorsal sinus [G *Dorsomediangefäß, dorsales Rumpfgefäß*]: In circulatory system, relatively short middorsal blood channel above pharynx and esophagus. Connected with genital sinus; conveys blood anteriorly to central sinus.

dorsal trunk nerve [G *Dorsalnerv, dorsaler Mediannerv*]: Relatively short nerve cord extending from collar ganglion to anus in metasome. Located middorsally and joined to ventral nerve cord at mesosome/ metasome (= collar/trunk) border by circumenteric connectives.

esophagus [G *Ösophagus, Oesophagus*]: Section of digestive tract between pharynx and stomach in metasome (= trunk).

genital sinus [G *Genitalsinus*]: In circulatory system, blood sinus surrounding each gonad in anterior region of metasome (= trunk); joined to dorsal sinus.

gill pore [G *Kiemenporus*]: In respiratory system, one of two dorsolateral pores on each side of anterior metasome (= trunk). Represents opening of gill slits and branchial sacs to exterior.

gill slit [G *Kiemenspalte*]: In respiratory system, one of two oval openings on each side of pharynx. Located dorsolaterally and opening to exterior via branchial sac and gill pore.

glomerulus [G *Glomerulus*]: In protosome, folded structure closely associated with ventral shield sinus under buc-

cal diverticulum (= stomochord). Considered to represent excretory organ.

gonad [G *Gonade*]: ♂ (testis) or ♀ (ovary) reproductive organ located anterodorsally in metasome (= trunk). Each gonad opens to exterior via gonoduct and gonopore. (elongated, oval, rounded, sacciform; single, paired)

gonoduct [G *Gonodukt*]: Relatively short, narrow section (vas deferens in ♂, oviduct in ♀) of reproductive system between gonad and gonopore.

gonopore [G *Gonoporus, Geschlechtsporus, Genitalporus*]: Opening of gonad to exterior. Located anterodorsally in metasome (= trunk), i.e., near mesosome/ metasome border. (single, paired)

heart vesicle (pericardial sac) [G *Herzblase, Perikard, Pericard*]: Muscular section of circulatory system more or less surrounding central sinus. Serves to pump blood posteriorly.

intestine [G *Mitteldarm*]: Elongate, relatively narrow section of digestive tract between stomach and rectum in metasome (= trunk). Curves anteriorly to open near mesosome/metasome border.

knob (terminal knob, glandular knob) [G *Endanschwellung*]: Modified, swollen tip of arm. Glandular, lacking tentacles. Elongated and rod-like distally in second arm pair of naked pterobranch.

lip: lower lip, upper lip

lower lip [G *Unterlippe*]: Slightly protruding body wall below mouth. Adjoined posteriorly by oral lamella. (See also upper lip)

mesentery [G *Mesenterium*]: One in a series of longitudinal partitions dividing body cavities (coeloms). According to position one may distinguish dorsal, ventral, and lateral mesenteries.

mesocoel (collar coelom) [G *Kragencoelom, Mesocoel, Mesosoma-Coelom*]: Body cavity (coelom) of mesosome (= collar). Paired, extending into arms and oral lamellae; may send pair of anterior outpocketings into protocoel. Opens to exterior via pair of collar canals. (See also metacoel, protocoel)

mesosome (collar) [G *Mesosoma, Kragen*]: Relatively short division of body between protosome (= buccal shield, cephalic shield, proboscis) and metasome (= trunk). Bears mouth ventrally and arms dorsally. (with one pair/five to nine pairs of arms) (See also mesocoel)

metacoel (trunk coelom) [G *Metacoel, Rumpfcoelom, Metasoma, Metasoma-Coelom*]: Body cavity (coelom) of metasome (= trunk). Basically paired, extending into stalk and occasionally sending pair of anteriorly directed outpocketings into mesocoel.

metasome (trunk) [G *Metasoma, Rumpf*]: Posterior and largest of three divisions (protosome, mesosome, metasome) of body. Divided into sac-shaped part (trunk sac) containing digestive tract and slender posterior stalk. (See also metacoel)

mouth [G *Mund, Mundöffnung*]: Anterior slit-shaped opening of digestive tract. Located ventrally in mesosome (= collar) and more or less covered by downturned protosome. Bordered by upper lip, lower lip and oral lamellae; opens into buccal tube.

muscle [G *Muskel*]: Any one of various circular, longitudinal, or radial muscles enabling body movement, retraction into tubes, closing of pores (sphincter muscles), or operation of circulatory system.

oral lamella [G *Orallamelle, Kragenfalte*]: In mesosome (= collar), fold of body wall extending from base of arms around each side to mouth. Right lamella may be larger than left lamella. Serves in feeding.

ovary [G *Ovar, Ovarium*]: gonad

oviduct [G *Ovidukt, Eileiter*]: gonoduct

peribuccal channel [G *Ringgefäß*]: In circulatory system, pair of blood channels surrounding digestive tract (buccal tube) and connecting ventral shield sinus with ventral trunk sinus.

pharynx [G *Pharynx*]: Section of digestive tract in anterior part of metasome (= trunk) between buccal tube and esophagus. May bear pair of gill slits; gives rise to dorsal outpocketing (dorsal recess). (with/without gill slits)

pigment stripe [G *Pigmentstreifen, Pigmentbinde*]: Transverse band of red pigment across face of protosome.

plume: arm

proboscis [G *Proboscis*]: protosome

progenitor [G *Stammindividuum*]: First individual of a colony. Gives rise to additional individuals (zooids) by budding.

protocoel [G *Protocoel, Kopfschildzölom, Prosoma-Coelom*]: Body cavity (coelom) of first division (protosome = buccal shield = cephalic shield = proboscis) of body. Basically unpaired, enclosing complex of organs including buccal diverticulum (= stomochord), central sinus, heart vesicle, and glomerulus. Opens to exterior via shield canals and shield pores. (See also mesocoel, metacoel)

protosome (buccal shield, cephalic shield, proboscis) [G *Protosoma, Mundschild, Kopfschild, Proboscis, Prosoma*]: Anterior of three divisions (protosome, mesosome, metasome) of body. Forms downturned, shield-like structure containing complex of organs including buccal diverticulum (= stomochord), central sinus, heart vesicle, and glomerulus. Bears transverse pigment stripe across outer face, bears glandular areas, and secretes tube or coenecium. (disciform, shield-shaped; with/without lateral notch)

rectal diverticulum [G *Enddarmdivertikel*]: In digestive tract of naked pterobranch, short outpocketing of rectum.

rectum [G *Enddarm*]: Posteriormost, somewhat expanded section of digestive tract between intestine and anus. May give rise to rectal diverticulum.

ring [G *Ring*]: In tube, one in a series of successive ring-like sections secreted by first division (protosome) of body.

seminal vesicle [G *Samenblase*]: Expanded distal section of ♂ reproductive system in which sperm are stored. May be separated from testis proper by constriction.

septum [G *Septum, Diaphragma*]: Transverse partition separating three body cavities (coeloms) from one another, i.e., protocoel from mesocoel and mesocoel from metacoel.

shield canal [G *Kopfschildpforte, Coelomodukt*]: In protosome cavity (protocoel), one of two canals leading to exterior via shield pore.

shield pore [G *Protocoelporus*]: One of two openings of protosome cavity (protocoel) to exterior. Each pore is located at base of anterior arm and joined to protocoel by shield canal.

sperm duct (vas deferens) [G *Samenleiter, Vas deferens*]: gonoduct

spicule [G *Stachel*]: In coenecium, one of numerous rigid, spine-like projections supporting branches and opening of zooids.

stalk [G *Stiel*]: Slender posterior part of metasome (= trunk). Hollow, muscular, and serving in attachment or movement.

stalk nerve [G *Stielnerv*]: ventral trunk nerve

stalk sinus [G *Stielgefäß*]: ventral trunk sinus

stomach [G *Magen*]: Enlarged section of digestive tract between esophagus and intestine; occupies greater part of metasome (= trunk).

stomochord (buccal diverticulum) [G *Stomochord, Kopfschilddarm, protosomales Darmdivertikel*]: Anteriorly directed, narrow outpocketing of digestive tract. Originates on roof of buccal tube and extends into protosome.

tentacle [G *Tentakel*]: One in a series of numerous projections along each side of arm.

testis [G *Hoden*]: gonad

trunk [G *Rumpf*]: metasome

tube [G *Rohr, Wohnrohr*]: One in a series of tube-like structures housing pterobranch individuals (zooids). Consists of series of successive rings secreted by glands on protosome. May be indistinguishable due to various degrees of fusion or reduction within coenecium.

upper lip [G *Oberlippe*]: Slightly protruding body wall above mouth. (See also lower lip)

ventral shield sinus [G *ventrales Kopfschildgefäß*]: In circulatory system, blood channel under buccal diverticulum (= stomochord) in protosome. Receives blood from central sinus and conveys it via peribuccal channels to ventral trunk sinus. Closely associated with glomerulus.

ventral sinus: ventral shield sinus, ventral trunk sinus

ventral trunk nerve [G *Ventralnerv, ventraler Rumpfnerv, Ventromedianband*]: Longitudinal nerve cord extending from behind mouth to tip of stalk (ventral stalk nerve), where it may continue forward on dorsal side (dorsal stalk nerve).

ventral trunk sinus [G *ventrales Rumpfgefäß, Ventromediangefäß*]: In circulatory system, midventral blood channel in metasome (= trunk). Extends from mesosome/metasome border to tip of stalk (ventral stalk sinus), where it may continue forward on dorsal side (dorsal stalk sinus).

zooid [G *Zoid*]: Single pterobranch individual. Consists of three body regions (protosome, mesosome, metasome). Either naked, encased in tube, or living within cavities of coenecium.

ENTEROPNEUSTA

acorn worm, enteropneust [G *Eichelwurm, Enteropneust*]

ENTEROPNEUSTA

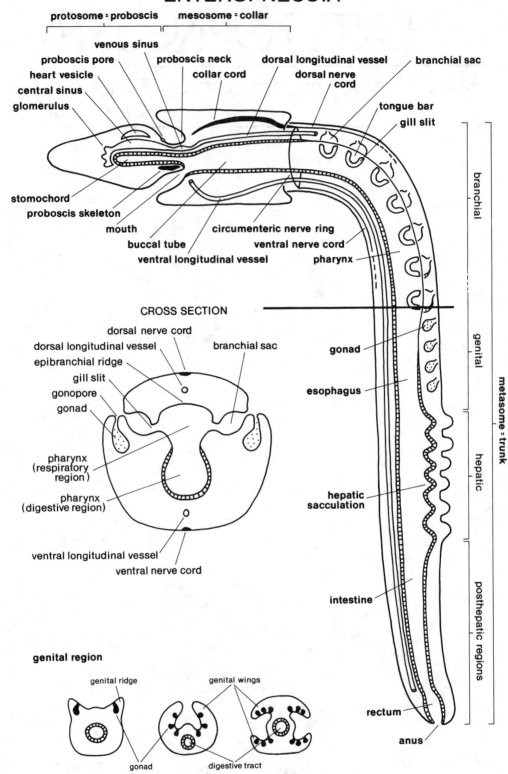

ENTEROPNEUSTA

acorn worm, enteropneust [G *Eichelwurm, Enteropneust*]

anterior nerve ring [G *vorderer Nervenring*]: Ring of nerve tissue surrounding posterior part of proboscis. Gives rise to dense network of anteriorly directed, longitudinal nerves.

anus [G *After*]: Posterior opening of digestive tract at end of trunk.

appendix [G *Frontalfortsatz*]: In certain acorn worms, slender, tubular extension of stomochord (buccal diverticulum); projects anteriorly far into proboscis.

branchial region (gill region) [G *Kiemenregion*]: Anteriormost region of trunk corresponding internally to gill slits; bears two dorsolateral series of gill pores externally. (See also branchiogenital region, hepatic region, posthepatic region)

branchial sac [G *Kiemensack, Kiementasche*]: In anterior region of trunk, one in a series of sac-like pouches into which each gill slit opens. Leads to exterior via gill pore.

branchiogenital groove (submedian line, sublateral line) [G *Branchiogenitalrinne*]: Longitudinal groove on each side of anterior trunk region. Contains series of gill pores and gonopores.

branchiogenital region (thorax) [G *Branchio-Genitalregion, Thorax*]: Term applied to anterior region of trunk if this is visibly modified externally (ridges, swellings, wings) by reproductive system.

buccal cavity [G *Mundhöhle*]: Term applied to anteriormost part (buccal tube) of digestive tract from which stomochord (= buccal diverticulum) originates.

buccal diverticulum: stomochord

buccal tube [G *mesosomaler Teil des Pharynx, Mundhöhle*]: Tube-like anterior section of digestive tract in collar between mouth and pharynx. Gives rise to buccal diverticulum (stomochord). Adjoined laterally by horns of proboscis skeleton.

cardiopericardial sac: heart vesicle

caudal region: posthepatic region

cauliflower organ (racemose organ) [G *blumenkohlähnliches Organ*]: Posterior region of each ventrolateral compartment of proboscis body cavity (protocoel); wall covered with outpocketings.

central sinus (heart) [G *Zentralsinus, zentraler Blutraum, Herz*]: Expanded part of circulatory system in posterior end of proboscis. Located above stomochord (buccal diverticulum) and below heart vesicle. Blood, coming from dorsal longitudinal vessel and venous sinus, is pumped via central sinus into glomerulus. (simple, bifurcate)

chondroid tissue [G *chondroides Gewebe*]: Rigid tissue formed at junction between body cavities (coeloms) of proboscis and collar. Located in proboscis stalk, above proboscis skeleton.

ciliated strip [G *Wimperstreifen, Wimperrinne, dorsolateraler Wimperstreifen*]: In digestive tract, dorsolateral strip or groove in intestine. Bears series of long cilia. (single, paired)

circumenteric nerve ring (pre-branchial nerve ring) [G *präbranchialer Nervenring, hinterer Nervenring*]: Ring of nerve tissue connecting ventral and dorsal longitudinal nerve cords at collar/trunk border.

coelom [G *Coelom*]: Epithelium-lined body cavity. Basically divided into protocoel of proboscis, mesocoel of collar, and metacoel of trunk.

collar (mesosome) [G *Kragen, Mesosom*]: Relatively short division of body between proboscis = protosome and trunk = metasome. Surrounds narrow posterior part (proboscis stalk) of proboscis and encloses anterior section (buccal tube) of digestive tract. (See also collarette)

collar canal [G *Kragenpforte, Coelomodukt*]: One of two dorsolateral canals leading from collar cavity (mesocoel) to exterior. Extends through collar/trunk border and opens to exterior via first branchial sac and gill pore. (See also collar pore)

collar cord (neurocord) [G *Kragenmark, Nervenrohr, Mesosomamark*]: Main longitudinal concentration of nerve tissue in collar. Represents infolded body wall and therefore contains either continuous longitudinal canal or series of cavities (lacunae). Associated with giant cells; continues posteriorly into trunk as more superficial dorsal nerve cord. (See also neuropore)

collarette [G *Kragentrichter*]: Funnel-shaped anterior part of collar. Encloses posterior part of proboscis and proboscis stalk.

collar pore [G *Kragenporus*]: Pore of canal (collar canal) leading from collar cavity (mesocoel). Refers either to pore which each canal makes when passing through collar/trunk border or to opening in branchial sac through which canal then opens to exterior.

digestive region [G *Nahrungsdarm, nutritiver Darm*]: In digestive tract,

ventral part of pharynx. Variously delimited from dorsal respiratory part.

dorsal longitudinal vessel [G *Dorsalgefäß, Dorsomediangefäß*]: In circulatory system, middorsal blood vessel extending along trunk and collar. Blood flows anteriorly through this vessel into venous and central sinuses. Collects blood from gill slits via series of branchial veins. (See also ventral longitudinal vessel)

dorsal nerve cord [G *Dorsalnerv, dorsaler medianer Nervenstreifen, dorsaler Nervenstamm*]: One of two main longitudinal nerve cords (dorsal, ventral) in trunk. Located middorsally and joined to ventral cord by circumenteric nerve ring. Extends into collar as collar cord.

efferent glomerular artery [G *abführendes Eichelgefäß*]: In circulatory system, one of two blood vessels extending posteriorly from excretory organ (glomerulus). Leads to ventral longitudinal vessel via peribuccal vessel.

epibranchial ridge [G *Epibranchialstreifen*]: In upper (respiratory) region of pharynx, narrow middorsal strip above lateral series of gill slits.

esophageal pore [G *Darmpforte, Ösophagusporus*]: In digestive tract, one in a series of pores extending to exterior from dorsal part of esophagus; typically paired. (paired, unpaired; with/without skeletal supports)

esophagus [G *Ösophagus, Oesophagus*]: In trunk, section of digestive tract between pharynx and intestine. May be variously regionated or bear pores (esophageal pores) opening to exterior. (with/without esophageal pores)

genital ridge [G *Genitalwulst*]: Longitudinal swelling along each side of anterior trunk region. Corresponds internally to series of gonads. (See also genital wing)

genital wing [G *Genitalflügel, Genitallamelle, Genitalfalte*]: Longitudinal, wing-like extension along each side of anterior trunk region. Contains gonads and may curve dorsally to almost meet opposite member.

giant cell [G *Riesenzelle*]: In nerve system, one of numerous large nerve cells typically associated with collar cord. Processes of giant cells extend posteriorly to dorsal nerve cord and lateral walls of trunk.

gill pore [G *Kiemenporus*]: In respiratory system, one in a series of dorsolateral pores on each side of anterior trunk region; each branchial sac opens to exterior via gill pore. Arranged within longitudinal groove (branchiogenital groove) or on longitudinal elevation. (elongate, oval, round)

gill region: branchial region

gill slit [G *Kiemenspalte*]: In respiratory system, one in a series of U-shaped openings on each side of pharynx. Located in dorsal region of pharynx; opens to exterior via branchial sac and gill pore. Divided medially from above by tongue bar. (with/without synapticules) (See also septum)

glomerulus [G *Glomerulus*]: In proboscis, mass of tubular structures closely associated with complex of organs including stomochord (buccal diverticulum), central sinus, and heart vesicle. Considered to represent excretory organ.

gonad [G *Gonade*]: One in a series of ♂ (testis) or ♀ (ovary) reproductive organs along each side of anterior trunk region; may or may not overlap with pores of gill slits. Position may be indicated externally by longitudinal swellings (genital ridges) or winglike extensions (genital wings). Each gonad opens to exterior via gonopore. (elongated, lobulated, sacciform)

gonopore [G *Gonoporus, Geschlechtsporus, Genitalporus*]: In reproductive system, opening of each gonad to exterior. Arranged in a series along each side of anterior trunk region, often within groove (branchiogenital groove).

heart: central sinus

heart vesicle (pericardial sac, cardiopericardial sac) [G *Herzblase, Perikard, Pericard*]: Muscular section of circulatory system in posterior part of proboscis. Located above central sinus; serves to pump blood anteriorly into excretory organ (glomerulus). (simple, bifurcate)

hepatic region [G *Leberregion, Leberzone*]: Term applied to midregion of trunk if this is visibly modified externally by outpocketings (hepatic sacculations) of anterior intestine. (See also branchial region, branchiogenital region, posthepatic region)

hepatic sacculation [G *Lebersack, Lebersäckchen, Mitteldarmdivertikel*]: One in a series of paired outpocketings of intestine. Visible externally as a series of bulges in midregion (hepatic region) of trunk.

horn [G *Crura, Schenkel*]: One of two posteriorly directed branches of proboscis skeleton. Extends backward into collar and closely adjoins wall of digestive tract. (See also median plate)

hypobranchial ridge [G *Hypobranchialstreifen*]: In digestive tract, narrow midventral strip of pharynx. Represents reduced digestive region of pharynx (as opposed to large dorsal respiratory region bearing gill slits). (See also epibranchial ridge)

intestine [G *Mitteldarm*]: In trunk, section of digestive tract between esophagus and rectum or anus. May be variously regionated into hepatic and posthepatic intestines. (with/without ciliated strip)

lateral longitudinal vessel [G *Lateralsinus, Lateralgefäß*]: In circulatory system, pair of longitudinal blood vessels, one on each side of digestive tract. Connected to dorsal longitudinal vessel.

lateral pharyngeal vessel [G *Grenzsinus, Grenzgefäß*]: In circulatory system, pair of longitudinal blood vessels, one on each side of pharynx. Located at border of upper (respiratory) and lower (digestive) parts of pharynx; supplies gill slits with series of branchial arteries.

lateral proboscis vein [G *Eichelvene, dorsolaterale Eichelvene*]: In circulatory system, pair of blood vessels on each side of proboscis. Extends posteriorly to venous sinus at beginning of collar.

median plate [G *Basalplatte*]: Main body of proboscis skeleton. Typically bears midventral keel and two posteriorly directed extensions (horns).

mesentery [G *Mesenterium, Scheidewand*]: One in a series of longitudinal partitions dividing body cavities (coeloms). According to position, one may distinguish dorsal, ventral, and lateral mesenteries.

mesocoel (collar coelom) [G *Kragencoelom, Mesocoel, Mesosoma-Coelom*]: Body cavity (coelom) of collar. Basically paired, opening to exterior via pair of collar canals. (See also metacoel, protocoel)

mesosome: collar

metacoel (trunk coelom) [G *Rumpfcoelom, Metacoel, Metasoma-Coelom*]: Body cavity (coelom) of trunk. Basically paired, being divided by mesenteries into lateral halves, occasionally additionally into dorsal and ventral compartments. Sends anterior outpocketings (peribuccal folds, perihemal spaces) into collar coelom. (See also mesocoel, protocoel)

metasome: trunk

middorsal proboscis artery [G *dorsale Eichelarterie, dorsales Eichelgefäß, dorsomedianes Eichelgefäß*]: In circulatory system, narrower dorsal blood vessel extending from excretory organ (glomerulus) into proboscis. (See also midventral proboscis artery)

midventral keel (keel) [G *Mediankiel, Kiel*]: In proboscis skeleton, longitudinal keel along underside of median plate.

midventral proboscis artery [G *ventrale Eichelarterie, ventrales Eichelgefäß, ventromedianes Eichelgefäß*]: In circulatory system, larger ventral blood vessel extending from excretory organ (glomerulus) into proboscis. (See also middorsal proboscis artery)

mouth [G *Mund, Mundöffnung*]: Anterior opening of digestive tract. Located ventrally at border of proboscis and collar. Opens into buccal tube.

muscle [G *Muskel*]: Any one of various longitudinal, circular, or radial muscles enabling body movement, closing pores (sphincter muscle), or operating circulatory system.

neurocord: collar cord

neuropore [G *Neuroporus*]: In type of collar cord with continuous longitudinal cavity (central canal), opening of canal to exterior. According to position one may distinguish an anterior and posterior neuropore.

operculum: In certain acorn worms, term occasionally applied to posterior part of collar which slightly overhangs trunk.

ovary: gonad

parabranchial ridge [G *Längsfalte, Grenzwulst*]: In digestive tract, one of two longitudinal ridges protruding into pharynx and dividing it into dorsal respiratory region with gill slits and ventral digestive region.

peribuccal fold [G *Peripharyngealraum*]: One of two flat outpocketings of trunk body cavity (metacoel). Projects into collar where both surround buccal tube.

peribuccal vessel [G *Ringgefäß*]: In circulatory system, pair of blood vessels surrounding digestive tract (buccal tube) in collar. Receives blood from efferent glomerular artery and conveys it to ventral longitudinal vessel.

pericardial sac: heart vesicle

perihemal canal: perihemal space

perihemal space (perihaema canal) [G *Perihämalraum, Perihaemalkanal, Perihaemalsinus*]: One of two anterior finger-like outpocketings of trunk body cavity (metacoel). Projects into collar, where it surrounds dorsal blood vessel.

pharynx [G *Pharynx*]: Section of digestive tract in anterior part of trunk between buccal tube and esophagus. Typically subdivided into dorsal respiratory region bearing gill slits and ventral digestive region. (See also epibranchial ridge, hypobranchial ridge, parabranchial ridge)

postbranchial canal [G *postbranchialer Kiemendarm*]: In digestive tract, short section of pharynx posterior to last gill slits; represents extension of dorsal respiratory region and is constricted from esophagus. Gives rise to postbranchial cecum.

postbranchial cecum [G *Blindtasche, Darmtasche*]: In digestive tract, short anterior extension of postbranchial canal.

posthepatic region (caudal region) [G *Abdominalregion*]: Posteriormost region of trunk when this is externally subdivided into branchiogenital, hepatic, and posthepatic regions.

prebranchial nerve ring: circumenteric nerve ring

preoral ciliary organ [G *präorales Wimperorgan*]: U-shaped, ciliated depression at proboscis/proboscis stalk border.

proboscis (protosome) [G *Eichel, Prosoma, Protosoma*]: Anterior of three divisions (proboscis, collar = mesosome, trunk = metasome) of body. Muscular, highly movable, and narrowing posteriorly to form proboscis stalk. Contains complex of organs including buccal diverticulum (stomochord), central sinus, heart vesicle, and glomerulus. (elongate, short; rounded, conical; cross section: circular, with middorsal groove) (See also protocoel)

proboscis canal [G *Eichelpforte*]: In proboscis cavity (protocoel), canal leading to exterior (via proboscis pore). Typically located on left side of proboscis stalk.

proboscis pore [G *Eichelporus*]: Opening of proboscis cavity (protocoel) to exterior. Typically located on left side of proboscis stalk; joined to protocoel by proboscis canal.

proboscis skeleton [G *Eichelskelett*]: Small skeletal element within slender posterior section (proboscis stalk) of proboscis. Consists of median plate bearing midventral keel and pair of posteriorly directed horns. Serves as site of muscle attachment. (See also chondroid tissue)

proboscis stalk [G *Eichelstiel, Eichelhals, Prosomastiel*]: Slender posterior region of proboscis. Surrounded by anterior part of collar and supported internally by proboscis skeleton.

protocoel (proboscis coelom) [G *Eichelcoelom, Procoel, Protocoel, Prosoma-Coelom*]: Body cavity (coelom) of proboscis. Basically unpaired, enclosing complex of organs including buccal diverticulum (stomochord), central sinus, heart vesicle, and glomerulus. (See also cauliflower organ)

protosome: proboscis

pygochord [G *Pygochord*]: In digestive tract, longitudinal band of cells associated with posterior (posthepatic) region of intestine. Located midventrally; extends from intestine wall to body wall. (hollow, solid)

racemose organ: cauliflower organ

rectum (end gut) [G *Enddarm*]: Posteriormost, variously differentiated section of digestive tract between intestine and anus.

respiratory region [G *Kiemendarm*]: In digestive tract, dorsal part of pharynx bearing series of gill slits. Variously delimited from ventral digestive part.

septum [G *Kiemenseptum, Septum, Hauptbogen*]: In upper (respiratory) region of pharynx, one in a series of partitions between successive gill slits. Supported internally by skeletal rod.

siphon [G *Nebendarm*]: In digestive tract of certain acorn worms, slender tube developing from and running parallel to anterior (hepatic) region of intestine.

skeletal rod [G *Skelettstab*]: In upper (respiratory) region of pharynx, one in a series of small skeletal elements supporting both tongue bars of gill slits and septa between slits. Each rod consists of two prongs, one extending into tongue bar, the second into adjoining septum.

stomochord (buccal diverticulum) [G *Stomochord, Eicheldarm, protosomales Darmdivertikel, Mundhöhlendivertikel*]: Anteriorly directed outpocketing of digestive tract. Originates in roof of buccal tube and extends anteriorly through proboscis stalk over proboscis skeleton. Divides into several branches. (with/without appendix)

synapticule [G *Synaptikel, Querbrücke, Querfortsatz*]: In upper (respiratory) region of pharynx, one in a series of minute bridges between tongue bars and adjoining walls of gill slits.

testis: gonad

thorax: branchiogenital region

tongue bar [G *Kiemenzunge, Zunge, Nebenbogen*]: In respiratory system, median extension of tissue into each gill slit (resulting in U-shaped configuration of latter). Typically supported by skeletal rods and joined to adjacent slit walls (septa) by series of bridges (synapticules).

tornaria [G *Tornaria*]: Free-swimming larval stage. Characterized by development of winding ciliary bands on outer surface.

trunk (metasome) [G *Rumpf, Metasom*]: Posterior and largest of three divisions (proboscis = protosome, collar = mesosome, trunk = metasome) of body. May be variously subdivided into branchial, genital (or branchiogenital), hepatic, and post-hepatic regions.

venous sinus [G *Venensinus, Sinus venosus*]: In circulatory system, somewhat enlarged section of dorsal longitudinal vessel at front end of collar. Opens anteriorly into central sinus; receives pair of lateral proboscis veins.

ventral longitudinal vessel [G *Ventralgefäß, Ventromediangefäß*]: In circulatory system, contractile midventral blood vessel extending from collar to end of trunk. Gives rise to series of branches to digestive tract. (See also dorsal longitudinal vessel)

ventral nerve cord [G *Ventralnerv, ventromedianer Nervenstreifen, Ventromedianstrang*]: One of two main longitudinal nerve cords (dorsal nerve cord, ventral nerve cord) extending through trunk. Joined to dorsal cord at collar/trunk border by circumenteric nerve ring.

POGONOPHORA

pogonophoran, beard worm [G *Bartwurm, Bartträger*]

POGONOPHORA

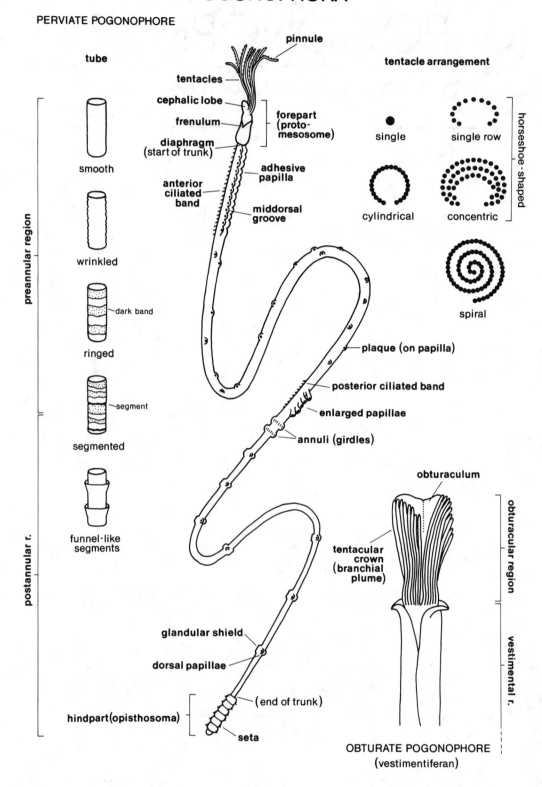

PERVIATE POGONOPHORE

tube

smooth

wrinkled

ringed — dark band

segmented — segment

funnel-like segments

preannular region

postannular r.

pinnule

tentacles

cephalic lobe

frenulum

diaphragm (start of trunk)

forepart (proto-mesosome)

adhesive papilla

anterior ciliated band

middorsal groove

tentacle arrangement

single

single row

cylindrical

concentric

horseshoe-shaped

spiral

plaque (on papilla)

posterior ciliated band

enlarged papillae

annuli (girdles)

obturaculum

tentacular crown (branchial plume)

obturacular region

vestimental r.

glandular shield

dorsal papillae

(end of trunk)

hindpart (opisthosoma)

seta

OBTURATE POGONOPHORE
(vestimentiferan)

POGONOPHORA

pogonophoran, beard worm [G *Bartwurm, Bartträger*]

adhesive papilla (metameric papilla, paired papilla) [G *paarige Papille, pseudomere Befestigungspapille*]: Term applied to voluminous, paired papillae serially lining middorsal groove on anterior section of preannular region; often provided with cuticular platelets or pyriform glands.

anchor: opisthosoma

annulus (girdle, belt) [G *Annulus, Gürtel*]: One of two to five typically elevated and dorsally interrupted rings of protrusible, denticulate platelets. Provides basis for division of trunk into preannular and postannular regions. Serves to anchor animal in tube.

anterior ciliated band: ciliated band

brain [G *Gehirn*]: Main concentration of intra- to basiepithelial nerve tissue (plexus), including a vague dorsal commissure; located ventrally in pretentacular lobe close to tentacle bases. Gives rise posteriorly to midventral nerve.

bridle: frenulum

cephalic lobe: pretentacular lobe

chaeta: seta

ciliated band (anterior ciliated band, ventral ciliated band) [G *ventrales Wimperband, präannulares Wimperband*]: Ventral band of cilia in preannular region of trunk; during development often subdivided into an elongate anterior portion and short posterior portion.

coelom [G *Coelom*]: One of four body cavities: an unpaired cavity in protosome, paired cavities in mesosome and metasome, and segmented cavities in telosome. (anterior coelom: corkscrew-shaped, horseshoe-shaped, sac-shaped; posterior coelom: paired, unpaired)

cuticle [G *Kutikula*]: Noncellular, variously secreted layer covering outer surface of body and tentacles; perforated by microvilli of underlying epidermis cells.

diaphragm [G *Diaphragma*]: Transverse septum marking boundary between forepart and trunk.

dorsal blood vessel [G *dorsales Blutgefäß*]: Less prominent of two major longitudinal blood vessels (dorsal and ventral blood vessels); expanded into heart at base of tentacles and extending into each tentacle. Blood flows anteriorly.

filament [G *Filament*]: Thin, extremely elongate structure at one end of spermatophore; used for attachment of spermatophore.

forepart (protomesosome) [G *Vorderende, Protomesosom, Vorderkörper*]: Anteriormost division of body (forepart, trunk, hindpart); equivalent to protosome and mesosome and delimited from trunk by diaphragm.

frenulomere (frenular section) [G *Mesosoma, Zügelabschnitt*]: Term applied to division of body bearing frenulum and corresponding to first paired coelomic body cavities; equivalent to mesosome.

frenulum (bridle) [G *Frenulum, Zügel*]: Elongate, comb-like, cuticular structures in anterior section of frenulomere. Obliquely arranged and representing a dorsally united, V-shaped configuration.

giant fiber (neurochord) [G *Riesenfaser*]: In nerve system, special axons of motor cells extending from brain to annulus; enables rapid contraction of longitudinal musculature.

girdle [G *Gürtel*]: Term applied to annulus on trunk and occasionally also to frenulum on forepart.

glandular shield [G *Drüsenschild, Drüsenfeld*]: Flattened elevations bearing pyriform glands; located ventrally on postannular region of trunk.

goblet cell (mucous cell) [G *Körnerzelle, Schleimdrüsenzelle*]: Unicellular, granule-producing gland often containing a single cilium; opens through cuticle to exterior via neck-like structure. Arranged over entire body but most numerous in forepart and preannular region.

gonad [G *Gonade, Keimdrüse*]: ovary, testis

gonadal region [G *Metasoma*]: Neutral term applied to division of body bearing gonads; equivalent to gonomere, metasome, or trunk.

gonomere (trunk) [G *Metasoma, Mittelkörper*]: Term applied to division of body bearing gonads (vs. frenulomere); equivalent to gonadal region, metasome, or trunk.

gonopore [G *Geschlechtsöffnung*]: ovary, testis

heart [G *Herz*]: Muscular enlargement of dorsal blood vessel in anterior region of forepart; blood is pumped anteriorly into tentacles.

hindpart (telosome, opisthosoma, anchor) [G *Anker, Hinterende, Telosoma, Opisthosoma*]: Short posteriormost division of body; consists of 5 to 23 segments, bears setae, and contains series of coelomic body cavities. (cavities: paired, unpaired)

lophophoral organ: obturaculum

mesosome (frenulomere) [G *Mesosoma*]: Externally indistinguishable division of body between tentacular region or protosome and trunk or metasome; corresponds to first paired coelomic body cavities. Protosome and mesosome together make up forepart. (See also vestimentum)

metameric papilla: adhesive papilla

metameric region [G *pseudometamere Region, metamere Region*]: Anterior region of trunk which takes on a "metameric" appearance (seriality) due to paired papillae (adhesive or metameric papillae) lining middorsal groove.

metasome [G *Metasoma*]: Elongate division of body corresponding to third paired coelomic body cavities; equivalent to gonadal region, gonomere, or trunk.

microvillus [G *Mikrovillus*]: One of numerous thin cytoplasmic extensions of epidermal cells; perforates overlying cuticle.

middorsal groove [G *Dorsalfurche, präannulare Längsfurche*]: Groove extending posteriorly from diaphragm on preannular region of trunk; lined by adhesive papillae.

midventral nerve cord [G *ventraler Nervenstrang*]: Longitudinal nerve arising from brain and extending to posterior end.

obturaculum (lophophoral organ) [G *Obturaculum*]: In certain pogonophorans, anterior region of body bearing tentacles. Equivalent to protosome.

opisthosoma: hindpart

ovary [G *Ovar*]: Section of ♀ reproductive system located in anterior half of trunk and typically containing large, yolky eggs.

oviduct [G *Ovidukt*]: Posterior section of ♀ reproductive system opening to exterior in middle of trunk.

papilla [G *Befestigungspapille*]: General term for variously elaborated and positioned, typically glandular elevations of body wall (adhesive papilla, postannular papilla).

pericardium [G *Perikard*]: Sac-like body cavity associated with ventral wall of heart.

pinnule [G *Pinnula*]: Highly elongated projection of epidermal cell protruding from tentacle in single or multiple rows; with U-shaped blood sinus.

plaque: More heavily cuticularized, oval elevation on trunk.

platelet (toothed platelet) [G *Plätchen*]: One in a series of mushroom-shaped, cuticularized structures with denticulate, oval surface; connected to annuli and adhesive (metameric) papillae.

plexus [G *Nervennetz, Plexus*]: Network of nerves principally consisting of intra- to basiepithelial nervous tissue; locally concentrated as brain, ventral nerve cord, or (incomplete) nerve rings below frenulum and annulus and above diaphragm.

postannular papilla [G *postannulare Papille, postannulare Befestigungspapille*]: Any papilla posterior to annuli; often arranged in transverse rows at regular intervals.

postannular region [G *postannularer Abschnitt*]: Region of trunk posterior to annuli.

posterior ciliated band: ciliated band

preannular region [G *prännularer Abschnitt*]: Region of trunk anterior to annuli. (See also postannular region)

pretentacular lobe (cephalic lobe) [G *Medianlobus, Kopflappen*]: Ventral lobe forming anteriormost region of forepart and bearing brain. (rounded, triangular)

protomesosome (forepart) [G *Protomesosoma, Vorderende*]: Term occasionally applied to anterior region of body comprising protosome and mesosome, especially if these two regions are not externally delimited by a constriction.

protosome (tentacular region) [G *Protosoma, Prosoma, Tentakelabschnitt*]: Anteriormost region of body comprising cephalic lobe (pretentacular lobe) and tentacles and corresponding to first unpaired coelomic body cavity.

pyriform gland [G *mehrzellige Drüse*]: Multicellular gland opening to exterior via cuticle-lined pore and aggregated in forepart behind bridle, in paired papillae on trunk, and in trunk (glandular shields). Functions in secretion of organic part of tube.

septate region (hindpart, setigerous section, anchor) [G *Anker, Opisthosoma, segmentierter Abschnitt, Telosoma, Hinterende*]: Neutral term for posteriormost segmented division of body. Equivalent to telosome. According to this terminology, body is divided into tentacular, frenular, gonadal, and septate regions.

seta (chaeta) [G *Seta, Borste*]: Bristle-shaped, chitinous structure associated with annuli on trunk as well as with each segment of hindpart (typically four setae per segment); short, rod-shaped, with oval head bearing short teeth.

spermatophore [G *Spermatophor*]: Packet of spermatozoa bearing attachment filament at one end. Shape is of taxonomic importance. (flattened, fusiform)

telosome: hindpart

tentacle [G *Tentakel*]: Long, ciliated, pinnule-bearing tentacle(s) arising from base of cephalic lobe on forepart. Number, configuration, and degree of fusion are of taxonomic importance.

tentacular region [G *Protosoma*]: Neutral term for anteriormost tentacle-bearing division of body. Equivalent to protosome. According to this terminology, body is divided into tentacular, frenular, gonadal, and septate regions.

testis [G *Hoden, Samensack*]: Elongate, paired section of ♂ reproductive system largely filling coelomic body cavity in posterior half of trunk.

trunk [G *Metasoma, Mittelkörper*]: Long middle division of body between forepart and setigerous hindpart; typically bears papillae and annuli. Equivalent to gonadal region, gonomere, or metasome; subdivided into preannular and postannular regions.

tube [G *Röhre*]: Extremely thin, chitinous tube inhabited by pogonophoran and secreted by epidermal glands; may be formed at both ends. (smooth, wrinkled; ringed, unringed; soft, stiff; segmented, unsegmented)

vas deferens (sperm duct) [G *Vas deferens, Samenleiter*]: Spermatophore-filled section of ♂ reproductive system leading anteriorly from testes and opening to exterior behind diaphragm.

ventral blood vessel [G *ventrales Blutgefäß*]: More prominent of two major longitudinal vessels (dorsal blood vessel, ventral blood vessel). Blood flows from tentacles and pretentacular lobe in a posterior direction.

ventral ciliated band: ciliated band

vestimentum [G *Vestimentum*]: In certain pogonophorans, second expanded region of body. Equivalent to mesosome. (See also obturaculum)

white cell [G *weiße Zelle*]: Whitish or yellowish, granule-containing cell occurring in bands or patches on forepart and trunk. Distribution is of taxonomic importance.

TUNICATA (UROCHORDATA)

ASCIDIACEA

ascidian, sea squirt [G *Ascidie, Seescheide*]

ASCIDIACEA

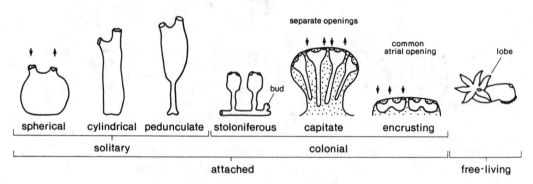

spherical   cylindrical   pedunculate | stoloniferous   capitate   encrusting   lobe

separate openings

common atrial opening

bud

|← solitary →|←——————— colonial ———————→|

|←———————————— attached ————————————→|← free-living →|

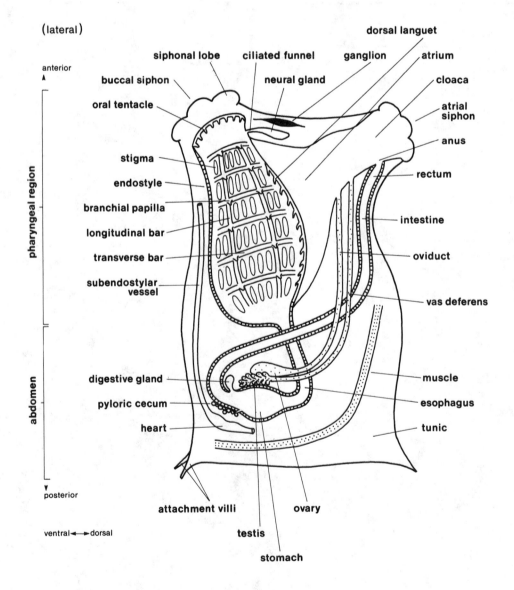

(lateral)

anterior

pharyngeal region

abdomen

posterior

ventral ←→ dorsal

buccal siphon
siphonal lobe
oral tentacle
ciliated funnel
neural gland
ganglion
dorsal languet
atrium
cloaca
atrial siphon
anus

stigma
endostyle
branchial papilla
longitudinal bar
transverse bar
subendostylar vessel

rectum
intestine
oviduct
vas deferens

digestive gland
pyloric cecum
heart

muscle
esophagus
tunic

attachment villi
testis
ovary
stomach

TUNICATA (UROCHORDATA)

ASCIDIACEA

ascidian, sea squirt [G *Ascidie, Seescheide*]

abdomen (abdominal region)
[G *Abdomen, Hinterleib*]: Basic division of body between pharyngeal region and postabdomen. Contains digestive tract posterior to pharyngeal basket and, if postabdomen is not developed, heart and reproductive organs.

anus [G *Anus, After*]: Posterior opening of digestive tract. Located alongside pharyngeal basket; opens to exterior via cloaca and atrial siphon.

appendicularia: tadpole larva

atrial languet [G *zungenartiger Fortsatz*]: Variously developed process overhanging atrial opening. (simple, lobed, trilobed)

atrial siphon (cloacal siphon, exhalent siphon) [G *Atrialsipho, Egestionssipho, Ausstromöffnung, Egestionsöffnung*]: Tube-like extension of body through which water exits after passing through pharyngeal basket. Rim may be elaborated into a number of lobes. (See also siphon)

atrium (atrial cavity) [G *Atrium, Atrialraum, Peribranchialraum*]: In pharyngeal region, space surrounding pharyngeal basket; basically divided into left and right halves by septa and traversed by strands (trabeculae). Water passes through slits (stig-

mata), enters atrium, and exits via cloaca and atrial siphon.

attachment villus [G *wurzelähnlicher Ausläufer, wurzelartiger Haftfortsatz*]: One of numerous root-like extensions of tunic at posterior end of body. Serves to attach ascidian to substrate.

blastozooid [G *Blastozoid*]: bud

branchial papilla [G *Papille, papillenartiger Forsatz*]: In pharyngeal basket, inwardly directed projection at intersection of transverse and certain longitudinal bars. Hollow, bearing blood vessels and frontal cilia. (simple, bifid)

branchial sac: pharyngeal basket

branchial tentacle: oral tentacle

brood sac (incubatory pouch)
[G *Bruttasche*]: Variously modified cloaca or atrium for brooding of embryos.

buccal siphon (branchial siphon, oral siphon) [G *Branchialsipho, Ingestionssipho, Einstromsipho, Ingestionsöffnung, Einstromöffnung*]: Tube-like extension of body through which water enters pharyngeal basket. Rim may be elaborated into a number of lobes. (See also atrial siphon, siphon)

bud (blastozooid) [G *Knospe, Blastozoid*]: In asexual reproduction, young ascidian developing from stolon, from epicardial tissue, or from wall of atrium. (enteroepicardial, epicardial, mesoblastic, peribranchial, stoloniferous)

ciliated funnel (dorsal tubercle) [G *Flimmergrube*]: Opening of neural organ; located inside buccal siphon between oral tentacles and dorsal lamina or dorsal languets.

circular muscle [G *Ringmuskel*]: One in a series of circular muscle bands typically underlying longitudinal muscles in body wall. Especially well developed around siphons.

cloaca [G *Kloake, Kloakalraum*]: Term applied either 1) in solitary ascidian to part of atrium near atrial siphon into which gonads and anus open or 2) in compound ascidian to common excurrent system of grouped individuals.

cloacal siphon: atrial siphon

coelom: epicardium

cupola organ: In nerve system, receptor structure projecting into atrium. Consists of dome-shaped aggregation of sensory cells covered by finger-shaped extension of tunic.

digestive gland (liver, hepatic cecum) [G *Leberanhang, Leberblindsack*]: In digestive tract, outpocketing of stomach.

dorsal abdominal sinus [G *Sinus cardiovisceralis*]: In circulatory system, blood channel originating from dorsal side of heart.

dorsal lamina [G *Dorsalfalte, Dorsalorgan*]: In digestive tract, middorsal organ running longitudinally along pharyngeal basket. Consists of continuous ridge which is curved to form gutter; serves to compact and convey food posteriorly to esophagus. (See also dorsal languet)

dorsal languet [G *Dorsalzunge, Zunge der Dorsalfalte, Läppchen*]: In digestive tract, one in a series of projections running longitudinally along middorsal wall of pharyngeal basket. Each languet is curved,

bears frontal cilia, and serves to compact and convey food to esophagus. (See also dorsal lamina)

dorsal tubercle: ciliated funnel

endostyle [G *Endostyl, Endostylrinne, Hypobranchialrinne*]: In digestive tract, deep groove running longitudinally along midventral wall of pharyngeal basket. Ciliated and glandular, producing mucus which is conveyed to dorsal side of basket. (See also retropharyngeal band)

epicardium [G *Epikard*]: In abdomen or postabdomen, cavity adjoining heart/pericardium complex. Arises as paired outpocketing of posterior end of pharyngeal basket and may more or less enclose digestive tract. Variously interpreted as excretory organ, as aiding pumping activity of circulatory system, and as representing body cavity (coelom).

esophagus (oesophagus) [G *Ösophagus, Oesophagus*]: Relatively short, narrow section of digestive tract between pharyngeal basket and stomach; lacks muscles. (See also gut loop)

exhalent siphon: atrial siphon

follicle [G *Hodenbläschen, Hodenfollikel*]: testis

frontal cilium [G *Frontalcilium*]: In pharyngeal basket, type of cilium along longitudinal bars and branchial papillae. Beats upward and transports mucus and food particles to dorsal lamina or dorsal languets. (See also lateral cilium)

ganglion [G *Ganglion, Zerebralganglion, Gehirnganglion*]: Main concentration of nerve tissue; located dorsally between buccal and atrial siphons. Adjoins neural gland. Gives rise anteriorly to nerves supplying buccal siphon (anterior siphonal nerves) and posteriorly to nerves supplying atrial siphon and other organs (posterior siphonal nerves). Lateral nerves arise from ganglion in compound ascidian.

genital cord [G *Genitalstrang, Dorsalstrang*]: Solid cord of tissue extending posteriorly from neural gland to gonads.

genital duct [G *Geschlechtsausfuhrgang, Geschlechtsgang, Gonodukt*]: oviduct, vas deferens

gill slit: stigma

gonad [G *Gonade*]: ♂ (testis) or ♀ (ovary) reproductive organ. (paired, unpaired; in a single row, scattered irregularly; in/below/beside gut loop; club-shaped, cylindrical, flask-shaped, U-shaped)

gut loop (digestive loop) [G *Darmschleife, Schleife des Verdauungstraktes*]: Part of digestive tract including esophagus, stomach, and early part of intestine; forms U-shaped curve or up to several loops due to coiled anterior intestine. (below/ to left of/to right of pharyngeal basket)

gutter: In digestive tract, channel formed by curved dorsal lamina along inner wall of pharyngeal basket. Serves to compact and convey food to esophagus.

heart [G *Herz*]: In circulatory system, main pumping organ located below pharyngeal basket in abdomen or postabdomen. Simple, tubular, and more or less enclosed by pericardium. Gives rise ventrally to subendostylar vessel, dorsally to dorsal abdominal sinus. May pump blood in either direction. (curved, U-shaped, V-shaped)

hepatic diverticulum: digestive gland

incubatory pouch: brood sac

inhalent siphon: buccal siphon

intestine [G *Intestinum, Dünndarm*]: In abdomen, narrow, elongate section of digestive tract between stomach and rectum; extends anteriorly in direction of atrial siphon. (coiled, looped) (See also gut loop)

languet: atrial languet, dorsal languet

lateral cilium [G *Lateralcilium*]: Type of cilium along walls of slits (stigmata) in pharyngeal basket. Draws water into basket and passes it through stigmata out atrial siphon. (See also frontal cilium)

liver: digestive gland

lobe [G *Siphonallappen, Lappen*]: One in a series of lobes forming outer rim of buccal siphon; may bear ocelli.

longitudinal bar (longitudinal vessel, internal longitudinal branchial bar) [G *Längsgefäß, primäres Längsgefäß*]: In pharyngeal basket, one in a series of anteroposteriorly directed ridges between slits (stigmata); contains blood vessels and bears frontal cilia. (with/without papillae) (see also branchial papilla, transverse bar)

longitudinal muscle [G *Längsmuskel*]: Anteroposteriorly directed muscle generally located outside of circular muscles in body wall; typically arranged in series of bands.

mantle [G *Mantel*]: tunic

mantle vessel [G *Mantelgefäß, Mantelkanal*]: Network of small blood channels extending through tunic.

median dorsal sinus [G *Dorsalsinus, mediodorsaler Sinus, Sinus viscerobranchialis*]: In circulatory system, longitudinal blood channel extending along pharyngeal basket (under dorsal lamina or dorsal languets). Joined to transverse bars of basket; divides posteriorly to supply digestive tract.

neural gland (subneural gland) [G *Neuraldrüse*]: Glandular organ adjoining ganglion between two siphons. Opens inside buccal siphon via short canal and ciliated funnel. (dorsal/lateral/ventral to ganglion)

ocellus [G *Ocellus*]: Term applied to one of several pigment spots along rim of siphons or between siphonal lobes. Not considered to represent light-sensitive organs.

oral siphon: buccal siphon

oral tentacle (branchial tentacle) [G *Tentakel*]: One in a circlet of short, inwardly directed projections within buccal siphon. Hollow, functioning as first filter for incoming food particles (i.e., rejecting too large particles). (branched, compound, simple)

ovary [G *Ovar, Ovarium, Eierstock*]: ♀ reproductive organ either adjoining gut loop or in anterior body wall at level of pharyngeal basket. Opens into cloaca via oviduct. (single, paired; compact, elongated and tubular, L-shaped, saccular, sinuous; dorsal to/posterior to testes)

oviduct [G *Ovidukt, Eileiter*]: Slender section of ♀ reproductive system between ovary and cloaca. Typically extends along intestine and parallels sperm duct (vas deferens).

parastigmatic bar (parastigmatic vessel) [G *parastigmatisches Gefäß*]: In pharyngeal basket, one in a series of longitudinal or transverse tissue strands over gill slits (stigmata). Arises from papilla-like projections and bears blood vessels.

pericardium [G *Perikard, Pericard*]: In circulatory system, tissue giving rise to and more or less surrounding heart. Space between pericardial wall and heart is termed pericardial cavity.

peripharyngeal ridge (peripharyngeal ciliated band) [G *Peripharyngealband, Flimmerbogen, Wimperbogen*]: In digestive tract, one of two (left and right) semicircular ciliated ridges around anterior end of pharyngeal basket. Represents extension of ventral groove (endostyle) and continues around to dorsal side to beginning of dorsal lamina or dorsal languets.

pharyngeal basket (branchial sac, pharynx) [G *Kiemenkorb, Kiemendarm, Pharynx*]: Expanded, sac-shaped anterior part of digestive tract below oral tentacles. Occupies and defines greatest part (pharyngeal region) of body. Perforated by series of slits (stigmata); bears midventral groove (endostyle) as well as middorsal structure (dorsal lamina or dorsal languets). (with/without folds, with/without longitudinal bars)

pharyngeal region (thorax) [G *Rumpf, Thorax*]: Anterior of three basic divisions (pharyngeal region, abdomen, postabdomen) of body. Contains pharyngeal basket.

pharynx: pharyngeal basket

postabdomen [G *Postabdomen*]: Posteriormost of three basic divisions (pharyngeal region, abdomen, postabdomen) of body. Contains heart and reproductive organs.

prepharyngeal region [G *präbranchiale Zone*]: Anterior region of pharynx between oral tentacles and first row of slits (stigmata). Ciliated funnel of neural gland opens here.

protostigma [G *Protostigma*]: One of several first-formed primary gill slits in young ascidian.

pyloric cecum (pyloric gland) [G *Pylorusdrüse, darmumspinnende Drüse, Darmdrüse*]: In digestive tract, branched structure forming network on stomach and early intestine. Opens into stomach via one or more ducts.

raphe [G *Herzraphe*]: In circulatory system, longitudinal seam formed by infolded body walls of pericardium/heart complex.

rectum [G *Enddarm*]: Posteriormost section of digestive tract between intestine and anus. Lies alongside pharyngeal basket.

renal sac [G *Speicherniere*]: In certain ascidians, relatively large, bean-shaped excretory organ on right side of body near heart.

retropharyngeal band [G *Retropharyngealband*]: In digestive tract, pair of ciliated bands at posterior end of pharyngeal sac. Represents continuation of endostyle and extends to esophagus.

septum [G *Septum*]: One of two longitudinal membranes extending through space (atrium) around pharyngeal basket and joining basket with body wall. Basically divides atrium into left and right halves. According to position one may distinguish a dorsal septum under dorsal lamina and ventral septum under endostyle.

siphon [G *Sipho, Siphon*]: One of two tubular extensions of body bearing incurrent opening (buccal siphon) and excurrent opening (atrial siphon). (similar, dissimilar; close together, widely separated)

sperm duct: vas deferens

spicule [G *Kalkkörperchen*]: In compound ascidian, one of numerous minute calcareous elements in tunic. (burr-like, irregular, rod-shaped, stellate)

standard measurements:
1. distance between two siphons (often as percent of body length)
2. number of rows of stigmata in pharyngeal basket (in adult)

stigma (gill slit) [G *Kiemenspalte*]: One of numerous slits in wall of pharyngeal basket; arranged into rows, each row containing series of stigmata. Water passes through stigmata and out atrial siphon. (coiled, spiral, straight, transverse) (See also standard measurements)

stolon [G *Stolon*]: In compound ascidian, slender, tubular structure connecting bases of individuals. Adheres to substrate and enables communication between zooids (e.g., circulatory system).

stomach [G *Magen*]: In abdomen, expanded section of digestive tract between esophagus and intestine; may bear digestive gland and pyloric cecum. (with/without folds; barrel-shaped, globular, ovoid, spindle-shaped; horizontal, longitudinal = vertical)

subendostylar vessel (subendostylar sinus) [G *Ventralgefäß, Hypobranchialgefäß*]: Main blood channel originating from ventral side of heart. Extends anteriorly along pharyngeal basket (under endostyle) and gives rise to series of branches into transverse bars of basket.

subneural gland: neural gland

tadpole larva (appendicularia) [G *Kaulquappen-Larve, Larve*]: Free-swimming larva characterized by head (bearing internal organs and adhesive papilla) and tail (with notochord and neural tube).

test: tunic

testis [G *Hoden*]: ♂ reproductive organ either adjoining gut loop or in anterior body wall at level of pharyngeal basket. Typically consists of cluster of small bodies (follicles) opening into cloaca via sperm ducts (vasa deferentia). (single, paired; elongate, ovoid, pear-shaped, forming rosette)

testis follicle [G *Hodenbläschen*]: testis

thorax [G *Thorax, Rumpf*]: pharyngeal region

trabecula [G *Trabekel*]: In pharyngeal region, one of numerous strands spanning across space (atrium) surrounding pharyngeal basket and thus connecting outer wall of basket with body wall. Bears blood vessels.

trabecular vessel [G *Trabekulargefäß, Dermatobranchialgefäß*]: In circulatory system, blood channels within strands (trabecula) spanning from pharyngeal basket to surrounding body wall.

transverse bar (transverse vessel) [G *Quergefäß, primäres Quergefäß*]: In pharyngeal basket, one in a series of circular ridges between rows of slits (stigmata). Contains blood vessels and may form branchial papillae at point of intersection with certain longitudinal bars.

tunic (test, mantle) [G *Tunica, Testa, Mantel*]: Variously developed layer enclosing body in solitary ascidian or forming matrix in which compound ascidians are embedded. Excreted by epidermis and permeated by blood channels. Considered as being synonymous with test and mantle or interpreted as consisting of outer tunic (test) and inner tunic (mantle). (semitransparent, translucent; cartilaginous, gelatinous, firm, soft; with/without spicules; coated/not coated with foreign bodies, mammillated, papillose, smooth, tessellated, wrinkled, with/without fibrillar processes, with branched spines)

vas deferens (sperm duct)

[G *Vas deferens, Samenleiter*]: Slender section of ♂ reproductive system between testis and cloaca. Typically extends along intestine and parallels oviduct. (uncoiled, with spiral turns)

vas efferens [G *Vas efferens*]: In type of ♂ reproductive system consisting of nu-

merous separate bodies (follicles), one of numerous canals leading to common sperm duct (vas deferens).

velum [G *Velum*]: In buccal siphon, inwardly projecting circular fold. Represents border of tunic within siphon and, if present, lies above (anterior to) oral tentacles.

visceral region: abdomen

zooid [G *Zoid*]: In compound ascidian, one of numerous individuals either joined basally or embedded in common matrix (tunic).

TUNICATA
(UROCHORDATA)

THALIACEA

thaliacean: doliolid, salp, pyrosome [G
Thaliacee: Tönnchensalpe, Salpe, Feuerwalze]

THALIACEA

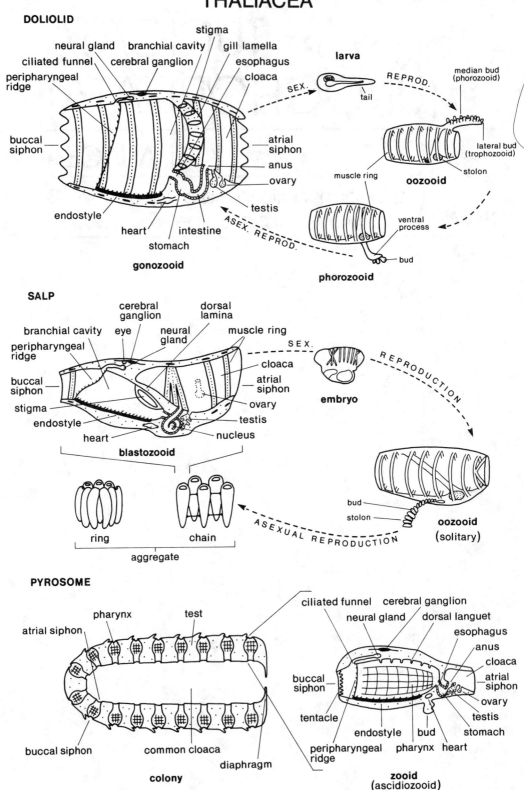

DOLIOLID

larva

neural gland · branchial cavity · stigma · gill lamella
ciliated funnel · cerebral ganglion · esophagus
peripharyngeal ridge · cloaca

REPROD.
SEX.

median bud (phorozooid)

buccal siphon

atrial siphon
anus
ovary
testis

endostyle
heart · intestine
stomach

gonozooid

tail

muscle ring

lateral bud (trophozooid)
stolon

oozooid

ventral process
bud

phorozooid

ASEX. REPROD.

SALP

cerebral ganglion · dorsal lamina
branchial cavity · eye · neural gland · muscle ring
peripharyngeal ridge

SEX.

REPRODUCTION

buccal siphon
stigma
endostyle
heart

cloaca
atrial siphon
ovary
testis
nucleus

blastozooid

embryo

ring · chain

aggregate

bud
stolon

oozooid
(solitary)

ASEXUAL REPRODUCTION

PYROSOME

pharynx · test
atrial siphon

ciliated funnel · cerebral ganglion
neural gland · dorsal languet
esophagus
anus
cloaca
atrial siphon

buccal siphon

ovary
testis
stomach

buccal siphon · common cloaca
diaphragm

colony

tentacle
endostyle · bud · heart
peripharyngeal ridge · pharynx

zooid
(ascidiozooid)

TUNICATA (UROCHORDATA)

THALIACEA

thaliacean: doliolid, salp, pyrosome [G
Thaliacee: Tönnchensalpe, Salpe, Feuerwalze]

anus [G *After*]: Posterior opening of digestive tract into cloaca. (in doliolid: median, on right side, subterminal)

ascidiozooid [G *Ascidiozoid*]: In pyrosome, one in a series of larvae produced by budding (asexual reproduction) of cyathozooid. Set free to form new colony. First set of ascidiozooids occasionally referred to as primary ascidiozooids, later individuals as secondary ascidiozooids.

atrial siphon (cloacal aperture, exhalent opening) [G *Egestionsöffnung*]: Opening through which water exits after passing through pharynx. Positioned opposite buccal siphon and equipped with sphincter muscle. Opens either to exterior or, in pyrosome, into common cloaca.

atrium [G *Peribranchialraum*]: In pyrosome, space surrounding pharynx; basically divided into left and right halves by septa and traversed by strands (trabeculae). Leads via cloaca to atrial siphon. Also employed as synonym for cloaca.

blastozooid [G *Blastozoid*]: General term for individual produced by budding (i.e., asexually produced individual). Represented in doliolid by gonozooid, phorozooid, and trophozooid, in salp by aggregate form,

and in pyrosome by ascidiozooid. (See also oozooid)

branchial basket: pharynx

branchial bar [G *Kiemengefäß*]: In pharynx, one in a series of longitudinal or transverse ridges between slits (stigmata); contains blood vessels.

branchial cavity [G *Pharyngealhöhle*]: Space within pharynx. Water in cavity passes through slits (stigmata) of pharynx and is expelled to exterior via atrium, cloaca, and atrial siphon.

buccal cavity (prepharyngeal region) [G *Schlundrohr, Präbranchialzone*]: In digestive tract of pyrosome, anterior region of pharynx between buccal siphon and first row of slits (stigmata); ciliated funnel of neural gland opens here.

buccal siphon (oral aperture, incurrent opening) [G *Ingestionsöffnung*]: Opening through which water enters pharynx. Positioned opposite atrial siphon and equipped with sphincter muscle.

bud [G *Knospe*]: Any asexually produced young thaliacean. Produced on stolon and variously transported to appropriate site on colony (in pyrosome) or cadophore (in doliolid). (on cadophore: lateral, median)

cadophore (dorsal stolon, dorsal process) [G *Dorsalstolo, Rückenfortsatz*]: In doliolid, posteriorly directed dorsal process of oozooid. Originates from and incorporates one or more posterior muscle bands. Asexually produced buds migrate from ventral stolon to cadophore, where they form lateral and median rows.

cecum [G *Blindsack, Magenblindsack*]: In digestive tract, single or paired (left and right) blind outpocketing of stomach. (single, paired; simple, branched)

cerebral ganglion (dorsal nerve ganglion) [G *Gehirnganglion, Zerebralganglion, Hirnganglion*]: Main concentration of nerve tissue located dorsally above pharynx. Adjoins neural gland and gives rise to numerous nerves innervating muscle bands and internal organs. (See also eye, statocyst)

ciliated funnel (dorsal tubercle) [G *Flimmergrube*]: Opening of neural gland; located dorsally in anterior region of pharynx (buccal cavity) and associated with peripharyngeal bands.

cloaca [G *Kloake, Kloakenhöhle*]: Cavity into which anus and gonads open and into which water enters after passing through slits (stigmata) of pharynx. In pyrosome, refers either to posterior part of atrium or to common cavity of colony into which atrial siphons of all individuals open.

cuirass: In doliolid, continuous sheet of muscle in body wall; formed by expansion and abutment of originally distinctly separate muscle rings.

cyathozooid [G *Cyathozooid*]: In pyrosome, rudimentary larva produced by sexual reproduction (i.e., via egg). Located within parent and buds (asexual reproduction) to form first larval individuals (ascidiozooids) of colony. Represents an oozooid.

diaphragm [G *Diaphragma*]: In pyrosome, ring-shaped membrane constricting opening of common cloaca to exterior.

Expulsion of water through restricted opening of diaphragm enables colony to swim.

dorsal blood vessel [G *Dorsalgefäß*]: In circulatory system, one or more longitudinal blood channels. Variously joined to ventral blood channel and leading posteriorly to digestive tract.

dorsal lamina (gill bar) [G *Kiemenbalken*]: In salp, middorsal strip of pharynx; runs obliquely from anterodorsal to posteroventral part of body. Contains blood vessel and bears single elongated slit (stigma) on each side.

dorsal languet [G *Rückenzapfen*]: In pyrosome, one in a series of projections along middorsal wall of pharynx. Ciliated, conveying food and mucus posteriorly to esophagus.

dorsal nerve ganglion: cerebral ganglion

dorsal stolon: cadophore

dorsal tubercle: ciliated funnel

eleoblast [G *Elaeoblast, Eläoblast*]: Sac-shaped cell aggregation in embryo of salp. Considered either to represent rudimentary notochord or to contain food reserves.

endostyle [G *Endostyl*]: In digestive tract, deep groove running longitudinally along midventral wall of pharynx. Ciliated and glandular, producing mucus which is variously conveyed to esophagus. (See also retropharyngeal band)

esophagus [G *Ösophagus, Oesophagus*]: Relatively short, narrow section of digestive tract between pharynx and stomach; linked to endostyle of pharynx via retropharyngeal band.

eye [G *Auge*]: Light-sensitive organ located dorsally and associated with cerebral ganglion. May be subdivided in salp.

gastrozooid [G *Gasterozoid*]: trophozooid

gill bar: dorsal lamina

gill lamella [G *Kiemenlamelle*]: In doliolid, posterior part of pharynx bearing series of slits (stigmata). Separates anterior branchial cavity from posterior cloaca.

gill slit: stigma

gonad: ovary, testis

gonozooid [G *Gonozoid, Geschlechtstier*]: In doliolid, one of three types of blastozooid (gonozooid, phorozooid, trophozooid). Develops from bud on ventral process of phorozooid.

heart [G *Herz*]: In circulatory system, main pumping organ located below pharynx between posterior end of endostyl and digestive tract. More or less enclosed by pericardium. Gives rise anteriorly to ventral blood vessel, posteriorly to vessel extending to digestive tract. May pump blood in either direction.

intestine [G *Mitteldarm*]: Narrow, more elongate section of digestive tract between stomach and anus. (coiled, straight)

longitudinal bar [G *Längsgefäß, Längsbalken*]: branchial bar

luminous organ [G *Leuchtorgan*]: Structure containing light-producing bacteria. Typically located laterally/dorsolaterally in buccal cavity (in pyrosome) or associated with digestive tract or lateral body wall (in salp).

mantle vessel [G *Mantelgefäß*]: Network of blood channels extending through test. Supplied with muscles and involved in movement of diaphragm in pyrosome.

muscle ring (muscle band) [G *Muskelring, Muskelreifen*]: One in a series of well-developed, circular muscle bands more or less surrounding body. Number, shape, orientation, and relative proportion of rings and interspaces between rings (especially in old nurses in doliolid) are of taxonomic importance. (complete, not complete; parallel, not parallel, converging/not converging dorsally, touching dorsally, in contact/not in contact laterally; bifurcating, not bifurcating ventrally; aclinous, amphiclinous, eurymyonic, holomyonic, stenomyonic) (See also cuirass)

myoplane: Position on body in which successive muscle bands become narrower both anteriorly and posteriorly.

neural gland [G *Neuraldrüse, Subneuraldrüse*]: Glandular organ adjoining cerebral ganglion. Located dorsally above pharynx; opens into buccal cavity via canal and ciliated funnel.

nucleus [G *Nucleus, Eingeweidenucleus, Darmknäuel*]: In salp, compact, ball-shaped elaboration of digestive tract (esophagus, stomach, intestine, cecum).

nurse [G *Amme, Ammentier*]: In doliolid, late stage of sexually produced (and asexually reproducing) individual (oozooid). Carries or has carried (old nurse) asexually produced buds (blastozooids) along dorsal process (cadophore).

ocellus: eye

old nurse [G *alte Amme*]: nurse, muscle ring

oozooid [G *Oozoid*]: General term for individual arising from fertilized egg (i.e., sexually produced individual). Reproduces asexually by budding along ventral stolon. (See also blastozooid, cyathozooid, nurse)

otocyst: statocyst

otolith: statocyst

ovary [G *Ovar, Ovarium, Eierstock*]: ♀ reproductive organ located ventrally behind digestive tract. Opens into cloaca via oviduct.

oviduct [G *Eileiter, Ovidukt*]: In ♀ reproductive system, short canal leading from ovary to cloaca.

pericardium [G *Perikard*]: In circulatory system, tissue giving rise to and more or less surrounding heart. Space between pericardial wall and heart is termed pericardial cavity.

peripharyngeal ridge (peripharyngeal band) [G *Flimmerbogen*]: In digestive tract, one of two (left and right) semicircular ridges around anterior end of pharynx. Extends from ventral groove (endostyle) to ciliated funnel dorsally.

pharynx (branchial basket)
[G *Kiemendarm, Pharynx*]: Expanded, sac-shaped anterior section of digestive tract between buccal siphon and esophagus. Variously perforated by slits (stigmata); bears midventral groove (endostyle). Serves as combined digestive/respiratory/locomotory organ.

phorocyte [G *Phorozyte, Phorocyte*]: Type of cell in test; star-shaped, serving in asexual reproduction to transport buds from site of production to definitive site.

phorozooid [G *Phorozoid, Pflegetier, Medianzoid, Mediansproß*]: In doliolid, one of three types of blastozooid (gonozooid, phorozooid, trophozooid). Develops from median row of buds along dorsal process (cadophore) of oozooid. Characterized by ventral process bearing buds which eventually give rise to gonozooids. (See also probud)

prepharyngeal region: buccal cavity

probud [G *Geschlechtsknospe*]: In doliolid, bud produced on stolon of oozooid and migrating to phorozooid on dorsal process (cadophore) of oozooid. Attaches to stalk (= later ventral process) of phorozooid and gives rise, by budding, to gonozooids.

raphe [G *Herzraphe*]: In circulatory system, longitudinal seam formed by infolded walls of pericardium/heart complex.

rectum [G *Enddarm*]: Term occasionally applied to posterior section of intestine.

retropharyngeal band [G *Retropharyngealband, Mundrinne*]: In digestive tract, pair of ciliated bands at posterior end of pharynx. Represents continuation of endostyle and extends to esophagus.

sperm duct (vas deferens)
[G *Samenleiter, Vas deferens*]: In ♂ reproductive system, short canal leading from testis to cloaca.

sphincter muscle [G *Sphinkter, Schließmuskel*]: Band of muscle around either buccal siphon or atrial siphon. Serves to close opening(s) and enables, in combination

with muscle rings (in doliolid and salp), swimming movement through expulsion of water.

statocyst [G *Statozyste*]: In doliolid, organ of equilibrium located on left side of body at level of pharynx. Innervated by cerebral ganglion; consists of sac-like structure enclosing solid element. Enclosed element or entire organ occasionally also termed otolith.

stigma (gill slit) [G *Kiemenspalte*]: One in a series of slits surrounding pharynx (in pyrosome), in posterior pharynx wall (gill lamella) (in doliolid), or along strip (dorsal lamina) in dorsal wall of pharynx (in salp).

stolon [G *Stolo prolifer*]: In oozooid, ventral structure posterior to endostyle. Contains tissue from several organs of oozooid and gives rise to buds. In salp, extends through test to exterior.

stomach [G *Magen*]: Expanded section of digestive tract between esophagus and intestine. May be developed as nucleus and may bear ceca.

tentacle [G *Tentakel, Mundtentakel*]: In pyrosome, single ventral or one in a circlet of short, inwardly directed projections within buccal siphon.

test [G *Mantel*]: Variously developed layer enclosing body of solitary thaliacean or forming matrix in which colonial forms are embedded. Transparent, containing blood vessels and cells. (cartilaginous, gelatinous, leathery; firm, flabby; barrel-shaped, cylindrical, fusiform = spindle-shaped, globular, prismatic)

testis [G *Hoden*]: ♂ reproductive organ typically located ventrally behind digestive tract and in front of ovary. May consist of numerous follicles; opens into cloaca via sperm duct. (sac-shaped, ball-shaped, bottle-shaped, elongate)

trabecula [G *Trabekel*]: In pyrosome, one of numerous strands of tissue spanning across space (atrium) surrounding pharynx. Connects outer wall of pharynx with body wall.

transverse bar [G *Quergefäß, Querbalken*]: branchial bar

trophozooid (gastrozooid) [G *Trophozoid, Gasterozoid, Nährtier, Lateralsproß*]: In doliolid, one of three types of blastozooid (gonozooid, phorozooid, trophozooid). Develops from two lateral rows of buds along dorsal process (cadophore) of oozoid. Serves in feeding oozoid and other attached buds.

vas deferens: sperm duct

ventral blood vessel [G *Ventralgefäß, Hypobranchialgefäß*]: In circulatory system, one of up to three longitudinal blood channels extending under endostyle. Arises from anterior end of heart; variously joined to dorsal blood vessel. May also refer to short ventral vessel arising from posterior part of heart and leading to digestive tract.

ventral process [G *Haftstiel*]: In doliolid, posteriorly directed ventral extension of phorozooid. Represents remnant of stalk by which phorozooid was attached to oozoid. Bears probud and, later, series of additional buds giving rise to gonozooids.

zooid [G *Zoid*]: General term for single thaliacean individual, either as solitary form or within colony. (See also blastozooid, gonozooid, oozooid, phorozooid, trophozooid)

APPENDICULARIA

appendicularian in house (complex type)

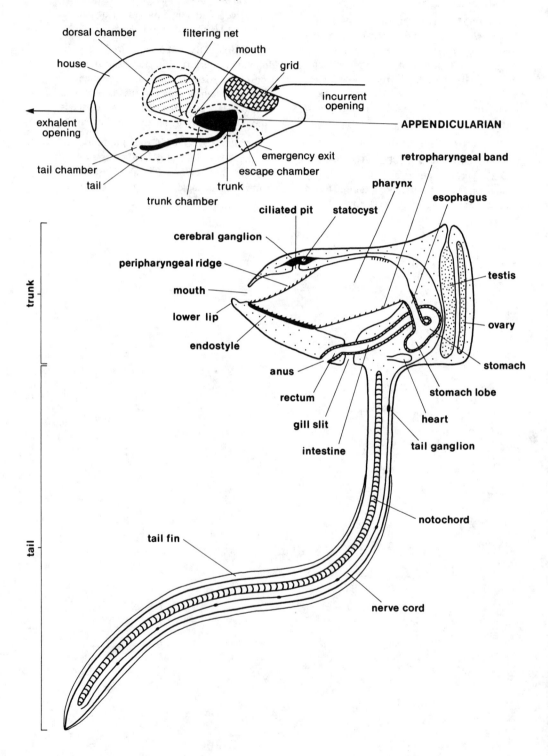

TUNICATA (UROCHORDATA)

APPENDICULARIA (LARVACEA)

appendicularian, larvacean
[G *Appendicularie, Larvacee*]

---

amphichordal cell [G *amphichordale Zelle*]: In certain appendicularians, one in a series of cells surrounding tip of tail.

anus [G *After, Anus*]: Posterior opening of digestive tract; located in midregion of trunk, either ventrally or somewhat to right side.

buccal gland [G *Munddrüse*]: Gland associated with midventral groove (endostyle) in pharynx. (oval, spherical)

cerebral ganglion [G *Gehirnganglion, Zerebralganglion*]: In trunk, main concentration of nerve tissue above anterior end of pharynx. Associated with statocyst and ciliated pit; gives rise to numerous nerves including posterior nerve cord extending into tail.

ciliated pit [G *Flimmergrube*]: Funnel-shaped, ciliated sensory organ associated with cerebral ganglion; opens into pharynx.

dorsal blood vessel [G *Mediodorsalsinus*]: In circulatory system, middorsal blood channel in trunk. Joined anteriorly to ventral blood vessel; extends posteriorly into tail as tail sinus.

dorsal chamber [G *Dorsalkammer, Rückenkammer*]: One of several chambers (dorsal, escape, tail, trunk chambers) which may be distinguished within complex type of house; refers to posterodorsal cavity surrounding filtering nets.

dorsal organ [G *Dorsalorgan*]: Term applied to ciliated middorsal strip of pharynx. Extends to esophagus and represents continuation of peripharyngeal ridges.

Eisen's oikoplast [G *Eisen'scher Oikoplast*]: One in a group of gland cells (oikoplasts) in outer body wall of trunk; located more posteriorly and secretes grid of house. (See also Fol's oikoplast)

emergency exit (escape hatch) [G *Fluchtöffnung*]: In complex type of house, anteroventral opening at end of escape chamber. Appendicularian may escape from house via membrane-covered exit.

endostyle [G *Endostyl*]: In trunk, ciliated groove along midventral wall of pharynx. Ciliated and glandular, producing mucus which is conveyed to esophagus. (See also retropharyngeal band)

escape chamber [G *Fluchtkammer*]: One of several chambers (dorsal, escape, tail, trunk chambers) which may be distinguished within complex type of house; refers to anteroventral cavity through which appendicularian can escape (via emergency exit) to exterior.

esophagus [G *Ösophagus, Speiseröhre*]: Narrow section of digestive tract between pharynx and stomach; heavily ciliated internally.

exhalent opening [G *Ausstromöffnung*]: In complex type of house, posterior opening through which water passes to exterior.

filtering net [G *Fangapparat*]: In complex type of house, one of two multilayered nets through which water passes before being expelled to exterior via exhalent aperture. Retained food is conveyed along trough at junction of two nets and is funneled into mouth of appendicularian. Secreted by Fol's oikoplasts.

Fol's oikoplast [G *Fol'scher Oikoplast*]: One in a group of gland cells (oikoplasts) in body wall of trunk; located more anteriorly and secretes filtering net of house. (See also Eisen's oikoplast)

germ of house: Surface of body in region of oikoplasts which secrete house.

gill slit [G *Kiemenspalte*]: One of two openings of pharynx. Water in pharynx passes through slits and is expelled to exterior. (short, slit-shaped; with/without tube; anterior/posterior to anus)

gonad: ovary, testis

grid (filtration grid, protective grid) [G *Schutzgitter, Gitterfenster, Einstromgitter*]: In complex type of house, grid-like structure covering each incurrent opening. Secreted by Eisen's oikoplasts; serves to prefilter oversized particles from incoming water.

heart [G *Herz*]: Pumping organ of circulatory system; located ventrally in trunk below digestive tract. May pump blood in either direction. (See also ventral blood vessel)

house [G *Gehäuse*]: Variously shaped gelatinous structure more or less surrounding appendicularian. Secreted by specialized glands (oikoplasts); serves in feeding and as protection. If fully developed, characterized by incurrent openings protected by grids, internal filtering nets, exhalent aperture through which water is passed to exterior, and several cavities (dorsal, escape, tail, trunk chambers). (enclosing/not enclosing body)

incurrent opening [G *Einstromöffnung*]: One of two dorsal openings at anterior end of complex type of house. Each opening is protected by grid.

intestine [G *Mitteldarm, Dünndarm*]: Narrow section of digestive tract between stomach and rectum. (with/without gland cells)

lip [G *Lippe*]: One of up to several lip-like processes surrounding mouth. Lower lip may be especially well developed.

lower lip [G *Unterlippe*]: lip

mouth [G *Mund*]: Anterior opening of digestive tract. Associated with filtering net of house; opens into pharynx. May be bordered by one or more processes (lips).

muscle [G *Muskel*]: Plate of muscle tissue forming heart or bands of muscle surrounding notochord in tail.

nerve cord [G *Neuralrohr, Neuralstrang*]: Longitudinal dorsal nerve cord arising from cerebral ganglion and extending posteriorly through trunk and into tail. May bear series of tail ganglia.

notochord [G *Chorda*]: Longitudinal supporting structure extending through tail; surrounded by muscle and adjoined by nerve cord and tail sinuses.

oikoplast [G *Oikoplast*]: One of numerous glands in body wall of trunk; secretes house. According to section of house secreted one may distinguish, for example, fields of Eisen's oikoplasts and Fol's oikoplasts.

ovary [G *Ovar, Ovarium*]: ♀ reproductive organ located posteriorly in trunk. Ruptures to release eggs.

pericardium [G *Perikard, Pericard, Perikardialsack*]: In circulatory system, saclike structure surrounding heart.

peripharyngeal ridge (peripharyngeal band) [G *Wimperbogen, Peripharyngealband, peripharyngeales Wimperband*]: In trunk, one of two (left and right) ciliated bands around anterior end of pharynx. Extends from ventral groove (endostyle) to dorsal organ.

pharynx [G *Kiemendarm, Pharynx*]: In trunk, expanded, sac-shaped anterior part of digestive tract between mouth and esophagus. If fully developed, bears midventral groove (endostyle), peripharyngeal and retropharyngeal bands, and dorsal organ. Alternately, may bear two series of dorsal processes and two series of ventral processes.

pylorus [G *Pylorus, Pylorusdarm*]: stomach

rectum [G *Enddarm*]: Heavily ciliated posteriormost section of digestive tract between intestine and anus.

retropharyngeal band [G *Retropharyngealband, retropharyngeales Wimperband*]: In trunk, pair of ciliated bands at posterior end of pharynx. Represents continuation of endostyle and may extend to esophagus.

sperm duct [G *Samenleiter*]: Short section of ♂ reproductive system between testis and exterior.

statocyst [G *Statozyste, Statolithenblase*]: In trunk, sac-shaped organ of equilibrium associated with cerebral ganglion. Contains solid element (statolith).

statolith [G *Statolith*]: Solid element within sac-shaped organ of equilibrium (statocyst).

stomach [G *Magen*]: Expanded section of digestive tract between esophagus and intestine. May be variously regionated into left and right lobes as well as pyloric region.

stomach lobe [G *Magen-Lappen, Magensack, Magentasche*]: In certain appendicularians, one of two (left and right) lobes of stomach. (sac-like, pentagonal)

subchordal cell (subnotochordal cell) [G *Subchordalzelle*]: In tail, one of up to numerous characteristic cells below notochord. (two cells, series of cells)

tadpole larva [G *Kaulquappen-Larve, Schwimmlarve*]: Free-swimming larval stage characterized by tail.

tail [G *Schwanz*]: One of two major divisions (tail, trunk) of body. Contains notochord, nerve cord, muscles, and blood sinuses. Serves in locomotion and to drive current within house. Length relative to trunk is of taxonomic importance. (forked, pointed, rounded, simple)

tail chamber [G *Schwanzkammer*]: One of several chambers (dorsal, escape, tail, trunk chambers) which one may distinguish within complex type of house; refers to ventral cavity surrounding tail.

tail fin [G *Flossensaum*]: Membranous extension along each side of of tail. (tip: bifurcated, square-cut)

tail ganglion [G *Schwanzganglion*]: One in a series of ganglia along dorsal nerve in tail. First ganglion at base of tail may be larger.

tail sinus [G *Schwanzsinus*]: In circulatory system, one of two longitudinal blood channels through tail.

test: house

testis [G *Hoden*]: ♂ reproductive organ located posteriorly in trunk. Opens to exterior via short sperm duct. (single, paired)

trunk [G *Rumpf*]: One of two major divisions (tail, trunk) of body. Encloses major organs (pharynx and remaining digestive tract, cerebral ganglion, heart, and gonads). (oval, tripartite)

trunk chamber [G *Rumpfkammer*]: One of several chambers (dorsal, escape, tail, trunk chambers) which one may distinguish within complex type of house; refers to anterior cavity surrounding trunk of appendicularian.

ventral blood vessel [G *Medioventralsinus*]: In circulatory system, blood channel arising from anterior end of heart. Extends below endostyle of pharynx and is joined anteriorly to dorsal blood vessel.

ACRANIA
(CEPHALOCHORDATA)

lancelet [G Lanzettfischchen]

ACRANIA

CROSS SECTIONS
(viewed posteriorly)

(lateral)

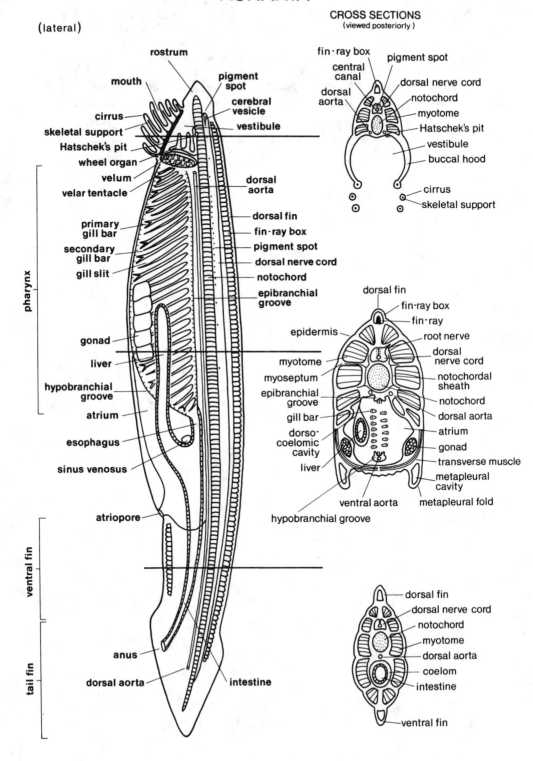

ACRANIA (CEPHALOCHORDATA)

lancelet [G *Lanzettfischchen*]

anus [G *After*]: Opening of digestive tract to exterior; located ventrally, posterior to opening of atrium (atriopore) in region of tail fin.

aorta: dorsal aorta, ventral aorta

atriopore [G *Atrioporus, Porus abdominalis*]: Unpaired opening of atrium to exterior; located ventrally at border of metapleural folds and ventral fin, anterior to anus.

atrium [G *Atrium, Peribranchialraum*]: Space surrounding pharynx. Water passing through gill slits of pharynx enters atrium and passes to exterior via atriopore.

branchial basket: pharynx

buccal cirrus (cirrus, tentacle) [G *Mundcirrus, Lippententakel*]: One in a series of elongate projections flanking each side of mouth. Originates from margin of buccal hood and contains skeletal element. Cirri form first sieve-like apparatus preventing large or indigestible particles from entering pharynx. (See also velar tentacle)

buccal hood (oral hood) [G *Wange*]: Anterolateral cheek-like fold forming each wall of vestibule; bears series of buccal cirri around mouth.

bulbillus [G *Bulbillus, Kiemenherz*]: In circulatory system, one in a series of contractile enlargements of ventral aorta in pharynx region. Pumps blood dorsally through gill bars.

cardinal vein [G *Cardinalvene*]: In circulatory system, one of four (two anterior and two posterior) vessels transporting deoxygenated blood from body wall and gonads to sinus venosus. (See also duct of Cuvier)

central canal [G *Zentralkanal, Canalis centralis*]: Slit-like canal extending throughout dorsal nerve cord; associated with pigment spots and somewhat expanded anteriorly to form cerebral vesicle.

cerebralvesicle [G *Ependymbläschen, Hirnbläschen, Stirnbläschen*]: Anterior tip of dorsal nerve cord with somewhat expanded central canal. Composed of several types of cells, including photosensitive pigment spots.

cirrus: buccal cirrus

coelom [G *Coelom*]: Epithelium-lined body cavity. According to position, the following coeloms have been distinguished: dorsocoelomic canal, gill bar coelom, subendostylar coelom, gonocoel, metapleural cavity, and intestinal coelom.

cyrtopodocyte (solenocyte) [G *Crytopodozyte, Solenozyte*]: In excretory system, one of several flagellum-bearing structures attached at one end to glomerulus and opening at other end into atrium. Flagellum, surrounded by series of elongate rods, lies in dorsocoelomic canal.

dorsal aorta [G *Dorsalaorta, Aorta dorsalis*]: In circulatory system, dorsal vessel in which blood is transported posteriorly; gives off branches into gut, gonads, and somatic muscles. Paired above pharynx (left and right aortas or radix aortae), fused posteriorly.

dorsal nerve cord (neural tube) [G *Rückenmark, Neuralrohr*]: Elongate nerve cord extending above notochord from base of rostrum to tail; contains slit-shaped central canal with pigment spots and gives rise to segmentally arranged root nerves. Canal is somewhat expanded anteriorly to form cerebral vesicle.

dorsocoelomic canal [G *Subchordalcoelom*]: coelom

duct of Cuvier [G *Ductus cuvieri*]: In circulatory system, one of two ducts transporting deoxygenated blood from cardinal veins to sinus venosus.

endostylar groove: hypobranchial groove

endostyle: hypobranchial groove

enterostome [G *Enterostom*]: Opening in velum through which vestibule and pharynx communicate. Formed by inner margins of velum and surrounded by velar tentacles.

epibranchial groove [G *Epibranchialrinne*]: Longitudinal, ciliated groove forming roof of pharynx; receives food particles trapped between gill bars and transports them into esophagus.

epidermis [G *Epidermis*]: Single layer of cells forming outer layer of body. Overlies cutis.

esophagus [G *Ösophagus*]: Relatively short section of digestive tract between pharynx and intestine.

excretory organ (protonephridium [G *Exkretionsorgan*]: One in a series of segmentally arranged excretory structures, one pair to each pair of gill slits. Each organ consists of expanded glomerulus and adjoining cyrtopodocytes which open into atrium.

external blood vessel [G *Aussengefäß, aüßeres Achsengefäß, Coelomgefäß*]: In circulatory system, outermost vessel extending through gill bar; transports blood dorsally from ventral to dorsal aorta. (See also somatic blood vessel, visceral blood vessel)

eye spot: pigment spot

fin [G *Flosse, Flossensaum*]: Thin median extension of body. According to position one may distinguish a dorsal fin extending from rostrum to tail, a ventral fin between atriopore and anus, and a tail fin. The former two are supported proximally by a series of noncellular fin-rays extending into fin-ray boxes.

fin-ray [G *Flossenstrahl*]: Noncellular material extending into each fin-ray box.

fin-ray box [G *Flossenkammer, Flossenkästchen, Flossenhöhle*]: One in a series of minute chambers forming basal part of dorsal and ventral fins, each box containing noncellular material (fin-ray). Three to five boxes are present per somatic muscle segment (myotome).

gill bar (pharyngeal gill bar) [G *Kiemenbogen, Bogen*]: One in a series of up to 150 oblique bars forming each wall of pharynx. One may distinguish primary and secondary (tongue) bars. The former contain three blood vessels (somatic, visceral, external) and a coelom, are interconnected by synapticulae, and are forked ventrally; the latter contain only two blood vessels, lack a coelom, and are not forked. Both types bear a skeletal rod as well as tufts of frontal and lateral cilia. Serve to trap food particles and, to a lesser extent, in respiration.

gill bar coelom [G *Kiemenbogencoelom, Branchialcoelom*]: coelom

gill slit (pharyngeal gill slit, pharyngeal cleft) [G *Kiemenspalte*]: In wall of pharynx, one in a series of slits between gill bars; interrupted by synapticulae between bars and largely filled by lateral cilia of adjoining bars.

glomerulus [G *Glomerulus*]: Expanded section of each excretory organ; located at end of secondary gill bar shortly before entrance of gill blood vessels into dorsal aorta (radix aortae). Cryptopodocytes are attached to glomerulus.

gonad [G *Gonade*]: ♂ (testis) or ♀ (ovary) reproductive organ segmentally arranged along wall of atrium. Eggs and sperm reach exterior through atriopore after rupture of gonad wall. (along one side of body, along both sides of body)

gonocoel [G *Gonadencoelom*]: coelom

Hatschek's pit [G *Hatschek'sche Grube, Geißelgrube*]: Ciliated sensory pit in roof of vestibule; associated with dorsal projection of wheel organ.

hepatic diverticulum: liver

hepatic reticulum [G *Leberpfortaderkreis*]: In circulatory system, network of fine capillaries around liver; deoxygenated blood is transported from hepatic reticulum to sinus venosus.

hindgut (rectum) [G *Enddarm*]: Short posteriormost section of digestive tract between intestine and anus.

hood: buccal hood

hypobranchial groove (endostyle, endostylar groove) [G *Hypobranchialrinne, Endostyl*]: Longitudinal, ciliated groove forming floor of pharynx; secretes mucus and transports food particles trapped between gill bars to epibranchial groove and intestine. Adjoined ventrally by supports (endostylar skeletal plates), ventral aorta, and subendostylar coelom.

intestinal coelom: coelom

intestinal reticulum [G *kapilläres Darmgeflecht, Darmsinus*]: In circulatory system, network of fine capillaries around intestine; from reticulum, blood enters subintestinal channel and is transported to hepatic reticulum.

intestine (midgut) [G *Darm, Nahrungsdarm*]: Elongate section of digestive tract between esophagus and hindgut. Gives rise to relatively large, anteriorly directed diverticulum (liver). Nonmuscular, inner wall lined with well-developed cilia.

liver (midgut diverticulum, midgut cecum) [G *Leberblindsack, Mitteldarmdrüse*]: Relatively large, anteriorly directed diverticulum originating at junction of esophagus and intestine. Extends far into anterior region of body along right side of pharynx. Serves in digestion.

metapleural cavity [G *Metapleuralcoelom, Metapleuralhöhle*]: One of two ventrolateral body cavities delimited by metapleural folds and transverse muscles.

metapleural fold [G *Metapleuralfalte*]: One of two longitudinal folds extending along ventral surface of body from mouth to atriopore. Encloses body cavity (metapleural cavity).

midgut: intestine

midgut cecum (midgut diverticulum): liver

mouth [G *Mund, Mundöffnung*]: Anterior opening of digestive tract under rostrum. Flanked on each side by series of buccal cirri; opens into vestibule. (See also enterostome)

muscle: somatic muscle, transverse muscle

myomere: myotome

myoseptum [G *Myoseptum*]: Partition of connective tissue separating individual muscle segments (myotomes). Spans from cutis to either notochordal sheath or wall of atrium and forms sheath-like structure enclosing myotome.

myotome (myomere) [G *Myotom, Myomer*]: One of up to 75 segmentally arranged bands of muscle (somatic muscle) along each side of body. Each myotome is V-shaped, with bend pointing anteriorly. Myotomes are separated by myosepta and those of left and right body walls show a staggered pattern.

nerve cord: dorsal nerve cord

neural tube: dorsal nerve cord

notochord [G *Chorda dorsalis*]: Elongate, cylindrical structure extending anteriorly from tip of rostrum to end of tail fin under dorsal nerve cord. Composed of densely packed vertical discs (muscle fibrils) and enclosed by notochordal sheath. Serves as skeletal support.

notochordal sheath [G *Chordascheide*]: Sheath of connective tissue surrounding and imparting rigidity to notochord. Continuous with myosepta between muscle segments (myotomes).

ocellus: pigment spot

olfactory pit [G *Kölliker'sche Grube, Geißelgrube*]: Small, ciliated pit above tip (cerebral vesicle) of dorsal nerve cord.

oral hood: buccal hood

ovary: gonad

pharyngeal gill slit: gill slit

pharynx (branchial basket) [G *Kiemendarm, Pharynx*]: Prominent food-straining and respiratory organ between velum and esophagus. Each side wall of pharynx consists of a series of alternating gill bars and gill slits which separate inner pharynx lumen from surrounding atrium. Food particles are collected by gill bars, transferred to epibranchial groove, and transported posteriorly into esophagus.

pigment spot (eye spot) [G *Pigmentfleck, Pigmentbecherocelle, Rückenmarksauge*]: One in a series of variously oriented photosensitive structures along dorsal nerve cord. Associated with central canal and concentrated at anterior end. Refers also to photoreceptor cells at expanded anterior tip (cephalic vesicle) of dorsal nerve cord.

primary gill bar [G *Hauptkiemenbogen, Hauptbogen*]: gill bar

protonephridium: excretory organ

radix aortae: aorta

root nerve (nerve root) [G *Nervenwurzel, Spinalnerv*]: One in a series of segmentally arranged nerves (alternating on left and right sides) originating on dorsal nerve cord. One may distinguish dorsal and ventral root nerves.

rostrum [G *Rostrum*]: Anteriormost part of body; projects over vestibule and contains tip of notochord.

secondary gill bar [G *Zwischenbogen, Zungenbogen*]: gill bar

sinus venosus [G *Sinus venosus*]: Contractile enlargement of circulatory system at junction of esophagus, intestine, and liver. Pumps deoxygenated blood anteriorly to ventral aorta and gill bars.

solenocyte: crytopodocyte

somatic blood vessel [G *Aussengefäß, äußeres Achsengefäß*]: In primary gill bar, median vessel extending from ventral to dorsal aorta. Blood is pumped dorsally by contractile ventral aorta or by bulbilli. (See also external blood vessel, visceral blood vessel)

somatic muscle (segmental muscle, longitudinal muscle) [G *somatischer Muskel*]: Well-developed muscles forming greater part of body wall. Subdivided into V-shaped muscle bands (myotomes) separated by myosepta. Restricted to dorsal and lateral sides anteriorly, but gradually extending to midventral line after atriopore. Serves in locomotion. (See also transverse muscle)

stomach: Term occasionally applied to somewhat expanded section of digestive tract at junction of esophagus, intestine, and liver.

subendostylar coelom [G *Endostylcoelom, Endostylarcoelom, Hypobranchialcoelom*]: coelom

subintestinal channel [G *Vena subintestinalis, Pfortader, Vena portae*]: In circulatory system, longitudinal vessel under intestine; receives blood from intestine and transports it anteriorly to liver capillaries (hepatic reticulum).

synapticule [G *Synaptikel*]: In pharynx, short interconnection between adjacent primary gill bars. Contains blood vessel.

tail [G *Schwanzregion*]: Part of body posterior to anus; contains segmented muscles (myotomes), dorsal nerve cord, and notochord and is extended dorsally and ventrally to form tail fin.

tail fin [G *Schwanzflosse, Schwanzflossensaum*]: Dorsal and ventral extension of body posterior to anus; lacks fin-ray boxes.

tentacle: buccal cirrus, velar tentacle

testis: gonad

tongue bar (secondary bar) [G *Zwischenbogen, Zungenbogen*]: gill bar

transverse muscle [G *Transversalmuskel, Viszeralmuskel, Quermuskel*]: Sheath of muscle extending ventrally from somatic musculature (myotomes) and surrounding atrium. Reduced posteriorly, gradually being replaced by somatic musculature posterior to atriopore. Water current produced by contraction of transverse muscles serves to unclog mouth region of particles.

velar tentacle [G *Velartentakel*]: One in a series of tentacle-like projections of inner margin of velum. Tentacles surround pore (enterostome) through which vestibule and pharynx communicate and form second sieve-like apparatus preventing large or indigestible particles from entering pharynx. (See also buccal cirrus)

velum [G *Velum*]: Inwardly directed, muscular fold of buccal hood; projects into and forms posterior wall of vestibule. Bears velar tentacles along inner margins and forms median pore (enterostome) through which vestibule and pharynx communicate.

vena hepatica [G *Vena hepatica, Lebervene*]: In circulatory system, vessel transporting deoxygenated blood from liver to sinus venosus.

ventral aorta [G *Aorta ventralis, Truncus arteriosus*]: In circulatory system, vessel running through floor (endostyle) of pharynx and transporting blood anteriorly from sinus venosus; bears series of contractile bulbilli and pumps blood through gill bars to dorsal aorta.

vestibule [G *Präoralhöhle, prävelarer Mundraum*]: Cavity between mouth opening and pharynx. Outer walls of vestibule formed by buccal hood; contains Hatschek's pit and wheel organ and is delimited posteriorly by velum.

visceral blood vessel [G *Innengefäß, inneres Achsengefäß*]: Innermost of two (in secondary gill bar) or three (in primary gill bar) vessels extending through gill bar from ventral to dorsal aorta. Blood is pumped dorsally by contractile ventral aorta or by bulbilli. (See also external blood vessel, somatic blood vessel)

wheel organ [G *Räderorgan*]: Band of elevated tissue around inner wall of buccal hood; located in front of velum at posterior end of hood. Bears highly ciliated, finger-like projections extending into vestibule and creates water current directed toward pharynx. Associated dorsally with Hatschek's pit.

ADJECTIVE SECTION

a- [G *a-*]: Prefix indicating absence of a structure or condition.

ab- [G *ab-*]: Directed away from or lying opposite a specified structure. (See also ad-)

abanal [G *abanal*]: Type of gill arrangement in chiton (polyplacophoran) in which gill size increases posteriorly, the last one in each row being the largest; gills cease some distance from anus. (See also adanal)

abapertural [G *abapertural*]: In snail (gastropod) shell, located on or slightly oblique to shell axis and directed or facing away from aperture. (See also adapertural)

abapical [G *abapical*]: Directed away from or lying opposite to apex. In snail (gastropod) shell, directed away from apex. (See also adapical)

abaxial [G *abaxial*]: Directed away from or lying opposite to axis. In snail (gastropod) shell, directed outward, away from axis. (See also adaxial)

abbreviated [G *verkürzt*]: Relatively short.

abfrontal [G *abfrontal*]: In mollusc, margin of gill lying opposite to inhalent current. (See also frontal)

aboral [G *aboral*]: Directed away from or lying opposite to mouth. (See also adoral)

aborted [G *abortiv*]: Not fully developed. In certain colonial hydrozoans, structure which may represent reduced or modified hydranth.

abranchiate [G *kiemenlos*]: Lacking gills, e.g., as in certain polychaetes.

absent [G *fehlend, fehlen*]: (See also present)

acantho- [G *akantho-, acantho-*]: Prefix indicating spiny nature of a structure, e.g., spiny elaboration of sponge spicule types.

acantho- [G *akantho-, acantho-*]: In tapeworm (cestode), prefix applied to type of plerocercoid (stage in tapeworm life cycle) whose scolex bears spines. (See also acetabulo-, bothridio-, bothrio-, culcito-, glando-, tentaculo-)

acanthopodous [G *acanthopod*]: In barnacle, type of thoracic appendage (cirrus) in which anterior margin bears few or no setae, while setae of posterior margin are stout, sharp, and restricted to region of articulation of segments. (See also ctenopodous, lasiopodous)

acanthorostello- [G *acanthorostello-*]: In tapeworm (cestode), prefix applied to type of cysticercoid (stage in tapeworm life cycle) with armed rostellum. (See also anacanthorostello-, arostello-, rostello-)

acanthostegous: In bryozoan, type of external brooding chamber ("ovicell") formed by extension of spines over frontal membrane.

acaudate [G *schwanzlos, acaudat*]: Lacking tail. In tapeworm (cestode), refers to absence of tail (cercomer) in procercoid, plerocercoid, or cystocercoid stages of life cycle. (See also caudate)

accessory [G *akzessorisch*]: Refers to presence of additional structure.

acetabulo- [G *acetabulo-*]: In tapeworm (cestode), prefix applied to type of plerocercoid (stage in tapeworm life cycle) bearing suckers (acetabula). (See also acantho-, bothridio-, bothrio-, culcito-, glando-, tentaculo-)

achaetous (asetigerous) [G *borstenlos*]: In polychaete, lacking bristles (chaetae, setae).

achelate [G *achelat*]: In crustacean, lacking pincer-like structure at end of appendage. (See also chelate, subchelate)

acicular [G *aciculär*]: In polychaete, type of short bristle (chaeta, seta) similar in thickness to aciculum supporting branch of parapodium.

aciculate [G *nadelförmig*]: Needle-shaped, e.g., slender, tapering to sharp point.

acleithral: In bryozoan, type of brood chamber (ooecium) not closed by lid (operculum) of parent zooid. (See also cleithral)

acline: orthocline

aclinous: Type of arrangement of muscle bands in certain thaliaceans (old nurses in doliolid); bands 2–8 do not form a series in which they gradually become broader and then narrower. (See also amphiclinous; eurymyonic, holomyonic, stenomyonic)

aclitellate [G *aclitellat*]: In oligochaete, lacking clitellum. Typically reflects sexual immaturity, occasionally also a condition after a period of sexual activity. (See also clitellate)

acoelous [G *acoel*]: Lacking body cavity (coelom).

acraspedote [G *acraspedot*]: In hydrozoan or scyphozoan medusa, lacking horizontal fold (velum) projecting inward from umbrella margin. In tapeworm (cestode), type of strobila in which each segment (proglottid) does not overlap immediately following proglottid. (See also craspedote)

acrembolic [G *akrembolisch*]: In snail (gastropod), type of proboscis which is completely invaginable. (See also pleurembolic).

actinodont [G *actinodont*]: In certain freshwater bivalves, type of hinge (dentition) composed of pseudocardinal and pseudolateral teeth.

acuminate [G *spitz, zugespitzt*]: Tapering or abruptly tapering structure, e.g.,

bristle (chaeta, seta) of polychaete or tooth on exoskeleton of crustacean.

acute (acute-angled) [G *spitzwinklig*]: Coming together at a narrow angle or forming a sharp point. (See also obtuse)

ad- [G *ad-*]: Directed toward or lying close to a specified structure. (See also ab-)

adanal [G *adanal*]: Located near anus. In chiton (polyplacophoran), type of gill arrangement in which series of gills continues up to anus; largest gill in each row is followed posteriorly by several smaller ones. (See also abanal)

adapertural [G *adapertural*]: In snail (gastropod) shell, directed toward or facing aperture. (See also abapertural)

adapical [G *adapical*]: Directed toward or facing apex. In snail (gastropod) shell, located on or slightly oblique to shell axis and directed toward apex. In sea urchin shell (test), positioned closer to apical plates surrounding anus, e.g., adapical suture. (See also abapical)

adaxial [G *adaxial*]: Directed toward or facing axis. In snail (gastropod) shell, directed inward toward shell axis. (See also abaxial)

adenal [G *adenal*]: In flatworm (turbellarian), type of structure (rhabdite) secreted in deeper layer of body wall (vs. epidermal rhabdite).

adeoniform [G *adeoniform*]: Type of bryozoan colony: erect, rigid, arborescent, bilaminar, with flattened branches, and firmly attached to hard substratum by calcareous plate.

adiverticulate [G *unverzweigt*]: Lacking a diverticulum or diverticula. In oligochaete, refers to spermatheca lacking diverticula. (See also diverticulate)

adnate: In bryozoan, type of growth in which one surface of individual (zooid) or colony (zoarium) is adherent to substratum.

adont [G *adont*]: Simple type of hinge in mussel shrimp (ostracod) carapace. Characterized by elongate ridge or bar along dorsal margin of one valve which fits into groove of opposite valve. (See also amphidont, archidont, entomodont, heterodont, lobodont, lophodont, merodont, schizodont)

adoral [G *adoral*]: Directed toward or located near mouth. (See also aboral)

adpressed [G *adpress*]: In snail (gastropod) shell, condition in which whorls overlap in such a manner that their outer surfaces converge very gradually. (See also disjunct, imbricate, immersed; appressed)

adradial [G *adradial*]: Relating to adradius of cnidarian; e.g., in tetramerous radial symmetry of medusa, positioned on or along one of eight sectors lying between each perradius and interradius. In sea urchin shell (test), positioned between ambulacral and interambulacral plates, e.g., adradial suture.

advolute [G *advolut*]: In snail shell, condition in which successive whorls do not distinctly overlap but rather just barely touch one another. (See also adpressed, disjunct, imbricate, immersed)

afferent [G *afferent, zuführend*]: Leading or conveying to an organ or structure, e.g., as in an afferent blood vessel supplying a particular organ. (See also efferent)

aggregate [G *gregat*]: In certain pelagic thaliaceans (salps), colonial form/phase consisting of circular or chain-like group of individuals (zooids).

ahermatypic [G *ahermatypisch*]: Type of growth which does not lead to reef formation. Also designates type of coral (anthozoan) which lacks symbiotic algae (zooxanthellae).

alate [G *flügelförmig ausgebreitet, alat*]: Bearing wing-like extension (ala). Type of snail shell bearing wing-like extension of outer lip. In bivalve, characterized by an extension of dorsal (hinge) region beyond main body (disc) of shell. In ostracod, shell (valve) with wing-like extension. (See also bialate, auriculate)

aliform [G *flügelförmig*]: Wing-shaped, e.g., in polychaete, aliform dorsal branch (notopodium) of appendage (parapodium).

alimbate: Lacking an extended margin, e.g., type of polychaete capillary seta not exhibiting a lateral wing.

alivincular [G *alivinkulär*]: In hinge of bivalve, flat type of ligament positioned centrally and having lamellar layer both anterior and posterior to central fibrous layer. (See also duplivincular, multivincular, paravincular)

alternate [G *alternierend, wechselständig*]: Arranged along an axis so as to be positioned on a different level than opposite member, e.g., type of polyp or branch arrangement in erect hydrozoan colony. (See also opposite, spiral, subopposite, whorled)

alveolar [G *alveolär*]: In sea star, type of pedicellaria partly sunk into endoskeletal depression (alveolus); may be further subdivided into bivalved, excavate, and spatulate types. (See also multilocular)

alveolate [G *wabig*]: Pitted and resembling a honeycomb, e.g., as in type of ornamentation on shell (valve) of mussel shrimp (ostracod).

ambulatory [G *Gang-*]: Serving in walking. (See also natatory)

amphiclinous: Type of arrangement of muscle bands in certain thaliaceans (old nurses in doliolid); bands 2–8 form a series in which they gradually become broader and then narrower. (See also aclinous; eurymyonic, holomyonic, stenomyonic)

amphidelphic [G *gegenständig*]: Type of uterus configuration in roundworm (nematode); uteri opposed in their origin. (See also opisthodelphic, prodelphic; didelphic, monodelphic, polydelphic, tetradelphic)

amphidetic [G *amphidet*]: In hinge of bivalve shell, type of ligament extending on both sides of (anterior and posterior to) beak. (See also opisthodetic, prosodetic).

amphidont [G *amphidont*]: Type of compound hinge in mussel shrimp (ostracod); consists of four elements, i.e., two terminal (anterior and posterior) elements as well as subdivided median element (into

shorter anteromedian and longer posteromedian elements). According to elaboration of these elements, four subtypes
(archidont, heterodont, lobodont, schizodont) may be distinguished. (See also adont,
merodont)

amphisternous [G *amphistern*]:
Type of region (plastron) on underside of irregular sea urchin shell (test); plastron in
which labrum is followed by two large plates
(See also meridosternous)

amphistome [G *amphistom*]: General category of larval (cercaria) or adult digenetic fluke (trematode); refers to forms
with anterior oral sucker and ventral sucker
(acetabulum) located at or near posterior end
of body. (See also distome, echinostome, gasterostome, holostome, monostome, prosostomatous)

amphithyridid [G *amphithyrid*]: In
lamp shell (brachiopod), position of pedicel
opening (foramen) relative to beak of ventral
valve: foramen shared by margin of delthyrium (ventral valve) and groove in beak of
dorsal valve. (See also epithyridid, hypothyridid, mesothyridid, permesothyridid,
submesothyridid)

amplete: In mussel shrimp (ostracod),
condition of carapace (or valve) in which
greatest height is in front of midlength. (See
also postplete, preplete)

anacantho-: In tapeworm (cestode),
prefix applied to cysticercus (stage in cestode
life cycle) lacking hooks. (See also heter-
acantho-, homacantho-)

anacanthorostello-: In tapeworm
(cestode), prefix applied to cysticercoid
(stage in tapeworm life cycle) with unarmed
rostellum. (See also acanthorostello-, aros-
tello-, rostello-)

anapolytic: In tapeworm (cestode),
type of species in which eggs are shed before
segments (proglottids) detach from worm
body.

anascan: Type of bryozoan in which
feeding individual (autozooid) lacks compensation sac (ascus).

anastomose [G *anastomosierend*]:
Forming a network.

anastrophic [G *anastroph*]: In snail
(gastropod) shell, growth form in which embryonic shell (protoconch) is coiled in opposite direction of remaining whorls
(teloconch) and nucleus is directed toward
base of shell. (See also heterostrophic)

anchorate: Bearing one or more processes resembling the flukes of an anchor.
Refers to certain types of sponge spicules,
e.g., anchorate isochelae.

angulate (angular) [G *kantig, ge-
schultert, getreppt*]: Forming an angle. In
snail (gastropod) shell, refers to type of whorl
which forms an angle rather than being
rounded.

aniso- [G *aniso-*]: Prefix indicating
unequal dimension or other dissimilarity.

anisodiametric [G *anisodiamet-
risch*]: Not of same diameter throughout,
e.g., butt of stinging cell (nematocyst) in
cnidarian. (See also isodiametric)

anisomyarian (heteromyarian)
[G *anisomyar, ungleichmuskelig*]: In bivalve, condition in which two adductor muscles differ significantly in size (anterior
muscle typically smaller than posterior). (See
also dimyarian, isomyarian = homomyarian, monomyarian)

annular: In oligochaete, type of clitellum that is complete ventrally and thus encircles body. (See also saddle-shaped)

annulate [G *geringelt*]: Consisting of
or characterized by ring-like structures due
to series of constrictions. May be used as synonym for multisegment anthropod appendage. (See also smooth)

anodont (edentulous) [G *ano-
dont, zahnlos*]: In bivalve, type of hinge lacking teeth. (See also dysodont)

anomphalous [G *nabellos*]: In snail
(gastropod), type of shell lacking umbilicus.

anoperculate [G *deckellos*]: Lacking a lid (operculum), e.g., in egg of tapeworm (cestode). (See also operculate)

anteclitellar: In oligochaete, position of male gonopore anterior to clitellum. (See also intraclitellar, postclitellar)

antenniform [G *antennenförmig*]: Shaped like an antenna.

antepenultimate [G *drittletzt*]: Positioned before the next to last (penultimate) element in a series, e.g., segment in arthropod appendage.

anteriad [G *anteriad*]: Extending toward anterior end.

anterior [G *vordere, Vorder-*]: Directed toward or lying in region defined by head.

antero- [G *antero-*]: Prefix indicating position or orientation involving anterior end, e.g., anterodorsal, anterolateral, anteromedian, anteroventral.

antidromous [G *antidrom*]: Type of reflexed ovary in roundworm (nematode): entire germinal and growth zone is reflexed against oviduct. Eggs wander with one pole toward reflection, with other pole toward vulva. (See also homodromous)

apertural [G *apertural*]: In snail (gastropod), pertaining to or on same side as opening (aperture) of shell. (See also abapertural, adapertural)

apharyngeate [G *apharyngeat, schlundlos*]: Lacking a pharynx. Refers to type of larva (cercaria) of digenetic fluke (trematode). (See also pharyngeate)

aphelenchoid [G *aphelenchoid*]: Type of three-part pharynx in roundworm (nematode); similar to tylenchoid, yet pharyngeal glands protrude posteriorly from pharynx. Also refers to type of caudal ala (bursa) in nematode; characterized by three or more pairs of genital papillae.

aphodal [G *aphodal*]: Type of choanocyte chamber in leuconoid sponge; characterized by short canal (aphodus) leading from chamber to excurrent canal. (See also eurypylous)

apical [G *apikal*]: Pertaining to or located at or near tip (apex).

apiculate [G *feinspitzig*]: Ending abruptly in small, distinct point.

apodous (apodal) [G *fußlos, gliedmaßenfrei*]: Lacking limb (in arthropod) or parapodium (in polychaete).

apolytic: In tapeworm (cestode), type of species in which segments (proglottids) detach from animal before eggs are shed.

apomorphic [G *apomorph*]: Derived, recently acquired structure differing from ancestral condition. (See also plesiomorphic)

aporose: imperforate

apotrichous [G *apotrich*]: Type of stinging cell (nematocyst) of cnidarian: only distal section of tubule bears spines. Refers to nematocyst type (isorhiza) which lacks butt and in which tubule is of same diameter throughout. (See also atrichous, basitrichous, heterotrichous, holotrichous, merotrichous, telotrichous)

appendiculate: Type of distome digenetic fluke (trematode) whose thin-walled, tail-like posterior end (ecsoma) can be withdrawn into thick-walled anterior end (soma). (See also nonappendiculate).

appressed (adpressed) [G *anliegend, angedrückt, angeschmiegt*]: Lying flat against a structure, e.g., spines along arm of brittle star (ophiuroid). (See also erect)

arakoderan: Type of caudal ala (bursa) in roundworm (nematode): term proposed for ala meeting anteriorly and posteriorly, i.e., forming a complete oval and thus surrounding cloacal area. (See also leptoderan, peloderan)

arbacioid [G *arbacioid*]: In shell (test) of sea urchin, type of compound ambulacral plate composed of median primary plate with demiplate adapical and adoral to it. (See also diadematoid, echinoid, echinothurioid)

arborescent [G *baumförmig, bäumchenartig verzweigt*]: Resembling a small tree with stem and branches.

archidont [G *archidont*]: Type of compound hinge in mussel shrimp (ostracod): one of four basic subtypes of amphidont hinge—anteromedian element is only poorly differentiated from posteromedian element (e.g., being only more coarsely crenulate, not developed as tooth). Anterior element is notched. Also used as synonym for entomodont by certain authors. (See also adont, amphidont, heterodont, lobodont, lophodont, merodont, schizodont)

arcticoid (cyprinoid): heterodont

arcuate [G *arcuat, bogenförmig, gebogen*]: Curved like a bow.

areolate [G *gegittert*]: Characterized by network with interspaces.

aristate [G *borstig*]: Bearing tuft of small hairs or bristles, e.g., as in type of seta in polychaetes.

armed [G *bewehrt, bewaffnet*]: Equipped with or bearing additional structures, typically sharp or pointed element(s) serving in defense, feeding (stylets on proboscis of nemertine), or reproduction (spines on copulatory organ).

arostello- [G *arostello-*]: In tapeworm (cestode), prefix applied to cysticercoid (stage in tapeworm life cycle) lacking rostellum. (See also acanthorostello-, anacanthorostello-, rostello-)

articulate [G *gegliedert*]: Consisting of series of elements united by joints. (See also multiarticulate, uniarticulate)

articulate [G *artikulat*]: Type of lamp shell (brachiopod) in which dorsal and ventral valves are hinged, i.e., held together posteriorly by both muscles and tooth-and-socket arrangement. (See also inarticulate)

asconoid [G *asconoid*]: Type of sponge in which incurrent canals lead directly into central cavity (spongocoel). (See also leuconoid, syconoid)

ascophoran: Type of bryozoan in which feeding individual (autozooid) bears compensation sac (ascus). (See also anascan)

aseptate [G *aseptat*]: Lacking partitions (septa) or elevated ridges, e.g., as in type of adhesive organ (opisthaptor) of fluke (trematode). (See also septate)

asetose (asetigerous): achaetous

astomatous [G *mundlos*]: Lacking a mouth, e.g., as in roundworm (nematode) either lacking mouth cavity (stoma) or lacking cheilostome yet retaining unexpanded posterior esophostome.

asulcal [G *asulcal*]: In anthozoan, side of polyp opposite flagellated groove (siphonoglyph) in pharynx (also considered to be dorsal). In anthozoan with two (diglyphic) siphonoglyphs (siphonoglyph, sulculus), side of polyp corresponding to occasionally smaller sulculus. (See also sulcal)

asymmetrical [G *asymmetrisch*]: Having dimensions of one side not being equal to those of other side, e.g., chewing elements (trophi) of pharynx of certain rotifers or appendage pairs of certain arthropods. (See also symmetrical)

atentaculate [G *tentakellos*]: Lacking tentacles. (See also tentaculate)

athecate (symnoblastic) [G *athekat*]: Type of hydrozoan either lacking periderm or with periderm extending only to base of hydranth. (See also thecate)

atokous [G *atok*]: In polychaete, nonsexual stage or unmodified, sexually reproductive individuals; may also refer to unmodified anterior end of heteronereis.

atrichous [G *atrich*]: Type of stinging cell (nematocyst) of cnidarian: tubule lacks visible (under light microscope) spines along its entire length. Refers to nematocyst type in which tubules lack butt and are either of same diameter throughout (isorhiza) or slightly dilated toward base (anisorhiza). (See also apotrichous, basitrichous, heterotrichous, holotrichous, merotrichous, telotrichous)

attenuate [G *verjüngt*]: Gradually becoming more slender.

aulodont [G *aulodont*]: In sea urchin, type of chewing apparatus (Aristotle's lantern) in which teeth are broadly U-shaped (i.e., no keel) and in which space (foramen magnum) between demipyramids is open above (i.e., not covered by fused epiphyses). (See also camarodont, stirodont)

auricular: auriform

auriculate [G *aurikulat*]: Bearing ear-like process. In bivalve, type of shell with anterior and/or posterior extensions of dorsal (hinge) region. (See also alate, bialate)

auriform [G *ohrförmig*]: Ear-shaped.

avicular [G *aviculär*]: Type of seta in polychaete; Z- or S-shaped seta resembling a bird's head.

awl-shaped: subulate

axial [G *axial*]: Running parallel to axis, e.g., orientation of sculpture with regard to axis of shell in gastropod.

basal [G *basal*]: Pertaining to or located at or near basis or point of attachment. (See also proximal)

basitrichous [G *basitrich*]: Type of stinging cell (nematocyst) of cnidarian: spines are restricted to proximal section of tubule. Refers to type of nematocyst (isorhiza) in which tubule lacks butt and is of same diameter throughout. (See also apotrichous, atrichous, heterotrichous, holotrichous, homotrichous, merotrichous, telotrichous)

beaded [G *perlschnurartig*]: In snail (gastropod), type of shell sculpture (rib, cord) which is broken up into series of bead-like units. (See also moniliform)

beak-shaped [G *schnabelförmig*]: rostriform

bell-shaped [G *glockenförmig*]

biacerate: bihamate

bialate [G *bialat*]: Bearing two wing-like processes. In bivalve, having extension of dorsal (hinge) region beyond main body (disc) of shell both anteriorly and posteriorly. (See also alate, auriculate)

biannulate [G *biannulat*]: In oligochaete, condition in which a segment is subdivided externally into two rings by a secondary furrow (annulus). (See also triannulate)

biarticulate [G *zweigliedrig*]: Consisting of two segments.

bichromatic: dichromatic

biconical [G *doppelkegelförmig*]: Resembling two cones placed base to base.

biconvex [G *bikonvex*]: Curved outward (convex) on both sides, e.g., as in shell of lamp shell (brachiopod).

bicuspid [G *zweispitzig, doppelspitzig*]: Ending in two points. (See also tricuspid)

bidentate [G *zweizähnig*]: Ending in two teeth. In sea urchin, modified tridentate type of pedicellaria with only two long, pointed jaws. (In sea urchin, see also biphyllous, tridentate)

bifascial: In crinoid, type of articulation between two endoskeletal elements (ossicles/columnals of stalk); face of ossicle characterized by median ridge separating two depressions accommodating ligaments.

bifid [G *bifid, zweispaltig, in zwei Teilen gespalten*]: Split into two equal parts (often lobe-like) by median cleft. Refers in echiuran to type of proboscis with two arms. (See also trifid)

biflagellate [G *biflagellat, zweigeißelig*]: Bearing two flagella, e.g., as in certain sperm cells. (See also monoflagellate)

bifoliate [G *bifoliat*]: Type of bryozoan colony: erect, with two layers of individuals (zooids) growing back to back away from multizooidal median wall. (See also trifoliate)

bifurcate [G *gabelförmig, gegabelt*]: Forking into two parts. (See also furcate, multifurcate)

bigeminate [G *bigeminat*]: In sea urchin shell (test), type of modified ambulacral plate (compound plate) bearing two pore pairs. (See also trigeminate; oligoporous, polyporous)

bihamate [G *bihamat, C-förmig*]: Bent at both ends into hook, e.g., as in skeletal element (spicule) of sea urchin. (See also hamate)

bilabiate [G *zweilippig*]: Bearing two lips. Refers, in bryozoan, to type of aperture (orifice): orifice bears two lip-like structures, one of which closes it.

bilaminar (bilaminate) [G *bilaminar*]: Consisting of two layers (laminae). Refers, in bryozoan, to type of colony with two layers of individuals (zooids) growing back to back. (See also plurilaminar, unilaminar)

bilateral [G *bilateral*]: Located on both sides or characterized by having right and left sides (bilateral vs. radial symmetry). (See also unilateral)

bilimbate [G *bilimbat*]: Type of bristle (seta) in polychaete: having two flattened margins or wing-like extensions along sides.

bilobate [G *zweilappig*]: Divided into two lobes. (See also multilobed, trilobate, unilobate)

bipartite [G *zweiteilig*]: Divided into two parts. (See also tripartite)

bipectinate [G *bipectinat, doppelfiedrig*]: Having series of tooth-like projections, as in comb, along two sides. (See also monopectinate = unipectinate, pectinate)

biphyllous: In sea urchin, modified triphyllous type of pedicellaria with only two short, broad jaws. (See also triphyllous)

bipinnate [G *doppelfiedrig*]: Twice pinnate, i.e., projections along each side of axis also bear projections along opposite sides.

bipyramidal [G *doppelpyramidförmig*]: Resembling two pyramids placed base to base.

biradial [G *biradial*]: Type of symmetry, as in ctenophore, in which radially arranged parts may be divided into equal left and right halves by median longitudinal plane.

biramous [G *zweiästig, biram*]: Consisting of two branches (rami), e.g., as in appendage of crustacean or parapodium of polychaete. (See also sesquiramous, subbiramous, uniramous)

biserial [G *biseriat, zweizeilig*]: Arranged into two rows or series, e.g., as in zooid arrangement in bryozoan colony or in skeletal elements (brachia) along arms of crinoid. (See also pluriserial, uniserial)

bisulcate [G *bisulkat*]: Bearing two grooves (sulci). Refers to type of sculpture in shell (valve) of mussel shrimp (ostracod): two sulci separate three rounded elevations (lobes). (See also sulcate, trisulcate, unisulcate)

bitesticular: In oligochaete, condition in which two pairs of testes are associated with one pair of chambers (atria) occurring in the most posterior testis-bearing segment. (Versus monotesticular)

biunguiculate: Condition in which tip of last segment (dactylus) of crustacean limb is produced into two claw-like structures (ungues). (See also triunguiculate, uniunguiculate)

bivalved [G *zweiklappig*]: Consisting of two valves, as in shell of bivalved mollusc. (See also univalve)

bivalved [G *klappenförmig*]: In sea star, type of alveolar pedicellaria in which two jaws are horizontally elongated and resemble a minute clam.

bladder-like [G *blasenförmig*]: Expanded and pouch-like; refers to type of intestine in rotifer.

blind [G *blind, blind endend*]: Closed or not communicating with other structures at one end, e.g., as in outpocketing of digestive tract.

boat-shaped: navicular

bothridio- [G *bothridio-*]: In tapeworm (cestode), prefix applied to describe type of plerocercoid (stage in tapeworm life cycle) which bears bothridia. (See also acantho-, acetabulo-, bothrio-, culcito-, glando-, tentaculo-)

bothrio- [G *bothrio-*]: In tapeworm (cestode), prefix applied to describe type of plerocercoid (stage in tapeworm life cycle) which has bothrium. (See also acantho-, acetabulo-, bothridio-, culcito-, glando-, tentaculo-)

bottlebrush [G *flaschenbürstenförmig*]: Type of anthozoan colony: numerous, crowded, short branches arise all around main stem.

bottle-shaped [G *flaschenförmig*]

bow-shaped [G *bogenförmig*]: arcuate

box-like [G *kästchenförmig*]: Refers, for example, to box-shaped (vs. cylindrical or vase-like) type of individual (zooid) in bryozoan colony.

brachyuran: Pertaining to crab-like decapod, i.e., with more or less dorsoventrally flattened (depressed) carapace and abdomen folded up under cephalothorax. (See also natant)

branched [G *verzweigt*]

branchial [G *branchial*]: Adjoining or serving as respiratory structure, e.g., as in region of carapace covering gills in decapod.

branchiferous [G *kiementragend*]: In polychaete, segment bearing gills (branchia).

branchycnemous: In hexamerous anthozoan, arrangement of septa in which all septa except directives form pairs composed of one macroseptum and one microseptum; only one pair of complete septa present. (See also macrocnemous)

brevicaudate: In tapeworm (cestode), condition in which tail (cercomer) of cysticercoid (stage in tapeworm life cycle) is short. (See also circumcaudate, longicaudate; acaudate, caudate)

brevifurcate: In fluke (trematode) larva with forked tail (furcocercous cercaria), condition in which stem of tail is long and forks are short. (See also longifurcate)

buccal [G *bukkal, buccal*]: Pertaining to mouth.

bulboid (oxyuroid): Type of pharynx in roundworm (nematode): characterized by posterior bulb.

bulbous: Expanded like a bulb. Type of pharynx in flatworm (turbellarian), as opposed to simple or plicate type. (In turbellarian, see also doliiform, rosulate)

bullate [G *blasenartig, aufgetrieben*]: Broadly expanded or blister-like.

bushy [G *buschig*]: Type of anthozoan colony: with many branches arising immediately above holdfast and not forming an obvious stem, as opposed to arborescent, bottlebrush, capitate, digitate, encrusting, lobate, pinnate, reticulate colonies, etc.

byssiferous [G *byssustragend*]: In bivalve, bearing attachment threads (byssus).

C-shaped [G *C-förmig*]: Refers, for example, to type of skeletal element in sea urchin.

caducous [G *abfallend*]: In bryozoan, type of feeding individual (autozooid) which is shed from colony once functional life is spent.

calcareous [G *kalkig*]: Consisting of or containing calcium carbonate. Refers, for example, to type of brachiopod shell (as opposed to chitinophosphatic shell).

calcinate: Resembling the calyx of a flower.

callous [G *kallös*]: In snail (gastropod) shell, condition in which area (parietal region) on one side of aperture is coated with thickened shelly material.

calyptoblastic: thecate

camarodont [G *camarodont*]: In sea urchin, type of chewing apparatus (Aristotle's lantern) in which teeth have longitudinal keel and in which space (foramen magnum) between demipyramids is closed above by fused epiphyses. (See also aulodont, stirodont)

camerated: Divided into series of chambers, as in type of seta in polychaete.

campylonemidan: Type of feeding apparatus (lophophore) in bryozoan: tentacles are "bent."

canaliculate [G *kanalikuliert*]: Characterized by series of fine canals. In snail (gastropod) shell, refers to type of border (suture) between whorls; suture forms fine channel coiling around shell. (In gastropod sutures, see also adpressed, channeled, flush, impressed)

cancellate (clathrate) [G *gegittert*]: Forming crisscrossed pattern due to intersection, generally at right angles, of elevated surface features, e.g., as in sculpture of gastropod or bivalve shell.

candelabriform: lyrate

caperate: Bearing fine wrinkles. Refers, in mussel shrimp (ostracod), to shell (valve/carapace) surface or to walls (muri) along valve surface.

capillary [G *kapillar*]: In polychaete, type of seta: long, slender, and tapering like a hair.

capilliform: capillary

capitate [G *geknöpft, pilzförmig, kapitat*]: With head-shaped distal expansion, e.g., as in type of tentacle in hydrozoan or colony shape in anthozoan.

cardate [G *cardat*]: Sucking type of trophi in anterior digestive tract of rotifer; suction is created not by muscular hypopharynx but by unci in conjunction with epipharynx. (See also forcipate, virgate)

cardiac: Pertaining either to heart or to stomach or part of stomach, e.g., as in region of carapace more or less covering stomach in decapod.

carinate [G *gekielt*]: Bearing a keel (carina).

carnose [G *fleischig*]: Fleshy; refers to type of bryozoan.

carpo-subchelate: Type of pincer-like structure (subchela) in crustacean limb: last (dactylus) and next to last (propodus) segments both bear down on enlarged carpus.

castellate [G *zinnenartig*]: Bearing series of squarish elevations along margin, like battlements of a castle wall, e.g., as in rim of hydrotheca surrounding polyp in hydrozoan.

catenicelliform: Type of bryozoan colony with erect, jointed branches.

caudad [G *caudad*]: Extending toward tail.

caudal [G *caudal*]: Directed toward or lying in region defined by tail. Strictly applied, refers to region between anus and posterior extremity, i.e., tip of tail or fin, and should not be applied to organisms having a terminal anus.

caudate [G *mit Schwanz*]: Bearing tail. In tapeworm (cestode), indicates presence of tail (cercomer) in certain larval stages (e.g., procercoid, plerocercoid). (See also acaudate)

celate: Obscured by overlapping structure. Refers to sculpture in shell (valve) of mussel shrimp (ostracod) in which outer layer of calcite overlaps underlying ornamentation.

celliporiform [G *celliporiform*]: Type of bryozoan colony characterized by the heaping of multilaminar masses of individuals (zooids); forms nodular, typically small structures.

cephalic [G *am Kopf gelegen, kopfwärts*]: Located on or directed toward head.

cervical [G *zervikal*]: Located on or representing part of body directly behind head, particularly if this is constricted and neck-like.

chaetomicrocnemous: Variation of small-tailed (microcnemous) cercaria larva of fluke (trematode): tail spinous.

chelate [G *scherentragend, chelat*]: In crustacean, bearing pincer-like structure at end of appendage. (See also achelate, carpo-subchelate)

cheliform [G *scherenförmig*]: Pincer-like, i.e., resembling claw (pincer) of crab.

chevroned (en chevron): Configuration characterized by two diagonal elements meeting at an angle and thus forming a V-shaped or inverted V-shaped structure. Refers to type of ligament grooves on shell of bivalve or to type of sclerite arrangement in octomerous anthozoan.

chitinophosphatic: In lamp shell (brachiopod), type of shell consisting predominantly of calcium phosphate and chitin. (See also calcareous)

chromodorid [G *chromodorid*]: Type of pharynx in roundworm (nematode): three-part pharynx consisting of corpus, isthmus, and posterior bulb.

ciliate [G *bewimpert*]: Bearing series of minute, movable, hair-like extensions (cilia).

cingulate [G *gegürtelt*]: In gastropod, sculpture or color band winding spirally around shell.

circular [G *kreisförmig*]: Resembling or forming a circle, e.g., as in muscle bands around body (vs. longitudinal muscles). Refers also to amphid type in roundworm (nematode). (See also sub-)

circumcaudate: In tapeworm (cestode), condition in which tail (cercomer) of cysticercoid (stage in tapeworm life cycle) surrounds scolex. (See also brevicaudate, longicaudate; acaudate, caudate)

circumenteric [G *darmumschlingend*]: Surrounding digestive tract, e.g., as in nerve ring originating from brain and extending around both sides of anterior digestive tract.

circumoral [G *circumoral*]: Surrounding mouth.

cirrate [G *cirrat*]: Bearing slender, cylindrical process(es). Refers also to type of cephalopod whose arms each bear two lateral rows of elongate, finger-shaped projections (cirri).

cirriferous [G *cirrentragend*]: In crinoid, type of endoskeleton element (ossicle, plate) bearing cirri. (See also noncirriferous)

cirriform [G *cirrenförmig, cirriform*]: Resembling an elongate, finger-shaped process (cirrus).

cirrigerous [G *cirrentragend*]: In polychaete with dorsal scales (elytra), parapodium bearing a dorsal cirrus. (See also elytrigerous)

clathrate: cancellate

clavate (club-shaped) [G *keulenförmig*]: Club-shaped, i.e., gradually becoming enlarged near distal end.

claviform [G *claviform*]: In sea urchin, type of pedicellaria consisting only of stem and three poison sacs (i.e., jaws reduced). May also refer to nail-shaped bryozoan colony.

cleithral: In bryozoan, type of brood chamber (ooecium) closed by lid (operculum) of parent zooid. (See also acleithral)

cleithridiate: Keyhole-shaped. In bryozoan, refers to keyhole-shaped aperture (orifice) of individual (zooid) or to similarly shaped ovicell.

clitellate [G *clitellat*]: In oligochaete, possessing a clitellum. (See also aclitellate)

closed [G *geschlossen*]: Refers, for example, to type of bivalve shell which does not gape along margins when valves are shut or, in sea urchin, to type of petals which contract at their ambital extremity. (In sea urchin petal, see also open)

club-shaped: clavate

clypeate (clypeiform) [G *schildförmig*]: Shaped like a shield.

coarse [G *rauh, grob, gerauht*]

coeloconoid: In snail (gastropod), type of shell shape which resembles cone yet has concave sides.

coiled: Refers to type of umbo in bivalve: umbo incurved to such a degree that it forms a spiral of a complete whorl or more. (See also opisthogyrate, orthogyrate, prosogyrate)

collabral: In snail (gastropod) shell, type of growth line conforming to shape of outer lip.

colonial [G *koloniebildend, kolonial*]: (See also compound, solitary)

comb-like: pectinate

commarginal: concentric

compact: In sea urchin, type of apical plate system with no separation between anterior and posterior plates.

complete [G *vollständig*]: Refers, in anthozoan, to type of mesentery which extends inward from body wall and reaches pharynx. Also termed perfect. (See also incomplete)

composite: compound

compound [G *zusammengesetzt*]: Composed of more than one element. Refers in polychaete, for example, to type of seta (also termed composite seta) consisting of two parts joined by joint.

compound [G *kolonial, koloniebildend*]: Type of ascidian, referring either to colonial ascidians in general or to colony in which individuals are embedded in common matrix and share common excurrent opening. (See also simple)

compressed [G *zusammengedrückt, seitlich zusammengedrückt*]: Flattened laterally, e.g., as in shell of bivalve or carapace of crustacean. (See also depressed)

concave [G *konkav*]: Curved inward, forming depression or hollow. (See also convex)

concentric [G *konzentrisch*]: Having a common center. Refers, in bivalve, to type (or direction) of shell sculpture (also termed commarginal sculpture) which parallels outer shell margin. (See also radial)

concrescent [G *verschmolzen*]: Refers to a growing together of originally separate elements. In barnacle, total fusion of certain plates.

confluent [G *zusammenlaufend, zusammenwachsend*]: Running together or meeting without distinct junction. In plate of sea urchin shell (test), refers to arrangement of depression (scrobicule) around primary tubercle: scrobicules have no intervening small secondary tubercles (scrobicular tubercles), i.e., rows of these small tubercles are lost due to partial overlap of adjoining scrobicula. (In sea urchin, see also contiguous)

conical [G *kegelförmig, konisch*]: Cone-shaped.

conispiral: In snail (gastropod), type of shell in which spire projects in a cone-shaped configuration.

conjugate [G *konjugiert, gejocht*]: In sea urchin shell (test), type of pore pair in ambulacral plate: pores of pair connected by a groove. (See also nonconjugate)

conjunct [G *konjunkt*]: Joined, united. In ventral valve of lamp shell (brachiopod), condition of deltidial plates in which these are in contact with each other anterodorsally of stalk (pedicle). (See also disjunct)

conjunctive: Type of spine on shell (valve) of mussel shrimp (ostracod): spine located at junction of walls (muri) or reticulate sculpture. (See also disjunctive)

connate: Firmly united or fused. In bryozoan, refers to arrangement of individuals (zooids) in which erect distal portions are fused in linear groups.

conoid (conoidal) [G *konoid*]: Approaching a cone in shape. Refers, for example, to type of nematode tail or gastropod shell. (See also cyrtoconoid)

constricted [G *eingeschnürt*]: Narrowed.

contiguous [G *sich berührend*]: Touching or adjoining. Refers, for example, to condition in which two eyes of isopod are fused medially. (See also discontiguous)

continuous [G *geschlossen*]: In sea urchin, type of perignathic girdle in which individual elements (apophyses, auricles) are fused with one another. (See also discontinuous)

contractile [G *kontraktil*]: Capable of becoming shorter or smaller. Refers, in anthozoan, to type of polyp that diminishes in size without introversion (i.e., as opposed to retractile polyp).

conulose: In sponge, type of surface characterized by cone-shaped elevations.

convex [G *konvex*]: Curved outward, forming rounded surface. (See also biconvex, concave)

convolute [G *konvolut*]: In snail (gastropod), type of shell in which last whorl completely conceals earlier ones. (See also involute)

convoluted [G *geschlängelt*]: Highly wound, coiled, or folded.

cordate (cordiform) [G *herzförmig*]: Heart-shaped.

coriaceous [G *lederartig*]: With tough, leather-like texture as, for example, in type of bryozoan colony.

cormidial: In bryozoan, type of orifice that is the joint product of more than one individual (zooid).

coronate: In gastropod shell, type of sculpture in which shoulder of whorls bears projections (tubercles or nodes).

corroded [G *korrodiert, erodiert*]: In gastropod or bivalve, condition in which certain parts (umbo, apex) of shell are frequently worn away.

cortical [G *kortikal*]: Lying within outer layer (cortex). In tapeworm, structure (ovary, testis, uterus, yolk glands) located exterior to outer longitudinal muscles.

corticate: In sponge, covered or lined with a cortex.

costate [G *gerippt*]: Bearing costae, i.e., prominent, round-topped axial (in gastropod) or either radial or concentric sculpture (in bivalve).

costellate [G *fein gerippt*]: Bearing small costae (costellae), e.g., as in sculpture on shell of bivalve or gastropod.

cotylocercous (cotylomicrocercous): Type of larva (cercaria) of distome digenetic trematode: tail cup-like, forming a sucker containing adhesive glands.

coupled [G *gegenüberstehend*]: In anthozoan, arrangement of septa in which each septum on one side of gastrovascular cavity has mate of same age on opposite side. (See also paired)

craspedote [G *craspedot*]: In hydrozoan or scyphozoan medusa, bearing horizontal fold (velum) projecting inward from umbrella margin. In tapeworm (cestode), type of strobila in which each segment (proglottid) overlaps immediately following proglottid.

crenate (crenulate) [G *krenuliert*]: Bearing series of rounded elevations along margin, as in condition of ribs on bivalve shell, outer lip of gastropod shell, or tubercle or spine socket (acetabulum) on shell plate of sea urchin. (In sea urchin, versus noncrenulate)

crescentic [G *halbmondförmig, sichelförmig*]: Crescent-shaped.

crispate [G *gekräuselt*]: With crinkled margin.

cristate: Bearing an elevated, sharp crest (crista).

crossed [G *gekreuzt*]: In sea star, type of stalked (pedunculate) pedicellaria in which jaws (valves) are curved and crossed basally. (See also straight)

cross-shaped: cruciform

cruciform [G *kreuzförmig*]: Cross-shaped.

cryptocystidean: Type of ascus-bearing (ascophoran) bryozoan in which frontal wall forms an interior partition below an overlying hypostegal coelom. (See also gymnocystidean)

cryptodicyclic (pseudomonocyclic) [G *cryptodizyklisch*]: In crinoid, type of dicyclic arrangement of calyx plates: one circlet (infrabasals) is reduced or lost in adults. (See also dicyclic, monocyclic)

cryptomphalous [G *verborgen*]: In snail (gastropod) shell, condition in which umbilicus is completely plugged. (See also hemiomphalous, phaneromphalous)

ctenodont [G *ctenodont*]: Type of hinge (dentition) in bivalve: with numerous short hinge teeth oriented transverse to margin.

ctenopodous: In barnacle, type of thoracic appendage (cirrus) in which setae along anterior margin are elongate and arranged in series of pairs. (See also acanthopodous, lasiopodous)

cuboidal [G *würfelförmig, kubisch*]: Approaching a cube in shape, as in bell of certain scyphozoans.

culcito-: In tapeworm (cestode), prefix applied to describe scolex of plerocercoid (stage in tapeworm life cycle) which bears pads or cushions. (See also acantho-, acetabulo-, bothridio-, bothrio-, glando-, tentaculo-)

cuneiform [G *keilförmig*]: Wedge-shaped, as in shell of bivalve or spicule of roundworm (nematode).

cultriform [G *messerförmig*]: Shaped like the blade of a knife.

cup-shaped: cyathiform

cuspidate [G *cuspidat*]: Ending in a point, e.g., as in tooth in pharynx of priapulid or type of seta. (In priapulid, see also pectinate)

cyathiform [G *becherförmig*]: Cup-shaped. Refers, in nematode, to amphid type.

cyclodont [G *cyclodont*]: Type of dentition in bivalve: hinge plate is lacking or reduced and teeth curve out from under umbones.

cylindrical [G *zylindrisch*]: Refers to structure having same diameter throughout. Not necessarily hollow (i.e., tubular).

cymose [G *cymös*]: Type of branching in hydroid in which budding zone below temporary terminal hydranth gives rise to new branches. (See also sympodial)

cyrtoconoid: Approaching a cone in shape, yet having convex sides.

cystocercous: Type of larva (cercaria) of distome digenetic fluke (trematode): base of large tail contains cavity into which main body can be retracted. (See also cystophorous, macrocercous)

cystophorous: In fluke (trematode), variation of cystocercous cercaria: tail consists of rounded vesicle bearing various long appendages.

D-shaped [G *D-förmig*]: In irregular sea urchin, shape of area (peristome) around mouth. (See also crescentic, oval, pentagonal)

dactylous: In sea urchin, type of pedicellaria with three to five relatively immovable, spoon-shaped jaws.

dagger-like [G *dolchförmig*]: Refers to type of cirrus in tardigrade.

decabrachiate [G *zehnarmig*]: Type of cephalopod with 10 arms.

decamerous [G *decamer*]: In anthozoan, modification of hexamerous condition characterized by 10 pairs of complete septa and numerous incomplete septa. (See also hexamerous, octomerous)

deciduous [G *abfallend*]: Characterized by tendency to fall off or detach, e.g., as in certain palps of polychaetes, the proboscis of certain echiurans, the periostracum of a bivalve shell, or type of spicule shed by sponges.

decollated: In snail (gastropod) shell, condition in which apex is broken off.

decussate [G *sich kreuzend, dekussiert*]: Intersecting to form X-shaped structure, e.g., as in sculpture on gastropod shell or crossed nerves.

deflexed [G *hinuntergebogen*]: Turning abruptly downward, as in type of rostrum shape in decapod.

dendritic [G *baumartig verzweigt*]: Branching like a tree, e.g., as in tentacle of sipunculid.

dendrobranchiate **(dendritic)** [G *dendrobranchiat*]: In decapod, type of gill characterized by highly subdivided branches along axis. (See also phyllobranchiate, trichobranchiate)

dentate [G *gezahnt*]: Bearing teeth.

denticulate [G *gezähnelt*]: Bearing small teeth (denticles).

dentiform [G *zahnförmig*]: Tooth-shaped.

dependent: In bryozoan, type of brood chamber (ooecium) which develops resting on distal individual (zooid). (See also independent)

depressed [G *flach gedrückt*]: Flattened dorsoventrally. In snail (gastropod), refers to shell which is low in proportion to its diameter. (See also compressed)

desmodont [G *desmodont*]: Type of hinge (dentition) in bivalve: two cardinal teeth are fused to form short, spoon-shaped process. Synonymous with adont hinge in ostracod valve.

deviated: In snail (gastropod) shell, condition of earliest whorls (protoconch) in which protoconch axis forms distinct angle with axis of later whorls (teloconch).

dextral [G *rechtsgewunden*]: Turning to the right, e.g., as in right-handed direction of coiling in snail (gastropod) shell or in certain tube-dwelling polychaetes. (See also sinistral)

diadematoid [G *diademoid*]: In shell (test) of sea urchin, type of compound ambulacral plate composed of three primary plates. (See also arbacioid, echinoid, echinothurioid)

diagenodont [G *diagonodont*]: Type of hinge (dentition) in bivalve: differentiated cardinal and lateral teeth present, yet not more than two lateral or three cardinal teeth per valve.

diagonal: In fluke (trematode), refers to arrangement of testes in which testes do not lie in same transverse plane or are not located one behind the other. (See also opposite, tandem)

diaulic [G *diaul*]: In reproductive system, for example of snail (gastropod), condition in which sperm duct and oviduct open separately to exterior. (See also monaulic, triaulic)

dibranch [G *dibranch*]: Type of cephalopod characterized, among other features, by two gills, two auricles, and two nephridia.

dichotomous [G *dichotom, dichotom verzweigt*]: Dividing, as a rule repeatedly, into two parts. Refers, for example, to type of branching in anthozoan colony; branching pattern is a repeated bifurcation.

dichromatic   (bichromatic) [G *dichromatisch*]: Exhibiting or having two colors. Refers, in crustacean, to type of pigment-bearing cell (chromatophore) with two color pigments. (See also monochromatic, polychromatic)

dicondylic [G *dikondyl*]: Type of joint in crustacean appendage in which adjoining segments articulate on two processes (condyles). (See also monocondylic)

dicyclic [G *dizyklisch*]: In crinoid, condition of plates forming skeleton (calyx) surrounding main body (crown): having three circlets of plates, i.e., two circlets (basals, infrabasals) proximal to radials. (See also cryptodicyclic = pseudomonocyclic, monocyclic)

didelphic [G *paarig*]: In roundworm (nematode), condition in which female reproductive system is doubled. (See also monodelphic, polydelphic)

digenetic [G *digenetisch*]: Type of fluke (trematode) whose life cycle involves two to four hosts; typically endoparasitic. (See also monogenetic)

digitate [G *handförmig, fingerförmig*]: Bearing finger-like processes or being finger-shaped. (See also multidigitate)

digitiform [G *fingerförmig*]: Shaped like a finger.

diglyphic [G *diglyph*]: Type of anthozoan with two flagellum-bearing grooves (siphonoglyphs) along pharynx. (See also monoglyphic)

dimorphic [G *dimorph*]: Characterized by the occurrence of two different types of individuals, e.g., as in anthozoan colonies with autozooids and siphonozoids. (See also monomorphic, polymorphic, trimorphic)

dimyarian [G *dimyar, zweimuskelig*]: In bivalve, condition in which two adductor muscles (anterior and posterior) span between valves. Visible as two muscle scars on inner surface of valves. (See also anisomyarian = heteromyarian, isomyarian = homomyarian, monomyarian)

dioecious [G *getrenntgeschlechtlich*]: Having separate sexes, i.e., no individuals having both male and female reproductive systems. (See also monoecious)

diorchic [G *paarig*]: Type of male reproductive system in roundworm (nematode): paired, typically opposed testes. (See also monorchic, proorchic)

diotocardian [G *diotokard*]: In snail (gastropod), type of heart basically consisting of two auricles and one ventricle. (See also monotocardian)

diplogasteroid: Type of three-part pharynx in roundworm (nematode); characterized by corpus, isthmus, and posterior bulb, with corpus being further divisible into procorpus and metacorpus.

discoidal [G *scheibenförmig*]: Disc-shaped, i.e., rounded and flattened. In snail (gastropod), type of shell in which whorls form a flat coil. (In gastropod, see also convolute, involute)

disconnected: Characterized by interruptions. In sculpture of shell (valve) of mussel shrimp (ostracod), refers to type of ridge (carina or ponticulus) consisting of a series of elements (clavae).

discontiguous: Not contiguous, e.g., as in coxal plates of amphipod.

discontinuous [G *diskontinuierlich, unterbrochen*]: In sea urchin, type of perignathic girdle in which individual elements (apophyses, auricles) are not fused with one another. (See also continuous)

disjunct [G *disjunkt*]: Characterized by separation. In bivalve, condition in which pallial line is not continuous but rather broken up into shorter muscle scars. In ventral valve of brachiopod, condition of deltidial plates in which these are not in contact with each other anterodorsally of stalk (pedicle). (In brachiopod, see also conjunct)

disjunctive: Type of spine on shell (valve) of mussel shrimp (ostracod); spine located in wall (murus) of reticulated sculpture, but not at their junctions. (See also conjunctive)

distal [G *distal*]: Pertaining to or located away from point of attachment or from center of body. Refers, in bryozoan, also to side away from ancestrula or origin of growth. (See also proximal)

distichous [G *distich*]: In caudofoveate or solenogaster, type of radula characterized by one pair of teeth in each transverse row. (See also monostichous, polystichous)

distome [G *distom*]: General category of larval (cercaria) or adult fluke (trematode); refers to forms with typical anterior oral sucker and ventral sucker (acetabulum) located at or near center of body. (See also amphistome, echinostome, gasterostome, holostome, monostome, prosostomatous)

divaricate [G *gegabelt, abstehend*]: Spreading apart widely. In bivalve shell, refers to type of sculpture composed of widely diverging elements (e.g., costules).

diverticulate [G *mit Divertikel, verzweigt*]: Bearing a diverticulum or diverticula. (See also adiverticulate)

docoglossate [G *docogloss*]: Type of radula in patelliform snail (gastropod): each transverse tooth row consists of few relatively strong teeth. Rachidian tooth may be missing and lateral teeth may be positioned one behind the other. (See also hystrichoglossate, ptenoglossate, rachiglossate, rhipidoglossate, taenioglossate, toxoglossate)

doliiform [G *doliiform*]: One of two main types (doliiform, rosulate) of bulbous pharynx in flatworm (turbellarian): barrel-shaped, oriented parallel to body axis.

dorsad [G *dorsad*]: Extending toward dorsal side.

dorsal [G *dorsal*]: Pertaining to or located on dorsum ("back"). Not uniformly applicable in invertebrates, often requiring differentiation between morphological dorsal side and upwardly = "dorsally" directed side as regards orientation of organism in the environment. Refers, for example, to hinge region in bivalve or to face of bryozoan colony (zoarium) opposed to orifice-bearing side. Replaced in certain invertebrate groups (cnidarians, echinoderms) by the preferred term aboral. (See also ventral)

dorsomedian [G *dorsomedian*]: Located near middle of dorsal side or extending along middle of dorsal side.

dorsoventral [G *dorsoventral*]: Extending from dorsal to ventral side.

dorylaimoid (two-part) [G *dorylaimoid*]: Type of pharynx in roundworm (nematode); characterized by cylindrical corpus and expanded postcorpus.

dumbbell-shaped [G *hantelförmig*]: Refers, for example, to spicule of sea urchin (echinoid) or statolith of nemertine.

duplivincular [G *duplivinkulär*]: In hinge of bivalve, type of ligament similar to multivincular one (alternating lamellar and fibrous components), yet with lamellar layer inserting in a series of chevron-shaped grooves in cardinal area. (See also alivincular, multivincular, paravincular)

dysodont [G *dysodont*]: In bivalve, term referring either to type of hinge lacking teeth or to hinge with small denticles close to beaks.

ear-shaped: auriform

echinoid [G *echinoid*]: In shell (test) of sea urchin, type of compound ambulacral plate composed of demiplate with primary plate adapical and adoral to it. (See also arbacioid, diadematoid, echinothurioid)

echinostome [G *echinostom*]: General category of larval (cercaria) or adult digenetic trematode; refers to distome forms whose anterior oral sucker is surrounded by spinous collar (head collar). (See also amphistome, distome, gasterostome, holostome, monostome, prosostomatous)

echinothurioid [G *echinothurioid*]: In shell (test) of sea urchin, type of compound ambulacral plate composed of primary plate with two included plates adoral to

it. (See also arbacioid, diadematoid, echinoid)

echinulate [G *dicht beborstet*]: With dense cover of small spines or bristles, i.e., reminiscent of a sea urchin.

ectal [G *ectal*]: Outermost or external. (See also ental)

ectocochliate (ectocochleate) [G *ectocochliat, ectocochleat*]: In cephalopod, type of shell lying outside mantle. (See also endocochliate)

ectoparasitic [G *ektoparasitisch*]: Refers to parasite living on outer surface of host. (See also endoparasitic)

edentulous: adont

efferent [G *efferent, wegführend*]: Leading to or conveying to an organ or structure, e.g., as in an efferent blood vessel. (See also afferent)

effuse [G *klaffend*]: In snail (gastropod) shell, condition of aperture when this is interrupted by short spout for siphonal outlet or when it extends from one end of shell to other and is open at both ends.

egg-shaped: ovate

elated [G *überglücklich*]: Condition of author upon completion of this book.

elevated: In snail (gastropod), type of shell in which spire is relatively high in proportion to its diameter.

ellipsoid [G *ellipsoid*]

elliptical [G *elliptisch*]

elongate [G *länglich, lang-*]: Lengthened; often used in combination with other adjective specifying basic shape, e.g., elongate-oval.

elongate-oval [G *länglich oval, lang-oval, länglich eiförmig*]

elytrigerous [G *elytrentragend*]: In polychaete, segment or parapodium bearing scales (elytra). (See also cirrigerous)

emarginate [G *eingekerbt*]: With notched or indented margin, e.g., as in outer lip of snail (gastropod).

embryonated [G *mit Embryo versehen*]: Having an embryo. Refers, in fluke (trematode), to egg that contains miracidium at time of oviposition. (Versus unembryonated)

enantiomorphic: Exhibiting a mirror-image relationship. In bryozoan, orientation of individual zooids such that each is a mirror image of its predecessor.

en chevron: chevroned

encrusting [G *inkrustierend*]: Forming a crust-like cover on substratum; typically applied to type of colony, as opposed to erect, stolonate growth, etc.

endocochliate (endocochleate) [G *endocochliat, endocochleat*]: In cephalopod, type of shell enclosed by mantle. (See also ectocochliate)

endocoelic [G *dem Zwischenfach zugewendet*]: In anthozoan, directed toward, facing, or located within endocoel; condition in which longitudinal retractor muscle band along one member of septum pair faces retractor of second member (i.e., both face into endocoel or space between that septum pair). (See also exocoelic)

endocyclic [G *endozyklisch*]: In crinoid, condition of mouth: mouth positioned in center of membrane (tegmen) covering crown. (See also exocyclic)

endogastric [G *endogastrisch*]: In snail (gastropod), typical condition of adult in which body is coiled so as to extend backward from aperture over extruded head-foot mass. Also applied to shell of certain cephalopods. (See also exogastric)

endogenous [G *endogen*]: Growing from or on inside. In tapeworm (cestode), type of budding in hydatid characterized by inward proliferation of germinal epithelium. Results in a brood cyst or daughter cyst. (See also exogenous)

endoparasitic [G *endoparasitisch*]: Refers to parasite living inside host. (See also ectoparasitic)

endopunctate (punctate): In lamp shell (brachiopod), type of shell bearing minute canals (endopunctae). (See also impunctate)

endotoichal [G *endotoichal*]: In bryozoan, type of brood chamber (ooecium) which appears immersed in distal individual (zooid); opens independently to exterior. (See also endozooidal)

endozooidal [G *endozoidal*]: In bryozoan, type of brood chamber (ooecium) which appears immersed in distal individual (zooid); opens below lid (operculum) of parent zooid. (See also endotoichal)

enoploid (one-part) [G *enoploid*]: Type of pharynx in roundworm (nematode): typically cylindrical or conoid in shape.

ensiform [G *lang scheidenförmig, schwertförmig*]: Shaped like a sword, typically with sharp edges and somewhat curved. Refers to type of bivalve shell.

ental [G *ental*]: Innermost or internal. (See also ectal)

entalophoriform: pustiliporiform

enteronephric [G *enteronephridisch*]: In oligochaete, condition in which metanephridium opens into digestive tract rather than through nephridiopore. (See also exonephric)

entire [G *ganzrandig*]: Smooth or uninterrupted, typically with regard to margin or edge lacking teeth or notches. Refers in snail (gastropod) shell to type of aperture uninterrupted by siphonal canal, in bivalve shell to pallial line lacking sinus.

entomodont [G *entomodont*]: Type of compound hinge in mussel shrimp (ostracod): subtype of merodont hinge in which terminal (anterior and posterior) elements are crenulate or lobed, while median part is either smooth or crenulated.

ephemeral [G *ephemer*]: Existing only briefly, as in ligament sacs of certain acanthocephalans. (See also persistent)

epi- [G *epi-*]: Prefix indicating position above or covering a structure, e.g., as in epibranchial and epigastric regions of decapod carapace.

epilobous (epilobic) [G *epilobisch*]: In oligochaete, type of prostomium in which posterior extension (tongue) encroaches onto peristomium dorsally, yet does not reach furrow separating peristomium (segment 1) from segment 2. Posterior margin of extension (tongue) may be open or delimited (closed) by transverse groove. (See also proepilobous, prolobous, tanylobous, zygolobous)

epithyridid: In articulate lamp shell (brachiopod), position of pedicle opening (foramen) relative to deltidial plates and umbo of ventral valve: foramen located wholly within ventral umbo, i.e., outside delthyrium and lacking contact with beak ridges. (See also amphithyridid, hypothyridid, mesothyridid, permesothyridid, submesothyridid)

epitokous [G *epitok*]: In polychaete, modified, sexually reproductive individual; may also refer only to modified posterior end of heteronereis.

equal [G *gleich groß*]: (See also subequal, unequal)

equilateral: In bivalve, condition in which part of shell anterior to beak is equal or almost equal in length to part posterior to beak.

equivalve [G *gleichklappig*]: In bivalve, type of shell in which both valves are of same shape and size.

erect [G *abstehend, aufrecht*]: Oriented upward or outward from a surface, e.g., as in colony growth form or arm spine position in brittle star (ophiuroid). (In ophiuroid, see also appressed)

eroded [G *erodiert*]: Condition in which a structure tends to be worn, e.g., as in shell of bivalve or carapace of decapod.

errant (errantiate) [G *errant*]: Category of polychaete basically composed of highly motile or free-swimming forms with well-developed appendages (parapodia). (See also sedentary)

ethmolytic (ethmolysian) [G *ethmolytisch*]: In sea urchin, type of apical plate system in which madreporite is shifted posteriorly (and no longer entirely surrounded by genital and ocular plates). (See also ethmophract)

ethmophract [G *ethmophraktisch*]: In sea urchin, type of apical plate system in which madreporite is enlarged and occupies central position (i.e., surrounded by genital and ocular plates). (See also ethmolytic)

eulamellibranch [G *eulamellibranch*]: In bivalve, type of gill (ctenidium) in which descending and ascending limbs of filaments are fused to each other and to filament before and behind it (except for small pores). (See also pseudolamellibranch)

eurymyonic: Type of arrangement of muscle bands in certain thaliaceans (old nurses in doliolid): bands are broader than one-half the interspaces. (See also holomyonic, stenomyonic; aclinous, amphiclinous)

eurypylous [G *eurypyl*]: Type of choanocyte chamber in leuconoid sponge: chamber connected directly to incoming (prosopyle) and outgoing (apopyle) opening. (See also aphodal)

euthyneurous [G *euthyneur*]: In snail (gastropod), condition of nerve system in which pleurovisceral connectives are not crossed. (See also streptoneurous)

eversible [G *umstülpbar, ausstülpbar*]: Capable of being turned inside out, e.g., as in proboscis of polychaete when extended or tube of nematocyst when triggered.

everted: In snail (gastropod) shell, type of outer lip whose edge is turned outward.

evolute: In snail (gastropod) shell, type of coiling in which whorls are out of contact.

excavate [G *ausgehöhlt*]: Characterized by a depression or cavity. In sea star, type of alveolate pedicellaria in which elongated jaws (valves) resemble pair of tongs and fit into depression (alveolus) of underlying ossicle when opened. (See also bivalved, spatulate)

excurrent (exhalent) [G *nach außen führend*]: Pertaining to structures involved in conveying water to exterior, e.g., as in opening or siphon in bivalve or ascidian. (See also incurrent)

excurvate [G *nach außen gekrümmt*]: Bent or curved outward.

exhalent: excurrent

exocoelic: In anthozoan, directed toward, facing, or located in exocoel; condition in which longitudinal retractor muscle band along each member of septum pair faces in opposite directions (i.e., into exocoel or space between septum pairs). (See also endocoelic)

exocyclic [G *exozyklisch*]: In crinoid, condition of mouth; mouth positioned near margin of membrane (tegmen) covering crown. (See also endocyclic)

exogastric [G *exogastrisch*]: In snail (gastropod), condition in which body is coiled so as to extend forward from aperture over front of extruded head-foot mass; pertains to early developmental stage before torsion. Also applied to type of shell coiling in cephalopod in which coil is held aloft dorsally and mantle and body space lie below opening in front. (See also endogastric)

exogenous [G *exogen*]: Growing from or on outside. In tapeworm (cestode), type of budding in hydatid characterized by proliferation of germinal epithelium outward through cyst wall. Results in a daughter cyst. (See also endogenous)

exonephric [G *exonephridisch*]: In oligochaete, condition in which metanephridium opens to exterior via nephridiopore on surface of body. (See also enteronephric)

explanate: In snail (gastropod) shell, condition in which outer lip spreads outward and becomes flat.

exsert [G *exsert*]: In sea urchin, condition of ocular plates in which these are not in contact with test margin (periproct) around anus. (See also insert)

extant [G *heute existierend, heute lebend*]: Living today, as opposed to extinct.

extracecal: Located outside of ceca. Refers in fluke (trematode) to position of internal organs (testes, vitellaria) lateral to each intestinal cecum. (See also intercecal)

extramural: In oligochaete, type of calciferous gland lying outside esophagus and opening into esophagus lumen via tube. (See also intramural)

extravascular: In tapeworm, located lateral to osmoregulatory canals. (See also intervascular)

extrazooidal (multizooidal): In bryozoan, structure arising from or shared by more than a single individual (zooid).

exumbrellar [G *exumbrellar*]: In medusa of hydrozoan or scyphozoan, positioned on or located closer to upper convex surface of body (umbrella). (See also subumbrellar)

falcate (falciform) [G *sichelförmig*]: Sickle-shaped.

fan-shaped [G *fächerförmig*]: flabellate

fascicled [G *gebündelt*]: Arranged or fused into bundles, e.g., as in type of stem in hydrozoan colony (vs. unfascicled) or arrangement of erect, peristomial portions of autozooids in certain byrozoans.

fasciculate [G *büschelförmig*]: Synonym for fascicled. In sea star, type of sessile pedicellaria consisting of cluster of movable spines on single underlying skeletal element (ossicle). (See also pectinate, spiniform, valvulate)

feather-shaped: plumose

fenestrate [G *gefenstert*]: Perforated by holes (fenestrae) or having transparent areas. Refers to type of bryozoan colony in which branches form reticulate pattern or to construction of endoskeletal elements in certain echinoderms.

fertile [G *fertil*]: In anthozoan, refers to mesentery with gonads. (See also infertile)

filamentous: Synonym for filiform. For type of gill in decapod, see trichobranchiate.

filibranch [G *filibranch*]: In bivalve, type of gill (ctenidium) in which descending and ascending limbs of each filament are connected to each other and loosely to filament before and behind it. (See also eulamelli-branch, protobranchiate, pseudolamelli-branch)

filiform [G *fadenförmig, filiform*]: Shaped like a thread, e.g., as in type of tentacle in hydrozoan or type of tail in roundworm (nematode). (In hydrozoan, see also capitate)

fimbriate [G *gefranst*]: Having a fringed margin, as a rule with slender processes. (See also laciniate)

finger-shaped: digitiform

fistulous [G *fistulös*]: In certain anthozoan colonies, branch with inrolled edges that may be fused partially or completely.

flabellate (flabelliform) [G *fächerförmig*]: Fan-shaped.

flagellate [G *begeißelt, mit Geißel versehen, flagellat*]: Having or bearing a flagellum.

flagelliform [G *geißelförmig*]: Whip-like, i.e., elongate, slender, and tapering. May refer, for example, to appendage of crustacean or to anthozoan colony type.

flanged: Type of bristle (seta) in polychaete: elongate with flattened edge or margin.

flattened [G *abgeflacht, abgeplattet*]: compressed, depressed

fleshy [G *fleischig*]

fluted [G *geriffelt, gerillt*]: Bearing series of narrow, parallel, rounded excavations, e.g., as in longitudinal markings along spine of sea urchin.

folded: plicate

foliaceous [G *blattförmig*]: Leaf-shaped. May refer to flattened appendage of crustacean.

foliate [G *blattformig, blätterig*]: Synonym for foliaceous. May also refer to structure formed by series of flattened, leaf-like elements.

follicular [G *follikulär*]: Condition of gonad, for example in flatworm (turbellarian)

or fluke (trematode), in which numerous small testes or ovaries are present.

forcipate [G *forcipat*]: Protrusile, grasping type of trophi in rotifer; rami are well developed and form pincer-like structures, usually bearing teeth.

forcipiform: straight

forficiform: crossed

fossorial [G *grabend*]: Modified for digging.

foveolate [G *kleingrubig, fein genarbt*]: Bearing minute pits. Refers, in ostracod for example, to shell (valve, carapace) or to wall (murus) forming reticulation on valve surface.

free: In male reproductive system of flatworm (turbellarian), condition in which prostatic vesicle is connected to ejaculatory duct by prostatic duct. (See also interpolated)

free-living [G *freilebend*]: Refers to nonparasitic organism.

fringed [G *gefranst*]

frondose [G *farnwedelartig*]: Resembling leaf of a palm or fern. Type of colony form in byrozoan.

frontal [G *frontal*]: Directed toward or lying in region defined as "front" (typically anterior end). In bryozoan, for example, refers to exposed or orifice-bearing side of zooid or colony.

fulcrate [G *fulcrat*]: Piercing type of trophi in rotifer; only three larger trophi, an elongated median piece (fulcrum) and an anteriorly expanded piece (manubrium) to each side, are developed. Several accessory elements may be present.

fungiform [G *pilzförmig, fungiform*]: Mushroom-shaped, as in growth form of anthozoan colony.

funnel-shaped [G *trichterförmig*]: (See also infundibular)

furcate [G *gabelförmig, gegabelt*]: Forked. In sea star, type of pedunculate, straight pedicellaria in which jaws (valves) have bifurctaed distal ends.

furcocerous: Type of larva (cercaria) of distome digenetic fluke (trematode); tail is forked.

fused [G *verschmolzen, verwachsen*]

fusiform [G *spindelförmig, fusiform*]: Spindle-shaped, i.e., tapering at both ends.

gaping [G *klaffend*]: In bivalve, condition in which margins of valves do not seal tightly when shell is closed.

gasterostome [G *gasterostome*]: General category of larval (cercaria) or adult digenetic fluke (trematode); refers to forms with mouth located at middle of body. (See also amphistome, distome, echinostome, holostome, monostome, prosostomatous)

gastric [G *gastrisch*]: Relating in origin or position to stomach. May also refer to region of carapace more or less covering stomach in decapod.

geniculate [G *gekniet, genikulierend*]: Bent like a knee, as in antennule of certain crustaceans or type of growth form in hydrozoan colony.

gibbose (gibbous) [G *aufgewölbt*]: Very rounded, as in whorls of certain snails (gastropods).

glabrous [G *kahl, unbehaart*]: Smooth and often glistening, i.e., lacking hairs.

glando-: In tapeworm (cestode), prefix applied to describe scolex of pleurocercoid (stage in tapeworm life cycle) which bears glands. (See also acantho-, acetabulo-, bothridio-, bothrio-, culcito-, tentaculo-)

globiferous [G *globifer, gemmiform*]: In sea urchin, type of pedicellaria with three long jaws typically bearing terminal tooth and associated with poison glands.

globose [G *kugelförmig*]: More or less spherical, as in shape of snail (gastropod) shell or sea urchin test.

globular [G *kugelförmig*]: More or less spherical.

glomerate: Type of colony in anthozoan: arborescent and sparsely branched, with numerous bundles of polyps crowded to form rounded bunches.

gnathostomatous [G *gnathostom*]: In copepod, type of mouth which is open and bears normally developed complement of mouthparts. (See also siphonostomatous)

gonochoristic: dioecious

gorgoderine: macrocercous

gradate [G *treppenförmig*]: Characterized by a step-like configuration. Refers to step-like arrangement of successive whorls in snail (gastropod) shell due to angular shoulders of whorls.

granulate (granulose) [G *granuliert, gekörnelt*]: Covered with minute protuberances (granules), e.g., as on exoskeleton of crustacean or shell of mollusc.

granuloreticulate: In ostracod shell, type of sculpture consisting of granules arranged in intersecting rows.

gravid [G *gravid*]: Full of eggs. Refers, for example, to condition of segment (proglottid) in tapeworm (cestode).

gymnoblastic: athecate

gymnocystidean: Type of ascus-bearing (ascophoran) bryozoan in which frontal wall is formed by the calcification of the outer body wall. (See also cryptocystidean)

hairy: hirsute, pilose, pubescent

hamate [G *hakenförmig*]: With hook-shaped end. Refers, for example, to spicule type in roundworm (nematode). (See also bihamate)

harpoon-shaped [G *harpunförmig*]

hastate [G *spießförmig*]: Shaped like a triangle or spear blade with widened base.

heart-shaped: cordate

hectocotylized [G *hectocotylisiert*]: In cephalopod, condition in which one or more arms are modified as spermatophore-transferring (copulatory) organs.

helmet-shaped [G *helmförmig*]: Refers to shape of nemertine larva or to type of bract in free-swimming colonial hydrozoan. (In hydrozoan, see also leaf-shaped, prismatic)

hemidapedont [G *hemidapedont*]: Type of dentition in bivalve; characterized by weak hinge and hinge plate and poorly developed cardinal teeth (lateral teeth often lacking).

hemigomph: Type of bristle (seta) in polychaete: with asymmetrical articulation nearly at right angles to long axis of shaft. (See also heterogomph, homogomph)

hemiomphalous: In snail (gastropod) shell, condition in which opening of umbilicus is partly plugged.

hemispherical [G *halbkugelig*]: Refers to three-dimensional structure (as opposed to semicircular, two-dimensional shape).

hepatic [G *hepatisch*]: Relating in origin or position to digestive gland ("liver"). May also refer to region of carapace in decapod.

hermaphroditic: monoecious

hermatypic [G *hermatypisch*]: In anthozoans, refers to reef-building corals containing symbiotic, unicellular algae within their tissues.

heteracanthus: In tapeworm (cestode), hook arrangement on tentacle; no chainette is present and hooks are in alternating oblique rows extending from the internal to the external surface. Hooks may be similar in size (homeomorphus) or dissimilar (heteromorphous). (See also homeoacanthus, poeciloacanthus)

heteracantho-: In tapeworm (cestode), prefix applied to cysticercus (stage in tapeworm life cycle) with two rings of dissimilar hooks. (See also anacantho-, homacantho-)

heterochelous (heterochelate) [G *heterochel*]: In crustacean, condition in which left and right pincers (chelae) differ in size and shape.

heterodont [G *heterodont*]: General term for type of hinge or dentition in bivalve; characterized by differently shaped teeth, i.e., cardinal and lateral teeth, if present, are distinctly different and generally few in number. According to elaboration, one may distinguish arcticoid, corbiculoid, and lucinoid subtypes. In ostracod, one of four subtypes of amphidont hinge: anteromedian element is differentiated into single strong tooth, with anterior and posterior teeth of opposite valve also in the form of a strong tooth.

heterogomph [G *heterogomph*]: Type of bristle (seta) in polychaete: with articulation clearly oblique to long axis of shaft. (See also hemigomph, homogomph)

heteromorphous [G *heteromorph*]: Differing in size or shape. In tapeworm (cestode), type of hooks in heteracanthus hook arrangement on tentacle: hooks differ in size and shape. (See also homeomorphous)

heteromyarian: anisomyarian

heteronomous [G *heteronom*]: Having segments differing in size or shape, e.g., as in tapeworm (cestode) proglottids. (See also homonomous)

heterorhabdic [G *heterorhabdisch*]: In bivalve gill (ctenidium), condition in which two types of filaments (numerous smaller and single enlarged filament) exist. (See also homorhabdic)

heterostrophic [G *heterostroph*]: In snail (gastropod) shell, condition in which earliest whorls (protoconch) appear to be coiled in opposite direction to those of later whorls (teloconch). (See also homeostrophic)

heterotrichous [G *heterotrich*]: Type of stinging cell (nematocyst) in cnidarian: spines on tubule are larger proximally than distally. Refers to type of nematocyst in which tubule lacks butt yet increases in diameter proximally (anisorhiza). (See also apotrichous, atrichous, basitrichous, holotrichous, homotrichous, merotrichous, telotrichous)

hexagonal [G *sechseckig, hexagonal*]: Having six angles and six sides. Refers, for example, to carapace shape in crab-like decapod. (See also sub-)

hexamerous [G *sechsteilig*]: General category of anthozoan in which polyp is not characterized by eight single complete septa, but typically by septa in pairs in cycles of six or a multiple of six. (See also octomerous, polymerous; biradial)

hirsute [G *steifhaarig, borstig*]: Covered with coarse, stiff, hair-like projections.

hispid [G *kurzborstig*]: With fur-like cover of short, stiff, hair- or spine-like projections.

holandric: In oligochaete, condition in which two pairs of testes are present, one in segment 10 and one in segment 11. (See also meroandric, metandric, proandric)

holobranchial [G *holobranchial*]: Type of gill arrangement in chiton (polyplacophoran) in which gills extend from anterior to posterior end of pallial groove along foot. (See also merobranchial; abanal, adanal)

hologonic: Type of ovary or testis in roundworm (nematode): germ cells are formed along entire length of above reproductive organs. (See also telogonic)

holomyonic: Type of arrangement of muscle bands in certain thaliaceans (old nurses in doliolid): bands 2–8 are united into continuous sheet (cuirass). (See also eurymyonic, stenomyonic; aclinous, amphiclinous)

holostomatous [G *holostom*]: In snail (gastropod) shell, type of aperture which lacks any interruption by notch or canal (siphonal canal). (See also siphonostomatous)

holostome [G *holostom*]: General category of larval (cercaria) or adult digenetic fluke (trematode); refers to forms having large adhesive organ (holdfast) in addition to anterior oral sucker and ventral sucker (acetabulum). (See also amphistome, distome, echinostome, gasterostome, monostome, prosostomatous)

holotrichous [G *holotrich*]: In type of nematocyst in which tubule lacks butt and is of same diameter throughout (isorhiza), condition in which tubule bears minute

spines along entire length. In nematocyst with long, distally dilated butt (macrobasic eurytele), condition in which butt bears spines along entire length. (See also apotrichous, atrichous, basitrichous, heterotrichous, homotrichous, merotrichous, telotrichous)

homacantho-: Prefix applied to cysticerus (stage in tapeworm life cycle) with only one ring of similar hooks.

homeoacanthus: In tapeworm (cestode), hook arrangement on tentacle: hooks are approximately alike in size and shape and are arranged in spirals or quincunxes. (See also heteracanthus, poeciloacanthus)

homeomorphous: Similar in size or shape. In tapeworm (cestode), type of hooks in heteracanthus hook arrangement: hooks are similar. (See also hetermorphous)

homeostrophic: In snail (gastropod) shell, condition in which earliest whorls (protoconch) are clearly coiled in same direction as later whorls (teloconch). (See also heterostrophic)

homodromous [G *homodrom*]: Type of reflexed ovary in roundworm: germinal and only part of growth zone are reflexed against oviduct. (See also antidromous)

homogomph [G *homogomph*]: Type of bristle (seta) in polychaete: with articulation distinctly and symmetrically at right angles to long axis of shaft.

homomyarian: isomyarian

homonomous [G *homonom*]: Having segments or appendages differing in size or shape. (See also heteronomous)

homorhabdic [G *homorhabdisch*]: In bivalve gill (ctenidium), condition in which only one type of filament exists; all are simple (not forming folds). (See also heterorhabdic)

homotrichous [G *homotrich*]: In type of nematocyst in which tubule lacks butt yet increases in diameter proximally (anisorhiza), condition in which tubule bears spines of equal length throughout. In nematocyst with short, distally expanded butt (microbasic eurytele), condition in which spines on butt are equal in size. (See also apotrichous, atrichous, basitrichous, heterotrichous, holotrichous, merotrichous, telotrichous)

hooded: Type of bristle (seta) in polychaete: covered distally by delicate chitinous envelope or guard.

hook-shaped [G *hakenförmig*]

horseshoe-shaped [G *hufeisenförmig*]

hydrorhizal (stolonate): Type of colony growth in hydrozoan; polyps originate singly and irregularly from stolons. (See also monopodial, sympodial)

hyperapolytic: apolytic

hyperstomial [G *hyperstomial*]: In bryozoan, type of brood chamber (ooecium) which rests on or is partly embedded in distal individual (zooid); opens above lid (operculum) of parent zooid.

hyperstrophic [G *hyperstroph*]: In snail (gastropod), condition in which shell coils to one side, but internal organs (especially gonads) are positioned as in specimens in which shell coils in opposite direction. (See also ultradextral, ultrasinistral)

hypothyridid [G *hypothyrid*]: In lamp shell (brachiopod), position of pedicle opening (foramen) relative to beak of ventral valve: foramen located below or on dorsal side of beak ridges, with umbo intact. (See also amphithyridid, epithyridid, mesothyridid, permesothyridid, submesothyridid)

hystrichoglossate [G *hystrichogloss*]: Type of radula in snail (gastropod): modified rhipidoglossate radula in which a large number of more lateral teeth bear terminal tuft of bristles. (See also docoglossate, ptenoglossate, rachiglossate, rhipidoglossate, taenioglossate, toxoglossate)

idiomorphic (automorphic): In bivalve, condition of shell or valve in which these are not deformed by crowding or attachment to other objects.

idmoneiform (idmidroneiform) [G *idmoneiform*]: Type of bryozoan colony:

creeping or semierect, arborescent with numerous and regular dichotomies. Branches relatively narrow, with feeding individuals (autozooids) opening on one face only and arranged in bundles alternating from right to left of branch.

illoricate: Type of rotifer without thickened outer layer (lorica). (See also loricate)

imbricate [G *dachziegelartig, imbrikiert, übereinandergreifend*]: Overlapping like tiles or shingles on a roof. Refers, for example, to arrangement of scale-like structures, to sculpture on outer surface of bivalve shell, or to type of plate arrangement in shell (test) of sea urchin.

immersed: In snail (gastropod) shell, condition in which initial whorls are concealed by later whorls.

imperfect: incomplete

imperforate [G *imperforat*]: Lacking holes (perforations). Refers, in plate of sea urchin shell (test), to type of spine-bearing elevation (tubercle) without depression on top. In bryozoan, type of frontal wall. In exoskeleton-producing, hexamerous anthozoan, condition in which skeletal cup (corallite) is solid and bears no perforations. (See also perforate; anomphalous, aporose)

impressed [G *eingedrückt*]: Type of suture in snail (gastropod) shell: sunk below surface due to inward turning of both adjoining whorl surfaces.

impunctate: In brachiopod, type of shell lacking minute canals (endopunctae). (See also punctate)

inarticulate [G *inartikulat*]: Type of lamp shell (brachiopod) in which dorsal and ventral valves are unhinged, i.e., held together posteriorly only by muscles (See also articulate)

incised [G *eingeschnitten*]: Bearing a cleft or cut (incision). May refer to margin of exoskeletal structure in crustacean.

incomplete (imperfect) [G *unvollständig*]: Refers, in anthozoan, to type of mesentery that extends inward from body wall and does not reach pharynx. (See also complete)

incudate [G *incudat*]: Protrusile, grasping type of mastax in rotifer; rami well developed and pincer-like as in forcipate type, yet protrusion entails a rotation of 90° to 180°.

incurrent (inhalent) [G *nach innen führend*]: Pertaining to structure involved in conveying water to interior, e.g., as in opening or siphon in bivalve or ascidian. (See also excurrent)

independent: Refers, in bryozoan, to type of brood chamber (ooecium) which develops independently of distal individual (zooid). (See also dependent)

indurate [G *verhärtet*]: Hardened and rigid, as in condition of certain appendages in crustacean.

inequilateral: In bivalve, condition in which part of shell anterior to beak and that posterior to beak differ appreciably in length. (See also equilateral)

inequivalve [G *ungleichklappig*]: In bivalve, type of shell in which one valve is larger than the other. (See also equivalve)

inferior [G *tieferstehend*]: Indicating position below (ventral to) a second structure. (See also superior)

infertile [G *steril*]: Refers, in anthozoan, to type of mesentery without gonads. (See also fertile)

inflated [G *bauchig, aufgeblasen, aufgetrieben, gebläht, gewölbt*]: Strongly convex or swollen. Refers, for example, to type of shell in snail or bivalve.

inflected [G *einwärts umgebogen*]: In gastropod shell, type of outer lip whose margin is turned inward. (See also everted)

infra- [G *infra-*]: Prefix indicating position below (occasionally also posterior to) a specified structure, e.g., as in infraorbital spine of decapod carapace. (See also supra-)

infundibular [G *trichterförmig*]: Funnel-shaped.

inhalent: incurrent

insert [G *insert*]: In sea urchin, condition of ocular plates in which these are in contact with test margin (periproct) around anus. (See also exsert)

integripalliate: In bivalve, condition in which pallial line on inner surface of valve lacks posterior, inwardly directed curve (sinus). (See also sinupalliate)

intercalated [G *eingeschaltet, eingefügt*]: Inserted between.

intercecal: Located between ceca; refers in fluke (trematode) to position of internal organs (testes, vitellaria) medial to each intestinal cecum. (See also extracecal)

intermediate [G *dazwischenliegend*]: Located between two structures, e.g., as in cirrus between notopodium and neuropodium in appendage (parapodium) of polychaete. May also refer to one of six valves between anterior and posterior valves in chiton (polyplacophoran).

interpolated: In male reproductive system of turbellarian, condition in which prostatic vesicle is continuous with and located directly between seminal vesicle and penis. (See also free)

interporiferous: In sea urchin shell (test), region of ambulacral plate which does not bear pore pairs.

interradial [G *interradial*]: In tetramerous radial symmetry of cnidarian, positioned on or along one of four sections lying between perradii. In sea urchin shell (test), type of suture between two rows of interambulacral plates.

interramal [G *interramal*]: In polychaete, located between dorsal branch or ramus (notopodium) and ventral branch (neuropodium) of appendage (paramodium).

intersegmental [G *intersegmental*]: In polychaete, located between segments.

intervascular: In tapeworm (cestode), located medial to osmoregulatory canals. (See also extravascular)

intraclitellar: In oligochaete, located in region of clitellum; refers, for example, to male pores. (See also preclitellar)

intramural: In oligochaete, type of calciferous gland consisting of series of folds within inner wall of esophagus. (See also extramural)

introversible [G *einstülpbar*]: Capable of being shortened by turning inside out, as opposed to shortening by mere contraction. (See also retractile)

involute [G *involut*]: In snail (gastropod), type of shell in which last whorl envelops earlier ones, the latter being more or less visible in umbilicus. (See also convolute)

iridescent [G *irisierend, schillerend*]: Showing rainbow colors in shifting patterns and hues according to angle of viewing.

irregular (exocyclic) [G *irregulär*]: Type of sea urchin or type of shell (test) of sea urchin in which anus and system of plates (periproct) around anus are not located within circlet of apical plates (genital and ocular plates). (See also regular)

isodiametric [G *isodiametrisch*]: Of the same diameter throughout, e.g., tubule or butt of stinging cell (nematocyst) in cnidarian. (See also anisodiametric)

isodont [G *isodont*]: Type of hinge (dentition) in bivalve; characterized by small number of relatively large, typically blunt and hook-shaped teeth alternating with deep pits (sockets).

isomyarian (homomyarian) [G *gleichmuskelig*]: In bivalve, condition in which two adductor muscles (anterior and posterior) are equal or almost equal in size (visible as two equal muscle scars). (See also anisomyarian = heteromyarian, dimyarian, monomyarian)

isostrophic [G *isotroph*]: In snail (gastropod), symmetrically sided shell coiled in one plane.

jointed: articulate

keeled: carinate

keyhole-shaped: cleithridiate

kidney-shaped: reniform

knee-shaped: geniculate

knife-shaped: cultriform

L-shaped [G *L-förmig*]: Refers, for example, to type of spicule in roundworm (nematode).

labial [G *labial*]: Pertaining to lip-like structure around mouth. In snail (gastropod) shell, refers to inner lip of aperture.

labiate [G *lippentragend*]: Bearing lip-like structure(s).

labral [G *labral*]: Pertaining to lower lip (labrum) below mouth or, in snail (gastropod) shell, to outer lip of aperture.

laciniate [G *geschlitzt, tief eingeschnitten*]: Having a fringed margin, as a rule with broader, irregular processes separated by deep, narrow clefts. (See also fimbriate)

lamellar [G *lamellös*]: Composed of or arranged in elongate, thin plates (lamellae). (For gill type in decapod, see phyllobranchiate)

lamellate (lamellose) [G *lamellös, lamellenförmig*]: Composed of or resembling an elongate, thin plate (lamella).

lamelliform [G *lamellenförmig, blattförmig*]: Resembling an elongate, thin plate (lamella).

lanceolate [G *lanzettförmig, lanzettlich, lanzeolat*]: Shaped like a lance head, i.e., pointed at one end, rounded at other.

lasiopodius: In barnacle, type of thoracic appendage (cirrus) in which setae along anterior margin are arranged in groups in region of articulation of articles. (See also acanthopodous, ctenopodous)

lateral [G *lateral, seitlich*]: Pertaining to or located on side.

leaf-shaped [G *blattförmig*]: Refers, in free-swimming hydrozoan colony for example, to type of bract. (See also foliaceous; in hydrozoan, see also helmet-shaped, prismatic)

lenticular [G *linsenförmig, lentikular*]: Shaped like a lentil or biconvex lens. Refers, for example, to bivalve shell shape.

lepralioid: cleithridiate

leptoderan: Type of caudal ala (bursa) in nematode: alae restricted to sides of body and do not surround or meet posterior to tip of tail. (See also arakoderan, peloderan)

leuconoid [G *leuconoid*]: Complex type of sponge in which numerous small choanocyte chambers are located between incurrent canals and excurrent canals; central cavity (spongocoel) is reduced or lacking. (See also asconoid, syconoid)

limbate [G *gerandet, gesäumt*]: In polychaete, type of bristle (seta) with flattened margin or wing.

linguiform [G *zungenförmig*]: Tongue-shaped.

lirate: Type of sculpture on snail (gastropod) shell: bearing lira (fine line or groove).

lobate [G *lappenförmig, lappig, gelappt*]: Resembling a lobe or bearing lobe-like (i.e., rounded) processes. (See also bilobate, multilobed)

lobodont: Type of compound hinge in ostracod. One of four basic subtypes of amphidont hinge: anteromedian element is well differentiated from posteromedian element (in the form of a lobed, tooth-like structure).

lobular [G *kleinlappig, lobulär*]: Resembling a small lobe (lobule) or bearing series of lobules.

loculate [G *in Fächer oder Kammern geteilt*]: Having or subdivided into a number of small cavities (loculae).

long-handled [G *langschäftig*]: In polychaete, type of hook-shaped seta (uncinus) with long basal supporting rod. (See also short-handled)

longicaudate: In tapeworm (cestode), condition in which tail (cercomer) of cysticercoid (stage in tapeworm life cycle) is long. (See also brevicaudate, circumcaudate; acaudate, caudate)

longifurcate: In fluke (trematode) larva with forked tail (furcocercous cercaria), condition in which stem of tail is short and forks are long. (See also brevifurcate)

longitudinal [G *longitudinal*]

loop-shaped: unispiral

lophocercous: Type of larva (cercaria) in digenetic fluke (trematode): body bears dorsal fin.

lophodont [G *lophodont*]: Type of compound hinge in mussel shrimp (ostracod): one of two basic subtypes (lophodont, entomodont) of merodont hinge—all three elements (anterior, median, posterior) are smooth.

loricate [G *lorikat*]: Type of rotifer with thickened outer layer (lorica). (See also illoricate)

loriferate: Type of scyphozoan medusa with long, slender oral arms bearing three rows of frills. (See also scapulate, tripterous)

lucinoid: heterodont

lumbricine (octochaetine) [G *lumbricid*]: In oligochaete, arrangement of bristles (setae) consisting of four pairs on each segment, two located ventrally and two lateroventrally. (See also perichaetine)

lunate [G *halbmondförmig*]: Shaped like a crescent.

lunulitiform [G *lunulitiform*]: Type of bryozoan colony; discoidal or conical, normally not anchored to (sandy or muddy) substratum. Feeding individuals (autozooids) restricted to convex face. Movements of vibracula stabilize community.

lyrate (lyriform) [G *leierförmig, lyraförmig*]: Resembling a lyre, i.e., with two parallel elements curving distally in opposite directions.

macrobasic: In cnidarian, type of stinging cell (nematocyst) with butt (heteronemes): condition in which butt is long (more than three times capsule length). (See also microbasic)

macrocercous (gorgoderine): In tapeworm (cestode), variation of cystocercous cercaria (stage in tapeworm life cycle): tail long, with or without vesicular swelling containing main body.

macrocnemous: In hexamerous anthozoan, arrangement of septa in which fourth and fifth septa on each side of dorsal directive are both macrosepta; three pairs of complete septa are present. (See also branchycnemous)

macrophreate: Type of crinoid: cavity in large aboral plate (centrodorsal) is large and deep, tegmen is naked, and 10 arms are typically present. (See also oligophreate)

macrurous [G *macrur*]: Type of decapod adapted either for swimming or with well-developed, cylindrical or laterally compressed tail; tail extends posteriorly and is not folded under cephalothorax. (See also brachyuran)

malleate [G *malleat*]: Complicated grinding or to a lesser extent grasping type of mastax in rotifer in which all trophi are relatively well developed and movable; rami untoothed, unci bearing several large prongs for chewing.

malleoramate [G *malleoramat*]: Slight modification of ramate type of mastax in rotifer: rami strongly toothed.

mammillated (mammilliform) [G *mammillär*]: Shaped like a breast or bearing nipple-shaped (i.e., bluntly rounded or wart-like) protuberance(s), e.g., as in certain papillae on sipunculan or sculpture on shell of snail (gastropod) or mussel shrimp (ostracod).

marginal [G *randständig, marginal*]: Relating to or located along edge, e.g., as in sensory bodies (rhopalia) along margin of scyphozoan medusa or groove along edge of decapod carapace.

marginate [G *mit Rand versehen*]: Provided with distinct margin, e.g., as in strengthed margin of outer lip in snail (gastropod) shell.

marmorate [G *marmoriert*]: Having veined or streaked markings or coloration as in marble stone.

massive [G *massiv, massig*]: Forming large, dense mass. Refers, for example, to type of colony growth in sponge or bryozoan.

masticatory [G *Kau-, kauende*]: Serving in the chewing of food.

maxilliform [G *maxilliform, maxillenförmig*]: In crustacean, development of an anterior appendage (e.g., maxilliped) to resemble and function as a mouthpart (not as a locomotory limb). (See also pediform)

mediad [G *mediad*]: Extending toward midline.

medial [G *medial*]: Located in or extending toward middle.

median [G *median*]: Located in the middle or lying in a plane dividing bilateral organism into left and right halves.

medusiform [G *medusenförmig*]: Bell- or umbrella-shaped like a jellyfish (i.e., like a medusa of a hydrozoan or scyphozoan).

membraniporiform [G *membraniporiform*]: Type of bryozoan colony: encrusting, usually unilaminar. Basal side partly or totally calcified.

membranous [G *membranös*]: Consisting of or resembling a membrane. In anthozoan, type of thin, soft colony growth covering substratum. (In anthozoan, see also encrusting)

meridosternous [G *meridostern*]: Type of region (plastron) on underside of irregular sea urchin shell (test): plastron in which lip is followed by single large plate (See also amphisternous)

mermithoid: Type of two-part pharynx in roundworm (nematode): highly elongate, tubular, and nonmuscular and often associated with stichosome.

meroandric: In oligochaete, condition in which only one pair of testes is retained. If anterior pair is present, the designation proandric is applied; if only posterior pair is present, then metandric.

merobranchial [G *merobranchial*]: Type of gill arrangement in chiton (polyplacophoran); gills occupy only posterior half of pallial cavity. (See also holobranchial; abanal, adanal)

merodont [G *merodont*]: Type of compound hinge in mussel shrimp (ostracod); consists of three elements, i.e., two terminal (anterior and posterior) elements as well as an undivided median element. According to whether the three elements are smooth or crenulate, two subtypes (leptodont, entomodont) may be distinguished. (See also adont, amphidont, archidont, heterodont, lobodont, lophodont, schizodont)

merotrichous [G *merotrich*]: Type of stinging cell (nematocyst) in cnidarian: spines are located on proximal section of butt, yet lacking on distal section. Refers to nematocyst type (macrobasic eurytele) with long, distally dilated butt. In nematocyst lacking butt and with tubule of same diameter throughout (isorhiza), condition in which spines are present only on intermediate section. (See also apotrichous, atrichous, basitrichous, heterotrichous, homotrichous, telotrichous)

mesial [G *mesial, in der Mittelebene gelegen*]: Lying on vertical longitudinal plane dividing bilaterally symmetrical organism into left and right halves.

mesostomate: Type of excretory system in larval stage (cercaria) of fluke (trematode): main excretory (collecting) ducts extend only from posterior excretory vesicle to midbody, where they join excretory tubules. (See also stenostomate)

mesothyridid [G *mesothyrid*]: In articulate lamp shell (brachiopod), position of pedicle opening (foramen) relative to deltidial plates and beak of ventral valve: foramen located partly in ventral umbo and partly in delthyrium, with beak ridges appearing to bisect foramen. (See also amphi-

thyridid, epithyridid, hypothyridid, permesothyridid, submesothyridid)

metandric: In oligochaete in which only one pair of testes is retained (meroandric forms), condition in which posterior testes are present (in segment 11). (See also holandric, proandric)

microbasic: In cnidarian, type of stinging cell (nematocyst) with butt (heteronemes): condition in which butt is short (less than three times capsule length). (See also macrobasic)

microcercous: Type of larva (cercaria) in distome digenetic fluke (trematode): tail very small. (See also obscuromicrocercous, sulcatomicrocercous)

monactinal: Having a single ray with fundamentally different ends, e.g., as in type of sponge spicule.

monaulic [G *monaulic*]: In reproductive system, for example in snail (gastropod), condition in which sperm duct and oviduct open to exterior via common pore. (See also diaulic, triaulic)

moniliform [G *perlschnurförmig, perlschnurartig, moniliform*]: Resembling a string of beads, e.g., as in arrangement of bulbs in nematode pharynx.

monobasal: In shell (test) of irregular sea urchin, type of apical system of plates in which five original genital plates are fused to form single large central plate. (See also tetrabasal)

monochromatic [G *monochrom*]: Exhibiting or having one color. Refers, in crustacean, to type of pigment-bearing cell (chromatophore) with one color pigment. (See also dichromatic, polychromatic)

monocondylic [G *monokondyl*]: Type of joint in crustacean appendage in which adjoining segments articulate on a single process (condyle). (See also dicondylic)

monocyclic [G *monozyklisch*]: In crinoid, condition of plates forming skeleton (calyx) surrounding main body (crown): having two circlets of plates, i.e., only one circlet (basals) proximal to radials. (See also cryptodicyclic, dicyclic)

monodelphic [G *unpaar*]: In roundworm (nematode), condition in which female reproductive system consists of single tract. (See also didelphic, polydelphic, tetradelphic)

monodisc [G *monodisk*]: In scyphozoan, type of medusa production (strobilation) in which one young medusa (ephyra) forms at a time. (See also polydisc)

monoecious [G *zwittrig, hermaphroditisch, monözisch*]: Containing gonads (ovary, testis) of both sexes within a single individual. (See also dioecious)

monoflagellate [G *monoflagellat, eingeißelig*]: Bearing one flagellum. (See also biflagellate)

monogenetic: Type of fluke (trematode) whose life cycle involves only a single host; typically ectoparasitic. (See also digenetic)

monoglyphic: In anthozoan, form with only one flagellum-bearing groove (siphonoglyph) along pharynx. (See also diglyphic)

monomorphic [G *monomorph*]: Characterized by occurrence of only one type of individual, e.g., as in anthozoan colony with autozooids only. (See also dimorphic, polymorphic)

monomyarian [G *einmuskelig*]: In bivalve, condition in which only one adductor muscle (typically posterior muscle) spans between valves (visible as a single muscle scar on inner surface of valve). (See also anisomyarian = heteromyarian, dimyarian, isomyarian = homomyarian)

monopectinate: unipectinate

monopodial [G *monopodial*]: Having a single main axis of growth but giving rise to series of lateral branches. Refers, in hydrozoan colony, to type of growth occurring either below permanent terminal hydranths or at terminal growing points. (See also sympodial)

monorchic [G *unpaar*]: Type of male reproductive system in roundworm (nematode): having only single testis. (See also diorchic, proorchic)

monosiphonic: Type of main stalk (hydrocaulus) of hydrozoan consisting of single element. (See also polysiphonic = fascicled)

monostichous [G *monostich*]: In caudofoveate or solenogaster, type of radula characterized by (functionally) single tooth in each transverse row. (See also distichous, polystichous)

monostome [G *monostom*]: General category of larval (cercaria) or adult digenetic trematode; refers to forms having only one (typically oral) sucker or lacking suckers. (See also amphistome, distome, echinostome, gasterostome, holostome, prosostomatous)

monostyliferous: In ribbon-worm (nemertine), condition in which proboscis has only one stylet. (See also polystyliferous)

monotocardian [G *monotokard*]: In snail (gastropod), type of heart basically consisting of one auricle and one ventricle. (See also diotocardian)

monozoic [G *monozoisch*]: Type of tapeworm (cestode) whose body is not divided into segments (proglottids). (See also polyzoic)

mucronate [G *stachelspitzig*]: Ending in an abrupt, sharp tip or process.

multiarticulate (multisegmented) [G *vielgliederig*]: Consisting of a long series of elements united by joints, e.g., as in flagellum of crustacean antenna. (See also articulate)

multibrachiate: Type of crinoid bearing more than 10 arms (brachia).

multibulbar: Type of pharynx in roundworm (nematode): characterized by series of bulbs posteriorly. (See also moniliform)

multicephalo-: In tapeworm (cestode), prefix applied to cysticercus (stage in tapeworm life cycle) with multiple scoleces.

(See also anacantho-, heteracantho-, homacantho-, strobilo)

multidigitate [G *vielfingerig*]: Bearing many finger-like processes. (See also digitate)

multifurcate: Forking into more than two parts, e.g., as in type of tubercle on shell of mussel shrimp (ostracod). (See also bifurcate, trifurcate)

multilobed (plurilobed) [G *mehrlappig*]: Bearing or consisting of many lobe-like processes.

multilocular [G *multilokulär*]: In tapeworm (cestode), type of hydatid (stage in tapeworm life cycle) with highly branched structure and typically with exogenous budding. (See also alveolar, unilocular)

multiplanar [G *in mehreren Ebenen*]: Refers, in anthozoan, to type of branched colony in which branches grow in several planes.

multiramous: polyramous

multisegmented: multiarticulate

multiserial [G *multiserial*]: Arranged in several to many series. Refers, in bryozoan, to type of colony in which individuals (zooids) are arranged in several rows. In snail (gastropod), arrangement of teeth in radula. (See also biserial, uniserial)

multispiral [G *mehrspiralig, vielspiralig*]: Refers, in snail (gastropod), to shell consisting of numerous whorls; in roundworm (nematode), type of amphid characterized by several coils.

multivincular [G *multivinkulär*]: In hinge of bivalve, type of ligament consisting of series of elements of alivincular type, i.e., succession of alternating lamellar and fibrous components. (See also alivincular, duplivincular, paravincular)

multizooidal: extrazooidal

muricate [G *stachelig*]: Spiny, i.e., covered with sharp, hard points as on surface of snail (gastropod) shell.

nacreous [G *perlmuttrig*]: Consisting of or resembling pearly or shiny layer (nacre) found in certain bivalves.

naked [G *nackt*]: Lacking typical external armature or covering. Refers, for example, to type of barnacle (cirriped) lacking plates.

natant [G *schwimmende*]: Category of decapod including shrimp-like forms, i.e., typically laterally flattened (compressed) body and more or less adapted for swimming. (See also brachyuran, reptant)

natatory [G *Schwimm-*]: Serving in swimming. (See also ambulatory)

navicular [G *bootförmig, kahnförmig*]: Shaped like or resembling a boat.

needle-shaped [G *nadelförmig*]: (See also spicular)

nemic: Of or related to roundworms (nematodes).

nodose [G *knotig*]: Bearing nodes, i.e., knob-like protuberances.

nodulose [G *knötchenförmig*]: Bearing nodules, i.e., small, knob-like protuberances.

nonappendiculate: Type of distome digenetic fluke (trematode) not regionated into anterior ecsoma and posterior soma. (See also appendiculate)

noncirriferous [G *cirrenlos*]: Type of crinoid lacking cirri. (See also cirriferous)

nonconjugate: In sea urchin shell (test), type of pore pair in ambulacral plate: pores distinctly separated. (See also conjugate)

nonoculate: Lacking photosensitive organs. Refers to type of larva (cercaria) of digenetic fluke (trematode). (See also oculate)

nonoperculate: Lacking a lid (operculum), e.g., as in certain tube-dwelling polychaetes. (See also operculate; in tapeworm, see also anoperculate)

nonporiferous: interporiferous

nonstalked [G *ungestielt*]: sessile

nonstrophic: In lamp shell (brachiopod), type of shell with posterior margin not parallel with hinge axis. (See also strophic)

notched [G *eingekerbt*]: With margin or edge characterized by V-shaped indentation(s).

nuchal [G *Nacken-*]: Relating to or located on back of "neck", i.e., constricted region posterior to head. Applied, in polychaete, to various sense organs at anterior end of body.

ob- [G *verkehrt, umgekehrt*]: Prefix applied to denote inverted (upside-down) orientation.

obconical [G *verkehrt kegelförmig*]: Shaped like an upside-down cone.

obcordate [G *verkehrt herzförmig*]: Shaped like an upside-down heart.

oblique [G *schief*]: Slanting, i.e., neither parallel nor perpendicular to main body axis or other specified structure. In snail (gastropod), refers to slanting aperture of shell; in bivalve, to structure extending in direction neither parallel nor perpendicular to hinge axis.

obliquely ovate [G *schief-oval, schrägoval*]: In bivalve, egg-shaped shell in which main body of valve grows at an angle to hinge axis.

oblong [G *länglich*]: Deviating from either square or circuluar shape due to elongation.

obovate [G *verkehrt eiförmig*]: Shaped like an upside-down egg, i.e., narrower end pointing downward or forming base or site of attachment. (See also ovate)

obpyruvate [G *verkehrt birnenförmig*]: Shaped like an upside-down pear, i.e., narrower end pointing downward or forming base or site of attachment. (See also pyriform)

obscuromicrocercous: Variation of small-tailed (microcercous) cercaria larva of fluke (trematode): tail indistinctly set off from body.

obsolescent [G *allmählich schwindend*]: Condition in which elements of a series gradually become smaller or disappear. Also used as synonym for obsolete.

obsolete [G *unvollkommen entwickelt, obsolet*]: Indistinct or poorly developed.

obtuse [G *stumpfwinklig*]: Coming together at a wide angle. (See also acute)

occipital [G *okzipital*]: Pertaining to back part of head. Refers, in polychaete, to posterodorsal part of prostomium.

occluded [G *verschlossen*]: Blocked or closed off.

octagonal [G *achteckig*]: Having eight sides or angles. Refers, for example, to carapace shape in decapod.

octobrachiate [G *octobrachiat, achtarmig*]: Type of cephalopod with eight arms. (See also decabrachiate)

octochaetine: lumbricine

octomerous [G *achtteilig, achtstrahlig, octomer*]: In anthozoan, condition in which polyp is characterized by 8 single complete septa and 8 pinnate tentacles. May also refer to modified hexamerous condition characterized by 8 or 16 pairs of complete septa and numerous incomplete septa. (See also polymerous; biradial)

ocular [G *Augen-*]: Pertaining to eyes, e.g., as in ocular peduncle (= eyestalk) of crustacean.

oculate: With photosensitive organs. Refers to type of larva (cercaria) of digenetic fluke (trematode). (See also nonoculate)

oligogyral: paucispiral

oligophreate: Type of crinoid: cavity in large aboral plate (centrodorsal) is small and shallow, tegmen bears numerous small plates, and more than 10 arms are typically present. (See also macrophreate)

oligoporous: In sea urchin shell (test), type of modified ambulacral plate (compound plate) bearing few (two or three) pore pairs. (See also polyporous; bigeminate, trigeminate)

omphalous (umbilicate) [G *genabelt*]: In snail (gastropod), type of shell with umbilicus. (See also anomphalous)

one-part: enoploid

one-segmented [G *eingliedrig, einteilig*]: uniarticulate

open [G *offen*]: Refers, in sea urchin, to type of petals which do not contract markedly at their ambital extremity. (See also closed)

operculate [G *deckeltragend, operculat*]: Having or forming a lid (operculum). Refers, for example, to type of egg in tapeworm (cestode), to snail shell with operculum, or to condition in which abdominal appendages (pleopods) are modified to form cover in isopod.

operculiform [G *deckelförmig, operculiform*]: Shaped like a lid (operculum).

ophicephalous [G *ophiocephal*]: In sea urchin, type of pedicellaria with short, concave, blunt-ended jaws that lock together when closed due to basal element.

ophthalmic [G *Augen-*]: Pertaining to eye.

opisthocline: In bivalve, type of shell growth in which body of valve slopes posteriorly, i.e., angle between hinge axis and line bisecting umbo is greater than 90°. Also refers to posteriorly sloping hinge teeth. (See also acline, proscoline)

opisthocline: In snail (gastropod) shell, type of growth line which leans backward (i.e., inclined adapically) with respect to growth direction of helicocone. (See also orthocline, prosocline)

opisthocyrt: In snail (gastropod) shell, type of growth line that arches backward with respect to growth direction of helicocone. (See also prosocyrt)

opisthodelphic [G *nach hinten gerichtet*]: Type of uterus configuration in roundworm (nematode): uteri are parallel and posteriorly directed. (See also amphidelphic, prodelphic)

opisthodetic [G *opisthodet*]: In hinge of bivalve shell, type of ligament extending only posterior to beak. (See also amphidetic, prosodetic)

opisthogoneate [G *opisthogoneat*]: Having the genital opening near posterior end of body. (See also progoneate)

opisthogyrate [G *opisthogyr*]: In bivalve shell, condition in which umbones are curved so that beaks point in posterior direction. (See also orthogyrate, prosogyrate, spirogyrate)

opposite [G *gegenständig*]: Refers, in hydrozoan, to type of polyp arrangement along branch of erect colony. (See also alternate, subopposite, whorled)

oral [G *oral*]: Directed toward mouth or representing side on which mouth is located. (See also aboral)

orbicular [G *kreisförmig, scheibenförmig, kugelförmig, orbicular*]: Circular and flat (i.e., more "two-dimensional" as in an outline) or spherical ("three-dimensional"). (See also sub-)

orbital [G *Orbital-*]: Surrounding or immediately adjoining eye.

orthocline: In bivalve, type of shell growth in which body of valve extends directly in ventral direction, i.e., angle between hinge axis and line bisecting umbo approaches 90°. Also refers to hinge tooth which is aligned perpendicular to hinge axis. (See also opisthocline, prosocline)

orthocline: In snail (gastropod) shell, type of growth line which is oriented at right angle with respect to growth direction of helicocone. (See also opisthocline, prosocline)

orthogyrate [G *orthogyr*]: In bivalve, condition in which umbones are curved directly toward other valve (beaks therefore not pointing anteriorly or posteriorly but at right angles to hinge axis). (See also opisthogyrate, prosogyrate, spirogyrate)

orthostrophic [G *orthostroph*]: In snail (gastropod), condition in which shell is normally coiled. (See also hyperstrophic)

ostiate [G *ostiumtragend*]: Type of heart bearing openings (ostia).

outstretched [G *ausgestreckt*]: In roundworm (nematode), type of female reproductive system: all parts of ovary and oviduct oriented in a single direction. (See also reflexed)

oval [G *oval, eirund, eiförmig*]: Resembling an egg in outline (i.e., more "two-dimensional"). Also used as synonym for ovate.

ovarian: Pertaining to an ovary. Refers, in oligochaete, to ovary-bearing segment of body. (See also testicular)

ovate [G *eiförmig*]: Egg-shaped (i.e., more "three-dimensional"). Also used as synonym for oval.

overlapping: imbricate

ovigerous [G *eitragend*]: Bearing eggs, e.g., as in certain appendages of crustacean.

oxy- [G *oxy-*]: Prefix indicating sharp or pointed elaboration.

oxyuroid: bulboid

pachydont [G *pachydont*]: Type of hinge (dentition) in bivalve: characterized by one to three asymmetrical, cone-shaped teeth fitting into deep recesses in opposite valve.

paired [G *paarig*]: (See also single = unpaired)

palatal [G *palatal*]: In snail (gastropod), pertaining to outer lip of shell.

palmate [G *handförmig*]: Resembling a hand with fingers spread apart.

papillate (papillose) [G *warzig, papillös*]: Bearing papillae, i.e., nipple-shaped protuberances. Also used as synonym for papilliform.

papilliform [G *warzenförmig, papillenförmig*]: Shaped like a papilla, i.e., like a nipple.

pappose: Refers to type of bristle (seta) with secondary projections (setules) arising on all sides of shaft. (See also plumose)

parasigmoidal [G *parasigmoid*]: Curved like a reverse "S."

parasitic [G *parasitisch*]: Type of organism metabolically dependent on a host to complete its life cycle and typically negatively affecting host. (See also ectoparasitic, endoparasitic, free-living)

paravincular [G *parivinkulär*]: In hinge of bivalve, elongate, cylindrical type of ligament posterior to beaks. (See also alivincular, duplivincular, multivincular)

parietal [G *parietal*]: Pertaining to wall of cavity or other structure. In snail (gastropod) shell, referring to broader upper portion of inner lip adjoining aperture.

patelliform [G *napfschneckenförmig*]: Limpet-shaped, i.e., resembling a flattened cone.

patulous [G *klaffend*]: Spread widely apart, e.g., as in expanded aperture of certain snail (gastropod) shells.

paucispiral [G *paucispiral*]: Type of lid (operculum) in snail (gastropod) shell: with relatively few whorls.

pearly: nacreous

pear-shaped: pyriform

pectinate [G *kammförmig*]: Comb-like, i.e., with series of projections along one side. In sea star, type of sessile pedicellaria in which series of opposing, movable spines are short and curved. (See also bipectinate; in sea urchin, see also fasciculate, spiniform, valvulate)

pedal [G *Fuß-, Pedal-*]: Located in or otherwise pertaining to foot-like structure.

pedate: General category of sea cucumber (holothurian) characterized by tube feet.

pedicellate [G *gestielt*]: Characterized or attached by stalk (pedicel).

pediform [G *beinförmig, pediform*]: Leg-like. Refers, in crustacean, to development of anterior appendage (e.g., maxilliped) to resemble or function as a walking limb (and not as a mouthpart).

pedigerous [G *beintragend*]: Bearing foot-like appendage, e.g., as in segment of crustacean body.

pedunculate [G *gestielt*]: Characterized or attached by stalk (peduncle). Refers to general category of barnacles, to type of protective individual (vicarious avicularium) in bryozoan colony, and, in sea star, to type of pedicellaria with short, stout stalk lacking internal calcareous support. (See also sessile; in sea star, see also alveolar, crossed, furcate, straight)

peloderan: Type of caudal ala (bursa) in roundworm (nematode): alae surround or meet at tip of tail. (See also arakoderan, leptoderan)

pellucid [G *durchsichtig, klar*]: Transparent or clear, e.g., as in shell of certain snails (gastropods).

peltate [G *schildförmig*]: Shield-shaped, e.g., as in type of tentacle in sea cucumber (holothurian).

pendent [G *herabhängend*]: Suspended or hanging down.

pendulous [G *hängend, frei schwebend, schwingend*]: Suspended so as to swing freely, e.g., as in type of manubrium in medusa.

penicillate [G *pinselförmig*]: Brush-like, i.e., resembling or tipped by tuft of hair-like projections. Type of bristle (seta) in polychaete or type of tube foot (podium) in phyllode of sea urchin. (See also pseudopenicillate)

pennoned: In polychaete, type of bristle (seta) with teardrop-shaped tip.

pentagonal [G *fünfeckig*]: Having five sides or angles. Refers, for example, to carapace shape in decapod. (See also subpentagonal)

pentamerous [G *fünfteilig, pentamer*]: Divided into or consisting of five parts. Refers, for example, to body plan of sea cucumber, where major organs are arranged in five sets.

pentaradiate (pentaradial) [G *fünfstrahlig*]: Divided into or consisting of five parts, especially if parts are arranged so as to radiate from a common center. Refers to symmetry of echinoderm.

penultimate (penult) [G *vorletzte*]: Next to last, e.g., as in next to last in a series of segments in crustacean appendage or body. In snail (gastropod) shell, whorl preceding last-formed whorl, i.e., preceding body whorl.

perfect: complete

perforate [G *durchlöchert, perforiert*]: Characterized by hole(s). Refers, for example, to type of frontal wall in bryozoan or condition of skeletal cup (corallite) in anthozoan. In plate of sea urchin shell (test), type of spine-bearing elevation (tubercle) with small depression on top for ligament. (See also imperforate)

perichaetine [G *perichätin*]: In oligochaete, arrangement of bristles (setae): numerous setae surround equator of each segment. (See also lumbricine)

peristomial: In bryozoan, type of brood chamber (ovicell) formed by enlarged orificial collar (peristome).

permesothyridid [G *permesothyrid*]: In lamp shell (brachiopod), position of pedicle opening (foramen) relative to deltidial plates and beak of ventral valve: foramen located mostly within ventral umbo, i.e., outside delthyrium yet in contact with beak ridges. (See also amphithyridid, epithyridid, hypothyridid, mesothyridid, submesothyridid)

perradial [G *perradial*]: In tetramerous radial symmetry of cnidarian, positioned on or along one of four major radial axes. In sea urchin shell (test), positioned along midline of plates forming each ambulacrum, e.g., as in perradial suture.

persistent [G *persistent*]: Refers to structure that is retained, e.g., as in ligament sacs of certain spiny-headed worms (acanthocephalans).

petaloid: At apex of irregular sea urchin shell (test), expanded, petal-shaped condition of each ambulacrum. (See also subpetaloid)

petraliiform [G *petraliiform*]: Type of bryozoan colony: unilaminar, only attached to substratum by chitinous rhizoids arising from isolated pores on dorsal face.

phaneromphalous: In snail (gastropod), shell with completely open umbilicus. (See also anomphalous, omphalous)

pharyngeate: Having pharynx. Refers to type of larva (cercaria) of digenetic fluke (trematode). (See also apharyngeate)

phyllobranchiate (lamellar) [G *phyllobranchiat*]: In decapod, type of gill characterized by broad and flattened, leaf-like branches along axis. (See also dendrobranchiate, trichobranchiate)

phyllopodous: In crustacean, refers to broad and flattened, leaf-like appendage (phyllopod).

pilose [G *haarig, behaart, dicht behaart*]: Velvety, i.e., covered with (usually soft) short hairs.

pinnate [G *gefiedert*]: Feather-shaped, i.e., with main stem and series of branches along each side. (See also bipinnate, pectinate)

planar [G *in einer Ebene*]: In anthozoan, type of colony in which branches grow more or less in one plane.

planispiral (planospiral) [G *flachgewunden, plan-spiralig*]: In snail (gastropod), type of shell coiled in one plane, typically with symmetrical sides yet also applied to shells with asymmetrical sides.

plate-like [G *tellerförmig, plattenförmig*]

plectolophous [G *plectoloph*]: Type of lophophore in lamp shell (brachiopod): median coiled arm developed in addition to two lateral arms. (See also ptycholophous, schizolophous, spirolophous, trocholophous, zugolophous)

plesiomorphic [G *plesiomorph*]: Primitive, ancestral structure. (See also apomorphic)

pleurembolic [G *pleurembolisch*]: In snail (gastropod), type of proboscis in which only basal part is invaginable; remainder is merely retracted into proboscis sheath. (See also acrembolic)

pleurolophocercous: Type of larva (cercaria) of distome digenetic fluke (trematode): tail bears longitudinal fin.

plexiform [G *netzgeflechtartig*]: Forming a network (plexus), e.g., as in system of nerves or blood vessels.

plicate [G *gefaltet*]: Folded or ridged lengthwise like a fan. May also refer to type of gill in snail (gastropod).

plumose [G *gefiedert, federartig*]: Feather-shaped, e.g., as in type of bristle (seta) with series of long, thin branches (setules) along each side. (See also pappose, subsetulate)

pluriarticulate: multiarticulate

plurilaminar: Type of colony growth in bryozoan: successive layers of individuals (zooids) are formed by frontal budding. (See also bilaminar, unilaminar).

plurilobed: multilobed

pluriserial: multiserial

poeciloacanthus: In tapeworm (cestode), hook arrangement on tentacle: chainette or longitudinal band of small hooks is present on externolateral surface. (See also heteracanthus, homeoacanthus)

polyaxial (polyaxonic): Having many axes, e.g., as in sponge spicule with more than four centered rays.

polychromatic [G *polychrom, mehrfärbig*]: Exhibiting or having many colors. Refers, in crustacean, to type of pigment-bearing cell (chromatophore) with three or four color pigments. (See also dichromatic, monochromatic)

polyclad [G *polyclad*]: In flatworm (turbellarian), condition in which digestive system consists of central tubular intestine and series of lateral branches (diverticula). (See also triclad)

polydelphic: In roundworm (nematode), condition in which female reproductive system consists of many tracts. (See also didelphic, monodelphic, tetradelphic)

polydisc [G *polydisk*]: In scyphozoan, type of medusa production (strobilation) in which series of young medusa (ephyrae) are formed at the same time. (See also monodisc)

polygastric: In certain floating colonial hydrozoans, condition in which units of different zooid types occur in series along stem of colony; distinguished from single detached units (eudoxids).

polygonal (multiangular) [G *polygonal, vieleckig*]: Having many sides or angles. May refer to areoles in cuticle of gordian worm (nematomorph) or to carapace shape in decapod.

polygyral: multispiral

polymerous [G *polymer, vielteilig*]: In anthozoan, condition in which polyp is neither hexamerous nor octomerous. Also refers to type of sea spider (pycnogonid) with more than four trunk segments (and thus more than eight legs).

polymorphic [G *polymorph, vielgestaltig*]: Characterized by the occurrence of several different types of individuals. (See also dimorphic, monomorphic, trimorphic)

polypharyngeal [G *polypharyngeal*]: In flatworm (turbellarian), condition in which several pharynges are present.

polypiferous [G *polypentragend*]: Polyp-bearing, e.g., as in upper region of erect anthozoan colony.

polyporous: In sea urchin shell (test), type of modified ambulacral plate (compound plate) bearing more than three pore pairs (usually five or more). (See also oligoporous; bigeminate, trigeminate)

polyramous (multiramous): With numerous branches. Refers to biramous crustacean limb (i.e., protopod bearing enodopod and exopod) having additional large processes (endites and epipod). (See also biramous, uniramous)

polysiphonic [G *polysiphonal*]: In hydrozoan, type of main stalk (hydrocaulus) consisting of a number of bundled elements. (See also monosiphonic)

polystichous [G *polystich*]: In caudofoveate or solenogaster, type of radula characterized by several to many teeth in each transverse row. (See also distichous, monostichous)

polystyliferous [G *polystilifer*]: In ribbon-worm (nemertine), condition in which proboscis bears numerous stylets. (See also monostyliferous)

polyzoic [G *polyzoisch*]: Type of tapeworm (cestode) whose body is divided into segments (proglottids). (See also monozoic)

ponticulate: Type of ridge (carina) on shell of mussel shrimp (ostracod): walls of carina perforated by series of openings resulting in bridge-like structure. (See also undercut)

porcelaneous [G *porzellanartig*]: Resembling porcelain, e.g., as in type of snail (gastropod) shell with enamel-like luster.

poriferous: Bearing pores. In sea urchin shell (test), region of ambulacral plate which bears pore pairs. (See also interporiferous)

porose [G *porös*]: Having pores.

post- [G *post-*]: Prefix indicating position behind a specified structure or region, e.g., postcephalic, postorbital, postrostral. (See also pre-)

postclitellar: In oligochaete, position of male gonopore posterior to clitellum. (See also anteclitellar, intraclitellar)

posterior [G *hintere, Hinter-*]: Directed toward or lying in region opposite head. (See also anterior)

postoral [G *postoral*]: Located behind mouth.

postplete: In mussel shrimp (ostracod), condition of carapace or valve in which greatest height is behind midlength. (See also amplete, preplete)

postseptal [G *postseptal*]: In polychaete and oligochaete, section of nephridial tubule located after septum, i.e., in same segment as that bearing nephridiopore. (See also preseptal)

postsetal [G *postsetal*]: In polychaete parapodium, located posterior to point of insertion of bristles (setae); typically refers to lamella-like lobe. (See also presetal)

posttesticular: In fluke (trematode), condition in which ovary is positioned posterior to testis. (See also pretesticular)

pre- [G *prä-*]: Prefix indicating position in front of specified structure or region, e.g., pregastric.

preclitellar: In oligochaete, located anterior to clitellum, e.g., as of gut diverticulae. (See also anteclitellar)

prehensile [G *Klammer-, zum Greifen geeignet*]: Adapted for grasping or clinging, e.g., as in antenna of certain copepods.

preoral [G *präoral*]: Located in front of mouth.

preplete: In mussel shrimp (ostracod), condition of carapace or valve in which greatest height is at or near midlength. (See also amplete, postplete)

present [G *vorhanden*]: One of two basic character states (present/absent). Few structures or organs are present in all representatives of a major taxonomic group (e.g., phylum); this term is therefore not listed as a character state in the first part of this glossary.

preseptal [G *präseptal*]: In polychaete or oligochaete, section of nephridial tubule located before septum, i.e., in segment anterior to that bearing nephridiopore. (See also postseptal)

presetal [G *präsetal*]: In polychaete parapodium, located anterior to point of insertion of bristles (setae); typically refers to lamella-like lobe. (See also postsetal)

pretesticular: In fluke (trematode), condition in which ovary is positioned anterior to testis. (See also posttesticular)

primary [G *primär*]: In snail (gastropod) shell, type of spiral sculpture appearing early in development. (See also secondary)

prionodont: taxodont

prismatic [G *prismatisch, prismaförmig*]: Prism-shaped or composed of prisms. In the former, refers to shape of buccal cavity in roundworm (nematode) or to shape of bract in free-swimming colonial hydrozoan. In the latter, refers to snail (gastropod) shell consisting of prisms of calcite or aragonite.

proandric: In oligochaete in which only one pair of testes is retained (meroandric forms), condition in which anterior testes are present (in segment 10). (See also holandric, metandric)

prosbosciform [G *rüsselförmig*]: Shaped like a proboscis, i.e., elongate, tubular, and capable of being extended and retracted. Refers, for example, to male copulatory organ in barnacle (cirriped).

prodelphic [G *nach vorne gerichtet*]: Type of uterus configuration in roundworm (nematode); uteri parallel and anteriorly directed. (See also amphidelphic, opisthodelphic; didelphic, monodelphic, polydelphic, tetradelphic)

proepilobous [G *proepilobisch*]: In oligochaete, combined pro- and epilobous condition of prostomium, i.e., prostomium is demarcated from peristomium by transverse groove, yet posterior extension (tongue) extends onto peristomium. (See also epilobous, prolobous, tanylobous, zygolobous)

progoneate [G *progoneat*]: Having genital opening near anterior end of body. (See also opisthogoneate)

prolobous [G *prolobisch*]: In oligochaete, type of prostomium which is demarcated from peristomium dorsally by simple transverse groove. (See also epilobous, proepilobous, tanylobous, zygolobous)

proorchic: Type of male reproductive system in roundworm (nematode): paired, anteriorly directed testes have undergone sex reversal. (See also diorchic, monorchic)

prosocline: In bivalve, type of shell growth in which main body of valve slopes anteriorly, i.e., angle between hinge axis and line bisecting umbo is less than 90°. Also refers to anteriorly sloping hinge teeth. (See also opisthocline, orthocline)

prosocline: In snail (gastropod), type of growth line which leans forward (i.e., inclined adapically) with respect to growth direction of helicocone. (See also opisthocline, orthocline)

prosocyrt: In snail (gastropod) shell, type of growth line which arches forward with respect to growth direction of helicocone. (See also opisthocyrt)

prosodetic [G *prosodet*]: In bivalve shell, located anterior to beak, e.g., as in ligament. (See also amphidetic, opisthodetic)

prosogyrate [G *prosogyr*]: In bivalve, condition in which umbones are curved so that beaks point in anterior direction (most typical condition). (See also coiled, opisthogyrate, orthogyrate, spirogyrate)

prosostomatous: General category of digenetic fluke (trematode); refers to forms with mouth at anterior end of body (as opposed to centrally located mouth = gasterostome). (See also amphistome, distome, echinostome, holostome, monostome)

protobranchiate [G *protobranchiat, protobranch*]: In bivalve, type of gill (ctenidium) consisting of series of flattened lamellae (rather than filaments).

protrusile [G *vorstreckbar*]: Capable of being protruded. (See also eversible)

proximal [G *proximal*]: Pertaining to or located at point of attachment or at center of body. Refers, in bryozoan, also to side toward ancestrula or origin of growth. (See also distal)

pseudocompound: In polychaete, type of bristle (seta) which superficially appears to be articulated.

pseudolamellibranch: In bivalve, type of gill (ctenidium) in which distal end of ascending limb of each filament in demi-branch has fused laterally with mantle (and mesially with base of foot). (See also eulamellibranch, filibranch, protobranchiate)

pseudomonocyclic: cryptodicyclic

pseudopenicillate: Type of bristle (seta) in polychaete: intermediate between penicillate forms (with terminal tuft) and those with tip projecting beyond tuft.

pseudosolitary [G *pseudosolitär*]: In anthozoan, colony consisting of single autozooid and numerous siphonozooids.

ptenoglossate [G *ptenogloss*]: Type of radula in snail (gastropod): each transverse tooth row consists of numerous hooked teeth increasing in size laterally. Central rachidian tooth is lacking. (See also docoglossate, hystrichoglossate, rachiglossate, rhipidoglossate, taenioglossate, toxoglossate)

ptycholophous: Type of lophophore in lamp shell (brachiopod): bearing four lobes due to indentation of simple, trocholophous type. (See also plectolophous, schizolophous, spiralophous, trocholophous, zugolophous)

pubescent [G *fein behaart, flaumhaarig*]: Covered with hair-like projections that are fine, short, and soft. Applied, for example, to claws or carapace of certain decapods.

punctate [G *punktiert*]: Minutely pitted, e.g., as in shell of bivalve or snail (gastropod) or cuticle of roundworm (nematode).

pustulate [G *pustulös*]: Shaped like or covered with blister-like protuberances (pustules), e.g., as in type of sculpture on shell of snail (gastropod) or bivalve.

pustuliporiform (entalophoriform) [G *pustuliporiform*]: Type of bryozoan colony: with erect, cylindrical branches formed by feeding individuals (autozooids) distributed radially and equally in all directions.

pyramidal [G *pyramidal, pyramidenförmig*]: Shaped more or less like a pyramid.

pyriform [G *birnenförmig*]: Pear-shaped. (See also obpyruvate)

quadrangular [G *vierseitig, viereckig*]: Having four angles and four sides.

quadrate [G *quadratisch*]: Square or almost square.

quadridentate [G *vierzähnig*]: Bearing or ending in four tooth-like projections, e.g., as in type of bristle (seta) in polychaete.

quadrifid [G *vierspaltig, vierteilig*]: Split into four equal (often lobe-like) parts, e.g., as in tip of spine in bryozoan.

quadrilobate [G *vierlappig*]: Divided into four lobes. May also refer to type of carapace (valve) of mussel shrimp (ostracod) bearing four lobes.

quadripartite [G *vierteilig*]: Consisting of or divided into four parts. Refers, for example, to mouth in certain scyphozoans.

quincuncial [G *quincunxial*]: Arrangement of five structures so that four are located at the corners of a rectangle and one is located in the center. Refers, for example, to arrangement of hooks on proboscis of spiny-headed worm (acanthocephalan).

racemose [G *traubenförmig, racemös*]: Resembling bunch of grapes, i.e., bushy with each branch ending in expanded, sac-like structure. Refers, for example, to type of gland and occasionally also applied in connection with type of branching in hydrozoan colony (growth below terminal hydranth).

rachiglossate: Type of radula in snail (gastropod): a transverse tooth row consists of three denticulate teeth—a central rachidian tooth flanked on each side by a lateral tooth. (See also docoglossate, hystrichoglossate, ptenoglossate, rhipidoglossate, taenioglossate, toxoglossate)

radial [G *radiär*]: Located on, arranged along, or otherwise pertaining to a radius. Refers to type of symmetry, e.g., as represented by five radii in echinoderm or variously elaborated in cnidarian. In bivalve, type of shell sculpture extending from beak to ventral margin.

ramate [G *ramat*]: Grinding type of trophi in rotifer: rami and unci are well developed, fulcrum and manubrium reduced. Unci bear parallel ridges for chewing.

ramose [G *verzweigt, verästelt*]: Branched.

raptorial [G *Greif-*]: Adapted to seizing prey.

rectangular [G *rechteckig*]: Having right angles and typically resembling an elongated square, e.g., as in carapace shape in certain crab-like decapods. (See also quadrate)

rectimarginate [G *rektimarginat*]: In lamp shell (brachiopod), type of commissure between dorsal and ventral valves: anterior commissure lies in one plane.

recumbent [G *zurückliegend, anliegend, fest anliegend*]: Refers to structure which tends to rest upon surface from which it extends. (See also erect)

recurved [G *zurückgebogen*]: Curved backward. Refers, in snail (gastropod) shell, to type of siphonal canal which curves away when shell is observed from side bearing opening (aperture).

reduced [G *reduziert*]: Poorly or incompletely developed. Virtually every structure is reduced in one or the other representative of a major group (e.g., phylum); this term is therefore not listed as a character state in the first part of this glossary.

reflected [G *zurückgebogen*]: Turned back, e.g., as in outer lip and columellar lip in snail (gastropod) shell or in gill element.

reflexed [G *umgeschlagen, flach zurückgebogen*]: Turned back upon itself, especially if this is abrupt. Refers, for example, to condition of female reproductive system in roundworm (nematode). (In nematode, see also antidromous, homodromous)

regular (endocyclic) [G *regulär*]: Type of sea urchin or type of sea urchin shell (test) in which system of plates (periproct) around anus is located within circlet of apical plates (genital and ocular plates). (See also irregular)

reniform [G *nierenförmig*]: Kidney-shaped.

reptant [G *kriechend*]: Category of decapod including crab- and lobster-like forms, i.e., typically dorsoventrally flattened (depressed) and more or less adapted for crawling. (See also natant)

reteporiform [G *reteporiform*]: Type of bryozoan colony: net-shaped.

reticulate [G *netzförmig, retikuliert*]: Forming or resembling a net, e.g., as in type of sculpture on snail (gastropod) shell. Also refers to type of anthozoan colony in which branches anastomose to form net-like structure.

retort-shaped [G *retortenförmig*]: Resembling a retort, i.e., with bulb-like section and elongate, tapering tube projecting away at an angle. Applied, for example, to gonad shape in ribbon-worm (nemertine).

retractile [G *retraktil, einziehbar, zu-rückziehbar*]: Capable of being drawn back or into (as into a sheath). Does not necessarily imply introversion but indicates more than mere shortening by contraction.

rhipidoglossate [G *rhipidogloss*]: Type of radula in snail (gastropod): a transverse tooth row consists of central rachidian tooth flanked on each side by five lateral teeth and a large number of narrow, increasingly smaller marginal teeth. (See also docoglossate, hystrichoglossate, ptenoglossate, rachiglossate, taenioglossate, toxoglossate)

rhomboid [G *rhombenförmig, rhomboid*]: Shaped like a rhombus, i.e., like a slanted square (equal, parallel sides yet no right angles). Refers, for example, to shape of bryozoan individual (zooid).

rhomboidal [G *rhomboidförmig, rhomboidal*]: Shaped like a rhomboid, i.e., like an elongated, slanted square (parallel, unequal sides yet no right angles).

rhopalocercous: Type of larva (cercaria) of distome digenetic fluke (trematode): tail heavy and broad.

rimate [G *spaltförmig*]: Consisting of narrow cavity or fissure, e.g., as in type of umbilicus in snail (gastropod) shell.

ringent [G *klaffend*]: Gaping, e.g., as in shells of certain bivalves. In polychaete, refers also to type of bristle (seta) with series of annular serrations on both prongs.

rod-like [G *stabförmig*]

rostello-: In tapeworm (cestode), prefix applied to cysticercoid (stage in tapeworm life cycle) with rostellum. (See also acanthorostello-, anacanthorostello-)

rostrad [G *rostrad*]: Extending toward rostrum, e.g., as in decapod.

rostrate [G *geschnäbelt, schnabelartig ausgezogen*]: Bearing a rostrum or pointed, beak-shaped end. Refers, for example, to type of bivalve shell with beak-like extension.

rostriform [G *schnabelförmig*]: Shaped like a rostrum or beak.

rosulate [G *rosettenartig geordnet*]: Arranged in the form of a rosette. Refers to type of pharynx in flatworm (turbellarian): spherical, oriented at right angle to body axis, and provided with conspicuous radial muscle fibers and glands. (In turbellarian, see also doliiform)

rounded [G *abgerundet*]

rudimentary [G *rudimentär*]: reduced

rugose [G *runzelig*]: Covered with wrinkles (rugae).

S-shaped [G *S-förmig*]

sacciform (saccular) [G *sackförmig, sackartig*]: Shaped like a sac or pouch.

saddle-shaped: In oligochaete, type of clitellum which is not complete ventrally and thus does not surround entire body. (See also annular)

sagittal [G *sagittal*]: Lying on or otherwise related to plane of symmetry dividing organism into left and right halves. Also refers to planes parallel to the above median plane.

scabrous [G *rauh*]: Rough, e.g., as in type of sculpture (ridge, costa) on bivalve shell which is additionally roughened by minute projections.

scalariform [G *leiterförmig, wendeltreppenförmig*]: Resembling a ladder. In snail (gastropod), refers to type of shell which is loosely coiled like a spiral staircase.

scale-shaped [G *schuppenförmig*]

scalloped [G *bogenförmig ausgebuchtet*]: With margin produced into series of rounded processes. Refers to type of bell margin in medusa or, in bivalve, to type of shell margin in which internal flutings correspond to external sculpture (costae).

scaly: squamous

scapulate: Type of scyphozoan medusa bearing fringed outgrowths (scapulets) on base of each oral arm. (See also loriferate, tripterous, two-/three-winged)

schizodont [G *schizodont*]: Type of hinge (dentition) in bivalve: median tooth of left valve is strong and bifid (and fits into correspondingly shaped set of teeth on right valve). Type of compound hinge in mussel shrimp (ostracod): one of four basic types of amphidont hinge—anterior and anteromedian elements are bifid.

schizolophous [G *schizoloph*]: Type of lophophore in lamp shell (brachiopod): bilobate due to median indentation of simple, trocholophous type. (See also plectolophous, ptycholophous, spiralophous, trocholophous, zugolophous)

schizoporellid: schizostomatous

schizostomatous: In bryozoan individual (zooid), type of aperture (orifice) characterized by median sinus or notch at proximal margin.

scutellate [G *plättchenförmig, schuppenartig*]: Flattened and rounded or oval in shape, often with expanded rim.

secondary: In snail (gastropod) shell, type of spiral sculpture appearing later in development. (See also primary)

sedentary [G *sedentär*]: Category of polychaete basically composed of less motile forms, often those confined to a tube. (See also errant)

segmented [G *segmentiert, gegliedert*]: Consisting of segments. Refers, in anthozoan colony, to type of axis consisting of alternating calcareous and horny stretches.

semicircular [G *halbkreisförmig*]: Refers to two-dimensional shape (as opposed to hemispherical, three-dimensional structure).

semiophoric: Type of stinging cell (nematocyst) in cnidarian: tubule is bent and whip-like and bears large, flat spine medially. Refers to category with short, distally dilated butt (microbasic eurytele).

septate [G *septiert, mit Scheidewänden versehen*]: With dividing walls or membranes. Refers, in fluke (trematode), to type of opisthaptor with elevated partitions between loculi.

serial [G *serial, reihenweise angeordnet*]: Arranged in a series. In sea star, type of gonad system in which numerous testes or ovaries are present in a row along each side of arm.

serrate [G *gesägt, gezackt*]: With evenly toothed margin resembling a saw. (See also entire)

serrulate [G *gezähnelt*]: With finely and evenly toothed margin. (See also triserrulate)

sesquiramous: In polychaete, type of parapodium in which dorsal branch (notopodium) is reduced to a dorsal cirrus, an aciculum, and sometimes one or two setae.

sessile [G *sessil, ungestielt, sitzend*]: Either more or less firmly attached (vs. free-swimming) or lacking a stalk. Refers to unstalked eye in crustacean, unstalked opisthaptor of fluke (trematode), or unstalked barnacle (cirriped). In bryozoan, type of protective individual (vicarious avicularium) adhering to or embedded in normal feeding individual (autozooid). In sea star, type of pedicellaria lacking stalk and consisting of groups of two or more movable spines attached to underlying skeletal elements (ossicles).

setaceous [G *borstig, borstenartig*]: Resembling or bearing bristles (setae). Refers to long, slender type of spiculum in roundworm (nematode).

setiferous [G *borstig*]: Bearing bristles (setae).

setigerous [G *borstentragend*]: Bearing or producing bristles (setae). In oligochaete, type of segment which bears setae; in polychaete, part (lobe) of notopodium or neuropodium which bears setae.

setose [G *borstig*]: Bearing bristles (setae).

setoselliniform [G *setoselliniform*]: Type of bryozoan colony: discoidal, encrusting, with spiral growth.

setulose [G *fein beborstet*]: Bearing small, hair-like bristles (setules).

shield-shaped: clypeate

short-handled [G *kurzschaftig*]: In polychaete, type of hook-shaped seta (uncinus) without long, rod-shaped support. (See also long-handled)

sickle-shaped [G *sichelförmig*]: Refers to type of spiculum in roundworm (nematode) or seta in polychaete.

sigmoid (sigmoidal) [G *sigmoid, sigmoidal*]: S-shaped or C-shaped.

simple [G *einfach*]: Refers, in polychaete, to type of bristle (seta) which consists of one piece (i.e., unjointed). Also, solitary sea squirt (ascidian) as opposed to compound type. In sea star, type of ampulla of tube foot which is not double or bilobed.

single (unpaired) [G *unpaar*]: (See also paired)

sinistral [G *linksgewunden*]: Turning to the left, e.g., as in left-handed direction of coiling in snail (gastropod) shell or in certain tube-dwelling polychaetes. (See also dextral)

sinuate [G *gebuchtet*]: With wavy margin. (See also trisinuate)

sinuous [G *geschlängelt, gewunden, sinuös*]: Characterized by series of S-shaped curves, e.g., as in type of intestine in ribbonworm (nemertine) (vs. straight), in ciliated bands of tornaria larva of acorn worm (enteropneust), or type of ovary in roundworm (nematode).

sinupalliate [G *sinupalliat*]: In bivalve, condition in which pallial line on inner surface of valve bears posterior, inwardly directed curve. (See also integripalliate)

siphonate [G *siphonat*]: General category of bivalve with siphons.

siphonostomatous [G *siphonostom*]: In snail (gastropod) shell, type of opening (aperture) which is notched or drawn out into a canal (siphonal canal) to accommodate a siphon. General category of copepod with tubular mouth adapted for sucking blood. (In snail, see also holostomatous; in copepod, see also gnathostomatous)

slit-like [G *schlitzförmig*]: Refers, for example, to type of amphid in roundworm (nematode) or to gill pore shape (vs. round) in acorn worm (enteropneust).

smooth [G *glatt, nackt*]

solitary [G *solitär*]: Living or growing alone. (See also colonial, compound, simple)

spatulate [G *spatelförmig*]: Shaped like a spatula, i.e., like a flattened spoon with blunt tip. Refers, for example, to type of head (cephalic lobe) shape in ribbon-worm (nemertine) or to type of bristle (seta) in polychaete. In sea star, type of alveolate pedicellaria in which jaws (valves) are elongate and broadened at the ends.

spear-shaped: hastate, lanceolate

spherical [G *kugelförmig, kugelig, sphärisch*]: (See also sub-)

spicate [G *spitz*]: Pointed, e.g., as in type of tail shape in roundworm (nematode).

spicular [G *nadelförmig, spikelförmig*]: Resembling a spicule (i.e., needle-shaped).

spiculate [G *mit Nadeln besetzt*]: Bearing needle-shaped projections.

spindle-shaped: fusiform

spiniform [G *stachelförmig*]: Spine-shaped. In sea star, type of sessile pedicellaria in which movable spines are long. (In sea star, see also fasciculate, pectinate, valvulate)

spinigerous: In polychaete, type of compound seta (spiniger): blade tapers to a fine point.

spinose [G *bestachelt, stachelig*]: Bearing spines.

spinulose (spinulate) [G *fein bestachelt, feindornig*]: Bearing small spines (spinules).

spiral [G *Spiral-, spiralig*]: Refers to type of amphid in roundworm (nematode) or, in snail (gastropod) shell, to type of sculpture winding continuously around whorls parallel to suture. (In gastropod, see also axial)

spiralophous [G *spiraloph*]: Type of lophophore in lamp shell (brachiopod): two lateral arms become elongate and coil into helicoidal spirals. (See also plectolophous, ptycholophous, schizolophous, trocholophous, zugolophous)

spirogyrate [G *spirogyr*]: In bivalve shell, condition in which umbones are coiled so as to form a spiral turning away from hinge axis. (See also coiled, opisthogyrate, orthogyrate, prosogyrate)

spiruroid: Type of two-part pharynx in roundworm (nematode): superficially similar to dorylaimoid pharynx, yet expanded posterior region consists of both corpus and postcorpus.

spoon-shaped [G *löffelförmig*]: (See also spatulate)

squamiform [G *schuppenförmig*]: Shaped like a scale, i.e., as in antenna of certain decapods.

squamous (squamose) [G *schuppig*]: Bearing or covered with scales. Often applied as synonym for squamiform.

squat [G *gedrungen*]: In snail (gastropod), type of shell which is relatively broad in proportion to height. Typically employed together with other descriptive terms, i.e., squat turbinate (vs. tall turbinate).

stalked [G *gestielt*]: (See also pedunculate)

star-shaped: stellate

stellate [G *sternförmig*]: Star-shaped.

stenomyonic: Type of arrangement of muscle bands in certain thaliaceans (old nurse in doliolid): bands are narrower than one-half the interspaces. (See also eurymyonic, holomyonic; aclinous, amphiclinous)

stenostomate: Type of excretory system in larval stage (cercaria) of fluke (trematode): main excretory ducts extend from posterior excretory vesicle to anterior part of body, where they turn posteriorly before joining excretory tubules. (See also mesostomate)

sterile [G *steril*]: Incapable of producing functional sex cells. In entoproct colony, section of stolon not bearing a stalk.

stirodont [G *stirodont*]: In sea urchin, type of chewing apparatus (Aristotle's lantern) in which teeth have longitudinal keel and in which space (foramen magnum) between demipyramids is open above (i.e., not covered by fused epiphyses). (See also aulodont, camarodont)

stolonate (stolonal) [G *stolonat*]: In anthozoan, type of colony consisting of several polyps interconnected by stolons. (See also hydrorhizal)

stoloniferous (stoloniform) [G *stoloniform*]: Type of colony in bryozoan: delicate, branching, with erect branches arising from creeping portion.

stomatoporiform [G *stomatoporiform*]: Type of bryozoan colony: linear, uniserial or pluriserial.

stout [G *gedrungen*]

straight (forcipiform) [G *gerade*]: In sea star, type of stalked (pedunculate) pedicellaria in which jaws (valves) are more or less straight and parallel and attached basally to basal ossicle.

streptoneurous [G *gekreuztnervig*]: In snail (gastropod), condition of nerve system in which, due to torsion, pleurovisceral connectives are crossed and form a figure eight. (See also euthyneurous)

streptospondylous (strepto-spondyline) [G *streptospondyl*]: Type of articulation of vertebral ossicles in arm of brittle star: projections of adjoining vertebrae intermesh such that three-dimensional movement of arms is possible. (See also zygospondylous)

striate [G *gestreift*]: Bearing very fine grooves (striae).

strobilo-: In tapeworm (cestode), prefix applied to cysticercus (stage in tapeworm life cycle) in which forebody initiates segmentation. (See also anacantho-, heteracantho-, homocantho-, multicephalo)

strophic: In lamp shell (brachiopod), type of shell in which true hinge line is parallel to hinge axis. (See also nonstrophic)

styliform [G *griffelförmig, stiletförmig*]: More or less parallel-sided, with sharp-pointed tip. Applied to describe shape of various crustacean appendages (as opposed to lamellate or fan-shaped).

sub- [G *sub-, annähernd, fast*]: Prefix meaning "almost" and combined with term specifying idealized shape or structure (e.g., subcircular, subdigitate, subglobose, subhexagonal, suborbicular, subpentagonal, subquadrate, subrectangular, subspherical, subtrigonal). Application is often a matter of judgment, as virtually no larger structures or organs describe a perfect geometric shape. Also indicates position or region below a specified structure or area (e.g., subdistal, submedian, suborbital).

subbiramous [G *subbiram*]: In polychaete, type of appendage (parapodium) which is neither completely uniramous nor biramous (i.e., dorsal branch = notopodium is reduced).

subchelate [G *subchelate*]: In crustacean, type of appendage ending in a subchela, i.e., type of pincer in which last segment (dactylus) bears down on next to last

segment (propodus). (See also achelate, carpo-subchelate, chelate)

subcordate [G *fast herzförmig*]: Almost heart-shaped.

subcylindrical [G *subzylindrisch*]: Almost cylindrical, e.g., as in carapace of shrimp-like decapod or crustacean appendage (as opposed to foliaceous appendage).

subdivided: multiarticulate

subequal [G *fast gleich*]: Almost equal, e.g., as in first pair of appendages in crustacean or in valves of mussel shrimp (ostracod). (In ostracod, see also unequal)

subhemispherical: Almost hemispherical. Refers, for example, to umbrella shape in medusa of hydrozoan or scyphozoan.

subjacent [G *unmittelbar darunterliegend*]: Lying directly under or below.

submalleate [G *submalleat*]: In rotifer, modified type of malleate trophi characterized by slightly longer fulcrum and manubria.

submesothyridid [G *submesothyrid*]: In lamp shell (brachiopod), position of pedicle opening (foramen) relative to deltidial plates and beak of ventral valve: foramen located partly in ventral umbo, but mainly in delthyrium. (See also amphithyridid, epithyridid, hypothyridid, mesothyridid, permesothyridid)

subopposite [G *fast gegenständig*]: Located almost opposite to one another. In hydrozoan, type of arrangement of polyps along branch of erect colony. (See also opposite)

subpentagonal: Almost pentagonal. Applied to carapace shape in certain crab-like decapods or to body shape of certain sea stars.

subpetaloid [G *subpetaloid*]: At apex of irregular sea urchin shell (test), feebly developed, petal-shaped condition of each ambulacrum. (See also petaloid)

subsetulate: Type of bristle (seta) bearing minute projections (subsetules). (See also plumose)

substomodeal [G *subpharyngeal*]: In comb jelly (ctenophore), lying closer to stomodeal (= sagittal = pharyngeal) plane, i.e., plane passing through flattened stomodeum. (See also subtentacular)

subtentacular [G *subtentakulär*]: In comb jelly (ctenophore), lying closer to tentacular plane, i.e., longitudinal plane through tentacle bulbs. (See also substomodeal)

subterminal [G *subterminal*]: Almost at end, e.g., as in position of mouth or genital pore on body or position of expanded section (head) on sponge spicule. (See also terminal)

subtriangular [G *subtriangulär, fast dreieckig*]

subulate [G *ahlenförmig, pfriemenförmig*]: Awl-shaped, i.e., elongate, very slowly tapering to a sharp point.

subumbrellar [G *subumbrellar*]: In medusa of hydrozoan or scyphozoan, positioned on or located closer to lower concave surface of body (umbrella). (See also exumbrellar)

sulcal [G *sulcal*]: In anthozoan, side of polyp bearing flagellated groove (siphonoglyph) in pharynx (also considered to be ventral). (See also asulcal)

sulcate [G *sulkat*]: Bearing furrow or groove (sulcus). Refers to type of lamp shell (brachiopod) with major depression extending radially from umbo or to type of dorsoventral sculpture on valve of mussel shrimp (ostracod).

sulcatomicrocercous: Variation of small-tailed (microcercous) cercaria larva of fluke (trematode): tail triangular with ventral groove (sulcus).

sulculular: asulcal

superior [G *höherstehend, obere*]: Indicating position above (dorsal to) a second structure. (See also inferior)

supra- [G *supra-*]: Prefix indicating position above a specified structure, e.g., as in supraorbital spine of decapod carapace.

surficial: In the absence of criteria to determine the dorsal and ventral sides of tapeworm (cestode), term applied to flat surface of body.

syconoid [G *syconoid*]: Type of sponge in which choanocyte chambers are more or less restricted to radial outpocketings of central cavity (spongocoel). (See also asconoid, leuconoid)

symmetrical [G *symmetrisch*]: Having dimensions of one side being equal to those of other side, e.g., as in chewing elements (trophi) of pharynx in most rotifers. (See also asymmetrical)

sympodial [G *sympodial*]: Type of colony growth in hydrozoan: stalk of terminal hydranth buds to form new stalk and hydranth (or new branch). Typically gives rise to series of alternate or opposite hydranths. (See also hydrorhizal, monopodial)

synarthrial: In crinoid, type of articulation between endoskeletal elements (ossicles/columnals) of stalk: faces of ossicles are bifacial, i.e., characterized by median ridge separating two depressions to accommodate ligaments. (See also synostosial)

synostosial: In crinoid, type of articulation between endoskeletal elements (ossicles/columnals) of stalk: faces of ossicles are closely fitting and broadly concave to accommodate ligaments. (See also synarthrial)

taenioglossate [G *taeniogloss*]: Type of radula in snail (gastropod): a transverse tooth row consists of seven teeth—a median rachidian tooth flanked on each side by one lateral tooth and two marginal teeth. (See also docoglossate, hystrichoglossate, ptenoglossate, rachiglossate, rhipidoglossate, toxoglossate)

tall [G *hoch*]: In snail (gastropod), type of shell which is relatively high in proportion to its width. Typically employed together with other descriptive terms, i.e., tall turbinate. (See also squat)

tandem (in tandem) [G *hinterein-ander angeordnet*]: Positioned one behind the other, e.g., as in type of testes arrangement in fluke (trematode) (as opposed to diagonal or opposite testes).

tanylobous (tanylobic) [G *tanylobisch*]: In oligochaete, type of prostomium in which posterior projection (tongue) extends dorsally across peristomium and reaches furrow separating peristomium (segment 1) from segment 2. (See also epilobous, proepilobous, prolobous, zygolobous)

taxodont [G *taxodont*]: Type of hinge (dentition) in bivalve: characterized by numerous parallel teeth which are arranged in a series along hinge margin. Teeth are small, short, similar in shape, and alternate with sockets.

telogonic: Type of ovary or testis in roundworm (nematode): germ cells are formed only at blind end of reproductive organs. (See also hologonic)

telotrichous: Type of stinging cell (nematocyst) in cnidarian: butt bears spines only distally. Refers to nematocyst type (macrobasic eurytele) with long, distally dilated butt.

tentaculate [G *tentakeltragend*]: Bearing tentacles.

tentaculiform [G *tentakelförmig*]: Tentacle-shaped.

tentaculo-: In tapeworm (cestode), prefix applied to type of plerocercoid (stage in tapeworm life cycle) which bears tentacles. (See also acantho-, acetabulo-, bothridio-, bothrio-, culcito-, glando-)

terete [G *zylindrisch*]: Cylindrical.

terminal [G *endständig, terminal*]: Located at end, e.g., as in position of mouth on body (as opposed to subterminal or ventral). In bivalve, type of beak forming most anterior or posterior point of valve.

tessellated [G *mosaikartig*]: Resembling a mosaic, i.e., having surface with network of grooves.

testicular: Pertaining to a testis. Refers, in oligochaete, to testis-bearing segment of. (See also ovarian)

tetrabasal: In shell (test) of irregular sea urchin, type of apical system of plates in which number of genital plates is reduced to four (from five). (See also monobasal)

tetrabranch [G *tetrabranch, tetrabranchiat*]: Type of cephalopod characterized, among other features, by four gills and four auricles. (See also dibranch)

tetradelphic: In roundworm (nematode), condition in which female reproductive system consists of four tracts. (See also didelphic, monodelphic, polydelphic)

tetramerous [G *tetramer, vierteilig*]: Characterized by four parts (or multiples of four). Refers to type of symmetry in medusa of hydrozoan or scyphozoan.

thecate [G *thekat*]: Type of hydrozoan in which periderm surrounds entire hydranth (as hydrotheca) and gonophore (as gonotheca). (See also athecate)

thin-shelled [G *dünnschalig*]

thorn-like [G *dornartig*]

three-lobed: trilobate

three-part: General category of pharynx in roundworm (nematode). (See also chromodorid)

three-segmented: triarticulate

three-winged (tripterous) [G *dreiflügelig*]: In medusa of scyhozoan, type of oral arm with three wing-like processes. (See also two-winged; loriferate, scapulate)

torted [G *mit Torsion*]: General category of mollusc which has undergone torsion, i.e., altered symmetry due to twisting of certain internal organs. (See also untorted)

toxoglossate [G *toxogloss*]: Type of radula in snail (gastropod): consists of series of elongate, often harpoon-like teeth which may be associated with poison sac. (See also docoglossate, hystrichoglossate, ptenoglossate, rachiglossate, rhipidoglossate, taenioglossate)

translucent [G *durchscheinend*]: Transmitting and partly diffusing light, i.e., so that structures on other side are somewhat unclear.

transparent [G *durchsichtig*]: Transmitting light without diffusion, i.e., so that structures on other side are entirely visible.

transverse [G *quer, querlaufend, transversal*]: Crossing or at right angles to longitudinal axis.

trapeziform [G *trapezähnlich, trapezförmig*]: With four straight sides, no two sides being parallel. With four straight sides, two of which are parallel (British usage).

trapezoidal [G *trapezoid, parallel-trapezförmig*]: With four straight sides, two of which are parallel. With four straight sides, no two sides being parallel (British usage).

tree-like: arborescent

triangular [G *dreieckig*]

triannulate: In oligochaete, condition in which a segment is subdivided externally into three rings by two secondary furrows (annuli), one on each side of setae. (See also biannulate)

triarticulate [G *dreigliedrig*]: Consisting of three segments, e.g., as in peduncle of antennae in certain crustaceans.

triaulic: In reproductive system, for example of snail (gastropod), condition in which three openings of reproductive system to exterior are present (i.e., vagina, male gonopore, female gonopore). (See also diaulic, monaulic)

tricarinate [G *dreifach gekielt*]: Bearing three keels (carinae).

trichobranchiate (filamentous): In decapod, type of gill characterized by undivided, filament-shaped branches along axis. (See also dendrobranchiate, phyllobranchiate)

trichocercous: Type of larva (cercaria) of distome digenetic fluke (trematode): tail beset with bristles.

triclad [G *dreiästig*]: In flatworm (turbellarian), condition of digestive system in which intestine basically consists of three branches, one extending from pharynx to anterior and two from pharynx to posterior end. (See also polyclad)

tricuspid [G *dreispitzig*]: Ending in three points. (See also bicuspid)

tridactyle: tridentate

tridentate (tridactyle) [G *tridentat, tridactyl*]: Ending in three teeth. In sea urchin, type of relatively large pedicellaria with three long, pointed jaws; lacks both terminal teeth and poison glands. (In sea urchin, see also bidentate, triphyllous)

trifid [G *dreispaltig*]: Split into three equal parts (often lobe-like) by clefts. (See also bifid)

trifoliate [G *trifoliat*]: Type of bryozoan colony; erect, with three median walls, each bearing two layers of individuals (zooids) growing back to back. (See also bifoliate)

trifurcate [G *dreigabelig, dreizackig, dreifach gegabelt*]: Forking into three parts. (See also bifurcate, multifurcate)

trigeminate: In sea urchin shell (test), type of modified ambulacral plate (compound plate) bearing three pore pairs. (See also bigeminate; oligoporous, polyporous)

trigonal [G *dreieickig, dreikantig*]: Three-cornered.

trilobate (trilobed) [G *dreilappig*]: Divided into three lobes. (See also bilobate, unilobate)

trimorphic [G *trimorph*]: Characterized by the occurrence of three types of individuals, e.g., as in hydrozoan colonies with gasterozooids, medusoids, and gonozooids. (See also dimorphic, monomorphic, polymorphic)

tripartite [G *dreiteilig*]: Divided into three parts. (See also bipartite)

triphyllous [G *trifoliat*]: In sea urchin, type of small pedicellaria with three short, broad jaws not hinged to one another. (See also biphyllous)

tripterous: three-winged

triradiate [G *dreistrahlig*]: Having three radiating parts or branches, e.g., as in cavity of rotifer mastax or sponge spicule.

triserrate: Type of bristle (seta) with three rows of projections. (See also triserrulate)

triserrulate: Type of bristle (seta) with three rows of minute projections. (See also serrulate, triserrate)

trisinuate: With wavy margin forming three rounded lobes.

trisulcate: Bearing three grooves (sulci). Refers to type of sculpture in shell (valve) of mussel shrimp (ostracod): three sulci separate four rounded elevations (lobes). (See also bisulcate, unisulcate)

triunguiculate: Condition in which tip of last segment (dactylus) of crustacean limb is produced into three claw-like structures (ungues). (See also biunguiculate, uniunguiculate)

trochoid [G *kreiselförmig, trochoid*]: Top-shaped. Applied to snail (gastropod) shell whose whorls are flat and taper rapidly from flat base to apex. (See also turbinate, turriform)

trocholophous [G *trocholoph*]: Type of lophophore in lamp shell (brachiopod): simple, disc-shaped, with relatively short, tentacle-bearing edge. (See also plectolophous, ptycholophous, schizolophous, spiralophous, zugolophous)

truncate [G *abgestutzt, gestutzt*]: With end being flat or straight, as if cut off. Refers, for example, to type of posterior telson margin in crustacean.

tuberculate [G *knotig, höckerig*]: Covered with small, knob-like protuberances (tubercles).

tubicolous [G *rohrenbewohnend*]: Living in self-constructed tube.

tubular [G *röhrenförmig*]: Elongate, hollow, typically with same diameter throughout. (See also cylindrical)

tubuliporiform [G *tubuliporiform*]: Type of bryozoan colony: branches flattened, more or less broad, little ramified, with all feeding individuals (autozooids) opening on face only.

tumid [G *geschwollen, aufgebaucht*]: Swollen or inflated.

turbinate [G *kreiselförmig*]: Top-shaped. Applied to snail (gastropod) shell whose whorls are rounded and taper rapidly from a broad, convex base to apex. (See also squat, tall; trochoid, turriform)

turriform [G *türmchenförmig*]: Type of snail (gastropod) shell shape: elongate, with somewhat rounded whorls tapering only gradually from base to pointed apex. (See also trochoid, turbinate)

two-part: dorylaimoid

two-valved: In bryozoan, type of brood chamber (ovicell) consisting of two hemispherical valves which represent modified spines or tubercles.

two-winged [G *zweiflügelig*]: In medusa of scyphozoan, type of oral arm with two wing-like processes. (See also three-winged = tripterous; loriferate, scapulate)

tylenchoid: Type of three-part pharynx in roundworm (nematode): similar to diplogasteroid pharynx, yet characterized by more slender anterior region.

U-shaped [G *U-förmig*]

ultradextral: In snail (gastropod), type of shell which appears to be sinistral (coiled to the left), yet animal is organized dextrally. (See also hyperstrophic, ultrasinistral)

ultrasinistral: In snail (gastropod), type of shell which appears to be dextral (coiled to the right), yet animal is organized sinistrally. (See also hyperstrophic, ultradextral)

umbellate [G *doldenartig*]: Type of colony growth in anthozoan: arborescent, with polyps bundled into umbel-like aggregates, all polyps thus being positioned on outer surface of colony.

umbilicate: omphalous

umbonuloid: Type of bryozoan in which frontal shield of feeding individual (autozooid) forms through calcification of lower side of fold of body wall, derived from proximal end of zooid.

unarmed [G *unbewehrt, unbewaffnet*]: Lacking additional structures such as sharp or pointed elements which might serve in defense, feeding (on proboscis of nemertine), or reproduction (on copulatory organ). (See also armed)

unbranched [G *unverzweigt*]

uncigerous: In polychaete, bearing uncini (hook-like seta).

uncinate [G *hakenförmig, gekrümmt*]: With tip bent like a hook. Refers also to type of trophi in rotifer. Voluminous (used to swallow prey), with trophi restricted to posterior section or mastax; rami and subunci well developed, fulcrum and manubria reduced, unci bearing several teeth.

undercut: Condition of tubercle (clava) or ridge (carina) on valve of mussel shrimp (ostracod): walls hollowed out, resulting in expanded upper margin. (See also ponticulate)

unequal [G *ungleich*]: In mussel shrimp (ostracod), condition in which two valves of carapace differ considerably (as opposed to being almost equal = subequal).

uniarticulate [G *eingliedrig, einteilig*]: Consisting of only one segment. Refers to condition of crustacean appendage. (See also articulate, multiarticulate)

unicolored (unicolorous) [G *einfärbig*]

unidentate [G *einzähnig, einspitzig*]: Ending in only a single tooth. (See also bidentate)

unilaminar: Type of colony growth in bryozoan: consisting of only one layer of individuals (zooids). (See also bilaminar, plurilaminar)

unilateral [G *einseitig*]: Located only on one side, e.g., as in condition of vitellaria in certain flukes (trematodes). (See also bilateral)

unilobate [G *einlappig*]: Bearing only one lobe. Refers to type of sculpture of shell (valve) in mussel shrimp (ostracod): surface is evenly elevated, without intervening grooves (sulci). (See also bilobate, trilobate)

unilocular [G *unilokulär*]: In tapeworm (cestode), type of hydatid (stage in tapeworm life cycle) with uniform outer wall and typically with endogenous budding. (See also multilocular)

unipectinate (monopectinate) [G *monopectinat*]: Type of gill in snail (gastropod): gill leaflets occur only on one side of septum. (See also bipectinate)

uniramous [G *einästig, uniram*]: Bearing only one branch (ramus). In polychaete, type of appendage (parapodium) in which one branch (typically dorsal branch = notopodium) is absent. In crustacean, appendage with only one branch (e.g., formed by endopod, with exopod reduced).

uniserial [G *einzeilig*]: Arranged in a single row (series). In bryozoan, type of colony in which individuals (zooids) bud in a single row. In crinoid, type of arm (brachium) supported by single row of endoskeletal elements (brachials). (See also biserial, pluriserial)

unispiral (loop-shaped): In roundworm (nematode), type of amphid characterized by one coil. (See also multispirial)

unisulcate [G *unisulcat*]: Bearing one groove (sulcus). Refers to type of sculpture in shell (valve) of mussel shrimp (ostracod) in which one sulcus separates two lobes. (See also bisulcate, trisulcate)

uniunguiculate: Condition in which tip of last segment (dactylus) of crustacean limb is produced into claw-like structure (unguis). (See also biunguiculate, triunguiculate)

univalve [G *einschalig*]: Type of mollusc whose shell is composed of a single element. (See also bivalved)

unpaired [G *unpaar*]: (See also paired)

unsegmented [G *ungegliedert*]: (See also articulate)

untorted [G *ohne Torsion*]: General category of mollusc which has not undergone torsion, i.e., has not altered its symmetry due to twisting of certain internal organs. (See also torted)

unwound [G *ungewunden*]: Type of amphid in roundworm (nematode) (as opposed to dorsally or ventrally wound).

V-shaped [G *V-förmig*]

valvulate (valvate) [G *muschelschalenähnlich*]: In sea star, type of sessile pedicellaria in which jaws (valves) resemble valve-like elements due to fusion of groups of spinelets. (See also fasciculate, pectinate, spiniform)

variegated [G *buntscheckig, buntgefleckt*]: With discrete markings of different colors.

vasiform [G *gefäßförmig*]: Vase-shaped, e.g., as in type of sponge colony.

ventral [G *ventral*]: Pertaining to or located on lower surface of body. Not uniformly applicable in invertebrates, often requiring differentiation between morphological ventral side and downwardly = "ventrally" directed side as regards orientation of organism in the environment. In bivalve, for example, refers to region of shell opposite hinge. (See also dorsal).

ventricose [G *stark aufgetrieben*]: Strongly inflated, e.g., as in whorl of snail (gastropod) or shell of bivalve.

vermiform [G *wurmförmig*]: Worm-shaped.

verticillate [G *wirtelig*]: In sea urchin, arrangement of secondary spines or thorns on larger primary spine in well-spaced whorls.

vesicular [G *blasig*]: Shaped like a small sac (vesicle). May be used as synonym for vesiculate.

vesiculate [G *blasenförmig*]: Containing small cavities (vesicles). May be used as synonym for vesicular.

vestigial [G *verkümmert, rudimentär*]: Comparatively imperfect in size or shape.

vicarious [G *vikariierend*]: In bryozoan, type of protective individual (avicularium) whose position in a colony replaces or substitutes that of a normal feeding (adventitious) individual (autozooid). Relatively large in size compared with autozooid.

vinculariiform [G *vinculariiform*]: Type of bryozoan colony: erect, rigid, arborescent, made up of subcylindrical branches divided dichotomously. Firmly attached to substratum by calcareous basis.

virgate: Piercing and sucking type of mastax in rotifer in which trophi are relatively well developed: fulcrum and manubria are elongated, unci have grasping function, rami reinforce mastax wall, and suction is created by hypopharynx in conjunction with epipharynx.

virgulate: Type of larva (cercaria) of digenetic fluke (trematode): bears mucus storage organ (virgula organ) in region of oral sucker.

wart-covered (warty) [G *warzig*]

wart-like (warty) [G *warzenförmig, warzenartig, warzenähnlich*]: (See also papilliform, tuberculate)

wedge-shaped [G *keilförmig*]: Refers, for example, to shape of endoskeletal elements (brachials) supporting arms in certain crinoids. (See also cuneiform)

whip-like: flagelliform

whorled [G *wirtelförmig*]: In hydrozoan, type of branching in which branches arise from stem in whorls, i.e., several branches radiate from a common point of stem (as opposed to alternate, opposite, or spiral).

wing-shaped [G *flügelförmig, flügelartig*]: (See also aliform)

wrinkled [G *runzlig*]: (See also rugose)

zigzag [G *zickzackförmig*]: Describes, for example, pattern formed by suture between plates in sea urchin shell (test).

zugolophous (zygolophous) [G *zygoloph*]: Type of lophophore in lamp shell (brachiopod): extended laterally to form two arms. (See also plectolophous, ptycholophous, schizolophous, spiralophous, trocholophous)

zygolobous [G *zygolobisch*]: In oligochaete, type of prostomium which is continuous with peristomium (not marked off by groove). (See also epilobous, proepilobous, prolobous, tanylobous)

zygophiuroid: zygospondylous

zygospondyline: zygospondylous

zygospondylous [G *zygospondyl*]: Type of articulation of vertebral ossicles in arm of brittle star: depressions and projections of one vertebra fit into projections and depressions in adjoining vertebra such that only sidewise (horizontal) movement of arm is possible. (See also streptospondylous)

INDEX